AF679446

Ken Ring's

NEW ZEALAND WEATHER ALMANAC

2025

BY THE SAME AUTHOR

Predict Weather Almanacs for New Zealand 1999-Present)
Predict Weather Almanacs for Australia (2006-Present)
Predict Weather Almanac for Ireland (2010-Present)
How to Get Your Kid to Like Maths
Predicting Weather by The Moon
New Meteorological Techniques
We Used to Call It Summer
The Romany's Apprentice
Moon and WeatherLore
Secrets of the Moon
Notes From the Pit
The Lunar Code
Fingermaths
Supertramp
A Fiery Tale

© Ken Ring 2024
All Rights Reserved

Paperback ISBN: 978-1-73862-303-7
Hardback ISBN: 978-1-73862-304-4
eBook ISBN: 978-1-73862-305-1

Email: enquiries@predictweather.com
Web: www.predictweather.com

The moral right of the author to the copyright of this work as regards concept and expression of the manuscript has been asserted. Reproduction is not allowed of any part by any means, including electronic and through ongoing distribution in media or print except by written consent of the author.

CONTENTS

DISCLAIMER

The forecasts in this work are best-endeavour opinions of the authors and associates. No guarantee or promise of certainty is claimed. The authors cannot be held accountable for decisions made on the basis of information in these pages.

POTENTIAL

There could be times that the rain maps only succeed predicting clouds that have a rain-look about them. It may rain if there is potential, or, with potential in place, for various reasons, it may not rain at all. But it will not rain without potential. Prediction is not of certainty of outcomes, crystal-ball gazing, or voodoo. Weather forecasting is an inexact science, like medicine, geology, psychology, and economics. In medicine, often a second 'opinion' might be sought. A political commentator may predict the outcome of an election, and it is informed opinion, nothing more. No one would call the authors of commercial tide tables 'satanic soothsayers', and yet tides will be found in real life to vary from previously published heights and times. Allow for a leeway of 1-3 days and as weather is general to an area of 50-80kms radius, that is the accepted range of applicability and error potential.

ABBREVIATIONS

RAIN, SUNSHINE AND TEMPERATURE TABLES

NORTH ISLAND

Ak = Auckland
Chat = Chateau
Co = Thames/Coromandel
CR = Cape Reinga
Dann = Dannevirke
Dar = Dargaville
Gis = Gisborne
Has = Hastings
Hml = Hamilton
Ker = Kerikeri
Kta = Kaitaia, Northland
Lev = Levin
Mast = Masterton
Mata = Matamata
MkH = Manukau Heads
Nap = Napier
NP = New Plymouth
Opo = Opotiki
Ot = Otorohanga
Pa = Paraparaumu/Kapiti
PN = Palmerston North
Rsl = Russell/Bay of Islands
Rt = Rotorua
Tai = Taihape
Taum = Taumarunui
Taup = Taupo
Taur = Tauranga
TeA = Te Awamutu
TeP = Te Puke
UH = Upper Hutt
Wang = Wanganui
Wgt = Wellington
Whg = Whangarei
Wkt = Whakatane

SOUTH ISLAND

Alex = Alexandra
Ash = Ashburton
Bal = Balclutha
BB = Bruce Bay/Haast
Ble = Blenheim
Ch = Christchurch
Crom = Cromwell
Dar = Darfield
Dun = Dunedin
Gore = Gore
Grey = Greymouth
Han = Hanmer
Hok = Hokitika
Inv = Invercargill
Kai = Kaikoura
LP = Lake Pukaki
LT = Lake Tekapo
Meth = Methven
Mot = Motueka
Nel = Nelson
Oam = Oamaru
Qu = Queenstown
Rak = Rakaia
Rang = Rangitata
Ta = Takaka
TeA = Te Anau
Tim = Timaru
Twi = Twizel
West = Westport/Buller
Wtn = Winton

MOON SYMBOLS

1st Q	Moon in first-quarter phase.
3rd Q	Moon in third-quarter phase.
^	Northern declination Moon at furthest point north for month.
A	Apogee Moon furthest from earth for the month.
F	Full moon.
N	New moon.
P	Perigee Moon at closest point to the earth out of the 27 day cycle.
P(n)	Priority closest e.g. P3 means third closest for year.
V	Southern declination Moon at furthest point south for the month.
XhN	Moon crossing the equator heading north.
XhS	Moon crossing the equator heading south.

RAIN SUMMARY TABLES

d	Dry most of the day.
h	Heavy rain.
l	Light, intermittent showers.
n	Fog, mist, dew, or non-recordable showers.
p	Passing Showers
r	Rain.
s	Showery.

SUNSHINE SUMMARY TABLES

F	Fine, sunny (8+ hours sunshine)
pc	Partly cloudy (4-7 hrs sunshine)
c	Cloudy (1-3 hrs sunshine)
o	Overcast (0 hrs sunshine)

TEMPERATURE SUMMARY TABLES

clr	Cooler than average temperatures.
a-c	Average to cooler temperatures.
av	Average temperatures.
a-w	Average to warmer temperatures.
wrm	Warmer than average temperatures.

RAIN POTENTIAL MAPS

For the frost/snow maps, one town per region has been selected as representative of approximate sub zero temperatures. These towns are:

REGION	REPRESENTATIVE TOWN	REGION	REPRESENTATIVE TOWN
Northland	Kerikeri	Wanganui-	Levin
Auckland	Henderson	Manawatu	Upper Hutt
Bay of Plenty	Tauranga	Wellington	Takaka
Waikato	Hamilton	Nelson-Marlborough	Westport
Taranaki	New Plymouth	West Coast	Christchurch
Gisborne	Gisborne	Canterbury	Twizel
Hawkes Bay	Napier	Otago	Gore
		Southland	

THE NATURE OF WEATHER

Weather is determined by astronomical and physical reasons which are under relatively fixed control of cycles. It is not determined by any activities by Man. Climate-scientists erroneously claim that the heavier-than-air "trace" gas, carbon dioxide, which is absent in 99.96% of the atmosphere; and methane, which is absent in 99.9999%, have a controlling effect on the climate.

Approximately 80-85% of the weather, in this author's opinion, is caused by the moon, changing and repeating its position above the earth. Planning for major droughts or floods is therefore possible, but not by measuring any emissions of a negligible gas.

Here is the weighting system. There is space weather, climatic weather and everyday weather.

1. **Changes to whole planet**:
   (Irrespective of geography)
   • Deviation of Earth's rotational axis to the vertical.
   • Variation of Earth's orbit.
   • Magnetosphere
   • Heliosphere

2. **Climate at location**:
   (long-term trends in weather-patterns)
   • Proximity to equator (angle of the Sun's rays – latitude)
   • Direction of prevailing winds
   • Topographic factors – shape of land, proximity to other land masses, elevation and distance from coast.

3. **Weather at location:**
   (trends and daily occurrences, observable and recordable)
   • Influence of Moon on tides (including air tides), cycle of rain from ocean, and back to ocean. (80-85%), also ocean cycles e.g. El Nino.
   • Level and intensity of cloud cover (keeps heat in, determines what, where, when, of rain (subject to ocean and air tides)
   • Solar winds, cycles of Sunspots (temperature of seasons), hemispherical and seasonal factors.

4. **Not in the mix at all, either for climate or weather**
   • Land use changes, deforestation and greenification of areas of the earth.
   • Volcanic activity ongoing.
   • Ice-reflectivity feedback, constant.
   • Industrial and commercial pollution.
   • other activities of Man e.g. diet.

Immediate weather is modified by cycles of solar activity (position and number of occurring Sunspots), and the cyclic position of the Moon relative to the earth. Solar influences determine floods and drought periods.

The weather has no brain, so it does not know that there is a small percentage of earth that is not covered in ocean; to the moon it is all ocean and just more rain falls in certain areas around the globe, and not others. From the height rain falls from, up to 12km, it is immaterial whether there is land to fall on, or ocean. That's why rainforests make no difference to rainfall amounts. Forests grow where it rains often, not vice versa. Have you ever wondered why Brazil has no deserts, nor suffers periodic droughts?

About 85% of all the weather is controlled, directly or indirectly by the Moon. If you live by the ocean, then the influence is more obvious and direct. If you don't live on or near the coast, then weather systems take a little longer to reach you. But you can see, in a broad sense, the influence of the moon. For example, on almost every full moon night, you will see the full moon in all its glory. And that's because, on the night of the full moon, the moon clears the sky of clouds close to midnight, by dragging more air underneath it. This pushes away the cold air of space and prevents condensation of water vapor which make the clouds visible. The moon controls the weather every night, however, it is just not as obvious.

When the moon is in the sky i.e. above the horizon, it is less likely to rain. This is a Rule. Rain times are more common, if rain is about, when the moon is underfoot (IC), overhead (MH), or on the horizon (ASC=rising or DES=setting). The in-between positions i.e. every three hours, also have rain potential. Sir Isaac Newton identified that when the moon reached a 45° angle to the horizon, a gravitational force corridor was established, and gravity was lessened, i.e. the moon effect was enhanced. Just before it rains, there is a noticeable chill in the air. There are also extra wind gusts in the hours of moonrise and moon set. This was first noticed and recorded by the Greek philosopher Theophrastus in 300 BC. It still applies today. When the trees suddenly rustle for no reason, check out where the moon is.

Jupiter, Saturn, Uranus and Neptune, in that order, affect the Sun, and the sun and moon together have an effect on the oceans. Venus and Mercury have an effect on the tides, but not as much as sun and moon. Jupiter and Saturn mainly, are responsible for what we call the sunspot cycle. That is when the sun gets overactive and seems to put out visible blotches which some cultures like the Chinese have been watching and tabulating for hundreds of years, for they coincide with extra radiation levels. This in turn translates into increased thunderstorms, and also growth, agriculturally.

The sunspot cycle is about 12 years, which is slightly more than the Jupiter cycle of 11.8 years although this can vary between 9 and 14 years. There are variations in the positions of the other planets, and it's the oceans that get affected first. Because the moon is the tiny player in the big team, it gets affected by everything that the big team does, in the same way that a baby is affected by the moods of the whole family. Because the moon is the closest planet to us and so to the oceans, everything in terms of planet aspects involves the moon. There is the timing of the tides, the height of the tides, the speed of the tides, all factors in how the ocean behaves, and all that gets imparted to the weather, because all weather comes from the ocean.

That is not a known viewpoint, or if it is, it hasn't reached the ears of meteorologists yet, who still look to the sky for signs of change. That's why, in my opinion, they're wasting their time talking about cloud-seeding, weather modification, and geoengineering, those are all scams. Those are all things that people do with the sky using research funds made available by western governments, thinking by mistake that the weather originates there. But by the time a weather event reaches the sky, where we on the land see it, it's already half-over, it's obsolete, and just the final dregs of the system.

ORIGINS OF A STORM SYSTEM

All adverse weather comes from *under the boat*. I have never heard a sailor say, aw, a violent storm was going on above me, whilst we were sailing *on a flat, calm sea*. Usually, the flat sea accompanies a high-pressure system. Higher air pressure pushes the sea flat. It is as if your hand was the size of a bath and pushed down on the bath's surface. But in reality the flatter sea arrives first and allows more stable air to gather and consolidate over it, exerting more pressure.

It has been further observed that the deeper a low-pressure system is in the air, the faster the wind blows, imparting more energy, and the longer the wind blows over the same area, the bigger are the swells, which are the strong underwater currents. It begins in the sea. If currents increase in the deep of the ocean, then waves will increase at the surface and disturb the air over long distances, enough to generate those faster winds.

Strong winds will always accompany a swell. That's because the waves, which arise from the swell, cause the wind, The etymology says it all. Wind, weather and wing are words that come from the word wave, not the other way around. Fishermen have noticed that the direction in which the wind is blowing matters. The word west comes from wind as well. So does the word weather. When there is an onshore wind coming off the sea, what's really happening is that underwater waves are being whipped up, which disturbs marine worms out of their burrows, dislodge shellfish such as mussels, limpets and cockles from their home on rocks and force small fish, crabs and other forms of marine life out of weed beds. The currents change first. How could a surface wind possibly affect a seabed? Surface currents do not transmit downwards to the depths of the ocean.

To prove this theory, put some sand along the bottom of a bath and then splash along the surface, and see if the sand moves. Now place floating things, like corks, on the surface and generate underwater currents with your hand and see if the floating things are affected. Which moves more? There is no question that the energy transmission in the ocean is always upwards. It means that winds are more likely to be created by currents and swells than vice versa. It would be reasonable to conclude that a clear day indicates little wind BECAUSE the sea is calm, indicating deeper currents have lessened. Deep sea fish may have a sense of impending still tides and may plan to be there when small creatures that they prey on emerge from nooks and crannies.

An experienced fisherman can tell when there is a storm coming. The ocean 'speaks' to him and the sound of the swell reaches him first. Deep ocean currents are a fact, but science does not link them to winds. Seismic activity generates ocean currents, and earthquakes are energy waves up to 20,000 mph through solid rock. It is not likely that ocean currents go downwards to generate earthquakes, and besides we already know that some earthquake epicentres originate about 400 or 500kms below the seafloor.

It is more likely that the wave regimes that earthquakes are part of, transmit through rock to vibrate the ocean floor enough to generate the ocean currents, and on into the air via oscillating waves to generate unstable airflows. The fact that larger earthquakes happen around full moons and what we call perigees, together with stronger currents and kingtides, attest to that, also cyclones that form around these times. The moon and larger gravitational forces are responsible for all three. So, when it comes to what happens at sea and in the air above the sea, it is more feasible that ocean currents are in charge, rather than the surface winds.

As soon as the temperatures reverse, the winds reverse. Sailors who monitor Sea Surface Temperatures (SSTs) know this to be more reliable than isobaric maps, because sea-surface temperatures change before changes in air pressure, which gives warning of storms before they happen. Just by watching the temperature of the water, you can tell what the weather will do, in the next few hours or longer.

SUMMARY

A typical storm begins with the flat sea, the Sun shining above, heating the surface, accompanied by non-mixing of waters at various depths, relatively warm sailing conditions, and SSTs (sea surface temperatures) starting off high. The surface of the sea is warm because the layer of water constituting the surface is not being replaced by cooler water. Layers of water closer to the surface are the warmest, being closer to the warmth of the Sun. Then, as mixing begins below the surface, cool water is brought up from below, causing surface cooling. Further mixing brings more cooler waters up from the depths and there is an interchange with the warm waters close to the surface. Due to the drop in temperature at the surface, clouds may form as cooler air mixes with the remainder of warm air. Because warm air always moves to cold, the air becomes mobile and the wind picks up. There is a storm brewing. Warmer air continues to supply the area which has now considerably cooled, and the surface is described as windy, and then as a gale. The clouds increase and darken with the load of water that they gather. Condensation accompanies cooling, supplied by the movement of air. The surface of the sea begins to ripple strongly as the swell increases. Each wave has a convex and concave side. The convex has a higher pressure and the concave has a lower pressure. Waves are created, with higher pressure moving up each crest. The waves are joined together, and big areas of high pressure are juxtaposed with big areas of low pressure; and strong winds generate which go from high pressure to low pressure. The storm keeps up as long as it is fed by the waves and currents, which are the dynamics of the moon, interacting with gravitational factors that produce underwater earthquakes and faster swells, e.g. new moon, full moon, perigees, and when the moon crosses the equator which is called the lunar equinoxes.

The storm dissipates in the reverse order. Winds lessen and the sea surface temperature rises. Mixing below the surface ceases gradually. The Sun comes out as clouds seem to dissipate and the ocean surface flattens. It is all controlled by swells. Sailors know that to escape a storm, you sail into the incoming wind, which is into the whitecaps. They have known it for thousands of years, and there are ancient accounts in ancient texts, but they may not have realised why.

QUESTIONS FOR SCIENCE

If we ignore the moon then how could a huge celestial body between a third and a quarter the size of Earth, and only 10 earth circumferences away, not have any effect on earth, sea and the air? Astronomers have let us down over the years. They say that the moon only affects the tides, but nothing else. If it affects something like tides on the earth, then why does it not affect other things that are on the earth too, like humans? And how could all three layers (air, sea and land) not be interconnected - considering they completely and perfectly interface over most of the surface of the planet? How could it be that only one could get away with having a tide, but not the other two, when all three are joined? Aren't joined things part of a joined system that acts together? Science should answer all questions, for that is the essence of scientific pursuit.

CYCLONES / HEAVY RAIN

Tropical cyclones are formed when the sun within 6° latitude of the equator warms the waters that have to be 26°C -28°C, over a wide ocean area. And that produces massive evaporation. The heat from a full moon day, especially when the moon is close, in certain areas can create a tropical cyclone. When the full moon's heating capacity diminishes, then so will the cyclone. The huge accumulation of water in the sky, ready to fall and cause destruction, can only come from the sea.

That is why TC's immediately begin to fall apart, the minute that full moon moves away. And that's why tropical cyclones don't form south of Fiji. The water is just not warm enough That's why you never get tropical cyclones occurring in say, Brazil. Cyclones travel from either north or south depending on which hemisphere they are formed in. They will very rarely cross the equator. Heat from warmer water can significantly modify an air mass over 35-40 km.

Why is there a cyclone season? Cyclones, hurricanes, and typhoons predominate during equinox months, which are mainly when the atmosphere undergoes a twisting action, through the plane of what we call the ecliptic, due to the increase in speed of the Sun relative to the orbit of the earth. If you can picture the midpoint of a pendulum, that happens in both March and September. Furthermore, with spring and autumn full and new moons, king-tides occur when the moon rises at or near the equator. This represents the midpoint of yet another pendulum, the Sun on the equator being the first, and now the Moon on the equator for the tides. Sun and moon are moving faster relative to earth around equinoxes, so atmospheric disturbance increases.

2025 starts with a weak perigee near northern declination on 8 January, and the upswing of a potential cyclonic system forms around 20-24 January, bringing rain to the South Island West Coast and up through the North Island. Exactly a month on, another rain system may strike the North Island (9th and 20th) and the west and south of the South Island, but mid-February may experience a heavy rain washout for Northland. Around 21st, an ex-tropical cyclone is expected to move away from Fiji. In March there may be heavy rain and thunderstorms around 23rd affecting all areas, but more so south of Auckland, and bringing localized flooding to the Wellington region.

April sees the top half of the North Island battered by strong winds and rain on 8-9 April. This will be possibly the most rain to fall this year. The band of heavy rain is due to stall over Northland bringing risk of flooding to the Coromandel and Bay of Plenty, including Auckland, with eastern districts from Whangarei to the Bay of Islands at risk, and also flooding in the Rotorua district.

As can be seen, although there is intense rain activity, it falls short of being labelled tropically cyclonic, but may be destructive, nevertheless. The end of 2025 is expected to reveal El Nino conditions, which typically suppress TC activity during cyclonic seasons. Currents along the equatorial band, between Tahiti and Darwin stall, allowing for warmth to build up, as it did in 1998. That was an El Nino year, along with 1982/3, 1986/7 and 2016. 1997/8 is still regarded as one of the most powerful El Nino-Southern Oscillation events in recorded history.

There are twice as many weak La Nina years as there are weak El Nino years. But there are about equal moderate to strong La Nina and El Nino years. La Nina can be viewed as the planks on a deck, with the El Nino akin to the gaps between them. La Nina years are 2x wetter than El Nino years.

The El Niño/La Nina index has been neutral since 2021. It stayed close to the neutral range in 2022, 2023 and 2024. In 2025 the index edges up into the El Nino episode range.

RAINFALL AND DROUGHT CYCLES

We return now to what makes more rainfall and what makes less, over a long period. First we must go back to the cycles of the sun. With each passage of Jupiter, when Jupiter and Saturn sit on one side of the sun in conjunction, there is usually a major drought in the area of earth that they call, say, eastern Australia. Those years were 1901, 1920, 1940, 1961, 1981, 2020, and they are about 20 years apart. Halfway into this decadal period, Jupiter and Saturn face each other on opposite sides of the sun, and we call that an opposition, and those become flood years. Flood years have higher tides than drought years, and that leads to more rainfall which affect the rivers, and everything falls into place accordingly from the oceans to the farmland. We can note that the conjunctions i.e. drought years, were 1862, 1901, and every 12, 24, and about 36 years, so the next worst was 1959, and then around 2020 (but actually 2016-2023), we call them the **solar minimum** years, they were years of world-wide droughts because of low rainfall and low agricultural yields.

1997/8 was followed by 1999, also an El Nino year, and then a gradual return to cyclone seasons, The equivalent may occur in 2027. La Nina conditions are typified by a return of cyclonic activity. It just so happens that Jupiter and Saturn are on opposite sides of the sun, for example 1974, 1993 and 2012 which were especially wet in the east of Australia and NZ, now jump forward 18.6 years and the next lot of flood years may be around 2030, but around late 2026/7 we may begin to see significantly wetter years.

If the reader cares to look at the graph of what the Sun does in terms of peaks and troughs of the solar cycle, they will notice that they are about a decade apart, and the solar maximums are 2014 and then 2026 and then 2037. The solar minimums are between those, so 2020, and 2031, and so on, but droughts can be expected every 5 years, at both the solar maximum and solar minimum years, for different reasons (in solar maximum years there's too much heat coming off the ground and that stops the rain from falling until it gets over the cooler ocean, so you get drought below: and with the solar minimum years the sun is at a low radiating point and there's not enough evaporation going on). So effectively the rainiest times are on, or coming up to the solar maximum years, in other words after 2026.

Solar cycles determine annual temperatures and amounts of annual rain. Moon cycles determine timing of weather events. Cycle #23 began in 1998, peaked around 2000/1 (solar maximum), and died off (solar minimum) from 2007-2009. After a 2-3 year period of relative inactivity Cycle #24 began around midway through 2010, and peaked in 2014. It "bottomed-out" in 2020-21. 2020/2021 saw the end of Sunspot cycle #24. Cycle #25 began in 2022, peaks in 2026 and bottoms out again in 2032.

By combining lunar and solar influences, temperature (solar) may lead to amounts of rainfall (due to evaporation rates), whereas timing of severe weather events may follow strict lunar cycles.

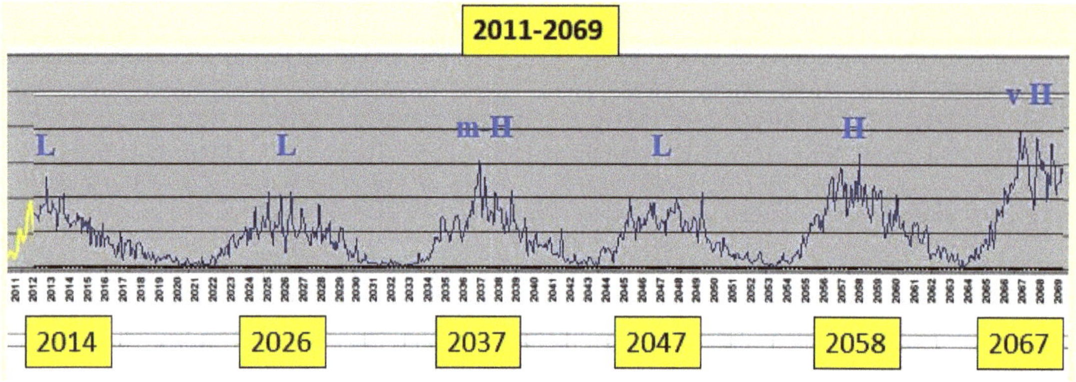

Contrary to global warming propaganda, globally warm periods have less and weaker hurricanes while cooler periods have more and stronger hurricanes. 2022 was such a year.

It is the Sun's heat that causes evaporation which brings rain. The amount of heat that the Sun puts out determines the amount of evaporation, which must then fall within about 7 days, near to the water source. Coastal locations are generally wetter than places that are some distance inland. Without a source of nearby water, a place quickly dries out. The *amount* of rain falling is due to the amount of solar radiation which causes evaporation. Therefore, rain amounts depend on the sunspot cycle. The amount of rain falling is also dependent on the extent of the air tide, which can be likened to volume of water in each sea tide. As the solar cycle of radiation is cyclic then so too are rain intensities.

Sea tide timing is predictable, subject to positions of Moon and Sun. The volume of water flowing into an estuary depends on wind direction subject to shape and size of surrounding hills, wind forces increased or confused by surface currents, and less visible factors like close proximity of underwater eruptions, fissures, earthquakes, and time of the day due to temperature gradients. Amounts of rain are influenced by Sun's warmth for evaporation and then how much cold air comes to the layers of the atmosphere containing clouds.

Sometimes colder air is at a higher or lower altitude than rain-filled clouds, and expected rains must wait for the marrying of the two factors. Until then a cloud may look like rain but heat from the ground is sufficient to buoy up the rain-laden cloud and prevent rain falling. Analyzing sunspot cycles is at least better than having nothing at all, by looking to see which previous years correspond to same points of the cycle.

THE ATMOSPHERE

Temperature, volume, and pressure became the focus of climate science a few centuries ago, when the Ideal Gas law was discovered. This states that the three factors are dependent, and if given a closed system, and played around with in the laboratory, then by changing one variable you can make others change as well. These principles were applied to the air in the atmosphere regarded as an enclosed gas. It is supposed to be enclosed because it is held here by gravity. This is where science broke down due to a fundamental misunderstanding.

This is not a laboratory, and the earth does not have a lid by which one can change one factor and watch the changes in the other two. In other words, the ideal gas law, which sounds very nice, cannot be applied to the atmosphere, which is open to space. Nor can it be shown to work in the laboratory without enclosing the gas and some container. Gravity is still working in the laboratory but is not doing much to the experiment other than showing that nothing works anymore when you remove the lid of your container. Therefore, we could arguably scrap the ideal gas law, as it applies to our environment, and the idea of temperature along with it.

What of volume and pressure? Well, except in isolated and controlled cases, we cannot watch the effects of air pressure on the air, because it is generated too locally close to the ocean surface to be of any use to the atmosphere as a whole. Besides, any wind is generated by currents in the sea that work their way to the surface, and create high and low-pressure areas, and wind between high and low pressure, which is only measured on the ocean's surface. Yet the surface of the sea is NOT where the air pressure is generated.

Which leaves us with volume. Science has not come up with a way of accurately quantifying the amount of air in the world, and therefore telling us that there is a danger of it running out. If every tree on the land was instantly destroyed, we would still have 4 million years' worth of oxygen in the atmosphere, because air contains 21% oxygen produced by photosynthesis on the biomass in the ocean. That means 99.9% of our vegetation, which produces oxygen for the survival of every species of life, operates mostly in the sea, and away from the air. That means CO2, or anything else introduced to the air, can have no effect.

The sun warms the earth, but the oceans cool it. Even on the hottest summer's day, a swim in the ocean will be refreshing because the water is always cooler than the air above it. What's more, it can never get hotter than the air above. Everything is in balance, and will remain so, and what we call the weather is what we perceive as the final outcome of planetary cycles that began thousands of years ago.

RAIN, SUN, AND TEMPERATURES 2024

RAIN NEW ZEALAND OVERVIEW

Most of New Zealand can expect around average rainfall for the year as a whole, with most regions that we monitor in the North Island within 10% of their normal precipitation, whilst many regions in the South Island will be well below average. Overall, July, August, and October will be wetter than normal while all other months will bring below average rainfall. Generally, winter and spring will see average to slightly more rainfall than normal, while summer and autumn will be drier than average.

NORTH ISLAND RAIN OVERVIEW

While Northland, the Greater Auckland region, Hawkes Bay and the Central Plateau regions can all expect rainfall amounts to average out at around 10% of yearly norm, Northland and Auckland will be slightly wetter overall, while the Hawkes Bay and Central Plateau slightly drier. The remaining North Island regions can expect up to 25-30% less rainfall than in a typical year. Winter will be the only season this year that will see above average rainfall, with the other seasons all being drier than normal. The Manawatu to the Lower North Island districts will see the lowest rainfall amounts this year, while Northland to the Central Plateau will have the highest rainfall.

The wettest periods of the year in the North Island are expected over the second half of July and second half of October, again during the last three weeks of August, as well as around the turn of the third and into the fourth weeks of January, third week of February, second week of May, second half of June, and around mid-September to the fourth week.

The driest periods of the year in the North Island are expected to be the first half of the second weeks of January, February, and June.

The months of July, August, and October will see above average rainfall, December average rainfall, and January, February, March, April, May, June, September, and November seeing below average rainfall. Overall, winter will see above average rainfall, while summer, autumn, and spring will be drier than average.

SOUTH ISLAND RAIN OVERVIEW

Overall, the South Island can expect around 18% less rainfall than normal as a whole, with all regions likely to experience below average rainfall in total, and only the Inland Lakes and Southland closer to their norms. All other regions can anticipate receiving only around 20-25% of their usual rainfall amounts. Autumn will be particularly drier than average, receiving the lowest amount of rainfall compared to the other seasons.

The wettest periods of the year in the South Island are expected over the October, from around the 10th to the end of the month, as well as around the turn of the first week into the second week in January and again over the fourth week of January, the second week of February, the last week of March, second and third weeks of May, June, July, and August, second week of September, and in December from around the 21st until the New Year.

The driest periods of the year in the South Island are expected to be during the over the third week of February, second week of March, fourth week of July, and first half of December.

The months of July and October will see above average rainfall, and all other months will see below average rainfall. Overall, all seasons will see below average rainfall with winter closest to the norm.

SUNSHINE HOURS

The North Island and South Island towns we monitor may both be sunnier than normal for the overall year but by only around 6% more on averageerage. Six months are expected to be sunnier than average for the country as a whole and six months with slightly less sunshine hours than normal for a year.

The North Island is expected to be the sunniest in November, followed in order of sunniest to cloudiest by February, March, January, April, December, October, September, May, August, June, then July with the least sunshine hours, with seven months with above average sunshine hours, one month with average sunshine hours, and four months below average sunshine.

In the North Island, the sunniest periods of the year are expected to be in January around the end of the second week and into the first half of the third week, over the second and third weeks of February, third week of March, first half of November and in the last two weeks of November.

The cloudiest periods of the year in the North Island can be expected during the second half of June, the third weeks of July and August, and in the week in the lead up to Christmas.

The monthly order of sunniest to least sunny in order for the South Island are November, December, March, February, September, April, October, January, May, June, August, and July with the least sunshine hours, with eight months of above average sunshine hours and four months with below average sunshine hours.

In the South Island, the sunniest periods of the year are expected over the second half of the second week of February until midway through the third week, over the first half of March, during the third week of September, and over the first three weeks of both November and December.

The cloudiest periods of the year in the South Island can be expected in the first week of May, first half of July, the second week of August and in the last few days before Christmas.

TEMPERATURES

Overall, maximum temperatures will be close to the average across the country as a whole based on the towns we monitor, with the North Island slightly warmer by an estimated 0.6°C above the average and the South Island slightly warmer but only by around 0.3°C. January, February, July, and October are all likely to be around 0.5-1°C cooler and January cooler by an estimated 1.8°C across the country as a whole. August, September, and November are all estimated to be average to slightly warmer, but within half a degree of the norm, and the remaining months of March, April, May, June, and December all warmer by an estimated 1-2°C above the average.

Temperature trends will be mostly the same across both the North and South Islands overall, except in July, when the North Island will be slightly warmer by around 0.6°C above the monthly average, and the South Island up to 1-1.5°C cooler than average, and in October, and in October, when the North Island will see maximum temperatures around the average, while the South Island will averge around 1-1.5°C below the average.

For the North Island, overall, the summer will be cooler than average by almost a degree compared to the norm, spring will come out average for the season, and autumn and winter will be warmer by around 1-1.5°C.

For the South Island, the summer season will overall be around 1-1.5°C cooler than average, while autumn sees a warmer than normal trend with overall maximum temperatures averaging around 1.5-2°C above the seasonal average, and winter and spring will both be close to the norm, with winter slightly milder than normal and spring slightly cooler but still within 0.5°C of the seasonal average.

The warmest period of the year will be late February to mid-March in the North Island, and the first half of March in the South Island. The second and third weeks of July will see the lowest maximum temperatures in the North Island, while the first, third, and fourth weeks of July will be have the lowest maximum temperature averages across the South Island.

REGIONS COMPARED TO AVERAGES

Northland, Auckland, Coromandel Peninsula, Bay of Plenty, Rotorua, Hamilton

January: Drier Dargaville to Waikato, wetter Northland, Bay of Plenty to Rotorua, cloudier Waikato to Bay of Plenty, average to sunnier elsewhere, cooler temperatures.

February: Wetter than average Northland, Tauranga, and Hamilton, drier elsewhere, average to sunnier, average to warmer temperatures Manukau Heads, and Bay of Plenty, average to cooler temperatures elsewhere.

March: Average to drier everywhere, significantly drier Northland, average to sunnier, average to warmer temperatures.

April: Drier Far North, Auckland, Waikato, Bay of Plenty, wetter elsewhere, average to sunnier, average to cooler temperatures Northland, average to warmer temperatures elsewhere.

May: Significantly drier than average, average to sunnier, warmer temperatures.

June: Significantly drier than average, sunnier than average, warmer temperatures.

July: Significantly wetter than average, average to cloudier, average to warmer temperatures.

August: Drier Northland except Kaitaia and Opotiki, wetter than average elsewhere, sunnier Auckland to Dargaville, average to cloudier elsewhere, average to warmer temperatures.

September: Wetter Far North, Whangarei to Warkworth, drier than average elsewhere, average-cloudier Waikato to Northland, sunnier-to-average Bay of Plenty to Rotorua, average to warmer temperatures Te Puke to Opotiki and Manukau Heads, average to cooler elsewhere.

October: Drier Northland, Tauranga to Whakatane, average to wetter elsewhere, average to sunnier Northland, average sunshine elsewhere, average to cooler temperatures Whakatane to Rotorua and Warkworth, average to warmer elsewhere.

November: Wetter Henderson, significantly drier elsewhere, average to cloudier Northland, sunnier elsewhere, average to warmer temperatures except cooler Warkworth.

December: Significantly wetter Northland and Coromandel, drier elsewhere, cloudier than average, cooler than average temperatures Warkworth, average to warmer temperatures elsewhere.

Western and Central North Island, not including Taupo

January: Wetter Wanganui, drier elsewhere, average to cloudier, cooler temperatures.

February: Drier than average, average to sunnier, average to warmer temperatures Chateau, average to cooler elsewhere.

March: Drier than average Stratford, wetter elsewhere, average to sunnier, warmer temperatures.

April: Drier than average, sunnier, average to warmer temperatures.

May: Drier than average, sunnier, warmer temperatures.

June: Average to drier rainfall, sunnier, average to warmer temperatures.

July: Wetter than average, cloudier, average to warmer temperatures.

August: Wetter than average, cloudier Wanganui, average sunshine elsewhere, average to warmer temperatures Taranaki, average to cooler temperatures Central Plateau.

September: Drier than average, cloudier Wanganui, average to sunnier elsewhere, average to cooler temperatures.

October: Drier New Plymouth, wetter elsewhere, significantly wetter Central Plateau, average to cloudier, average to warmer temperatures Taranaki, cooler temperatures Central Plateau.

November: Drier than average, average to sunnier, average to warmer temperatures Taranaki, average to cooler temperatures Central Plateau.

December: Wetter Waiouru, average to drier elsewhere, average sunshine, average temperatures Taumaranui, warmer temperatures elsewhere.

Taupo

January: Average rainfall and sunshine, cooler temperatures.
February: Drier than average, sunnier, average to cooler temperatures.
March: Wetter than average, sunnier, warmer temperatures.
April: Drier than average, sunnier, warmer temperatures.
May: Drier than average, sunnier, warmer temperatures.
June: Significantly drier than average, sunnier, warmer temperatures.
July: Average rain, cloudier, average to warmer temperatures.
August: Wetter than average, average sun, average temperatures.
September: Drier than average, sunnier, average temperatures.
October: Drier than average, average sunshine, average to cooler temperatures.
November: Significantly drier than average, sunnier, average to cooler temperatures.
December: Average rain, cloudier, warmer temperatures.

Gisborne/Hawkes Bay

January: Drier than average, average sunshine but cloudier Waipukurau, cooler temperatures.
February: Average rain Hastings, drier elsewhere, average sunshine, cooler temperatures.
March: Average rain Hastings, drier elsewhere, sunnier, warmer temperatures except cooler Waipukurau.
April: Drier than average, sunnier, warmer Gisborne, cooler Waipukurau, average temperatures elsewhere.
May: Significantly drier than average, sunnier, warmer temperatures.
June: Wetter than average, slightly sunnier, cooler temperatures Waipukurau, warmer elsewhere.
July: Significantly wetter than average, cloudier, slightly cooler Waipukura, warmer elsewhere.
August: Drier than average, average sunshine, average to cooler temperatures.
September: Wetter than average, sunnier, slightly warmer temperatures Waipukurau, cooler Napier, average elsewhere.
October: Drier Napier, wetter Gisborne, average elsewhere, average sunshine, average temperatures.
November: Drier than average, sunnier, warmer temperatures.
December: Wetter than average, average sun, average to cooler temperatures Waipukarau, warmer elsewhere.

Lower North Island

January: Drier than average, average to cloudier, average to cooler temperatures.
February: Wetter than average Palmerston North to Levin, drier elsewhere, average to sunnier, average to warmer temperatures Masterton to Wellington, average to cooler elsewhere.
March: Wetter Palmerston North to Levin, drier elsewhere, sunnier, warmer temperatures.
April: Significantly drier than average, sunnier, average to warmer temperatures.
May: Significantly drier than average, average to sunnier, warmer temperatures.
June: Drier than average, sunnier, warmer temperatures.
July: Average to wetter, cloudier, cooler temperatures Dannevirke, average-to warmer elsewhere.
August: Drier Masterton, average to wetter elsewhere, average to cloudier, average to warmer temperatures.
September: Drier than average, average to sunnier, average to warmer temperatures Dannevirke to Masterton, average to cooler elsewhere.
October: Average to wetter, average to cloudier, average to warmer temperatures Palmerston to Levin, average to cooler temperatures elsewhere.
November: Wetter Levin, drier elsewhere, average to sunnier, cooler temperatures Paraparaumu, average to warmer elsewhere.
December: Drier than average, sunnier Palmerston North, average to cloudier elsewhere, warmer.

Nelson and Marlborough

January: Drier Blenheim and Motueka, wetter elsewhere, cloudier, cooler temperatures.
February: Drier than average, average to cloudier, average temperatures except cooler Blenheim.
March: Drier than average, sunnier, warmer temperatures.
April: Significantly drier than average, average to sunnier, average to warmer temperatures.
May: Drier Motueka, average to wetter elsewhere, average to cloudier, warmer temperatures.
June: Wetter Nelson, drier elsewhere, sunnier, warmer temperatures.
July: Average to wetter, average to cloudier, warmer Nelson, average to cooler elsewhere.
August: Average to drier, average sunshine, average temperatures.
September: Drier than average, sunnier, average to warmer temperatures.
October: Wetter than average, average to cloudier, cooler temperatures.
November: Average to drier, sunnier, average to cooler Motueka to Takaka, warmer temperatures elsewhere.
December: Wetter Nelson, average to drier elsewhere, average to cloudier, warmer temperatures.

Canterbury

January: Drier Central Canterbury, wetter elsewhere, cloudier, cooler temperatures.
February: Wetter Twizel, drier elsewhere, average to cloudier, average to cooler temperatures.
March: Wetter Timaru to Twizel, drier elsewhere, sunnier, warmer temperatures.
April: Drier than average, sunnier, average to warmer temperatures.
May: Wetter Timaru to Twizel, drier elsewhere, average to sunnier, warmer temperatures.
June: Wetter Twizel, drier elsewhere, average to sunnier, warmer temperatures.
July: Drier than average North and South Canterbury, wetter elsewhere, average to cloudier, average to cooler temperatures.
August: Wetter Christchurch and Ashburton to Twizel, drier elsewhere, cloudier Timaru, average to sunnier elsewhere, warmer temperatures Hamner Springs, average to cooler temperatures elsewhere.
September: Wetter Twizel, drier elsewhere, sunnier, slightly cooler temperatures inland Lakes, average to warmer temperatures elsewhere.
October: Drier Rakaia to Methvan, wetter elsewhere, average Ashburton, average to cloudier elsewhere, average to cooler temperatures.
November: Drier than average, sunnier, average to warmer temperatures Christchurch to Timaru, average to cooler temperatures elsewhere.
December: Drier than average, average to sunnier, average to warmer temperatures.

Inland Otago

January: Wetter Ranfurly, drier elsewhere, cloudier, average to cooler temperatures.
February: Drier than average, sunnier, average temperatures.
March: Drier than average, average to sunnier, warmer temperatures.
April: Drier than average, sunnier, average to warmer temperatures.
May: Average to drier, sunnier, warmer temperatures.
June: Average rain Cromwell to Queenstown, drier elsewhere, average sunshine Queenstown, sunnier elsewhere, average to warmer temperatures.
July: Drier Queenstown, wetter elsewhere, average to cloudier, cooler temperatures.
August: Drier Cromwell and Ranfurly, wetter elsewhere, cloudier, average to cooler temperatures.
September: Drier than average, sunnier, average to warmer temperatures.
October: Wetter than average, average to cloudier, average to cooler temperatures.
November: Drier than average, sunnier, cooler temperatures Cromwell, average to warmer temperatures elsewhere.
December: Wetter Ranfurly, drier elsewhere, sunnier, warmer temperatures.

Coastal Otago, Dunedin

January: Wetter Balclutha, drier elsewhere, cloudier, cooler temperatures.
February: Drier than average, average to sunnier, cooler temperatures.
March: Drier than average, sunnier, warmer temperatures.
April: Drier than average, sunnier, average to warmer temperatures.
May: Average to drier, sunnier, warmer temperatures.
June: Average to drier, average to sunnier, warmer temperatures.
July: Drier Dunedin, wetter elsewhere, sunnier Dunedin, cloudier elsewhere, average to cooler.
August: Wetter than average, average to cloudier, average to cooler temperartures.
September: Drier than average, sunnier, average to warmer temperatures.
October: Average to wetter, average sunshine to cloudier, average to cooler temperatures.
November: Drier than average, average to sunnier, cooler temperatures Balclutha, warmer elsewhere.
December: Drier than average, sunnier, average to warmer temperatures.

Southland

January: Average rainfall to drier, cloudier, cooler temperatures.
February: Drier than average, sunnier Invercargill, cloudier elsewhere, average to slightly cooler.
March: Wetter than average, average to sunnier, average to warmer temperatures.
April: Drier than average, average to sunnier, average to warmer temperatures.
May: Drier than average, sunnier, warmer temperatures.
June: Average to wetter, average to sunnier, warmer temperatures.
July: Wetter, sunnier, average to cooler temperatures.
August: Wetter Invercargill, drier elsewhere, cloudier, average to slightly warmer temperatures.
September: Average rainfall to wetter, sunnier, average to warmer temperatures.
October: Average to wetter, average sunshine, average to cooler temperatures.
November: Average rainfall to drier, average to sunnier, average to cooler temperatures.
December: Drier, sunnier, warmer temperatures.

Buller, West Coast, Fiordland

January: Average to drier, cloudier, average to warmer temperatures West Coast, average to cooler temperatures Southland.
February: Wetter Westport and Fiordland, drier elsewhere, average to sunnier, average to warmer temperatures West Coast, average to cooler temperatures Southland.
March: Drier Bruce Bay, wetter elsewhere, sunnier, average to warmer temperatures.
April: Average rainfall Fiordland, significantly drier elsewhere, sunnier, average to warmer.
May: Wetter Fiordland, average rainfall to drier elsewhere, average sunshine, warmer temperatures.
June: Wetter Fiordland, average to drier elsewhere, cloudier Westport, sunnier elsewhere, average to warmer temperatures.
July: Drier Fiordland to Bruce Bay, wetter elsewhere, cloudier Westport, sunnier elsewhere, average to warmer temperatures West Coast, cool temperatures elsewhere.
August: Wetter Fiordland, average to drier elsewhere, cloudier, warmer temperatures West Coast, average temperatures elsewhere.
September: Wetter Fiordland, drier elsewhere, average to sunnier, average to warmer temperatures Southland, average to cooler temperatures elsewhere.
October: Drier Bruce Bay, wetter elsewhere, average to cloudier, warmer West Coast, cooler Southland to Fiordland.
November: Wetter than average Fiordland, drier elsewhere, sunnier, average temperatures West Coast, average to cooler elsewhere.
December: Wetter Hokitika, average to drier elsewhere, average sun Westport, slightly sunnier elsewhere, average to warmer temperatures.

REGIONAL RAIN SUMMARIES

The tables below show, from wettest to driest months, estimated monthly departures from locations' long-term averages. e.g. Northland is estimated to receive 19% more than average in January therefore receiving estimated 119% of the January average (100-77). All wetter-than-average months shown below in bold.

NORTHLAND		AUCKLAND		BAY OF PLENTY		WAIKATO		ROTORUA/TAUPO	
JUL	**119%**	JUL	**87%**	JUL	**57%**	AUG	**72%**	AUG	**30%**
APR	**96%**	AUG	**64%**	JAN	**22%**	OCT	**22%**	JUL	**26%**
DEC	**87%**	OCT	**55%**	AUG	**5%**	JUL	**18%**	JAN	**9%**
FEB	**24%**	APR	**51%**	OCT	-6%	FEB	-6%	DEC	-9%
SEP	**6%**	DEC	**12%**	APR	-8%	MAR	-16%	APR	-10%
JAN	**2%**	JAN	-17%	DEC	-27%	JAN	-16%	OCT	-14%
AUG	-1%	NOV	-20%	MAY	-28%	NOV	-33%	MAR	-19%
OCT	-25%	SEP	-23%	SEP	-44%	MAY	-34%	FEB	-49%
NOV	-40%	MAR	-25%	FEB	-45%	DEC	-39%	MAY	-50%
JUN	-54%	MAY	-40%	MAR	-56%	SEP	-39%	SEP	-55%
MAR	-66%	FEB	-48%	JUN	-70%	APR	-44%	JUN	-65%
MAY	-76%	JUN	-53%	NOV	-86%	JUN	-55%	NOV	-80%
YEAR	**6%**	YEAR	**7%**	YEAR	-24%	YEAR	-13%	YEAR	-24%

GIS / HAWKES BAY		CENTRAL PLATEAU		TARANAKI		MANAWATU		LOWER NI	
JUL	**104%**	OCT	**68%**	AUG	**45%**	OCT	**37%**	OCT	**73%**
DEC	**102%**	AUG	**61%**	JUL	**31%**	JUL	**17%**	AUG	**5%**
SEP	**59%**	MAR	**46%**	OCT	**17%**	AUG	**13%**	JUL	**0%**
OCT	**24%**	JUL	**26%**	JAN	-19%	MAR	-14%	JAN	-30%
JUN	**22%**	DEC	-11%	MAR	-19%	FEB	-14%	MAR	-37%
JAN	-29%	JUN	-12%	JUN	-23%	NOV	-30%	FEB	-44%
AUG	-31%	NOV	-30%	DEC	-41%	DEC	-39%	DEC	-52%
FEB	-34%	FEB	-31%	FEB	-47%	JAN	-40%	NOV	-52%
APR	-54%	JAN	-40%	MAY	-55%	SEP	-53%	MAY	-56%
MAR	-56%	SEP	-41%	SEP	-64%	JUN	-56%	APR	-60%
NOV	-66%	MAY	-55%	NOV	-65%	APR	-60%	JUN	-64%
MAY	-82%	APR	-60%	APR	-68%	MAY	-65%	SEP	-70%
YEAR	-5%	YEAR	-5%	YEAR	-23%	YEAR	-25%	YEAR	-31%

SI WEST COAST		TOP OF SI		CANTERBURY		STH CANTERBURY		COASTAL OTAGO	
OCT	**32%**	OCT	**77%**	OCT	**67%**	JAN	**31%**	AUG	**78%**
JUL	**19%**	JUL	**12%**	JAN	-7%	OCT	**4%**	JUL	**33%**
MAR	-2%	JAN	**4%**	JUL	-7%	AUG	**1%**	OCT	**12%**
JUN	-13%	MAY	**1%**	DEC	-29%	MAY	**0%**	JAN	-1%
DEC	-17%	DEC	-9%	MAY	-37%	JUL	-10%	MAR	-32%
JAN	-19%	AUG	-19%	AUG	-41%	MAR	-21%	JUN	-38%
MAY	-20%	JUN	-24%	APR	-44%	DEC	-40%	APR	-41%
AUG	-21%	FEB	-53%	FEB	-45%	SEP	-51%	DEC	-43%
FEB	-35%	SEP	-72%	NOV	-53%	NOV	-53%	NOV	-48%
SEP	-47%	MAR	-72%	MAR	-59%	APR	-53%	FEB	-55%
NOV	-57%	APR	-74%	SEP	-64%	FEB	-63%	SEP	-57%
APR	-57%	NOV	-89%	JUN	-70%	JUN	-67%	MAY	-57%
YEAR	-19%	YEAR	-23%	YEAR	-32%	YEAR	-26%	YEAR	-22%

SI LAKES	
OCT	**152%**
JUL	**45%**
AUG	**6%**
JAN	-1%
MAY	-11%
MAR	-16%
JUN	-21%
SEP	-24%
FEB	-30%
NOV	-45%
APR	-64%
DEC	-69%
YEAR	-4%

INLAND OTAGO	
OCT	**61%**
JUL	**30%**
AUG	**2%**
JAN	-13%
JUN	-18%
MAY	-23%
MAR	-26%
SEP	-40%
DEC	-42%
NOV	-58%
FEB	-61%
APR	-68%
YEAR	-22%

SOUTHLAND	
AUG	**43%**
MAR	**33%**
OCT	**32%**
JUL	**5%**
JUN	**2%**
SEP	**1%**
NOV	-5%
FEB	-10%
MAY	-14%
APR	-27%
JAN	-31%
DEC	-36%
YEAR	-2%

NORTHLAND	
Winter	**22%**
Summer	-5%
Autumn	-16%
Spring	-18%
Overall	**6%**

AUCKLAND	
Winter	**32%**
Spring	**3%**
Autumn	-4%
Summer	-28%
Overall	**7%**

BAY OF PLENTY	
Winter	-5%
Summer	-18%
Autumn	-31%
Spring	-44%
Overall	-24%

WAIKATO	
Winter	**11%**
Summer	-13%
Spring	-16%
Autumn	-32%
Overall	-13%

ROTORUA/TAUPO	
Winter	-5%
Summer	-14%
Autumn	-29%
Spring	-48%
Overall	-24%

GIS / HAWKES BAY	
Winter	**36%**
Spring	**7%**
Summer	-20%
Autumn	-66%
Overall	-5%

CENTRAL PLATEAU	
Winter	**25%**
Spring	**0%**
Summer	-18%
Autumn	-28%
Overall	-5%

TARANAKI	
Winter	**17%**
Summer	-25%
Spring	-34%
Autumn	-49%
Overall	-23%

MANAWATU	
Winter	-9%
Spring	-13%
Summer	-19%
Autumn	-48%
Overall	-25%

LOWER NI	
Spring	-12%
Winter	-20%
Summer	-21%
Autumn	-52%
Overall	-31%

WEST COAST	
Winter	-5%
Spring	-21%
Autumn	-26%
Summer	-26%
Overall	-19%

TOP OF SI	
Winter	-11%
Spring	-21%
Summer	-30%
Autumn	-45%
Overall	-23%

CANTERBUR	
Summer	-1%
Spring	-14%
Winter	-38%
Autumn	-47%
Overall	-32%

STH CANTERBURY	
Summer	**8%**
Winter	-24%
Autumn	-25%
Spring	-31%
Overall	-26%

COASTAL OTAGO	
Winter	**21%**
Summer	-14%
Spring	-29%
Autumn	-44%
Overall	-22%

SI LAKES	
Spring	**34%**
Summer	**15%**
Winter	**11%**
Autumn	-30%
Overall	-4%

INLAND OTAGO	
Winter	**4%**
Spring	-8%
Summer	-13%
Autumn	-39%
Overall	-22%

SOUTHLAND	
Winter	**15%**
Spring	**10%**
Autumn	-3%
Summer	-5%
Overall	-2%

NI OVERALL		SI OVERALL		NZ OVERALL	
JUL	59%	OCT	47%	OCT	198%
AUG	27%	JUL	13%	JUL	43%
OCT	23%	AUG	-4%	AUG	15%
DEC	0%	JAN	-11%	JAN	-13%
APR	-1%	MAR	-16%	DEC	-13%
JAN	-14%	MAY	-18%	APR	-23%
FEB	-23%	JUN	-25%	MAR	-24%
SEP	-31%	DEC	-29%	FEB	-30%
MAR	-31%	FEB	-40%	SEP	-36%
NOV	-47%	SEP	-44%	JUN	-39%
JUN	-47%	NOV	-51%	MAY	-40%
MAY	-55%	APR	-52%	NOV	-49%
YEAR	-1%	YEAR	-18%	YEAR	-8%

NI OVERALL		SI OVERALL		NZ OVERALL	
Winter	13%	Winter	-5%	Winter	6%
Spring	-17%	Spring	-13%	Spring	5%
Summer	-17%	Summer	-14%	Summer	-16%
Autumn	-31%	Autumn	-28%	Autumn	-30%
Overall	-1%	Overall	-18%	Overall	-8%

WEATHER GLOSSARY

To get the most out of the forecasts, it helps to be able to read a weather map and to understand the language used. A high air pressure – a fair-weather sign, is when the isobaric values increase inwards towards the center. A low-pressure zone or depression, indicates worsening conditions, and when isobaric values decrease towards the center. Closer isobars = stronger winds.

TERM	MEANING	WEATHER
Apogee	Moon furthest from Earth for month	Winds generally light. Calm conditions.
Perigee	Moon closest to Earth for month	Spells trouble Earthquake risk in some countries.
Cold front	Cold air advancing, pushing away warm air	Changeable
Depression	Area of low barometric pressure	Unpredictable, bad weather, strong winds
Ridge of high pressure	Edge of anticyclone	Good weather approaching, cooler nights in winter
Stationary front	Slow-moving weather	No change for 2-3 days
Northern declination	Moon at northernmost point for month. Moon rises in the north.	Slow moving weather system, northerlies, rain in north, colder conditions
Southern declination	Moon at southernmost point for month	Slow moving system, southerlies, precipitation in south, generally milder.
Moon crossing equator	Faster moving weather system, moonrise is due east.	Turbulent, changeable, often electrical storms, gales or winds expected.

Thanks

The almanacs are not possible without the considerable help and support of Glenys and Victoria Dreyer in our central North Island office. My huge thanks, also to Vicki in dispatch. We also acknowledge the NOAA for ongoing permission to adapt archival MSL maps, and we are grateful to NIWA for historical archival data.

We hope that the information provided here will help in your planning.

Ken Ring

© Ken Ring 2024

GRAPHS

The graphs show estimations (light bars) and their deviations from the approximately 30-year long term average (dark bars).

RAINFALL ESTIMATES 2025

The graph below shows estimated total rainfall amounts for 2025 for the North Island towns we monitor, as compared to their average.

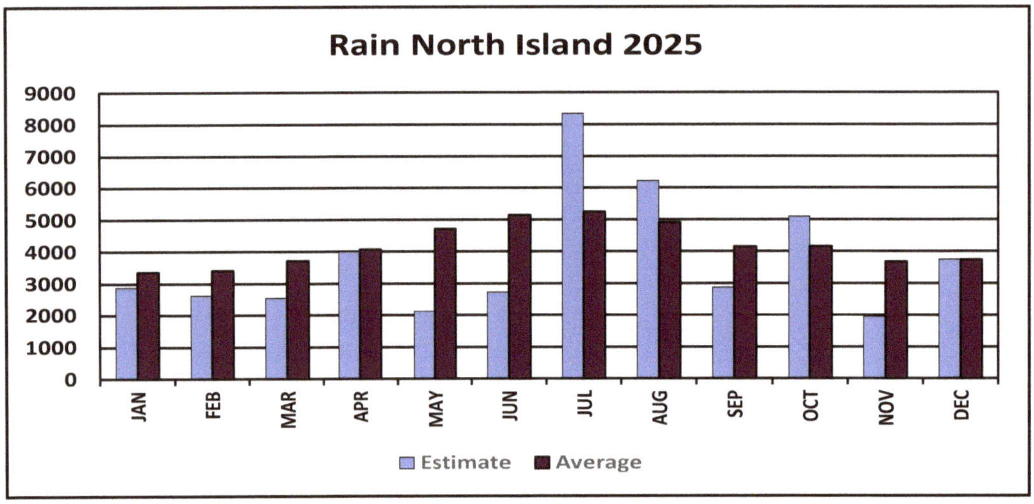

Overall, each season except for winter, is expected to bring average to below average rainfall overall this year, with winter averaging around 15% wetter than normal, while summer and spring will be drier by a similar variant, but autumn likely to only see around two-thirds of the usual amount of rain for that time of year. Despite this, July is likely to see up to 60% more rainfall than is typical for that time of year, August around a third wetter, and October around 25% wetter. Only April will have average rainfall, and all other months will be drier than normal, with the months of May, June, and November all seeing serious rainfall decifits. Overall, the North Island will balance out with its rainfall expections, with close to the norm amounts for the year.

The graph below shows estimated total rainfall amounts for 2025 for the South Island towns we monitor, as compared to their average.

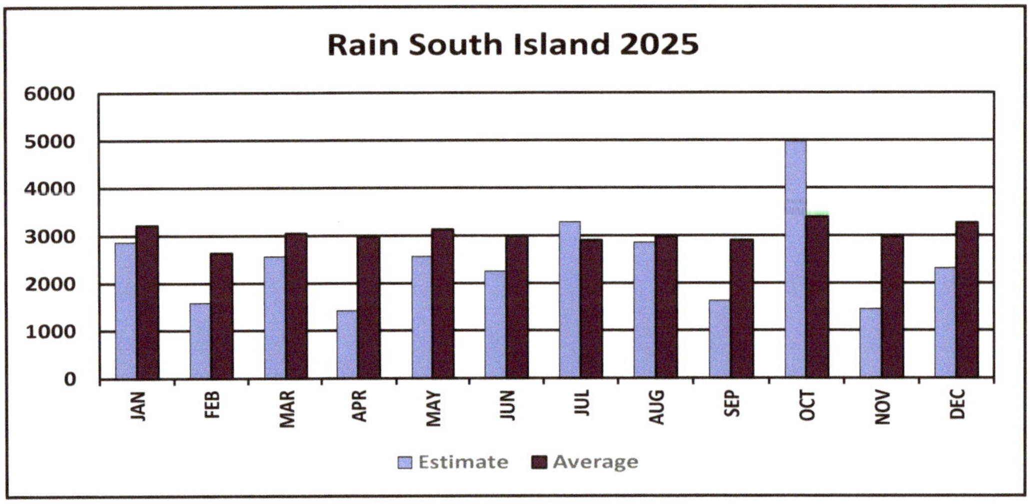

Overall, each season is expected to bring average to below average rainfall overall this year, with only the months of July and October likely to be wetter than average. All other months will see below average rainfall, with autumn up to a third drier than average, spring and summer around 15% drier than average, and only winter closer to the norm overall, but still slightly below average at around 5% drier overall. By the end of 2024, the South Island will be drier overall with only around 80% of their usual rainfall likely.

The graph below shows estimated total rainfall amounts for 2025 for both islands, as compared to their average.

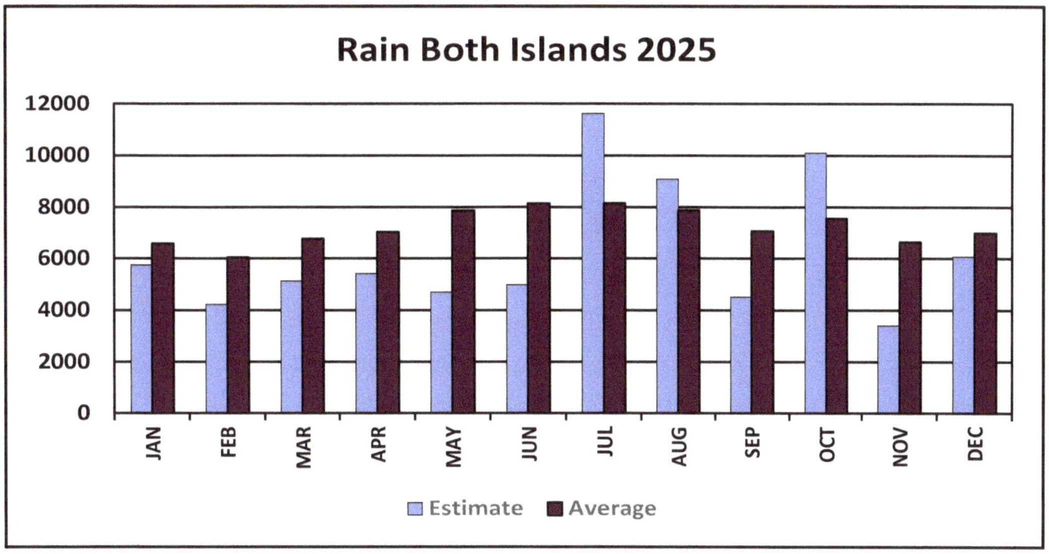

Overall, winter and spring are all expected to see around average rainfall for the country as a whole, with summer and autumn both drier than normal, with the overall expectation of around 8% less rainfall than normal for the year as a whole. The wettest months will be July, August, and October, and all other months will see below average rainfall.

HYDROLAKE LEVELS, 2025

Overall for 2025 the southern hydrolakes (Lake Pukaki and Tekapo) may receive the following:

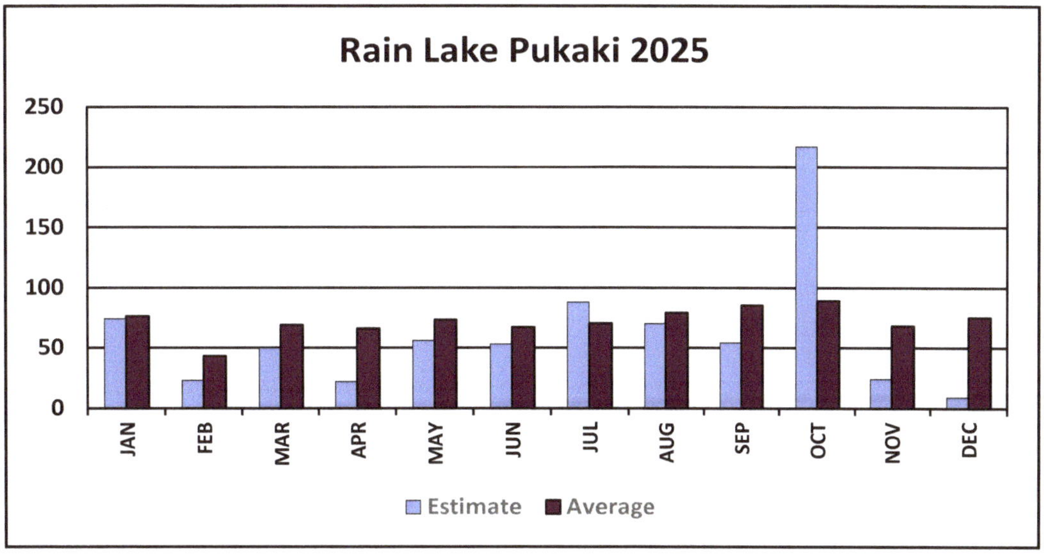

Lake Pukaki is estimated to receive around 86% of its yearly average rainfall, which is around 14% below the norm.

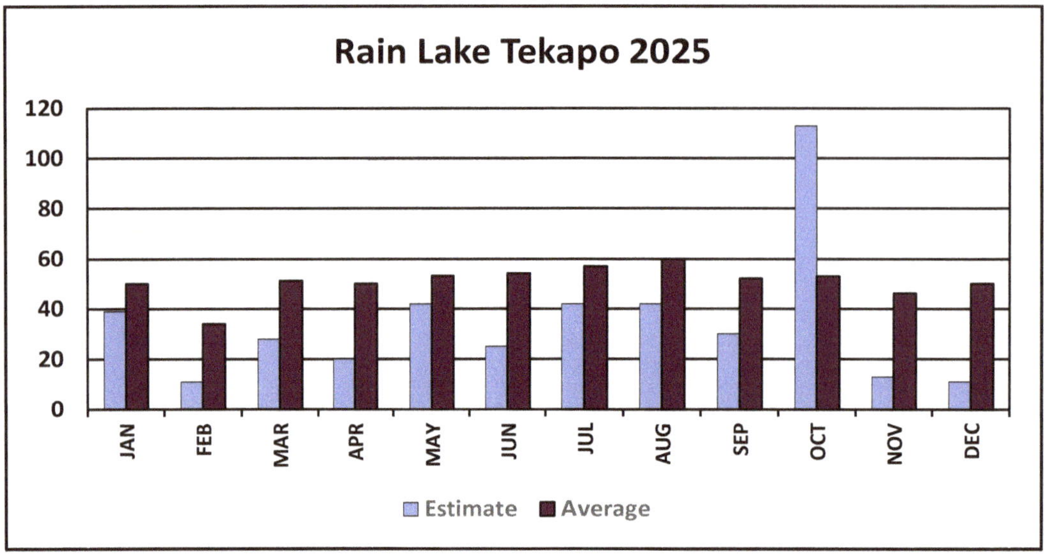

Lake Tekapo overall is estimated to receive around 32% less rain than average, which is well below the norm.

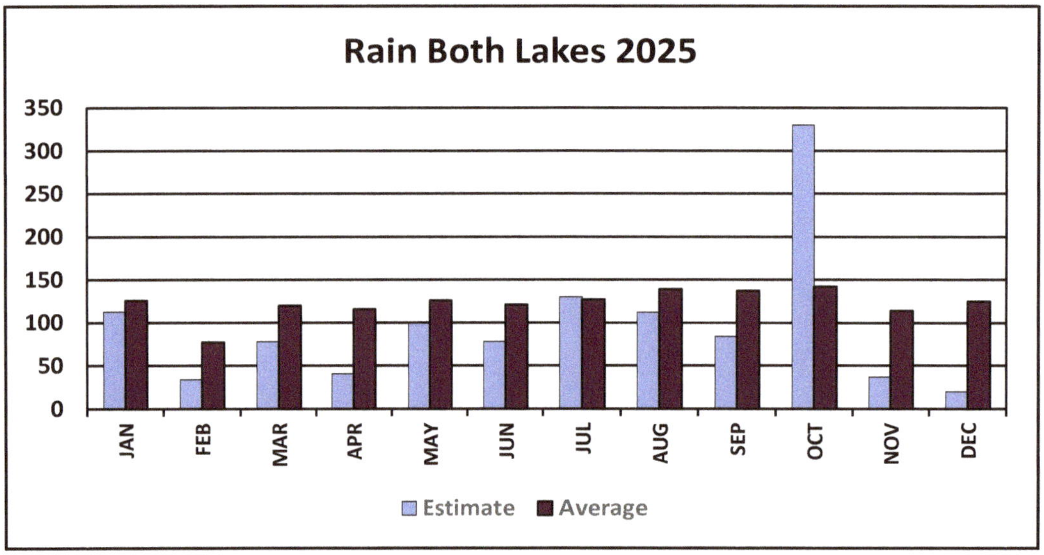

Regarded together overall both lakes are estimated to receive around 21% less rainfall than average for the year as a whole.

SUNSHINE ESTIMATES 2025

The graph below shows estimated sunshine hours for 2025 for the North Island towns we monitor, as compared to their average.

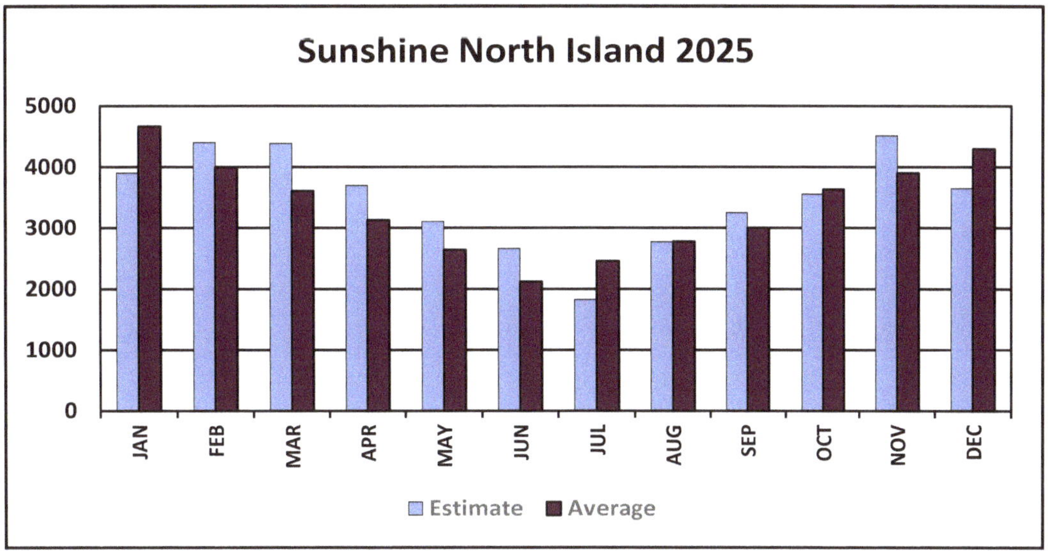

Overall this represents around 4% more sunshine than average, in other words in the range of an average year for the North island.

The graph below shows estimated sunshine hours for 2025 for the South Island towns we monitor, as compared to their average.

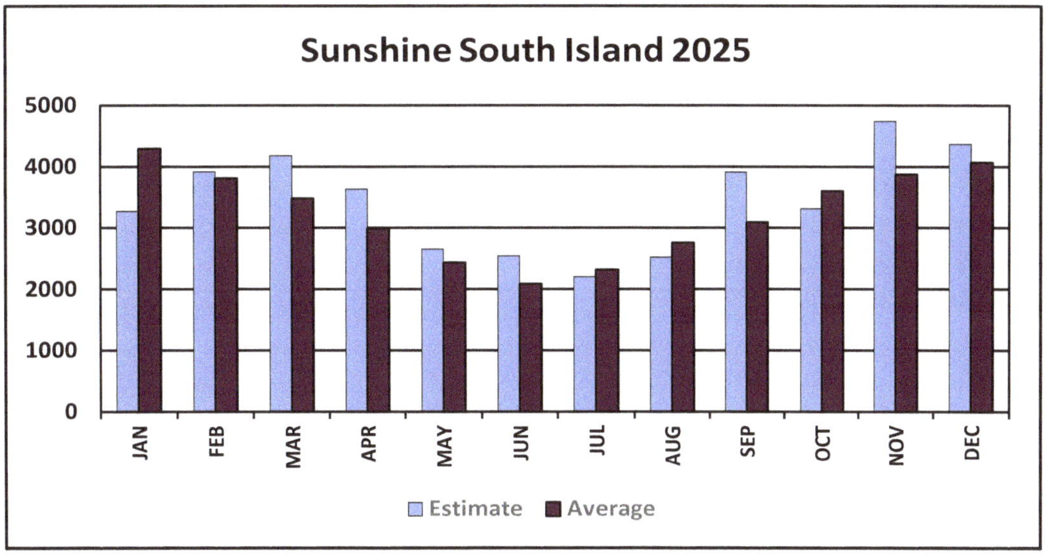

Overall this represents 6% more sunshine than average, in other words in the range of an average year for the South island.

The graph below shows estimated sunshine hours for 2025 for the combined islands' towns we monitor, as compared to their average.

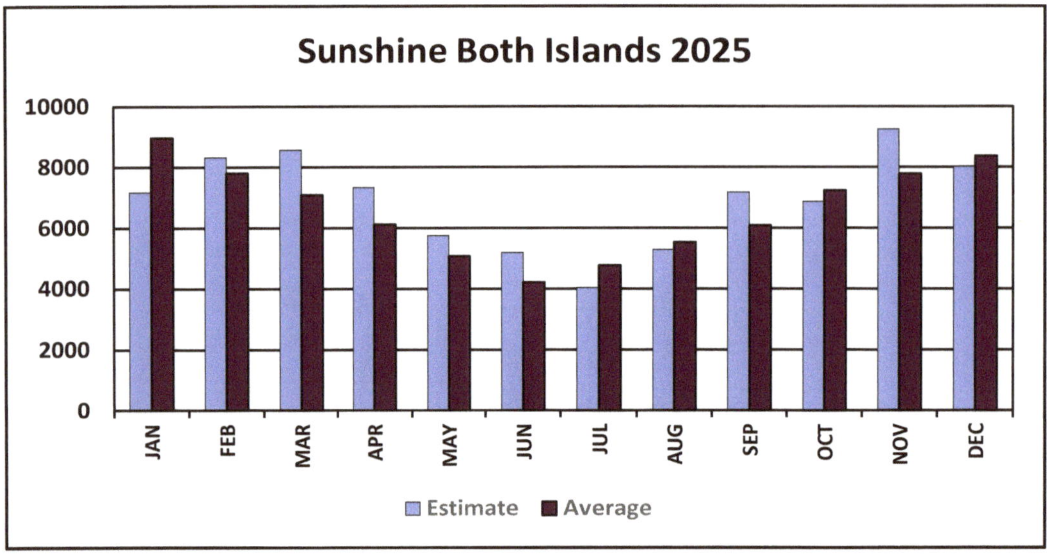

Overall this represents around 5% more sunshine hours than the average, in other words in the range of an average year for the country as a whole.

TEMPERATURE ESTIMATES 2025

The graph below shows estimated maximum temperatures for 2025 for the North Island towns we monitor as compared to their average.

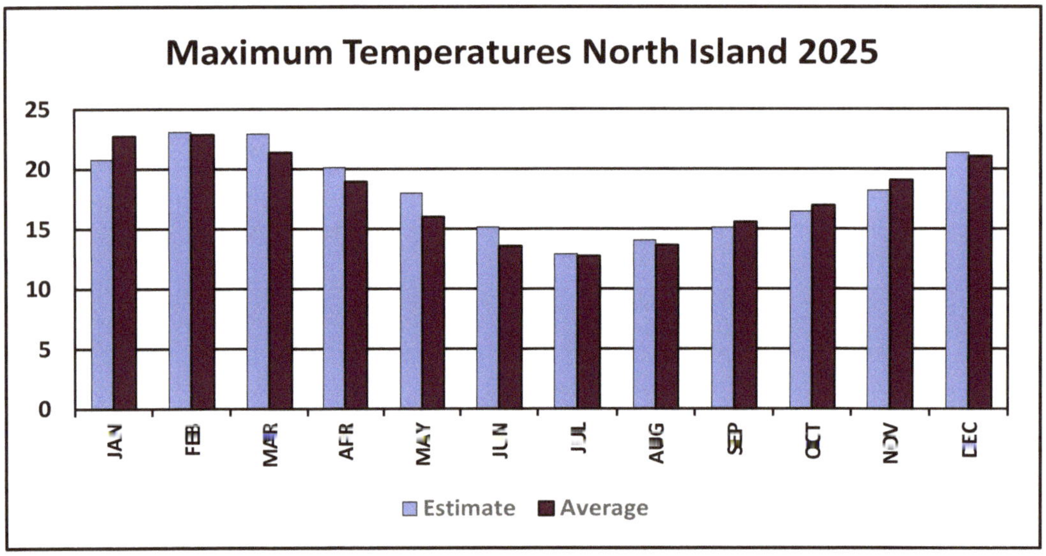

Overall 2025 is expected to be slightly warmer than normal across the North Island as a whole, but still within a half a degree of the norm, with the autumn season warmer than normal compared to the other seasons by up to 1-2°C more above the norm, while January will be much cooler than normal by a similar variant.

The graph below shows estimated maximum temperatures for 2025 for the South Island towns we monitor as compared to their average.

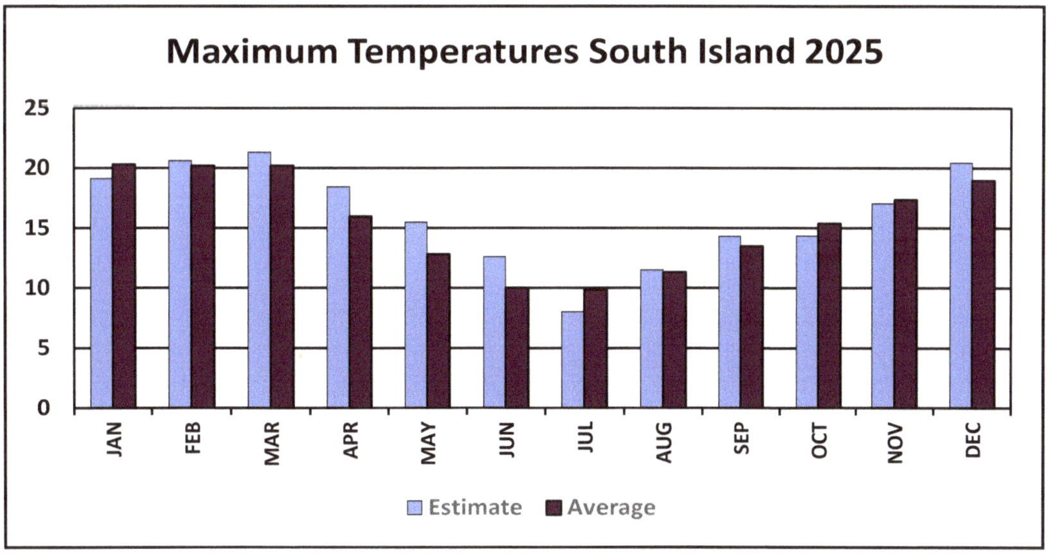

Overall 2025 is expected to be slightly warmer than normal across the South Island as a whole by just over a half a degree of the norm, with autumn up to 2-3°C warmer than the norm, while January, July and October all cooler by around 1-2°C below their average.

The graph below shows estimated maximum temperatures for 2025 for the combined islands towns we monitor as compared to their average.

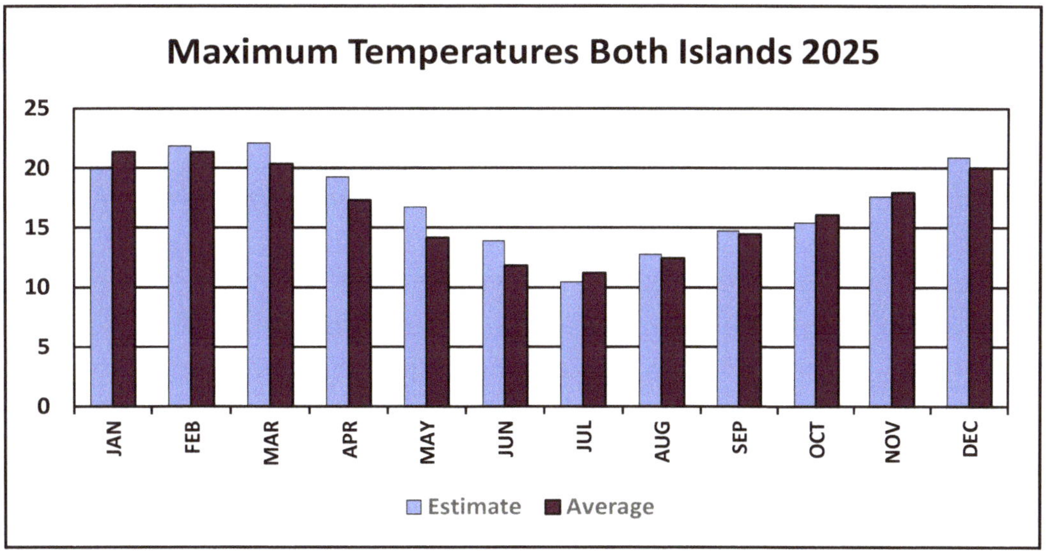

By the end of 2025, overall maximum temperatures will be slightly above the average but still within 1°C of the norm, averaging around 0.6°C warmer than normal across the country as a whole. The greatest variations for temperatures will be in January, when temperatures will be up to 1-1.5°C cooler than average, and in the autumn months and June up to 2-2.5°C warmer than average.

The graph below shows estimated minimum temperatures for 2025 for the North Island towns we monitor as compared to their average.

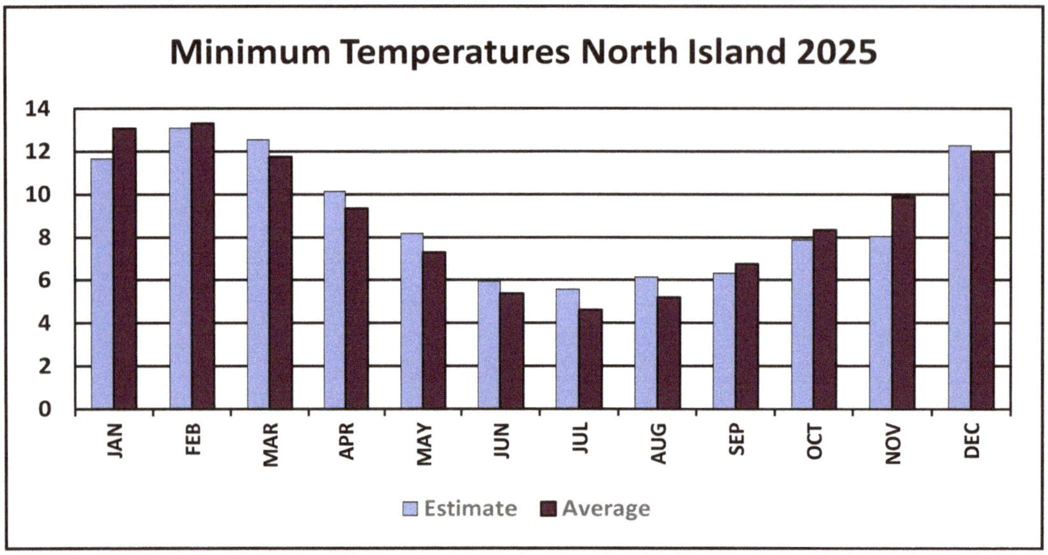

Overall, minimum temperatures across the North Island as a whole will be slightly warmer for 2025 but still within half a degree of the norm, with January and November around 1.5-2° cooler than average, and the autumn months, July and August all slightly warmer by around 0.5-1C above the average.

The graph below shows estimated minimum temperatures for 2025 for the South Island towns we monitor as compared to their average.

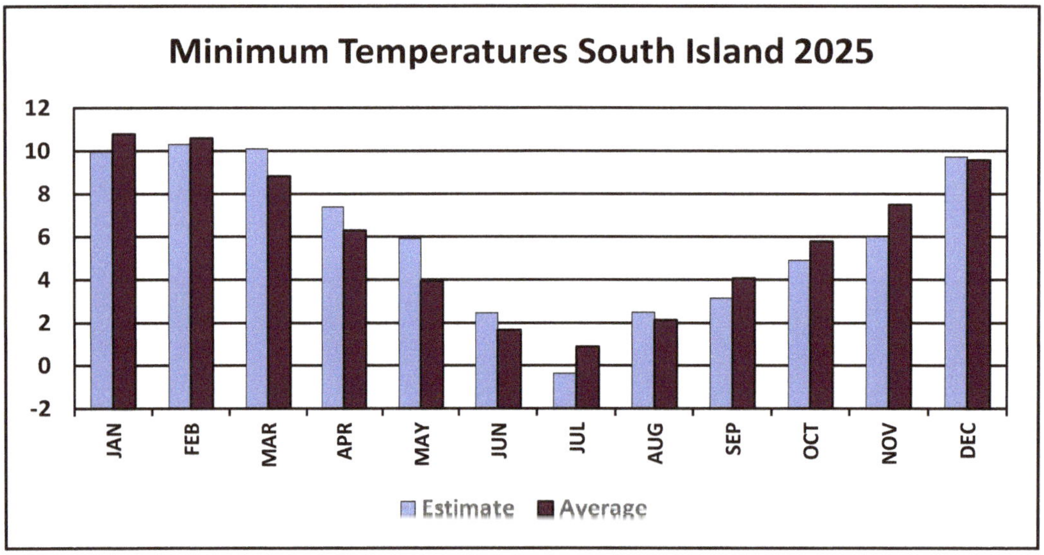

Overall, minimum temperatures across the South Island as a whole will be average for 2025, with the autumn months around 1-2°C warmer than average, and the spring months and July around 1-1.5°C cooler than average.

The graph below shows estimated minimum temperatures for 2025 for the combined islands towns we monitor as compared to their average.

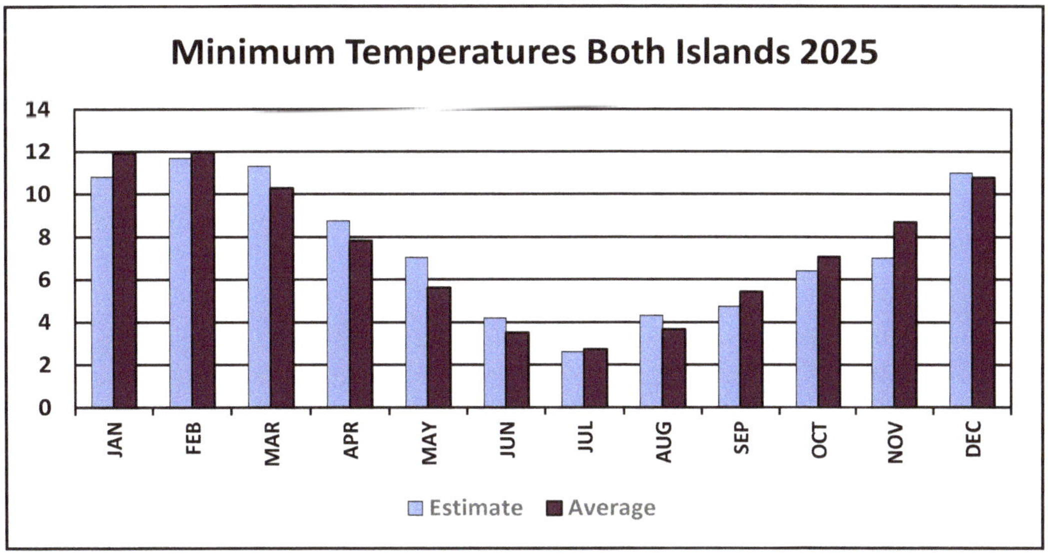

Overall, minimum temperatures across the country as a whole will be average for 2025. The months with the biggest variations from the norm will be January, September, October, and November all between 1-2°C below the average, while the months of March, April, May, June, and August all around 0.5-1°C above the norm for the year. All other months will have minimum temperatures closer to their average or within half a degree of the monthly norm.

MINIMUM TEMPS, RAIN & SNOW

The following graphs show estimated daily minimum temperature trends plus estimated rain. This is so that frosts and snow times may be better predicted. The reason is the estimation of potential below average temperatures and the necessity for farmers to protect young stock or vines at crucial times. The blue line represents minimum temperatures, with the scale in °C on the right. The black lines are estimated rain times and amounts are displayed in the left index. Snow is formed in temperatures 0-2°C. However whether it may be snow, flurries, hail, cold rain, sleet, or frost is often subject to conditions on the day and local factors.

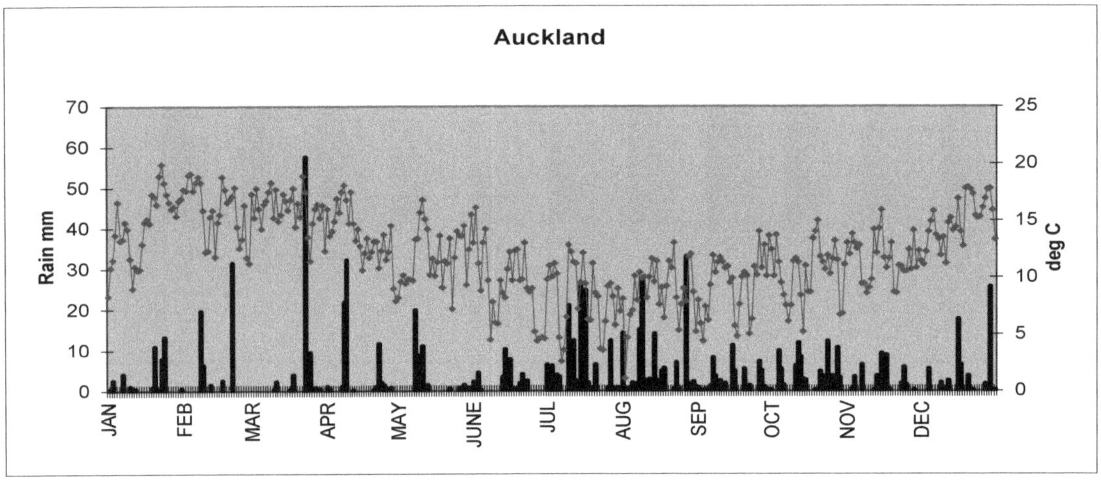

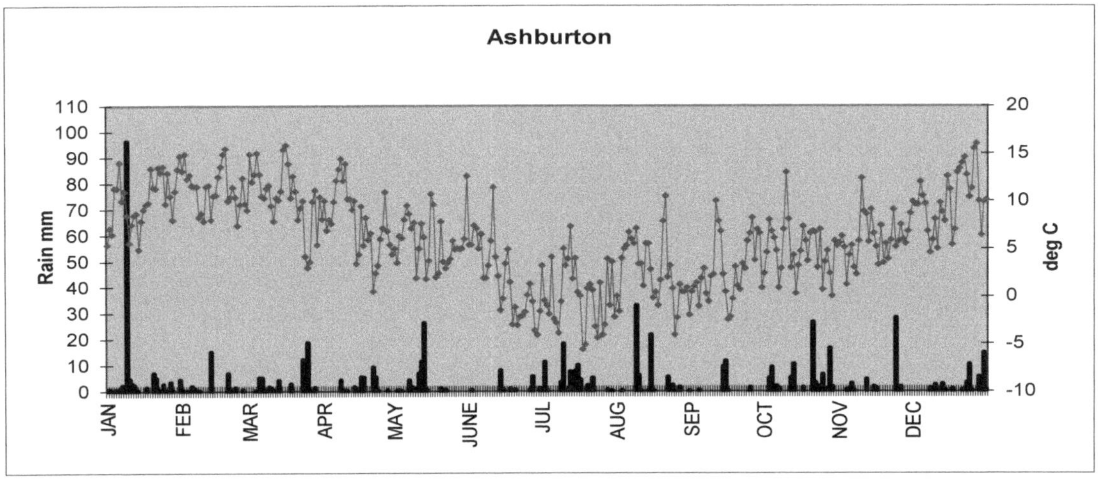

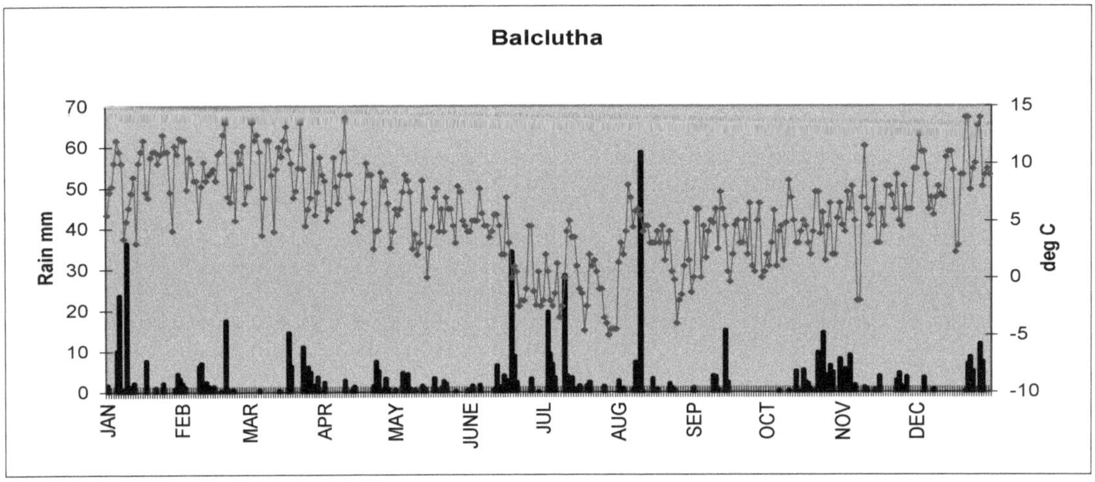

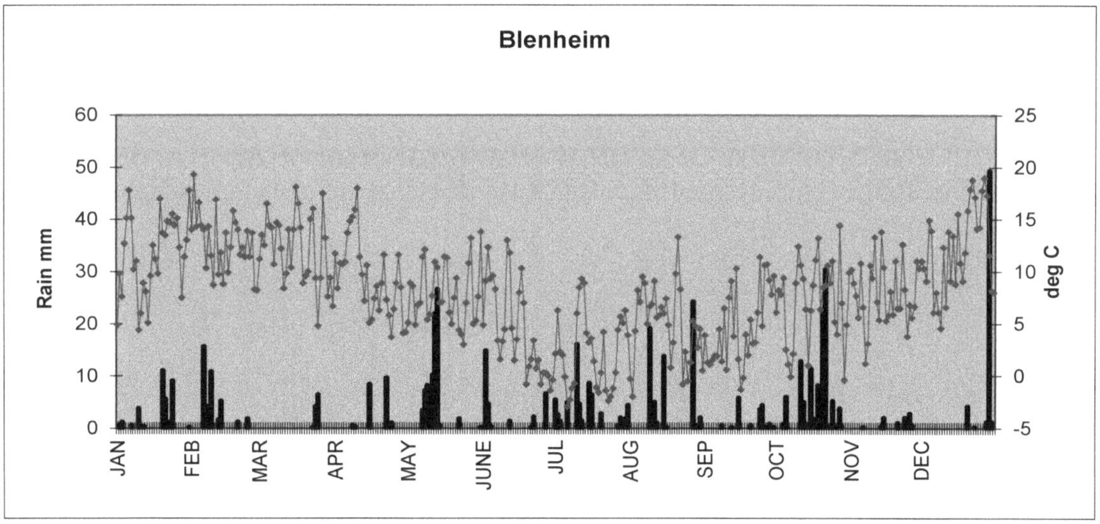
Blenheim
Rain mm
deg C
60
50
40
30
20
10
0
25
20
15
10
5
0
-5
JAN
FEB
MAR
APR
MAY
JUNE
JUL
AUG
SEP
OCT
NOV
DEC

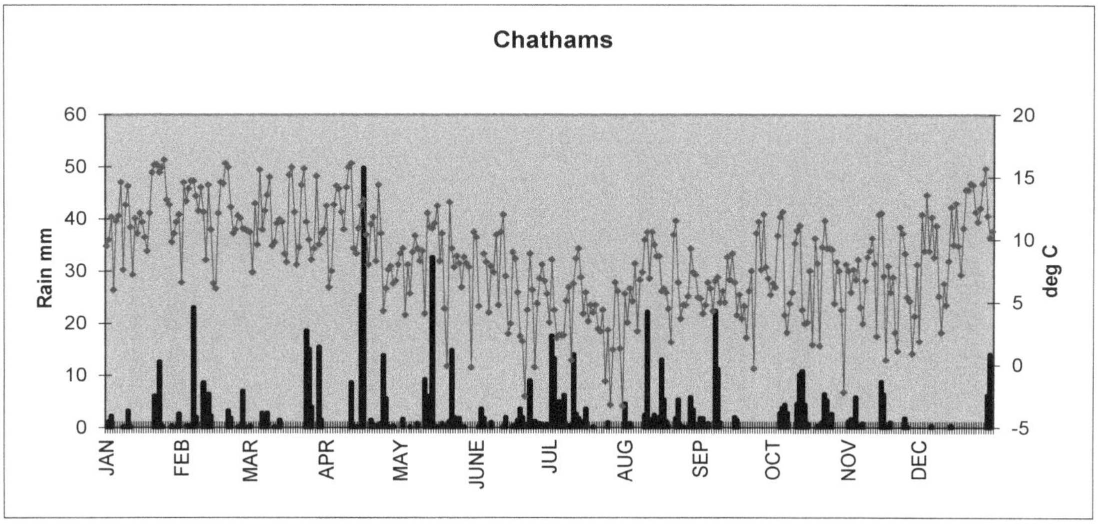
Chathams
Rain mm
deg C
60
50
40
30
20
10
0
20
15
10
5
0
-5
JAN
FEB
MAR
APR
MAY
JUNE
JUL
AUG
SEP
OCT
NOV
DEC

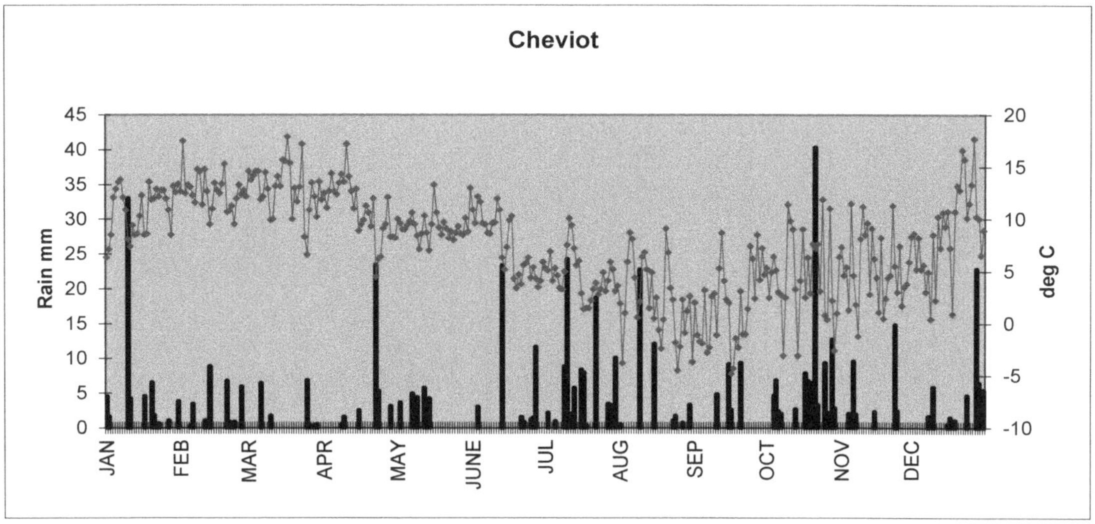
Cheviot
Rain mm
deg C
45
40
35
30
25
20
15
10
5
0
20
15
10
5
0
-5
-10
JAN
FEB
MAR
APR
MAY
JUNE
JUL
AUG
SEP
OCT
NOV
DEC

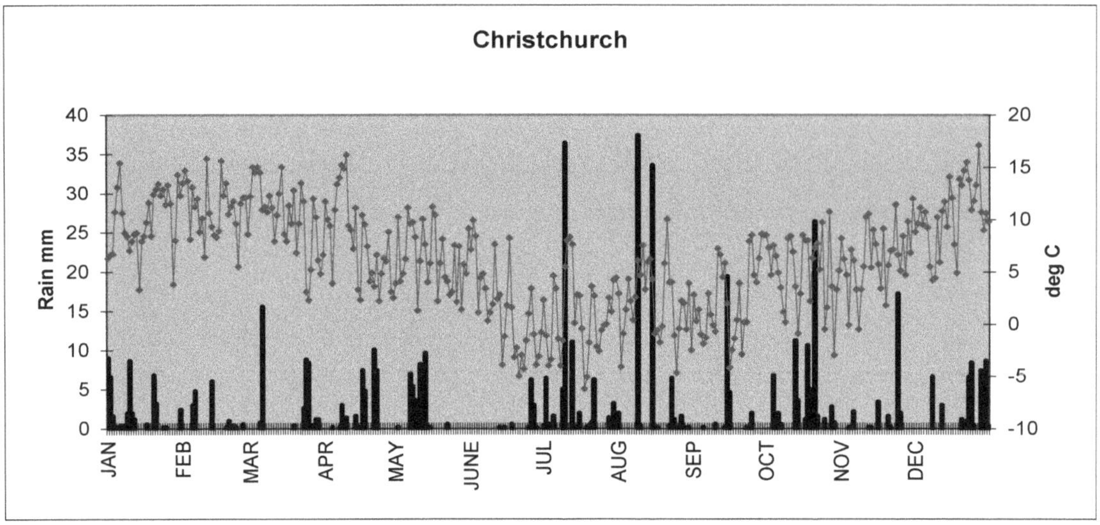
Christchurch
Rain mm
deg C
40
35
30
25
20
15
10
5
0
20
15
10
5
0
-5
-10
JAN
FEB
MAR
APR
MAY
JUNE
JUL
AUG
SEP
OCT
NOV
DEC

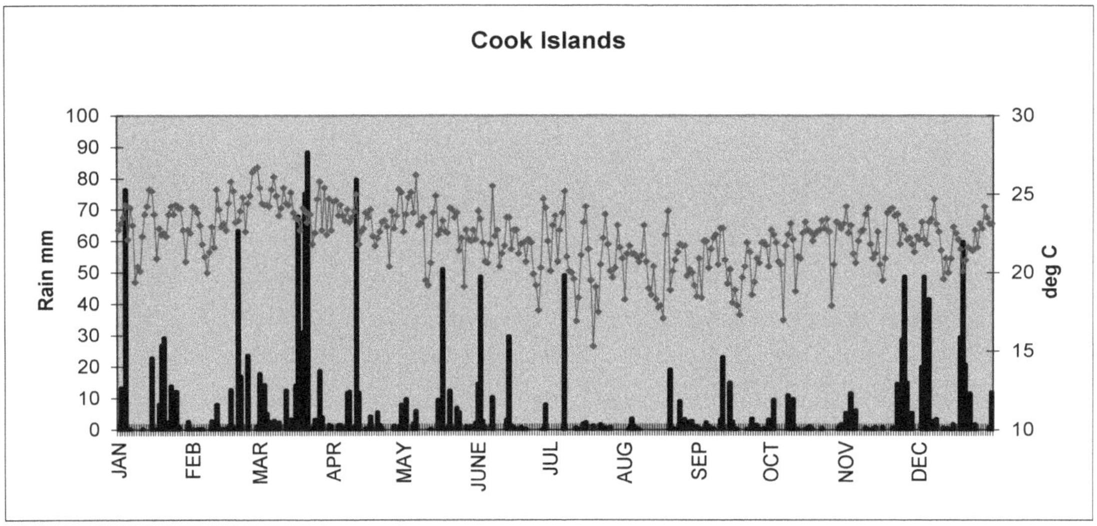
Cook Islands
Rain mm
deg C
100
90
80
70
60
50
40
30
20
10
0
30
25
20
15
10
JAN
FEB
MAR
APR
MAY
JUNE
JUL
AUG
SEP
OCT
NOV
DEC

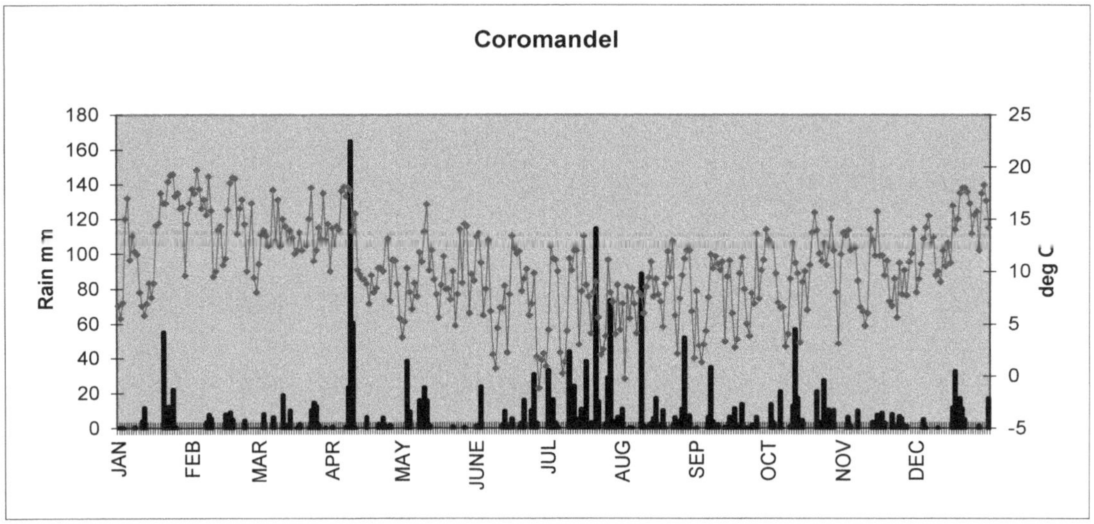
Coromandel
Rain mm
deg C
180
160
140
120
100
80
60
40
20
0
25
20
15
10
5
0
-5
JAN
FEB
MAR
APR
MAY
JUNE
JUL
AUG
SEP
OCT
NOV
DEC

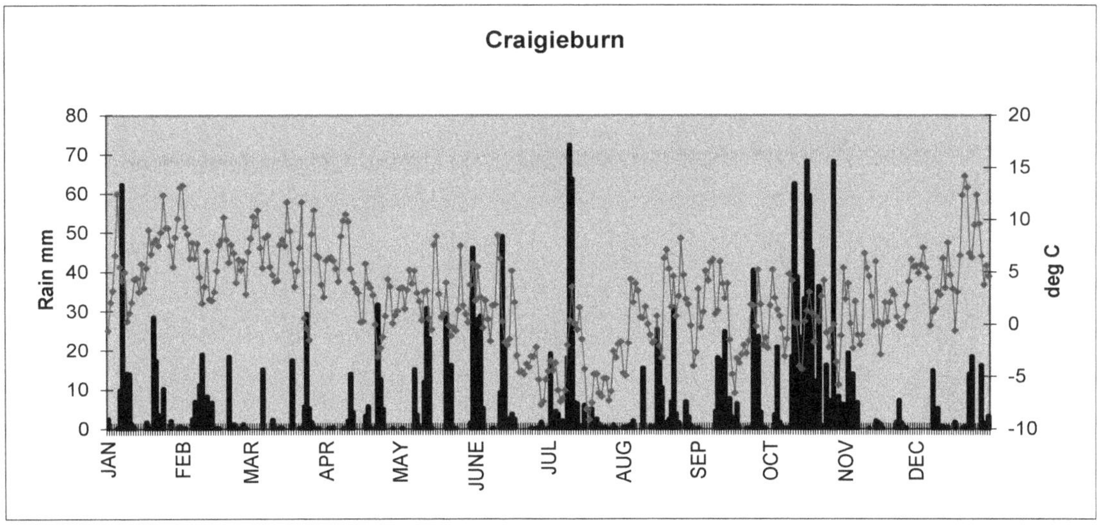
Craigieburn
Rain mm
deg C
80
70
60
50
40
30
20
10
0
20
15
10
5
0
-5
-10
JAN
FEB
MAR
APR
MAY
JUNE
JUL
AUG
SEP
OCT
NOV
DEC

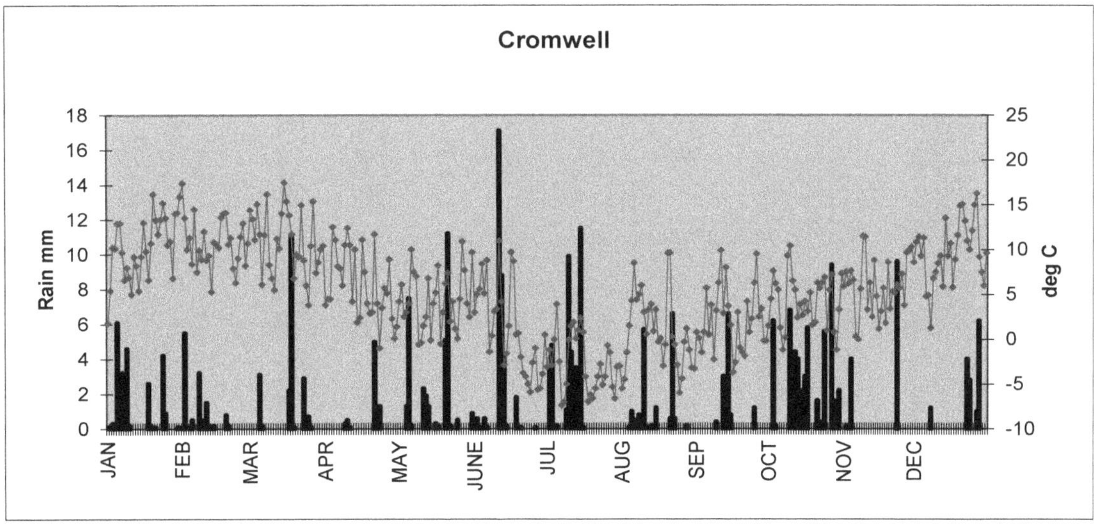
Cromwell
Rain mm
deg C
18
16
14
12
10
8
6
4
2
0
25
20
15
10
5
0
-5
-10
JAN
FEB
MAR
APR
MAY
JUNE
JUL
AUG
SEP
OCT
NOV
DEC

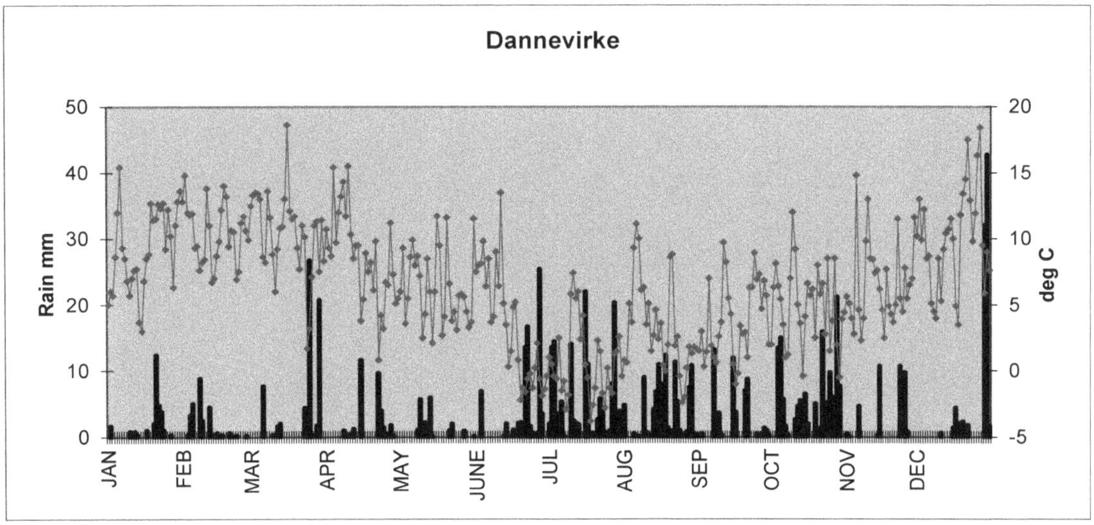
Dannevirke
Rain mm
deg C
50
40
30
20
10
0
20
15
10
5
0
-5
JAN
FEB
MAR
APR
MAY
JUNE
JUL
AUG
SEP
OCT
NOV
DEC

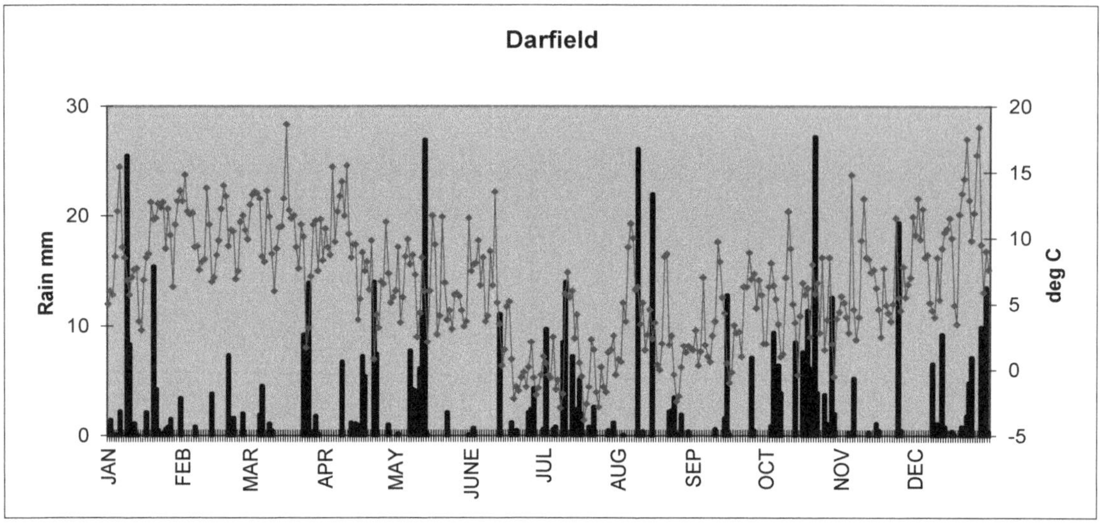
Darfield
Rain mm
deg C
JAN
FEB
MAR
APR
MAY
JUNE
JUL
AUG
SEP
OCT
NOV
DEC

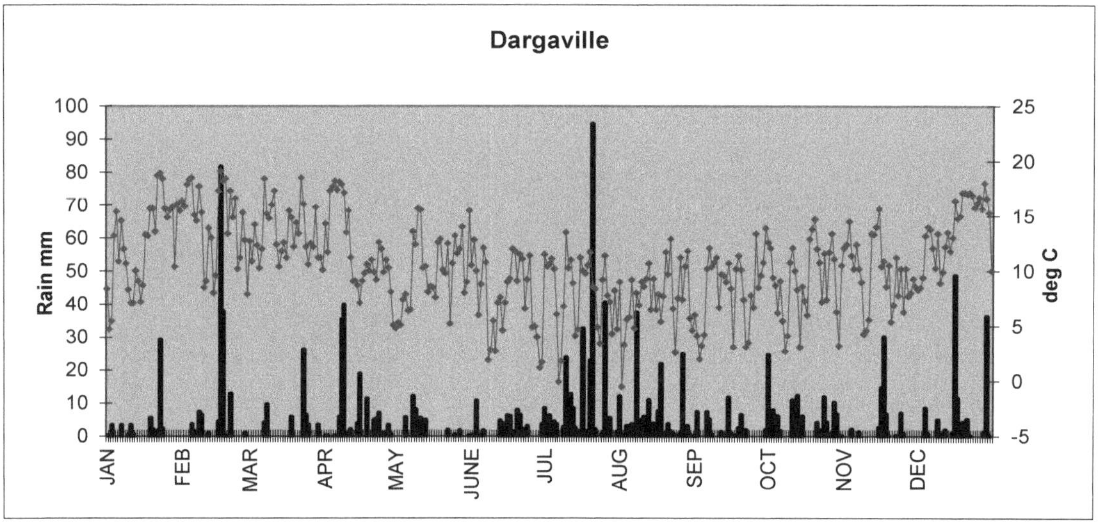
Dargaville
Rain mm
deg C
JAN
FEB
MAR
APR
MAY
JUNE
JUL
AUG
SEP
OCT
NOV
DEC

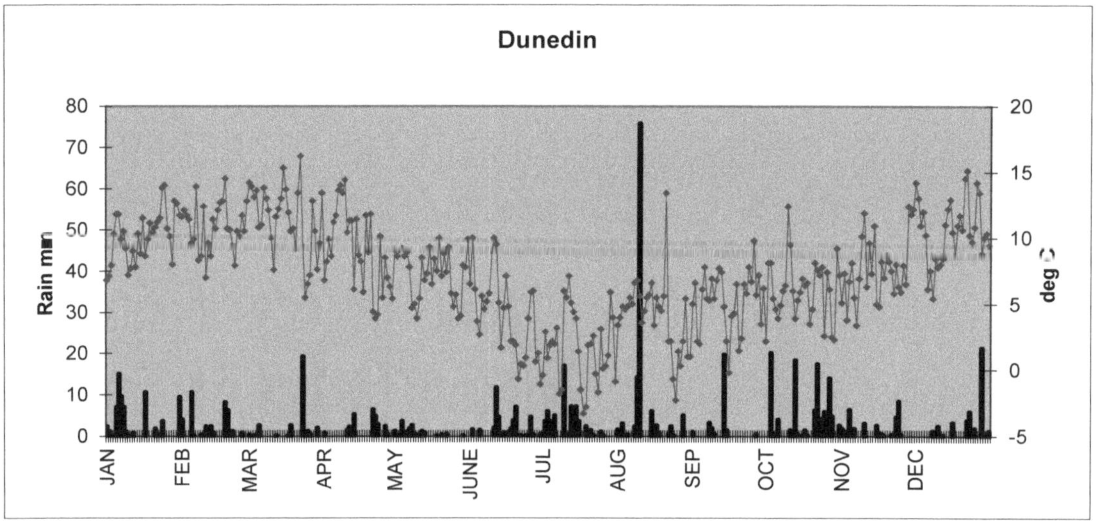
Dunedin
Rain mm
JAN
FEB
MAR
APR
MAY
JUNE
JUL
AUG
SEP
OCT
NOV
DEC

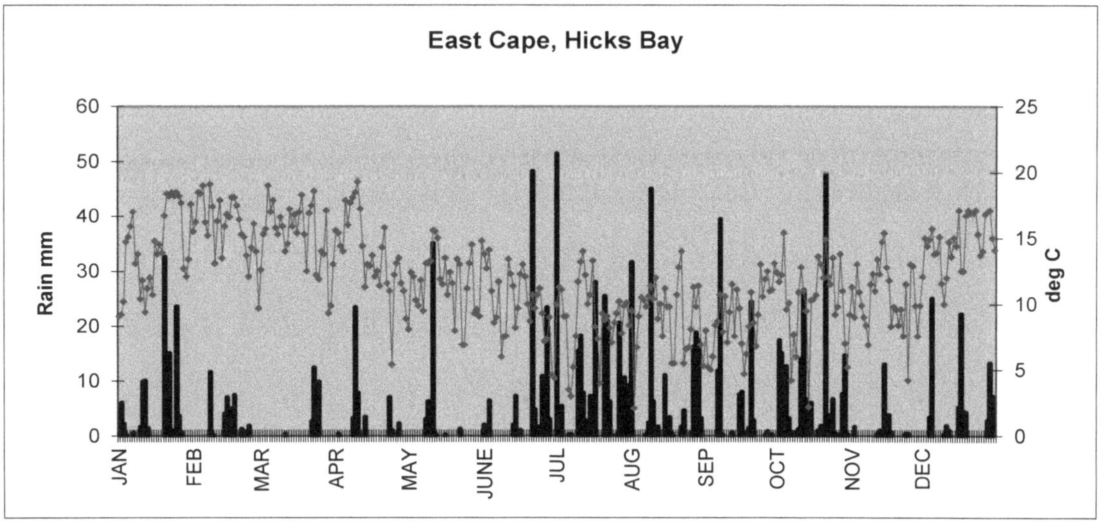
East Cape, Hicks Bay
Rain mm
deg C
60
50
40
30
20
10
0
25
20
15
10
5
0
JAN
FEB
MAR
APR
MAY
JUNE
JUL
AUG
SEP
OCT
NOV
DEC

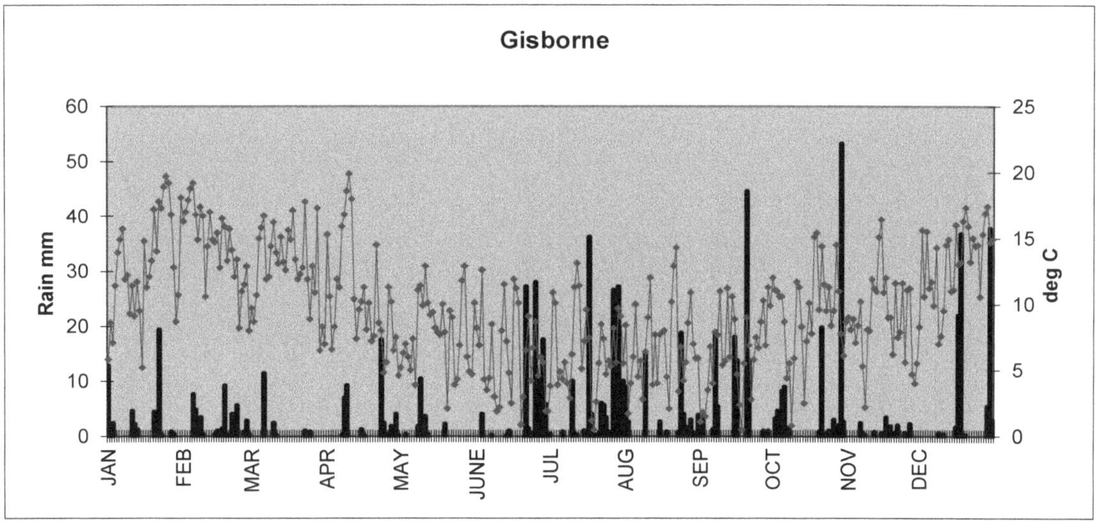
Gisborne
Rain mm
deg C
60
50
40
30
20
10
0
25
20
15
10
5
0
JAN
FEB
MAR
APR
MAY
JUNE
JUL
AUG
SEP
OCT
NOV
DEC

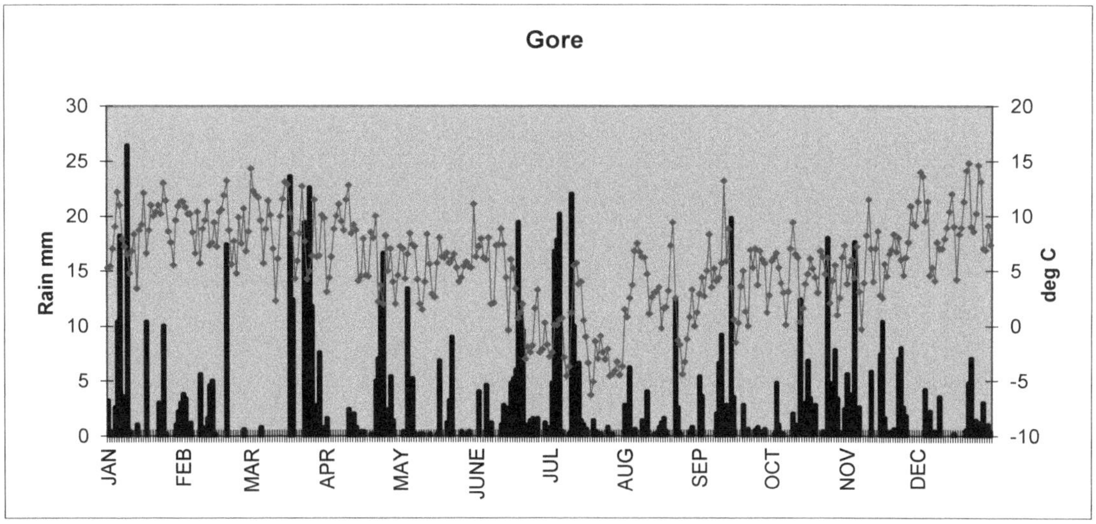
Gore
Rain mm
deg C
30
25
20
15
10
5
0
20
15
10
5
0
-5
-10
JAN
FEB
MAR
APR
MAY
JUNE
JUL
AUG
SEP
OCT
NOV
DEC

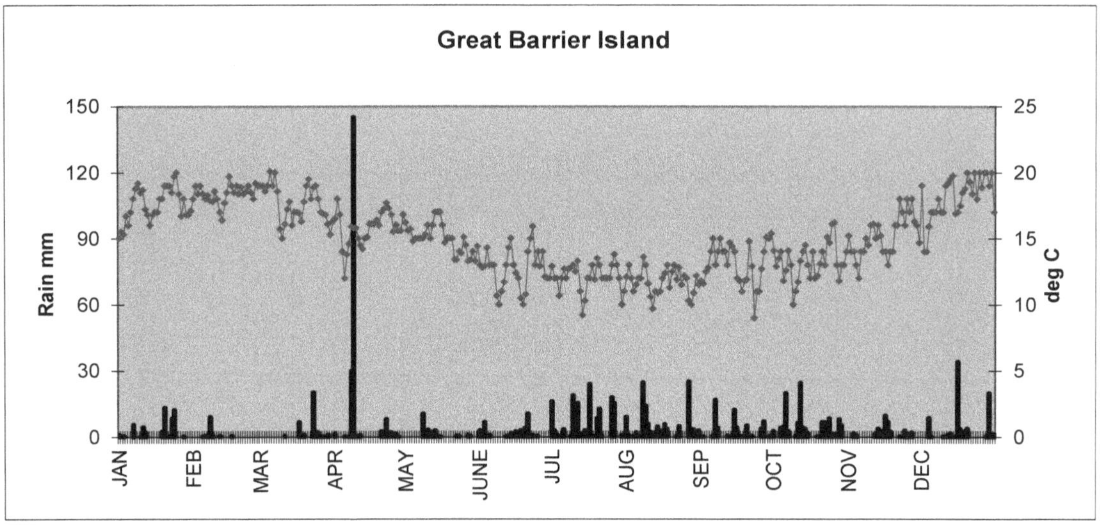
Great Barrier Island
Rain mm
deg C
150
120
90
60
30
0
25
20
15
10
5
0
JAN
FEB
MAR
APR
MAY
JUNE
JUL
AUG
SEP
OCT
NOV
DEC

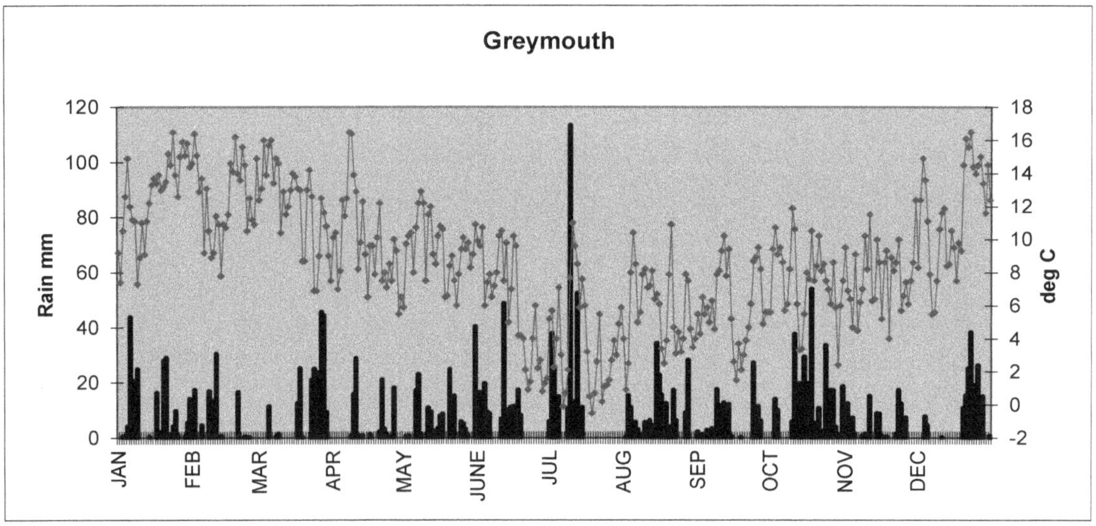
Greymouth
Rain mm
deg C
120
100
80
60
40
20
0
18
16
14
12
10
8
6
4
2
0
-2
JAN
FEB
MAR
APR
MAY
JUNE
JUL
AUG
SEP
OCT
NOV
DEC

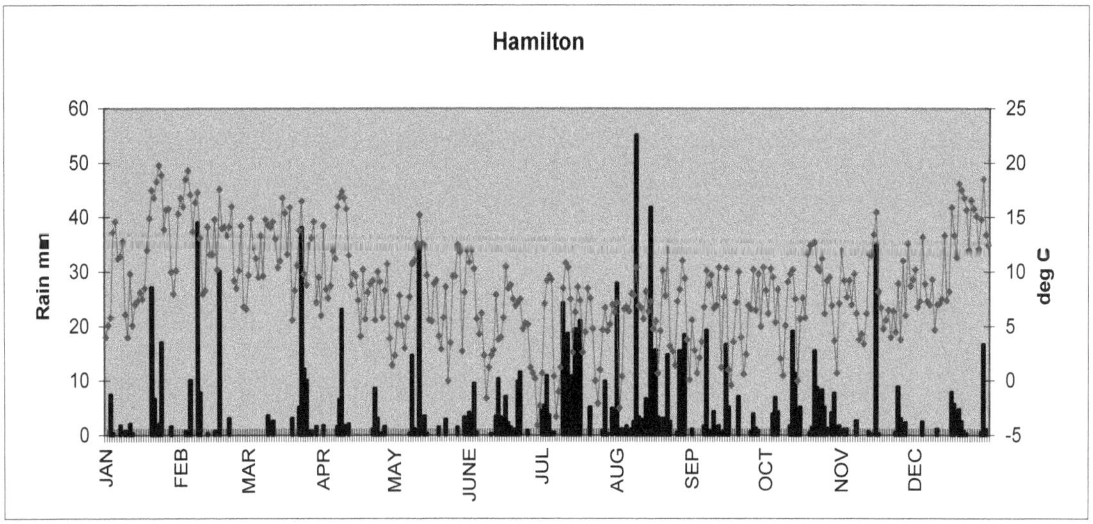
Hamilton
Rain mm
deg C
60
50
40
30
20
10
0
25
20
15
10
5
0
-5
JAN
FEB
MAR
APR
MAY
JUNE
JUL
AUG
SEP
OCT
NOV
DEC

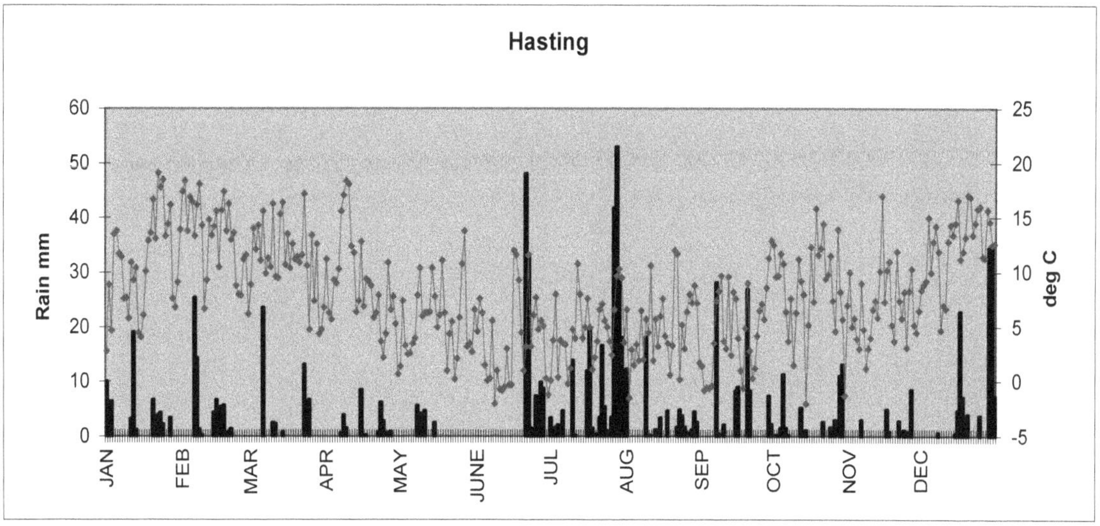
Hasting
Rain mm
deg C
60
50
40
30
20
10
0
25
20
15
10
5
0
-5
JAN
FEB
MAR
APR
MAY
JUNE
JUL
AUG
SEP
OCT
NOV
DEC

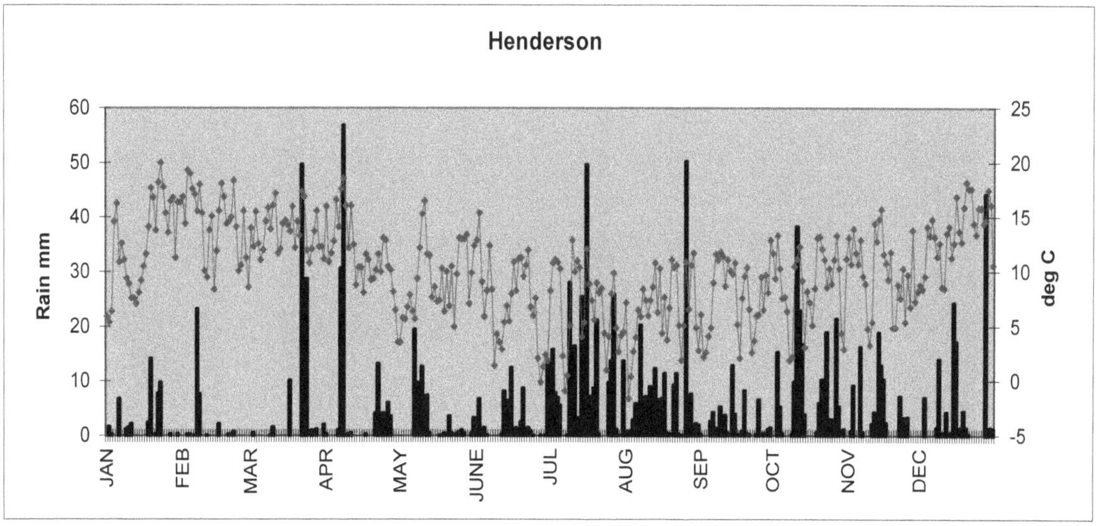
Henderson
Rain mm
deg C
60
50
40
30
20
10
0
25
20
15
10
5
0
-5
JAN
FEB
MAR
APR
MAY
JUNE
JUL
AUG
SEP
OCT
NOV
DEC

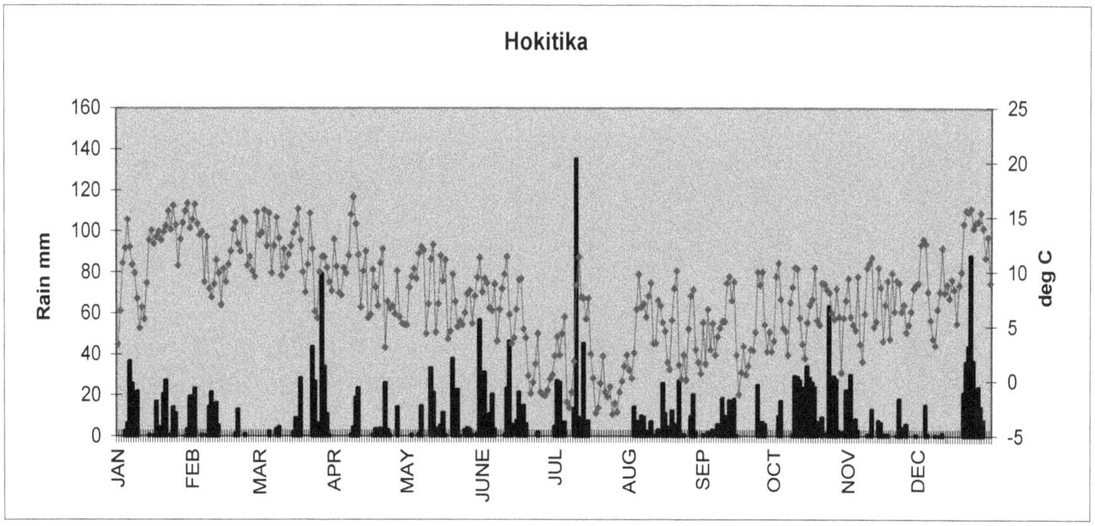
Hokitika
Rain mm
deg C
160
140
120
100
80
60
40
20
0
25
20
15
10
5
0
-5
JAN
FEB
MAR
APR
MAY
JUNE
JUL
AUG
SEP
OCT
NOV
DEC

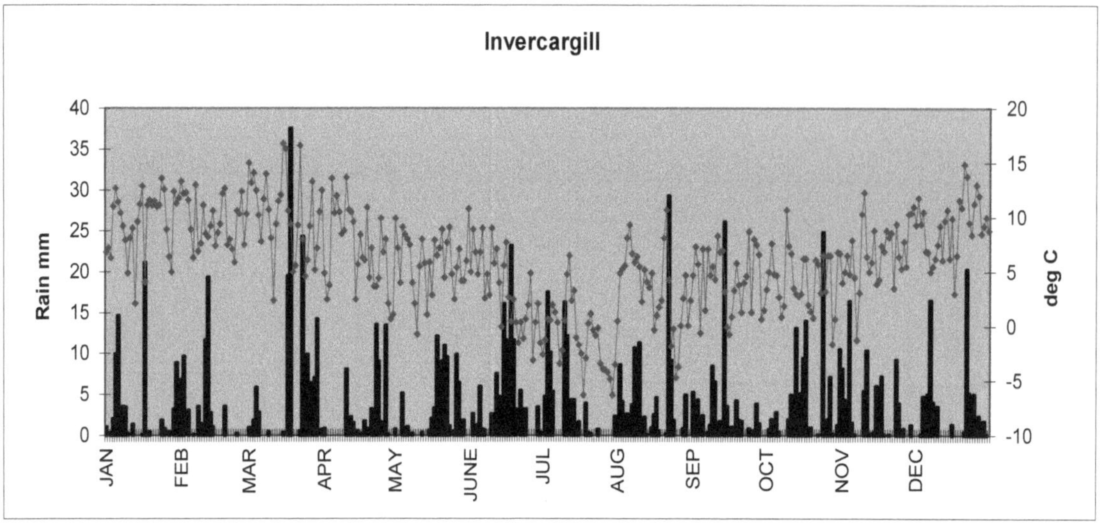
Invercargill
Rain mm
deg C
JAN
FEB
MAR
APR
MAY
JUNE
JUL
AUG
SEP
OCT
NOV
DEC

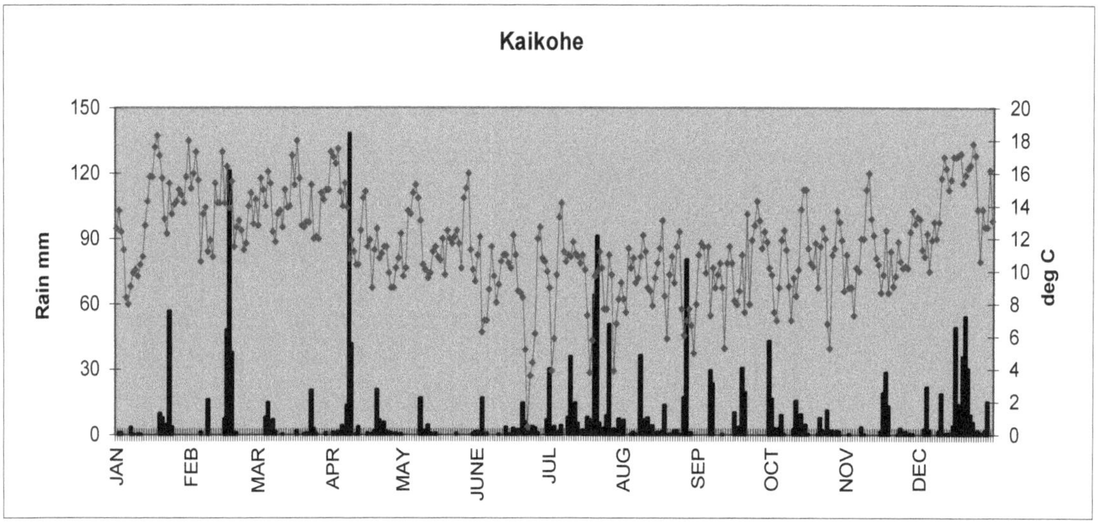
Kaikohe
Rain mm
deg C
JAN
FEB
MAR
APR
MAY
JUNE
JUL
AUG
SEP
OCT
NOV
DEC

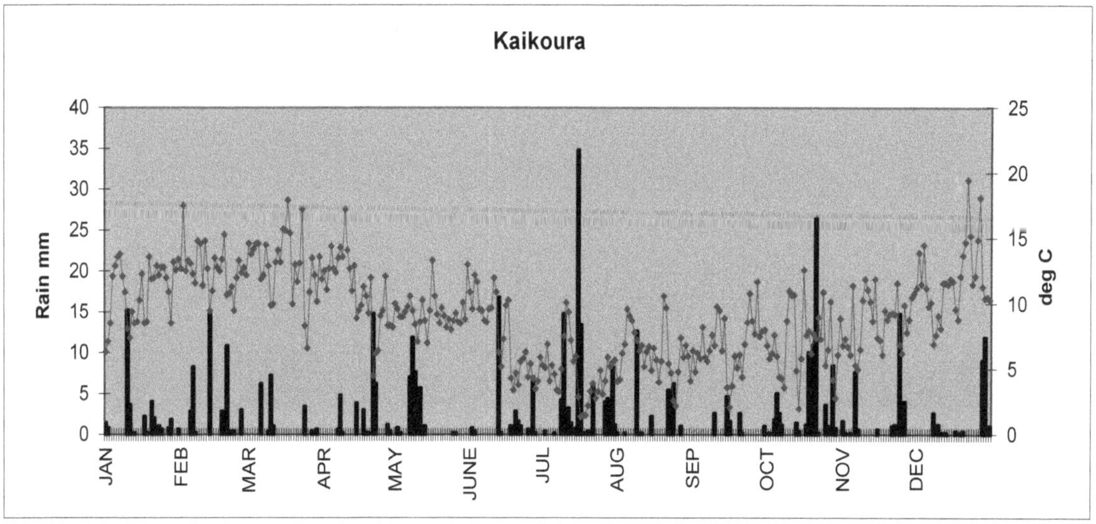
Kaikoura
Rain mm
deg C
JAN
FEB
MAR
APR
MAY
JUNE
JUL
AUG
SEP
OCT
NOV
DEC

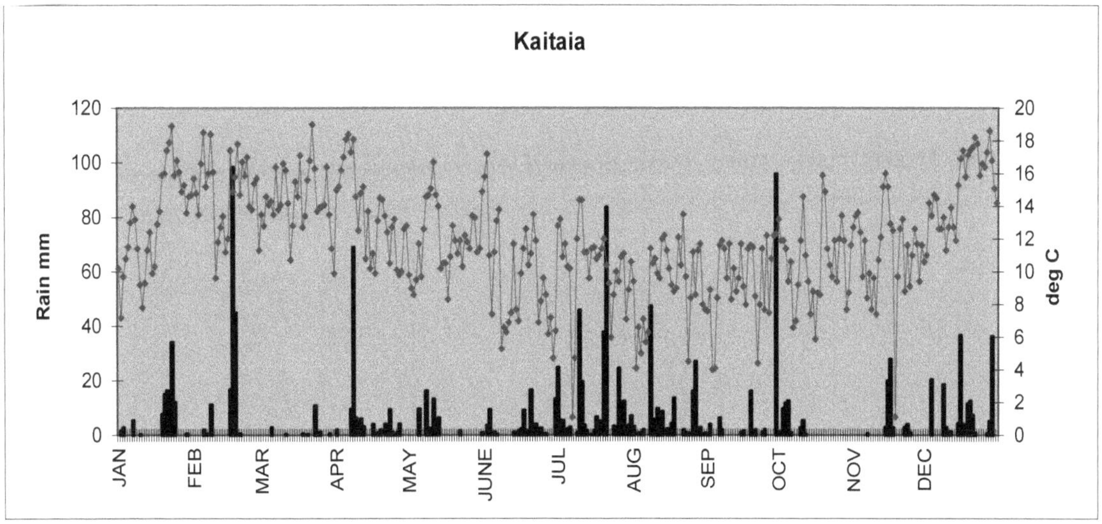
Kaitaia
Rain mm
deg C
JAN FEB MAR APR MAY JUNE JUL AUG SEP OCT NOV DEC

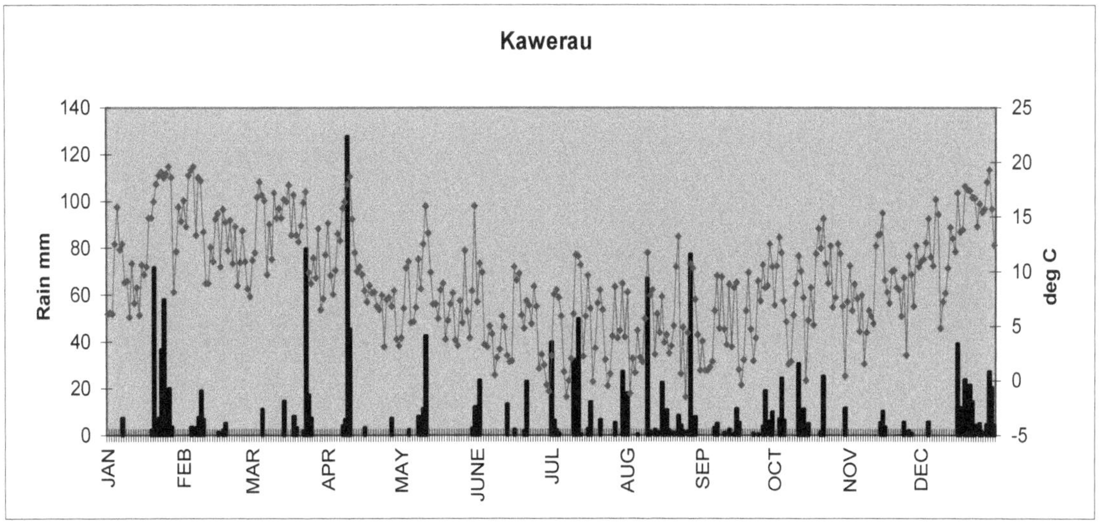
Kawerau
Rain mm
deg C
JAN FEB MAR APR MAY JUNE JUL AUG SEP OCT NOV DEC

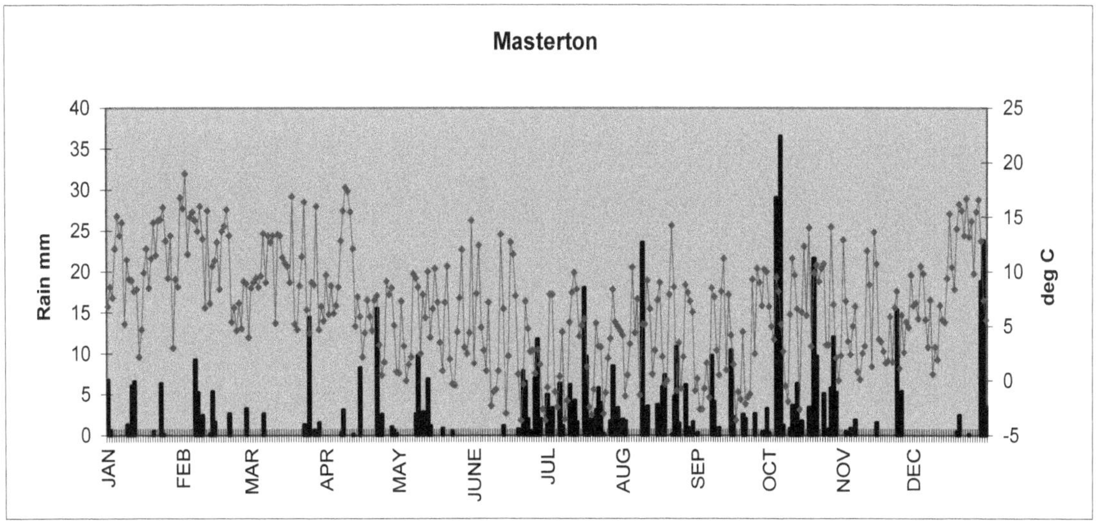
Masterton
Rain mm
deg C
JAN FEB MAR APR MAY JUNE JUL AUG SEP OCT NOV DEC

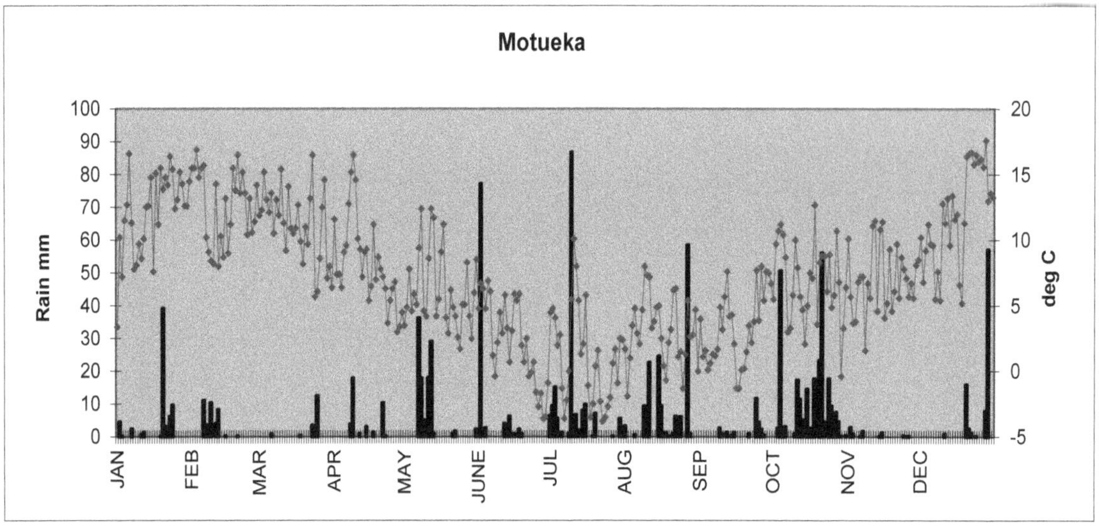
Motueka
Rain mm
deg C
100
90
80
70
60
50
40
30
20
10
0
20
15
10
5
0
-5
JAN
FEB
MAR
APR
MAY
JUNE
JUL
AUG
SEP
OCT
NOV
DEC

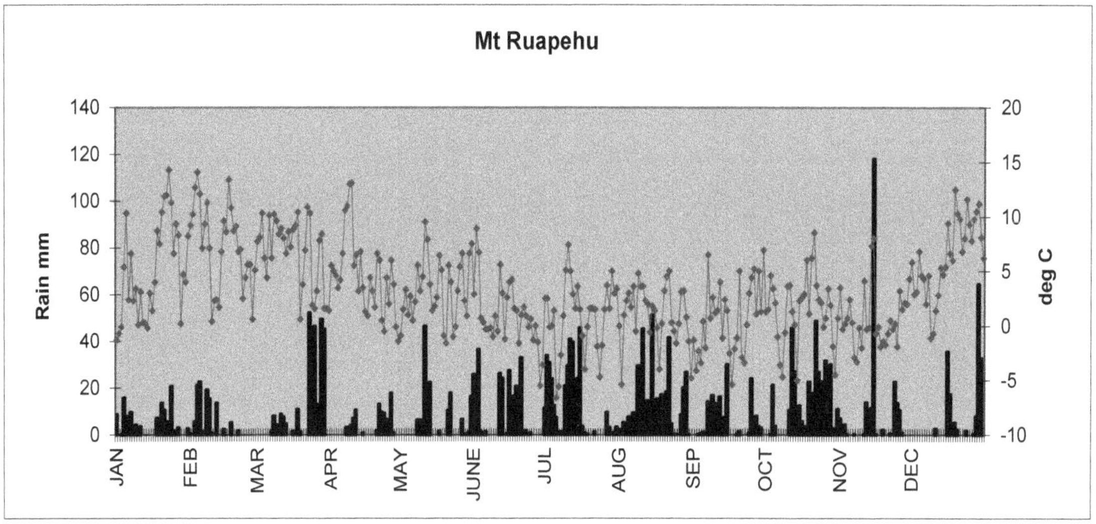
Mt Ruapehu
Rain mm
deg C
140
120
100
80
60
40
20
0
20
15
10
5
0
-5
-10
JAN
FEB
MAR
APR
MAY
JUNE
JUL
AUG
SEP
OCT
NOV
DEC

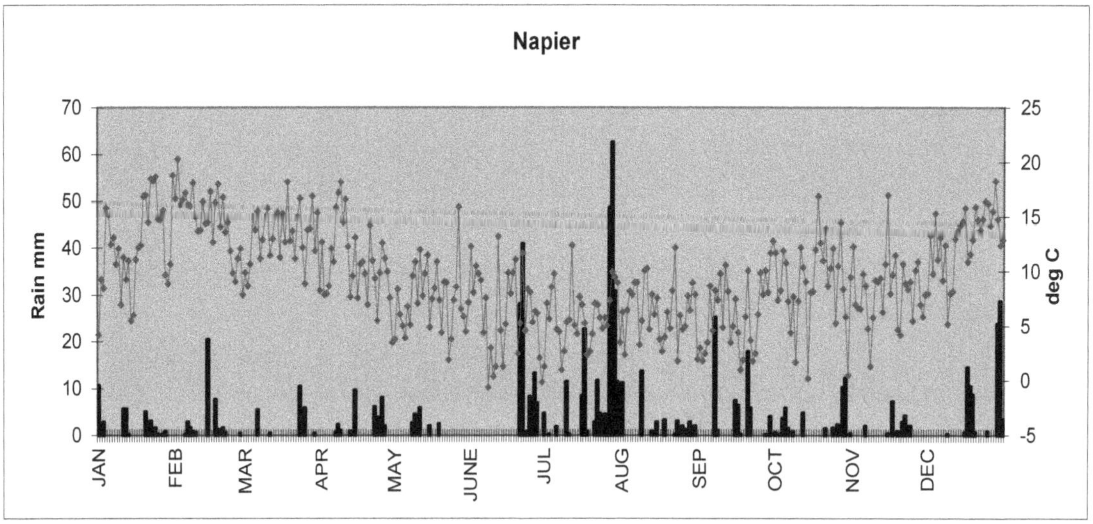
Napier
Rain mm
deg C
70
60
50
40
30
20
10
0
25
20
15
10
5
0
-5
JAN
FEB
MAR
APR
MAY
JUNE
JUL
AUG
SEP
OCT
NOV
DEC

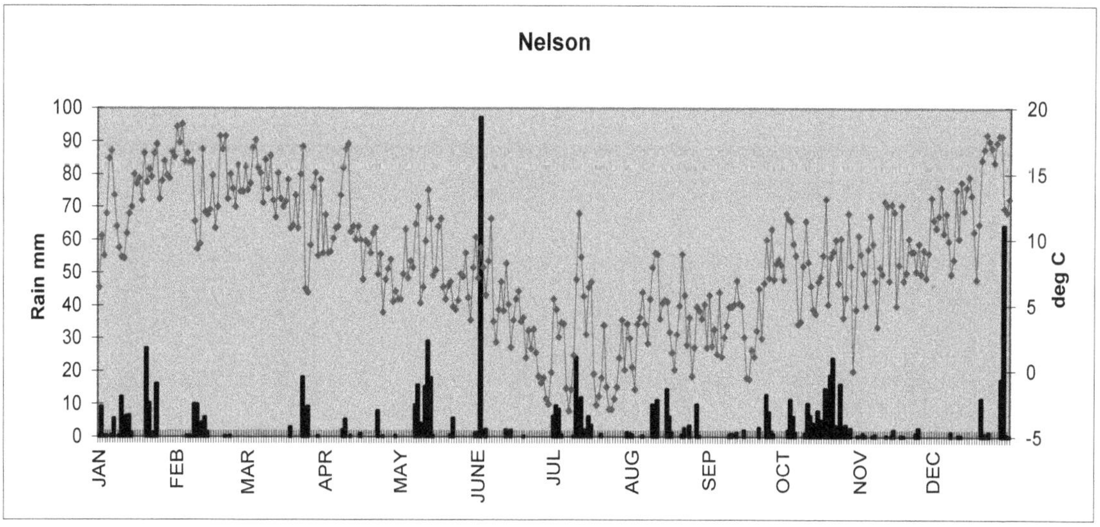

Nelson
Rain mm
deg C
100
90
80
70
60
50
40
30
20
10
0
20
15
10
5
0
-5
JAN
FEB
MAR
APR
MAY
JUNE
JUL
AUG
SEP
OCT
NOV
DEC

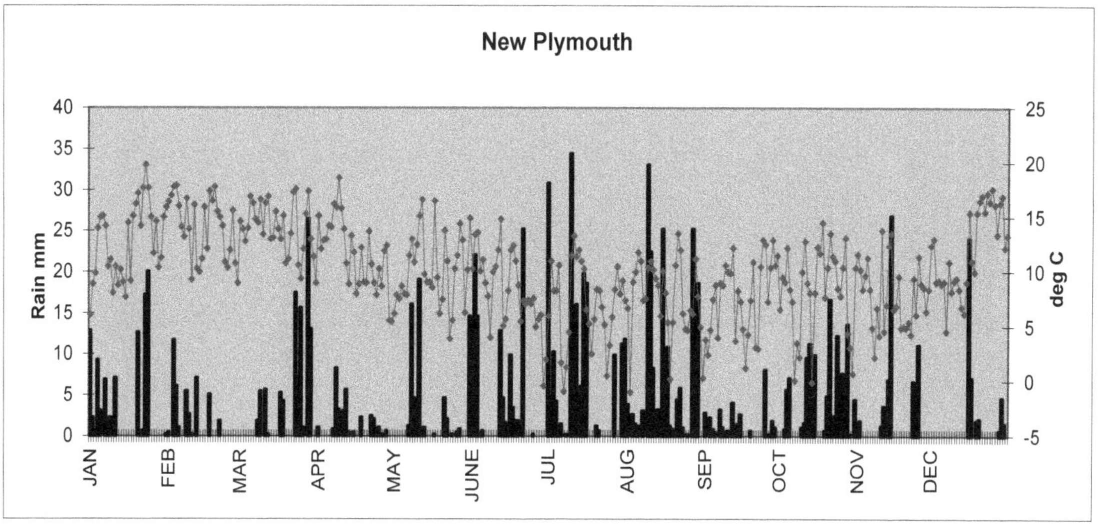

New Plymouth
Rain mm
deg C
40
35
30
25
20
15
10
5
0
25
20
15
10
5
0
-5
JAN
FEB
MAR
APR
MAY
JUNE
JUL
AUG
SEP
OCT
NOV
DEC

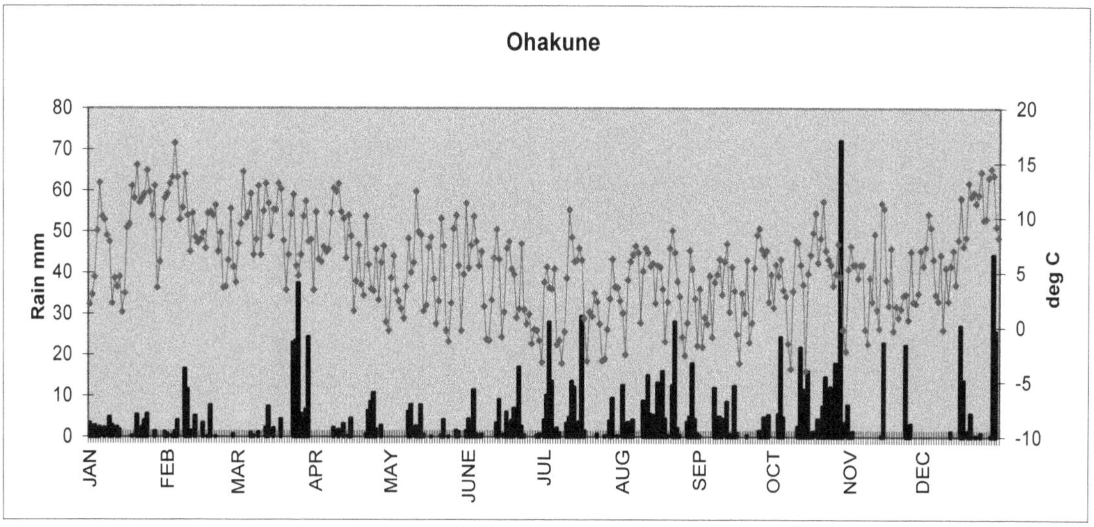

Ohakune
Rain mm
deg C
80
70
60
50
40
30
20
10
0
20
15
10
5
0
-5
-10
JAN
FEB
MAR
APR
MAY
JUNE
JUL
AUG
SEP
OCT
NOV
DEC

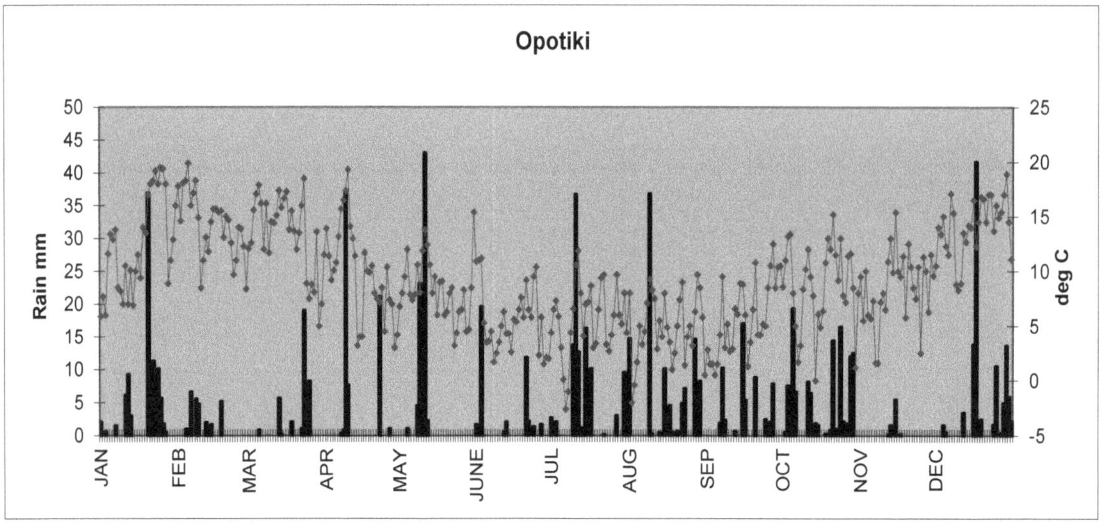
Opotiki
Rain mm
deg C
50
45
40
35
30
25
20
15
10
5
0
25
20
15
10
5
0
-5
JAN
FEB
MAR
APR
MAY
JUNE
JUL
AUG
SEP
OCT
NOV
DEC

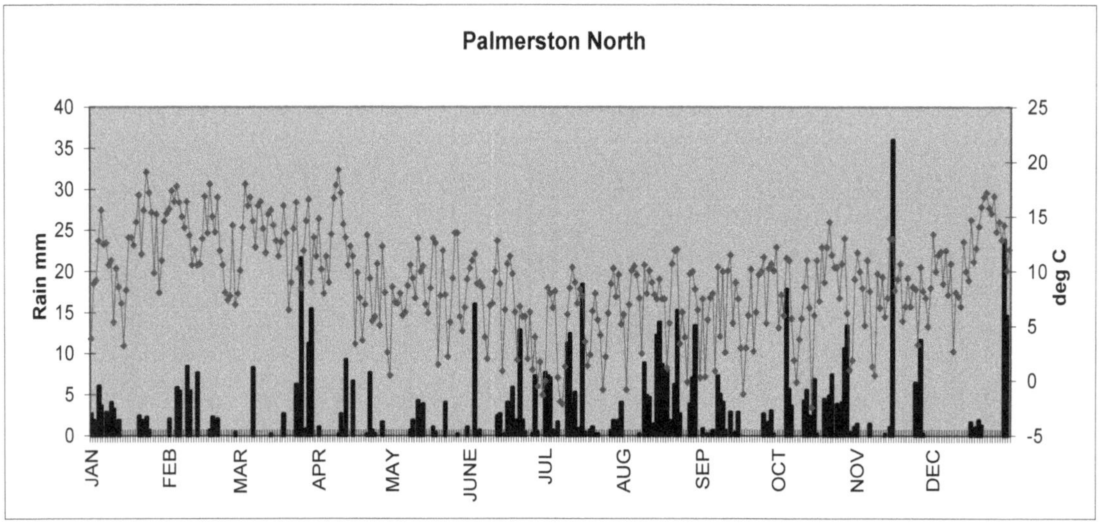
Palmerston North
Rain mm
deg C
40
35
30
25
20
15
10
5
0
25
20
15
10
5
0
-5
JAN
FEB
MAR
APR
MAY
JUNE
JUL
AUG
SEP
OCT
NOV
DEC

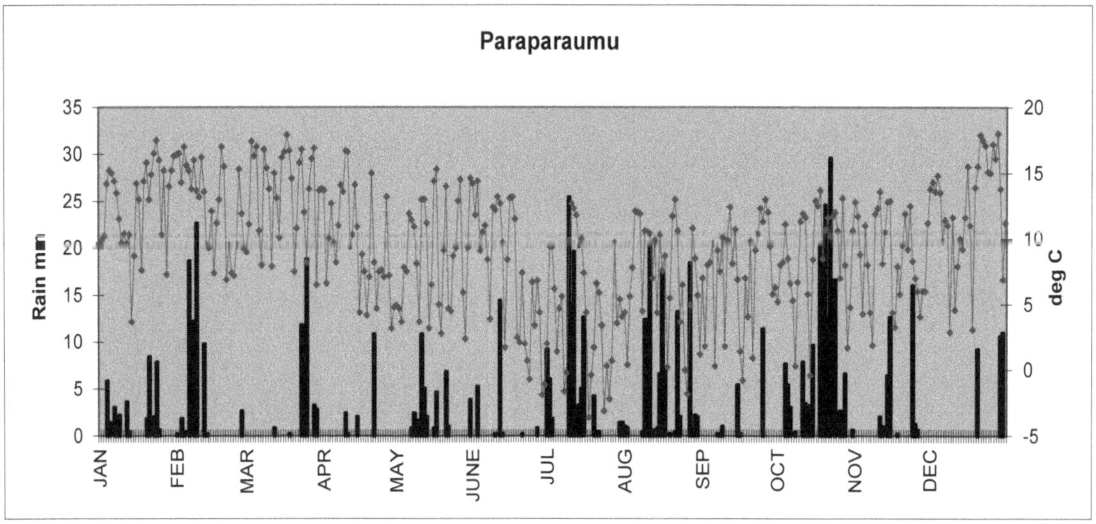
Paraparaumu
Rain mm
deg C
35
30
25
20
15
10
5
0
20
15
10
5
0
-5
JAN
FEB
MAR
APR
MAY
JUNE
JUL
AUG
SEP
OCT
NOV
DEC

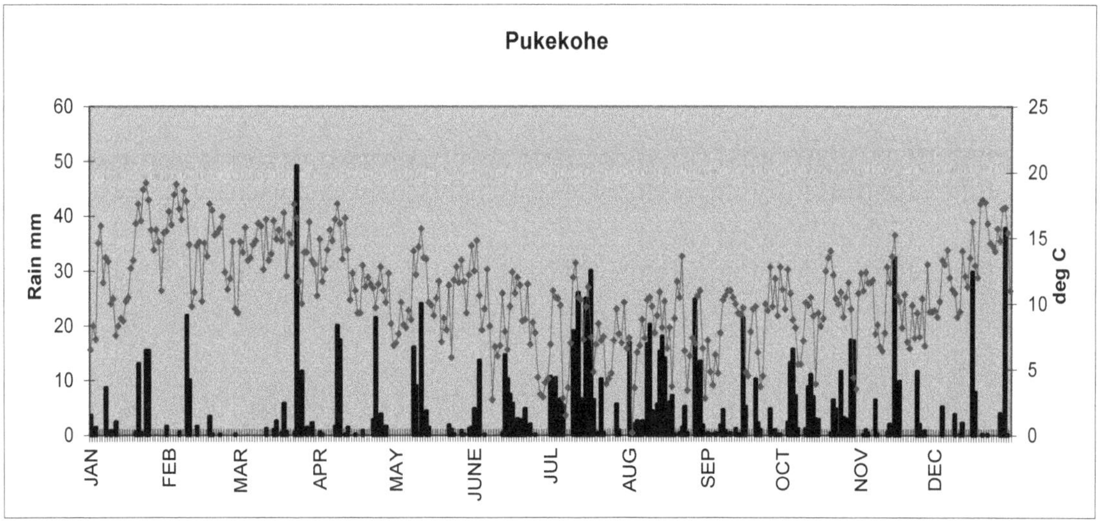
Pukekohe
Rain mm
deg C
60
50
40
30
20
10
0
25
20
15
10
5
0
JAN
FEB
MAR
APR
MAY
JUNE
JUL
AUG
SEP
OCT
NOV
DEC

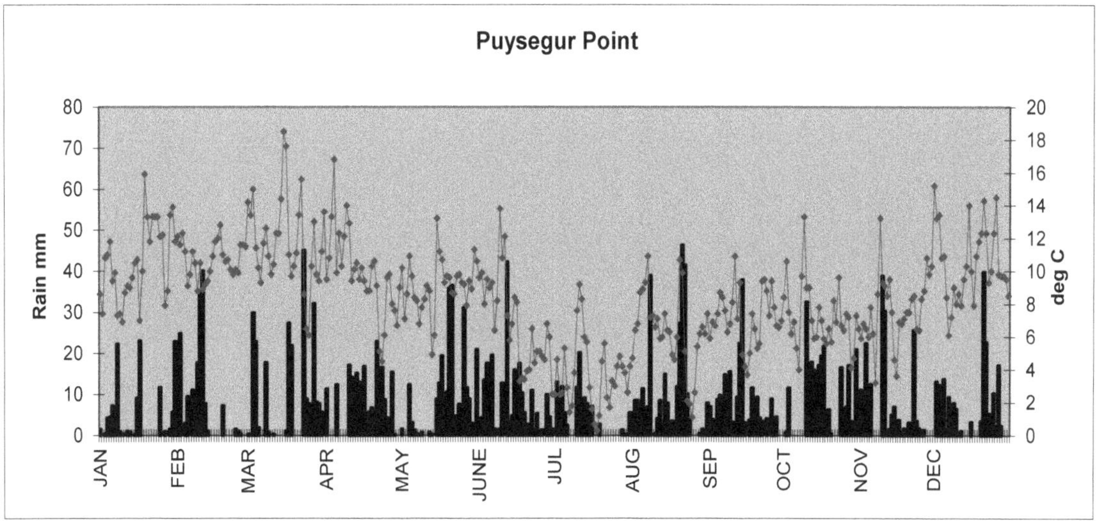
Puysegur Point
Rain mm
deg C
80
70
60
50
40
30
20
10
0
20
18
16
14
12
10
8
6
4
2
0
JAN
FEB
MAR
APR
MAY
JUNE
JUL
AUG
SEP
OCT
NOV
DEC

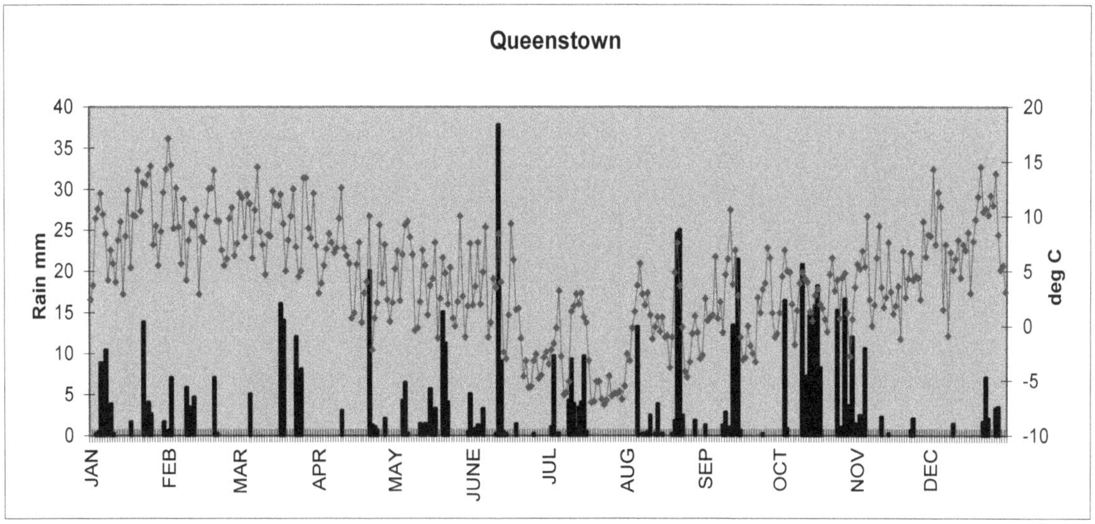
Queenstown
Rain mm
deg C
40
35
30
25
20
15
10
5
0
20
15
10
5
0
-5
-10
JAN
FEB
MAR
APR
MAY
JUNE
JUL
AUG
SEP
OCT
NOV
DEC

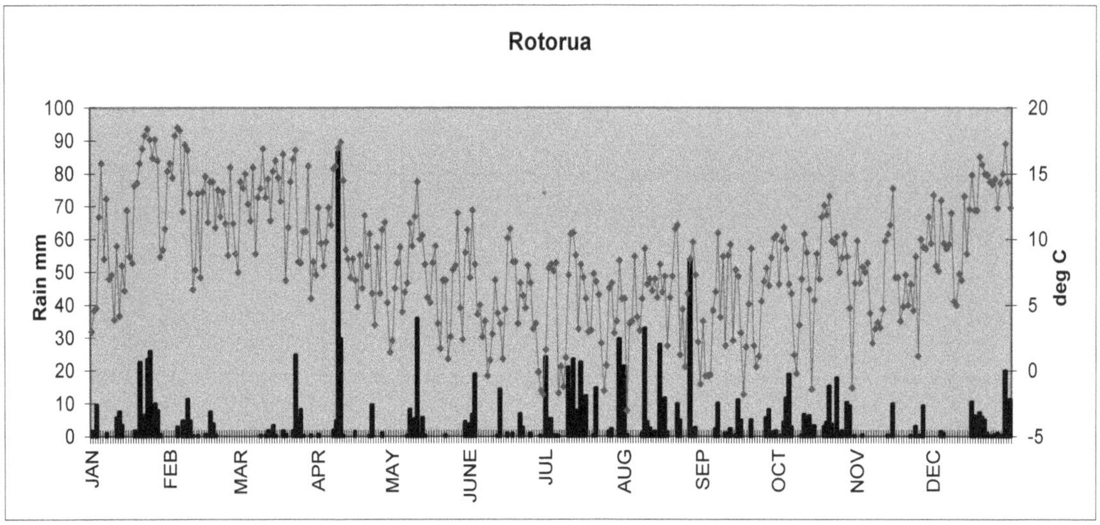
Rotorua
Rain mm
deg C
100
90
80
70
60
50
40
30
20
10
0
20
15
10
5
0
-5
JAN
FEB
MAR
APR
MAY
JUNE
JUL
AUG
SEP
OCT
NOV
DEC

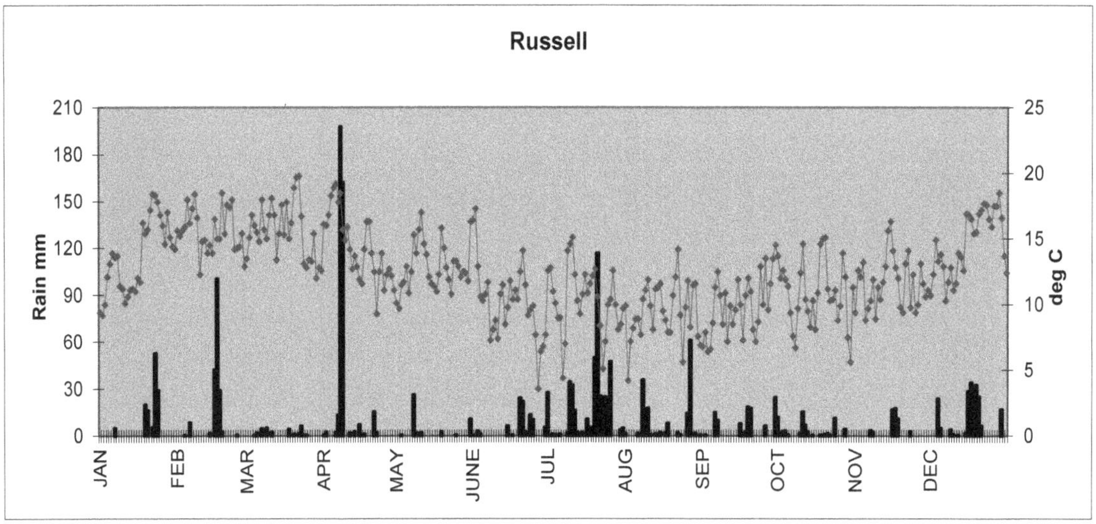
Russell
Rain mm
deg C
210
180
150
120
90
60
30
0
25
20
15
10
5
0
JAN
FEB
MAR
APR
MAY
JUNE
JUL
AUG
SEP
OCT
NOV
DEC

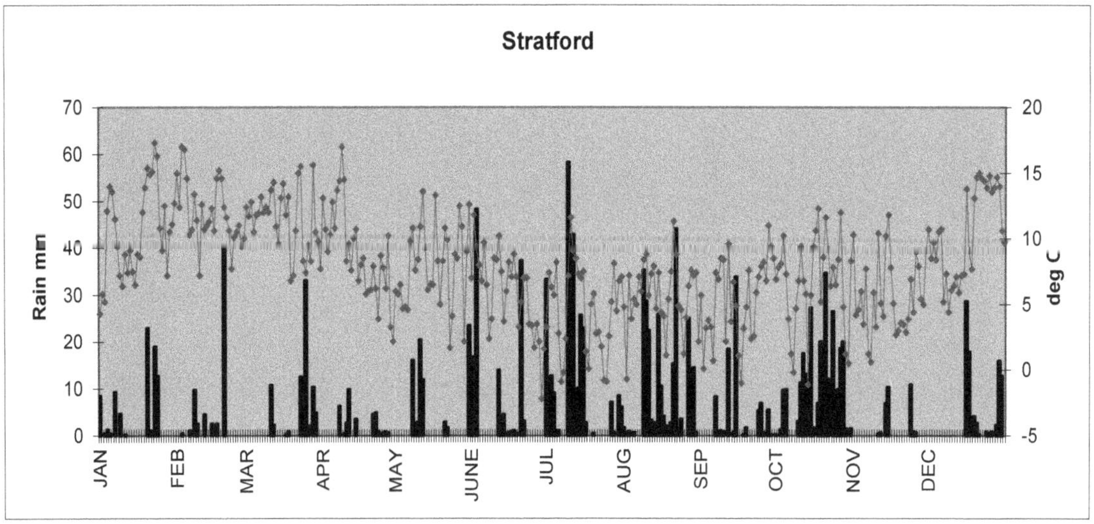
Stratford
Rain mm
deg C
70
60
50
40
30
20
10
0
20
15
10
5
0
-5
JAN
FEB
MAR
APR
MAY
JUNE
JUL
AUG
SEP
OCT
NOV
DEC

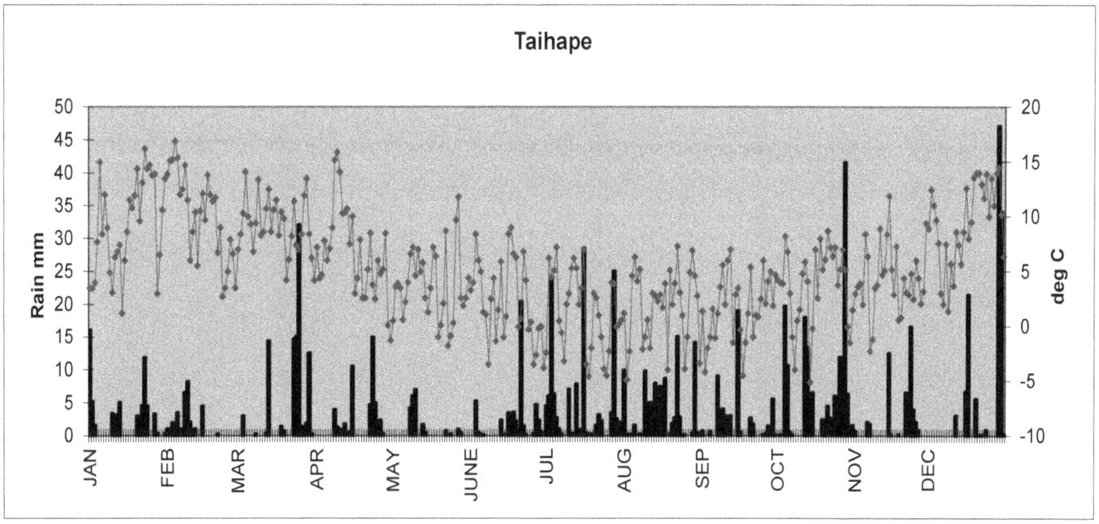
Taihape
Rain mm
deg C
50
45
40
35
30
25
20
15
10
5
0
20
15
10
5
0
-5
-10
JAN
FEB
MAR
APR
MAY
JUNE
JUL
AUG
SEP
OCT
NOV
DEC

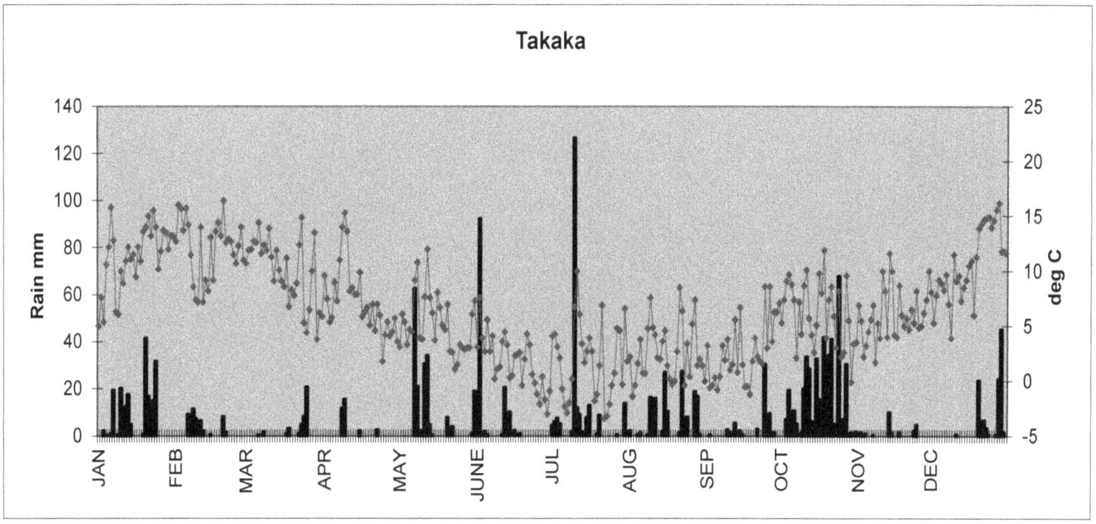
Takaka
Rain mm
deg C
140
120
100
80
60
40
20
0
25
20
15
10
5
0
-5
JAN
FEB
MAR
APR
MAY
JUNE
JUL
AUG
SEP
OCT
NOV
DEC

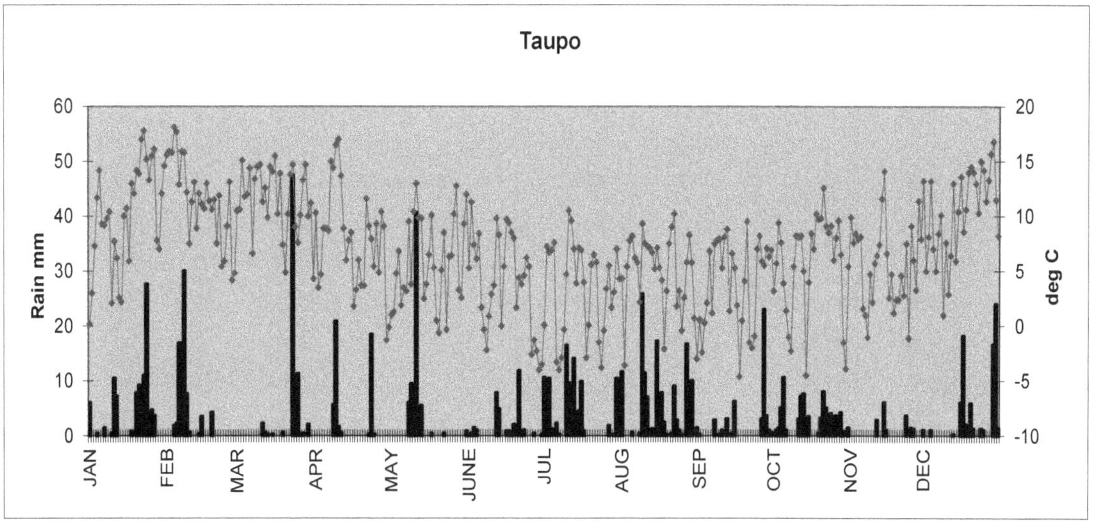
Taupo
Rain mm
deg C
60
50
40
30
20
10
0
20
15
10
5
0
-5
-10
JAN
FEB
MAR
APR
MAY
JUNE
JUL
AUG
SEP
OCT
NOV
DEC

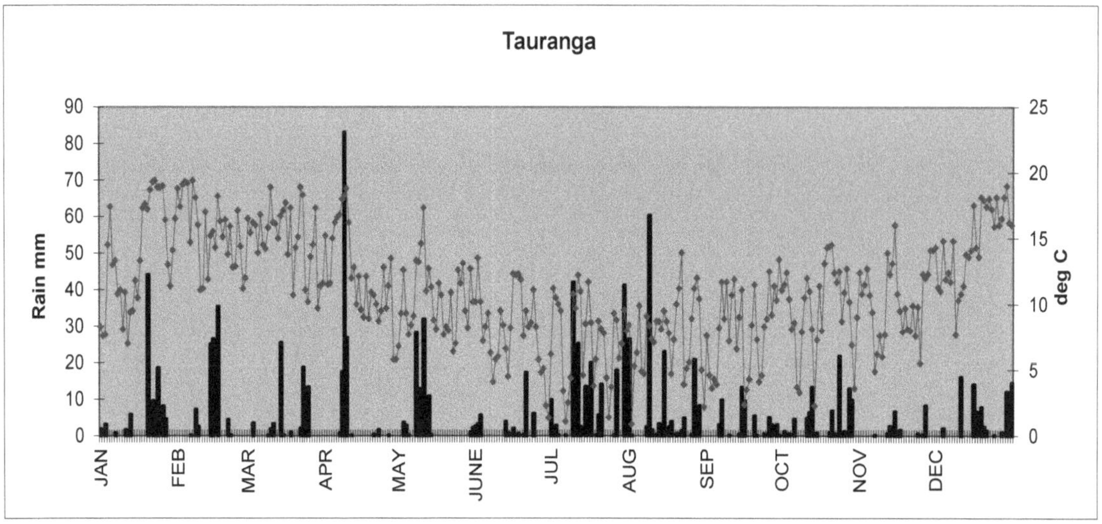
Tauranga
Rain mm
deg C
0
10
20
30
40
50
60
70
80
90
0
5
10
15
20
25
JAN
FEB
MAR
APR
MAY
JUNE
JUL
AUG
SEP
OCT
NOV
DEC

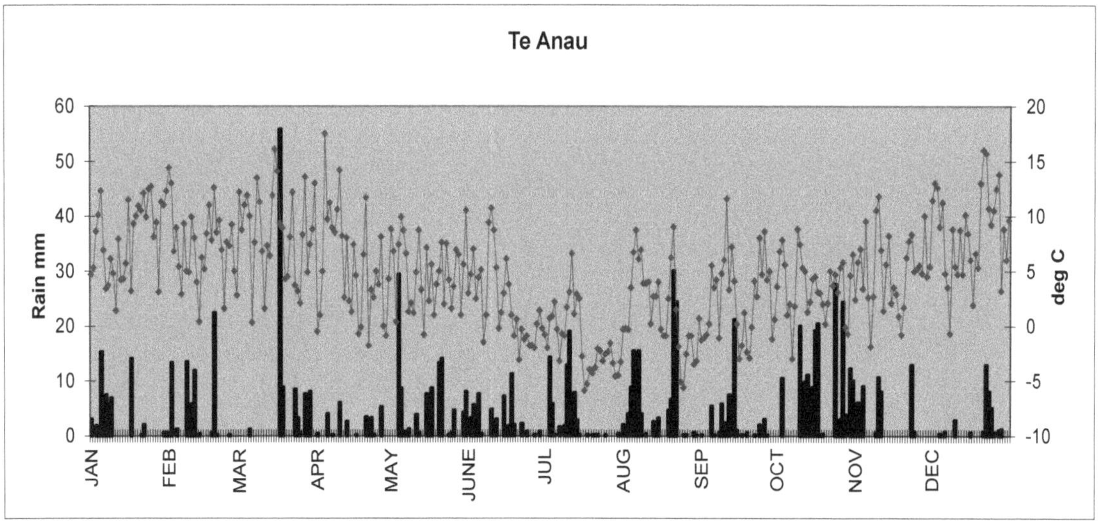
Te Anau
Rain mm
deg C
0
10
20
30
40
50
60
-10
-5
0
5
10
15
20
JAN
FEB
MAR
APR
MAY
JUNE
JUL
AUG
SEP
OCT
NOV
DEC

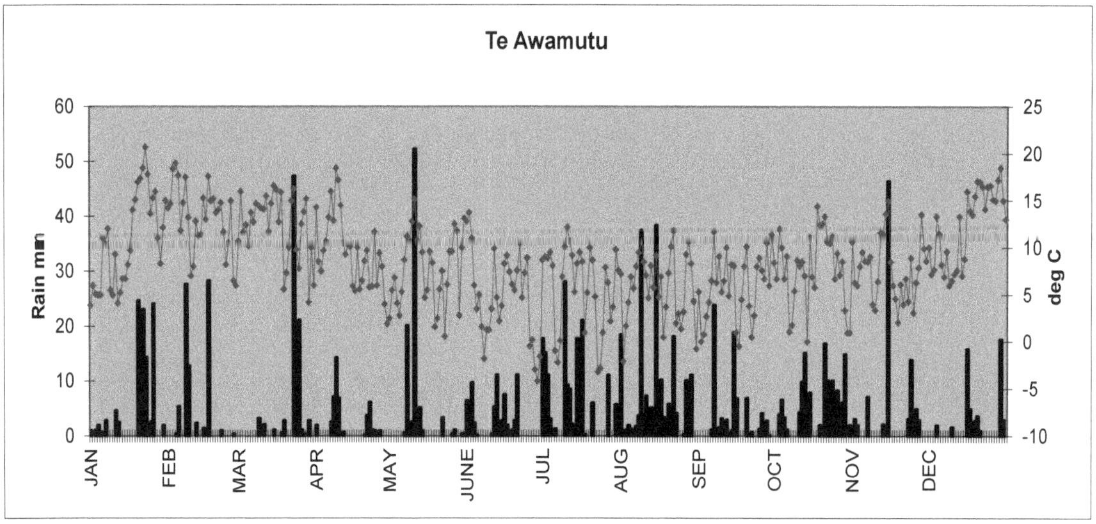
Te Awamutu
Rain mm
deg C
0
10
20
30
40
50
60
-10
-5
0
5
10
15
20
25
JAN
FEB
MAR
APR
MAY
JUNE
JUL
AUG
SEP
OCT
NOV
DEC

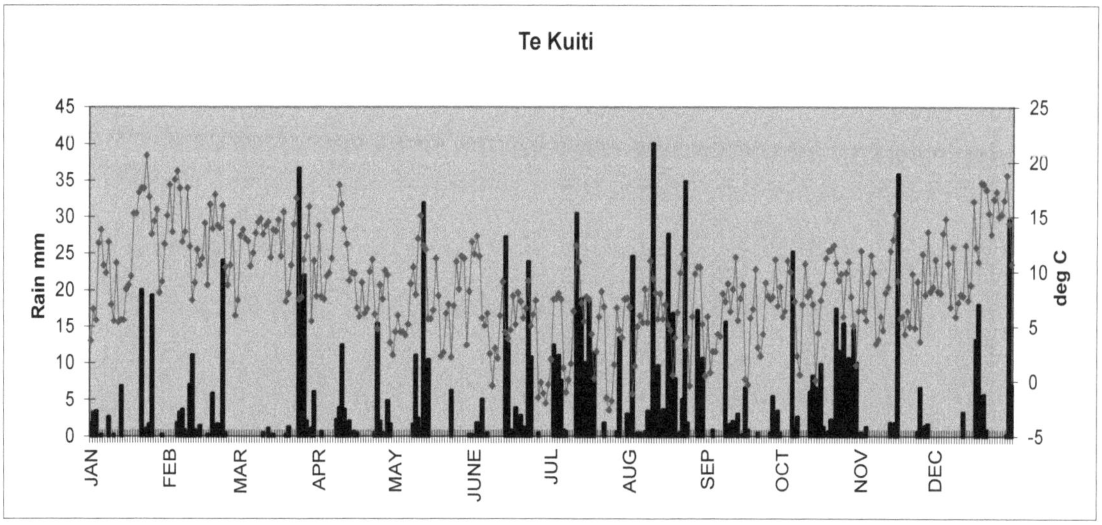
Te Kuiti
Rain mm
deg C
45
40
35
30
25
20
15
10
5
0
25
20
15
10
5
0
-5
JAN
FEB
MAR
APR
MAY
JUNE
JUL
AUG
SEP
OCT
NOV
DEC

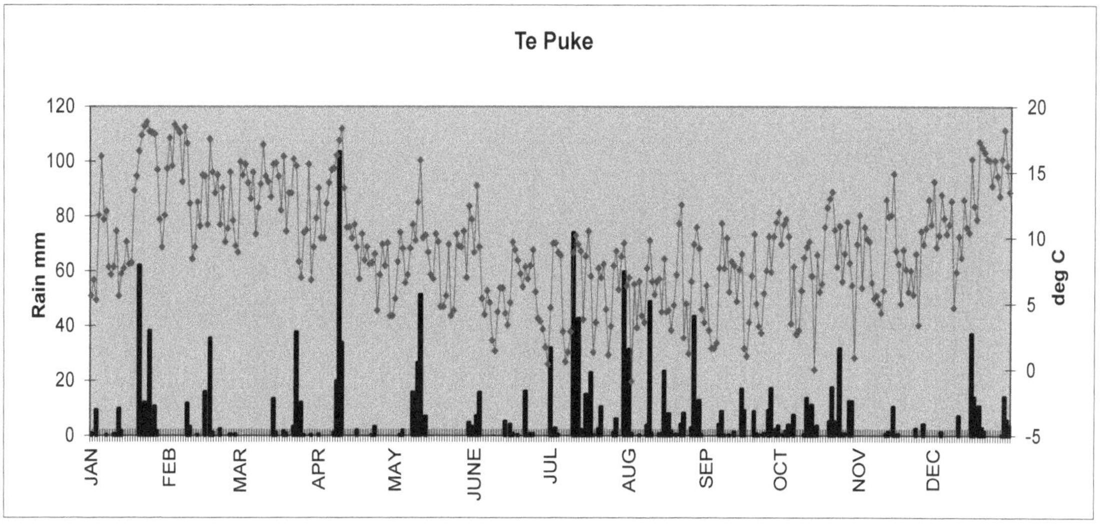
Te Puke
Rain mm
deg C
120
100
80
60
40
20
0
20
15
10
5
0
-5
JAN
FEB
MAR
APR
MAY
JUNE
JUL
AUG
SEP
OCT
NOV
DEC

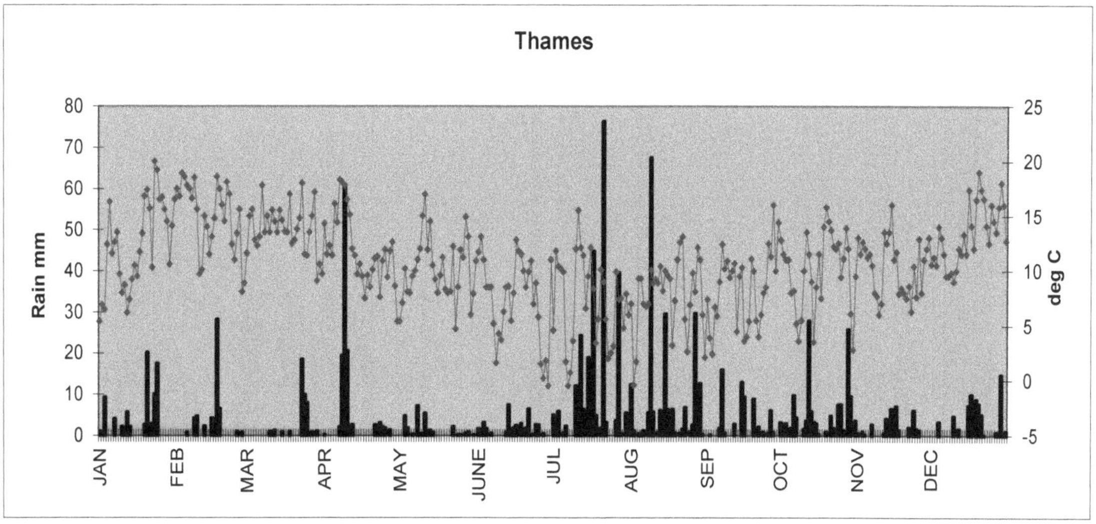
Thames
Rain mm
deg C
80
70
60
50
40
30
20
10
0
25
20
15
10
5
0
-5
JAN
FEB
MAR
APR
MAY
JUNE
JUL
AUG
SEP
OCT
NOV
DEC

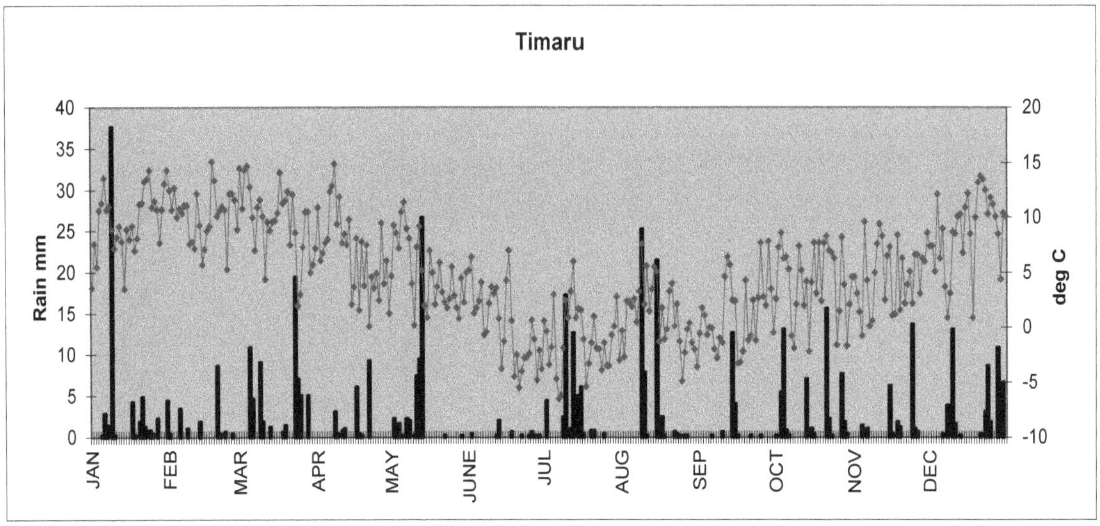
Timaru
Rain mm
deg C
40
35
30
25
20
15
10
5
0
20
15
10
5
0
-5
-10
JAN
FEB
MAR
APR
MAY
JUNE
JUL
AUG
SEP
OCT
NOV
DEC

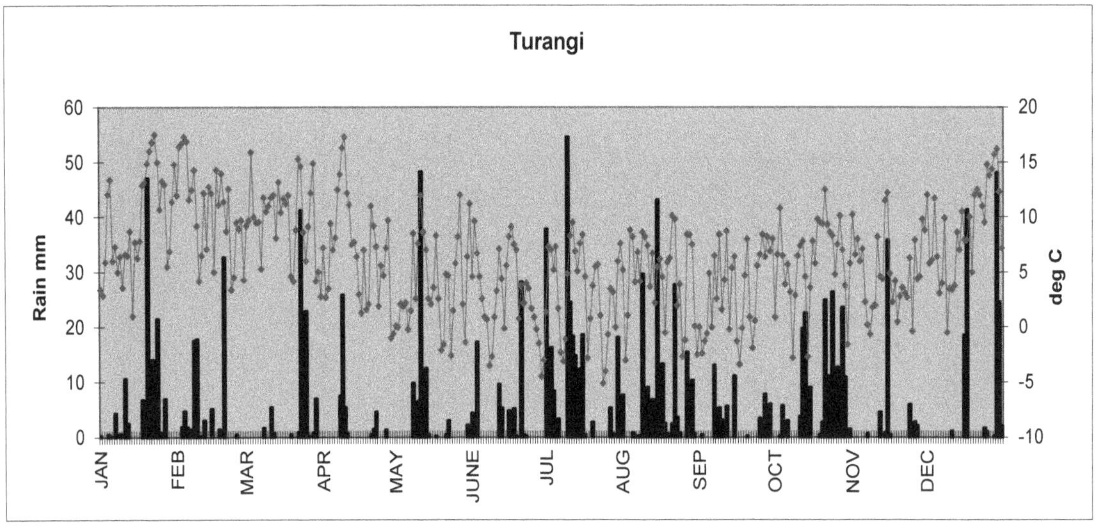
Turangi
Rain mm
deg C
60
50
40
30
20
10
0
20
15
10
5
0
-5
-10
JAN
FEB
MAR
APR
MAY
JUNE
JUL
AUG
SEP
OCT
NOV
DEC

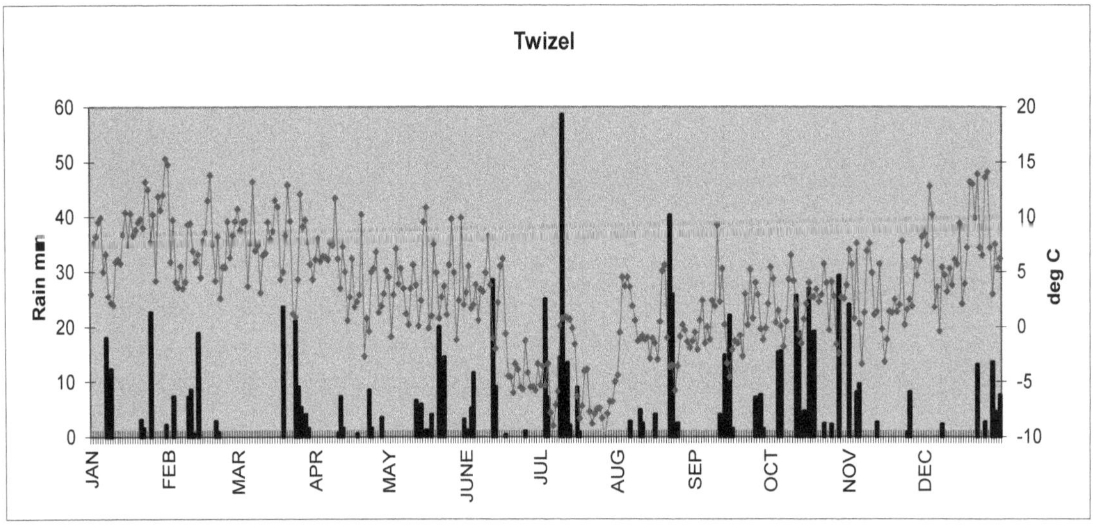
Twizel
Rain mm
deg C
60
50
40
30
20
10
0
20
15
10
5
0
-5
-10
JAN
FEB
MAR
APR
MAY
JUNE
JUL
AUG
SEP
OCT
NOV
DEC

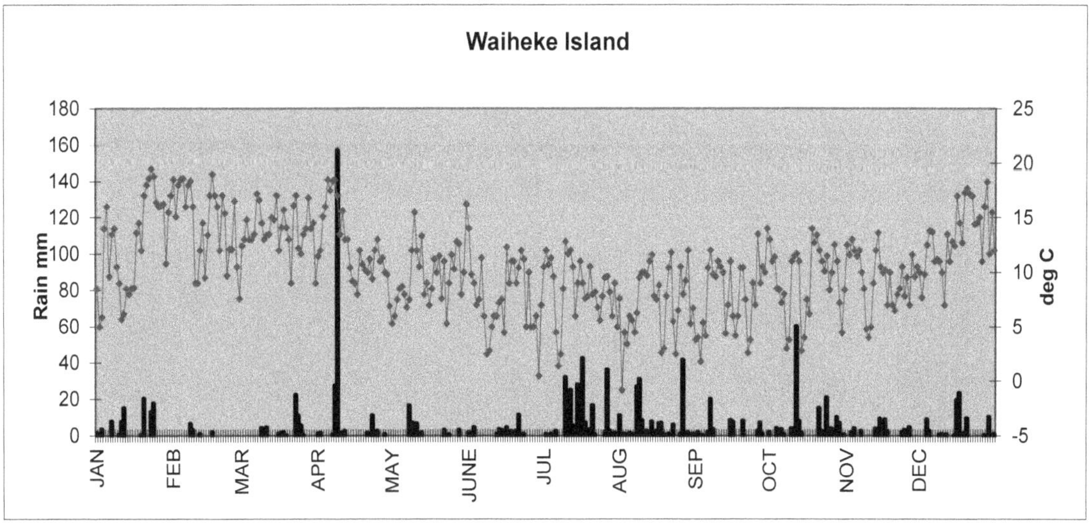
Waiheke Island
Rain mm
deg C
180
160
140
120
100
80
60
40
20
0
25
20
15
10
5
0
-5
JAN
FEB
MAR
APR
MAY
JUNE
JUL
AUG
SEP
OCT
NOV
DEC

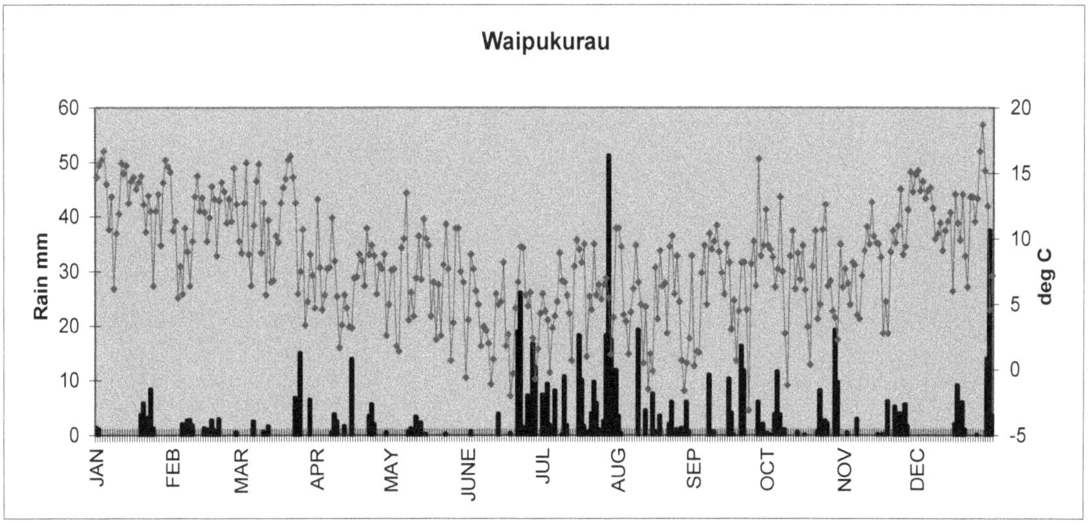
Waipukurau
Rain mm
deg C
60
50
40
30
20
10
0
20
15
10
5
0
-5
JAN
FEB
MAR
APR
MAY
JUNE
JUL
AUG
SEP
OCT
NOV
DEC

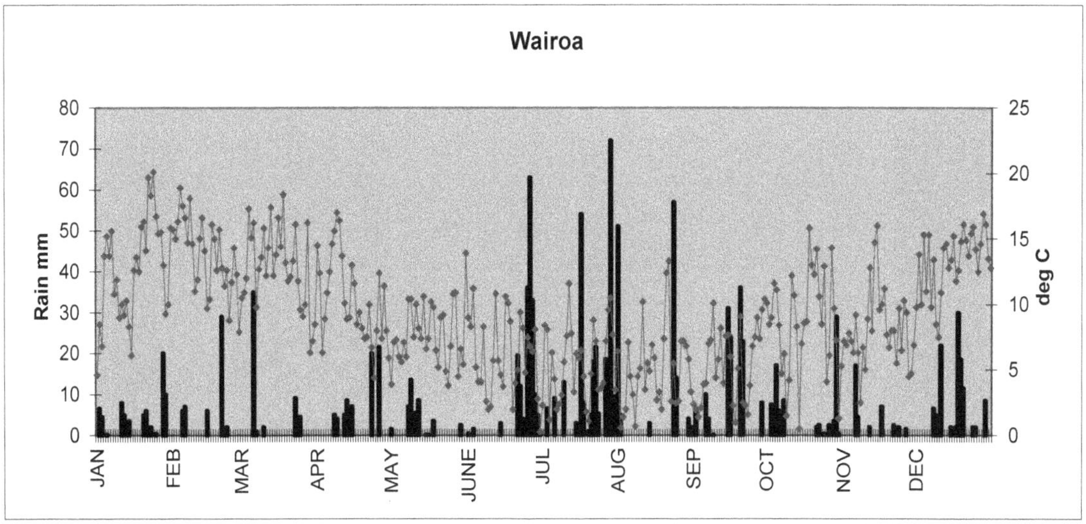
Wairoa
Rain mm
deg C
80
70
60
50
40
30
20
10
0
25
20
15
10
5
0
JAN
FEB
MAR
APR
MAY
JUNE
JUL
AUG
SEP
OCT
NOV
DEC

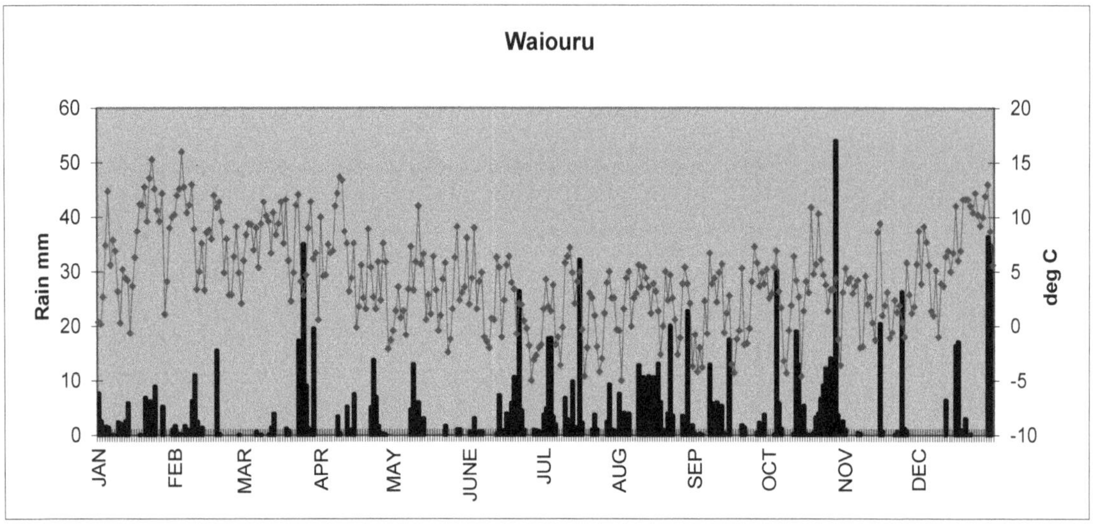
Waiouru
Rain mm
deg C
60
50
40
30
20
10
0
20
15
10
5
0
-5
-10
JAN
FEB
MAR
APR
MAY
JUNE
JUL
AUG
SEP
OCT
NOV
DEC

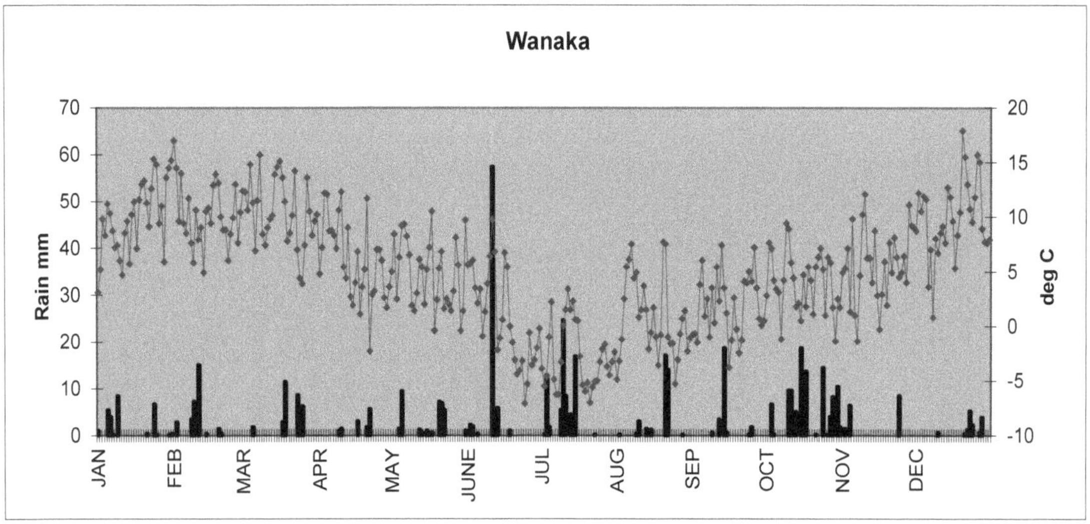
Wanaka
Rain mm
deg C
70
60
50
40
30
20
10
0
20
15
10
5
0
-5
-10
JAN
FEB
MAR
APR
MAY
JUNE
JUL
AUG
SEP
OCT
NOV
DEC

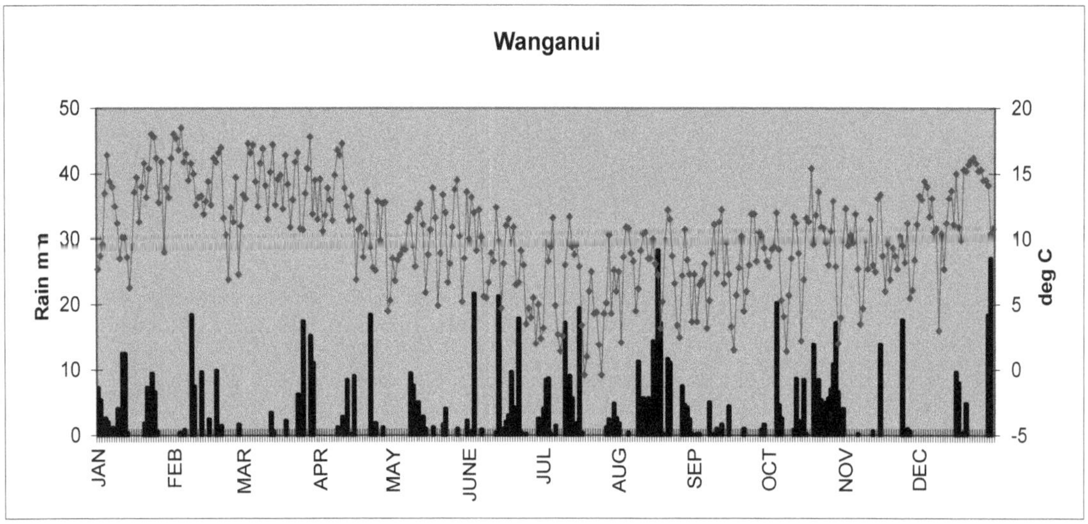
Wanganui
Rain mm
deg C
50
40
30
20
10
0
20
15
10
5
0
-5
JAN
FEB
MAR
APR
MAY
JUNE
JUL
AUG
SEP
OCT
NOV
DEC

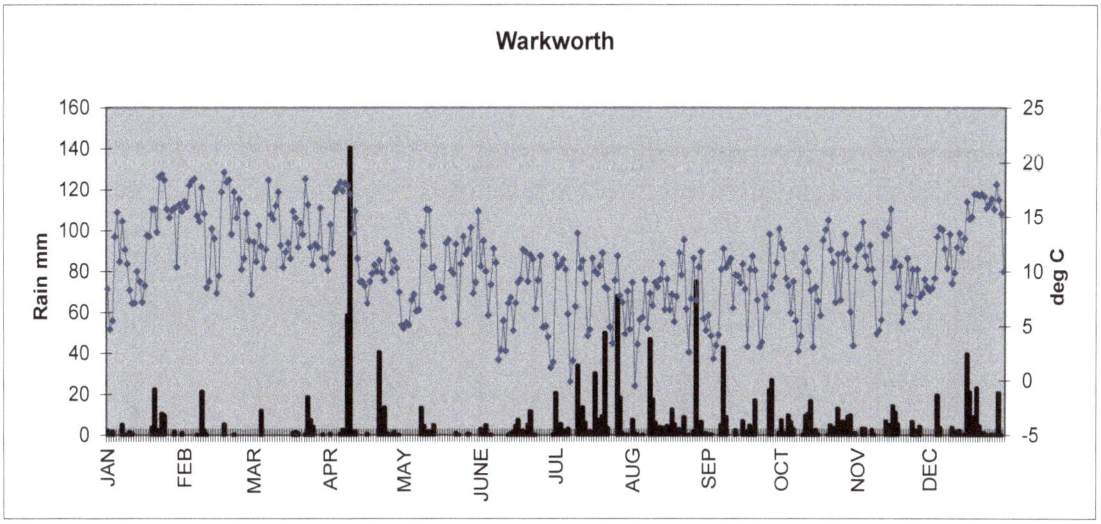
Warkworth
Rain mm
deg C
0
20
40
60
80
100
120
140
160
-5
0
5
10
15
20
25
JAN
FEB
MAR
APR
MAY
JUNE
JUL
AUG
SEP
OCT
NOV
DEC

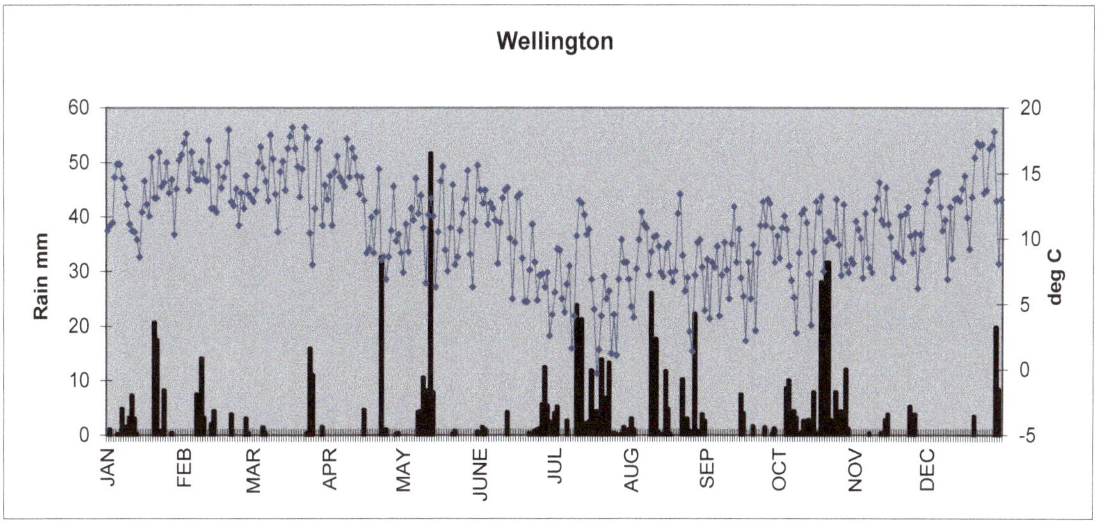
Wellington
Rain mm
deg C
0
10
20
30
40
50
60
-5
0
5
10
15
20
JAN
FEB
MAR
APR
MAY
JUNE
JUL
AUG
SEP
OCT
NOV
DEC

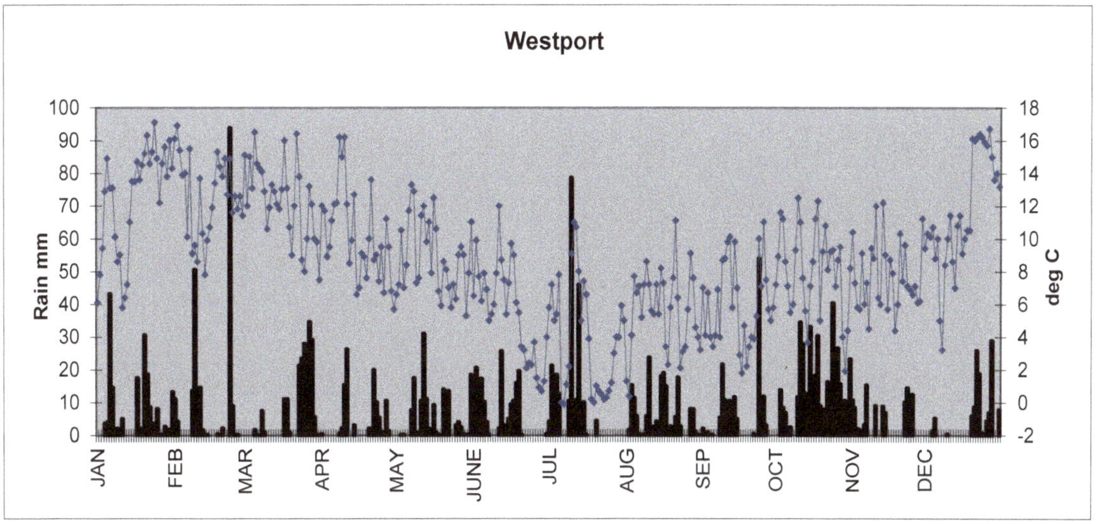
Westport
Rain mm
deg C
0
10
20
30
40
50
60
70
80
90
100
-2
0
2
4
6
8
10
12
14
16
18
JAN
FEB
MAR
APR
MAY
JUNE
JUL
AUG
SEP
OCT
NOV
DEC

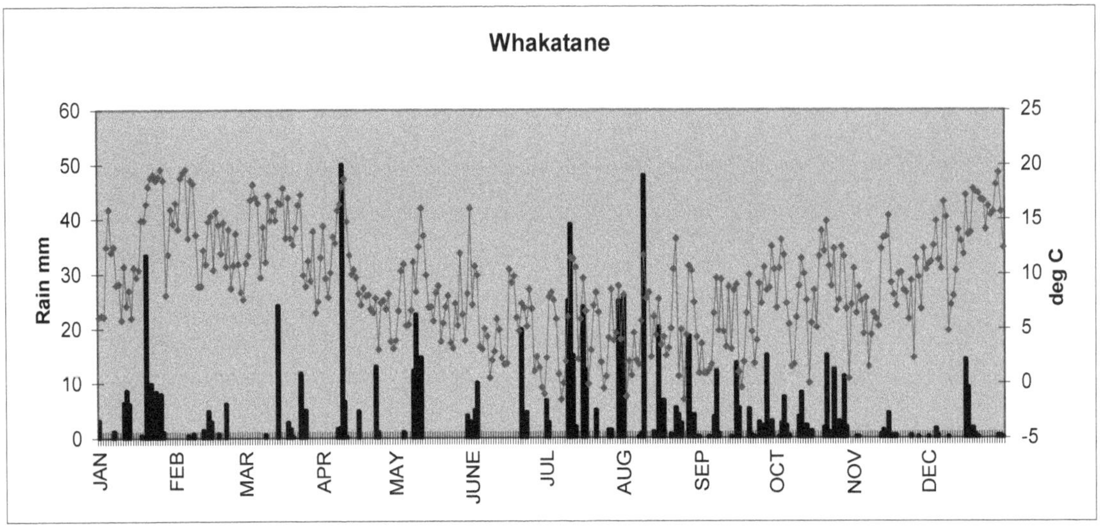
Whakatane
Rain mm
deg C
60
50
40
30
20
10
0
25
20
15
10
5
0
-5
JAN
FEB
MAR
APR
MAY
JUNE
JUL
AUG
SEP
OCT
NOV
DEC

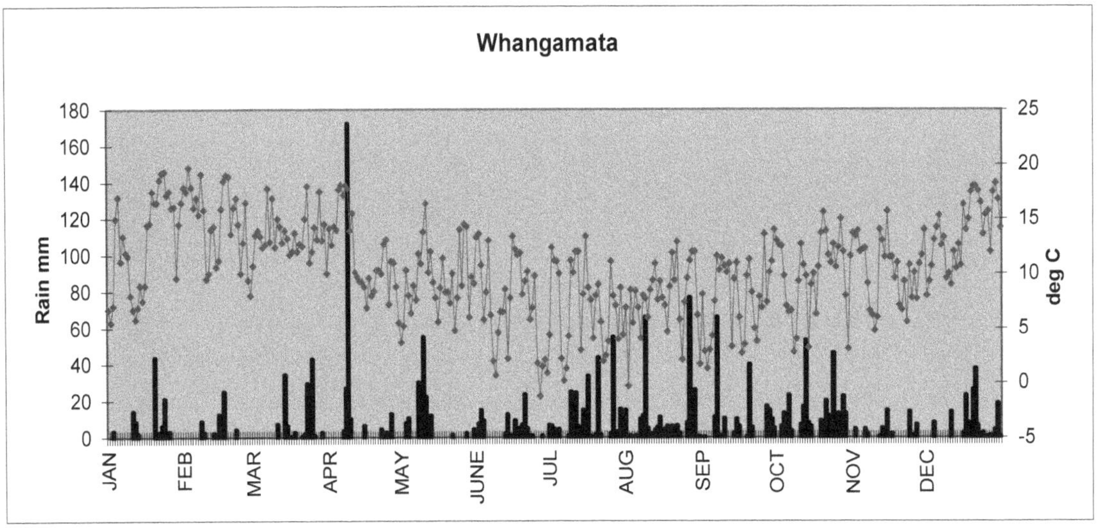
Whangamata
Rain mm
deg C
180
160
140
120
100
80
60
40
20
0
25
20
15
10
5
0
-5
JAN
FEB
MAR
APR
MAY
JUNE
JUL
AUG
SEP
OCT
NOV
DEC

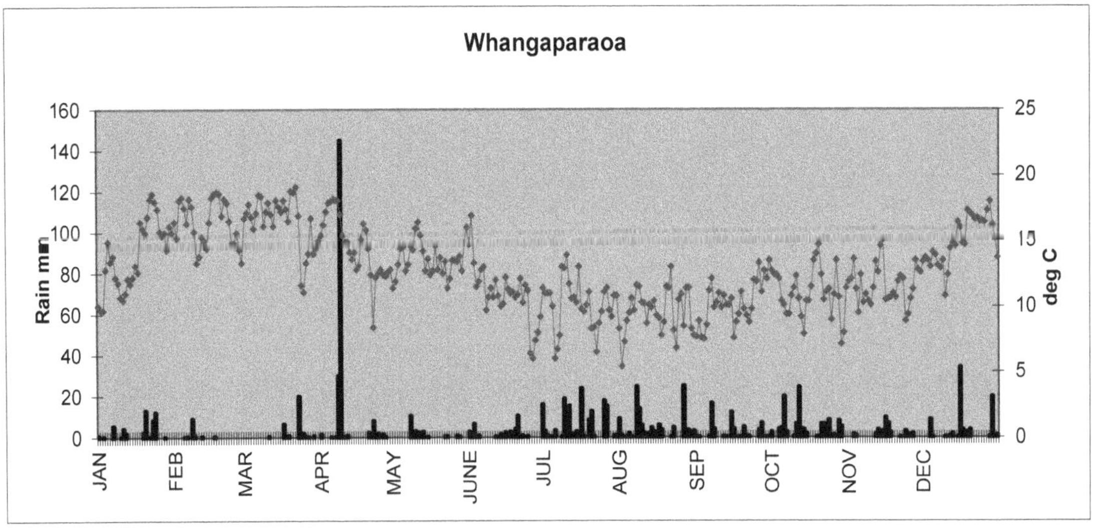
Whangaparaoa
Rain mm
deg C
160
140
120
100
80
60
40
20
0
25
20
15
10
5
0
JAN
FEB
MAR
APR
MAY
JUNE
JUL
AUG
SEP
OCT
NOV
DEC

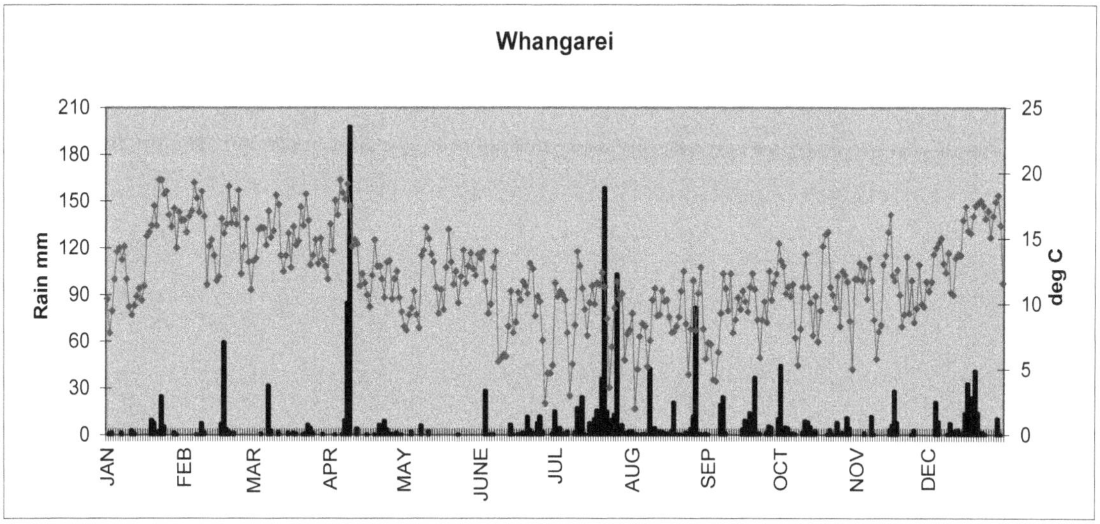
Whangarei
Rain mm
deg C
210
180
150
120
90
60
30
0
25
20
15
10
5
0
JAN
FEB
MAR
APR
MAY
JUNE
JUL
AUG
SEP
OCT
NOV
DEC

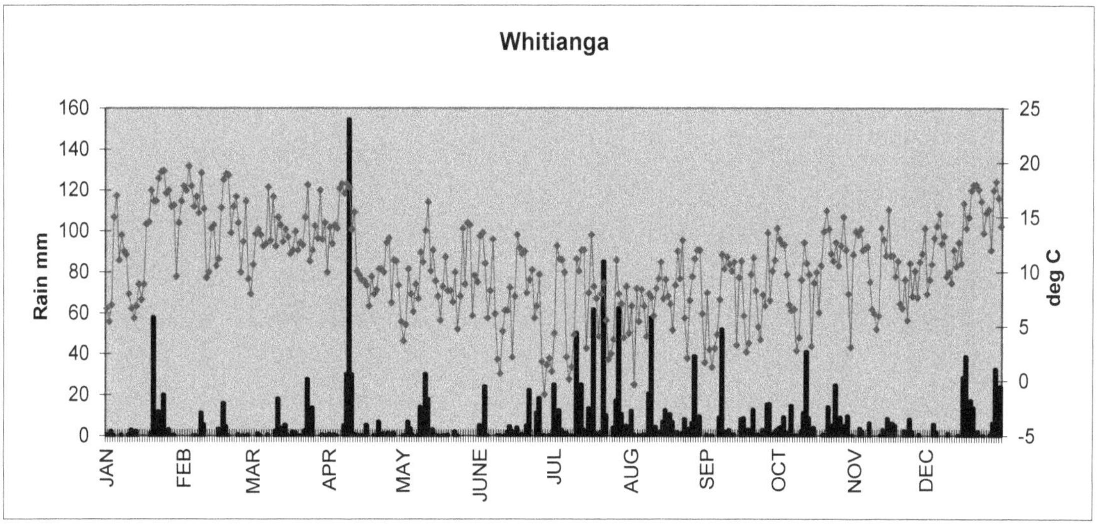
Whitianga
Rain mm
deg C
160
140
120
100
80
60
40
20
0
25
20
15
10
5
0
-5
JAN
FEB
MAR
APR
MAY
JUNE
JUL
AUG
SEP
OCT
NOV
DEC

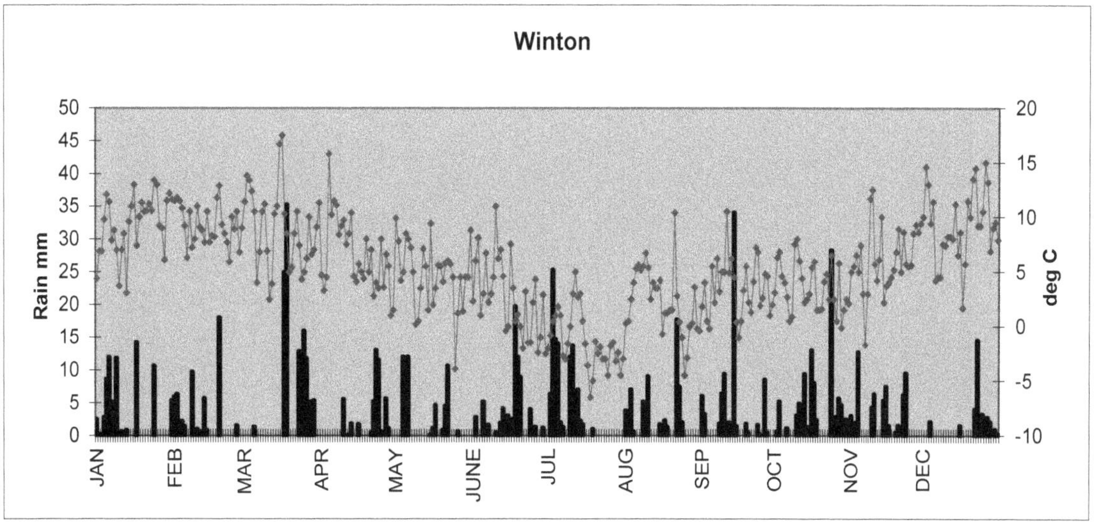
Winton
Rain mm
deg C
50
45
40
35
30
25
20
15
10
5
0
20
15
10
5
0
-5
-10
JAN
FEB
MAR
APR
MAY
JUNE
JUL
AUG
SEP
OCT
NOV
DEC

JANUARY

PHASES OF THE MOON

5th	Crossing Equator
6th	First Quarter
8th	Perigee #13
12th	Northern Declination
13th	Full Moon
19th	Crossing Equator
21st	Third Quarter
21st	Apogee
26th	Southern Declination
29th	New Moon

1	2	3	4	5	6	7	8	9	10	11	12	13	14	15	16	17	18	19	20	21	22	23	24	25	26	27	28	29	30	31
Wed	Thu	Fri	Sat	Sun	Mon	Tue	Wed	Thu	Fri	Sat	Sun	Mon	Tue	Wed	Thu	Fri	Sat	Sun	Mon	Tue	Wed	Thu	Fri	Sat	Sun	Mon	Tue	Wed	Thu	Fri
				XhN	1Q		P13				^	F						XhS		3Q A					V			N		

MONTHLY SUMMARY

4th-9th Wet front crosses the South Island.
4th-5th Chance of light fog and drizzle patches in the Christchurch region.
4th-6th Chance of heavy rain on the West Coast. Strong and blustery winds in the Wellington region.
8th-9th The passage of a wet front crosses southern parts of Canterbury, which may cause some localised flooding, including in the Ashburton district. Colder temperatures may also generate some out-of-season sleet or light snowfalls around the Porter's Pass area.

Over the next four weeks high pressure systems will dominate over the southern half of the country, while troughs of low pressure frequent over central New Zealand. As a result, southeasterly winds will prevail more often than is typical for this time of year over the South Island while frequent westerlies will dominate much of this outlook period over the north of the country. Rainfall will be below average in many regions, especially in eastern parts and including much of Auckland, the Waikato, Bay of Plenty and eastern parts of the country from the Hawkes Bay to Marlborough and down through Canterbury to Otago, where only around half the normal rainfall for this time of year is anticipated. It will also be drier than average in the Wairarapa, Westland and Southland. In contrast, Northland, the Bay of Plenty, Taupo to the East Cape, and Wellington to Nelson can expect to be wetter than average, with some parts possibly seeing up to twice as much recorded rainfall than normal. It will be sunnier by around 10% more than average in Southland while around 10% cloudier than normal in most other regions. Temperatures will be below average in the east of the South Island during this outlook period, while above average in most other parts of the country. Some parts of Wellington, Marlborough, and Canterbury may be up to 0.5-1°C below average, while the Gisborne, Hawkes Bay, the West Coast and Southern Lakes districts of the South Island will be warmer than normal. Most other districts will be closer to their norms for this time of year.

11th Very unsettled and blustery winds in the lower North Island including chance of gale gusts in the Wellington area.
12th Chance of hail in the northwest Waikato region, possibly affecting crops in the Te Puke area.
14th Unusually cold in the Waiouru area.
14th-16th Chance of fog in the Christchurch districts and scattered through Canterbury.
16th Heavy rain likely, particularly in the South Island from Fiordland and up into the West Coast.
18th Chance of fog patches in the lower North Island including in the Wellington region.
19th Chance of heavy rain in the Coromandel to the Bay of Plenty.
19th-24th Widespread wet spell across most of the North Island.
20th Chance of pockets of flooding in the Te Puke district.
20th-21st Wet front brings heavy rain in the West Coast and top of the South Island.
21st-24th Chance of light fog or misty patches-drizzle in the Auckland to the Waikato districts as well as in the Canterbury and Christchurch regions.
22nd The passage of a wet front is likely to bring high rainfall amounts to the north and east of the North Island, with particularly heavy rain in parts of Northland, the Coromandel, and Bay of Plenty.
22nd-23rd Chance of fog patches in the Wellington region.
23rd A wet front brings further heavy rain into the West Coast, particularly around South Westland.
25th-26th Further fog likely in the Christchurch districts as well as down the eastern parts of the South Island.
30th Chance of heavy rain in the West Coast to Southland.

January 2025

Monday	Tuesday	Wednesday	Thursday	Friday	Saturday	Sunday
30	31 Moon Sign	1 Capricorn Conjunct Sun Trine Uranus	2 Aquarius Opposition Mars Conjunct Plugo	3 Aquarius Trine Jupiter	4 Pisces Conjunct Venus	5 Pisces Sextile Sun Square Jupiter Conjunct Saturn
6 XhN Pisces Square Mercury Trine Mars Conjunct Neptune Sextile Pluto	7 1Q Aries Square Sun	8 P13 Aries Trine Mercury Square Mars Square Pluto	9 Taurus Trine Sun Sextile Saturn	10 Taurus Sextile Mars Opposition Uranus Sextile Neptune Trine Pluto Square Ascendant	11 Gemini Square Venus Conjunct Jupiter Square Saturn	12 ^ Gemini Opposition Mercury Square Neptune Square Ascendant
13 Cancer Opposition Mercury	14 FM Cancer Opposition Sun Conjunct Mars Sextile Uranus Trine Neptune Opposition Pluto	15 Leo Sextile Jupiter Square Ascendant	16 Leo Square Uranus Trine Ascendant	17 Virgo Trine Mercury Opposition Venus Square Jupiter	18 Virgo Trine Mercury Oppostion Venus Opposition Saturn Trine Uranus	19 XhS Libra Trine Sun Opposition Neptune Trine Pluto
20 Libra Square Mercury Trine Jupiter	21 A Libra Square Sun Square Mercury Square Mars Square Pluto	22 3Q Scorpio Square Sun Square Pluto	23 Scorpio Sextile Mercury Trine Venus Trine Mars Trine Saturn Opposition Uranus	24 Sagittarius Sextile Sun Neptune Sextile Pluto	25 Sagittarius Opposition Jupiter Square Saturn	26 Sagittarius Square Venus Square Neptune Square Ascendant
27 V Capricorn	28 Capricorn	29 Capricorn Conjunct Mercury Sextile Venus Opposition Mars Sextile Neptune Conjunct Pluto	30 NM Aquarius Conjunct Sun Conjunct Mercury Trine Jupiter Conjunct Pluto	31 Aquarius Square Uranus	1	2
3	4					

NORTH ISLAND RAINFALL

JANUARY 2025

d=mainly dry, n= fog, mist, dew or nonrecordable, l=light showers, p=passing showers, s=significant showers, r=rain, h=heavy falls

		Northland							Auckland					Bay Of Plenty				Waikato			
		CR	Kta	Ker	Rsl	Dar	Whg	Man	Wrk	Hen	MkH	Ak	Co	Tau	TeP	Wkt	Opo	Hml	TeA	Mata	Ot
1st	Wed	d	d	d	d	d	d	l	l	l	s	d	d	l	l	l	l	d	l	l	l
2nd	Thur	d	l	d	d	d	l	l	d	d	l	d	n	d	d	d	d	d	d	d	l
3rd	Fri	l	l	d	d	l	l	d	l	l	s	l	d	l	r	d	l	s	l	d	n
4th	Sat	d	d	d	d	d	d	d	d	d	d	d	d	d	d	d	d	n	l	d	l
5th	Sun	d	d	d	d	d	d	d	d	d	d	d	d	d	d	d	d	d	l	d	n
6th	Mon	d	d	d	d	d	d	d	d	d	d	d	d	d	d	d	d	d	d	d	d
7th	Tue	l	s	l	s	l	l	l	s	s	s	s	d	l	d	l	l	l	l	l	l
8th	Wed	d	d	d	d	d	d	d	d	d	d	d	l	d	d	d	d	d	d	d	l
9th	Thur	d	d	d	d	d	d	d	d	d	l	d	d	d	d	d	d	l	d	d	d
10th	Fri	d	d	d	d	l	d	l	l	l	l	l	d	d	n	d	d	d	d	d	l
11th	Sat	d	d	d	d	l	l	d	l	l	l	l	s	l	n	s	s	l	s	l	l
12th	Sun	l	d	d	d	l	l	l	d	l	s	n	r	l	r	s	r	n	l	d	l
13th	Mon	d	d	d	d	d	d	d	d	d	d	d	d	s	l	s	l	d	d	d	d
14th	Tue	d	d	d	d	d	d	d	d	d	d	d	d	d	d	d	d	d	d	d	d
15th	Wed	d	d	d	d	d	d	d	d	d	d	d	d	d	d	d	d	d	d	d	d
16th	Thur	d	d	d	d	d	d	d	d	d	d	d	d	d	d	d	d	d	d	d	d
17th	Fri	d	d	d	d	d	d	d	d	d	d	d	d	d	d	d	d	d	d	d	d
18th	Sat	d	d	d	d	d	d	d	d	d	d	d	d	d	d	n	d	d	d	d	d
19th	Sun	h	s	h	h	s	r	s	s	l	l	l	d	d	d	d	d	d	d	d	d
20th	Mon	l	r	h	r	l	s	h	h	r	s	r	h	h	h	h	h	h	h	h	r
21st	Tue	l	h	d	d	d	d	d	l	d	l	l	d	d	d	d	d	s	d	d	l
22nd	Wed	d	s	s	s	l	l	l	l	d	d	d	r	r	r	r	r	l	h	l	l
23rd	Thur	h	h	h	h	h	h	h	r	s	r	s	s	s	r	s	n	l	r	l	l
24th	Fri	l	r	r	h	l	s	h	r	r	h	r	h	h	h	s	r	h	l	h	h
25th	Sat	d	d	d	d	d	d	d	d	d	d	d	l	d	l	l	s	d	l	d	d
26th	Sun	d	d	d	d	d	d	d	d	d	d	d	d	s	r	s	l	d	h	l	d
27th	Mon	d	d	d	d	d	d	d	d	d	l	d	d	s	l	l	n	d	d	d	d
28th	Tue	d	d	n	d	d	l	d	l	d	n	d	d	d	d	d	d	l	d	d	l
29th	Wed	d	n	d	d	d	d	d	d	d	d	d	d	d	d	d	d	d	d	d	d
30th	Thur	d	d	d	d	d	d	d	d	d	d	d	d	d	d	d	d	d	l	d	d
31st	Fri	l	d	d	d	d	d	d	l	n	l	n	d	d	d	d	d	d	d	d	n
estimate:		63	99	122	127	51	59	80	64	50	71	42	111	103	158	92	89	67	106	52	61
average:		63	95	110	93	68	67	96	102	100	64	70	69	87	103	90	81	92	76	75	98
trend:		av	av	wtr	wtr	drr	av	drr	drr	drr	av	drr	wtr	wtr	wtr	av	av	drr	wtr	drr	drr

		Rot/Tau		Hawkes Bay			C Plateau			Taranaki			Manawatu			Lower North Island					
		Rt	Taup	Gis	Nap	Has	Taum	Chat	Waio	NP	Str	Wang	Pa	Dann	Lev	PN	Mast	UH	Wgt	all	Moon
1st	Wed	l	s	r	r	r	l	s	s	r	s	s	d	d	d	l	s	n	d	117	
2nd	Thur	d	d	l	l	l	l	d	l	l	d	s	d	l	l	l	l	l	l	33	
3rd	Fri	r	d	l	l	s	d	d	d	l	n	l	d	d	l	l	d	d	d	67	
4th	Sat	d	n	d	d	d	l	r	l	r	l	l	s	d	h	s	d	l	d	79	
5th	Sun	d	d	d	d	d	l	s	l	l	n	l	l	d	s	s	d	d	d	30	
6th	Mon	d	d	d	d	d	d	d	d	d	d	d	d	d	d	d	d	d	d	1	XhN
7th	Tue	l	l	d	d	d	l	r	d	s	r	l	l	d	s	l	d	s	s	112	1Q
8th	Wed	d	d	d	d	d	d	n	d	d	d	d	d	d	d	d	d	s	d	6	P13
9th	Thur	d	d	d	d	d	l	s	l	l	s	s	l	d	r	s	l	s	l	47	
10th	Fri	d	n	d	d	d	d	l	l	d	d	l	d	l	d	l	n	s	l	29	
11th	Sat	s	r	s	s	l	r	l	d	s	d	r	d	n	d	d	s	l	s	123	
12th	Sun	s	s	l	s	h	d	d	l	d	d	r	s	l	d	l	s	r	l	142	^
13th	Mon	l	d	l	d	l	d	d	s	d	d	n	n	n	d	d	d	d	d	30	
14th	Tue	d	d	d	d	d	d	d	d	d	d	d	d	d	d	d	d	d	d	0	F
15th	Wed	d	d	d	d	d	d	d	d	d	d	d	d	d	d	d	d	d	d	0	
16th	Thur	d	d	d	d	d	d	d	d	d	d	d	d	d	d	d	d	d	d	0	
17th	Fri	d	d	d	d	d	d	d	d	d	d	d	d	l	d	d	d	d	d	1	
18th	Sat	l	l	d	d	d	l	s	d	d	d	d	d	d	d	d	d	d	d	12	
19th	Sun	d	d	d	d	d	d	d	d	d	d	d	d	d	d	d	d	d	d	107	XhS
20th	Mon	h	s	s	s	s	r	r	s	r	h	l	l	l	l	l	l	s	h	612	
21st	Tue	l	r	d	d	d	l	r	d	d	l	s	s	r	l	l	d	h	h	126	A
22nd	Wed	s	l	h	l	s	n	s	s	l	d	s	d	s	d	d	d	d	d	163	3Q
23rd	Thur	h	r	d	l	s	s	s	l	h	h	r	l	s	l	l	s	l	l	439	
24th	Fri	h	h	d	l	l	r	h	s	h	r	s	s	l	l	l	d	r	s	470	
25th	Sat	d	l	d	d	d	d	l	d	d	d	l	l	d	n	d	d	d	d	15	
26th	Sun	r	s	d	d	d	d	d	d	d	d	d	d	d	d	d	d	d	d	69	
27th	Mon	s	s	l	d	l	d	l	s	d	d	d	d	n	d	d	d	d	n	32	V
28th	Tue	l	d	n	l	d	d	d	d	d	d	d	d	d	d	d	d	d	d	8	
29th	Wed	d	d	d	d	d	d	d	d	d	d	d	d	d	d	d	d	d	d	1	
30th	Thur	d	d	d	d	d	d	d	d	d	d	d	d	d	d	d	d	d	d	2	N
31st	Fri	d	d	d	d	d	d	l	l	d	d	d	d	d	l	d	d	d	d	10	
estimate:		129	92	49	38	63	62	119	58	95	80	81	37	29	69	34	28	84	69	2882	
average:		110	92	70	72	70	101	209	86	110	139	67	71	73	81	75	62	92	77	3356	
trend:		wtr	av	drr	drr	av	drr	drr	drr	drr	drr	wtr	drr	drr	drr	drr	drr	av	av	drr	

SOUTH ISLAND RAIN

JANUARY 2025

d=mainly dry, n= fog, mist, dew or nonrecordable, l=light showers, p=passing showers, s=significant showers, r=rain, h=heavy falls

		West Coast				Top South Island				Canterbury				South Canterbury				Coastal Otago			Lakes			Inland Otago				Southland						
		West	Grey	Hok	BB	Ta	Mou	Nel	Ble	Dar	Ham	Kai	Ch	Rak	Meth	Ash	Tim	Oam	Dun	Bal	Twi	LP	LT	Crom	Qstn	Ran	Alex	TeA	Wtn	Gore	Inv	Puy Pt	all	Moon
1st	Wed	d	d	d	d	d	d	d	d	l	l	l	r	d	l	l	d	n	l	l	d	h	n	d	d	d	d	l	l	l	l	l	56	
2nd	Thur	d	d	d	d	d	s	r	l	l	s	l	s	l	l	l	d	d	l	d	d	d	d	d	d	d	d	d	d	n	d	d	35	
3rd	Fri	d	l	d	d	l	d	d	l	d	d	d	l	d	d	d	d	d	d	d	d	d	d	n	d	d	d	l	d	d	n	n	9	
4th	Sat	s	d	l	d	d	d	d	d	d	l	d	d	d	l	l	d	d	d	d	d	l	d	d	n	d	d	d	l	l	l	s	23	
5th	Sun	s	s	s	r	d	d	d	d	d	d	d	d	d	d	d	d	l	s	r	d	d	l	s	s	s	s	r	s	r	r	s	121	
6th	Mon	h	h	h	h	l	d	n	d	l	d	d	n	d	d	l	l	r	r	h	d	d	l	l	l	r	s	l	r	h	r	s	322	XhN
7th	Tue	r	h	h	h	h	l	s	l	d	l	d	n	d	d	l	l	r	r	l	h	h	s	l	r	h	l	s	s	s	l	l	261	1Q
8th	Wed	l	l	l	l	d	d	d	d	d	d	d	d	d	d	d	d	d	s	d	s	d	d	d	d	s	d	s	l	l	l	h	61	P13
9th	Thur	l	h	h	r	d	d	d	d	h	d	d	l	h	h	h	h	r	l	h	r	d	h	s	s	h	s	s	r	h	l	l	463	
10th	Fri	d	d	d	d	h	d	r	s	s	r	r	s	l	s	s	d	d	d	l	d	d	l	d	d	d	d	d	d	n	d	n	100	
11th	Sat	s	d	d	d	l	l	s	d	l	s	s	l	d	d	l	d	d	d	d	d	d	l	d	d	d	d	d	l	d	d	d	27	
12th	Sun	d	d	d	d	r	l	s	n	l	l	d	l	l	l	l	d	n	l	l	d	d	d	d	d	l	d	d	d	d	l	l	39	^
13th	Mon	d	d	d	d	h	d	s	d	d	s	d	d	d	l	l	d	d	d	d	d	d	d	d	d	d	d	d	l	l	d	l	37	
14th	Tue	d	n	n	d	s	d	l	d	d	d	d	d	d	d	d	d	d	d	d	d	d	d	d	d	d	d	d	d	d	d	d	7	F
15th	Wed	d	d	d	l	d	d	d	d	d	d	d	d	d	d	d	d	d	d	d	d	d	d	d	d	d	d	d	d	d	d	d	1	
16th	Thur	d	d	d	l	d	d	d	d	d	d	d	d	d	d	d	d	d	d	d	d	d	d	d	d	d	d	d	d	d	d	r	10	
17th	Fri	h	h	h	h	d	d	d	d	l	s	l	l	l	s	l	s	l	r	s	d	d	d	d	l	l	l	r	r	r	h	h	201	
18th	Sat	l	l	s	d	d	d	d	d	d	d	d	d	d	d	d	d	d	d	d	d	d	d	l	d	d	d	d	d	d	d	d	11	
19th	Sun	l	l	d	d	n	d	d	d	d	d	d	d	d	d	d	d	d	d	d	d	d	d	d	d	d	d	d	d	d	n	d	5	XhS
20th	Mon	h	h	h	l	h	h	h	r	r	h	s	s	r	r	s	l	l	d	d	d	d	s	d	d	d	d	d	d	d	d	d	287	
21st	Tue	h	h	h	h	h	l	r	s	s	l	l	l	d	l	s	s	s	l	l	l	r	l	d	d	l	d	d	d	d	d	d	197	A
22nd	Wed	s	d	d	l	d	d	l	d	l	l	n	d	d	l	l	l	d	n	d	l	d	l	d	r	d	s	l	d	l	d	d	44	3Q
23rd	Thur	s	l	l	d	r	s	l	l	d	l	l	d	l	l	n	d	l	d	d	d	d	d	d	d	d	d	d	d	d	d	d	34	
24th	Fri	l	s	r	h	h	r	h	9	d	d	l	d	l	l	l	l	l	s	l	d	d	l	s	s	s	l	d	r	r	l	d	190	
25th	Sat	s	r	r	l	d	d	d	d	l	d	d	d	d	d	n	n	n	d	d	h	d	d	l	l	s	l	d	d	d	l	r	81	
26th	Sun	d	l	d	d	d	d	d	d	l	d	d	d	d	n	d	d	d	d	d	d	d	d	d	d	d	d	d	d	d	d	d	3	
27th	Mon	d	d	d	d	d	d	d	d	l	d	l	d	d	l	l	l	d	d	d	d	d	d	d	d	d	d	d	d	d	n	l	13	V
28th	Tue	l	d	d	l	d	d	d	d	d	l	l	d	l	l	d	d	d	d	d	d	d	d	d	d	d	d	d	d	d	d	d	12	
29th	Wed	l	l	d	d	d	d	d	d	d	d	d	d	d	d	d	d	d	d	l	d	d	d	d	d	d	d	d	d	l	l	l	9	
30th	Thur	l	s	l	s	d	d	d	d	d	d	d	d	l	l	d	d	d	d	s	d	d	l	d	l	l	l	l	d	l	s	s	49	N
31st	Fri	r	r	h	h	d	d	d	d	l	d	l	l	d	l	s	s	s	r	l	l	s	l	d	n	s	l	l	s	l	s	h	163	
estimate:		186	210	209	272	182	67	102	34	69	68	35	45	69	93	136	62	58	70	95	68	74	39	24	50	81	34	59	76	98	80	121	2866	
average:		182	212	240	445	165	77	72	55	57	69	55	52	72	78	63	62	62	87	77	57	76	50	42	80	50	45	127	102	96	100	207	3214	
trend:		av	av	drr	drr	wtr	av	wtr	drr	wtr	av	drr	av	av	wtr	wtr	av	av	drr	wtr	wtr	av	drr	drr	drr	wtr	drr	drr	drr	av	drr	drr	drr	

NORTH ISLAND SUNSHINE ESTIMATES
JANUARY 2025

F=fine (8-12 hours of sunshine), pc= partly cloudy (4-7 hours), c=cloudy (1-3 hours), o=overcast (0 hours)

		Kta	Whg	Dar	Ak	Hml	Tau	Rt	Taum	NP	Wang	Taup	Gis	Nap	Wai	Dann	PN	Mast	Pa	UH	Wgt	ALL	Moon
1st	Wed	F	F	F	F	F	F	F	pc	F	F	F	F	F	F	pc	F	pc	pc	pc	pc	186	
2nd	Thur	pc	pc	pc	pc	pc	pc	pc	pc	pc	o	pc	c	c	c	c	o	c	c	o	c	74	
3rd	Fri	pc	pc	pc	pc	pc	F	F	F	pc	c	F	F	F	pc	F	c	F	pc	c	F	140	
4th	Sat	F	F	pc	pc	o	o	o	o	o	o	o	pc	F	pc	F	o	F	o	c	pc	85	
5th	Sun	F	F	F	F	pc	pc	pc	F	o	c	pc	F	F	pc	pc	c	pc	o	o	o	123	
6th	Mon	c	c	c	c	c	pc	pc	c	pc	pc	pc	pc	o	c	F	pc	F	c	c	o	72	XhN
7th	Tue	pc	pc	F	F	F	F	F	c	pc	pc	F	F	F	F	pc	pc	pc	pc	c	F	161	1Q
8th	Wed	pc	pc	pc	pc	F	F	F	F	pc	pc	F	F	F	F	F	pc	F	F	pc	F	182	P13
9th	Thur	F	pc	F	F	F	F	F	F	F	pc	F	F	F	pc	pc	pc	pc	F	pc	F	158	
10th	Fri	F	pc	pc	pc	c	F	pc	pc	pc	pc	pc	pc	F	F	o	pc	o	c	o	c	97	
11th	Sat	F	F	F	F	pc	pc	c	pc	F	o	c	o	c	c	o	o	o	c	o	o	80	
12th	Sun	F	F	F	F	F	F	F	pc	pc	o	F	c	c	o	pc	o	pc	pc	o	o	123	^
13th	Mon	F	F	F	F	F	F	F	F	F	pc	F	c	c	c	o	pc	o	F	pc	pc	161	
14th	Tue	F	F	F	F	F	F	F	F	F	F	F	F	F	F	pc	F	pc	F	F	F	245	F
15th	Wed	pc	F	F	F	F	F	F	F	F	c	F	F	F	F	F	c	F	pc	pc	pc	196	
16th	Thur	F	pc	o	o	F	F	F	F	pc	c	F	F	F	F	pc	c	pc	o	pc	F	162	
17th	Fri	F	F	F	F	pc	F	F	F	F	F	F	pc	c	c	F	F	F	pc	c	pc	156	
18th	Sat	o	o	pc	pc	o	pc	c	o	F	o	c	F	F	pc	F	o	F	c	c	pc	90	
19th	Sun	o	o	o	o	o	o	o	o	o	o	o	o	c	o	F	o	F	o	o	o	24	XhS
20th	Mon	F	F	c	c	c	pc	pc	c	c	c	pc	F	F	pc	o	c	o	o	o	o	74	
21st	Tue	F	pc	pc	pc	o	pc	c	pc	F	F	c	pc	F	F	c	F	c	F	pc	F	120	A
22nd	Wed	o	o	o	o	o	o	o	o	o	o	o	o	o	o	c	o	c	c	o	pc	10	3Q
23rd	Thur	o	o	o	o	o	o	o	o	c	c	o	c	c	c	F	c	F	o	c	c	38	
24th	Fri	F	F	F	F	c	pc	pc	pc	c	o	pc	F	F	F	c	o	c	c	c	F	131	
25th	Sat	F	F	F	F	pc	c	pc	F	F	pc	pc	c	pc	F	F	pc	F	pc	F	F	144	
26th	Sun	F	pc	pc	pc	c	c	pc	c	pc	F	pc	F	pc	pc	pc	F	pc	F	F	F	143	
27th	Mon	o	o	c	c	c	o	c	pc	F	pc	c	o	F	o	F	pc	F	F	c	c	84	V
28th	Tue	pc	F	F	F	F	F	F	F	F	F	F	F	F	F	F	F	F	F	F	F	233	
29th	Wed	F	F	pc	pc	F	F	F	F	pc	o	F	F	F	pc	F	o	F	c	F	F	170	
30th	Thur	F	F	c	c	pc	F	c	pc	pc	c	c	F	F	pc	F	c	F	o	pc	c	108	N
31st	Fri	F	F	F	F	F	F	F	pc	c	o	F	F	F	pc	F	o	F	o	o	o	131	
est. hours:		259	213	207	207	169	235	211	199	194	132	211	224	241	200	212	132	212	147	122	177	3901	
av. hours:		230	229	227	232	232	253	245	202	246	244	227	241	242	233	211	208	232	239	242	246	4661	
trend:		more	av	av	less	less	av	less	av	less	less	av	av	av	less	av	less	av	less	less	less	less	

SOUTH ISLAND SUNSHINE ESTIMATES
JANUARY 2025

F=fine (8-12 hours of sunshine), pc= partly cloudy (4-7 hours), c=cloudy (1-3 hours), o=overcast (0 hours)

		Ta	Nel	Ble	Mou	West	Grey	Hok	Ham	Kai	Ch	Ash	Tim	Crom	Qstn	Bal	Dun	Wnt	Gore	Inv	Alex	ALL	Moon
1st	Wed	F	F	F	F	F	F	F	c	c	c	pc	F	F	F	o	c	c	c	o	F	145	
2nd	Thur	c	c	pc	pc	F	F	F	o	o	o	c	c	F	F	c	c	F	F	F	F	127	
3rd	Fri	pc	pc	F	pc	o	o	o	pc	pc	pc	c	c	pc	pc	c	o	o	o	o	pc	57	
4th	Sat	F	F	F	F	o	o	c	pc	pc	pc	pc	pc	c	c	o	pc	o	o	o	c	80	
5th	Sun	F	o	o	c	c	c	o	o	o	o	o	o	c	c	pc	o	o	o	c	c	35	
6th	Mon	F	c	pc	pc	c	c	c	pc	pc	pc	pc	c	pc	c	c	c	pc	pc	pc	pc	80	XhN
7th	Tue	c	F	F	F	F	F	F	F	F	F	F	F	F	pc	pc	pc	c	c	pc	F	163	1Q
8th	Wed	c	F	F	F	F	pc	c	F	F	F	pc	c	c	c	pc	pc	o	o	o	c	110	P13
9th	Thur	pc	F	F	F	F	F	F	o	o	pc	c	pc	c	c	c	pc	pc	pc	pc	c	107	
10th	Fri	o	pc	pc	pc	pc	F	F	o	o	c	c	pc	pc	pc	c	F	pc	pc	c	pc	102	
11th	Sat	c	c	pc	c	F	F	F	pc	pc	pc	pc	pc	F	F	o	c	o	o	c	F	94	
12th	Sun	F	pc	pc	F	F	F	F	o	o	pc	c	o	F	F	pc	pc	pc	pc	pc	F	152	^
13th	Mon	F	F	pc	F	F	F	F	pc	pc	F	F	c	F	F	F	F	F	F	F	F	211	
14th	Tue	o	o	F	c	o	pc	pc	F	F	F	F	F	F	F	F	F	F	F	pc	F	186	F
15th	Wed	F	F	F	F	c	c	o	F	F	F	F	F	F	F	F	F	F	F	F	F	193	
16th	Thur	o	F	F	F	pc	c	c	F	F	F	pc	c	o	c	c	c	pc	pc	pc	o	105	
17th	Fri	o	o	c	o	o	o	pc	o	o	c	c	o	F	pc	F	F	pc	pc	pc	F	69	
18th	Sat	pc	F	F	F	o	pc	pc	F	F	F	F	pc	c	o	o	pc	c	c	c	c	124	
19th	Sun	pc	o	o	o	o	o	c	c	c	c	pc	c	c	pc	o	o	pc	pc	F	c	55	XhS
20th	Mon	o	o	o	o	pc	o	o	o	o	o	o	o	o	o	o	o	o	o	o	o	9	
21st	Tue	pc	F	F	F	pc	c	pc	c	c	o	o	pc	pc	pc	o	c	pc	pc	pc	pc	93	A
22nd	Wed	F	pc	pc	pc	F	F	F	c	c	o	c	o	F	F	pc	o	F	F	F	F	133	3Q
23rd	Thur	pc	o	o	o	o	o	o	o	o	o	o	o	pc	F	F	F	F	F	F	pc	72	
24th	Fri	c	pc	pc	F	c	o	o	pc	pc	pc	c	c	c	c	c	c	pc	pc	pc	c	60	
25th	Sat	pc	F	F	F	pc	pc	F	pc	pc	F	pc	c	F	F	pc	o	F	F	F	F	169	
26th	Sun	pc	pc	F	F	pc	pc	F	pc	pc	pc	pc	pc	F	F	F	F	pc	pc	pc	F	153	
27th	Mon	c	o	o	c	o	F	F	o	o	o	o	o	F	F	F	pc	c	c	F	F	98	V
28th	Tue	o	c	F	pc	o	pc	pc	F	F	F	F	F	pc	pc	c	c	c	c	c	pc	114	
29th	Wed	F	pc	F	F	o	pc	pc	F	F	F	pc	pc	F	F	o	o	o	o	c	F	122	
30th	Thur	F	o	c	c	o	o	o	c	c	c	o	o	o	o	o	o	o	o	o	o	23	N
31st	Fri	F	pc	c	pc	o	o	o	o	o	c	o	o	c	o	o	o	o	o	o	c	30	
est. hours:		176	172	223	197	145	168	176	150	150	168	138	121	205	206	120	120	132	132	164	205	3267	
av. hours:		237	266	259	264	212	194	211	209	233	230	198	192	251	229	175	178	173	178	180	225	4294	
trend:		less	less	less	less	less	less	less	less	less	less	less	less	less	av	less	less	less	less	av	av	less	

JANUARY **1** WEDNESDAY

High pressure ridges over the west of New Zealand while low pressure moves away to the southeast of the country. South to southwest winds and occasional shower activity prevails in most districts predominately overnight, except in the Far North, and more isolated in the west of the South Island.

Auckland:	Mostly dry, mild and sunny, chance of brief overnight showers
Hamilton:	Sunny, dry, and mild, cold overnight
Wellington:	Partly cloudy, cool, dry, unsettled winds at times
Greymouth:	Very sunny, dry, pleasant breezes
Christchurch:	Windy, cloudy, cool, rain
Dunedin:	Cloudy, cool, brief showers, strong breezes at times

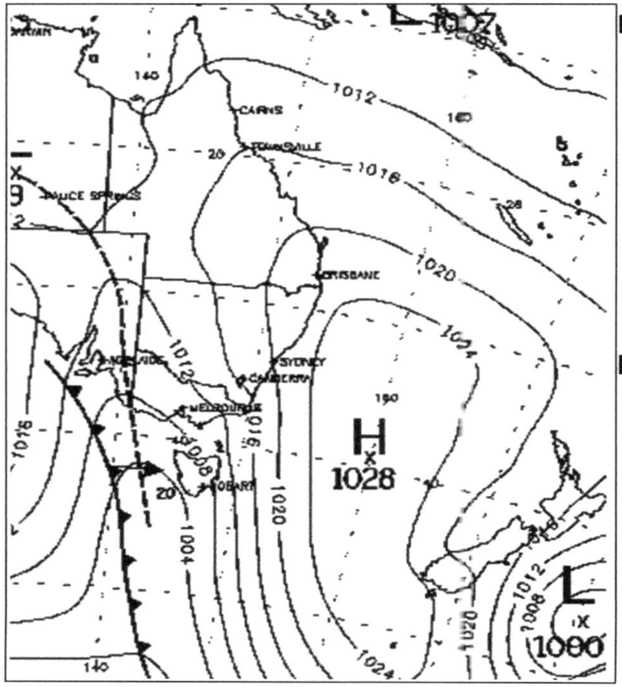

RAIN POTENTIAL

FROST/SNOW

JANUARY **2** THURSDAY

The high pressure ridge to the west of New Zealand weakens as does the low pressure to the southeast of the South Island. The southeast flow over the country weakens to light, with shower activity clearing in the south of the South Island and becoming more isolated and drizzly elsewhere.

Auckland:	Partly cloudy, mostly dry, mild, chance of isolated overnight showers
Hamilton:	Partly cloudy, dry, light air
Wellington:	Cloudy, cool, isolated showers, fluctuating winds
Greymouth:	Sunny, dry, mild day cool night, light breezes
Christchurch:	Overcast, scattered showers, moderate winds, cool overnight
Dunedin:	Cloudy, light showers, moderate breezes

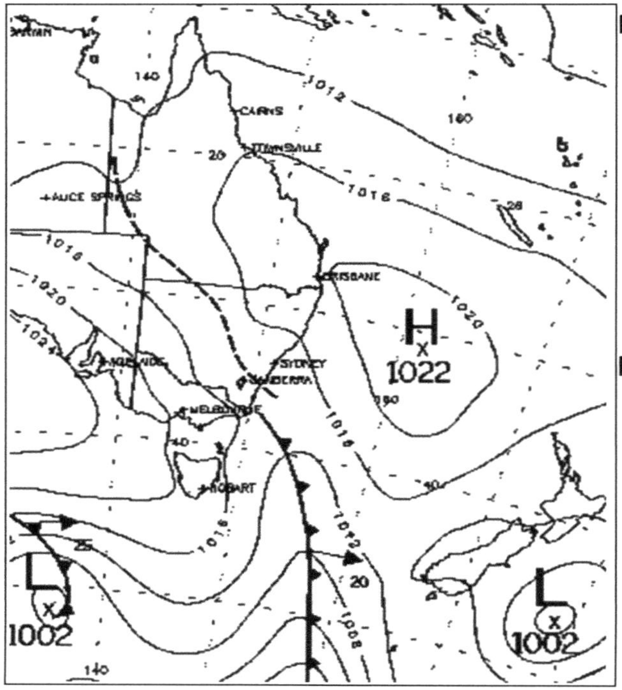

RAIN POTENTIAL

FROST/SNOW

JANUARY **3** FRIDAY

The anticyclonic ridge crosses eastwards to the north of the North Island, while a weak high pressure system forms in the southern Tasman Ocean to the west of the South Island. Winds remain light throughout and shower activity clears across the South Island while remaining light to isolated over the North island. A weak front approaches Fiordland bringing increased winds in the southeast of the country, while fog patches may briefly affect sheltered parts mostly in the east.

Auckland: Partly cloudy, spotty showers, fresh breezes
Hamilton: Changeable, mostly sunny, overnight showers, moderate breezes
Wellington: Sunny, dry, slightly warmer, moderate winds
Greymouth: Overcast, isolated showers, light winds
Christchurch: Cloudy, fluctuating breezes, brief showers, chance of fog patches
Dunedin: Overcast, dry, threats of rain, slightly warmer, fluctuating breezes

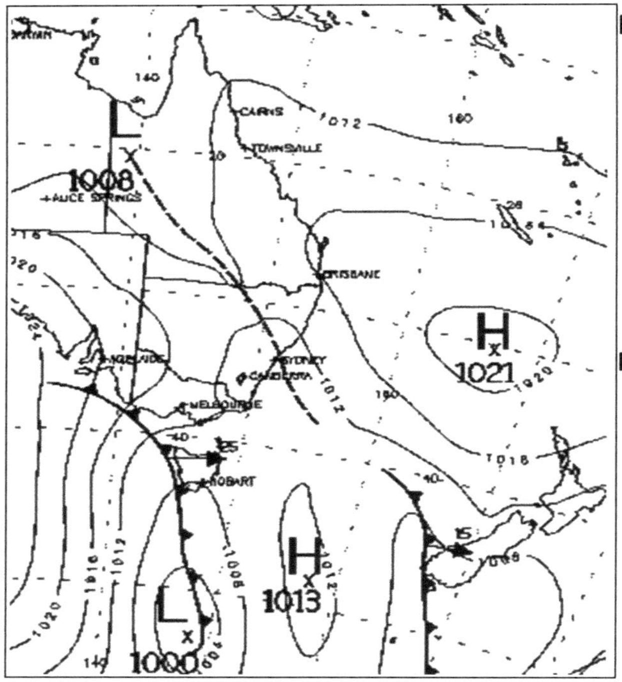

RAIN POTENTIAL

FROST/SNOW

JANUARY **4** SATURDAY

High pressure to the north of New Zealand slowly strengthens as it drifts to the east of the country, while a weak trough to the south of the ridge covers much of the southern half of the North Island and throughout the South Island. Mainly dry in the north and east of the North Island with pockets of rain in the west and south, and some misty showers prevail over much of the South Island. Chance of fog in the east of the South Island.

Auckland: Dry, cloudy spells, warmer, some breezy spells
Hamilton: Overcast, rain threatening, odd drizzle patch, light breezes, warmer
Wellington: Cloudy, warmer, mostly dry, odd threat of rain, moderate breezes
Greymouth: Overcast, dry with increasing threats of rain, cooler, mostly calm air
Christchurch: Mostly sunny and dry, warmer, calm air, possible misty patches
Dunedin: Partly cloudy, dry, warmer, gentle breezes to mostly calm air

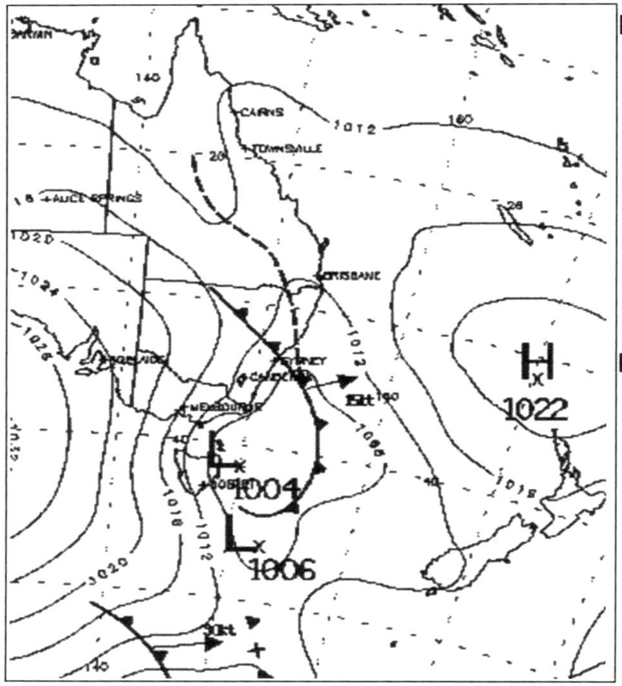

RAIN POTENTIAL

FROST/SNOW

JANUARY **5** SUNDAY

A complex of low pressure troughs form into a depression over the southern Tasman Sea and directs a freshening west to northwest flow over the South Island while high pressure weakens over the North Island. Mostly dry conditions continue in the Far North as well as around the Bay of Plenty to the East Coast and in the Wellington to Marlborough districts. Some showers elsewhere over the North Island, and rain developing over the south of the South Island with showers elsewhere.

Auckland:	Sunny and dry, light breezes
Hamilton:	Partly cloudy, dry, warmer, light breezes freshening later in the day
Wellington:	Dull, mostly dry, chance of drizzle, warmer, freshening breezes
Greymouth:	Cloudy, occasional showers, warmer, freshening winds
Christchurch:	Overcast, warmer, dry, unsettled and changeable winds
Dunedin:	Overcast, scattered showers, unsettled winds

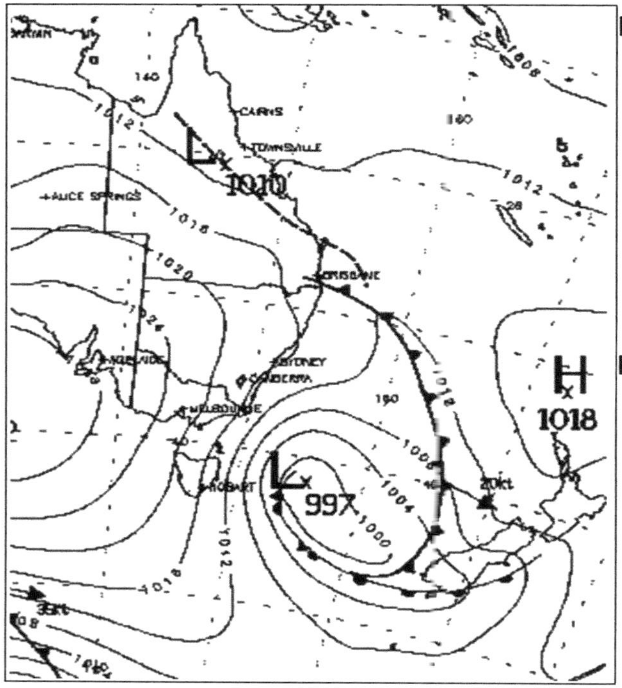

RAIN POTENTIAL

FROST/SNOW

JANUARY **6** MONDAY

A depression crosses over the south of the country bringing widespread rain throughout most of the South Island except for remaining mostly dry in the Marlborough region. Mostly dry over the North Island and very hot around East Cape. Winds lighter in the southern half of the South Island, while a more stronger and blustery west to northwest flow affects most other parts of New Zealand, especially through the Cook Strait.

Auckland:	Mostly cloudy, dry, odd threat of rain, light breezes freshening later
Hamilton:	Windy, warm, cloudy, mostly dry with odd threat of rain
Wellington:	Overcast, mostly dry, threats of rain, unsettled winds at times
Greymouth:	Stormy air, cloudy, heavy rain
Christchurch:	Partly cloudy, warmer, isolated showers, blustery at times
Dunedin:	Windy, cloudy, rain

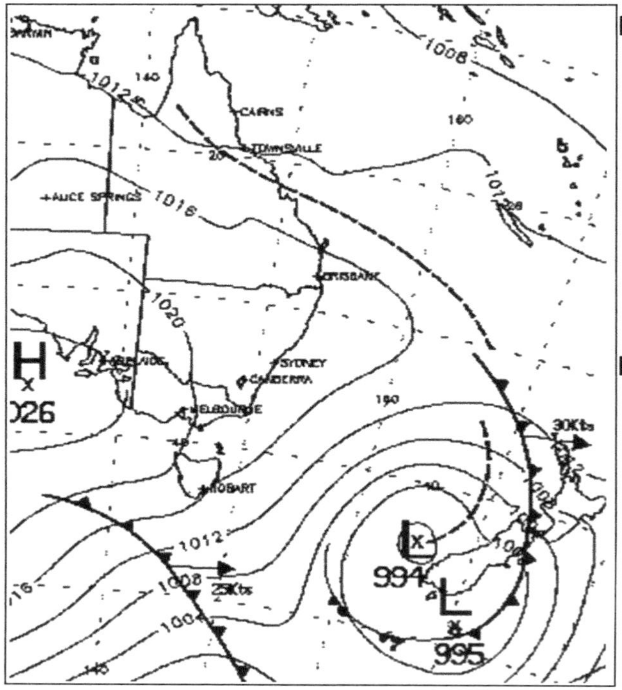

RAIN POTENTIAL

FROST/SNOW

JANUARY **7** TUESDAY

The depression to the south of New Zealand drops further southwards changing prevailing winds to a breezy southwest flow over most of the country. Wet fronts sweep rain activity up through the South Island and into the North Island, with heavy rain likely to affect the west and south of the South Island, and with occasional showers elsewhere. Mostly dry conditions however, are still likely to prevail in the east of the North Island.

Auckland: Mostly sunny, some brief showers, blustery at times
Hamilton: Sunny, warm, dry, chance of brief overnight showers, fresh breezes
Wellington: Overnight showers, sunny and mostly dry day, strong winds
Greymouth: Heavy overnight rain, sunny day, blustery winds at times
Christchurch: Sunny, hot, mostly dry, humid, chance of overnight showers
Dunedin: Cloudy, rain, cooler, windy

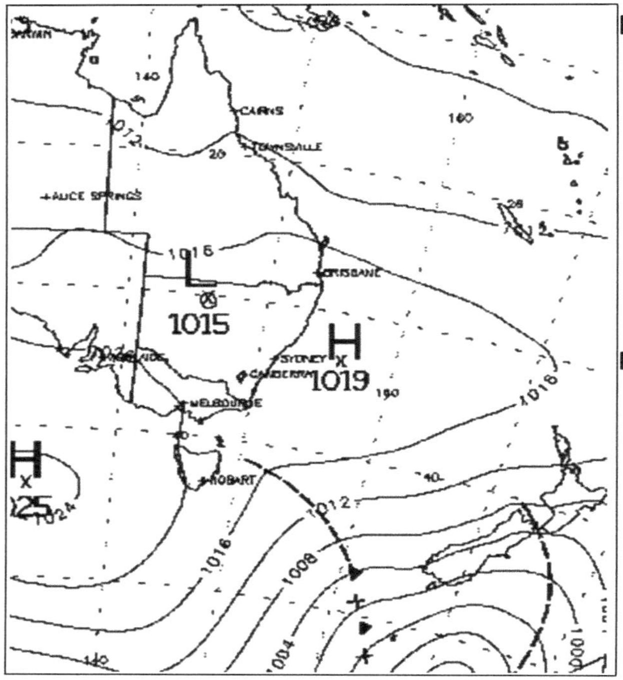

RAIN POTENTIAL

FROST/SNOW

JANUARY **8** WEDNESDAY

A strong high pressure system to the west of New Zealand ridges over the country clearing rain throughout the North Island and the northern half of the South Island, while the low pressure to the south of the country continues to bring some outbreaks of showers in the southern half of the South Island. Winds generally light in a south to southwest flow.

Auckland: Changeable, mostly dry, cloudy spells, chance of isolated showers
Hamilton: Very sunny, dry, warm, gentle breezes
Wellington: Very sunny, dry, mild, fluctuating winds
Greymouth: Partly cloudy, light showers, moderate winds
Christchurch: Sunny, dry, slightly cooler, fresh breezes
Dunedin: Mix of sunny spells and showers, warmer, fluctuating winds

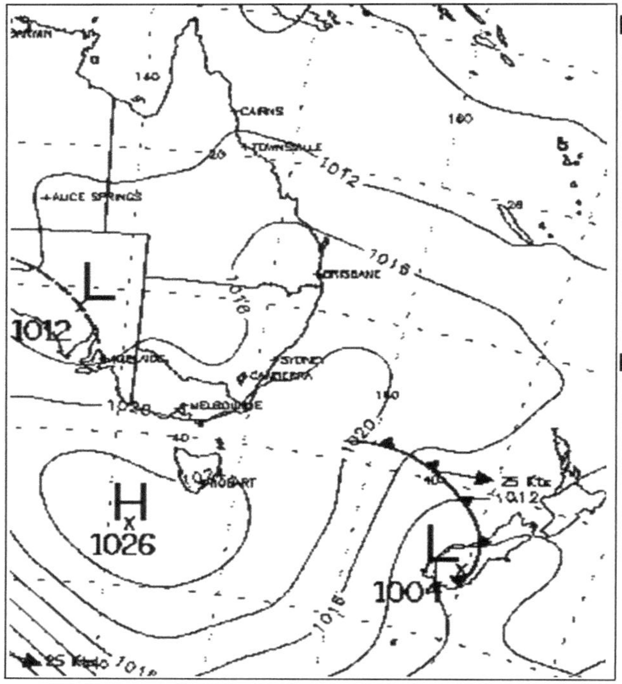

RAIN POTENTIAL

FROST/SNOW

JANUARY 9 THURSDAY

The weak depression to the south of New Zealand extends a rain-embedded front up through the country while the anticyclonic system to the west of the country in the southern Tasman Sea remains almost stalled as it builds in pressure. While it remains mostly dry in Northland and the east of the North Island, widespread rain spreads up through the southern half of the South Island, with scattered shower activity elsewhere.

Auckland: Mostly sunny, dry, cooler, chance of morning showers
Hamilton: Mostly sunny, dry, chance of morning showers, unsettled breezes
Wellington: Sunny, strong and blustery winds, isolated showers
Greymouth: Changeable, sunny day, heavy overnight rain, fluctuating winds
Christchurch: Partly cloudy, light showers, warm day cool night, unsettled winds
Dunedin: Partly cloudy, cooler, brief showers, fresh breezes

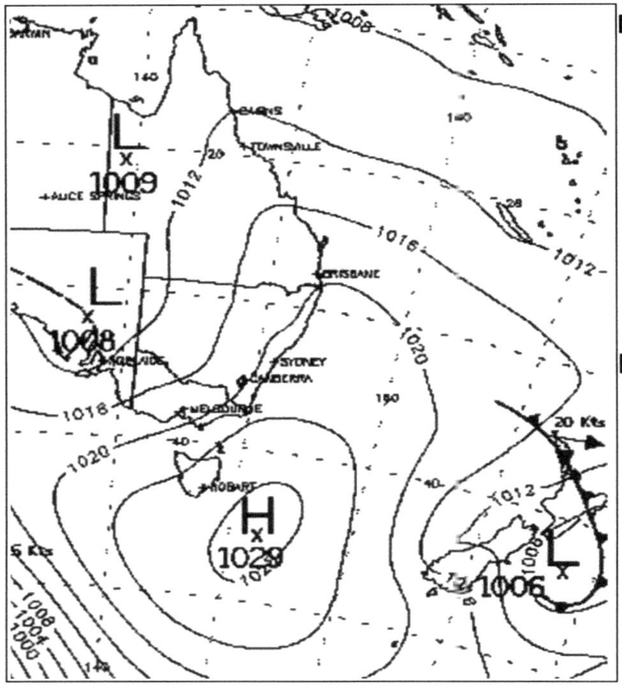

RAIN POTENTIAL

FROST/SNOW

JANUARY 10 FRIDAY

The large anticyclonic system to the west of New Zealand continues to slowly linger in the southern Tasman Sea, while low pressure maintains its presence over the country, maintaining a light south to southwest flow over the North Island and a changeable south to southeast flow in the South Island. Mostly dry in the Far North and east of the North Island and clearing to mostly dry in the southern half of the South Island. Outbreaks of heavy rain likely in the northern half of the South Island and some lingering showers elsewhere.

Auckland: Partly cloudy, isolated showers, light breezes, mostly calm air
Hamilton: Cloudy, dry, cooler, light breezes
Wellington: Cloudy, cooler, occasional showers, windy spells
Greymouth: Very sunny, dry, mild day cool night, shifting breezes
Christchurch: Colder, cloudy, spells of rain, fluctuating winds
Dunedin: Sunny and mostly dry, cooler, gentle breezes

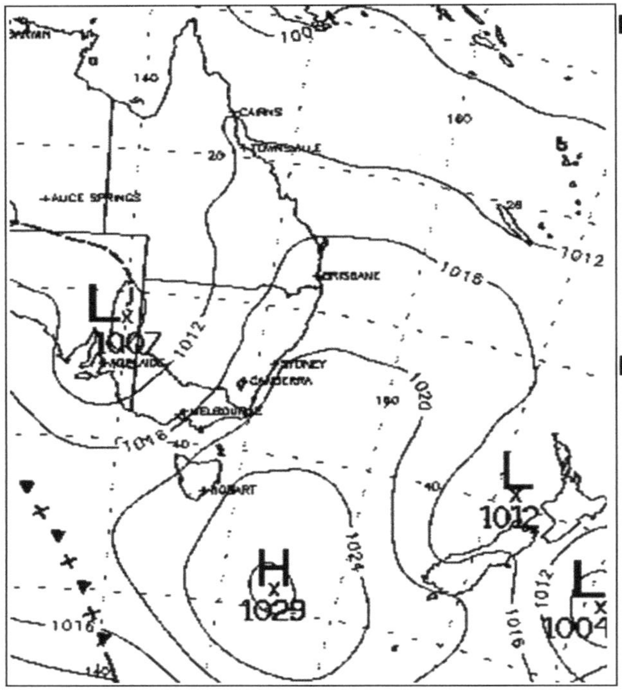

RAIN POTENTIAL

FROST/SNOW

JANUARY 11 SATURDAY

A large anticyclone to the southwest of the South Island ridges over the country while a weak depression drifts away to the east, resulting in a light south to southeaster flow over most districts and bringing some pockets of fog over the northern half of the North Island. Mostly dry in the southern half of the South Island with occasional showers elsewhere.

Auckland: Sunny, gentle breezes, chance of misty drizzles patches overnight
Hamilton: Partly cloudy, light showers, mostly calm air, chance of misty patches
Wellington: Overcast, scattered showers, cool, windy at times
Greymouth: Mostly sunny, dry, gentle breezes
Christchurch: Mostly sunny, light showers, cool, light breezes
Dunedin: Cloudy, dry, mild day cool night, light breezes

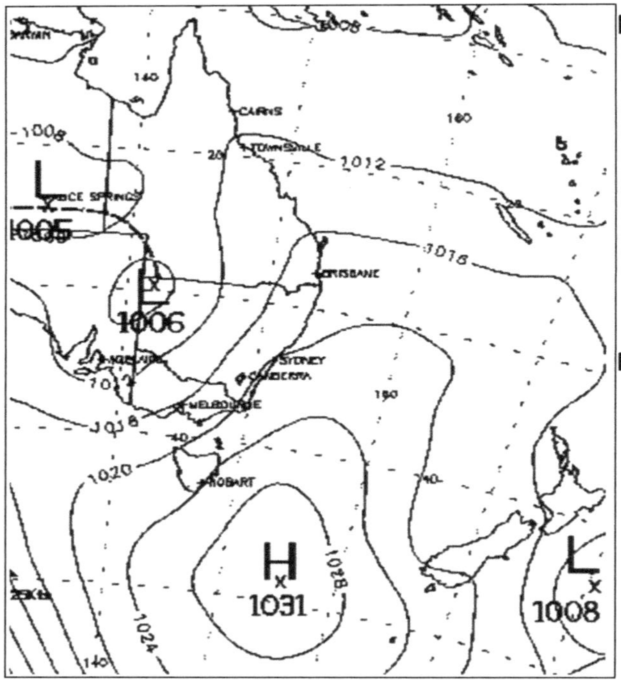

RAIN POTENTIAL

FROST/SNOW

JANUARY 12 SUNDAY

The large high to the southwest of New Zealand strengthens as it slowly builds over the country directing a light southerly flow throughout. Mostly dry in the Far North, while isolated showers prevail over most of the South Island and occasional showers elsewhere.

Auckland: Overnight showers, mostly sunny and dry day, moderate breezes
Hamilton: Sunny, milder, mostly dry, chance of misty patches
Wellington: Unsettled and breezy, widespread cloud, light showers
Greymouth: Bright and sunny, dry, light breezes
Christchurch: Cloudy, light showers, slightly warmer, moderate breezes
Dunedin: Partly cloudy, isolated showers, moderate breezes

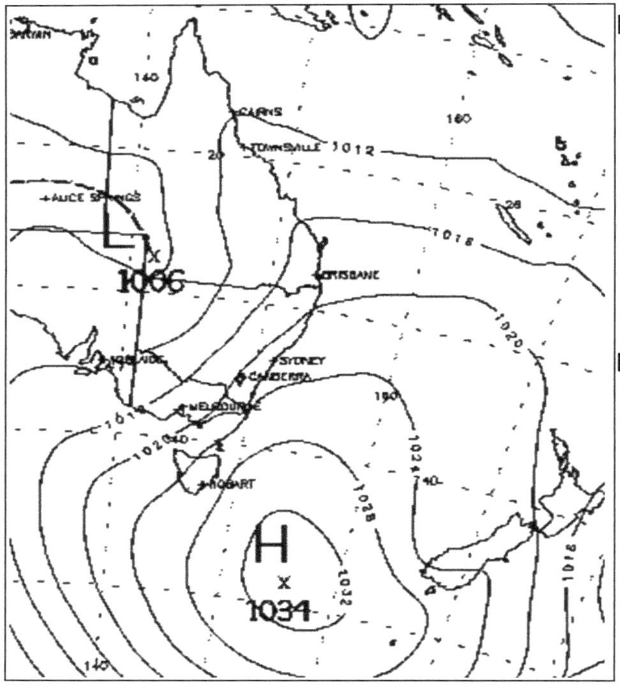

RAIN POTENTIAL

FROST/SNOW

JANUARY **13** MONDAY

A large anticyclonic system to the southwest of New Zealand remains stalled in the southern Tasman Sea, maintaining a gentle southwest flow over the country. Mostly sunny everywhere with some lingering showers here and there over the North Island and into the top of the South Island. Dry in the Far North again and dry with patches of fog over central and southern parts of the South Island.

Auckland: Very bright and sunny, dry, mostly calm air
Hamilton: Calm, dry, sunny, pleasant air
Wellington: Partly cloudy, mostly dry with odd threat of rain, fresh breezes
Greymouth: Very sunny, dry, mostly calm air
Christchurch: Dry, mild, sunny, mostly calm air, chance of fog patches
Dunedin: Sunny, dry, mild day cool night, mostly calm air

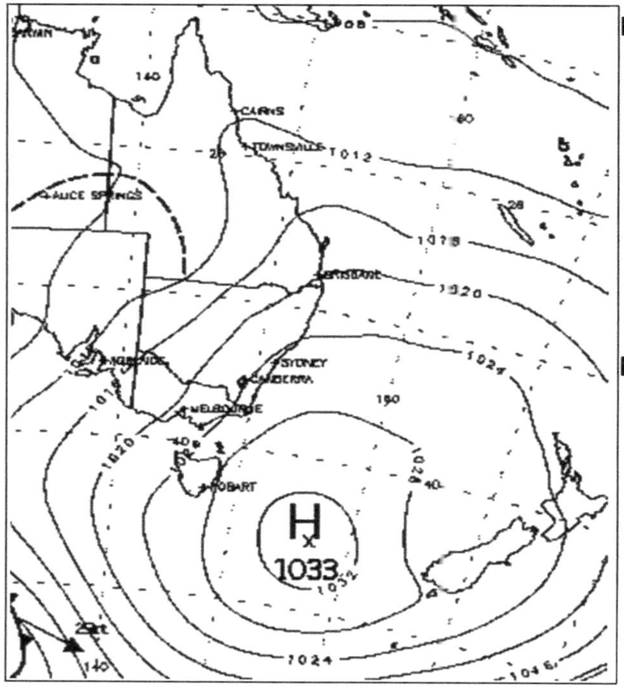

RAIN POTENTIAL

FROST/SNOW

JANUARY **14** TUESDAY

The large high pressure system continues to sit to the west of New Zealand with a light to mostly calm southwesterly air prevailing. Mostly sunny and dry everywhere, except for plenty of cloud in the West Coast and cloudy with isolated showers lingering around the top of the South Island. Chance of fog patches once more in the east of the South Island.

Auckland: Very sunny, dry, calm and serene air
Hamilton: Sunny, dry, warmer, mild overnight, calm and serene atmosphere
Wellington: Bright and sunny, dry, warmer day cooler night, light breezes
Greymouth: Partly cloudy, warmer, mostly dry, light breezes
Christchurch: Very bright and sunny, dry, cold overnight, calm air, fog likely
Dunedin: Dry, mild, sunny, light breezes

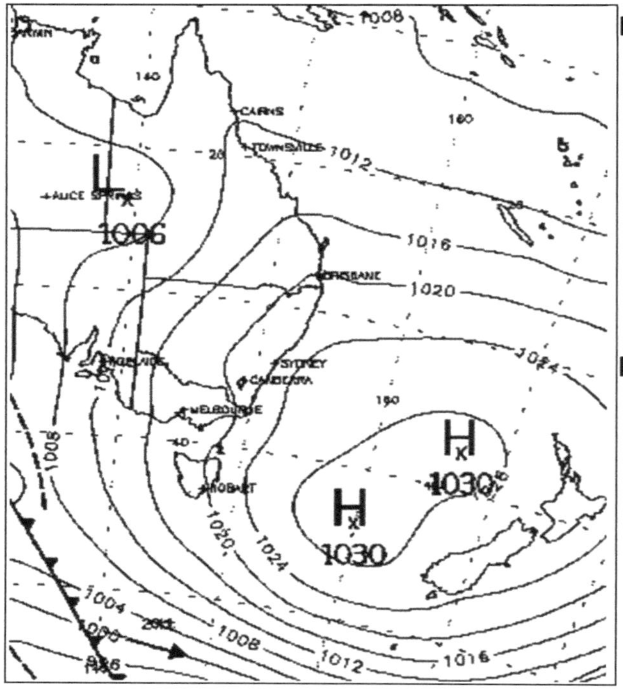

RAIN POTENTIAL

FROST/SNOW

JANUARY 15 WEDNESDAY

The large anticyclonic system that has prevailed to the west of New Zealand over the past few days slowly drifts eastward across the country bringing warmer, sunnier, and dry conditions throughout. Winds generally calm over the North Island while light northwesterly winds build over the South Island.

Auckland: Hot, dry, sunny, calm atmosphere
Hamilton: Warmer, bright and sunny, dry, pleasant air, mostly calm air
Wellington: Partly cloudy, dry, warmer, moderate breezes
Greymouth: Cloudy and dry, milder, light breezes
Christchurch: Very sunny, dry, warmer, light breezes, chance of misty patches
Dunedin: Sunny, dry, mild day cool night, fresh breezy spells

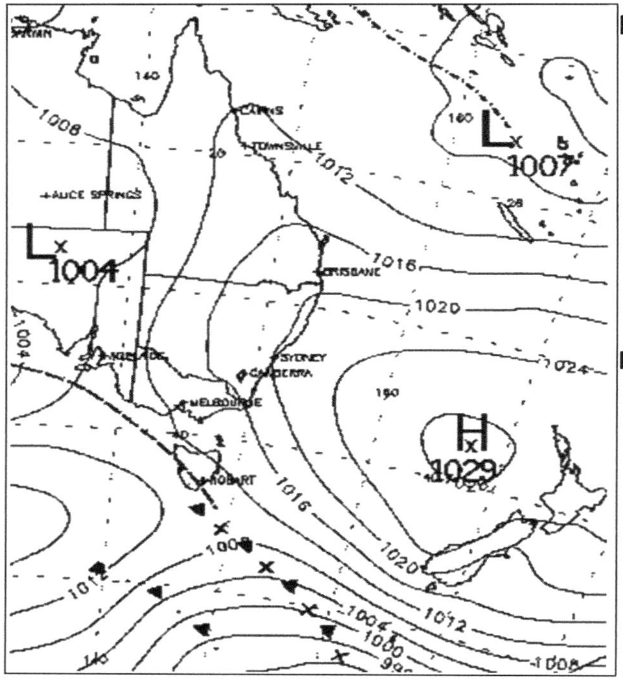

RAIN POTENTIAL

FROST/SNOW

JANUARY 16 THURSDAY

High pressure prevails over the north and west of the country while low pressure flows to the southeast. Generally, a hot northwesterly flow moves over central New Zealand bringing high temperatures across the country, while fresher southwest breezes prevail over the south of the South Island and northern half of the North Island. A weak front brushes Fiordland bringing some isolated showers to the region, while mainly dry and mostly sunny conditions prevail elsewhere.

Auckland: Cloudy, humid and dry, very warm, moderate breezes at times
Hamilton: Bright and sunny, dry, gentle breezes
Wellington: Unsettled and breezy, dry, sunny, warmer
Greymouth: Cloudy, dry with increasing threats of rain, fresh breezes at times
Christchurch: Sunny, warmer, dry, freshening breezes at times
Dunedin: Cloudy, warmer, dry with increasing threats of rain, windy at times

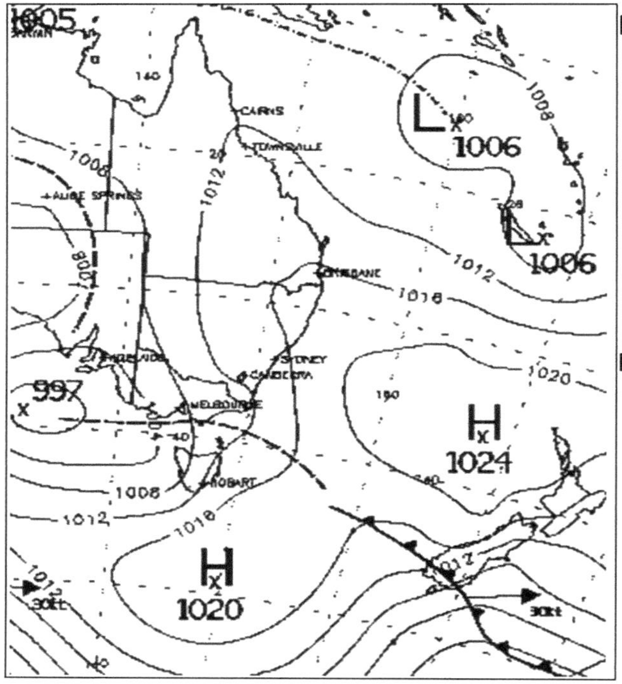

RAIN POTENTIAL

FROST/SNOW

JANUARY **17** FRIDAY

A decaying anticyclonic system continues to drift eastward over the country with cloud increasing in many districts and some rain likely in most parts of the South Island, heaviest in the West Coast to Fiordland regions. Mostly dry and sunnier in the North Island. Winds generally light to mostly calm throughout, and temperatures remaining high, particularly in the east of both islands.

Auckland: Mostly sunny, dry, gentle shifting breezes
Hamilton: Mostly sunny, dry, milder, mostly calm air
Wellington: Partly cloudy, warm, dry shifting breezes
Greymouth: Overcast, heavy rain, light breezes
Christchurch: Very hot, cloudy, isolated showers, light shifting breezes
Dunedin: Mix of sunny spells and rain, warmer, light winds

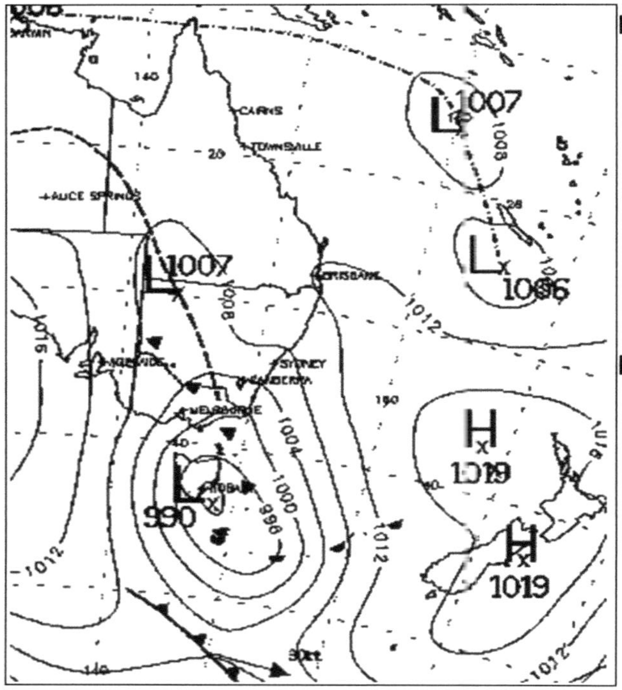

RAIN POTENTIAL

FROST/SNOW

JANUARY **18** SATURDAY

Low pressure moves over New Zealand from a weak depression to the north of the country, another depression moving eastward through the southern Tasman Sea and a third depression to the south. Overall, a west to northwest flow prevails, fresher over the South Island and lighter to the north. Mostly dry with temperatures remaining high over the North Island except for some spotty showers in central parts, while cooler over the South Island with rain clearing everywhere to isolated showers mostly confined to the west.

Auckland: Partly cloudy, dry, odd threat of rain, warm, shifting breezes
Hamilton: Very hot, dry, mostly overcast and humid, moderate breezes
Wellington: Mostly cloudy, dry, cooler, fluctuating breezes
Greymouth: Partly cloudy, fresh breezes, light showers
Christchurch: Cooler, sunny, dry, fresh breezes
Dunedin: Unsettled and breezy air, partly cloudy, dry, cooler

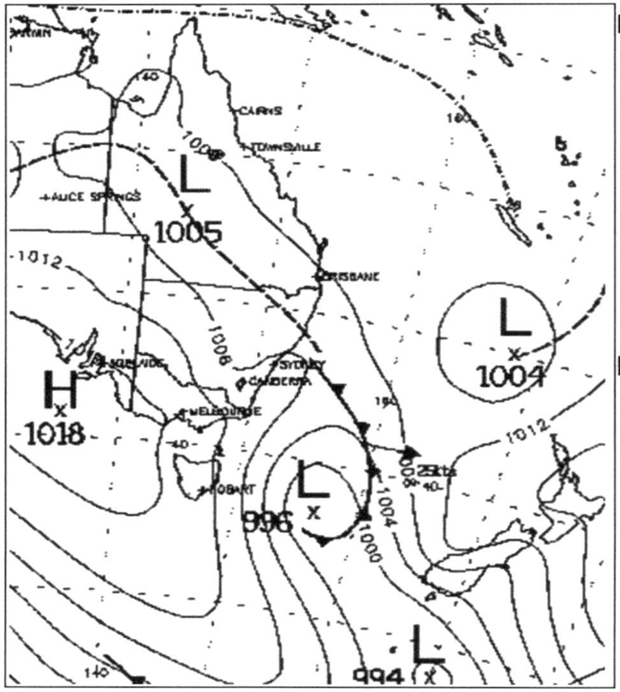

RAIN POTENTIAL

FROST/SNOW

JANUARY **19** SUNDAY

Low pressure continues to dominate the country with widespread cloud and warm and humid conditions prevailing almost everywhere. Some rain likely in Northland, isolated showers in the West Coast, and mainly dry elsewhere with increasing threats of rain. Winds generally light in a northeasterly flow.

Auckland: Overcast, drizzly showers, warmer, humid, fresh breezes at times
Hamilton: Overcast, hot, dry with increasing threats of rain, static air
Wellington: Overcast, warmer, mostly dry, threats of rain, breezy at times
Greymouth: Overcast, light showers, gentle breezes
Christchurch: Cloudy, dry with increasing threats of rain, hot, mostly calm air
Dunedin: Overcast, hot and dry, light shifting breezes

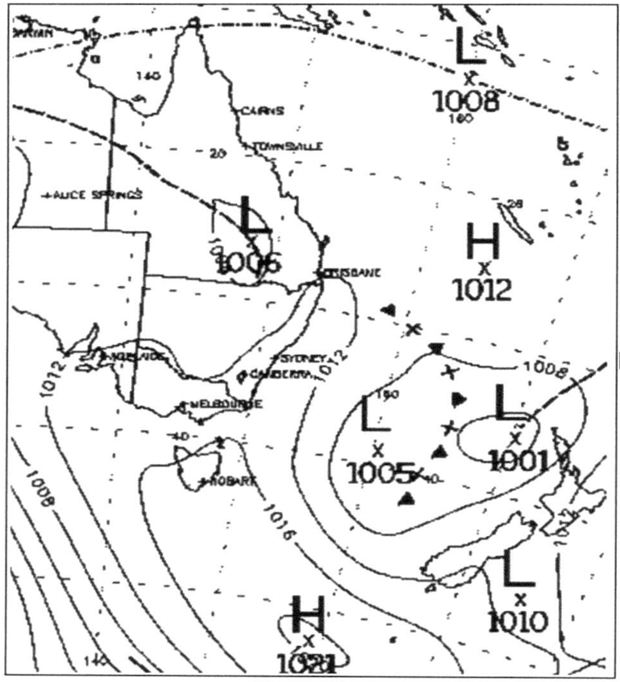

RAIN POTENTIAL

FROST/SNOW

JANUARY **20** MONDAY

A weak depression crosses New Zealand in a moderate northwest flow over the North Island, northeast over the South Island. Widespread cloud and rain activity throughout the North Island, including the chance of flooding in the Bay of Plenty to Waikato region. Heavy rain also over the northern half of the South Island, while mostly dry and cloudy in the south.

Auckland: Stormy air, cloudy, heavy rain, chance of thunderstorms
Hamilton: Cloudy, heavy rain and frequent drizzles, humid, shifting breezes
Wellington: Stormy air, overcast, heavy rain
Greymouth: Stormy air, overcast, heavy rain, fluctuating winds
Christchurch: Overcast, scattered showers, cooler, sea-breezes
Dunedin: Overcast, cooler, dry with passing threats of rain, moderate winds

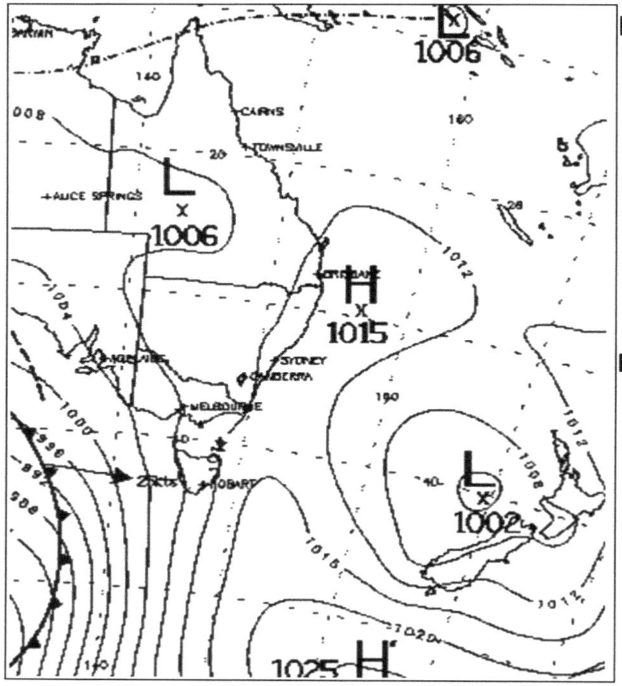

RAIN POTENTIAL

FROST/SNOW

JANUARY 21 TUESDAY

Low pressure continues to prevail over New Zealand with winds dropping to mostly calm over the northern half of the country, and shifting northerlies, later backing southeasterlies over southern districts. Some sunny spells in most districts during the day and remaining mostly dry in Southland and around the East Cape, while widespread scattered rain activity prevails elsewhere by evening. Rain may be heavy in the west of the South Island, while parts of Auckland to the Waikato and Canterbury districts may see some patches of fog.

Auckland: Warmer, humid, mix of showers and sunny spells, possible mist
Hamilton: Overcast, drizzly showers, mostly calm air, hot and humid
Wellington: Sunny, heavy evening showers, light breezes
Greymouth: Heavy rain, cloudy, cooler, strong breezes at times
Christchurch: Cooler, overcast, light drizzly showers, calm air
Dunedin: Cloudy, colder, light showers, fluctuating breezes

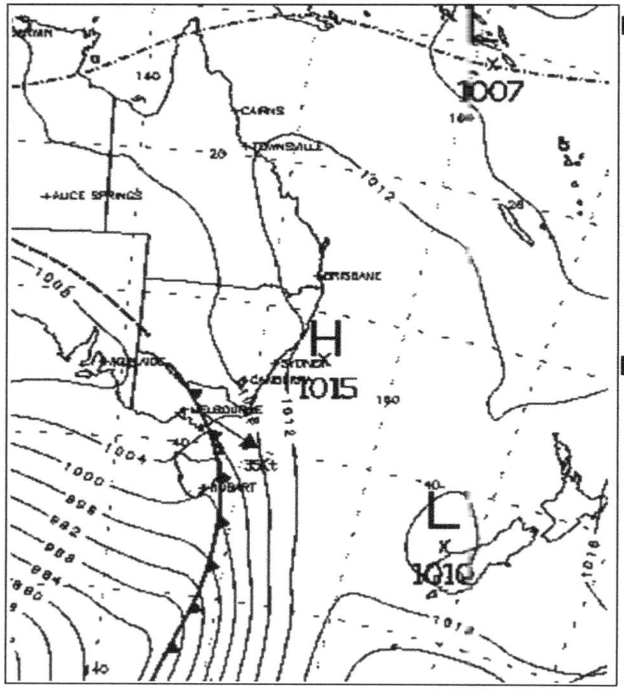

RAIN POTENTIAL

FROST/SNOW

JANUARY 22 WEDNESDAY

Low pressure shifts to the southeast of the country while weak high pressure ridges to the north. Very hot and cloudy with scattered showers and drizzle patches throughout most of the North Island, while rain clears to occasional showers and some sunny spells over the South Island. Winds generally very light in a southwesterly flow, with chance of fog in the Wellington region, Waikato, and Canterbury districts.

Auckland: Humid, overcast, patches of showers, light breezes to mostly calm air
Hamilton: Hot, humid, overcast, isolated showers, light breezes
Wellington: Cloudy, dry, warmer, mostly calm air, chance of overnight fog patches
Greymouth: Sunny, dry, mostly calm air
Christchurch: Dull and gloomy, dry, calm and hazy air
Dunedin: Overcast and mild, isolated showers, mostly calm air

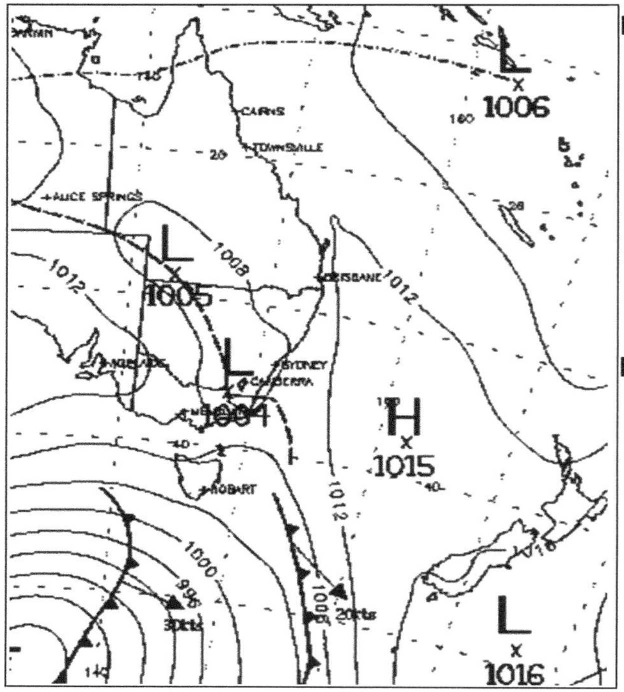

RAIN POTENTIAL

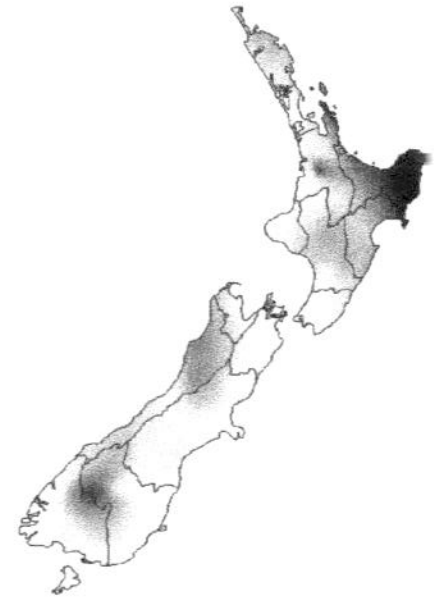

FROST/SNOW

JANUARY **23** THURSDAY

A weak depression forms to the west of the North Island, with a low pressure trough to the east of New Zealand, with a light northeasterly flow prevailing throughout. Hot and humid with widespread cloud and rain activity throughout most of the North Island, and into the northern half of the South Island. Mostly dry with good sunny spells in central and southern parts of the South Island. Chance of fog the Wellington and Canterbury districts.

Auckland: Overcast, outbreaks of rain, windy at times, chance of thunderstorms
Hamilton: Overcast, brief drizzly showers, humid, light shifting breezes
Wellington: Cloudy, passing drizzle patches, possible fog, gentle breezes
Greymouth: Overcast, light showers, mostly calm air
Christchurch: Overcast, warmer, dry, gentle breezes
Dunedin: Sunny and dry, mostly calm air

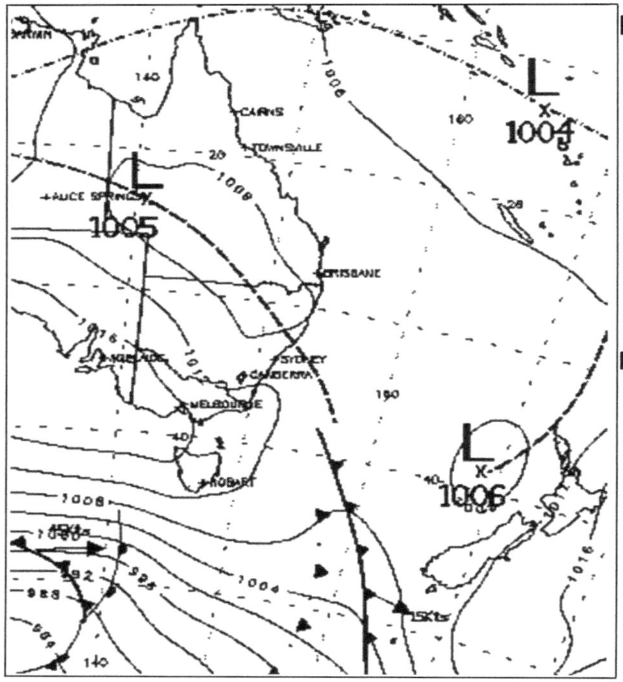

RAIN POTENTIAL

FROST/SNOW

JANUARY **24** FRIDAY

A dominance of low pressure over New Zealand continues to bring hot, humid, and wet conditions throughout most of the country, although some outbreaks of morning sunshine can be expected, particularly in the Waikato to the Bay of Plenty. A wet front moving up the country will bring some heavy rain to the west of the country and into the northern half of the North Island. Winds fresher in a westerly flow, backing southerly later over the North Island.

Auckland: Warm, sunny day, overnight rain, heavy in places, light breezes
Hamilton: Cloudy, warm and humid, some windy spells, heavy rain at times
Wellington: Changeable, partly cloudy, showers, warm, humid, unsettled breezes
Greymouth: Widespread cloud, scattered showers, humid, fresh winds at times
Christchurch: Freshening breezes, some cloudy spells, mostly dry, odd threat of rain
Dunedin: Cloudy, warmer, light showers, unsettled breezes, occasional gusts

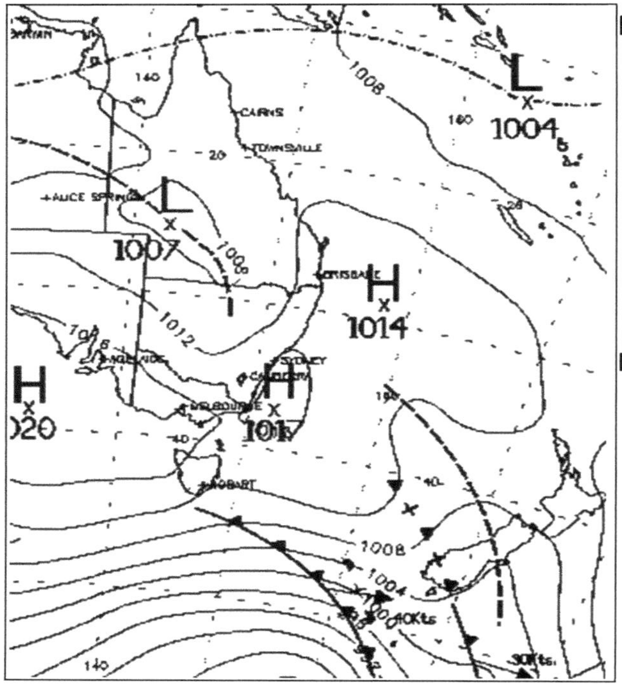

RAIN POTENTIAL

FROST/SNOW

JANUARY 25 SATURDAY

High pressure builds in the Tasman Sea and ridges over the west of New Zealand directing a light southwest flow over the country, with patches of fog likely in some eastern parts of the South Island. Warmer and sunnier and with rain clearing almost everywhere and very high maximas likely in the North Island East Coast region. Some overnight showers likely in the Waikato, and in the western half of the South Island.

Auckland: Mostly dry, sunny, hot, humid, chance of evening showers
Hamilton: Cloudy, dry, hot, mostly calm and humid air
Wellington: Sunny, dry, moderate breezes
Greymouth: Mix of sunny spells and outbreaks of rain, fresh breezes at times
Christchurch: Mostly sunny and dry, warmer, chance of misty patches
Dunedin: Overcast, dry, moderate breezes

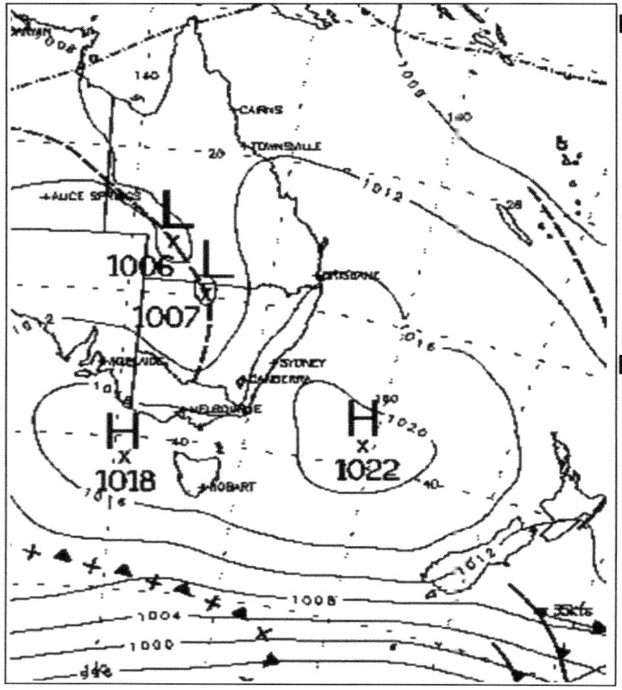

RAIN POTENTIAL

FROST/SNOW

JANUARY 26 SUNDAY

A large high pressure system continues to strengthen as it lingers in the Tasman Sea, while a low pressure trough moves off the country to the east of the North Island. Some showers in the Bay of Plenty to the Waikato, while mainly dry and mostly sunny everywhere else. Very hot in most districts. A light southwest airflow prevails, with possible fog patches observed in some eastern parts of the South Island.

Auckland: Partly cloudy, dry, warm, humid, fluctuating breezes
Hamilton: Cloudy, dry, hot, gentle shifting breezes
Wellington: Warmer day cooler night, dry, sunny, fresh breezes
Greymouth: Occasional showers, some sunny spells, moderate breezes
Christchurch: Very hot, dry, passing cloudy spells, moderate breezes
Dunedin: Fresh and cooling breezes, mostly sunny, dry

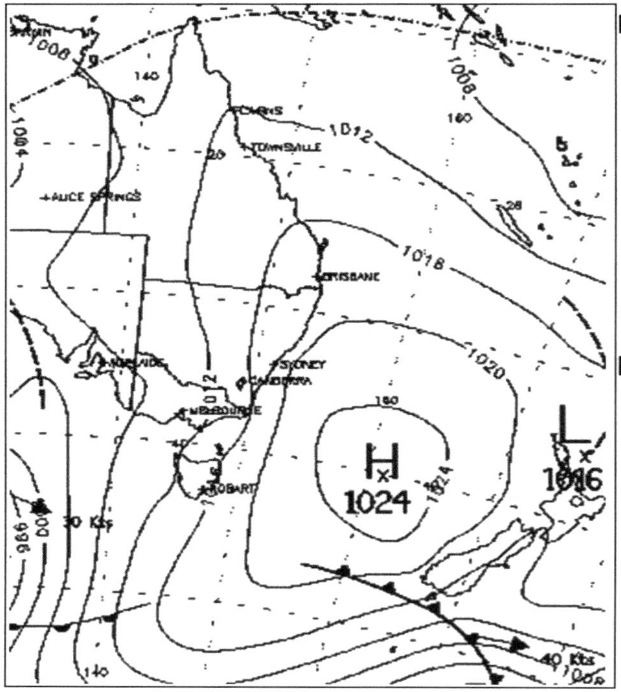

RAIN POTENTIAL

FROST/SNOW

JANUARY 27 MONDAY

A large anticyclonic system moves over the country in a light south to southeast flow. Warm throughout the country, with increased cloudy spells in the northern half of both islands. Some isolated showers likely in central parts of the North Island from the Bay of Plenty, through the Waikato, Hawkes Bay and down into the Manawatu, and chance of isolated showers in Canterbury. Mostly dry everywhere else.

Auckland: Cloudy, mostly dry with occasional threats of rain, moderate breezes
Hamilton: Cloudy, milder, dry, gentle breezes
Wellington: Cloudy, warm, mostly dry, chance of drizzle, humid
Greymouth: Sunny, dry, light breezes
Christchurch: Cooler, dry, mostly cloudy, fresh shifting breezes at times
Dunedin: Partly cloudy, dry, light shifting breezes

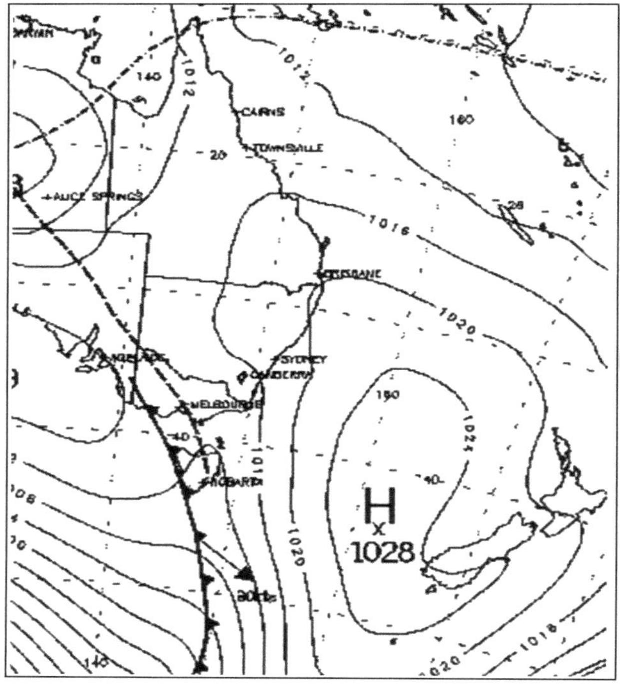

RAIN POTENTIAL

FROST/SNOW

JANUARY 28 TUESDAY

A large anticyclonic system slowly passes over the country with light westerly to mostly calm air prevailing throughout. Some isolated overnight showers likely in northern parts of the South Island, mainly dry elsewhere. Generally hot with some good sunshine hours expected throughout the North Island and slightly cooler in the South Island, mostly cloudy in Southland and the far northwest of the South Island, and partly cloudy elsewhere.

Auckland: Sunny, mostly dry, chance of overnight showers, gentle breezes
Hamilton: Overnight showers, sunny day, milder, gentle shifting breezes
Wellington: Very bright and sunny, cooler, dry, light shifting breezes
Greymouth: Scattered cloudy spells, dry, gentle breezes
Christchurch: Sunny, dry, colder, fresh breezes
Dunedin: Cloudy, colder, dry, some threats of rain, light breezes

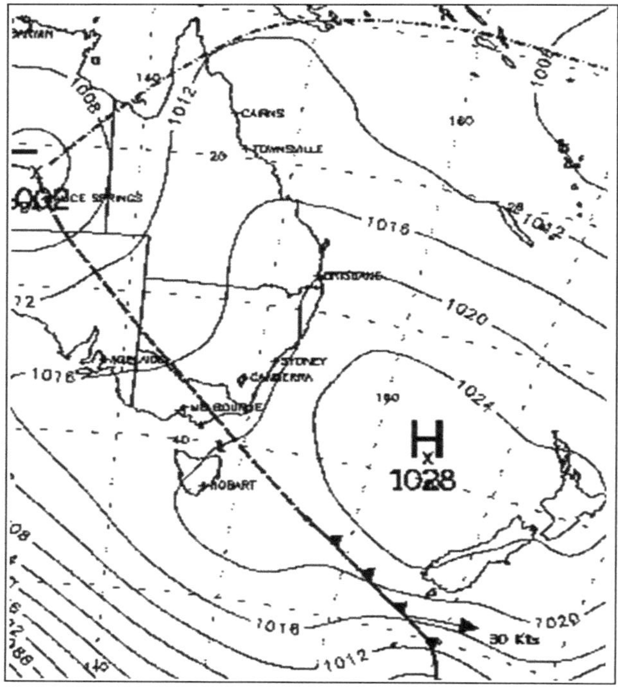

RAIN POTENTIAL

FROST/SNOW

JANUARY **29** WEDNESDAY

Anticyclonic. Winds generally light to mostly calm air in a southwesterly flow. Dry and warmer throughout, with good sunshine hours everywhere except for the chance of some spotty showers in Southland overnight.

Auckland: Cloudy, dry, serene atmosphere, warm and humid
Hamilton: Very bright and sunny, dry, warm day cool night, serene air
Wellington: Sunny, warmer, dry, light breezes
Greymouth: Mix of sunny spells and spotty showers, light breezes, warmer
Christchurch: Very sunny, dry, mild, light air with gentle breezes
Dunedin: Overcast, warmer, dry with passing threats of rain, mostly calm air

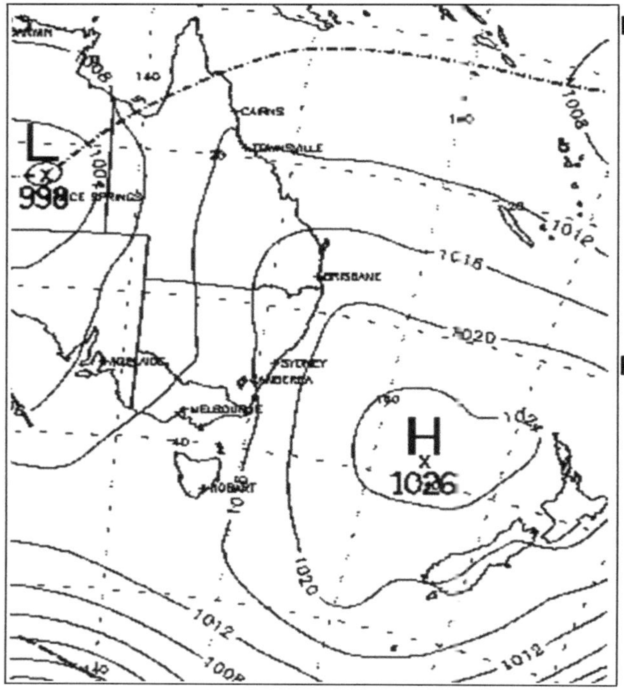

RAIN POTENTIAL

FROST/SNOW

JANUARY **30** THURSDAY

The large anticyclonic system prevailing over the country ridges over the North Island bringing a mostly sunny, warmer, and dry day throughout. To the south however, a wet front nears Southland with moderate northwesterly breezes prevailing and some showers spreading from the southern coasts and up into the West Coast overnight. Temperatures will be hot in the eastern half of the South Island.

Auckland: Cloudy, dry, gentle breezes, chance over overnight drizzle, humid
Hamilton: Mostly cloudy and dry, moderate breezes
Wellington: Cloudy, dry, moderate breezes
Greymouth: Overcast, breezy spells, occasional showers, static air
Christchurch: Very hot and dry, widespread cloud, static air, moderate breezes
Dunedin: Overcast, cooler, light breezes, dry with increasing threats of rain

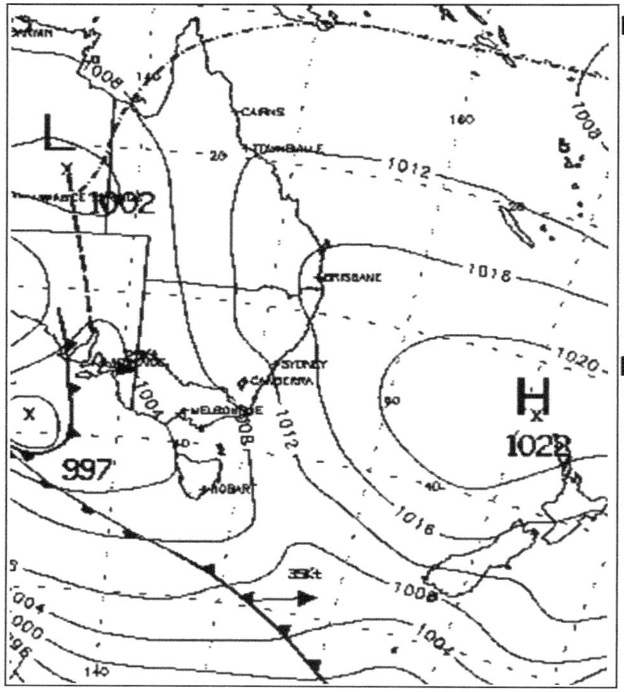

RAIN POTENTIAL

FROST/SNOW

JANUARY **31** FRIDAY

Anticyclonic to the north, low pressure to the south of the country with a southwesterly flow continuing over the North Island and a light northwesterly flow over the South Island. Very hot in the east of the North Island with milder temperatures elsewhere. Cloudy with rain spreading up through most of the South Island, with heavy rain likely in the West Coast. Mostly dry with sunnier spells over the North Island.

Auckland:	Mostly sunny, light overnight showers, milder, moderate breezes
Hamilton:	Mostly sunny, cooler, dry, fresh breezes at times
Wellington:	Overcast, dry with passing threats of rain, strong breezes
Greymouth:	Overcast, rain, light breezes
Christchurch:	Cloudy, cooler, light showers, gentle breezes
Dunedin:	Overcast, rain, light breezes

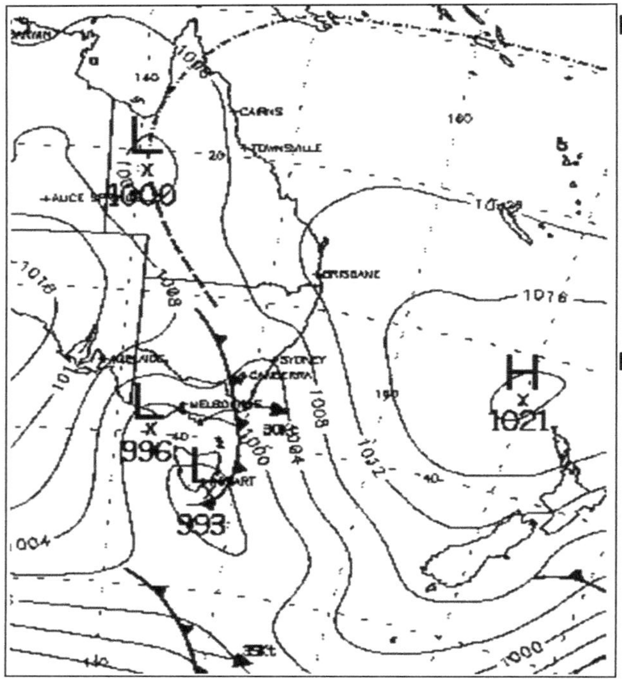

RAIN POTENTIAL

FROST/SNOW

FEBRUARY

PHASES OF THE MOON

2nd	Perigee #10
2nd	Crossing Equator
5th	First Quarter
8th	Northern Declination
12th	Full Moon
15th	Crossing Equator
18th	Apogee
20th	Third Quarter
22nd	Southern Declination
28th	New Moon

1 Sat	2 Sun	3 Mon	4 Tue	5 Wed	6 Thu	7 Fri	8 Sat	9 Sun	10 Mon	11 Tue	12 Wed	13 Thu	14 Fri	15 Sat	16 Sun	17 Mon	18 Tue	19 Wed	20 Thu	21 Fri	22 Sat	23 Sun	24 Mon	25 Tue	26 Wed	27 Thu	28 Fri
	P10 XhN			1Q			^				F			XhS			A		3Q		V						N

MONTHLY SUMMARY

1st Feb Unusually hot day in the Napier-Hastings region, as well as very hot in many other eastern districts from Gisborne to Otago. Heavy rain likely in the West Coast to Fiordland. Strong winds may cause some disruptions in Otago.
3rd-4th Possible fog in Christchurch.
8th-9th Heavy rain likely in the West Coast in the Buller region.
10th Heavy rain in the west of the South Island from Fiordland to South Westland and up into the Southern Alps. Strong winds possibly to gale force in the Wellington region. Chance of shallow fog patches in the Christchurch area.

Over the next four weeks, similar weather patterns to last month are likely to continue, with anticyclonic systems building more frequent to the southwest of the South Island. The accompanying high-pressure ridges will direct a frequent southeasterly flow over the South Island for the second month in a row, with westerly winds prevailing over the North Island. Rainfall will be up to 25-50% below average for this time on year in most parts of the country except in Northland, where up to twice as much rain as normal for this time of year may see some localised flooding, mostly due to heavy rain over a three-day period over the first week of this outlook period. Drought conditions may affect many eastern parts of the country from Gisborne to Central Otago, as well as parts of Auckland, the Waikato, eastern Bay of Plenty, Whanganui, the Manawatu, Wellington and even Nelson, and with severe soil moisture deficits in many other North Island regions. The Lake Tekapo region may not see any rainfall over the next four weeks. Accompanying the below average rainfall throughout most of the country will be much sunnier than average conditions over most of the country, with sunshine hours particularly higher than average by around 10% more in the west of the South Island, and temperatures also around 0.5°C above average in parts of Westland and the Southern Lakes region, Nelson and Buller. In contrast, however, many northern and eastern districts, including Gisborne to North Canterbury and Northland, will be cooler by a similar variant than what is typical for this time of year, and the Kairkoura Coast region possibly only seeing around 90% of their usual sunshine hours for this time of year.

11th Stormy air with possible gale winds, scattered thunderstorms and showers of hail likely in the Wellington region.
15th-18th Heavy rain in Northland to the Waikato which may cause some localised flooding in places, including in the areas from Kaitaia to Kerikeri, as well as around Dargarville and Whatawhata.
15th-19th Chance of fog in the Canterbury region, including in Christchurch.
17th Very hot in the Tasman-Marlborough region.
18th Heavy rain in Fiordland.
20th Chance of heavy rain and thunderstorms in the northern half of the North Island including in Auckland and the Waikato.
20th-23rd Chance a cyclone may form near Fiji and intensify while travelling in a southwesterly tract and pass close to the north of the North Island.
22nd Heavy rain likely in Buller.
27th-28th Possible fog in Christchurch.

February 2025

Monday	Tuesday	Wednesday	Thursday	Friday	Saturday	Sunday
27	28	29	30	31 **Moon Sign**	**1** **Pisces** Square Jupiter Conjunct Saturn	**2** **P10 XhN** **Pisces** Conjunct Venus Trine Mars Conjunct Saturn Sextile Uranus Conjunct Neptune
3 **Aries** Sextile Mercury Sextile Jupiter	**4** **Aries** Square Mars	**5** **1Q** **Taurus** Square Mercury	**6** **Taurus** Square Sun Conjunct Uranus Square Ascendant	**7** **^** **Gemini** Conjunct Jupiter Trine Pluto Square Ascendant	**8** **Gemini** Trine Sun Trine Mercury Square Saturn Square Uranus Square Ascendant	**9** **Cancer** Square Venus Square Ascendant
10 **Cancer** Conjunct Mars Trine Saturn Sextile Uranus Square Ascendant	**11** **Leo** Trine Venus Trine Neptune Opposition Pluto Square Ascendant	**12** **Leo** Opposition Sun Opposition Mercury Square Uranus Square Ascendant	**13** **FM** **Leo** Opposition Sun Opposition Mercury	**14** **Virgo** Square Jupiter Opposition Saturn	**15** **Virgo** Opposition Saturn Trine Uranus Trine Pluto	**16** **XhS** **Libra** Opposition Venus Trine Jupiter Trine Pluto
17 **Libra** Square Mars	**18** **A** **Scorpio** Trine Sun Trine Mercury Square Pluto	**19** **Scorpio** Trine Mercury Trine Mars Trine Saturn	**20** **Scorpio** Square Sun Trine Saturn Opposition Uranus Trine Neptune	**21** **3Q** **Sagittarius** Square Mercury Trine Venus Opposition Jupiter	**22** **Sagittarius** Square Saturn	**23** **V** **Capricorn** Sextile Sun Square Venus Square Ascendant
24 **Capricorn** Sextile Mercury Opposition Mars Sextile Saturn Trine Uranus	**25** **Aquarius** Conjunct Pluto	**26** **Aquarius** Trine Jupiter Square Uranus	**27** **Pisces** Conjunct Sun	**28** **NM** **Pisces** Conjunct Sun Conjunct Mercury Trine Mars Conjunct Saturn	1	2
3	4					

NORTH ISLAND RAINFALL
FEBRUARY 2025

d=mainly dry, n= fog, mist, dew or nonrecordable, l=light showers, p=passing showers, s=significant showers, r=rain, h=heavy falls

		Northland							Auckland					Bay Of Plenty				Waikato			
		CR	Kta	Ker	Rsl	Dar	Whg	Man	Wark	He	MkH	Ak	Co	Tau	TeP	Wkt	Opo	Hml	TeA	Mata	Ot
1st	Sat	d	d	d	d	d	d	d	d	d	d	d	d	d	d	d	d	d	d	d	d
2nd	Sun	d	d	d	d	d	d	d	d	d	d	d	d	d	d	d	d	d	d	d	d
3rd	Mon	d	d	d	d	d	d	d	d	d	l	d	d	d	d	d	d	l	d	l	d
4th	Tue	d	d	d	d	d	d	d	d	d	d	d	d	d	d	d	l	n	n	d	l
5th	Wed	l	l	l	d	s	d	d	d	d	l	d	d	d	d	d	d	r	s	s	l
6th	Thur	d	n	l	s	l	d	l	d	d	d	d	d	d	d	n	s	d	d	d	l
7th	Fri	d	d	d	d	d	d	d	d	d	d	d	l	d	d	d	d	d	d	d	l
8th	Sat	d	r	l	d	s	s	r	h	h	r	h	s	s	r	l	s	h	h	h	h
9th	Sun	d	d	d	d	s	l	n	l	s	h	s	s	l	l	d	s	s	r	s	s
10th	Mon	d	d	d	d	d	d	d	d	d	d	d	d	d	d	d	d	d	d	d	l
11th	Tue	d	d	d	d	d	d	d	d	d	d	d	d	d	d	d	d	d	d	l	d
12th	Wed	d	d	d	d	l	d	d	d	d	s	l	d	d	d	l	l	n	l	d	l
13th	Thur	l	d	d	d	d	d	d	d	d	n	d	d	d	d	d	l	d	d	l	d
14th	Fri	d	d	d	l	d	d	d	d	d	d	d	d	h	d	s	l	d	d	d	d
15th	Sat	d	d	d	d	d	d	d	d	d	l	d	s	h	r	l	d	l	l	d	s
16th	Sun	l	h	h	h	s	s	d	d	d	d	d	l	d	d	d	d	d	d	d	d
17th	Mon	h	h	h	h	h	h	r	s	l	l	l	r	h	h	d	d	h	h	r	s
18th	Tue	h	h	h	h	h	s	l	d	d	d	d	s	d	l	l	s	d	d	d	d
19th	Wed	d	d	n	l	d	l	n	d	d	d	d	d	d	d	d	d	d	d	d	d
20th	Thur	d	n	d	d	d	d	d	d	d	d	d	d	d	d	d	d	d	d	d	s
21st	Fri	d	d	d	d	r	l	s	n	d	l	h	d	s	l	s	d	l	d	d	l
22nd	Sat	l	d	d	d	d	d	d	d	d	l	d	d	d	d	d	d	d	l	d	l
23rd	Sun	l	d	d	d	d	d	d	d	l	d	d	s	d	d	d	d	d	d	d	d
24th	Mon	d	d	d	d	d	d	d	d	d	d	d	d	d	d	d	d	d	d	d	d
25th	Tue	d	d	d	n	d	d	r	d	d	d	d	d	d	n	d	d	d	d	d	s
26th	Wed	d	d	d	d	d	d	d	d	d	l	d	d	d	d	d	d	d	d	d	d
27th	Thur	d	d	d	d	l	d	d	d	d	l	d	d	d	n	d	d	d	n	d	d
28th	Fri	d	d	d	d	d	d	d	d	d	d	d	d	d	d	d	d	d	d	d	d
estimate:		158	173	151	184	158	83	44	30	34	45	61	44	101	70	17	27	92	79	62	73
average:		67	85	98	97	82	205	135	103	74	63	85	82	87	117	107	78	80	79	79	87
trend:		wtr	wtr	wtr	wtr	wtr	drr	drr	drr	drr	drr	drr	drr	wtr	drr	drr	drr	wtr	av	drr	drr

		Rot/Tau		Hawkes Bay			C Plateau			Taranaki			Manawatu			Lower North Island					
		Rt	Taup	Gis	Nap	Has	Taum	Chat	Waio	NP	Str	Wang	Pa	Dann	Lev	PN	Mast	UH	Wgt	all	Moon
1st	Sat	d	d	d	d	d	d	n	l	n	d	d	d	d	r	l	d	d	d	14	
2nd	Sun	d	d	d	d	d	d	d	d	d	d	d	d	d	l	d	d	l	d	1	P10 XhN
3rd	Mon	d	d	d	d	d	s	s	n	r	n	n	l	n	l	d	d	l	d	32	
4th	Tue	l	l	d	d	d	l	h	d	s	d	n	n	l	l	s	d	d	d	49	
5th	Wed	d	l	s	d	d	s	h	l	l	d	l	d	s	d	s	d	d	d	82	1Q
6th	Thur	s	h	s	l	h	l	l	l	d	l	d	h	d	l	d	r	l	s	119	
7th	Fri	l	d	d	l	r	n	n	d	d	d	d	d	d	n	d	s	d	d	27	
8th	Sat	r	h	l	l	l	r	h	s	s	r	h	r	s	h	s	l	h	r	480	^
9th	Sun	s	s	n	l	d	r	r	r	l	l	s	h	l	h	s	l	r	l	231	
10th	Mon	d	l	d	d	d	d	l	l	d	d	d	d	d	d	d	d	d	d	5	
11th	Tue	d	d	d	d	d	d	d	d	d	d	d	d	d	d	d	d	d	d	1	
12th	Wed	d	d	d	d	d	l	r	l	s	s	r	r	s	r	s	d	r	l	95	
13th	Thur	d	d	d	d	d	d	d	d	d	d	d	d	d	d	d	s	l	s	13	F
14th	Fri	d	d	n	h	s	d	d	d	d	d	d	d	n	d	d	l	d	d	60	
15th	Sat	l	l	l	d	s	r	l	d	d	l	l	d	l	d	d	d	d	d	94	
16th	Sun	d	d	l	d	d	d	d	d	d	d	d	d	d	d	d	d	d	d	134	XhS
17th	Mon	s	d	l	s	s	l	d	d	s	l	d	d	n	d	l	d	d	d	729	
18th	Tue	s	d	r	l	s	l	s	r	d	d	r	d	d	d	l	d	d	d	244	A
19th	Wed	n	s	d	d	d	s	d	d	d	d	d	d	d	d	d	d	d	d	17	
20th	Thur	d	d	d	l	l	d	d	d	d	h	l	d	l	l	l	l	l	s	62	
21st	Fri	d	d	s	l	l	n	l	d	l	d	d	d	d	d	d	d	d	d	79	3Q
22nd	Sat	d	d	d	d	d	d	d	d	d	d	d	d	d	d	d	d	d	d	4	
23rd	Sun	d	d	s	d	d	d	d	d	d	d	d	d	d	d	d	d	d	d	14	V
24th	Mon	d	d	d	d	d	d	d	d	d	d	d	d	d	d	d	d	d	d	0	
25th	Tue	d	d	d	d	d	l	d	d	d	d	d	d	d	d	d	d	d	d	21	
26th	Wed	d	d	l	d	d	l	d	d	d	d	d	d	d	d	d	d	l	l	9	
27th	Thur	d	d	l	n	d	d	d	d	d	d	l	l	d	l	n	l	d	n	16	
28th	Fri	d	d	n	d	d	d	d	d	d	d	d	d	d	d	d	d	d	d	0	N
estimate:		37	67	43	38	65	59	108	42	41	63	52	68	27	77	40	30	55	38	2634	
average:		105	97	80	75	67	92	150	60	102	119	75	61	71	68	75	67	74	77	3405	
trend:		drr	drr	drr	drr	av	drr	drr	drr	drr	drr	drr	av	drr	av	drr	drr	drr	drr	drr	

SOUTH ISLAND RAIN
FEBRUARY 2025

d=mainly dry, n= fog, mist, dew or nonrecordable, l=light showers, p=passing showers, s=significant showers, r=rain, h=heavy falls

		West Coast				Top South Island				Canterbury				South Canterbury				Coastal Otago			Lakes			Inland Otago				Southland						
		West	Grey	Hok	BB	Ta	Mou	Nel	Ble	Dar	Ham	Kai	Ch	Rak	Meth	Ash	Tim	Oam	Dun	Bal	Twi	LP	LT	Crom	Qstn	Ran	Alex	TeA	Wtn	Gore	Inv	Puy Pt	all	Moon
1st	Sat	r	s	s	s	d	d	d	d	d	d	d	d	d	d	l	n	l	s	l	d	d	d	d	l	l	d	n	s	s	s	h	77	
2nd	Sun	s	h	h	h	d	d	d	d	d	d	d	d	l	d	d	d	d	d	l	d	d	d	s	s	d	l	r	s	l	r	h	178	P10 XhN
3rd	Mon	d	d	d	d	d	d	d	d	d	d	d	d	d	l	d	d	d	d	d	s	d	d	d	d	d	d	d	d	d	d	d	10	
4th	Tue	d	d	d	d	d	d	d	d	d	d	d	d	d	d	d	d	d	d	d	d	d	d	d	d	d	d	l	l	l	l	l	11	
5th	Wed	d	s	l	d	d	d	d	d	d	d	l	l	s	l	l	l	l	r	d	d	d	d	l	d	d	s	d	l	d	d	r	55	1Q
6th	Thur	d	d	d	d	9	r	d	r	l	h	s	s	d	l	l	d	n	d	d	d	d	d	d	d	d	d	d	d	d	d	d	69	
7th	Fri	d	d	d	d	d	l	d	d	d	d	d	d	d	d	d	d	d	d	d	d	d	d	d	d	d	d	d	d	d	d	r	15	
8th	Sat	r	h	r	h	r	l	r	s	d	l	d	d	d	l	d	l	l	d	s	d	d	l	l	s	l	l	r	r	s	l	s	157	^
9th	Sun	h	r	h	r	s	r	r	r	d	l	d	d	d	d	d	d	d	n	s	s	s	l	l	l	d	l	s	d	l	l	h	192	
10th	Mon	l	d	l	l	d	d	d	d	d	d	d	d	d	d	d	d	d	d	d	s	s	d	d	d	d	d	s	l	l	l	h	61	
11th	Tue	r	h	r	h	s	s	s	d	d	l	d	d	d	d	d	d	d	l	l	l	d	s	l	s	d	l	r	l	l	r	h	197	
12th	Wed	l	n	s	d	l	s	s	l	d	d	d	d	l	l	d	d	d	d	l	d	d	d	d	d	d	d	d	d	s	h	s	61	
13th	Thur	n	l	d	d	d	d	l	s	s	l	r	s	d	s	r	l	d	l	l	h	r	d	d	d	d	d	n	s	s	l	l	105	F
14th	Fri	d	d	d	d	d	d	d	d	d	d	d	d	d	d	d	d	d	l	l	d	d	d	d	d	d	d	d	l	d	l	d	5	
15th	Sat	d	d	d	d	n	d	d	d	d	d	d	d	d	d	d	d	d	d	d	d	d	d	d	d	d	d	d	d	d	d	d	1	
16th	Sun	d	d	d	d	d	d	d	d	d	d	d	d	d	d	d	d	d	d	d	d	d	d	d	d	d	d	d	d	d	d	d	0	XhS
17th	Mon	d	d	d	d	d	d	d	d	d	d	d	d	d	l	d	d	d	d	d	d	d	d	d	d	d	d	d	d	d	d	d	1	
18th	Tue	n	d	d	d	d	d	d	d	d	d	l	d	d	d	d	d	d	d	d	d	d	d	d	d	d	d	l	d	d	d	d	4	A
19th	Wed	d	d	l	r	d	d	d	d	d	l	l	d	d	d	d	d	l	s	h	d	d	d	l	s	l	l	h	h	h	l	s	123	
20th	Thur	l	h	r	d	s	d	d	l	s	h	r	l	s	s	s	s	l	s	n	l	d	l	d	d	s	l	d	d	d	d	d	127	
21st	Fri	d	d	d	d	l	d	d	d	d	d	d	d	d	l	n	n	n	d	d	l	d	d	d	d	d	d	d	d	d	d	d	4	3Q
22nd	Sat	d	d	d	d	d	d	d	d	l	l	n	n	l	l	d	d	d	l	l	d	d	d	d	d	d	d	d	d	d	d	d	9	
23rd	Sun	h	n	l	l	d	d	d	d	d	d	n	d	d	l	l	l	d	d	d	d	d	d	d	d	d	d	d	d	d	d	d	100	V
24th	Mon	r	n	d	d	d	d	d	l	d	d	d	d	d	d	d	d	d	d	d	d	d	d	d	d	d	d	d	d	d	d	l	13	
25th	Tue	d	n	d	d	d	d	d	d	d	d	d	d	d	d	d	d	l	d	d	d	d	d	d	d	l	d	d	d	d	d	l	3	
26th	Wed	d	d	d	d	d	d	d	d	l	l	l	l	l	l	l	n	d	l	d	d	d	d	d	d	d	d	d	l	l	d	n	17	
27th	Thur	d	d	d	d	d	d	d	d	d	n	d	d	d	l	d	d	d	d	d	d	d	d	d	d	d	d	d	d	d	d	d	2	
28th	Fri	d	d	d	d	d	d	d	d	d	d	d	d	d	d	d	d	d	d	d	d	d	d	d	d	d	d	d	d	d	d	d	0	N
estimate:		203	110	99	147	46	41	32	40	16	47	44	16	16	33	28	17	9	37	41	46	23	11	13	29	13	15	74	53	45	64	189	1595	
average:		152	187	195	328	123	93	70	52	55	69	56	45	64	69	59	62	62	70	62	38	43	34	32	72	36	37	80	71	88	75	158	2637	
trend:		wtr	drr	drr	drr	drr	drr	drr	drr	drr	drr	drr	drr	drr	drr	drr	drr	drr	drr	drr	av	drr	drr	drr	drr	drr	drr	av	drr	drr	drr	wtr	drr	

NORTH ISLAND SUNSHINE ESTIMATES
FEBRUARY 2025

F=fine (8-12 hours of sunshine), pc= partly cloudy (4-7 hours), c=cloudy (1-3 hours), o=overcast (0 hours)

		Kta	Whg	Dar	Ak	Hml	Tau	Rt	Taum	NP	Wang	Taup	Gis	Nap	Wai	Dann	PN	Mast	Pa	UH	Wgt	ALL	Moon
1st	Sat	F	F	F	F	pc	F	F	F	F	pc	F	F	F	F	F	pc	F	c	c	F	172	
2nd	Sun	F	F	F	F	pc	F	pc	pc	o	o	pc	F	F	pc	c	o	c	o	c	c	110	P10 XhN
3rd	Mon	F	c	c	c	o	o	o	o	o	o	o	F	0.7	c	c	o	c	c	c	pc	43	
4th	Tue	pc	F	pc	pc	pc	pc	pc	c	c	o	pc	pc	pc	pc	F	o	F	c	pc	F	103	
5th	Wed	pc	o	c	c	pc	c	c	c	F	c	c	pc	c	c	F	c	F	pc	pc	pc	78	1Q
6th	Thur	F	pc	F	F	F	F	F	F	F	c	F	pc	pc	pc	o	c	o	F	c	pc	140	
7th	Fri	pc	F	F	F	F	c	F	pc	F	pc	F	F	F	F	o	pc	o	pc	c	pc	135	
8th	Sat	pc	F	F	F	pc	F	F	pc	F	F	F	F	F	pc	F	F	F	F	F	F	190	^
9th	Sun	F	F	F	F	F	F	F	pc	F	F	F	F	F	F	F	F	F	F	F	F	228	
10th	Mon	F	F	F	F	F	F	F	F	F	F	F	F	F	F	F	F	F	F	F	F	243	
11th	Tue	F	F	F	F	c	F	F	pc	c	o	F	F	pc	c	o	o	o	o	o	c	110	
12th	Wed	F	pc	F	F	F	F	F	F	F	F	F	F	F	F	F	F	F	F	F	F	201	
13th	Thur	pc	pc	F	F	pc	F	F	F	F	F	F	c	c	pc	o	F	o	pc	c	pc	139	F
14th	Fri	F	pc	F	F	F	o	o	o	pc	pc	o	pc	o	c	F	pc	F	F	pc	F	111	
15th	Sat	pc	pc	pc	pc	pc	c	pc	pc	pc	F	pc	F	F	F	F	F	F	F	F	F	173	
16th	Sun	o	o	pc	pc	F	pc	F	F	F	F	F	pc	pc	F	pc	F	pc	F	F	F	168	XhS
17th	Mon	o	o	F	F	F	pc	F	F	F	F	F	pc	F	F	F	F	F	F	F	F	167	
18th	Tue	F	pc	pc	pc	pc	F	F	F	F	F	F	F	F	F	F	F	F	F	F	F	194	A
19th	Wed	pc	F	F	F	pc	F	F	F	F	F	F	F	F	F	F	F	F	F	F	F	196	
20th	Thur	pc	pc	pc	pc	F	pc	pc	pc	F	pc	pc	c	o	o	c	pc	c	F	o	F	106	
21st	Fri	F	pc	c	c	F	F	F	pc	pc	F	F	c	c	pc	pc	F	pc	F	F	F	164	3Q
22nd	Sat	pc	pc	o	o	F	o	o	pc	F	c	o	c	pc	pc	o	c	o	F	c	c	66	
23rd	Sun	F	pc	pc	pc	F	F	F	F	F	F	F	pc	F	F	F	F	F	F	F	F	217	V
24th	Mon	F	F	F	F	F	F	F	F	F	F	F	F	F	pc	c	F	c	F	F	F	190	
25th	Tue	F	F	F	F	pc	F	F	F	F	pc	F	F	F	F	F	pc	F	F	pc	pc	202	
26th	Wed	pc	pc	F	F	F	F	pc	pc	F	pc	pc	pc	c	pc	pc	pc	pc	F	pc	pc	129	
27th	Thur	F	F	F	F	c	F	F	F	F	pc	F	pc	F	pc	F	pc	F	F	F	F	200	
28th	Fri	F	F	F	F	F	F	F	F	F	F	F	F	F	F	F	F	F	F	F	F	229	N
est. hours:		231	205	229	229	211	238	233	222	257	202	233	205	210	207	207	202	207	240	189	243	4401	
av. hours:		197	180	179	200	197	211	205	182	222	213	200	204	200	199	186	190	202	209	209	209	3994	
trend:		more	more	more	more	av	more	more	more	more	av	more	av	av	av	more	av	av	more	av	more	average	

SOUTH ISLAND SUNSHINE ESTIMATES
FEBRUARY 2025

F=fine (8-12 hours of sunshine), pc= partly cloudy (4-7 hours), c=cloudy (1-3 hours), o=overcast (0 hours)

		Ta	Nel	Ble	Mou	West	Grey	Hok	Ham	Kai	Ch	Ash	Tim	Crom	Qstn	Bal	Dun	Wnt	Gore	Inv	Alex	ALL	Moon
1st	Sat	pc	F	F	F	pc	o	o	pc	pc	pc	pc	pc	pc	F	pc	pc	c	c	c	pc	112	
2nd	Sun	c	F	F	F	c	pc	F	pc	pc	pc	c	c	F	F	pc	pc	c	c	pc	F	135	P10 XhN
3rd	Mon	pc	F	F	F	c	F	pc	o	o	o	pc	pc	F	F	pc	F	pc	pc	c	F	120	
4th	Tue	o	F	F	pc	F	F	F	c	c	c	o	o	pc	F	o	o	pc	pc	F	pc	107	
5th	Wed	pc	pc	c	pc	pc	pc	F	o	o	o	o	c	F	F	F	F	F	F	F	F	122	1Q
6th	Thur	F	pc	pc	c	pc	F	F	pc	pc	pc	o	c	pc	F	F	F	F	F	F	pc	160	
7th	Fri	F	c	pc	o	c	c	c	pc	pc	F	pc	c	c	o	o	c	c	c	c	c	64	
8th	Sat	pc	pc	F	pc	pc	c	c	pc	pc	F	F	F	pc	pc	pc	pc	F	F	F	pc	133	^
9th	Sun	F	F	F	F	F	F	F	F	F	F	F	F	F	F	F	F	F	F	F	F	234	
10th	Mon	c	F	F	F	F	o	o	pc	pc	pc	F	pc	pc	c	pc	c	F	F	pc	pc	123	
11th	Tue	F	pc	c	pc	pc	F	F	F	F	F	F	F	F	F	pc	F	c	c	o	F	145	
12th	Wed	F	F	F	F	pc	F	F	c	c	pc	pc	c	F	c	pc	pc	pc	pc	pc	F	141	
13th	Thur	F	pc	pc	c	F	F	F	F	F	F	F	F	F	F	pc	pc	pc	pc	pc	F	165	F
14th	Fri	pc	F	F	F	F	F	F	F	F	F	F	F	F	pc	F	F	F	F	F	F	204	
15th	Sat	F	F	F	F	F	F	F	F	F	F	F	F	F	F	F	F	F	F	F	F	228	
16th	Sun	F	F	F	F	F	F	F	F	F	F	F	F	F	F	c	F	pc	pc	pc	F	203	XhS
17th	Mon	o	pc	F	pc	F	F	F	F	F	F	F	pc	F	F	c	pc	pc	pc	F	F	167	
18th	Tue	pc	F	F	F	F	F	F	pc	pc	pc	F	pc	F	pc	pc	pc	o	o	o	F	131	A
19th	Wed	o	F	F	F	pc	c	pc	o	o	c	c	o	pc	c	o	o	o	o	F	pc	66	
20th	Thur	pc	pc	o	pc	c	F	F	o	o	o	o	o	pc	F	o	o	c	c	pc	pc	76	
21st	Fri	o	F	pc	pc	c	pc	F	o	o	o	o	o	F	pc	pc	pc	c	c	pc	F	75	3Q
22nd	Sat	F	o	o	o	o	pc	F	o	o	c	o	o	F	F	F	pc	F	F	F	F	115	
23rd	Sun	pc	pc	F	F	c	pc	F	pc	pc	c	c	c	F	F	F	F	pc	pc	c	F	131	V
24th	Mon	c	pc	o	c	F	F	F	pc	pc	F	F	F	F	F	pc	pc	c	c	F	F	142	
25th	Tue	pc	F	F	F	F	pc	F	o	o	c	pc	F	o	pc	c	pc	o	o	c	o	93	
26th	Wed	F	F	pc	F	F	F	F	c	c	pc	pc	pc	F	F	F	pc	pc	pc	F	F	170	
27th	Thur	F	pc	F	pc	F	F	F	F	F	F	F	F	F	F	F	F	pc	pc	pc	F	199	
28th	Fri	F	F	F	F	pc	pc	c	F	F	F	F	F	F	F	c	pc	c	c	o	F	158	N
est. hours:		192	225	227	217	188	220	242	156	156	184	186	167	255	236	154	174	149	149	186	255	3917	
av. hours:		217	229	229	227	188	172	184	178	195	196	171	171	223	206	163	152	169	163	164	212	3809	
trend:		less	av	av	av	av	more	more	less	less	av	av	av	more	more	av	more	less	av	more	more	average	

FEBRUARY 1 SATURDAY

High pressure will linger to the north of the North Island over the next few days directing a very warm west to northwest flow over the country. Very sunny and mostly dry over the North Island except for increased cloud and some isolated showers around the Kapiti Coast. Low pressure prevails over the South Island in a stronger westerly flow with a wet front bringing outbreaks of rain through the south and west of the South Island, but remaining mostly dry in the east.

Auckland: Mostly sunny, warm and dry
Hamilton: Partly cloudy, warm, dry, breezy at times
Wellington: Sunny and dry, windy at times
Greymouth: Overcast, rain, blustery winds with strong gusts
Christchurch: Warm, dry, partly cloudy, windy at times
Dunedin: Partly cloudy, occasional showers, unsettled winds

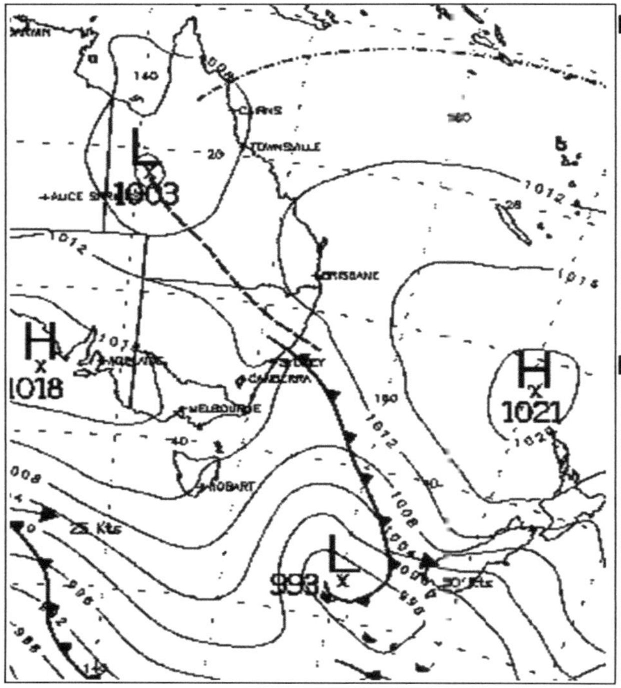

RAIN POTENTIAL

FROST/SNOW

FEBRUARY 2 SUNDAY

While weak high pressure prevails over New Zealand, a wet trough crosses up through the South Island bringing heavy rain, mostly overnight, through the West Coast and Fiordland with showers into Southland and Otago. The east of the South Island and the North Island, however, can expect a dry and partly cloudy day. Temperatures will be high everywhere, under a light northwesterly flow.

Auckland: Sunny, dry, warm, moderate breezes
Hamilton: Moderate breezes, dry, scattered cloudy spells, warm
Wellington: Cloudy, warmer, dry, moderate breezes
Greymouth: Stormy air easing, heavy overnight rain, mostly sunny day, warmer
Christchurch: Very hot and dry, scattered cloudy spells, light breezes, chance of fog
Dunedin: Moderate breezes, very hot, dry, mostly sunny

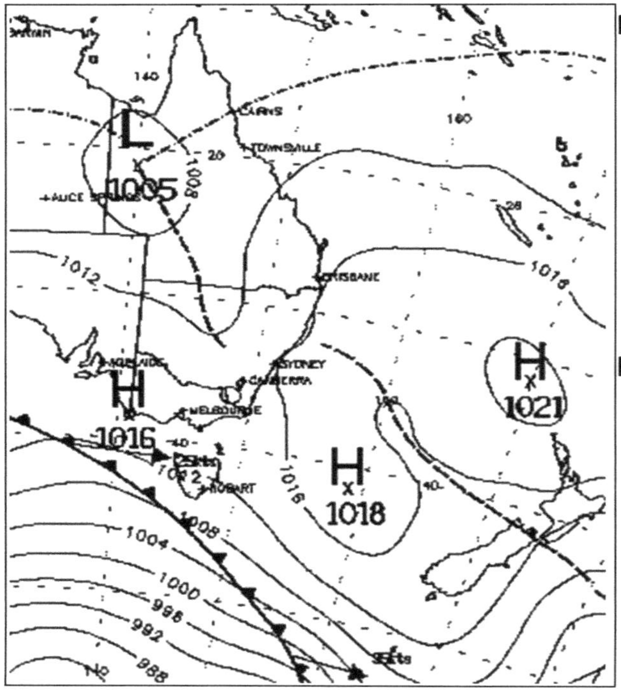

RAIN POTENTIAL

FROST/SNOW

FEBRUARY **3** MONDAY

A weak high pressure centre maintains its influence over the northern half of the country, directing a very hot, humid, and gentle northwesterly flow and widespread cloud throughout the North Island. To the south, a depression deep in the Southern Ocean directs a freshening northwest flow over the southern half of the South Island, with some good sunny spells in the southern districts. Some isolated showers likely in South Canterbury, and in the lower half of the North island from Taranaki to the Manawatu. Mostly dry elsewhere.

Auckland: Warmer, humid, mostly dry, cloudy, chance of drizzle overnight
Hamilton: Overcast, isolated showers, warm and humid, calm atmosphere
Wellington: Light shifting breezes, cloudy, dry, cooler
Greymouth: Partly cloudy, dry, mild temperatures, fluctuating breezes
Christchurch: Overcast, dry, threat of rain, gentle breezes, chance of misty patches
Dunedin: Sunny, dry, cooler, blustery at times

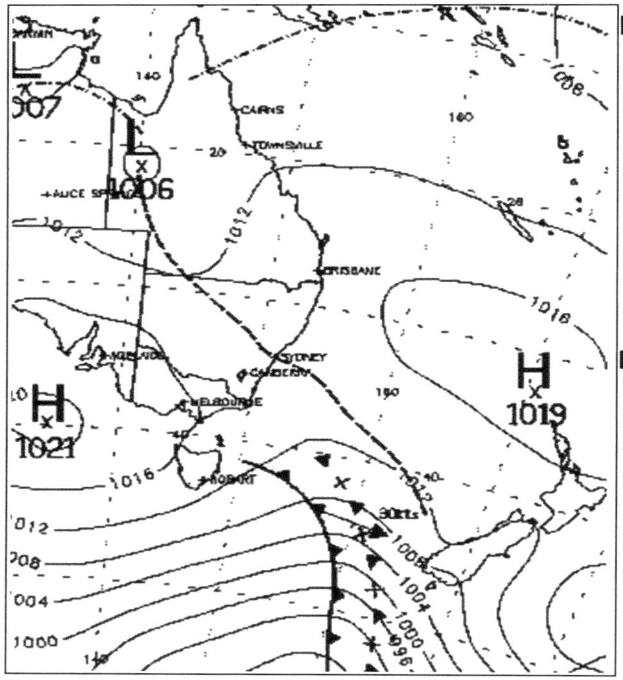

RAIN POTENTIAL

FROST/SNOW

FEBRUARY **4** TUESDAY

A very gentle southwest air flows over the country, with a mix of high pressure and low pressure centres to the west in the Tasman Sea and a weak low pressure flow to the southeast of the country. A couple of wet troughs bring some early rain activity through the west of the North Island and isolated showers in Southland with the remaining parts of the country mostly dry. Generally very hot and humid everywhere, sunniest in the west of the South Island and east and north of the North Island.

Auckland: Mostly sunny, dry, hot, calm and humid air
Hamilton: Partly cloudy, mostly dry, chance of overnight showers, light breezes
Wellington: Mostly sunny, dry, warmer, light breezes
Greymouth: Sunny, dry, mostly calm air
Christchurch: Cloudy, dry, odd threat of rain, cooler, moderate breezes
Dunedin: Overcast, dry, increasing threat of rain, moderate breezes

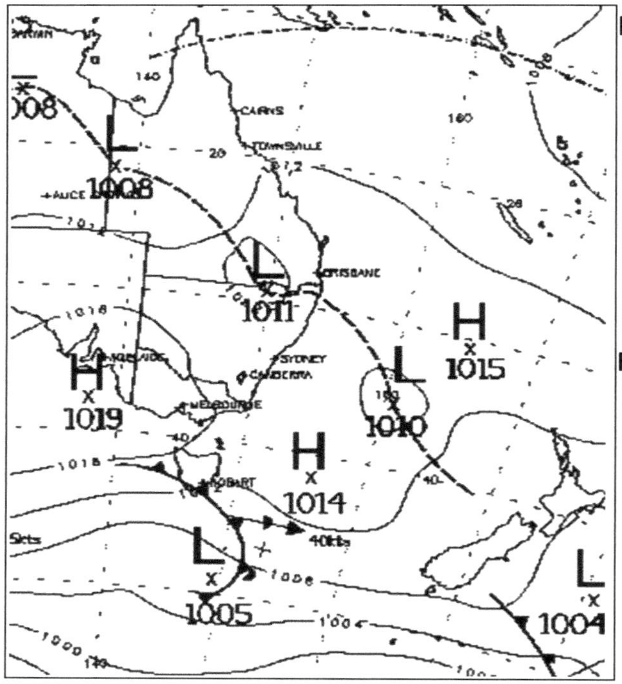

RAIN POTENTIAL

FROST/SNOW

FEBRUARY **5** WEDNESDAY

Low pressure prevails over the country with a very light easterly flow over the North Island; light northwesterly air over the South Island. Very hot and mostly cloudy with outbreaks of rain in the North Island, while slightly milder and sunnier over most of the South Island with light showers in the eastern and southern regions.

Auckland: Very hot, humid, mostly cloudy and dry, chance of overnight showers
Hamilton: Cloudy, hot, humid, rain mostly calm air
Wellington: Cloudy, warm and dry, increasing threats of rain, light air
Greymouth: Partly cloudy, overnight showers, gentle breezes
Christchurch: Mostly calm air, overcast and dull, light drizzly showers
Dunedin: Mostly sunny, overnight rain, cooler, calm atmosphere

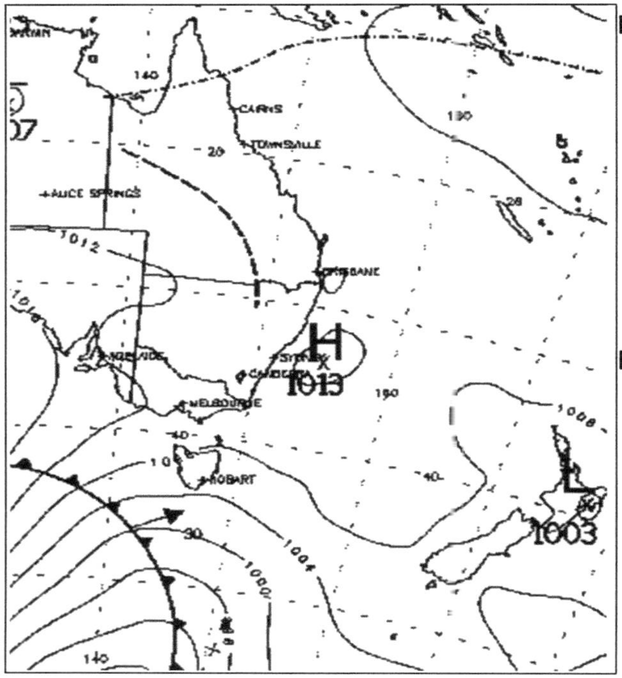

RAIN POTENTIAL

FROST/SNOW

FEBRUARY **6** THURSDAY

Mostly calm conditions continue throughout as a weak low pressure trough slowly crosses the country, with fresher northwesterly winds building over the southern half of the South Island ahead of a depression crossing through the Southern Ocean. Mostly dry in the southern half of the South Island and Northland to Auckland, with mostly morning rain activity either side of the Cook Strait to the central parts of the North Island, clearing to a mostly sunny day. Hot conditions continue over the North Island, milder to the south.

Auckland: Sunny, calm and dry air, hot
Hamilton: Sunny, dry and hot, calm and serene air
Wellington: Mix of sunny spells and showers, gentle breezes freshening later
Greymouth: Sunny, cooler, dry, light breezes freshening later in the day
Christchurch: Partly cloudy, occasional showers, light breezes freshening later
Dunedin: Sunny, dry, mild day cool night, freshening breezes

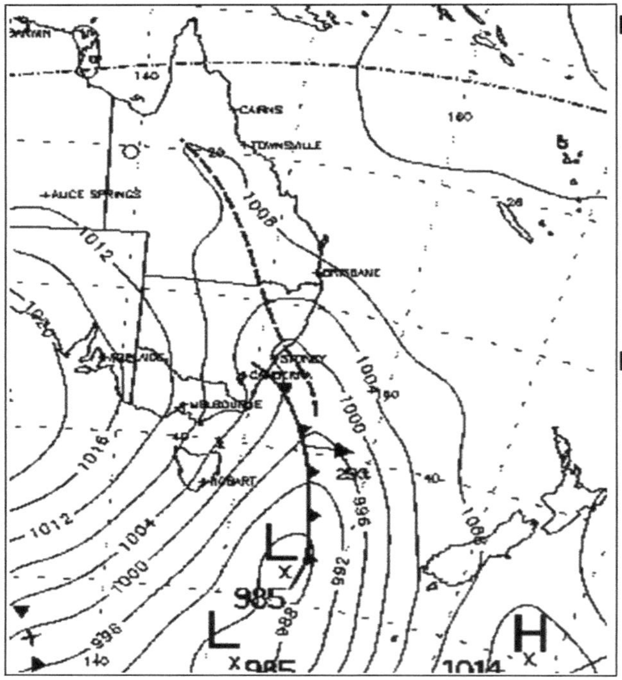

RAIN POTENTIAL

FROST/SNOW

FEBRUARY **7** FRIDAY

A windy low pressure flow crosses the country directing freshening west to northwest winds throughout. Warmer everywhere over the next couple of days. Very cloudy in the southern half of the South Island with sunnier spells elsewhere. Some showers likely in the southwest of the South Island and southeast of the North Island, while mostly dry with increasing threats of rain everywhere else.

Auckland:	Mostly dry and sunny, increasing threats of rain, freshening breezes
Hamilton:	Sunny, hot, humid, freshening winds, rain threatening later in the day
Wellington:	Partly cloudy, dry and windy, cooler
Greymouth:	Cloudy, blustery winds developing, dry with increasing threats of rain
Christchurch:	Mostly sunny and dry, warmer, strong breezes with occasional gusts
Dunedin:	Cloudy, mild, dry, strong and unsettled winds

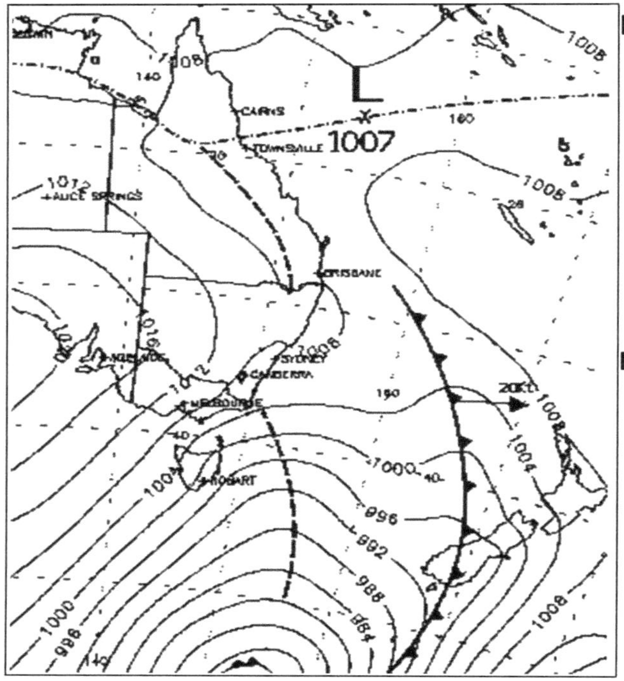

RAIN POTENTIAL

FROST/SNOW

FEBRUARY **8** SATURDAY

A hot west to northwest flow prevails over the country directed from a low pressure system crossing to the south of the South Island. Early morning scattered rain activity in most districts, except Canterbury, with heavy rain likely in the west of the South Island and through central parts of the North Island up into Northland. Some good sunny spells likely later in the day.

Auckland:	Changeable, mix of sunny spells and downpours, very hot and humid
Hamilton:	Heavy rain, some sunny breakthroughs, hot and humid, breezy spells
Wellington:	Changeable, overnight rain, sunny day, warm, blustery winds at times
Greymouth:	Mostly cloudy, heavy rain, blustery winds, chance of thunderstorms
Christchurch:	Sunny, hot day cool night, dry, fresh northwesterly breezes at times
Dunedin:	Dry and blustery, scattered cloudy spells

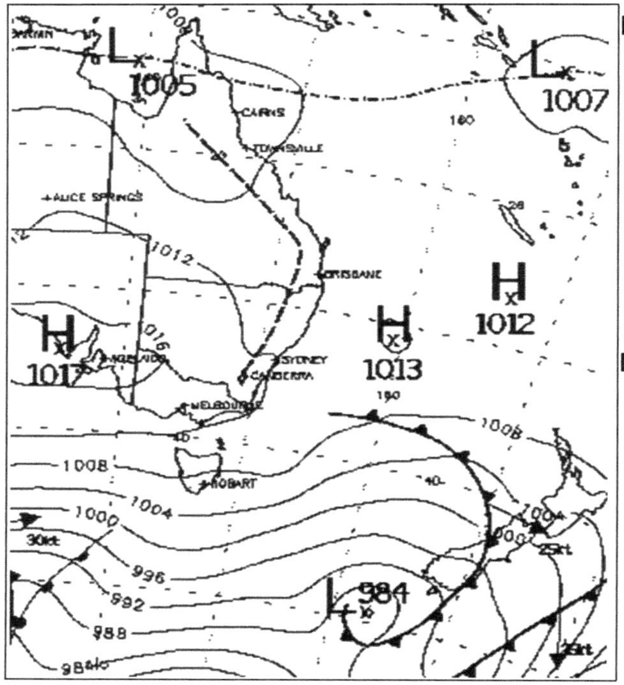

RAIN POTENTIAL

FROST/SNOW

FEBRUARY 9 SUNDAY

High pressure to the northeast of New Zealand holds a low pressure trough over the country for a time, allowing for widespread rain activity in most districts, sometimes accompanied by thunderstorm activity, particularly over the northern half of the North Island. A moderate southwest flow prevails, with high temperatures in the North Island and milder temperatures in the South Island. Generally changeable conditions with a mix of sunny spells and outbreaks of rain.

Auckland: Overnight rain, warm, humid, long sunny spells, strong winds at times
Hamilton: Overnight showers, sunny and warm day, fresh breezes at times
Wellington: Morning showers, sunny day, cooler, fresh to strong breezes at times
Greymouth: Overnight rain, sunny day, cooler, fresh breezes
Christchurch: Bright and sunny, dry, warm, moderate breezes
Dunedin: Unsettled air, very sunny, mostly dry, chance of isolated showers

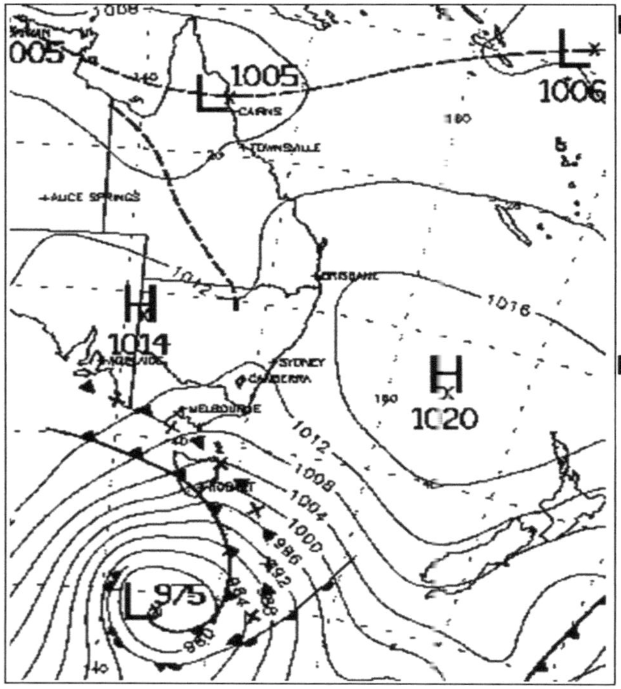

RAIN POTENTIAL

FROST/SNOW

FEBRUARY 10 MONDAY

High pressure crosses the North Island, while low pressure advances onto the South Island. Rain activity clears to isolated showers in the North Island, with high temperatures, long sunny spells, and moderate westerly breezes prevailing. In contrast, strong west to northwest winds build over the South Island, with heavy rain likely to continue around Fiordland, but clearing to isolated showers elsewhere. Winds may reach gale force for a time around the Cook Strait. Chance of fog briefly in some Canterbury districts.

Auckland: Sunny, dry, cooler, light shifting breezes
Hamilton: Warm day cool night, very sunny, dry, moderate breezes
Wellington: Dry, warm, sunny, unsettled winds freshening later in the day
Greymouth: Overcast, dry with passing threats of rain, freshening winds
Christchurch: Partly cloudy, dry, mild day cool night, unsettled windy spells
Dunedin: Cloudy, dry, strong and blustery winds

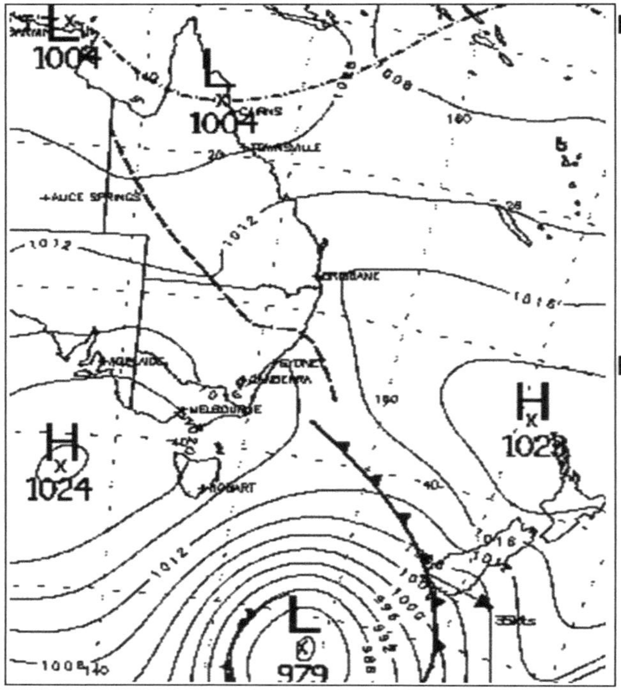

RAIN POTENTIAL

FROST/SNOW

FEBRUARY **11** TUESDAY

A depression to the south of the country directs a wet front up over the South Island in a fresh southwest flow. A stormy air with gales and thunderstorms likely over the west and south of the South island as well as the chance of hail showers and possible brief snow pellets in the higher reaches, despite the warm daytime temperatures and occasional sunny spells. Mostly dry, warm, and sunny in a lighter westerly flow throughout the North Island.

Auckland:	Very sunny and dry, mild temperatures, moderate breezes
Hamilton:	Cloudy, dry, warm day cool night, windy spells
Wellington:	Mostly dry, threats of rain, chance of thunderstorms, gusty winds
Greymouth:	Heavy rain, possible thunderstorms, sunny day, blustery at times
Christchurch:	Hot, sunny, dry, fresh breezes at times
Dunedin:	Sunny, light showers, warm day cool night, strong winds at times

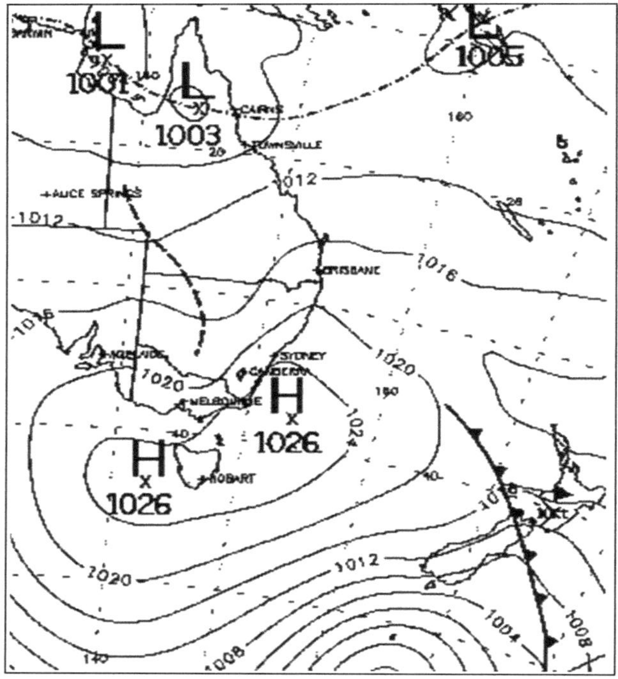

RAIN POTENTIAL

FROST/SNOW

FEBRUARY **12** WEDNESDAY

A large anticyclonic system crossing the Tasman Sea ridges over the country, with moderate southwest breezes prevailing. Rain activity clearing in the South Island while spreading into the North Island, but remaining mostly dry in the east of both islands, and very hot around the East Cape. Occasional sunny spells in the North Island, and sunnier in the South Island. Chance of very cold overnight with the chance of light snow in the higher reaches of the central North Island.

Auckland:	Light showers, sunny day, moderate southerly breezes
Hamilton:	Sunny, mostly dry, chance of morning showers, moderate breezes
Wellington:	Light showers, sunny day, cooler, chance of thunderstorms
Greymouth:	Sunny, mostly dry, chance of some isolated showers, fresh breezes
Christchurch:	Partly cloudy and dry, hot day cool night, gentle breezes
Dunedin:	Partly cloudy, cooler, dry, winds easing to light

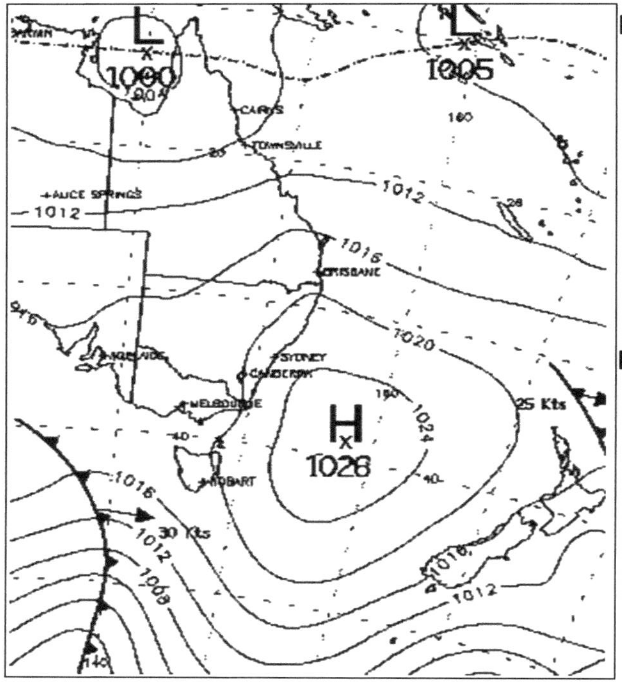

RAIN POTENTIAL

FROST/SNOW

FEBRUARY **13** THURSDAY

A large high pressure system crosses the country bringing a light northwest to mostly calm flow throughout. Shower activity clears in the west of both islands while spreading through the eastern half of the South Island and contracting to the south of the North Island. Mostly dry elsewhere with good sunshine in most parts of the country and remaining very hot in the North Island and northeast of the South Island.

Auckland: Dry, sunny, mild, mostly calm air, chance of brief misty patches
Hamilton: Partly cloudy, dry, milder, light air
Wellington: Changeable, mix of sunny spells and light showers, mostly calm air
Greymouth: Sunny, brief spotty showers, gentle breezes to mostly calm air
Christchurch: Mostly calm air, sunny, cooler, chance of fog patches or light showers
Dunedin: Mostly calm air, cooler, mix of sunny spells and light showers

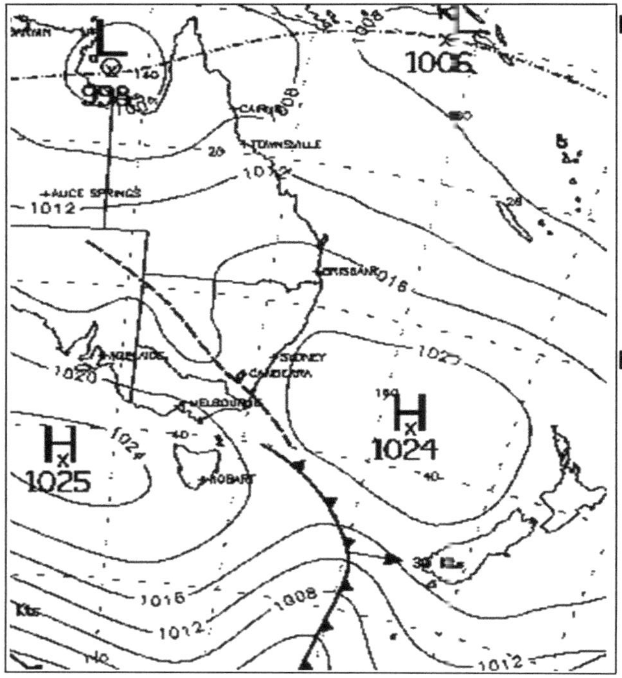

RAIN POTENTIAL

FROST/SNOW

FEBRUARY **14** FRIDAY

A wet front between two high pressure systems continues to move up through the country with rain activity clearing to mostly dry in the South Island while intensifying in the east of the North Island. Westerly breezes at first back northeasterly later in the day. Mostly sunny throughout the South Island and in the north of the North Island while dull and gloomy in the Bay of Plenty to the East Coast and central North Island. Slightly cooler.

Auckland: Sunny and mild, dry, mostly calm air
Hamilton: Sunny, dry, mostly calm air
Wellington: Sunny, cooler, dry, gentle breezes
Greymouth: Sunny, dry, light shifting breezes
Christchurch: Very sunny, dry, mild day cool night, calm air, fog patches possible
Dunedin: Very sunny, isolated evening showers, mostly calm air

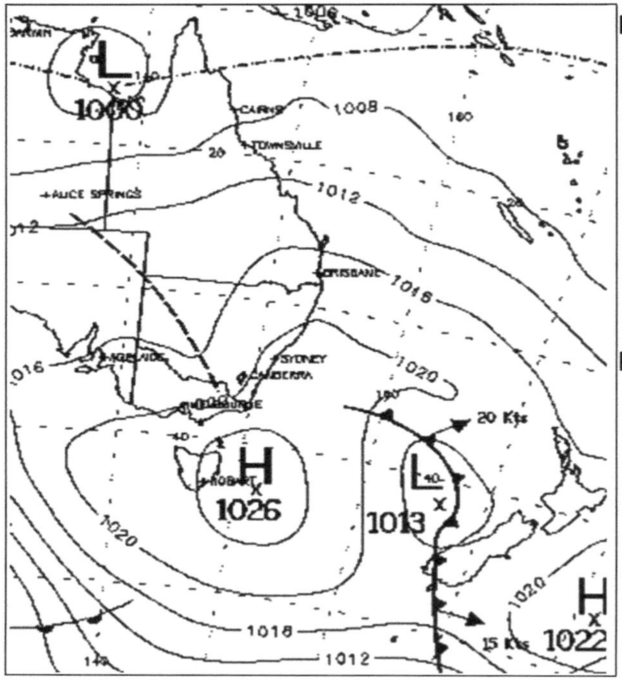

RAIN POTENTIAL

FROST/SNOW

FEBRUARY **15** SATURDAY

High pressure drifts to the southeast of New Zealand while a weak depression advances closer to the north of the country. Mostly dry, warmer and sunny throughout the South Island in a gentle to mostly calm easterly flow, with the chance of some patches of fog in Canterbury for a time. In contrast, while also being very warm and with moderate easterly breezes prevailing, a wet trough brings scattered rain activity throughout most of the North Island except in the far south.

Auckland: Changeable, mix of sunny spells and spotty showers, light breezes
Hamilton: Light breezes, cooler, scattered cloud, chance of isolated showers
Wellington: Very sunny, warmer, dry, light breezes
Greymouth: Very sunny, dry, mild temperatures, gentle breezes
Christchurch: Sunny, dry, warm day cool night, calm serene air, fog possible
Dunedin: Calm air, sunny and dry

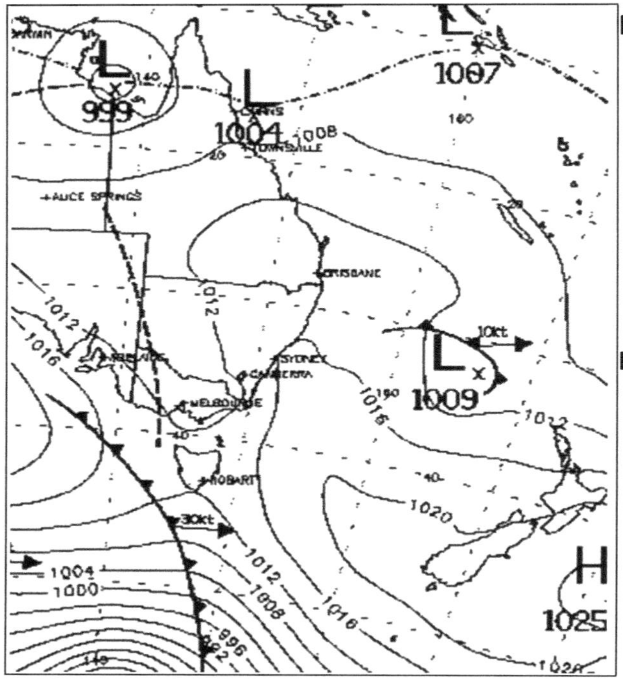

RAIN POTENTIAL

FROST/SNOW

FEBRUARY **16** SUNDAY

The high pressure system to the southeast of New Zealand weakens as it moves off the country, while the weak depression to the north of the North Island continues to deepen. A very warm easterly flow prevails with a mostly dry, sunny and hot day throughout the South Island. Mostly dry and hot across the North Island, except for increased cloudy spells in the west, and some heavy rain expected in the Far North, that may bring some pockets of localised flooding in the region over the next couple of days.

Auckland: Partly cloudy, mostly dry, breezy spells, rain threatening later
Hamilton: Sunny, warmer, dry, freshening breezes
Wellington: Bright and sunny, dry, warmer, light breezes
Greymouth: Warmer, bright and sunny, dry, mostly calm air
Christchurch: Sunny, dry, mild day cool night, gentle breezes at times, chance of fog
Dunedin: Light breezes, dry, very sunny

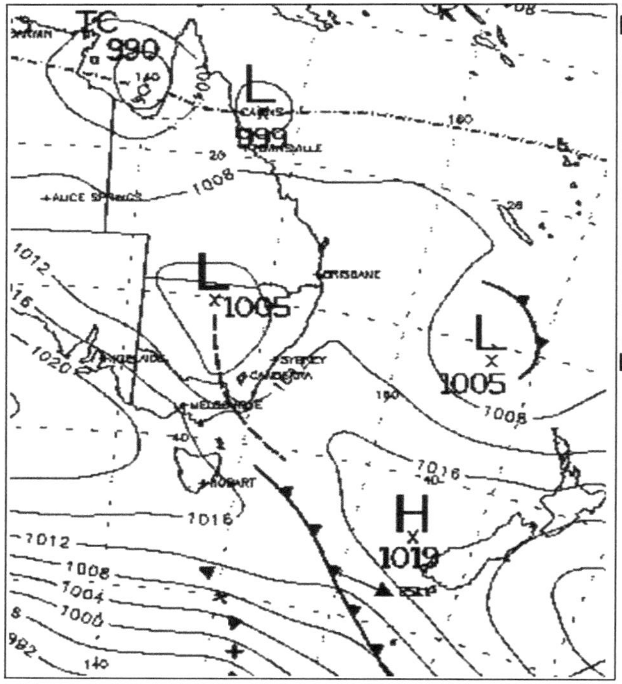

RAIN POTENTIAL

FROST/SNOW

FEBRUARY **17** MONDAY

The high pressure to the southwest of New Zealand maintains dry, hot and mostly sunny conditions throughout the South Island under a very light easterly flow, with the chance of some patches of morning fog in some eastern regions. In contrast, heavy rain associated with the weak depression north of the North Island extends showers into the central regions of the island, with the chance of pockets of flooding not just in Northland but also into the Waikato/Bay of Plenty region. Very hot and humid everywhere.

Auckland:	Overnight showers, sunny day, hot and humid, moderate breezes
Hamilton:	Heavy rain, sunny day, hot and humid, moderate breezes
Wellington:	Very hot and sunny, dry, light breezes
Greymouth:	Very sunny, warm, dry, gentle breezes
Christchurch:	Mild and dry, sunny, mostly calm air, chance of fog patches
Dunedin:	Partly cloudy, dry, warmer, calm atmosphere

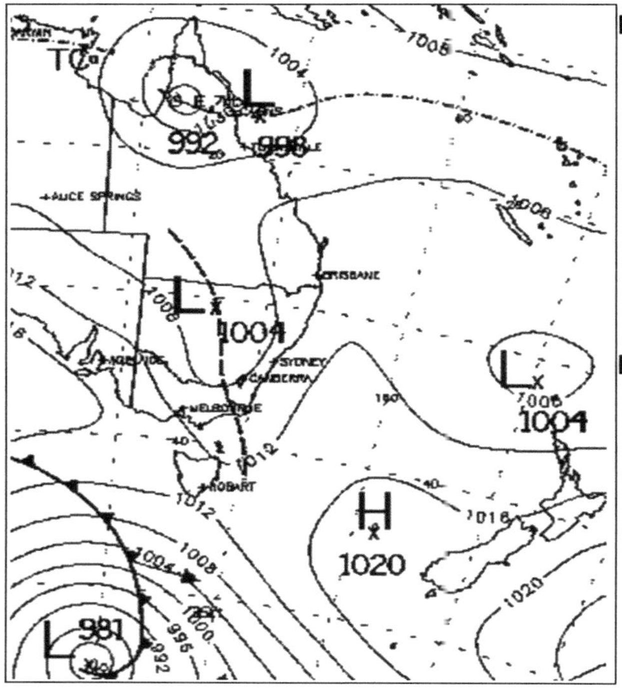

RAIN POTENTIAL

FROST/SNOW

FEBRUARY **18** TUESDAY

A weak high pressure flow continues to bring a mostly dry and very hot day over the South Island as the anticyclone drifts away to the east of the country, directing light northeasterly breezes. Low pressure over the North Island continues as the depression also moves away to the east, with heavy rain clearing to occasional showers in most districts, except in the Far North, where flooding may still be a risk. Very hot and humid throughout the country.

Auckland:	Partly cloudy, hot, humid, mostly dry, chance of isolated showers
Hamilton:	Partly cloudy, dry with passing threats of rain, gentle breezes
Wellington:	Sunny, cooler, dry, light shifting breezes
Greymouth:	Sunny, dry, hot, calm and serene air
Christchurch:	Partly cloudy, warmer, dry, gentle breezes
Dunedin:	Pleasant and serene air, dry, scattered cloudy spells, warm

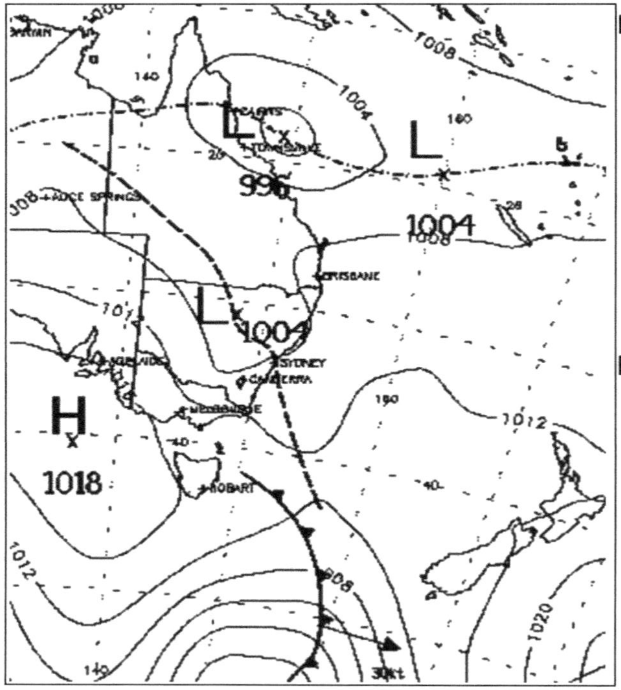

RAIN POTENTIAL

FROST/SNOW

FEBRUARY 19 WEDNESDAY

A wet front ahead of a large high in the southern Tasman Sea crosses the country. Rain clears to isolated showers and long sunny spells in the North Island, while heavy rain advances onto the south and west of the South Island. Winds generally very light in a gentle easterly flow over the North Island while backing fresher southwesterlies over the South Island. Very hot and humid everywhere.

Auckland: Sunny, hot, dry, calm and humid air
Hamilton: Cloudy, hot, humid, calm air
Wellington: Warm, sunny, dry, gentle shifting breezes freshening later in the day
Greymouth: Cloudy, shifting breezes, warm, dry, increasing threats of rain
Christchurch: Cloudy, warm, dry with increasing threats of rain, fluctuating breezes
Dunedin: Overcast, scattered showers, moderate breezes

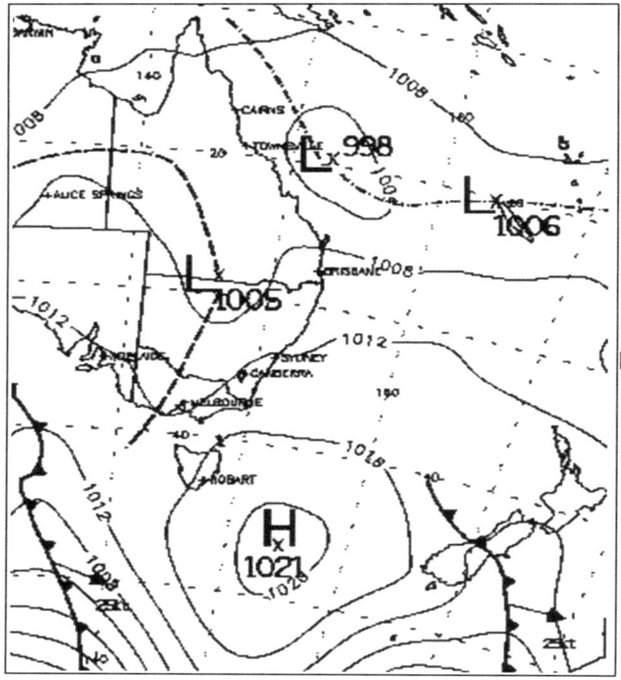

RAIN POTENTIAL

FROST/SNOW

FEBRUARY 20 THURSDAY

A large high builds in the southern Tasman Sea and extends its ridge over the country, clearing rain in the south of the South Island while pushing the wet front northwards. Scattered rain and thunderstorm activity can be expected throughout the northern half of the South Island and the lower half of the North Island while mostly dry Northland to Auckland. High temperatures ease in the south and west of the South Island while continue elsewhere.

Auckland: Very hot, humid, mostly sunny, dry, static air at times
Hamilton: Sunny, very hot and humid, dry, static air, light breezes
Wellington: Sunny, warm, some evening showers, fluctuating breezes
Greymouth: Sunny, warm, light breezes, heavy evening rain
Christchurch: Overcast, mild temperatures, shifting breezes, brief showers
Dunedin: Overcast, cooler, occasional showers, light shifting breezes

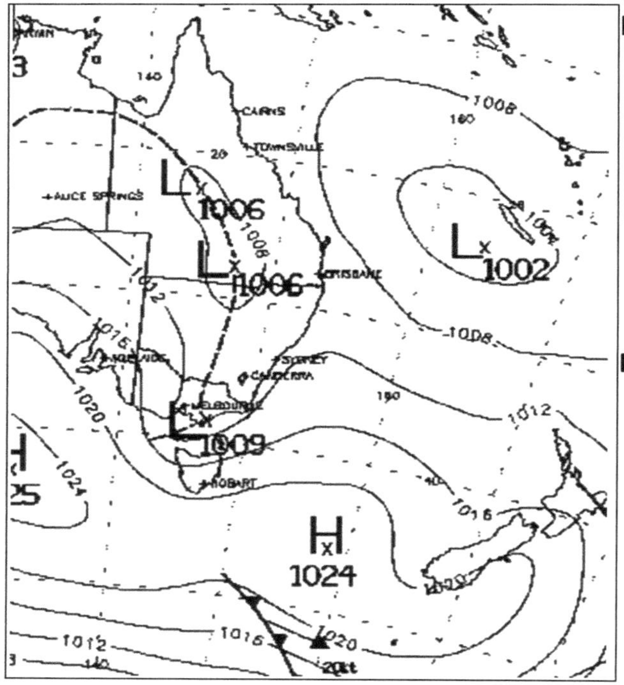

RAIN POTENTIAL

FROST/SNOW

FEBRUARY 21 FRIDAY

Anticyclonic, cooler and dry throughout the South Island with a very light westerly flow prevailing. North of New Zealand, a cyclonic system is likely to have formed near Fiji and although threatens New Zealand for a time, it is expected to move eastward away from the country. Despite this, the easterly flow will bring very high temperatures, and a band of afternoon thunderstorm activity over the northern half of the North Island. Mostly dry elsewhere.

Auckland:	Hot, humid, cloudy, chance of thunderstorms and heavy rain in places
Hamilton:	Sunny, hot, mostly dry, evening thunderstorms and showers possible
Wellington:	Very sunny and dry, cooler, moderate breezes
Greymouth:	Partly cloudy, dry, light shifting breezes
Christchurch:	Overcast, dry with odd threat of rain, cooler, calm air
Dunedin:	Partly cloudy, dry, cooler, light shifting breezes

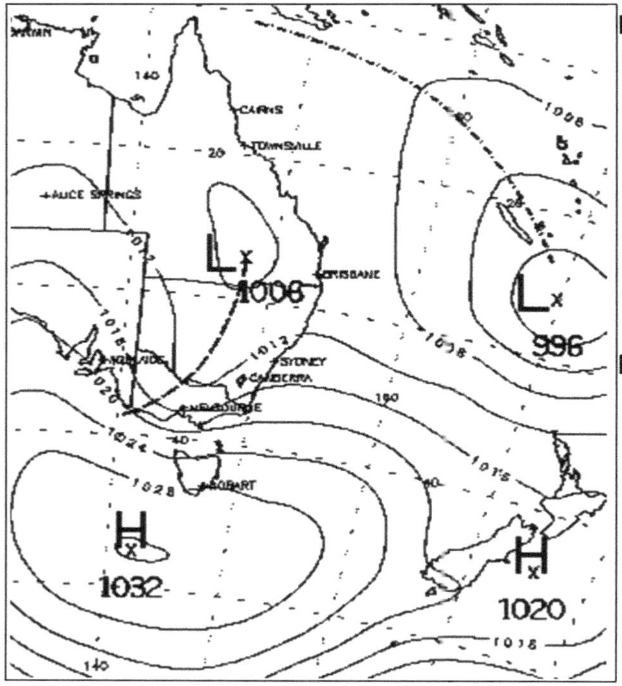

RAIN POTENTIAL

FROST/SNOW

FEBRUARY 22 SATURDAY

The high temperatures of the recent days ease around this time as a large anticyclonic system extends its ridge over the country, directing a light southwest flow over the South Island, while low pressure north of New Zealand maintains a moderate easterly flow, with the cyclonic system moving off the Fiji region and away from New Zealand. Some isolated showers linger around Canterbury and the Waikato, with mainly dry conditions elsewhere. Very sunny in Southland, Otago, and the central west of the North Island, while cloudier elsewhere.

Auckland:	Overcast, cooler, breezy, threats of rain, chance of showers, humid
Hamilton:	Very sunny, hot and humid, dry, moderate breezes
Wellington:	Cloudy, dry, mild temperatures, fresh breezes at times
Greymouth:	Partly cloudy, dry, gentle shifting breezes
Christchurch:	Cloudy, isolated showers, mild temperatures, gentle shifting breezes
Dunedin:	Changeable, partly cloudy, light showers, cool, light shifting breezes

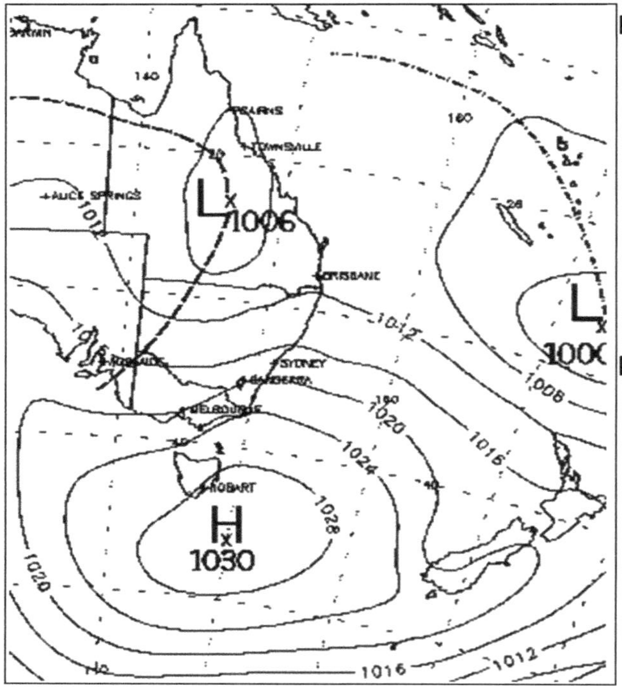

RAIN POTENTIAL

FROST/SNOW

FEBRUARY 23 SUNDAY

A large anticyclonic system to the west of New Zealand extends a ridge over the country with a very light southwest flow prevailing over the South Island, backing southeasterly over the northern half of the North Island. Milder temperatures and mostly sunny throughout. Some overnight rain likely in the northwest of the South Island and showers around the East Cape. Mainly dry elsewhere.

Auckland: Partly cloudy, cooler, some spotty showers, moderate breezes
Hamilton: Sunny, cooler, dry, mostly calm air
Wellington: Very sunny, mild and dry, mostly calm air
Greymouth: Partly cloudy, mostly dry, calm air, chance of evening showers
Christchurch: Mostly cloudy, dry, and calm air, cooler
Dunedin: Sunny, dry, mild day cool night, gentle breezes

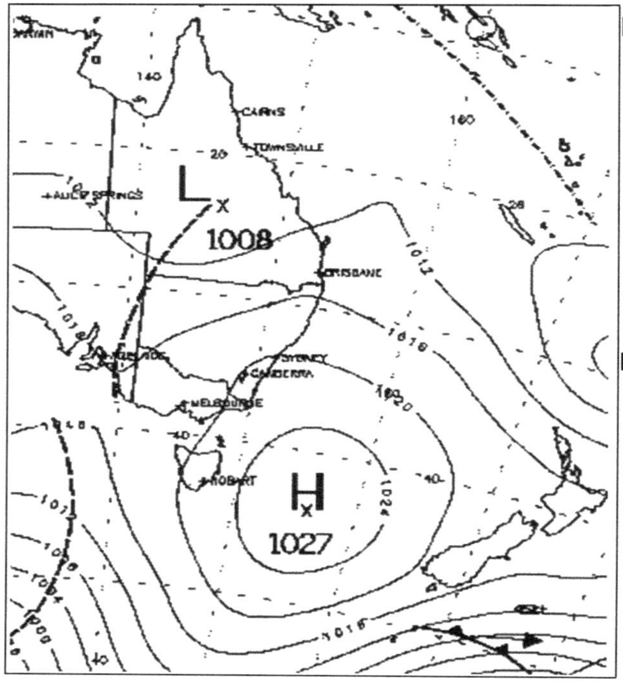

RAIN POTENTIAL

FROST/SNOW

FEBRUARY 24 MONDAY

Anticyclonic. Winds generally light in a southwesterly flow, turning southeasterly over the east of the North Island. Some lingering showers likely overnight in the northwest of the South Island, while dry, warm, and mostly sunny elsewhere.

Auckland: Sunny, dry, light breezes, warmer
Hamilton: Sunny, warmer day, cool night, dry, mostly calm air
Wellington: Sunny, warmer, dry, light shifting breezes
Greymouth: Sunny, mostly dry, mostly calm air, chance of evening showers
Christchurch: Sunny, dry, mild day cool night, calm and serene air
Dunedin: Partly cloudy, warmer, dry, gentle breezes to mostly calm air

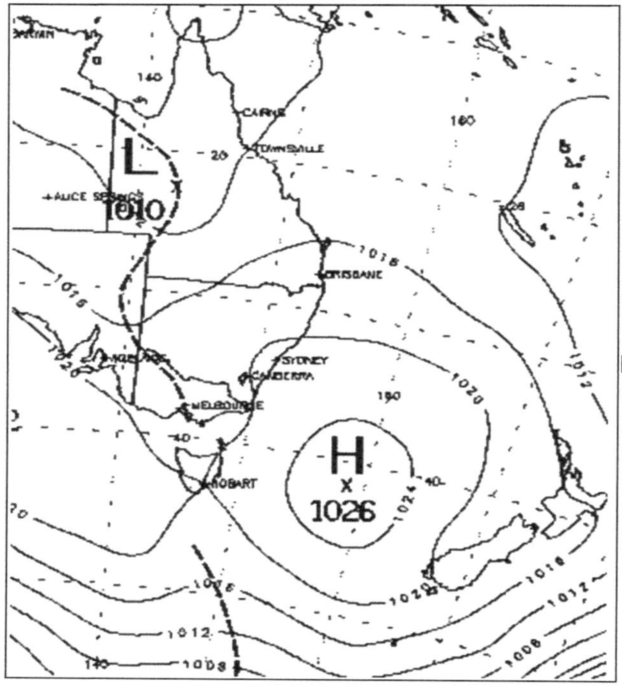

RAIN POTENTIAL

FROST/SNOW

FEBRUARY 25 TUESDAY

Anticyclonic. A gentle southwest flow prevails throughout with higher temperatures returning, cloudy in Canterbury to Otago, chance of some showers around Auckland and mostly dry and sunny elsewhere.

Auckland: Warm, sunny and dry, mostly calm air
Hamilton: Partly cloudy, dry, warm, calm and serene air
Wellington: Partly cloudy, cooler, dry, gentle breezes fresher at times
Greymouth: Partly cloudy, isolated showers, gentle breezes
Christchurch: Cloudy, warmer, dry, gentle breezes
Dunedin: Light breezes, partly cloudy, warm and dry

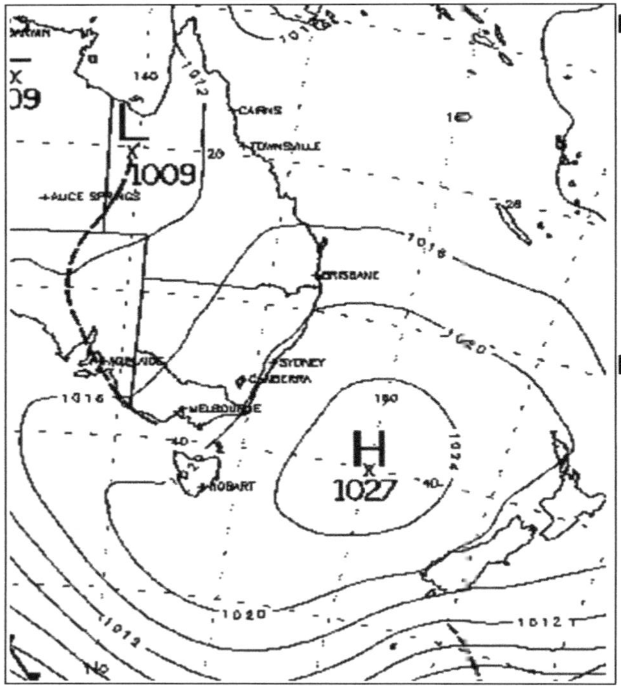

RAIN POTENTIAL

FROST/SNOW

FEBRUARY 26 WEDNESDAY

A large high to the west of New Zealand sits almost stationery in the central Tasman Sea as it strengthens. The high pressure ridge directs a light to moderate southwest flow over the country, high temperatures, and very sunny conditions almost everywhere. Some isolated overnight showers likely in the Marlborough to Canterbury region and through the Cook Strait to Wellington. Mostly dry elsewhere.

Auckland: Mostly sunny, dry, warm, fresh breezes, chance of evening showers
Hamilton: Warm, sunny, dry, light breezes
Wellington: Changeable, overnight showers, occasional sunny spells
Greymouth: Sunny, dry, slightly cooler, gentle shifting breezes
Christchurch: Changeable, partly cloudy, isolated showers, cooler, light breezes
Dunedin: Partly cloudy, isolated showers, warmer, breezy spells

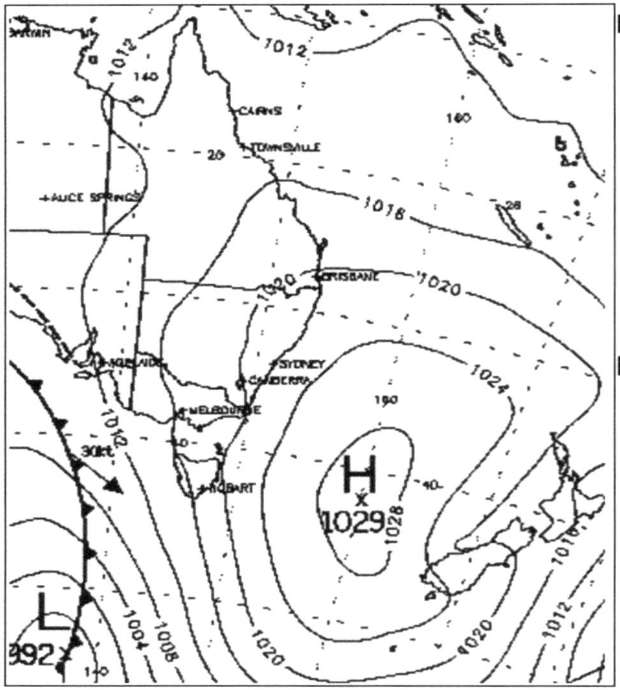

RAIN POTENTIAL

FROST/SNOW

FEBRUARY 27 THURSDAY

Anticyclonic. A large high over New Zealand directs a gentle westerly flow backing southeasterly later in the day. Mostly sunny and warm throughout. Some isolated overnight showers likely mostly in the easterly half of the North Island, while remaining mostly dry elsewhere.

Auckland: Very sunny, dry, fresh breezes at times, chance of evening showers
Hamilton: Cloudy, dry, warm day cool night, gentle breezes
Wellington: Very sunny, mostly dry, chance of overnight showers, light breezes
Greymouth: Sunny, dry, gentle breezes
Christchurch: Sunny, mild, dry, calm air, chance of overnight fog patches
Dunedin: Sunny, cooler, dry, mostly calm and serene air

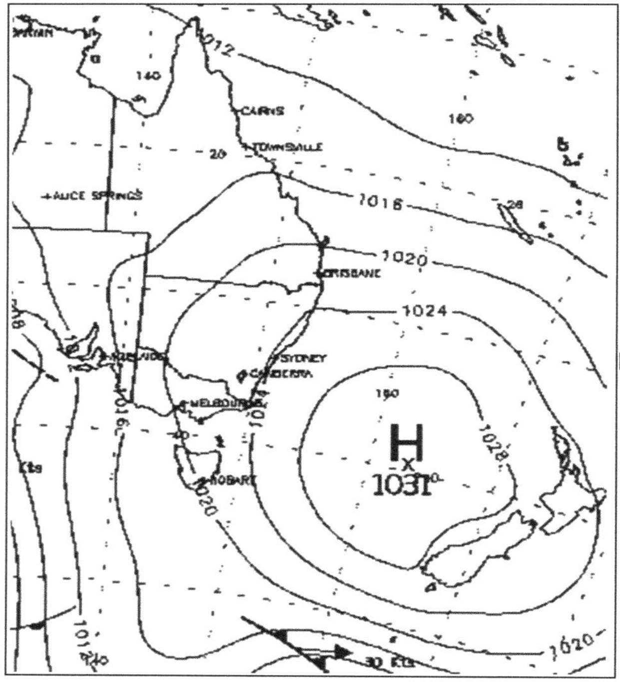

RAIN POTENTIAL

FROST/SNOW

FEBRUARY 28 FRIDAY

A large high pressure system remains stalled west of New Zealand in the central Tasman Sea, maintaining a gentle southwesterly flow over the country. Very hot, dry, and mostly sunny throughout, and particularly hot in the south and east of the South Island, and in the central North Island to Northland.

Auckland: Very sunny, dry, mild, calm and serene air
Hamilton: Sunny and dry, warm day cold overnight, calm and serene air
Wellington: Sunny, dry, mild, gentle breezes
Greymouth: Partly cloudy, warmer, dry, mostly calm air
Christchurch: Mostly calm air, sunny, warm day cool night, dry
Dunedin: Light breezes, dry, warmer, passing cloudy spells

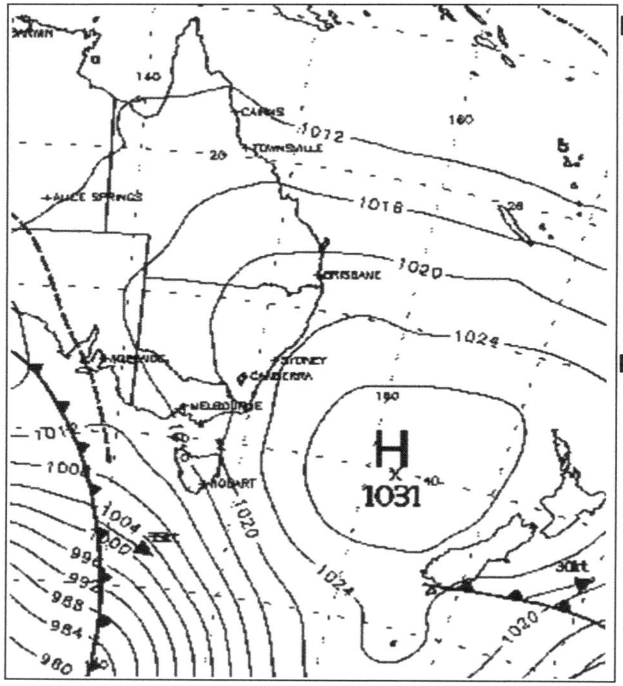

RAIN POTENTIAL

FROST/SNOW

MARCH

PHASES OF THE MOON

1st	Perigee #7
6th	First Quarter
7th	Northern Declination
14th	Full Moon
14th	Crossing Equator
17th	Apogee
22nd	Third Quarter
22nd	Southern Declination
28th	Crossing Equator
29th	New Moon
30th	Perigee #4

1	2	3	4	5	6	7	8	9	10	11	12	13	14	15	16	17	18	19	20	21	22	23	24	25	26	27	28	29	30	31
Sat	Sun	Mon	Tue	Wed	Thu	Fri	Sat	Sun	Mon	Tue	Wed	Thu	Fri	Sat	Sun	Mon	Tue	Wed	Thu	Fri	Sat	Sun	Mon	Tue	Wed	Thu	Fri	Sat	Sun	Mon
P7 XhN					1Q	^							F XhS			A					3Q V						XhN	N	P4	

MONTHLY SUMMARY

4th Misty patches possible in the Christchurch area.
6th Unusually cold overnight in the Manapouri region with chance of pockets of frost.
7th Very blustery in the south of the South Island.
8th Windy at times in the Christchurch region.

Over the next four weeks anticyclonic systems, previously prevailing more to the south and southwest of the country, will now predominately pass to the east of the North Island. This will see a change to more frequent northwesterly airflows prevailing than is typical for this time of year. The prevalence of anticyclonic systems and winds often from the north will result in more sunshine and warmth than is normal for this coming outlook period, with many days likely to feel more like a continuation of summer than the beginning of autumn, particularly in the east of the country. In a similar rainfall pattern to the previous four weeks, while much of the country will see below average rainfall in the coming four weeks, Northland is likely to be affected by very high rainfall and possible flooding again. Rainfall amounts in this look-ahead period will be around 25% less than normal in eastern parts from Gisborne to Marlborough and into the Nelson district, with below average rainfall also in most other eastern parts of the country. The Wellington and Central Otago districts may only see around half their usual rainfall. In contrast, above average rainfall of up to two to three times more than normal for this time of year will fall in Northland, and above average rainfall in parts of Auckland, Thames-Coromandel, the Central Plateau, and in parts of Buller, Westland and Southland. Most districts can expect to be sunnier than average, especially in eastern districts, as well as warmer than is typical for this time of year, with some parts of Canterbury and Otago possibly warmer by up to 1.5°C above their norm.

11th-15th Very hot spell in Central Otago.
14th-15th Chance of fog patches in Christchurch.
15th Unusually high maximas likely in the Darfield region.
16th-27th A period of high rainfall in the west and south of the South Island.
17th Chance of gale gusts in the Wellington region.
17th-18th Heavy rain in Southland.
19th-22nd Scattered fog patches likely in the Christchurch region.
21st Very hot in Central Otago.
22nd Chance of patches of shallow fog in the Wellington region.

23rd-24th A band of thunderstorms crossing central New Zealand may see lightning strikes and hail causing some damage parts and power outages in the lower North Island including around Wellington. Chance of heavy rain during this spell bringing some localised flooding in the Manawatu to Kapiti Coast region, particularly around the Paraparaumu district. Chance of light snow in the ranges around Wellington including on the Rimutaka Hills. Gale winds likely in Wellington.
23rd-25th Very strong westerly winds possibly to gale force for a time in the Auckland region that may cause some local damage, as well as the chance of strong winds causing possible damage in the Marlborough district.
23rd-29th Wet front in the West Coast brings a period of widespread heavy rain into the region and showers into the Southland districts. Occasional showers elsewhere in the South Island.
24th Stormy air in the Christchurch region that may bring gale gusts and hail showers.
24th-25th Very cold overnight with the chance of sub-zero frosts overnight in the central parts of the South Island including around the Mt Cook region and possible snowfall around Lake Rotoiti and frosts in the central plateau districts of the North Island.
27th Heavy rain in the Westport region may cause localised flooding and landslips.
28th Scattered thunderstorms around Canterbury may see heavy rain, hail, lightning strikes, power outages, and strong winds causing some disruptions in the Christchurch area.
29th Chance of thunderstorms in the Wellington region.
30th-31st Chance of fog in the Christchurch region.
31st Unstable winds across Taranaki may cause some damage as well as bring the chance of forming tornadoes. Chance of hail showers in the Wellington region.

March 2025

Monday	Tuesday	Wednesday	Thursday	Friday	Saturday	Sunday
24	25	26	27	28 **Moon Sign**	**1 XhN Pisces** Conjunct Sun Conjunct Mercury Trine Mars Square Jupiter Conjunct Saturn	**2 P7 Aries** Conjunct Mercury Conjunct Venus Conjunct Neptune Sextile Pluto
3 Aries Conjunct Venus Square Mars	**4 Taurus** Square Pluto	**5 Taurus** Sextile Mars Sextile Saturn Conjunction Uranus Square Ascendant	**6 Gemini** Sextile Mercury Conjunct Jupiter Trine Pluto Square Ascendant	**7 1Q Gemini** Square Sun Conjunct Jupiter Square Saturn Square Ascendant	**8 ^ Cancer** Square Mercury Square Neptune Square Ascendant	**9 Cancer** Trine Sun Square Venus Conjunct Mars Trine Saturn Square Ascendant
10 Cancer Trine Neptune Opposition Pluto Trine Ascendant	**11 Leo** Trine Mercury Trine Venus Sextile Jupiter Square Ascendant	**12 Leo** Square Uranus	**13 Virgo** Square Jupiter	**14 FM Virgo** Opposition Sun Sextile Mars Opposition Saturn Trine Uranus Opposition Neptune	**15 XhS Libra** Opposition Mercury Opposition Venus Opposition Neptune Trine Pluto	**16 Libra** Opposition Mercury Square Mars Trine Jupiter
17 Libra Square Pluto	**18 A Scorpio**	**19 Scorpio** Trine Sun Trine Mars Trine Saturn Opposition Uranus Trine Neptune	**20 Sagittarius** Trine Sun Trine Mercury Trine Venus Trine Neptune Sextile Pluto	**21 Sagittarius** Opposition Jupiter Square Saturn	**22 V Sagittarius** Square Sun Square Venus Square Neptune Square Ascendant	**23 3Q Capricorn** Opposition Mars
24 Capricorn Opposition Mars Sextile Saturn Trine Uranus Sextile Neptune Conjunct Pluto	**25 Aquarius** Trine Jupiter Conjunct Pluto	**26 Aquarius** Square Uranus	**27 Pisces** Square Jupiter	**28 Pisces**	**29 XhN Pisces** Conjunct Mercury Conjunct Venus Trine Mars Conjunct Saturn Sextile Uranus Conjunct Neptune	**30 NM P4 Aries** Conjunct Sun
31 Aries Square Mars	1					

NORTH ISLAND RAINFALL
MARCH 2025

d=mainly dry, n= fog, mist, dew or nonrecordable, l=light showers, p=passing showers, s=significant showers, r=rain, h=heavy falls

Date	Day	Northland CR	Kta	Ker	Rsl	Dar	Whg	Man	Auckland Wark	He	MkH	Ak	Co
1st	Sat	d	d	d	d	d	d	d	d	d	d	d	d
2nd	Sun	d	d	d	d	d	d	d	d	d	d	d	d
3rd	Mon	d	d	d	d	d	d	d	d	l	d	d	s
4th	Tue	d	d	d	n	d	l	r	r	d	d	d	d
5th	Wed	s	l	h	l	d	d	l	d	d	d	d	d
6th	Thur	d	d	d	d	d	d	d	d	d	d	d	d
7th	Fri	d	d	h	s	s	h	s	d	d	d	d	s
8th	Sat	d	d	l	d	r	l	l	d	d	d	d	d
9th	Sun	d	d	d	s	d	d	d	d	c	l	d	d
10th	Mon	d	d	d	d	d	d	d	d	c	l	d	d
11th	Tue	d	d	d	l	d	l	d	d	l	l	l	h
12th	Wed	d	d	d	d	d	d	d	d	c	d	d	l
13th	Thur	d	d	d	d	d	d	d	d	c	d	d	d
14th	Fri	d	d	d	d	d	d	d	d	c	d	d	r
15th	Sat	d	d	d	d	d	l	d	d	c	l	d	d
16th	Sun	d	d	d	d	d	d	d	d	c	d	d	d
17th	Mon	d	d	l	d	d	l	l	l	c	l	n	n
18th	Tue	n	n	d	s	s	l	d	l	[illegible]	l	s	l
19th	Wed	d	d	d	d	d	d	d	l	d	s	d	d
20th	Thur	d	d	l	l	d	d	d	d	d	d	d	d
21st	Fri	d	d	d	d	d	n	d	d	d	d	d	d
22nd	Sat	d	d	l	l	d	l	d	n	d	d	d	l
23rd	Sun	s	r	s	s	h	s	s	h	[illegible]	h	h	r
24th	Mon	d	l	d	n	s	s	d	s	s	s	s	r
25th	Tue	d	l	l	n	l	l	l	s	[illegible]	r	r	r
26th	Wed	d	d	d	d	d	d	d	d	d	n	d	l
27th	Thur	d	d	d	d	d	d	d	d	l	n	l	d
28th	Fri	d	d	d	d	d	d	d	d	d	d	d	d
29th	Sat	l	n	d	d	l	l	d	n	l	l	n	d
30th	Sun	d	d	d	d	d	d	d	d	d	d	d	d
31st	Mon	d	d	d	d	d	d	d	d	d	d	d	d
estimate:		13	16	62	26	59	54	31	47	100	68	79	88
average:		82	70	126	105	84	177	123	148	[illegible]	94	77	96
trend:		drr	drr	drr	drr	drr	drr	drr	drr	av	drr	av	av

Date	Day	Bay Of Plenty Tau	TeP	Wkt	Opo	Waikato Hml	TeA	Mata	Ot	Rot/Tau Rt	Taup	Hawkes Bay Gis	Nap	Has
1st	Sat	d	d	d	d	d	d	d	d	d	d	d	d	d
2nd	Sun	d	d	d	d	d	d	d	d	d	d	d	d	d
3rd	Mon	l	d	d	d	d	d	d	d	d	d	d	d	d
4th	Tue	d	d	d	d	d	d	d	d	d	d	d	d	d
5th	Wed	d	d	d	l	d	d	d	d	d	d	d	d	d
6th	Thur	d	d	d	d	d	d	d	d	d	d	r	s	h
7th	Fri	d	d	d	d	d	d	d	d	d	d	d	d	d
8th	Sat	d	d	d	d	d	d	d	l	d	d	d	d	d
9th	Sun	d	d	l	d	l	l	l	s	d	d	d	d	d
10th	Mon	l	d	d	d	d	d	d	d	d	d	l	d	l
11th	Tue	l	d	d	d	l	l	d	l	n	l	d	n	l
12th	Wed	d	d	d	d	d	d	d	d	l	l	d	d	d
13th	Thur	d	d	d	s	d	d	d	d	d	d	d	d	d
14th	Fri	h	r	h	d	d	d	d	d	l	d	d	d	l
15th	Sat	d	l	d	d	d	l	s	d	d	d	d	d	d
16th	Sun	d	d	d	d	d	d	d	d	d	d	d	d	d
17th	Mon	d	d	d	d	d	d	d	d	d	d	d	d	d
18th	Tue	l	l	l	l	d	l	l	l	l	d	d	d	d
19th	Wed	d	l	l	d	l	l	l	s	l	l	d	d	d
20th	Thur	d	d	d	d	d	d	d	d	d	d	d	d	d
21st	Fri	d	d	d	d	d	d	d	d	d	d	d	d	d
22nd	Sat	l	l	d	l	s	d	d	d	l	d	d	d	d
23rd	Sun	h	h	r	h	h	h	h	h	h	h	l	r	r
24th	Mon	l	l	l	d	r	r	s	r	s	r	d	d	d
25th	Tue	r	r	s	s	r	h	s	h	s	r	l	s	s
26th	Wed	d	d	d	d	l	l	d	l	d	d	d	d	d
27th	Thur	d	d	d	d	l	n	d	n	d	n	d	d	d
28th	Fri	d	d	d	d	d	d	d	l	d	d	d	d	d
29th	Sat	d	d	d	d	l	l	d	s	n	l	d	n	d
30th	Sun	d	d	d	d	d	d	d	d	d	d	d	d	d
31st	Mon	d	d	d	d	d	d	d	d	d	d	d	d	d
estimate:		72	72	47	37	77	96	47	109	48	74	16	22	48
average:		95	209	77	136	72	112	108	99	90	60	75	72	50
trend:		drr	drr	drr	drr	av	drr	drr	av	drr	wtr	drr	drr	av

Date	Day	C Plateau Taum	Chat	Waio	Taranaki NP	Str	Wang	Manawatu Pa	Dann	Lev	Lower North Island PN	Mast	UH	Wgt	all	Moon
1st	Sat	d	d	d	d	d	d	d	d	d	d	d	d	d	0	XhN
2nd	Sun	d	d	d	d	d	d	d	d	d	d	d	d	d	0	P7
3rd	Mon	d	d	d	d	d	d	d	d	d	d	d	d	d	12	
4th	Tue	d	d	d	d	d	d	d	d	d	d	d	d	d	28	
5th	Wed	d	d	d	d	d	d	d	d	l	d	d	d	l	36	
6th	Thur	d	d	l	d	d	d	d	s	h	s	l	l	n	80	
7th	Fri	d	l	d	d	d	d	d	d	d	d	d	d	d	75	1Q
8th	Sat	d	s	d	l	d	d	d	d	d	d	d	d	d	26	^
9th	Sun	l	s	d	s	d	d	d	d	d	d	d	d	d	33	
10th	Mon	d	d	d	d	d	d	d	n	d	d	d	d	d	8	
11th	Tue	r	s	d	s	r	d	d	d	d	d	d	d	d	86	
12th	Wed	l	s	l	d	l	l	l	l	l	d	d	d	d	24	
13th	Thur	l	s	s	d	d	l	d	l	d	d	d	d	d	19	
14th	Fri	d	d	d	d	d	d	d	d	d	d	d	d	d	77	F
15th	Sat	d	d	d	d	d	d	d	d	d	d	d	d	d	14	XhS
16th	Sun	d	l	d	d	d	d	d	d	d	d	d	d	d	1	
17th	Mon	d	d	d	s	d	d	d	d	d	d	d	d	d	14	
18th	Tue	l	r	l	s	l	l	d	d	d	l	d	l	d	70	A
19th	Wed	l	l	l	d	d	d	d	d	d	d	d	d	d	24	
20th	Thur	d	d	d	d	d	d	d	d	d	d	d	d	d	4	
21st	Fri	d	d	d	d	d	d	d	d	d	d	d	d	d	0	
22nd	Sat	d	d	d	d	d	d	d	d	d	d	d	d	d	17	V
23rd	Sun	h	h	h	h	r	s	r	s	h	s	l	d	n	742	3Q
24th	Mon	r	h	r	s	s	l	l	l	s	l	l	s	r	221	
25th	Tue	h	h	h	r	h	h	h	h	h	h	r	h	r	539	
26th	Wed	l	r	r	l	l	d	d	d	d	d	d	d	d	31	
27th	Thur	l	s	l	d	d	d	d	d	d	l	l	d	d	13	
28th	Fri	r	h	l	h	r	r	l	l	h	r	d	h	d	175	
29th	Sat	h	h	h	r	s	r	l	h	s	r	l	l	l	184	XhN
30th	Sun	d	d	d	d	d	d	d	d	d	d	d	d	d	0	N P4
31st	Mon	d	d	d	d	d	d	d	d	d	d	d	d	d	0	
estimate:		120	287	105	103	81	58	38	67	99	69	22	65	31	2553	
average:		88	190	72	90	155	55	74	86	77	60	55	96	85	3714	
trend:		wtr	wtr	wtr	wtr	drr	av	drr	drr	wtr	av	drr	drr	drr	drr	

SOUTH ISLAND RAIN

MARCH 2025

d=mainly dry, n=fog, mist, dew or nonrecordable, l=light showers, p=passing showers, s=significant showers, r=rain, h=heavy falls

		West Coast				Top South Island				Canterbury				South Canterbury				Coastal Otago			Lakes			Inland Otago				Southland						
		West	Grey	Hok	BB	Ta	Mou	Nel	Ble	Dar	Ham	Kai	Ch	Rak	Meth	Ash	Tim	Oam	Dun	Bal	Twi	LP	LT	Crom	Qstn	Ran	Alex	TeA	Wtn	Gore	Inv	Puy Pt	all	Moon
1st	Sat	d	d	d	d	d	d	d	d	d	d	d	d	d	d	d	d	d	n	d	d	d	d	d	d	d	d	d	d	d	l	d	2	XhN
2nd	Sun	d	d	d	d	d	d	d	d	d	d	d	d	d	d	d	d	d	d	d	d	d	d	d	d	d	d	d	d	d	d	d	0	P7
3rd	Mon	d	d	d	d	d	d	d	d	d	d	d	d	d	n	d	d	d	d	d	d	d	d	d	d	d	d	d	d	d	l	h	32	
4th	Tue	d	d	d	d	d	d	d	d	d	d	d	d	d	l	d	d	d	d	d	d	d	d	d	d	d	d	d	d	d	s	h	29	
5th	Wed	l	r	l	s	d	d	d	d	l	d	d	l	s	l	s	r	s	l	l	d	d	d	l	s	s	s	l	l	l	l	l	89	
6th	Thur	d	d	d	d	d	l	d	d	s	s	s	r	l	r	s	s	l	d	d	d	r	d	d	d	l	d	d	d	d	d	d	68	
7th	Fri	d	d	d	n	d	d	d	d	d	d	d	d	d	d	n	d	d	d	d	d	d	d	d	d	d	d	d	d	d	d	d	1	1Q
8th	Sat	s	l	l	h	l	d	d	d	d	d	d	d	d	d	d	d	d	d	d	d	d	d	d	d	d	d	d	d	d	d	h	52	^
9th	Sun	l	l	s	d	d	d	d	d	l	d	n	d	d	l	l	r	l	d	d	d	d	d	d	d	d	d	d	d	d	n	l	24	
10th	Mon	d	d	d	d	d	d	d	d	l	l	s	d	l	s	l	l	d	d	d	d	d	d	d	d	d	d	d	d	d	d	d	20	
11th	Tue	d	d	d	d	d	d	d	d	d	n	l	d	d	l	d	d	d	d	d	d	d	d	d	d	d	d	d	d	d	d	d	3	
12th	Wed	d	d	d	d	d	d	d	d	d	d	d	d	d	d	d	d	d	d	d	d	d	d	d	d	d	d	d	d	d	d	d	0	
13th	Thur	d	d	d	d	d	d	d	d	d	d	d	d	d	d	s	l	d	d	n	d	d	n	d	d	d	d	d	d	d	d	d	6	
14th	Fri	d	d	d	d	d	d	d	d	d	d	d	d	d	d	d	d	d	d	d	d	d	d	d	d	d	d	d	d	d	d	d	0	F
15th	Sat	d	d	l	d	d	d	d	d	d	d	d	d	d	d	d	d	d	d	d	d	d	d	d	d	d	d	d	d	d	n	d	2	XhS
16th	Sun	d	d	s	l	d	d	d	d	d	d	d	d	d	d	d	d	d	d	d	d	d	d	d	d	d	d	d	d	d	d	l	10	
17th	Mon	r	r	l	r	l	d	d	d	d	d	d	d	d	d	d	d	d	d	r	d	d	d	l	h	d	l	h	h	h	h	h	224	
18th	Tue	r	h	h	h	l	n	l	d	d	d	d	d	d	l	l	l	l	l	s	d	d	r	r	r	r	h	s	h	r	h	h	308	A
19th	Wed	l	n	d	d	d	d	d	d	d	d	d	n	d	l	d	l	d	d	d	h	d	d	d	d	d	d	d	d	d	d	d	27	
20th	Thur	d	d	d	d	d	d	d	d	d	d	d	d	d	d	d	d	d	d	d	d	d	d	d	d	d	d	d	d	d	d	d	0	
21st	Fri	d	d	d	d	d	d	d	d	d	d	d	d	d	d	d	d	d	d	d	d	d	d	d	d	d	d	d	d	d	d	d	0	
22nd	Sat	d	d	d	d	l	d	d	d	d	d	d	d	d	d	d	d	d	d	d	d	d	d	d	d	d	d	d	d	d	l	d	1	V
23rd	Sun	h	h	h	h	s	l	h	d	r	d	d	l	r	r	r	h	h	h	r	d	h	r	l	r	r	r	s	r	h	h	h	498	3Q
24th	Mon	h	h	h	h	s	l	l	s	s	s	l	s	s	s	r	s	s	d	d	h	l	s	d	l	d	d	l	s	l	r	s	232	
25th	Tue	h	h	s	s	h	r	r	s	r	r	d	s	r	s	h	s	d	l	s	r	s	s	l	s	d	d	l	h	h	r	s	279	
26th	Wed	r	r	s	n	d	d	d	d	d	d	d	d	d	d	n	d	d	l	s	s	d	d	d	d	d	d	d	r	r	s	s	77	
27th	Thur	h	h	h	h	d	d	d	d	n	d	n	n	d	d	d	d	d	d	d	d	d	d	d	d	d	d	s	l	l	s	h	298	
28th	Fri	h	h	h	h	d	d	d	d	l	d	d	l	l	l	l	s	l	d	l	s	d	d	d	d	d	d	s	s	l	s	s	176	
29th	Sat	s	r	r	h	d	d	d	d	d	d	l	l	d	d	d	d	l	l	s	l	l	d	d	d	d	d	s	s	s	r	s	97	XhN
30th	Sun	d	d	n	l	d	d	d	d	d	d	d	d	d	d	d	d	d	d	d	d	d	d	d	d	d	d	d	d	d	d	s	8	N P4
31st	Mon	d	d	d	d	d	d	d	d	d	d	d	d	d	d	d	d	d	d	d	d	d	d	d	d	d	d	d	d	l	l	d	2	
estimate:		190	232	254	319	40	19	31	11	41	26	19	39	45	46	65	66	44	29	49	64	50	28	21	56	35	32	98	121	106	148	244	2566	
average:		162	180	227	447	146	103	72	42	71	97	84	52	84	89	62	45	45	75	60	49	69	51	45	75	44	30	88	85	84	90	191	3044	
trend:		w tr	w tr	w tr	drr	drr	drr	drr	drr	drr	drr	drr	drr	drr	drr	av	w tr	av	drr	drr	w tr	drr	drr	drr	drr	av	av	av	w tr	w tr	w tr	w tr	drr	

NORTH ISLAND SUNSHINE ESTIMATES
MARCH 2025

F=fine (8-12 hours of sunshine), pc= partly cloudy (4-7 hours), c=cloudy (1-3 hours), o=overcast (0 hours)

		Kta	Whg	Dar	Ak	Hml	Tau	Rt	Taum	NP	Wang	Taup	Gis	Nap	Wai	Dann	PN	Mast	Pa	UH	Wgt	ALL	Moon
1st	Sat	F	pc	F	F	pc	F	F	F	F	F	F	pc	F	F	F	F	F	F	F	F	206	XhN
2nd	Sun	F	F	F	F	F	pc	F	F	pc	F	F	F	F	F	F	F	F	F	F	F	204	P7
3rd	Mon	pc	pc	pc	pc	F	pc	pc	F	pc	c	pc	F	F	F	F	c	F	o	o	c	125	
4th	Tue	pc	F	F	F	c	pc	F	F	pc	pc	F	F	F	F	F	pc	F	c	pc	c	146	
5th	Wed	F	F	F	F	pc	F	F	F	pc	o	F	pc	F	F	c	o	c	c	o	o	133	
6th	Thur	F	F	F	F	c	pc	pc	pc	F	F	pc	pc	c	pc	c	F	c	F	F	F	146	
7th	Fri	F	c	F	F	o	F	F	pc	F	pc	F	F	F	F	F	pc	F	pc	F	F	175	1Q
8th	Sat	F	F	F	F	pc	pc	pc	c	F	pc	pc	F	F	F	F	pc	F	pc	F	F	165	^
9th	Sun	F	pc	pc	pc	F	F	F	F	F	c	F	pc	F	pc	c	c	c	pc	c	pc	135	
10th	Mon	pc	pc	o	o	F	o	o	pc	pc	F	o	pc	o	o	c	F	c	F	c	F	85	
11th	Tue	F	F	F	F	o	o	c	o	o	c	c	F	pc	pc	pc	c	pc	c	pc	pc	90	
12th	Wed	F	F	F	F	F	F	pc	c	pc	c	pc	pc	F	pc	F	c	F	F	pc	F	151	
13th	Thur	F	F	F	F	pc	F	F	pc	pc	pc	F	c	c	c	F	pc	F	F	pc	F	154	
14th	Fri	F	F	pc	pc	o	o	o	pc	F	F	o	pc	pc	F	c	F	c	F	F	F	131	F
15th	Sat	F	F	F	F	pc	F	F	F	F	F	F	F	F	F	F	F	F	F	F	F	200	XhS
16th	Sun	F	F	F	F	F	F	F	F	pc	pc	F	F	F	F	F	pc	F	F	F	F	186	
17th	Mon	pc	c	c	c	o	c	c	c	c	c	c	F	F	F	pc	c	pc	pc	pc	pc	87	
18th	Tue	F	F	F	F	pc	pc	pc	pc	F	F	pc	F	F	pc	F	F	F	F	F	F	162	A
19th	Wed	F	F	F	F	F	F	F	F	F	F	F	F	F	F	F	F	F	F	pc	F	195	
20th	Thur	F	F	F	F	pc	F	F	F	F	F	F	F	F	F	F	F	F	F	F	F	199	
21st	Fri	c	c	pc	pc	pc	pc	pc	F	F	F	pc	F	F	F	F	F	F	F	F	F	166	
22nd	Sat	o	o	o	o	c	c	o	o	c	c	o	pc	c	c	pc	c	pc	c	c	c	39	V
23rd	Sun	pc	pc	pc	pc	pc	F	pc	pc	pc	c	pc	F	F	c	pc	c	pc	c	c	o	95	3Q
24th	Mon	F	pc	pc	pc	pc	pc	pc	c	o	o	pc	F	pc	pc	o	o	o	c	c	c	74	
25th	Tue	F	pc	F	F	F	F	pc	pc	pc	F	pc	F	F	F	F	F	F	F	F	F	171	
26th	Wed	pc	pc	pc	pc	pc	F	c	F	pc	pc	c	F	F	pc	F	pc	F	pc	F	F	131	
27th	Thur	c	o	o	o	pc	pc	c	c	o	o	c	pc	o	pc	o	o	o	o	o	o	33	
28th	Fri	c	c	c	c	pc	pc	o	pc	pc	pc	o	pc	pc	pc	pc	pc	pc	pc	pc	pc	86	
29th	Sat	F	pc	F	F	F	F	F	F	F	F	F	F	F	F	F	F	F	F	F	F	199	XhN
30th	Sun	F	pc	pc	pc	F	F	F	F	F	F	F	F	F	pc	F	F	F	F	F	F	184	N P4
31st	Mon	F	F	pc	pc	pc	F	F	F	pc	c	F	F	F	pc	pc	c	pc	pc	pc	pc	134	
est. hours:		246	212	218	218	185	223	199	205	202	198	199	263	255	236	229	198	229	222	211	238	4386	
av. hours:		186	153	170	176	184	197	191	156	202	186	176	185	194	181	163	170	165	190	190	191	3606	
trend:		more	more	more	more	av	more	av	more	av	av	more	more	more	more	more	more	more	more	more	more	more	

SOUTH ISLAND SUNSHINE ESTIMATES
MARCH 2025

F=fine (8-12 hours of sunshine), pc= partly cloudy (4-7 hours), c=cloudy (1-3 hours), o=overcast (0 hours)

		Ta	Nel	Ble	Mou	West	Grey	Hok	Ham	Kai	Ch	Ash	Tim	Crom	Qstn	Bal	Dun	Wnt	Gore	Inv	Alex	ALL	Moon
1st	Sat	F	F	F	F	F	F	F	F	F	F	F	F	F	F	F	F	F	F	pc	F	203	XhN
2nd	Sun	F	F	F	F	F	F	F	F	F	F	F	F	F	F	F	F	F	F	F	F	216	P7
3rd	Mon	F	F	F	F	o	pc	F	pc	pc	F	pc	F	F	pc	c	c	o	o	o	F	121	
4th	Tue	pc	F	pc	F	F	c	o	pc	pc	F	pc	o	F	F	o	c	o	o	c	F	101	
5th	Wed	F	F	o	F	F	F	F	o	o	o	o	o	F	F	F	o	c	c	F	F	116	
6th	Thur	F	F	pc	F	F	F	F	pc	pc	pc	c	c	F	F	pc	F	F	F	F	F	176	
7th	Fri	c	F	F	F	F	pc	c	F	F	F	F	F	pc	pc	c	pc	c	c	o	pc	127	1Q
8th	Sat	F	F	F	F	c	pc	F	F	F	F	F	F	F	F	F	F	F	F	F	F	201	^
9th	Sun	F	F	c	F	F	F	F	c	c	pc	c	pc	F	pc	pc	c	pc	pc	pc	F	133	
10th	Mon	pc	pc	o	pc	c	F	F	o	o	o	o	o	F	pc	F	F	F	F	F	F	113	
11th	Tue	F	F	F	F	pc	F	F	F	F	F	F	F	F	F	pc	F	pc	pc	c	F	179	
12th	Wed	F	F	F	F	F	pc	pc	c	c	o	c	o	c	c	o	c	o	o	o	c	87	
13th	Thur	F	F	F	F	F	F	F	F	F	F	F	F	F	F	pc	F	F	F	F	F	198	
14th	Fri	c	F	F	pc	F	F	pc	pc	pc	pc	pc	pc	pc	pc	c	pc	c	c	pc	pc	123	F
15th	Sat	F	F	F	F	F	F	F	F	F	F	F	F	pc	F	pc	pc	pc	pc	pc	pc	181	XhS
16th	Sun	F	F	F	F	F	c	pc	F	F	F	F	F	o	c	o	o	o	o	o	o	122	
17th	Mon	F	c	c	pc	pc	pc	c	pc	pc	pc	pc	pc	c	c	o	o	o	o	o	c	64	
18th	Tue	c	F	F	F	c	c	F	pc	pc	F	pc	c	pc	F	pc	pc	c	c	c	pc	120	A
19th	Wed	o	F	F	F	F	F	F	c	c	F	F	F	F	F	pc	F	F	F	F	F	167	
20th	Thur	F	pc	F	F	F	F	F	F	F	F	F	pc	F	F	F	F	F	F	F	F	199	
21st	Fri	c	F	F	F	pc	F	F	F	F	F	F	F	pc	pc	pc	F	pc	pc	pc	pc	158	
22nd	Sat	F	c	c	c	c	o	o	c	c	pc	c	pc	c	pc	c	c	c	c	o	c	52	V
23rd	Sun	pc	pc	c	F	c	pc	pc	o	o	o	c	o	c	pc	pc	pc	pc	pc	pc	c	75	3Q
24th	Mon	c	o	pc	c	c	c	pc	c	c	c	o	c	pc	c	pc	c	c	c	pc	pc	52	
25th	Tue	pc	F	F	F	c	c	pc	F	F	F	F	pc	pc	pc	o	c	o	o	o	pc	106	
26th	Wed	F	F	F	F	o	o	o	F	F	F	F	pc	pc	pc	F	pc	pc	pc	c	pc	133	
27th	Thur	F	c	o	o	o	o	o	F	F	F	pc	F	F	F	F	pc	pc	pc	F	F	120	
28th	Fri	pc	pc	pc	pc	pc	pc	pc	pc	pc	pc	F	F	F	pc	pc	pc	pc	pc	pc	F	126	
29th	Sat	F	F	F	F	F	c	pc	F	F	F	F	F	F	F	F	F	pc	pc	c	F	172	XhN
30th	Sun	F	F	F	F	F	F	F	F	F	F	F	pc	F	F	pc	F	pc	pc	c	F	170	N P4
31st	Mon	F	pc	pc	pc	o	o	o	pc	pc	pc	c	c	c	pc	pc	o	c	c	c	c	67	
est. hours:		245	271	264	264	191	189	219	213	213	230	221	189	230	216	167	174	151	151	150	230	4178	
av. hours:		221	212	215	210	161	152	170	171	175	183	160	160	218	189	142	139	141	136	136	187	3478	
trend:		more	more	more	more	more	more	more	more	more	more	more	more	av	more	more	more	av	more	av	more	more	

MARCH **1** SATURDAY

A large anticyclonic system sits over the country with a light northwest flow over the South Island turning south to southeasterly over the North Island. Mostly sunny, very hot, and dry everywhere.

Auckland:	Sunny, dry, pleasant and mostly serene air
Hamilton:	Partly cloudy, dry, hot day cool night, light breezes to mostly calm air
Wellington:	Very sunny, mild and dry, gentle breezes
Greymouth:	Sunny, dry, pleasant air, gentle breezes
Christchurch:	Sunny, hot day cool night, dry, gentle shifting breezes
Dunedin:	Very hot, sunny, mostly dry, chance of isolated showers, light air

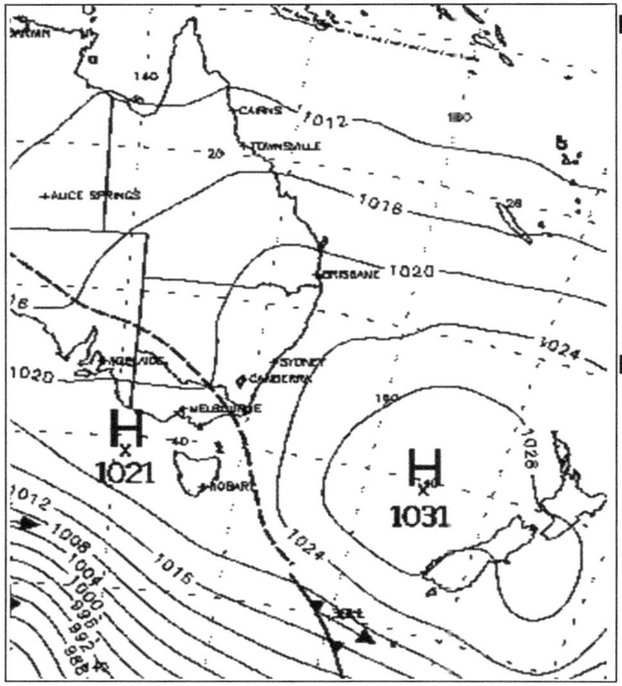

RAIN POTENTIAL

FROST/SNOW

MARCH **2** SUNDAY

Anticyclonic conditions continue to prevail over the country with an almost calm air throughout. Very hot, dry and sunny everywhere.

Auckland:	Hot, dry, sunny, calm atmosphere
Hamilton:	Sunny, warmer, dry, calm and serene air
Wellington:	Sunny, warm, dry, calm and serene air
Greymouth:	Very sunny, warm and dry, mostly calm air
Christchurch:	Sunny, dry, warm, light to moderate breezes
Dunedin:	Sunny, cooler, dry, gentle breezes to mostly calm air

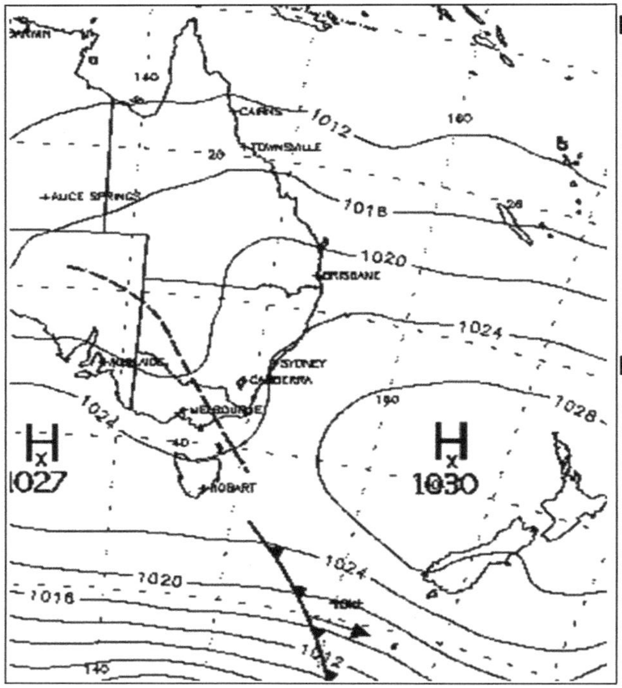

RAIN POTENTIAL

FROST/SNOW

MARCH **3** MONDAY

A large high pressure system remains almost stalled to the east of New Zealand maintaining a very hot and mostly calm air over the country. A wet front moves up through the Southern Ocean brushing the south of the South Island bringing some overnight rain activity mostly in Fiordland. Mainly dry and mostly sunny everywhere else.

Auckland: Partly cloudy, hot and humid, spotty showers, calm air
Hamilton: Very sunny, hot and dry, calm atmosphere
Wellington: Cloudy, hot and dry, calm atmosphere
Greymouth: Pleasant and serene air, dry, scattered cloudy spells
Christchurch: Mostly sunny and dry, warm, light shifting breezes
Dunedin: Cloudy, cooler, dry, occasional breezy spells

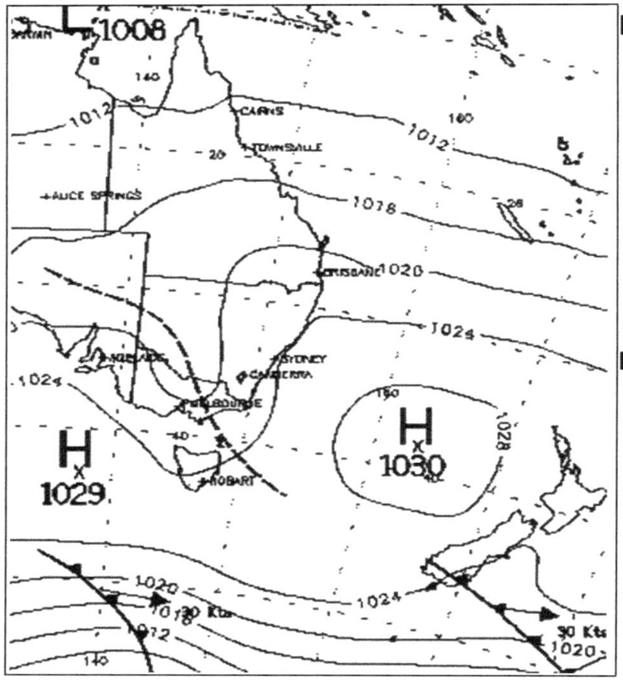

RAIN POTENTIAL

FROST/SNOW

MARCH **4** TUESDAY

The prevailing high pressure influence continues as it begins to weaken. Mostly calm and very high temperatures continue throughout, although some moderate westerly breezes are likely in the far south of the South Island. Overnight rain is likely in Fiordland and around the southern coastal fringes, as a wet front continues to brush the south of the South Island, and chance of some showers in Northland to north Auckland districts. Dry and mostly sunny elsewhere.

Auckland: Overnight rain, sunny and dry day, slightly cooler, serene air
Hamilton: Cloudy, slightly cooler, dry, calm atmosphere
Wellington: Cloudy, milder, dry, some fresh breezes at times
Greymouth: Cloudy, dry, light breezes
Christchurch: Partly cloudy, dry, gentle breezes
Dunedin: Cloudy and dry, occasional breezy spells

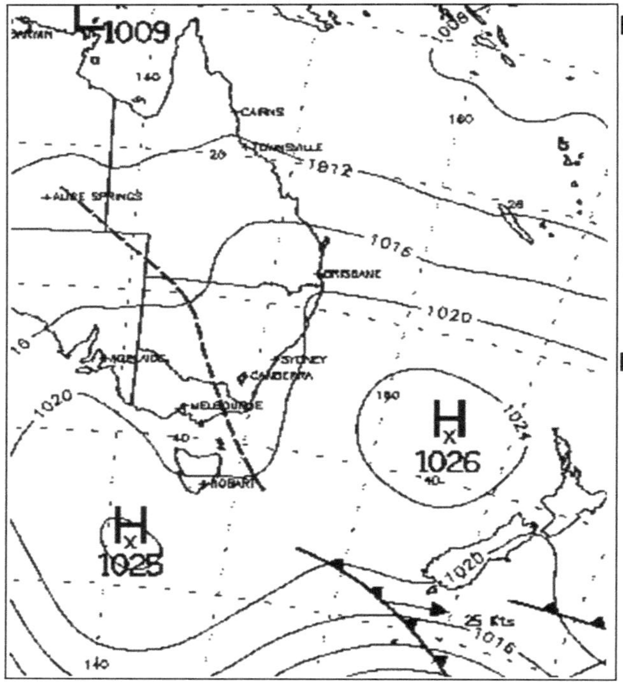

RAIN POTENTIAL

FROST/SNOW

MARCH 5 WEDNESDAY

Very hot and mostly sunny conditions continue throughout the country as high pressure continues to dominate. A wet front continues to slowly drift up the South Island bringing some outbreaks of showers in most South Island districts, as well as brief showers in Northland. Mainly dry everywhere else.

Auckland: Sunny, warm, and dry, light pleasant breezes
Hamilton: Partly cloudy, dry, warm day cool night, light breezes
Wellington: Overcast, milder, brief spotty showers, fresh breezes
Greymouth: Changeable, overnight rain, sunny day, cooler, fluctuating breezes
Christchurch: Fluctuating breezes, overcast, isolated showers
Dunedin: Overcast, brief showers, shifting breezes

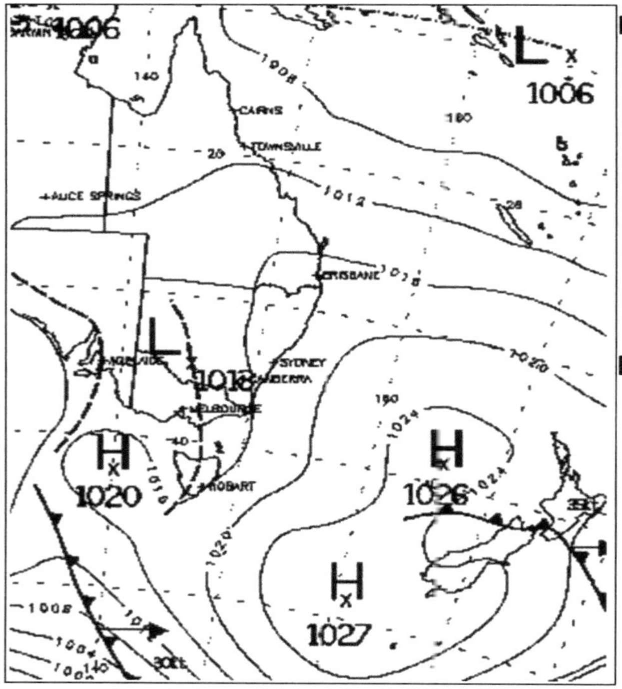

RAIN POTENTIAL

FROST/SNOW

MARCH 6 THURSDAY

The high pressure influence of the past week finally slips to be centred to the southeast of the country with a weak depression moving closer to the west of the South Island. Winds remain light in an easterly backing northerly flow, with scattered rain activity, mostly overnight, contracting to the eastern half of both islands, clearing to dry in the west. Temperatures remain hot over the North Island, but ease over central and southern districts of the South Island. Good sunny spells in most districts.

Auckland: Sunny, dry, warm, light shifting breezes
Hamilton: Hot and humid, cloudy, dry, light breezes
Wellington: Moderate breezes, sunny, cooler, mostly dry, chance of brief showers
Greymouth: Sunny, dry, mild temperatures, light shifting breezes
Christchurch: Overnight rain, cooler, some sunny breakthroughs, light breezes
Dunedin: Very sunny, cooler, dry, moderate breezes

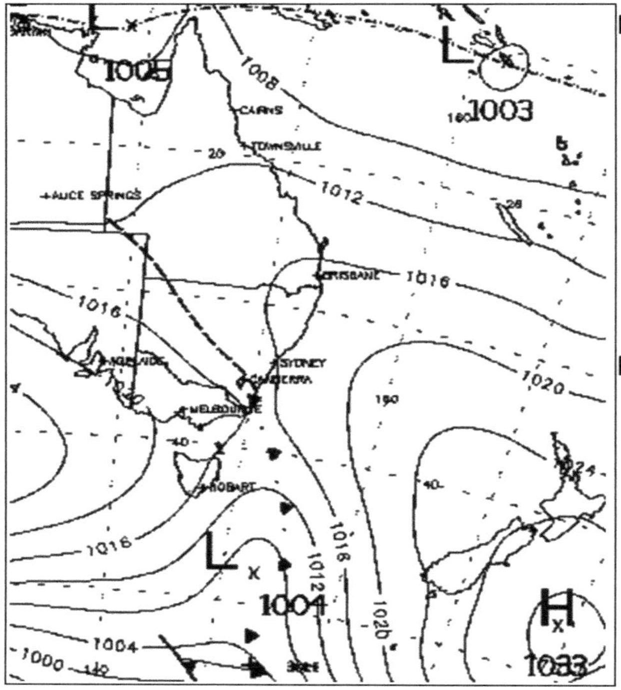

RAIN POTENTIAL

FROST/SNOW

MARCH 7 FRIDAY

A north to northwest flow prevails over the country with high pressure moving away to the east followed by a depression in the Southern Ocean bringing strengthening west to northwest winds over the southern half of the South Island. Widespread high temperatures return throughout the country and with good sunny spells in most districts except in the far south. Mostly dry everywhere except for the chance of some showers possible in the Northland region.

Auckland: Sunny, warm, mostly dry, shifting breezes, chance of brief showers
Hamilton: Hot day cool night, dry, widespread cloud,occasional threats of rain
Wellington: Mostly sunny, dry, warmer, fresh breezes at times
Greymouth: Partly cloudy, dry, moderate breezes
Christchurch: Sunny, dry, slightly warmer, moderate breezes
Dunedin: Partly cloudy, warmer, dry, fresh winds with occasional gusts

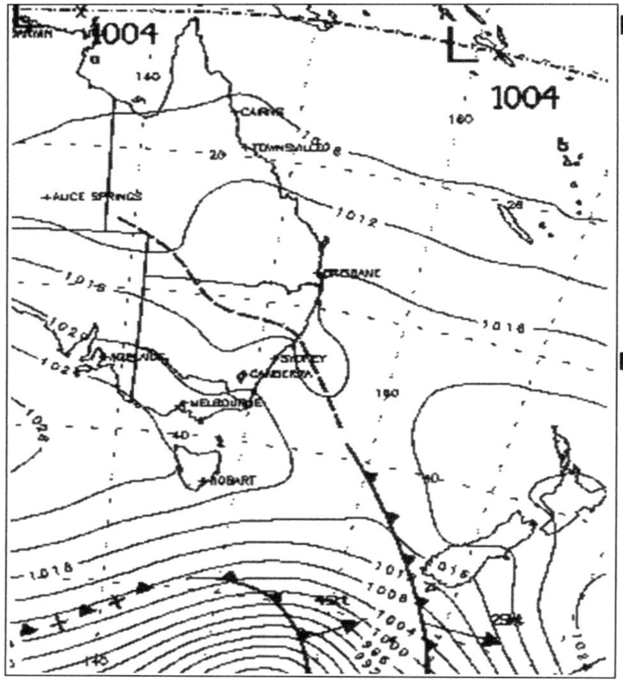

RAIN POTENTIAL

FROST/SNOW

MARCH 8 SATURDAY

High pressure crosses to the north of the North Island directing a light southwesterly flow, while a depression in the Southern Ocean directs a wet front up over the South Island in a fresh to strong southwesterly air. Rain in the west of the South Island, isolated showers in the north of the North Island and some spotty showers around the central districts of the North Island. Mainly dry elsewhere. Very hot in all districts with good sunny spells.

Auckland: Sunny, hot, dry, gentle breezes
Hamilton: Partly cloudy, hot and dry, light breezes
Wellington: Sunny, warm and dry, fresh breezes
Greymouth: Partly cloudy, isolated showers, windy spells
Christchurch: Unsettled and breezy, sunny, dry, hot day cool night
Dunedin: Mostly sunny, dry, warmer, windy spells

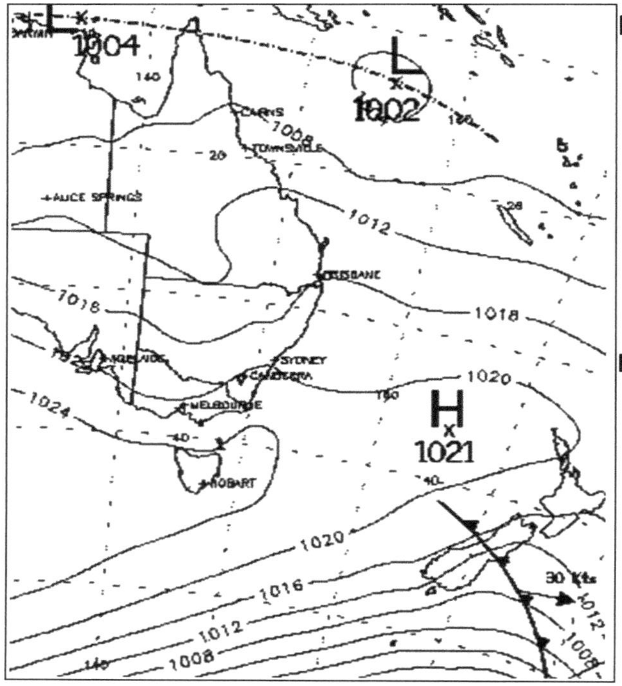

RAIN POTENTIAL

FROST/SNOW

MARCH **9** SUNDAY

A large high crossing to the south of Tasmania extends a high pressure ridge over New Zealand with light to moderate southwesterly winds prevailing. Very hot everywhere with good sunny spells. Some isolated morning showers likely in the West Coast to Canterbury districts as well as in the west of the North Island. Mostly dry elsewhere.

Auckland: Partly cloudy, mostly dry, chance of isolated showers, breezy spells
Hamilton: Hot, humid, light morning showers, sunny day, gentle breezes
Wellington: Partly cloudy, dry, warm, light breezes
Greymouth: Sunny, light morning showers, gentle breezes
Christchurch: Partly cloudy, dry, winds easing to light, very hot
Dunedin: Cloudy, cooler, dry, moderate breezes

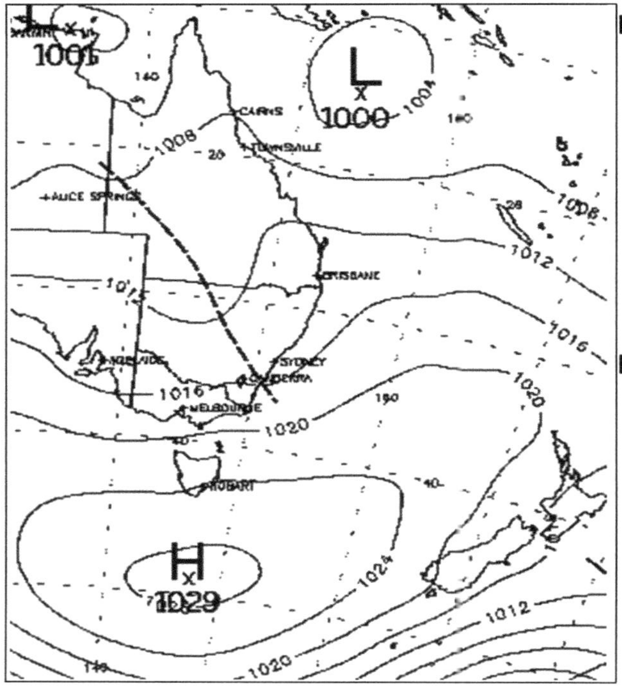

RAIN POTENTIAL

FROST/SNOW

MARCH **10** MONDAY

A large anticyclonic system dominates over New Zealand under a gentle west to southwest flow. Very hot temperatures continue over the North Island but ease in the South Island. Mostly cloudy with isolated showers contracting to the eastern fringes from the East Cape to South Canterbury while dry and mostly sunny conditions prevail elsewhere.

Auckland: Overcast, isolated showers with rain threatening, gentle breezes
Hamilton: Sunny, dry, gentle breezes, chance of overnight drizzle patches
Wellington: Sunny, cooler, dry, mostly calm air
Greymouth: Calm and serene air, mostly sunny, dry, slightly cooler
Christchurch: Overcast, cooler, dry, some threats of rain, calm atmosphere
Dunedin: Cooler, dry, sunny, pleasant and serene air

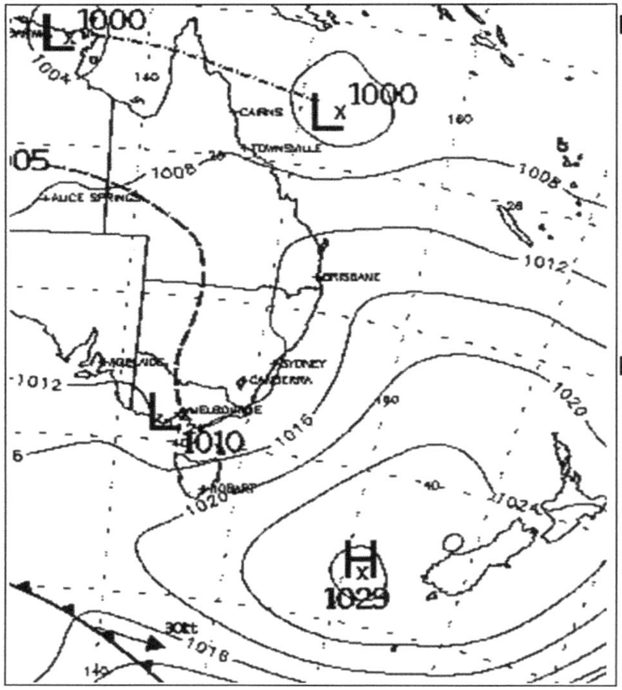

RAIN POTENTIAL

FROST/SNOW

MARCH **11** TUESDAY

High pressure continues to slowly drift over New Zealand with a slightly cooler and light westerly flow prevailing throughout. Widespread cloud and some isolated showers over the northern half of the North Island except in Northland, while mostly sunny and dry elsewhere.

Auckland:	Pockets of rain, cooler, occasional sunny spells, mostly calm air
Hamilton:	Overcast, light showers, calm and serene air
Wellington:	Partly cloudy, dry, cooler, light breezes, fresher at times
Greymouth:	Sunny, dry, gentle breezes
Christchurch:	Sunny, cooler, dry, gentle breezes
Dunedin:	Sunny, dry, gentle breezes

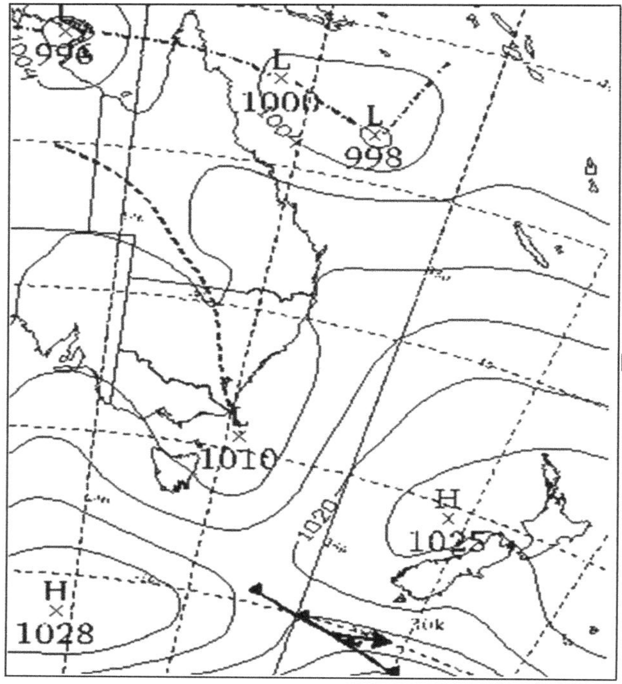

RAIN POTENTIAL

FROST/SNOW

MARCH **12** WEDNESDAY

A large anticyclonic system remains almost stalled over the country bringing a light east to northeast flow down through the west of New Zealand, backing southwesterly in the east. Warmer everywhere, and very hot in Canterbury. Some isolated showers likely from Taranaki to the central North Island, while mainly dry elsewhere. Very dull in the south and east of the South Island and in Taranaki, with good sunshine in the Waikato to the East Coast.

Auckland:	Sunny, mostly dry, possible overnight showers, light air
Hamilton:	Sunny, dry, calm and pleasant air
Wellington:	Sunny, warmer, dry, pleasant breezes
Greymouth:	Mostly sunny, dry, gentle breezes
Christchurch:	Warmer, widespread cloud, dry, light shifting breezes
Dunedin:	Cloudy, dry, gentle shifting breezes

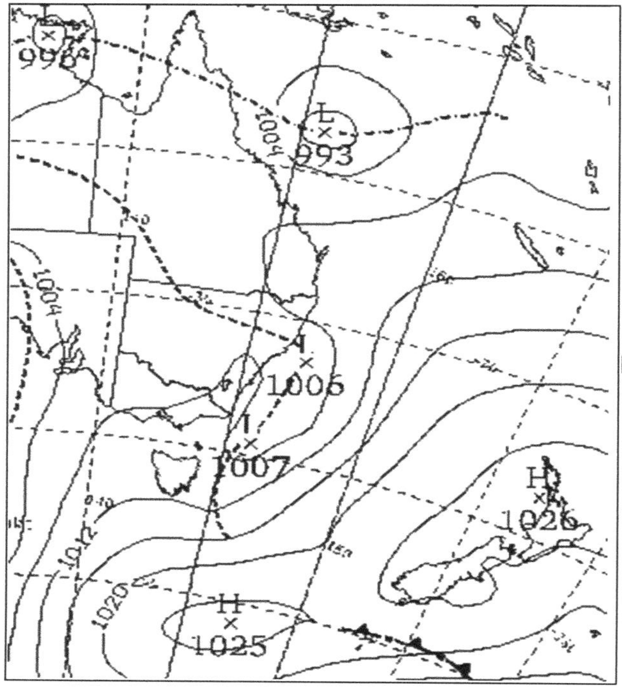

RAIN POTENTIAL

FROST/SNOW

MARCH **13** THURSDAY

The anticyclonic system stalled over New Zealand strengthens. A mostly calm air prevails, and hot and sunny everywhere. Chance of some spotty showers in the central parts of the North Island and possibly misty drizzles for a time along the coastal fringes of Canterbury. Mainly dry elsewhere.

Auckland:	Sunny, warm, humid, dry, calm atmosphere
Hamilton:	Partly cloudy, hot, dry, humid, serene air
Wellington:	Sunny, dry, gentle shifting breezes
Greymouth:	Sunny, dry, calm and pleasant air
Christchurch:	Sunny, dry, gentle breezes, chance of overnight fog patches
Dunedin:	Sunny and dry, calm and pleasant air

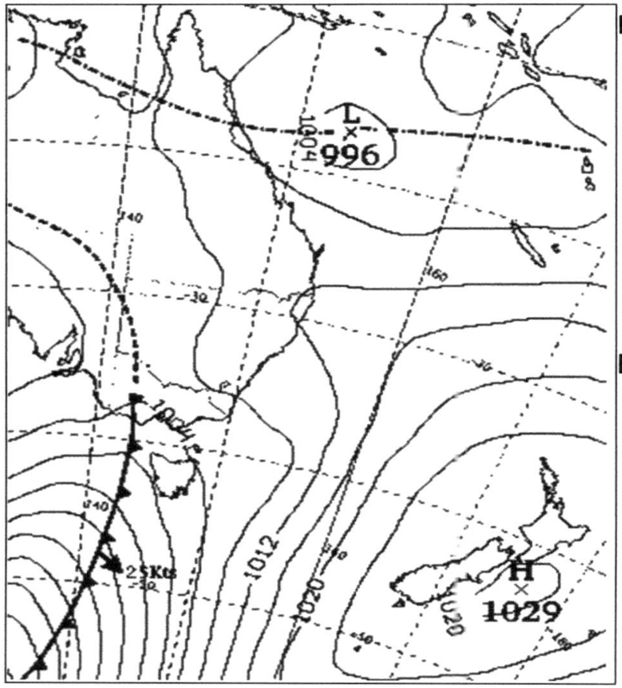

RAIN POTENTIAL

FROST/SNOW

MARCH **14** FRIDAY

The high pressure system which had dominated the past few days contracts up over the North Island allowing a freshening north to northwest flow associated with a depression to the south of New Zealand to build over the south of South Island while winds remain mostly calm over the North Island to Canterbury, with possible patches of fog around Christchurch. High maximum temperatures and mostly dry almost everywhere, except for some showers around the Bay of Plenty.

Auckland:	Partly cloudy, hot, humid, mostly dry, chance of morning rain
Hamilton:	Overcast, dry with odd threat of rain, hot, calm and humid air
Wellington:	Sunny, dry, slightly cooler, occasional breezy spells
Greymouth:	Partly cloudy, dry, light sea breezes
Christchurch:	Partly cloudy, warmer, dry, light shifting breezes
Dunedin:	Partly cloudy, dry, freshen breezes with occasional gusts

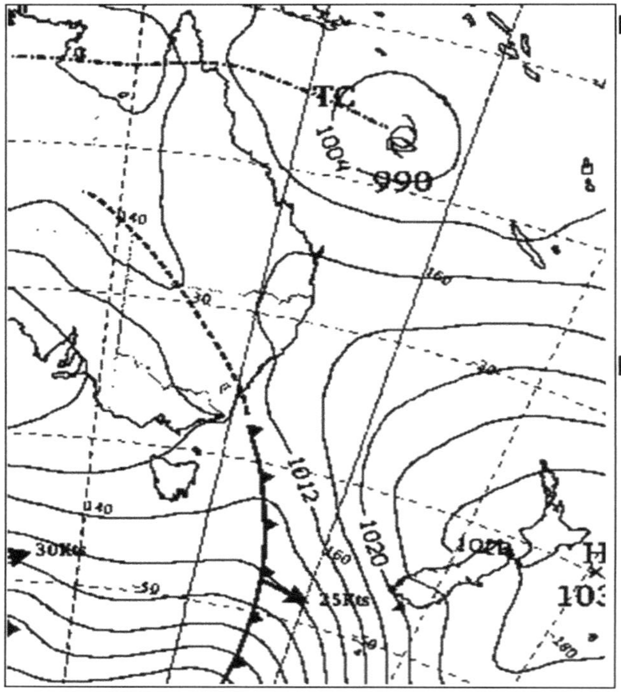

RAIN POTENTIAL

FROST/SNOW

MARCH 15 SATURDAY

High pressure remains stalled over the North Island with a mostly calm air prevailing, while a windier northwest flow associated with a low pressure trough covers the southern half of the South Island. Mostly dry, hot, and sunny everywhere, except for the chance of some isolated showers possibly lingering around the Bay of Plenty, and fog patches in Canterbury.

Auckland: Sunny, hot, mostly dry, chance of isolated showers, serene air
Hamilton: Calm air, partly cloudy, cooler, dry
Wellington: Dry, sunny, warm, moderate breezes
Greymouth: Sunny, moderate breezes, dry
Christchurch: Sunny, dry, mild day cool night, gentle breezes
Dunedin: Partly cloudy, warmer, dry, fresh breezes at times

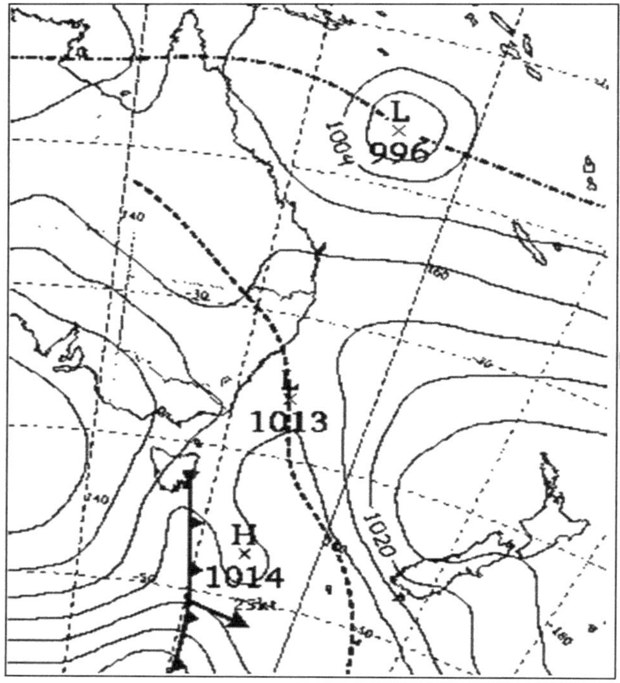

RAIN POTENTIAL

FROST/SNOW

MARCH 16 SUNDAY

A very hot westerly flow prevails throughout the country, gentle over the North Island and breezier over the South Island. Anticyclonic, hot, dry, and sunny throughout the North Island. Low pressure moves up over the South Island, with very high maximum temperatures and mostly sunny and dry conditions prevailing especially in Canterbury, but becoming cloudy in Southland and Otago, and chance of some isolated showers in the West Coast.

Auckland: Sunny, hot and dry, pleasant light breezes
Hamilton: Sunny, hot and dry, light air
Wellington: Very sunny, warm and dry, fresh breezes at times
Greymouth: Cloudy and dry, mild temperatures, unsettled windy spells
Christchurch: Very hot day cool night, dry, sunny, fresh northwesterly breezes likely
Dunedin: Very hot day cool night, cloudy, mostly dry with threats of rain

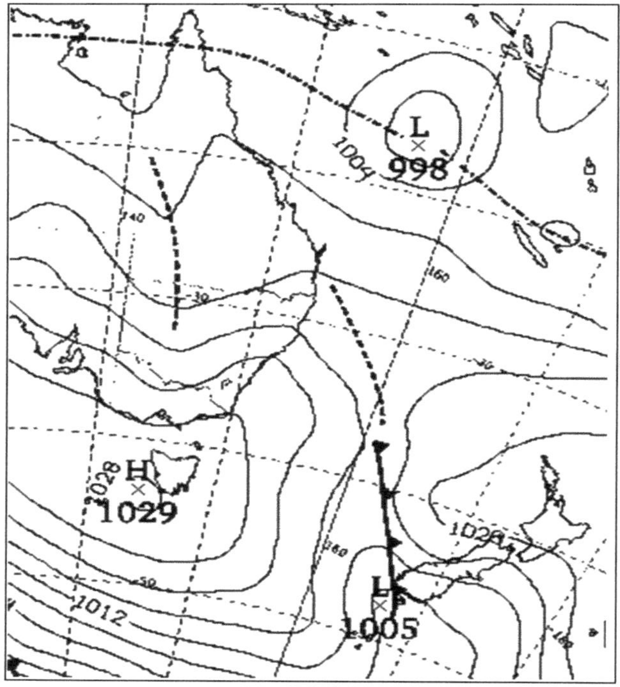

RAIN POTENTIAL

FROST/SNOW

MARCH 17 MONDAY

A weak depression moves up over the South Island while a high pressure ridge continues to linger over the North Island. Very hot throughout the country under a strong westerly flow over the South Island, gentle breezes in the North Island. Storms with heavy rain move into the south and west of the South Island, while the east remains dry with increasing threats of rain. Mainly dry and sunny throughout the North Island at first, with increasing cloud and threats of rain later in the day.

Auckland:	Cloudy, hot and humid, brief showers likely, light breezes
Hamilton:	Overcast, hot, humid, dry with passing threats of rain, mostly calm air
Wellington:	Partly cloudy, dry, warm, breezy with occasional gusts
Greymouth:	Cloudy, rain, unsettled winds with chance of gale gusts
Christchurch:	Very hot and dry, partly cloudy, unsettled winds
Dunedin:	Overcast, hot and dry, fluctuating and unsettled winds

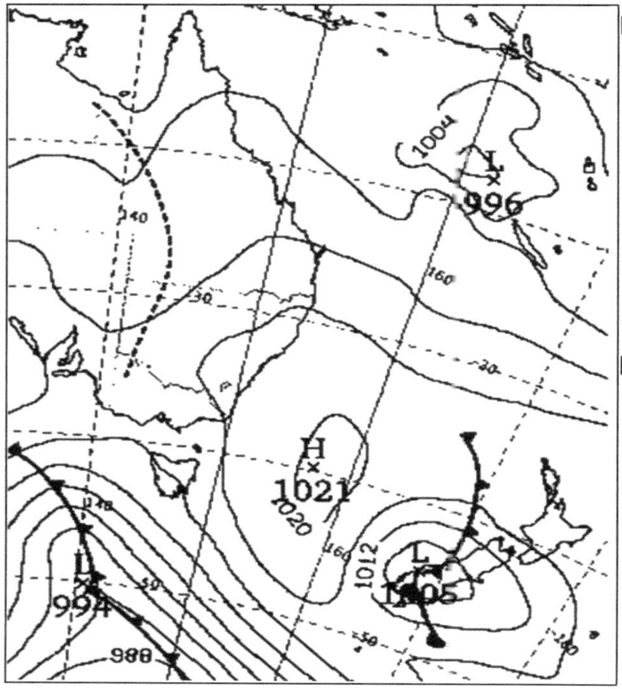

RAIN POTENTIAL

FROST/SNOW

MARCH 18 TUESDAY

A wet front moves up through the country bringing widespread heavy rain into the west and south of the South Island and lighter showers in most other parts of the country, mostly in the afternoon, although remaining mostly dry in the east from the East Cape to coastal Canterbury. Southwest winds prevail, gusty for a time particularly through the Cook Strait. Cooler in the south of the South Island, but remaining hot elsewhere. Good sunny spells in many districts.

Auckland:	Overnight rain, sunny day, hot, humid, fresh breezes at times
Hamilton:	Partly cloudy, hot and dry, freshening breezes
Wellington:	Dry and warm, sunny, fresh breezes
Greymouth:	Strong winds easing, cloudy, heavy rain
Christchurch:	Hot, mostly sunny, dry, light breezes, chance of fog patches
Dunedin:	Partly cloudy, cooler, brief showers, winds easing to light

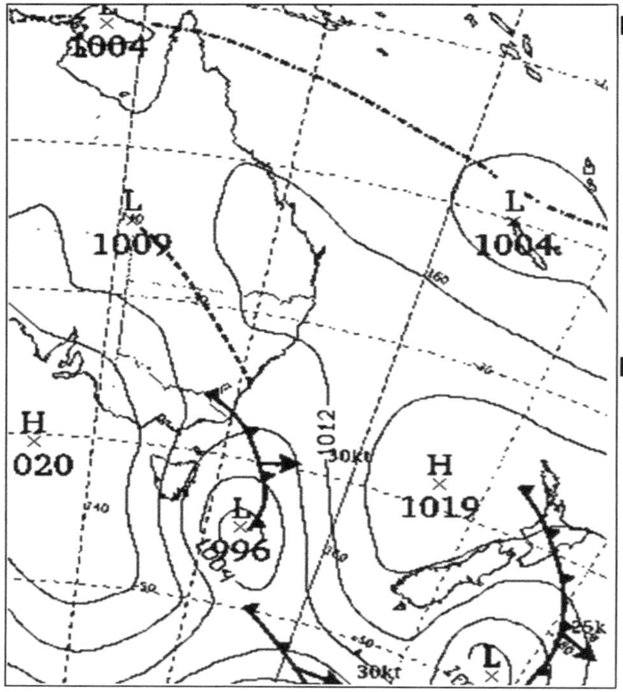

RAIN POTENTIAL

FROST/SNOW

MARCH **19** WEDNESDAY

A very weak high pressure system clears the depression away from New Zealand while re-establishing a gentle to mostly calm air throughout. Mostly dry and sunny everywhere except for some isolated showers lingering around the northern half of the North Island and in South Canterbury. Still hot in the North Island while milder in the South Island. Chance of fog patches in Canterbury.

Auckland: Brief showers, sunny day, calm and humid atmosphere
Hamilton: Sunny, brief evening showers, cold overnight, calm atmosphere
Wellington: Sunny, dry, warm, gentle breezes
Greymouth: Sunny, isolated showers, cooler, light breezes
Christchurch: Mostly sunny, dry, cooler, calm air, possible misty fog patches
Dunedin: Sunny, dry, mild, gentle breezes

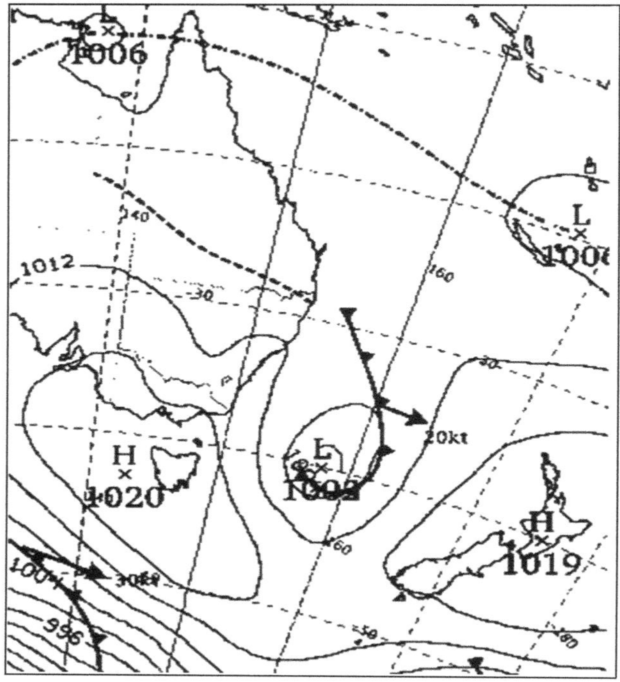

RAIN POTENTIAL

FROST/SNOW

MARCH **20** THURSDAY

High pressure builds over New Zealand as it directs a mostly calm to light northeasterly flow throughout. Mainly dry, hotter and sunny everywhere. Chance of fog in sheltered areas, particularly in the Canterbury region.

Auckland: Sunny, dry, hot, humid, gentle breezes
Hamilton: Partly cloudy, hot day cool night, dry, calm atmosphere
Wellington: Very sunny, warm and dry, gentle breezes
Greymouth: Sunny and dry, mild day cool night, gentle breezes
Christchurch: Sunny, dry, mild day cool night, gentle breezes, chance of fog
Dunedin: Very sunny, warmer, dry, gentle breezes

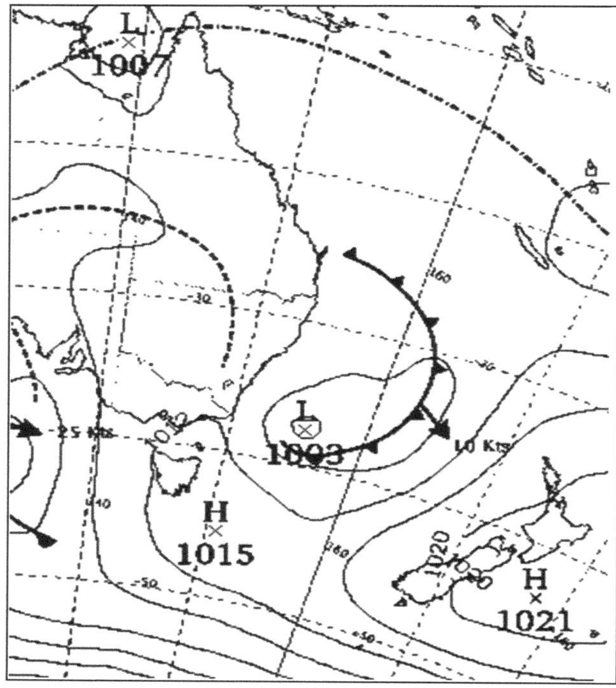

RAIN POTENTIAL

FROST/SNOW

MARCH **21** FRIDAY

High pressure continues to maintain its hot, dry, and mostly sunny influence throughout the country. A northeasterly air prevails over the North Island, backing northwesterly over the South Island. Chance of further fog patches in the Christchurch region.

Auckland: Calm and serene air, hot and humid, dry, increasing scattered cloud
Hamilton: Very hot day, cool night, dry, scattered cloudy spells, dry, calm air
Wellington: Sunny, dry, warmer, light breezes freshening later in the day
Greymouth: Sunny, warmer, dry, light breezes
Christchurch: Sunny, dry, cool overnight, gentle breezes
Dunedin: Sunny, warmer, dry, gentle breezes to mostly calm air

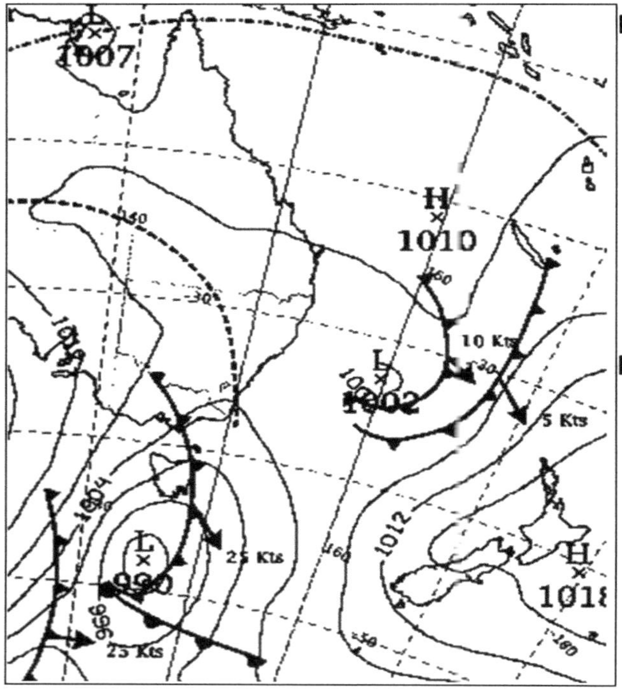

RAIN POTENTIAL

FROST/SNOW

MARCH **22** SATURDAY

A depression to the southwest of the country extends an unstable wet front up over the country in a fresh to strong northerly flow, lingering northeasterly over the north of the North Island. Very hot and with widespread cloud almost everywhere with rain threatening and some isolated showers developing later in the day around the Bay of Plenty.

Auckland: Hot and humid, overcast, isolated showers, freshening breezes
Hamilton: Unsettled breezes, cloudy, hot, scattered showers
Wellington: Very hot, cloudy, dry, passing threats of rain, windy, occasional gusts
Greymouth: Overcast, warmer, dry, unsettled winds developing
Christchurch: Partly cloudy, dry, strengthening winds later in the day
Dunedin: Cloudy, warmer, dry, increasing threats of rain, unsettled winds

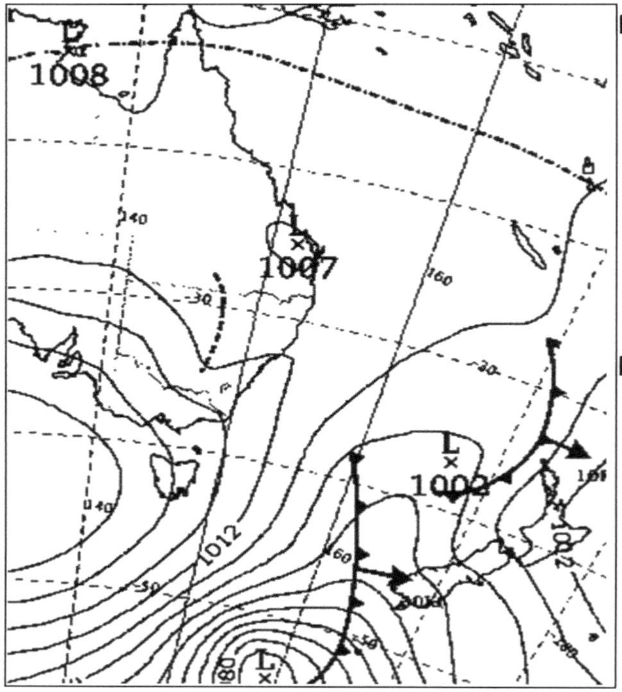

RAIN POTENTIAL

FROST/SNOW

MARCH 23 SUNDAY

A depression moves up over the country bringing widespread stormy conditions throughout. Blustery southwest to westerly winds prevail with gales likely around the Cook Strait. Heavy rain accompanies bands of thunderstorms across the country, which may see outbreaks of flooding in vulnerable areas, hail, power outages, and possible damage to local infrastructures. Chance of light snow in the higher reaches including around the Wellington hills and in Canterbury.

Auckland: Stormy air, heavy rain, some sunny spells, thunderstorms possible
Hamilton: Partly cloudy, heavy rain, stormy, winds possibly to gale force
Wellington: Overcast, isolated showers, chance of gales and thunderstorms
Greymouth: Heavy rain, some sunny spells, chance of thunderstorms and gales
Christchurch: Overcast, light squalls, blustery, hot, possible thunderstorms and hail
Dunedin: Cloudy, hot day cool night, heavy rain, fresh winds

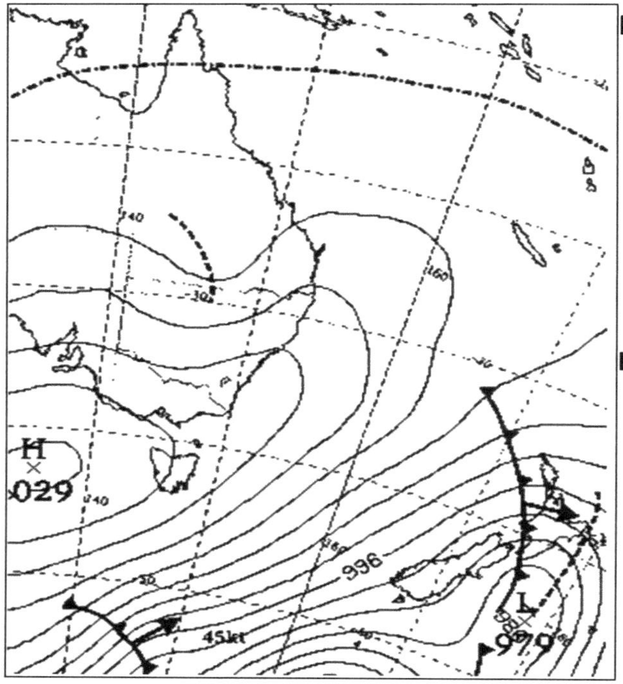

RAIN POTENTIAL

FROST/SNOW

MARCH 24 MONDAY

An intense depression brings a widespread stormy air throughcut the country with gale winds, rain, scattered thunderstorms, hail, and possible power outages and damage to infrastructure. Rain easing to showers in the south of the South Island and from the East Cape to Wellington, and with heavy falls still likely in most other districts. Storm conditions may see the chance of snow on the higher reaches in both the South Island and central North Island.

Auckland: Mix of some sunny spells and rain, winds possibly to storm force
Hamilton: Stormy air, gale gusts likely, rain, brief sunny breakthroughs, cooler
Wellington: Thunderstorms, gloomy, gale winds, widespread cloud, rain and hail
Greymouth: Cloudy, heavy rain, strong winds, thunderstcrms and gales likely
Christchurch: Cloudy, colder, rain, gusty winds, chance of overnight snow flurries
Dunedin: Strong and blustery winds, cloudy, mostly dry, colder

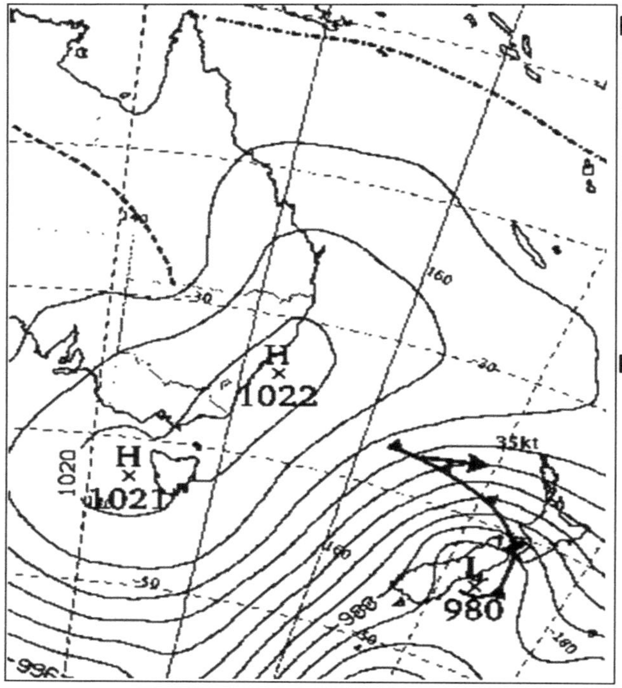

RAIN POTENTIAL

FROST/SNOW

MARCH **25** TUESDAY

Wet fronts associated with a Southern Ocean depression continue to bring stormy conditions throughout the country. Strong southwest winds prevail everywhere. Rain and possible thunderstorms in all districts, with heavy rain from Taranaki to the top of the South Island to central Canterbury. Cooler everywhere with possible pockets of frost appearing in central parts of both islands, and chance of snow in the higher reaches.

Auckland:	Stormy air, heavy rain, some sunny spells, strong gusty winds, cooler
Hamilton:	Changeable, sunny, cooler, evening rain, blustery winds with gusts
Wellington:	Changeable, sunny day, evening rain, cooler, blustery winds at times
Greymouth:	Blustery, cloudy, cooler, heavy rain, static air
Christchurch:	Changeable, sunny, blustery at times, showers, very cold overnight
Dunedin:	Cloudy, cool, light showers, strong winds, chance of gales and gusts

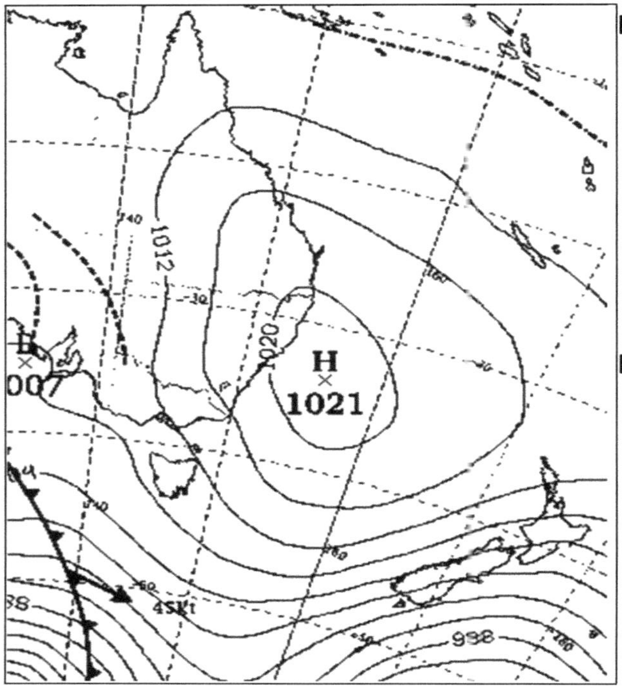

RAIN POTENTIAL

FROST/SNOW

MARCH **26** WEDNESDAY

The depression that had brought severe storm conditions over the past few days is pushed away as a high pressure system north of New Zealand ridges over the North Island, with west to southwest winds easing but still remaining fresh almost everywhere. Rain eases to mostly dry and sunny in the east of both islands, while outbreaks of rain continue in western regions from Taranaki to the West Coast as well as in Otago and Southland. Cooler everywhere.

Auckland:	Partly cloudy, spotty showers, strong winds with occasional gusts
Hamilton:	Partly cloudy, brief showers, fresh breezes
Wellington:	Sunny, mild and dry, fresh to strong winds with occasional gusts
Greymouth:	Overcast, rain, windy, static air
Christchurch:	Sunny, warmer, dry, breezy spells
Dunedin:	Partly cloudy, isolated showers, mild day cool night, fresh breezes

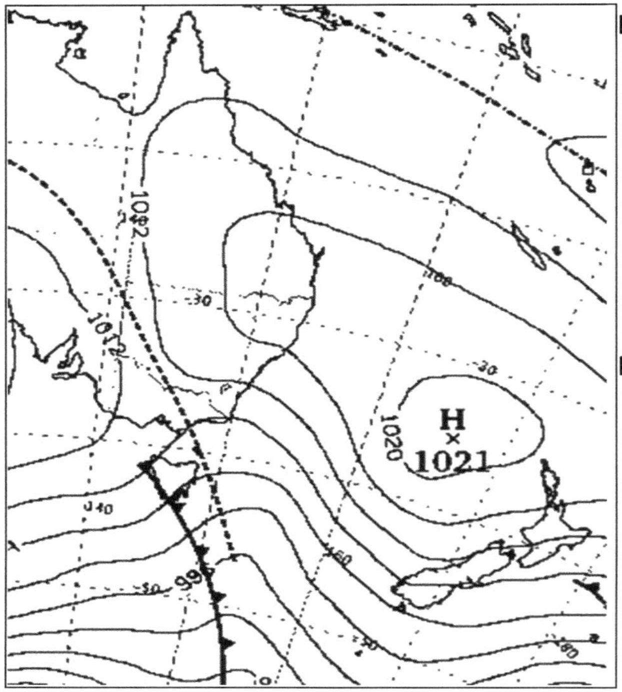

RAIN POTENTIAL

FROST/SNOW

MARCH **27** THURSDAY

A complex of high and low pressure systems affect New Zealand with high pressure ridging over the North Island and low pressure wet fronts crossing the South Island. Winds generally fresh to strong west to southwesterlies throughout the South Island and into the central North Island, while lighter northwesterlies prevail elsewhere. Heavy rain in the west of the South Island may see some localised flooding, and isolated showers likely in Southland, Otago, and Taranaki to the Manawatu. Mostly dry elsewhere. Warmer.

Auckland:	Overcast, squally drizzle patches, blustery winds at times
Hamilton:	Partly cloudy, spotty showers, strong winds with occasional gusts
Wellington:	Overcast and blustery, dry with passing threats of rain, warmer
Greymouth:	Overcast, heavy rain, strong gusty winds, chance of gales
Christchurch:	Unsettled winds, sunny, cooler, mostly dry, chance of drizzle patches
Dunedin:	Partly cloudy, warmer, dry, fluctuating breezes

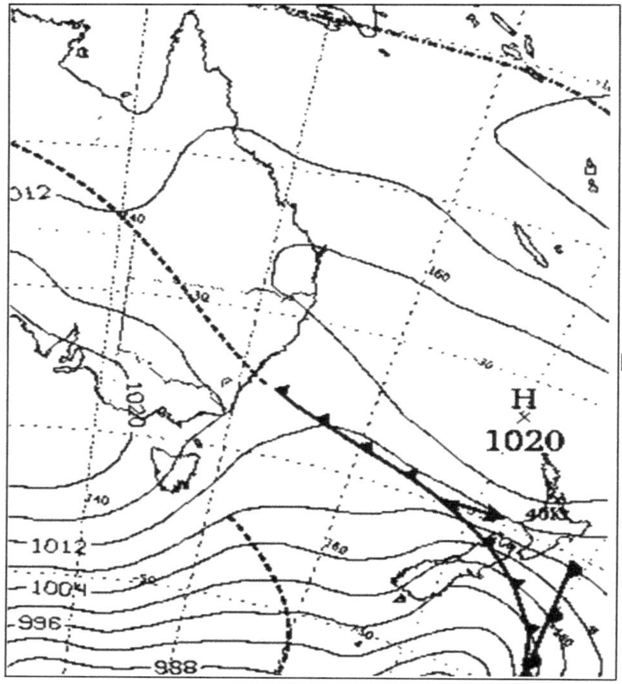

RAIN POTENTIAL

FROST/SNOW

MARCH **28** FRIDAY

A strong west to southwest flow moves up over the country pushing the high pressure ridge away. Rain throughout the west and south of both islands, with very heavy rain around the West Coast to Taranaki, with the chance of thunderstorms with hail, lightning strikes, power outages and gale gusts. Mostly dry Northland to the Bay of Plenty and East Coast, Marlborough, with light showers elsewhere. Warmer in all districts, particularly in Canterbury.

Auckland:	Cloudy, blustery, possible gales for a time, mostly dry
Hamilton:	Cloudy, dry, strong gusty winds
Wellington:	Partly cloudy, cooler, dry, blustery, possible gale gusts at times
Greymouth:	Partly cloudy, heavy overnight rain, strong gusty winds
Christchurch:	Very hot, cloudy spells, brief showers, strong winds with gusts
Dunedin:	Partly cloudy, dry, strong winds at times

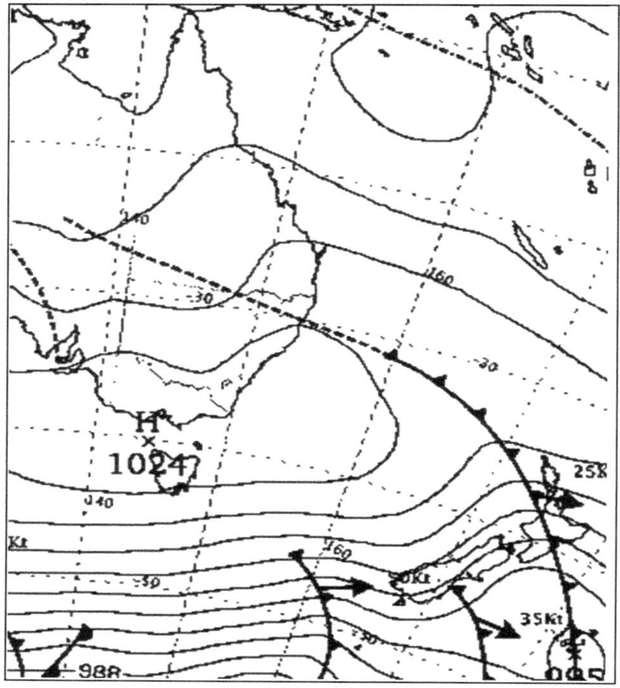

RAIN POTENTIAL

FROST/SNOW

MARCH 29 SATURDAY

High pressure ridges over the northeast of New Zealand once more pushing the low pressure back down over the South Island with strong westerly winds prevailing over the South Island, and lighter southwesterlies over the North Island. Heavy overnight rain in Taranaki to the Manawatu and down the West Coast, while mostly dry in the East Coast of the North Island and in Marlborough to Canterbury, with showery spells elsewhere. Mostly sunny everywhere.

Auckland: Sunny day, light evening showers, strong winds easing
Hamilton: Sunny, light overnight showers, moderate breezes
Wellington: Sunny and dry day, overnight showers, warmer, blustery at times
Greymouth: Cloudy, rain, strong winds easing later in the day, cooler
Christchurch: Sunny, colder, light showers, windy at times
Dunedin: Sunny, cooler, brief overnight showers, strong breezes

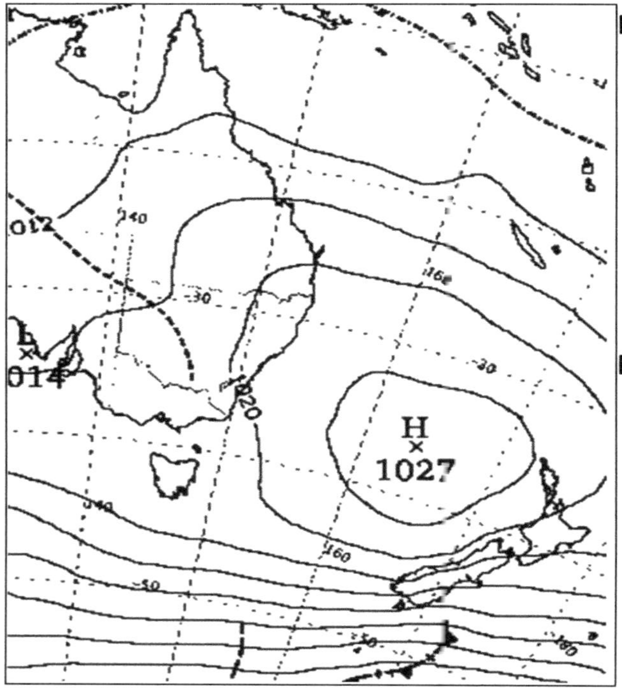

RAIN POTENTIAL

FROST/SNOW

MARCH 30 SUNDAY

A large anticyclonic system crosses the country in a light west to southwest flow, clearing rain to isolated showers in the far southwest of the South Island and around Taranaki. Mainly dry elsewhere. Warmer and with good sunny spells throughout.

Auckland: Partly cloudy, cooler, mostly dry, threats of rain, moderate breezes
Hamilton: Sunny, dry, mild day cool night, light breezes to mostly calm air
Wellington: Sunny, dry, fresh breezes easing to light
Greymouth: Sunny, dry, gentle breezes
Christchurch: Sunny, cool, dry, mostly calm air, chance of overnight fog patches
Dunedin: Sunny, warmer, dry, winds easing to light

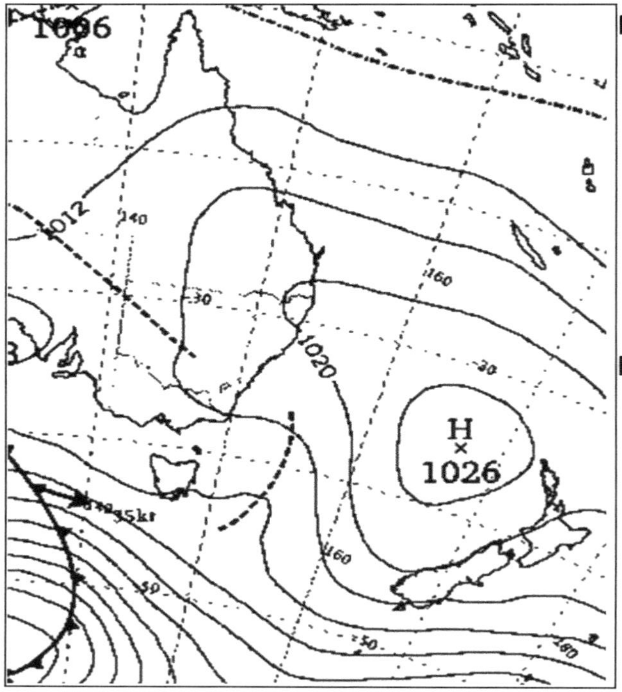

RAIN POTENTIAL

FROST/SNOW

MARCH 31 MONDAY

Anticyclonic over the North Island in a gentle westerly air, dry, warm, and mostly sunny. Low pressure moves up over the South Island in a moderate westerly flow, fresher through southern parts. Mainly dry and warm throughout the South Island with good sunny spells in the Nelson-Marlborough region and Otago, cloudier elsewhere.

Auckland: Partly cloudy, dry, gentle breezes
Hamilton: Gentle breezes, partly cloudy, dry, cool overnight
Wellington: Partly cloudy, dry, fluctuating breezes
Greymouth: Overcast, dry with occasional threats of rain, breezy spells
Christchurch: Partly cloudy, warmer, moderate breezes
Dunedin: Overcast, dry with passing threats of rain, windy spells

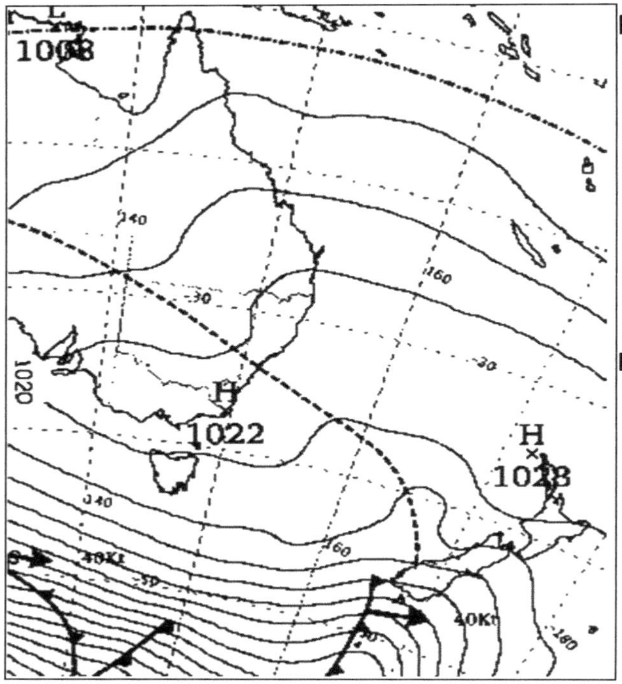

RAIN POTENTIAL

FROST/SNOW

APRIL

PHASES OF THE MOON

3rd	Northern Declination
5th	First Quarter
11th	Crossing Equator
12th	Full Moon
13th	Apogee
18th	Southern Declination
20th	Third Quarter
25th	Crossing Equator
27th	New Moon
27th	Perigee #3

1 Fri	2 Sat	3 Sun	4 Mon	5 Tue	6 Wed	7 Thur	8 Fri	9 Sat	10 Sun	11 Mon	12 Tue	13 Wed	14 Thur	15 Fri	16 Sat	17 Sun	18 Mon	19 Tue	20 Wed	21 Thur	22 Fri	23 Sat	24 Sun	25 Mon	26 Tue	27 Wed	28 Thur	29 Fri	30 Sat
N	XhN						A ^	1Q							XhS	F			P10		V	3Q						XhN	

MONTHLY SUMMARY

2nd-5th Light air in the east of the South Island may bring patches of fog in the Christchurch area.
3rd Unusually warm day in the east of the South Island from northern parts of Canterbury to Central Otago.
7th-10th A band of heavy rain stalls over Northland bringing another risk of further flooding in some northern districts to the Coromandel and Bay of Plenty and including in the Auckland region, with the eastern districts from Whangarei to the Bay of Islands particularly at risk. Heavy rain in the Rotorua district may also cause some brief localised flooding.
8th-9th Chance of fog in Christchurch.
10th-11th Very strong and blustery winds in the Wellington region likely and heavy rain in the West Coast of the South Island.

Over the next four weeks, anticyclonic systems will direct more frequent southerly airflows over the country than normal, as the high pressure systems will cross through lower latitudes south of Australia and Tasmania resulting in the arrival of typical autumn conditions. Generally, below average rainfall and above average sunshine continues to be the trend throughout most of the country over the next four weeks, with particularly high sunshine hours likely to be recorded in some parts of Northland, and in contrast, some parts in the east and south of the South Island likely to see below average sunshine. The districts of Northland, Taranaki, Gisborne, the Horowhenua and most of the western half of the South Island can expect rainfall levels below what is typical for this time of year with some places seeing only around half their usual amount of rain and with the remaining districts all likely to be around 25% drier than normal. Significant soil moisture deficits are likely to continue through the next four weeks in the Hawkes Bay, Marlborough, and Central Otago. Temperatures will be fairly normal for this time of year over the South Island, but around 0.5°C less than average in many North Island districts.

11th Unusually warm day in the East Coast, particularly in Hastings.
15th-16th Chance of shallow fog in the Waikato to Auckland districts and in the Canterbury and Christchurch regions.
17th-18th Chance of fog in the Christchurch area.
19th Blustery winds in Wellington.
20th Heavy rain in Fiordland.
20th-21st Possible patches of fog in Auckland, Waikato and Christchurch regions.

22nd Very cold in the Tara Hills, with severe frosts likely. Wet front crosses the South Island bringing heavy rain at times in the West Coast and Southland, and outbreaks of rain through northern and eastern parts including into Canterbury.
22nd-23rd Very strong southwest winds over the northern half of the North Island, particularly in the Auckland region, possible reaching gale force at times. Blustery in the Wellington region.
23rd Blustery in Christchurch.
24th Chance of snow in the Lake Rotoiti region.
26th-27th Possible fog in Christchurch.
30th Shallow fog likely in the Christchurch area.

April 2025

Monday	Tuesday	Wednesday	Thursday	Friday	Saturday	Sunday
31 **Moon Sign**	1 **Taurus**	2 **Taurus** Sextile Mercury Sextile Venus Sextile Saturn Conjunction Uranus Sextile Neptune Trine Pluto Square Ascendant	3 **Gemini** Sextile Sun Conjunct Jupiter Square Ascendant	4 **^** **Gemini** Square Mercury Square Venus Square Saturn Square Neptune Square Ascendant	5 **1Q** **Cancer** Square Sun Square Ascendant	6 ***DLS Ends*** **Cancer** Trine Mercury Trine Venus Conjunct Mars Trine Saturn Sextile Uranus Trine Neptune Square Ascendant
7 **Leo** Opposition Pluto Square Ascendant	8 **Leo** Trine Sun Square Uranus	9 **Virgo**	10 **Virgo** Opposition Venus Square Jupiter Opposition Saturn	11 **XhS** **Virgo** Opposition Mercury Opposition Venus Sextile Mars Opposition Saturn Trine Uranus Opposition Neptune Trine Pluto	12 **Libra** Trine Jupiter	13 **FM** **Libra** Opposition Sun Square Mars Trine Jupiter
14 **A** **Scorpio** Square Pluto	15 **Scorpio** Trine Venus Opposition Uranus	16 **Scorpio** Trine Mercury Trine Venus Trine Mars Trine Saturn Opposition Uranus Trine Neptune Sextile Pluto	17 **Sagittarius** Opposition Jupiter	18 **Sagittarius** Trine Sun Square Mercury Square Venus Opposition Jupiter Square Saturn Square Neptune Square Ascendant	19 **V** **Capricorn** Square Mercury Square Ascendant	20 **Capricorn** Trine Uranus
21 **3Q** **Aquarius** Square Sun Sextile Mercury Opposition Mars Conjunct Pluto	22 **Aquarius** Trine Jupiter	23 **Pisces** Sextile Sun	24 **Pisces** Square Jupiter	25 **XhN** **Aries** Conjunct Mercury Conjunct Venus Trine Mars Conjunct Saturn Conjunct Neptune Sextile Pluto	26 **Aries** Conjunct Mercury Sextile Jupiter	27 **Taurus** Conjunct Sun Square Mars Square Pluto
28 **NM P3** **Taurus** Conjunct Sun Square Mars Square Pluto	29 **Taurus** Conjunct Uranus Square Ascendant	30 **Gemini** Sextile Mars Trine Pluto Square Ascendant	1	2	3	4
5	6					

NORTH ISLAND RAINFALL

APRIL 2025

d=mainly dry, n= fog, mist, dew or nonrecordable, l=light showers, p=passing showers, s=significant showers, r=rain, h=heavy falls

		Northland							Auckland					Bay Of Plenty				Waikato				Rot/Tau		Hawkes Bay			C Plateau			Taranaki			Manawatu			Lower North Island					
		CR	Kta	Ker	Rsl	Dar	Whg	Man	Wark	Hen	MkH	Ak	Co	Tau	TeP	Wkt	Opo	Hml	TeA	Mata	Ot	Rt	Taup	Gis	Nap	Has	Taum	Chat	Waio	NP	Str	Wang	Pa	Dann	Lev	PN	Mast	UH	Wgt	all	Moon
1st	Tue	l	l	l	n	d	d	d	l	l	l	l	l	d	n	d	d	l	l	d	l	l	d	d	d	d	d	d	d	l	d	d	d	d	r	l	d	d	d	32	
2nd	Wed	l	d	d	l	d	d	l	d	n	l	n	d	d	d	d	d	d	d	d	d	d	d	d	d	d	d	d	d	d	d	d	d	d	d	d	d	d	d	8	
3rd	Thur	d	d	s	d	d	d	d	d	d	d	d	d	d	d	d	d	d	d	d	d	d	d	d	d	d	d	d	d	d	d	d	d	d	d	d	d	d	d	5	
4th	Fri	d	d	d	d	d	d	l	d	d	d	d	d	d	d	d	d	d	d	d	d	d	d	d	d	d	d	d	d	d	d	d	d	d	d	d	d	d	d	1	^
5th	Sat	l	d	l	d	d	d	l	l	d	d	d	d	d	d	d	d	d	d	d	d	d	d	d	d	d	d	d	d	d	d	d	d	d	d	d	d	d	d	4	IQ
6th	Sun	d	d	d	l	d	n	d	l	d	d	d	d	d	d	d	d	d	d	d	d	d	d	d	d	d	d	d	d	d	d	d	d	d	d	d	d	d	d	5	
7th	Mon	h	r	r	r	s	r	n	l	l	l	l	l	l	l	l	n	l	l	r	l	l	d	d	l	l	d	d	d	l	d	d	d	d	d	d	n	d	d	107	
8th	Tue	r	h	h	h	h	h	h	h	h	h	h	h	h	h	l	l	s	s	h	r	s	s	s	l	s	s	l	l	s	s	l	d	l	d	d	l	d	d	976	
9th	Wed	l	s	h	h	h	h	h	h	h	h	h	h	h	h	h	h	h	r	r	h	h	h	r	l	l	s	l	n	l	d	d	d	n	d	d	d	d	d	1591	
10th	Thur	d	d	d	d	n	d	d	d	d	d	d	h	h	h	s	s	l	s	l	s	h	l	d	d	d	l	s	d	d	n	l	l	n	s	l	d	l	d	206	
11th	Fri	d	s	d	d	l	d	d	l	n	l	d	n	d	d	d	d	d	d	d	l	d	l	d	d	d	d	s	d	l	l	l	d	d	l	l	d	l	d	31	XhS
12th	Sat	l	l	s	l	l	s	l	d	l	l	d	d	d	d	d	d	l	l	d	d	d	d	d	d	d	l	r	s	s	r	s	d	l	l	r	d	l	d	81	
13th	Sun	d	d	d	d	d	d	d	d	d	d	d	d	d	d	d	d	d	d	d	d	d	d	d	l	d	d	d	d	n	d	d	d	d	d	d	d	d	d	1	F
14th	Mon	d	d	d	l	d	d	d	d	d	d	d	d	d	d	d	d	d	d	d	d	d	d	d	d	d	d	d	l	d	d	d	d	d	d	d	d	d	d	5	A
15th	Tue	d	d	d	d	s	d	d	d	d	d	d	d	d	d	d	d	d	d	d	l	d	d	l	r	s	d	l	s	n	l	9	l	r	s	s	s	s	s	88	
16th	Wed	d	s	r	s	h	d	d	l	d	d	d	s	d	l	s	d	d	d	d	d	l	d	d	d	d	d	d	d	d	d	d	d	d	d	d	d	d	d	60	
17th	Thur	d	d	d	d	d	d	d	d	d	l	d	d	d	d	d	d	d	d	d	d	d	d	d	d	d	d	d	d	d	d	d	d	d	d	d	d	d	d	1	
18th	Fri	l	l	l	l	d	d	d	d	d	d	d	d	d	d	d	d	d	d	d	d	d	d	d	d	d	d	d	d	l	d	d	d	d	d	d	d	d	d	8	
19th	Sat	l	l	l	d	r	d	n	d	d	d	d	d	d	d	d	d	d	d	d	d	d	d	d	d	d	d	d	d	d	d	d	d	d	d	d	d	d	d	18	V
20th	Sun	s	d	d	d	d	d	d	d	d	d	d	d	d	d	d	d	d	d	d	d	d	d	d	d	d	d	d	d	d	d	d	d	d	d	d	d	d	d	7	
21st	Mon	r	s	l	d	d	s	l	h	d	d	d	l	n	d	d	d	d	d	d	l	d	d	d	d	d	l	l	d	d	d	d	d	d	n	d	d	d	d	73	3Q
22nd	Tue	l	l	d	r	s	s	r	r	s	l	l	d	d	d	d	d	l	s	l	l	d	d	d	d	l	l	r	s	l	s	h	r	r	h	s	r	h	h	228	
23rd	Wed	l	r	s	l	s	s	r	r	r	r	r	s	l	l	r	h	s	s	l	h	r	h	h	s	s	l	r	r	l	s	l	d	s	d	l	l	l	l	273	
24th	Thur	l	l	l	d	s	l	l	l	l	s	l	l	d	d	l	d	l	l	d	l	d	d	l	s	l	l	r	s	l	l	l	d	l	d	d	l	l	l	72	
25th	Fri	d	d	d	d	l	l	n	l	s	s	l	d	d	d	d	d	n	l	d	l	d	d	d	d	d	l	s	l	l	n	d	d	l	d	d	d	d	d	24	XhN
26th	Sat	l	l	d	d	l	l	l	d	l	l	l	l	d	d	d	d	n	n	l	l	d	d	n	s	l	l	n	d	d	l	d	d	d	d	d	d	d	d	23	
27th	Sun	l	s	s	d	l	d	n	l	s	l	n	d	d	d	d	l	l	l	d	l	l	d	l	l	l	l	h	n	d	l	l	d	l	d	l	d	d	d	59	
28th	Mon	d	d	d	d	l	l	d	d	l	l	l	d	d	d	d	d	d	d	d	l	d	d	d	d	d	n	l	d	l	l	d	d	d	d	d	l	d	d	14	N P3
29th	Tue	d	d	d	d	l	d	d	d	d	l	d	d	d	d	d	d	d	d	d	d	d	d	s	d	d	d	d	d	d	d	d	d	d	d	d	l	d	n	6	
30th	Wed	d	d	n	d	d	d	d	d	d	l	d	d	d	d	d	d	d	d	d	d	d	d	d	d	d	d	d	d	d	d	d	d	d	d	d	d	d	d	1	
estimate:		76	121	384	406	142	322	270	275	125	80	75	268	130	163	79	68	51	46	68	67	135	47	44	34	25	26	86	46	32	35	46	15	32	48	30	33	35	40	4006	
average:		109	122	142	121	101	152	129	136	116	54	95	105	125	132	107	113	95	96	113	110	112	92	85	75	63	114	198	78	117	162	70	77	86	77	82	67	96	97	4061	
trend:		drr	av	wtr	wtr	wtr	wtr	wtr	wtr	av	drr	drr	wtr	av	wtr	drr	drr	drr	drr	drr	drr	wtr	drr	drr	drr	drr	drr	drr	drr	drr	drr	drr	drr	drr	drr	drr	drr	drr	drr	av	

SOUTH ISLAND RAIN
APRIL 2025

d=mainly dry, n= fog, mist, dew or nonrecordable, l=light showers, p=passing showers, s=significant showers, r=rain, h=heavy falls

		West Coast				Top South Island				Canterbury				South Canterbury				Coastal Otago			Lakes			Inland Otago				Southland						
		West	Grey	Hok	BB	Ta	Mou	Nel	Ble	Dar	Ham	Kai	Ch	Rak	Meth	Ash	Tim	Oam	Dun	Bal	Twi	LP	LT	Crom	Qstn	Ran	Alex	TeA	Wtn	Gore	Inv	Puy Pt	all	Moon
1st	Tue	n	d	d	r	d	d	d	d	d	l	d	d	d	l	d	d	d	l	l	d	d	d	d	d	d	d	n	d	l	l	r	32	
2nd	Wed	d	d	d	d	d	d	d	d	d	d	d	d	d	d	d	d	d	d	d	d	d	d	d	d	d	d	d	d	d	d	d	0	
3rd	Thur	d	d	d	d	d	d	d	d	d	d	d	d	d	d	d	d	d	d	d	d	d	d	d	d	d	d	d	d	d	d	d	0	
4th	Fri	d	d	d	d	d	d	d	d	d	d	d	d	d	d	d	d	d	d	d	d	d	d	d	d	d	d	d	d	d	d	d	0	^
5th	Sat	d	d	d	d	d	d	d	d	d	d	d	d	d	d	d	d	d	d	d	d	d	d	d	d	d	d	s	d	d	d	r	16	1Q
6th	Sun	d	d	d	d	d	d	d	d	d	d	d	d	d	d	d	d	d	d	d	d	d	d	d	d	d	d	d	d	d	d	d	0	
7th	Mon	d	d	d	d	d	d	d	d	d	d	l	d	d	d	d	d	l	d	d	d	d	d	d	d	d	d	d	d	d	d	d	3	
8th	Tue	l	d	n	l	r	s	l	l	s	l	s	l	d	l	s	l	l	d	d	d	d	d	d	d	d	d	d	d	d	d	d	45	
9th	Wed	l	l	s	s	r	h	s	n	d	d	d	d	d	d	d	n	d	d	d	d	d	d	n	d	d	d	d	d	d	d	d	51	
10th	Thur	r	r	h	h	d	d	d	d	d	d	d	l	d	d	n	d	d	d	l	l	l	l	l	l	d	l	s	s	l	s	h	126	
11th	Fri	h	h	h	h	d	d	n	d	d	d	d	d	d	d	n	l	n	l	d	s	d	r	d	d	d	d	d	d	n	l	r	140	XhS
12th	Sat	d	d	d	d	d	l	d	d	l	d	d	d	d	l	d	l	d	l	d	l	r	d	d	d	d	d	d	d	l	l	s	31	
13th	Sun	d	l	d	d	d	d	d	d	d	d	d	d	d	d	d	d	l	d	n	d	d	d	d	d	d	d	l	l	l	l	r	24	F
14th	Mon	l	l	d	s	d	d	d	d	l	d	d	l	d	l	l	d	d	s	l	d	d	d	d	d	l	d	d	d	d	d	r	35	A
15th	Tue	d	d	d	d	l	l	l	s	l	l	s	d	l	d	d	d	d	d	d	d	d	d	d	d	d	d	d	d	n	l	s	30	
16th	Wed	d	d	d	d	d	d	d	d	d	d	d	d	d	d	d	d	d	d	d	d	d	d	d	d	d	d	d	l	n	d	h	19	
17th	Thur	d	l	l	l	d	d	d	d	s	d	d	s	s	r	s	s	s	d	d	d	l	l	d	d	s	d	d	d	d	d	l	63	
18th	Fri	d	d	l	l	d	l	d	d	s	l	l	s	l	l	s	n	n	d	d	l	d	d	d	d	d	d	d	d	d	l	s	40	
19th	Sat	d	d	d	d	d	d	d	d	d	d	d	d	d	l	d	d	d	d	d	d	d	d	d	d	d	d	d	d	d	l	s	9	V
20th	Sun	l	d	l	d	d	d	d	d	d	d	d	d	d	d	d	d	d	d	d	d	d	d	d	d	d	d	d	d	d	d	s	12	
21st	Mon	l	l	l	h	d	d	d	d	d	d	d	d	d	d	d	d	d	d	d	d	d	d	s	h	d	d	l	l	s	l	h	91	3Q
22nd	Tue	h	h	h	l	l	r	s	r	r	h	r	r	r	r	r	r	l	s	l	d	l	s	d	l	s	d	l	s	s	l	s	252	
23rd	Wed	r	l	l	s	d	d	d	l	s	r	s	s	l	s	s	d	l	s	s	s	d	d	l	l	l	l	l	r	r	r	h	165	
24th	Thur	s	l	n	d	d	d	d	l	d	l	d	d	d	d	d	d	d	l	s	l	d	d	d	l	d	l	d	r	h	r	s	69	
25th	Fri	l	n	d	d	d	d	d	d	d	d	d	d	d	d	d	d	d	d	n	d	d	d	d	d	d	l	d	l	l	l	s	11	XhN
26th	Sat	d	d	d	d	d	d	d	d	d	d	d	d	d	d	d	d	d	d	d	d	d	d	d	d	d	d	d	d	d	d	l	1	
27th	Sun	r	h	r	r	d	d	d	d	d	d	d	d	l	d	d	d	d	l	l	d	d	d	d	l	d	d	s	s	s	r	r	105	
28th	Mon	l	d	d	d	d	d	d	d	l	d	l	d	d	l	n	d	h	l	l	l	d	d	d	d	d	d	d	l	l	d	d	41	N P3
29th	Tue	d	d	d	d	d	d	n	d	d	d	n	d	d	d	d	d	l	d	d	d	d	d	d	d	d	d	d	d	d	d	d	3	
30th	Wed	d	d	d	d	d	d	d	d	d	d	d	d	d	d	d	d	d	d	d	d	d	d	d	d	d	d	d	d	d	d	d	0	
estimate:		**100**	**95**	**98**	**116**	**31**	**37**	**16**	**21**	**45**	**43**	**35**	**36**	**27**	**43**	**32**	**21**	**44**	**27**	**25**	**23**	**22**	**20**	**7**	**28**	**17**	**7**	**26**	**47**	**59**	**61**	**206**	**1414**	
average:		175	195	230	349	158	122	77	50	62	90	84	50	74	82	61	45	45	70	49	61	66	50	36	72	39	35	86	78	74	102	207	2974	
trend:		drr	drr	drr	drr	drr	drr	drr	drr	drr	drr	drr	drr	drr	drr	drr	drr	**av**	drr	drr	drr	drr	drr	drr	drr	drr	drr	drr	drr	drr	drr	**av**	drr	

NORTH ISLAND SUNSHINE ESTIMATES
APRIL 2025

F=fine (8-12 hours of sunshine), pc= partly cloudy (4-7 hours), c=cloudy (1-3 hours), o=overcast (0 hours)

		Kta	Whg	Dar	Ak	Hm	Tau	Rt	Taum	NP	Wang	Taup	Gis	Nap	Wai	Dann	PN	Mast	Pa	UH	Wgt	ALL	Moon
1st	Tue	o	o	pc	pc	F	F	F	F	F	F	F	F	F	F	F	F	F	F	F	F	154	
2nd	Wed	pc	pc	F	F	F	F	F	F	F	F	F	F	F	F	F	F	F	F	F	F	205	
3rd	Thur	F	pc	pc	pc	pc	pc	F	F	F	F	F	F	F	F	F	F	F	F	F	F	184	
4th	Fri	c	o	c	c	c	c	c	c	F	F	c	F	F	F	F	F	F	F	F	F	118	^
5th	Sat	pc	c	pc	pc	c	pc	pc	F	F	F	pc	pc	pc	F	F	F	F	F	F	F	144	1Q
6th	Sun	o	o	pc	pc	pc	pc	pc	F	pc	F	pc	pc	pc	pc	pc	F	pc	F	pc	pc	111	
7th	Mon	o	o	o	o	o	o	o	o	o	o	o	o	o	o	o	o	o	o	o	o	0	
8th	Tue	o	c	o	o	o	o	o	o	c	o	o	o	o	o	o	o	o	o	o	o	7	
9th	Wed	F	F	pc	pc	o	o	o	o	pc	pc	o	pc	pc	pc	F	pc	F	F	pc	F	101	
10th	Thur	F	pc	pc	pc	F	F	F	F	pc	c	F	F	F	pc	F	c	F	c	c	pc	126	
11th	Fri	pc	pc	c	c	c	c	c	F	pc	pc	c	pc	F	pc	pc	pc	pc	pc	pc	F	105	XhS
12th	Sat	pc	pc	F	F	F	F	F	pc	F	pc	F	pc	pc	pc	pc	pc	pc	F	F	F	157	
13th	Sun	F	pc	F	F	F	F	F	pc	pc	c	F	F	F	F	F	c	F	F	F	F	167	F
14th	Mon	F	pc	F	F	F	F	F	pc	pc	c	F	F	F	pc	pc	c	pc	c	c	c	122	A
15th	Tue	pc	c	pc	pc	pc	pc	pc	pc	F	F	pc	c	c	pc	pc	F	pc	F	F	F	133	
16th	Wed	F	F	F	F	F	F	F	F	F	F	F	pc	F	F	F	F	F	F	F	F	189	
17th	Thur	pc	pc	o	o	pc	pc	pc	pc	pc	pc	pc	pc	pc	pc	c	pc	c	pc	pc	pc	89	
18th	Fri	pc	pc	pc	pc	pc	pc	F	F	F	F	F	F	F	pc	pc	F	pc	F	F	pc	152	
19th	Sat	pc	F	F	F	F	F	F	F	F	F	F	F	F	pc	pc	F	pc	F	F	F	185	V
20th	Sun	pc	pc	pc	pc	pc	F	pc	pc	pc	o	pc	F	F	pc	F	o	F	o	c	F	114	
21st	Mon	F	pc	F	F	pc	F	pc	c	pc	c	pc	F	F	c	c	c	c	pc	c	pc	108	3C
22nd	Tue	pc	c	c	c	pc	c	o	o	c	o	o	pc	o	o	c	o	c	pc	c	pc	42	
23rd	Wed	pc	pc	c	c	pc	pc	pc	pc	pc	o	pc	pc	c	o	c	o	c	o	o	c	62	
24th	Thur	pc	c	c	c	c	F	F	c	F	pc	F	F	F	pc	F	pc	F	F	F	F	128	
25th	Fri	pc	pc	pc	pc	pc	F	pc	pc	pc	pc	pc	F	pc	F	F	pc	F	F	pc	F	131	XhN
26th	Sat	F	pc	pc	pc	pc	pc	F	c	F	c	F	pc	c	F	F	c	F	F	pc	c	120	
27th	Sun	F	pc	pc	pc	pc	F	F	pc	F	c	F	F	F	pc	F	c	F	F	pc	pc	139	
28th	Mon	F	pc	pc	pc	F	F	F	pc	pc	o	F	c	pc	pc	pc	o	pc	pc	c	c	108	N P3
29th	Tue	pc	pc	F	F	F	F	F	F	F	c	F	pc	c	pc	c	c	c	pc	c	pc	130	
30th	Wed	c	F	F	F	F	F	F	F	F	F	F	F	F	F	F	F	F	pc	c	pc	170	
est. hours:		169	142	169	169	178	203	200	186	218	159	200	207	200	179	188	159	188	210	174	200	3699	
av. hours:		158	145	137	164	164	175	168	128	173	153	157	156	163	191	140	139	146	160	158	155	3130	
trend:		av	av	more	av	av	more	more	more	more	av	more	more	more	av	more	more	more	more	av	more	more	

SOUTH ISLAND SUNSHINE ESTIMATES
APRIL 2025

F=fine (8-12 hours of sunshine), pc= partly cloudy (4-7 hours), c=cloudy (1-3 hours), o=overcast (0 hours)

		Ta	Nel	Ble	Mou	West	Grey	Hok	Ham	Kai	Ch	Ash	Tim	Crom	Qstn	Bal	Dun	Wnt	Gore	Inv	Alex	ALL	Moon
1st	Tue	F	F	F	F	F	F	F	F	F	F	F	F	F	F	c	F	F	F	F	F	182	
2nd	Wed	F	F	F	F	F	F	F	F	F	F	F	pc	F	F	F	F	F	F	F	F	196	
3rd	Thur	F	F	F	F	F	pc	pc	pc	pc	F	F	F	F	F	F	F	F	F	F	F	187	
4th	Fri	o	F	F	F	F	F	F	F	F	F	F	F	F	F	pc	F	pc	pc	F	F	169	^
5th	Sat	o	F	F	F	F	F	pc	pc	pc	F	pc	pc	o	o	o	pc	o	o	o	o	92	1Q
6th	Sun	c	F	c	F	F	F	F	o	o	o	c	c	F	pc	c	pc	o	o	c	F	90	
7th	Mon	pc	o	o	o	o	c	c	o	o	o	o	o	F	F	F	pc	pc	pc	pc	F	70	
8th	Tue	F	o	o	o	o	o	c	o	o	o	o	o	o	pc	pc	o	pc	pc	F	o	54	
9th	Wed	F	pc	F	pc	F	pc	o	F	F	F	F	pc	c	c	c	pc	c	c	c	c	103	
10th	Thur	F	c	pc	c	o	pc	pc	F	F	F	F	F	pc	pc	pc	pc	c	c	pc	pc	120	
11th	Fri	c	F	F	F	pc	F	F	pc	pc	pc	F	F	pc	c	c	o	c	c	o	pc	105	XhS
12th	Sat	pc	F	F	F	F	F	F	pc	pc	F	c	pc	F	F	c	c	pc	pc	pc	F	136	
13th	Sun	F	F	F	F	F	c	o	F	F	pc	pc	pc	pc	c	c	c	pc	pc	c	pc	111	F
14th	Mon	pc	c	c	c	pc	F	F	c	c	pc	pc	F	F	F	pc	F	o	o	pc	F	103	A
15th	Tue	c	F	F	F	F	F	F	F	F	F	F	F	F	F	F	F	F	F	F	F	181	
16th	Wed	pc	F	F	F	F	F	F	F	F	F	F	F	pc	c	o	o	o	o	c	pc	116	
17th	Thur	F	pc	F	pc	F	o	o	o	o	o	o	o	F	c	pc	pc	pc	pc	pc	F	81	
18th	Fri	pc	F	pc	pc	F	F	F	F	F	F	F	F	F	F	pc	pc	pc	pc	c	F	153	
19th	Sat	F	F	F	F	F	F	F	pc	pc	pc	pc	F	F	pc	pc	F	pc	pc	c	F	142	V
20th	Sun	c	F	F	F	c	c	c	F	F	F	F	F	pc	pc	pc	pc	c	c	c	pc	117	
21st	Mon	F	pc	pc	F	c	o	o	c	c	c	c	c	c	c	c	c	c	c	c	c	59	3Q
22nd	Tue	F	pc	pc	pc	F	F	F	F	F	F	F	F	c	pc	c	c	pc	pc	c	c	126	
23rd	Wed	c	pc	o	pc	c	pc	F	F	F	F	F	F	c	o	pc	c	o	o	c	c	95	
24th	Thur	F	F	F	F	o	c	pc	F	F	F	F	F	pc	pc	o	pc	c	c	c	pc	112	
25th	Fri	pc	F	F	F	pc	c	c	pc	pc	pc	pc	pc	F	pc	c	pc	c	c	c	F	103	XhN
26th	Sat	F	c	F	c	F	c	pc	F	F	F	F	F	pc	pc	pc	pc	pc	pc	pc	pc	128	
27th	Sun	F	F	pc	F	c	pc	F	c	c	pc	0.1	pc	F	F	pc	c	c	c	pc	F	98	
28th	Mon	F	F	pc	F	pc	F	pc	pc	pc	pc	F	pc	pc	pc	c	pc	pc	pc	pc	pc	115	N P3
29th	Tue	pc	F	F	F	F	F	F	pc	pc	F	F	F	F	pc	F	F	F	F	F	F	169	
30th	Wed	F	F	F	F	F	c	pc	F	F	F	F	pc	pc	pc	o	c	o	o	o	pc	122	
est. hours:		202	226	225	215	201	172	189	182	182	203	215	199	192	172	130	153	124	124	138	192	3635	
av. hours:		187	188	188	191	144	133	142	144	166	161	147	148	173	141	117	121	122	113	111	147	2984	
trend:		av	more	more	more	more	more	more	more	av	more	more	more	more	more	more	more	av	av	more	more	more	

APRIL **1** TUESDAY

A large anticyclonic system centred in the Tasman Sea extends a gentle northwest flow over the country, backing southeast over Northland. Warm, mostly sunny and dry everywhere during the day. Chance of overnight light showers in the Auckland region, around the Kapiti Coast, North Canterbury, down through the West Coast to Fiordland, and around coastal parts of Otago and Southland.

Auckland:	Changeable, mix of sunny spells and light showers, gentle breezes
Hamilton:	Light overnight showers, sunny day, mostly calm and serene air
Wellington:	Sunny, dry, warm, gentle breezes
Greymouth:	Partly cloudy, dry, mild temperatures, calm air
Christchurch:	Sunny, dry, calm air, chance of fog
Dunedin:	Sunny, isolated showers, mostly calm air

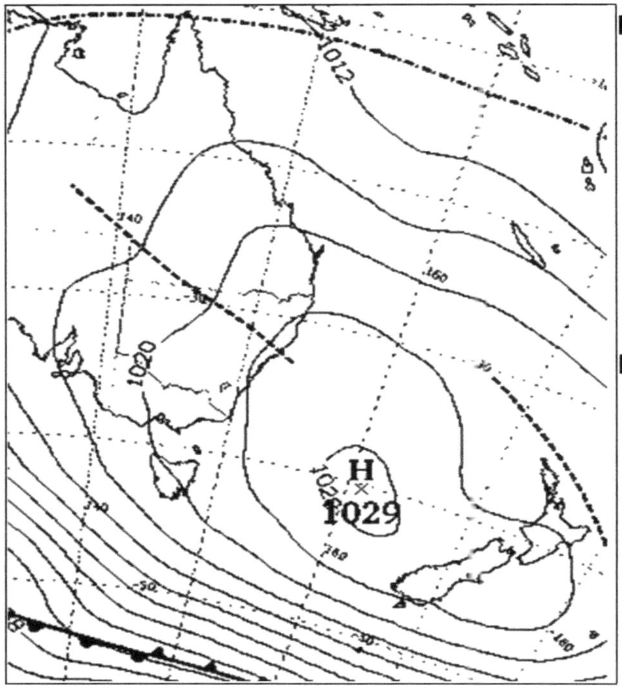

RAIN POTENTIAL

FROST/SNOW

APRIL **2** WEDNESDAY

Anticyclonic conditions strengthens over New Zealand with a gentle easterly air prevailing over the North Island, westerly over the South Island. Dry, warm, and mostly sunny throughout. Chance of fog in sheltered parts, particularly in the Christchurch area.

Auckland:	Sunny, mostly dry, some overnight showers, calm atmosphere
Hamilton:	Hot day cool night, sunny, dry, pleasant and serene air
Wellington:	Very sunny, dry, slightly cooler, gentle breezes to mostly calm air
Greymouth:	Sunny, dry, gentle breezes
Christchurch:	Sunny and dry, calm air, fog likely
Dunedin:	Sunny, mild, dry, light shifting breezes

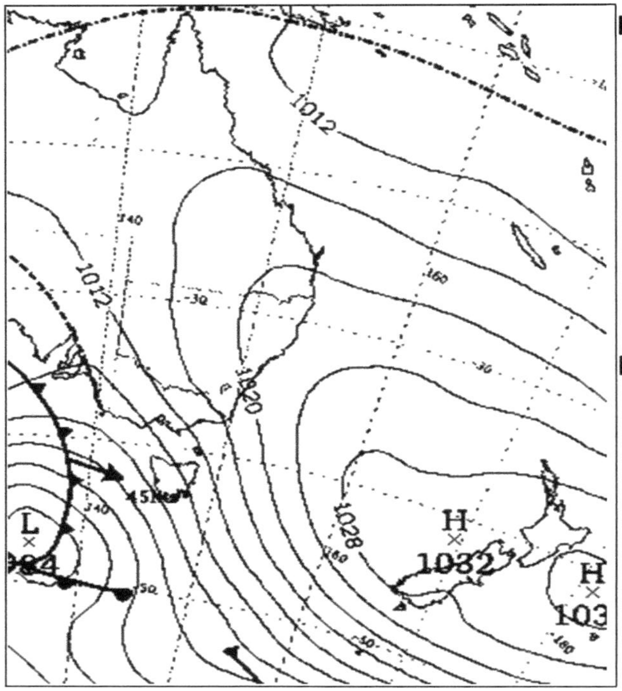

RAIN POTENTIAL

FROST/SNOW

APRIL 3 THURSDAY

High pressure maintains a warm and pleasant air throughout the country with high temperatures building in Canterbury and over the northern half of the North Island. Dry and mostly sunny in all districts. Winds generally easterly over the North Island backing westerly over the South Island. Chance of fog patches once more in the Canterbury region.

Auckland: Partly cloudy, warmer, dry, gentle breezes to mostly calm air
Hamilton: Very hot, dry, partly cloudy, light breezes
Wellington: Sunny, warmer, dry, gentle breezes
Greymouth: Partly cloudy, dry, mild day cool night, gentle breezes
Christchurch: Mostly sunny, warmer, dry, mostly calm air, possible fog patches
Dunedin: Sunny, warmer, dry, gentle breezes

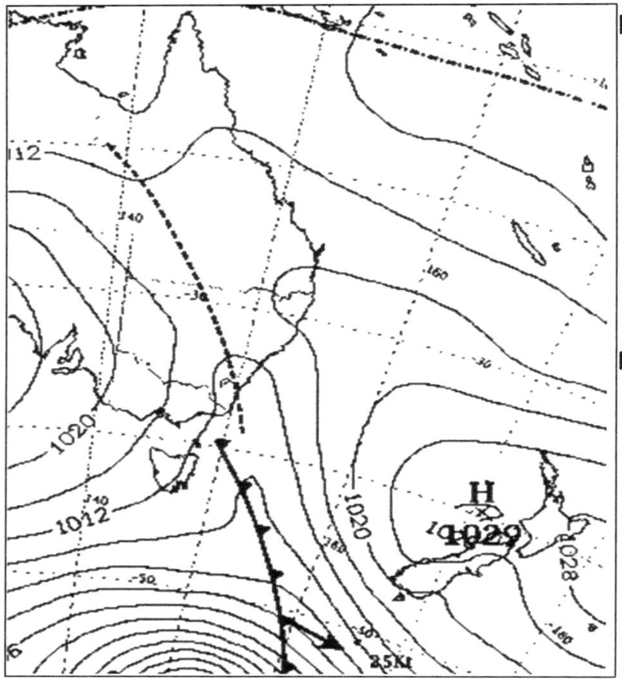

RAIN POTENTIAL

FROST/SNOW

APRIL 4 FRIDAY

High pressure begins to slip away to the east of New Zealand with light easterly breezes prevailing over the North Island backing west to northwesterly through central parts and the South Island. Dry everywhere with high maximum temperatures in many districts, especially in the east of the South Island. Mostly cloudy over the north of the North Island while good sunshine hours elsewhere. Fog patches still likely around Canterbury.

Auckland: Cloudy and dry, warm, gentle breezes
Hamilton: Cloudy, cooler, dry, light shifting breezes
Wellington: Unsettled breezes, dry, sunny
Greymouth: Sunny, cooler, dry, fluctuating breezes
Christchurch: Sunny, warm day cold night, mostly dry, gentle shifting breezes
Dunedin: Sunny, dry, mild day cool night, light breezes

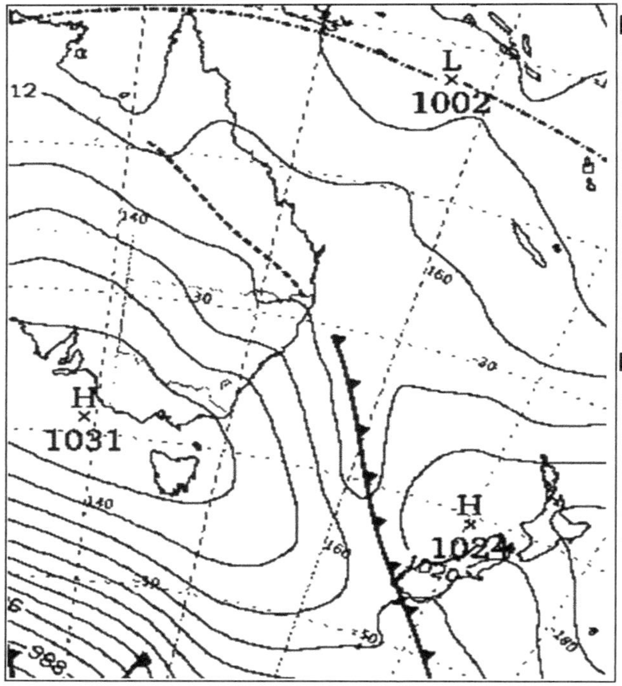

RAIN POTENTIAL

FROST/SNOW

APRIL **5** SATURDAY

A weak depression to the southwest of New Zealand brings widespread cloud over the south of the South Island and briefly extends a showery front into southern parts of the West Coast to Fiordland, while a weak high pressure flow brings hot, dry, and mostly sunny elsewhere. A moderate northeast flow prevails bringing much higher temperatures especially through the eastern districts of both islands.

Auckland: Partly cloudy, warm, mostly dry, chance of morning showers, light air
Hamilton: Mostly cloudy, dry and warm, gentle breezes
Wellington: Mostly sunny, dry, light breezes
Greymouth: Sunny, warmer, dry, fresh breezes at times
Christchurch: Dry, sunny, very hot day cool night, mostly calm air
Dunedin: Partly cloudy, warmer, dry, gentle breezes to mostly calm air

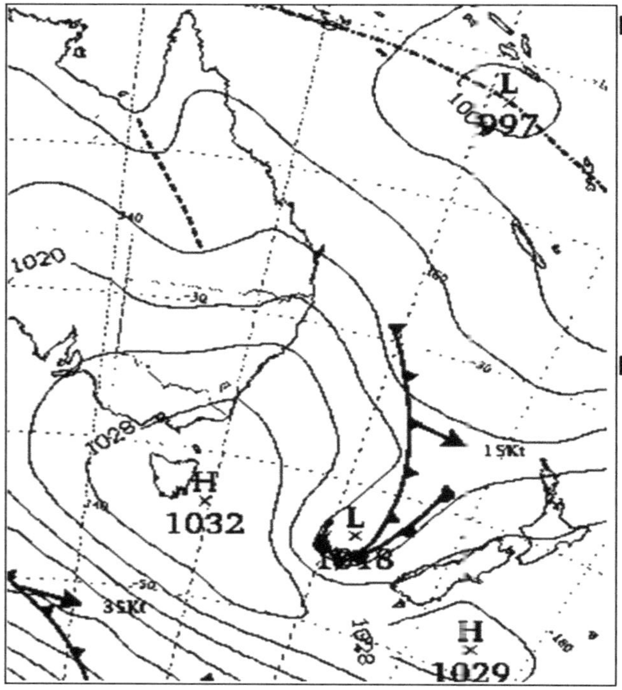

RAIN POTENTIAL

FROST/SNOW

APRIL **6** SUNDAY

High pressure system to the southeast of New Zealand directs a light to moderate east to northeast flow over the country. Generally dry everywhere with widespread cloud in the east and south of the South Island, and sunnier conditions elsewhere. High temperatures continue to prevail over the North Island, while temperatures ease to moderate over the South Island.

Auckland: Hot, humid, mostly cloudy and dry, chance of morning showers
Hamilton: Partly cloudy, dry, warm, moderate breezes
Wellington: Cloudy, cooler, dry, fresh breezes
Greymouth: Sunny, warmer, dry, moderate breezes
Christchurch: Cooler, overcast, mostly dry, odd threat of rain, moderate breezes
Dunedin: Mostly cloudy, dry, cooler, gentle breezes

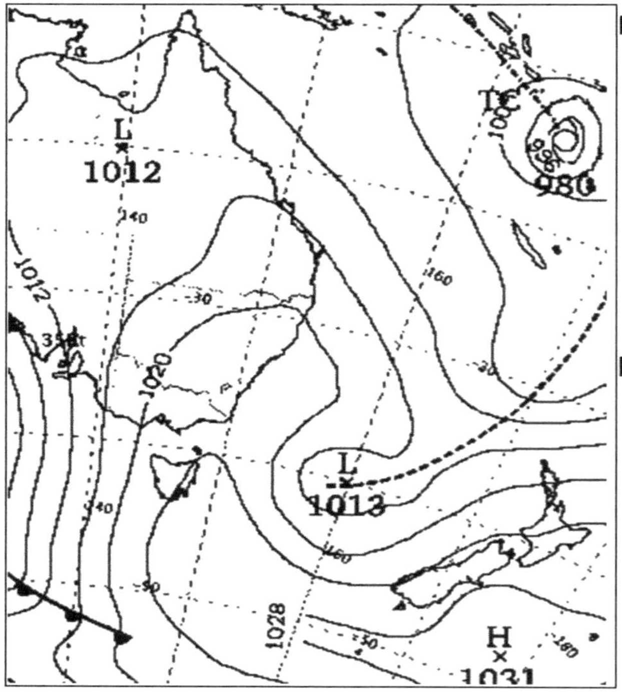

RAIN POTENTIAL

FROST/SNOW

APRIL **7** MONDAY

As the high pressure ridge moves away from New Zealand a low pressure trough moves into the west of the country spreading widespread dull and gloomy skies over the North Island and northern half of the South Island. Winds freshen in a northeasterly flow throughout. A wet front moves onto the north of the country bringing some showers from Northland to Auckland, and while mainly dry elsewhere, increasing threats of rain prevail.

Auckland: Overcast, fresh breezes, light showers, warm and humid
Hamilton: Overcast, drizzly showers, fresh breezes
Wellington: Overcast, dry, odd threat of rain, fresh breezes, occasional gusts
Greymouth: Cloudy, dry, warmer, moderate breezes
Christchurch: Overcast, mostly dry, chance of drizzle or fog patches, gentle breezes
Dunedin: Partly cloudy, dry, fresh breezes

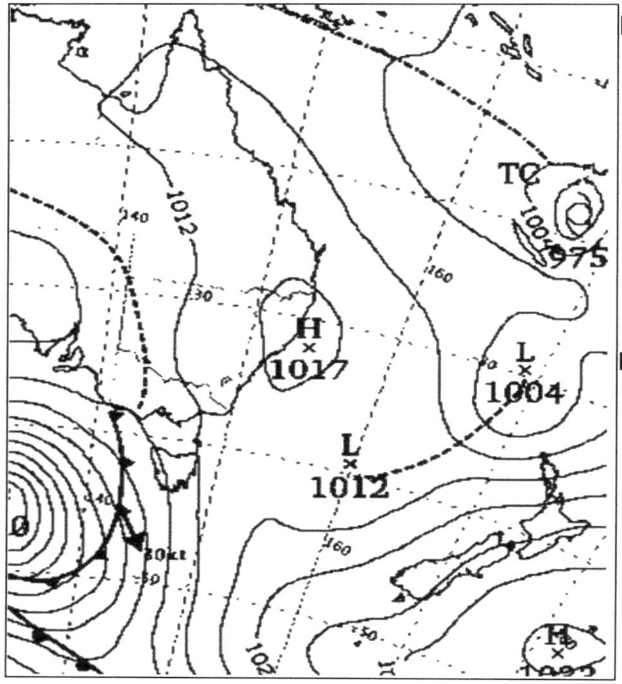

RAIN POTENTIAL

FROST/SNOW

APRIL **8** TUESDAY

A depression to the west of New Zealand advances onto central New Zealand bringing widespread cloudy conditions almost everywhere and strengthening northeasterly winds throughout. Outbreaks of rain spread southward from Northland to the central parts of the South Island, with chance of heavy rain in parts of Northland to the Bay of Plenty that may bring some pockets of localised flooding. Cooler.

Auckland: Stormy, overcast, heavy rain, chance of gale winds and flooding
Hamilton: Overcast, cooler, blustery, squally showers
Wellington: Overcast, mild temperatures, dry, passing threats of rain, blustery
Greymouth: Overcast, dry, mild temperatures, fresh breezes
Christchurch: Overcast, cooler, blustery, brief squalls, gusty winds, gales possible
Dunedin: Overcast and dry, odd threat of rain, fresh breezes

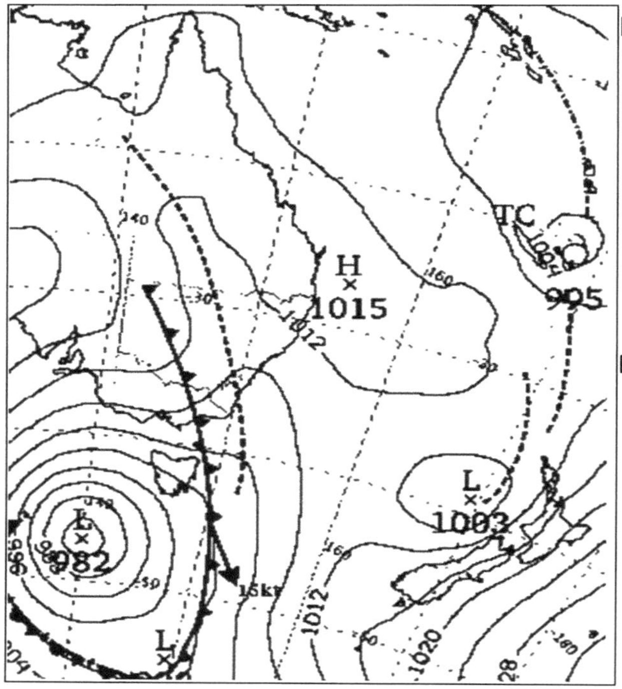

RAIN POTENTIAL

FROST/SNOW

APRIL **9** WEDNESDAY

A low pressure trough maintains frequent cloud throughout most of the country and with an easing north to northeasterly flow prevailing over the North Island, backing northwesterly over the South Island. Mostly dry in the east and south of the South Island and in the far south of the North Island, with rain elsewhere. Heavy rain may cause some flooding in parts of Northland to the Bay of Plenty.

Auckland:	Widespread heavy rain, chance of flooding, some sunny spells
Hamilton:	Overcast, heavy rain, moderate breezes
Wellington:	Sunny, dry, light breezes
Greymouth:	Changeable, partly cloudy, isolated showers, occasional breezy spells
Christchurch:	Mostly sunny, warmer, dry, fresh breezes
Dunedin:	Partly cloudy, dry, moderate breezes

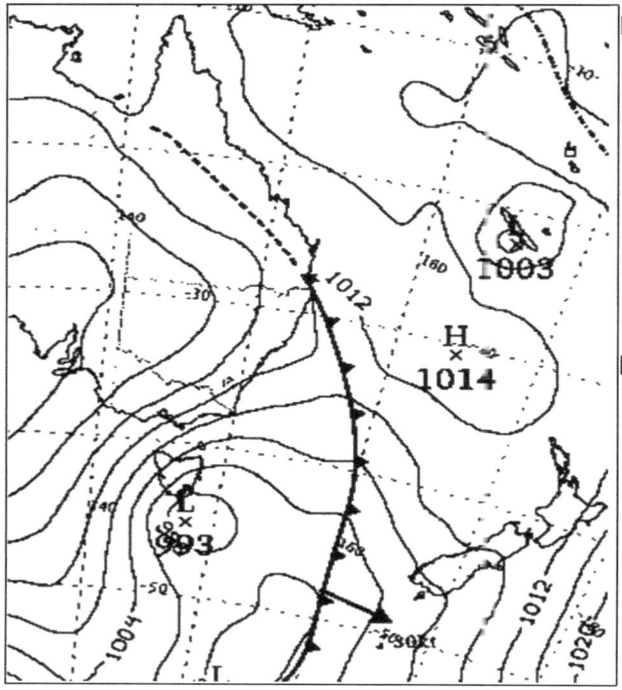

RAIN POTENTIAL

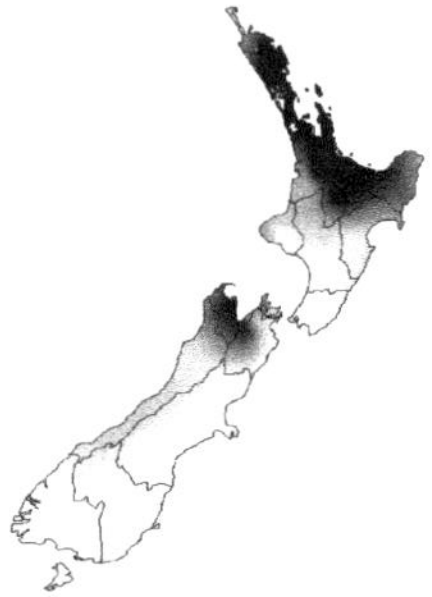

FROST/SNOW

APRIL **10** THURSDAY

A depression south of New Zealand directs a warmer and generally westerly flow over the country, strong and blustery at times particularly over the South Island. Scattered outbreaks of rain in most districts, with some heavy falls still likely around the Bay of Plenty as well as developing in the west of the South Island. Mostly dry in Northland and the east of the South Island. Warmer everywhere with occasional sunny spells.

Auckland:	Partly cloudy, pockets of heavy rain, fresh winds easing
Hamilton:	Sunny, mostly dry, chance of morning showers, moderate breezes
Wellington:	Very hot, partly cloudy, dry, strong winds at times
Greymouth:	Cloudy, warm, rain, blustery air
Christchurch:	Sunny, warmer, brief showers, blustery air
Dunedin:	Partly cloudy, warmer, dry, windy spells

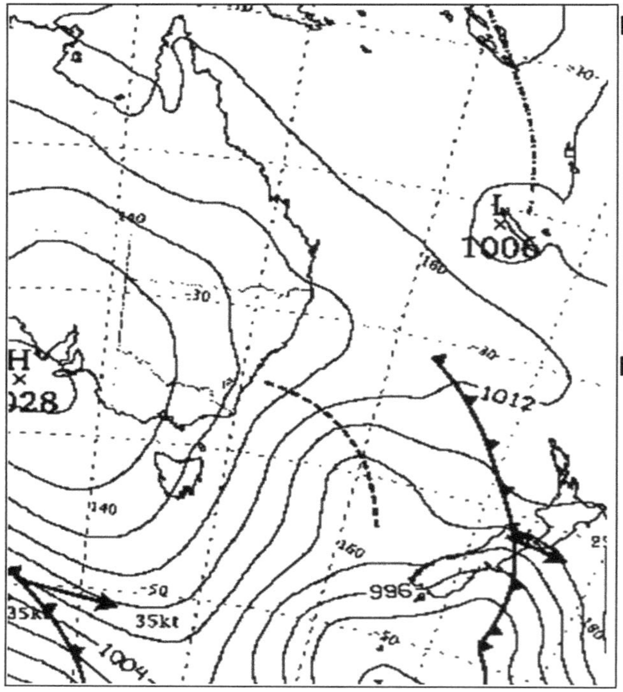

RAIN POTENTIAL

FROST/SNOW

APRIL 11 FRIDAY

A large high pressure system ridges near the west of New Zealand while low pressure moves away to the east. Southwest winds prevail over the South Island, while a mostly calm to light westerly air prevails over the North Island with fog patches likely here and there. A wet trough ahead of the high brings widespread and often heavy rain into the western half of the South Island, with isolated showers into the west of the North Island. Mostly dry in the east of the country. Warm temperatures continue.

Auckland: Mostly cloudy, isolated showers, light breezes
Hamilton: Cloudy, dry, gentle breezes
Wellington: Mostly sunny, dry, cooler, fresh winds, chance of gale gusts
Greymouth: Heavy overnight rain, sunny day, cooler, winds easing to light
Christchurch: Very hot and dry day, scattered cloudy spells, fresh breezes
Dunedin: Overcast, light showers, moderate breezes

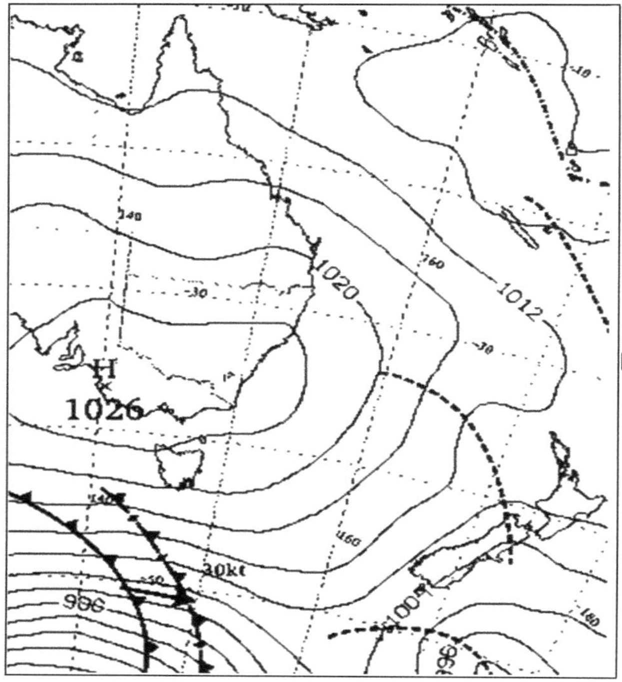

RAIN POTENTIAL

FROST/SNOW

APRIL 12 SATURDAY

A large anticyclonic system crossing the southeast of Australia extends a high pressure ridge over New Zealand directing a very light to mostly calm west to southwest flow throughout the country, slightly cooler in the South Island, while remaining warm in the North Island. A mixed day with sunny spells and the odd shower in most districts, with more persistent shower activity in the west of the North Island, while rain activity clears in the South Island.

Auckland: Sunny, mild, mostly dry, chance of isolated showers, gentle breezes
Hamilton: Sunny, mostly dry, chance of brief morning showers, gentle breezes
Wellington: Sunny and dry, winds easing to light
Greymouth: Mostly sunny, dry, light air
Christchurch: Sunny, cooler, dry, calm and pleasant air
Dunedin: Cloudy, cooler, brief showers, light breezes to mostly calm air

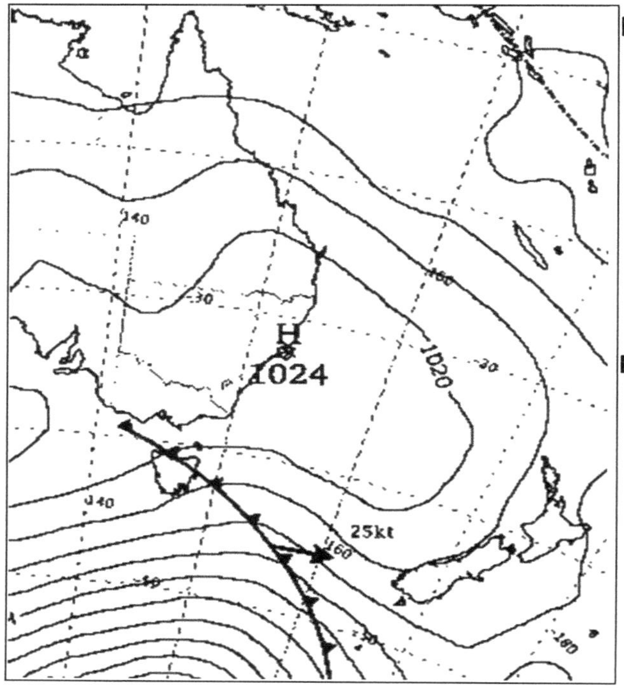

RAIN POTENTIAL

FROST/SNOW

APRIL **13** SUNDAY

High pressure prevails over the North Island while a wet front moves onto the South Island. A light to moderate westerly flow prevails in most districts, backing southeasterly in the northern half of the North Island later in the day. Some showers expected in the southwest of the South Island, while mostly dry and sunny conditions can be expected elsewhere. Temperatures remain warm over the North Island but slightly cooler in the South Island.

Auckland:	Very sunny, cooler, dry, light breezes
Hamilton:	Sunny, dry, warm day cool night, mostly calm air
Wellington:	Sunny, dry, cooler, gentle breezes to mostly calm air
Greymouth:	Cloudy, isolated showers, gentle breezes
Christchurch:	Partly cloudy, dry, cooler, light air
Dunedin:	Cloudy, dry, odd threat of rain, light breezes

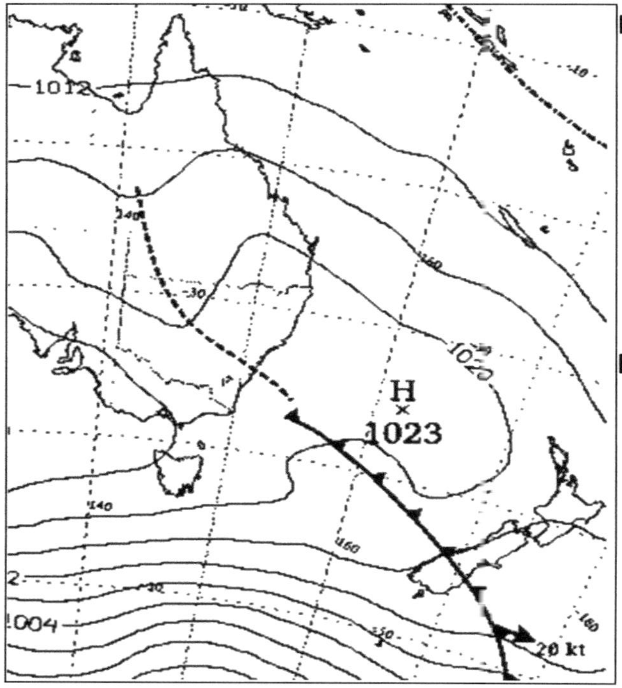

RAIN POTENTIAL

FROST/SNOW

APRIL **14** MONDAY

A weak wet front crosses the South Island bringing a mix of sunny spells and some showers into most districts with heavier showers in Fiordland. Mostly dry in the North Island with good sunny spells. Milder temperatures everywhere with the chance of overnight frost patches developing in the central North Island higher reaches.

Auckland:	Sunny and dry, fluctuating breezes
Hamilton:	Sunny, dry, warm day cool night, moderate breezes
Wellington:	Cloudy, dry with increasing threats of rain, warmer, light breezes
Greymouth:	Sunny, isolated showers, light shifting breezes to mostly calm air
Christchurch:	Partly cloudy, slightly warmer, light showers, fluctuating breezes
Dunedin:	Mix of sunny spells and showers, cool overnight, fluctuating breezes

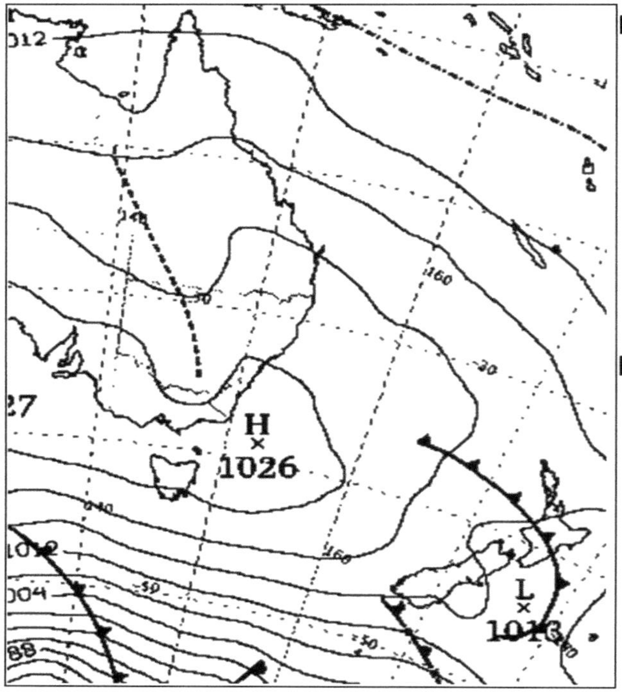

RAIN POTENTIAL

FROST/SNOW

APRIL 15 TUESDAY

A large high to the west of New Zealand directs a cooler west backing southeast flow over the country. The passage of yesterday's wet front moves showers up through the South Island and into the southern half of the North Island, with showers clearing to mostly dry in the east of the South Island, where colder overnight night temperatures may see some frosts appearing in Canterbury to Otago. Mostly dry but cloudy in the northern half of the North Island with increasing threats of rain.

Auckland: Dry and calm air, scattered cloudy spells
Hamilton: Partly cloudy, dry, mild day cool night, gentle breezes, chance of fog
Wellington: Changeable, sunny, overnight showers, cooler, gentle breezes
Greymouth: Sunny, dry, mostly calm air
Christchurch: Calm air, sunny, cooler, dry, chance of shallow fog
Dunedin: Sunny, dry, mild temperatures, gentle breezes

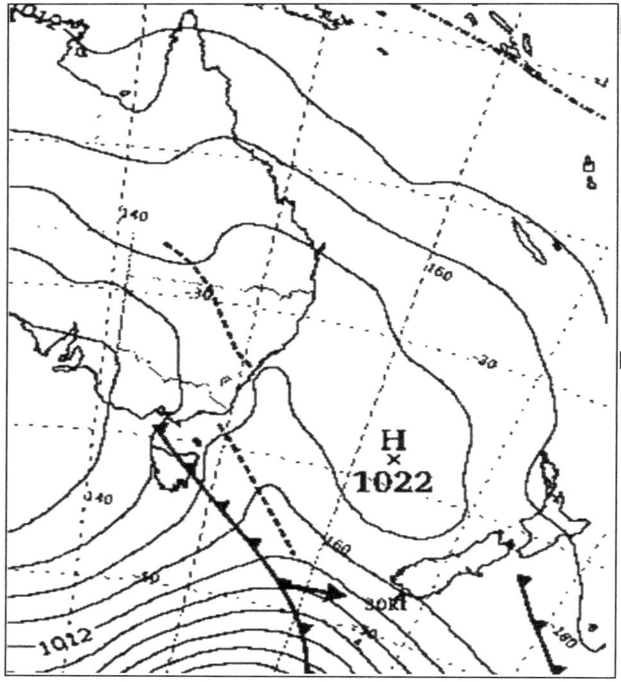

RAIN POTENTIAL

FROST/SNOW

APRIL 16 WEDNESDAY

A weak high pressure air prevails over most of the country ir a light to mostly calm south to southwest flow prevailing. The last of the showers associated with the wet front that has crossed the country the past couple of days clears Northland with most of the country seeing a dry and mostly sunny day with patches cf fog in sheltered parts. Another wet front nears the southern coasts of the South Island bringing some showers once more into Fiordland and Southland.

Auckland: Sunny, cooler, mostly dry, chance of fog, possible evening showers
Hamilton: Sunny, dry, cool overnight, calm air, fog patches possible
Wellington: Sunny, dry, milder, mostly calm air
Greymouth: Partly cloudy, dry, mild day cool night, light shifting breezes
Christchurch: Calm air, sunny, mild day cold night, fog possible
Dunedin: Overcast, dry with passing threats of rain, light breezes

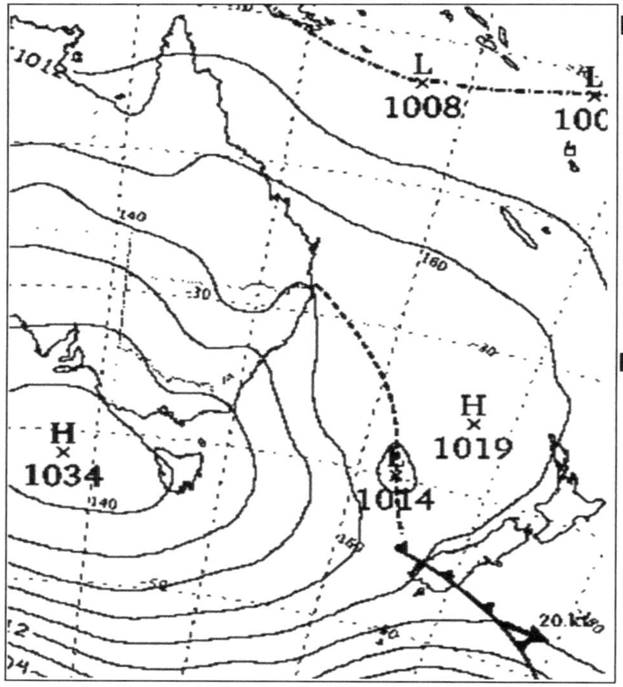

RAIN POTENTIAL

FROST/SNOW

APRIL **17** THURSDAY

A weak depression forms between two high pressure systems, with an intense high over Tasmania, and a weaker high moving off to the east of New Zealand, with low pressure moving into the west of the country. Mainly dry, calm and mostly sunny throughout the North Island, while a wet trough brings some showers here and there over the central and southern parts of the South Island. Milder temperatures with chance of fog in Canterbury.

Auckland:	Overcast, mostly dry, odd threat of rain, calm air, chance of drizzle
Hamilton:	Partly cloudy, dry, cool overnight, serene air, chance of fog patches
Wellington:	Partly cloudy, warmer day cool night, calm and serene air
Greymouth:	Overcast, light showers, light shifting breezes
Christchurch:	Overcast, scattered showers, milder, calm air, possible fog
Dunedin:	Cloudy, colder, dry, some threats of rain, calm atmosphere

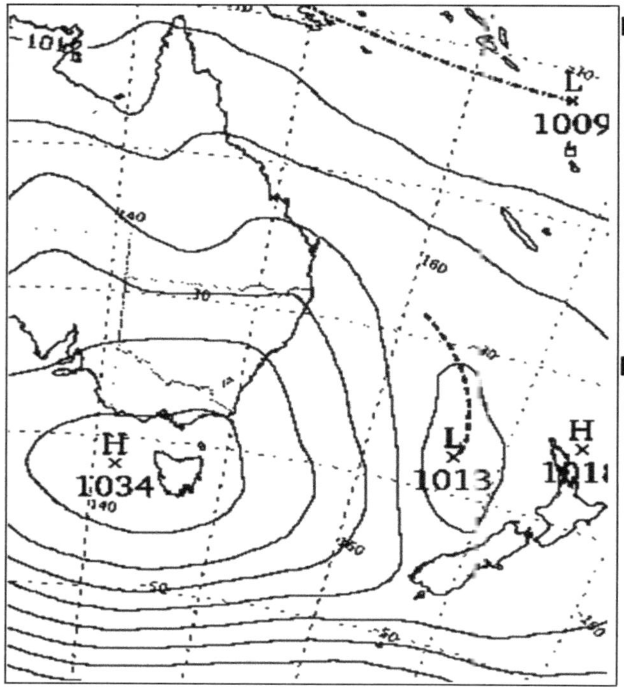

RAIN POTENTIAL

FROST/SNOW

APRIL **18** FRIDAY

A large high situated over the southeast of Australia extends its ridge onto the southwest of New Zealand, while a low pressure trough crosses northern parts of the country. A gentle southwest flow prevails over the South Island and gentle shifting breezes affect the northern half of the North Island. Some isolated showers persist in central and southern South Island districts, while mostly dry and sunny conditions prevail elsewhere. Chance of fog in sheltered parts, and possible frosts in the south of the South Island.

Auckland:	Partly cloudy, dry, mild, light shifting breezes
Hamilton:	Cooler day milder night, dry, scattered cloudy spells, mostly calm air
Wellington:	Mostly calm and serene air, partly cloudy, cooler, dry
Greymouth:	Sunny, dry, cooler, gentle breezes
Christchurch:	Cooler, mix of sunny spells and showers, mostly calm air
Dunedin:	Partly cloudy, dry, cool, mostly calm air

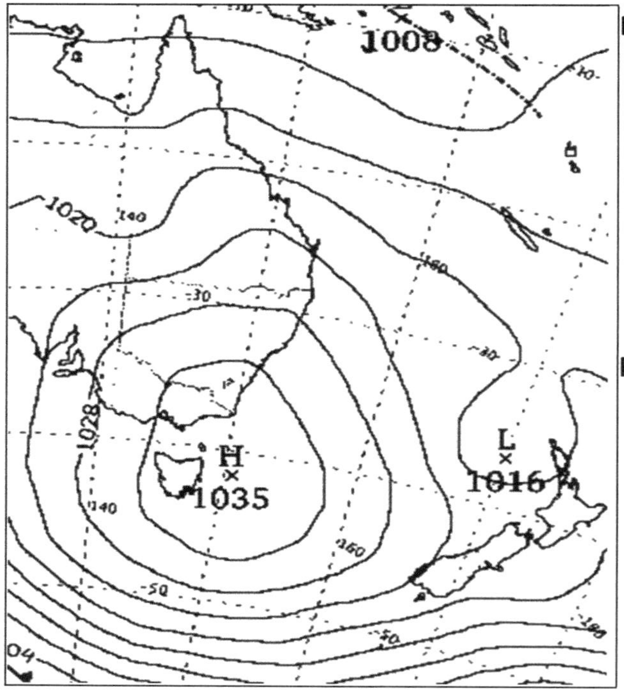

RAIN POTENTIAL

FROST/SNOW

APRIL 19 SATURDAY

A large anticyclonic system dominates the country directing a mostly calm and serene air throughout. Mild, mostly dry and sunny everywhere except for some showers possible in the far north of the North Island and in Fiordland. Chance of frosts in Otago and the Wellington area, and fog patches in the Auckland to Waikato districts and around Christchurch.

Auckland: Sunny, warmer, dry, calm and serene air
Hamilton: Sunny, dry, mild day cool night, calm and serene air
Wellington: Sunny, dry, calm and pleasant air
Greymouth: Sunny, warmer day cooler night, dry, gentle breezes
Christchurch: Partly cloudy, warmer, dry, calm air, fog likely
Dunedin: Mostly sunny, warmer, dry, mostly calm and pleasant air

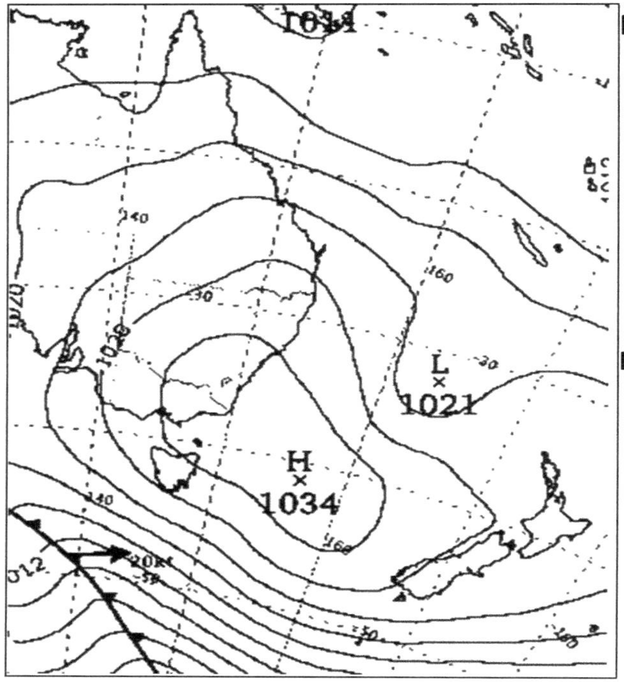

RAIN POTENTIAL

FROST/SNOW

APRIL 20 SUNDAY

Anticyclonic, mostly sunny, dry, and mild conditions prevail over the North Island with the chance of fog in the Auckland to Waikato regions. To the south however, a blustery wet front approaches the south of the South Island strengthening a westerly flow over the island and bringing increased cloud and evening showers into the western and southern districts. Chance of heavy rain overnight in Fiordland.

Auckland: Partly cloudy, warmer, dry, mostly calm air, chance of fog
Hamilton: Partly cloudy, dry, warmer, mostly calm air, chance of overnight fog
Wellington: Dry, mild temperatures, sunny, light breezes freshening later
Greymouth: Cloudy and dry, freshening breezes
Christchurch: Sunny, dry, mild day cool night, light breezes freshening later
Dunedin: Partly cloudy, dry, mild day cool night, freshening breezes

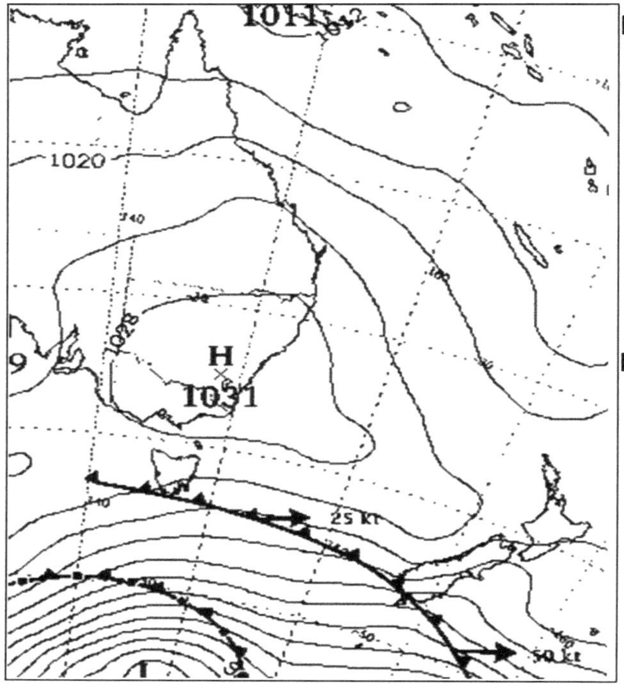

RAIN POTENTIAL

FROST/SNOW

APRIL 21 MONDAY

An unstable wet front associated with an intense depression crossing through the Southern Ocean brings widespread and heavy rain into the south and west of the South Island overnight, while low pressure north of the country brings some showers into the Northland region. Mostly dry everywhere else with increasing cloud and passing threats of rain. A moderate easterly flow prevails over the north of the North Island while a warmer westerly air prevails elsewhere.

Auckland:	Unsettled breezes, sunny, mostly dry, chance of evening rain
Hamilton:	Partly cloudy, cooler, dry, fresh breezes
Wellington:	Partly cloudy, dry, increasing threats of rain, strengthening winds
Greymouth:	Strong winds, overcast, squalls, chance of thunderstorms and gales
Christchurch:	Unsettled blustery winds, cloudy, dry, chance of gales
Dunedin:	Cloudy, mild, dry, increasing threats of rain, unsettled blustery winds

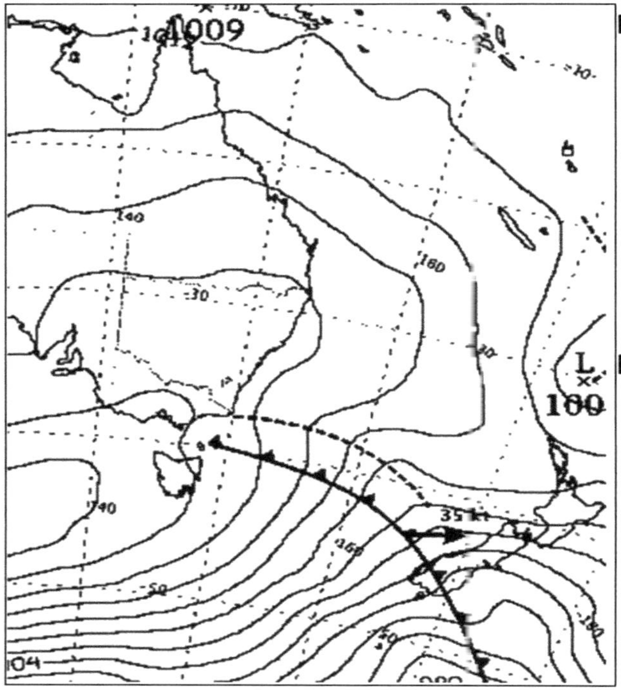

RAIN POTENTIAL

FROST/SNOW

APRIL 22 TUESDAY

An intense depression south of New Zealand brings widespread stormy conditions over much of the country by evening, although it will remain mostly dry in the Bay of Plenty to the East Cape. All other parts of the country can expect some rain activity to develop with heavy falls in the northern half of the South Island and across the Cook Strait into the lower half of the North Island. Very cold west to southwest winds prevail throughout, which may bring wintry showers into the higher reaches of both islands.

Auckland:	Cloudy, scattered showers, blustery winds, possible gales at times
Hamilton:	Mix of showers and occasional sunny breakthroughs, fresh winds
Wellington:	Partly cloudy, heavy rain, chance of thunderstorms and gale gusts
Greymouth:	Heavy rain overnight, sunny day, thunderstorms and hail possible
Christchurch:	Sunny day, warm day cool night, evening rain, unsettled winds
Dunedin:	Stormy air, cloudy, squalls, chance of gales, thunderstorms and hail

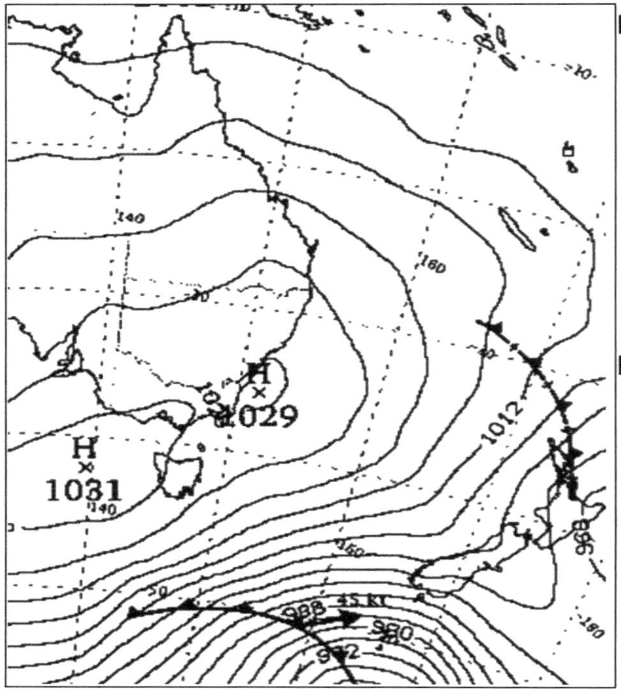

RAIN POTENTIAL

FROST/SNOW

APRIL 23 WEDNESDAY

A stormy wet front moves up through the country in a blustery southwest flow, with chance of gales through the Cook Strait. Widespread cloud and rain throughout and colder temperatures in the south of the South Island may see wintry precipitation at times, particularly in the higher reaches where some snowfalls in the Alps may be observed.

Auckland:	Cooler, cloudy, rain, strong winds possibly to storm force for a time
Hamilton:	Mix of showers and sunny spells, cooler, strong winds, gales possible
Wellington:	Static air, cloudy, isolated showers, gusty winds, gales possible
Greymouth:	Unsettled and blustery, static air, partly cloudy, light showers, cooler
Christchurch:	Cooler, sunny day, evening showers, blustery, chance of gales
Dunedin:	Stormy, cloudy, showers, chance of gales, thunderstorms and hail

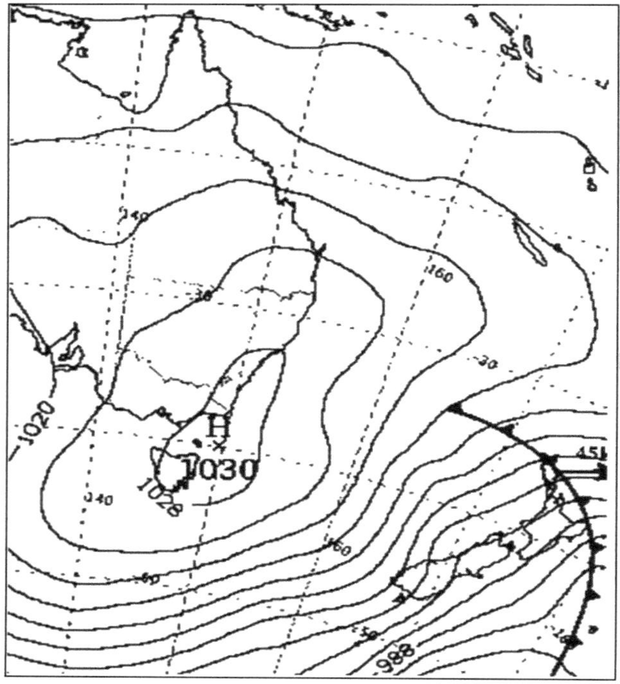

RAIN POTENTIAL

FROST/SNOW

APRIL 24 THURSDAY

While a high pressure system west of New Zealand ridges near the west of the country, a blustery southwest flow associated with the depression to the south continues to bring stormy conditions over most of the country, but with rain activity clearing to occasional showers in most districts. Temperatures colder than of recent, with frosts likely in the Bay of Plenty and the east of the South Island and chance of snow in the Southern Alps and higher parts of the Bay of Plenty to the Central Plateau.

Auckland:	Unsettled blustery air, cloudy, squally showers
Hamilton:	Cloudy, light squally showers, strong winds, possible gales for a time
Wellington:	Unsettled strong winds, sunny, mostly dry, chance of brief showers
Greymouth:	Cloudy, windy, light showers
Christchurch:	Sunny, colder, dry, strong breezes easing later in the day
Dunedin:	Mostly cloudy, blustery, squalls, colder

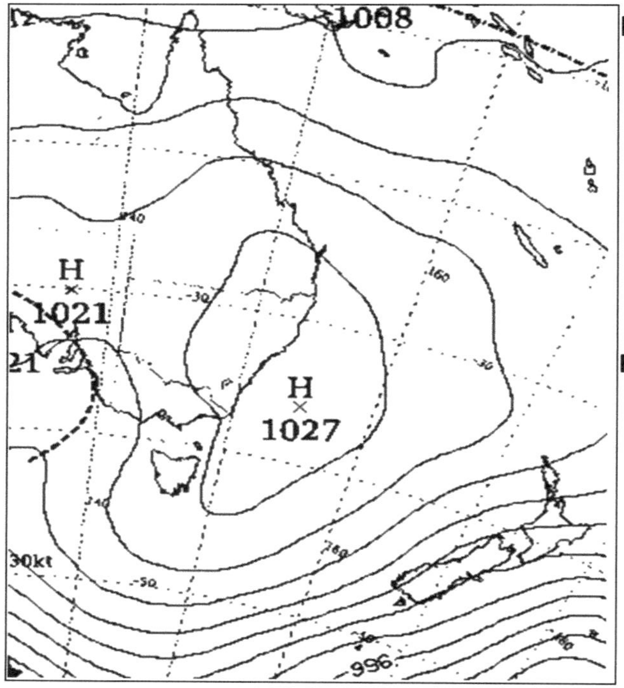

RAIN POTENTIAL

FROST/SNOW

APRIL **25** FRIDAY

A large anticyclonic system moves onto the country with southwest winds prevailing, but easing to moderate throughout. Rain clears overnight to mostly dry with increasing sunny spells in most districts. Milder everywhere. Chance of fog patches in the Auckland to Waikato region and isolated showers lingering around the west of the North Island to the Central Plateau and in Fiordland and Southland.

Auckland: Cloudy, occasional showers, fresh to strong winds
Hamilton: Cloudy, mostly dry, chance of isolated evening showers, windy spells
Wellington: Sunny, dry, mild day cool night, fresh to strong winds
Greymouth: Cloudy, isolated showers, winds easing
Christchurch: Partly cloudy, warmer, dry, light breezes, chance of misty patches
Dunedin: Cloudy, cool, dry with lingering threats of rain, winds easing

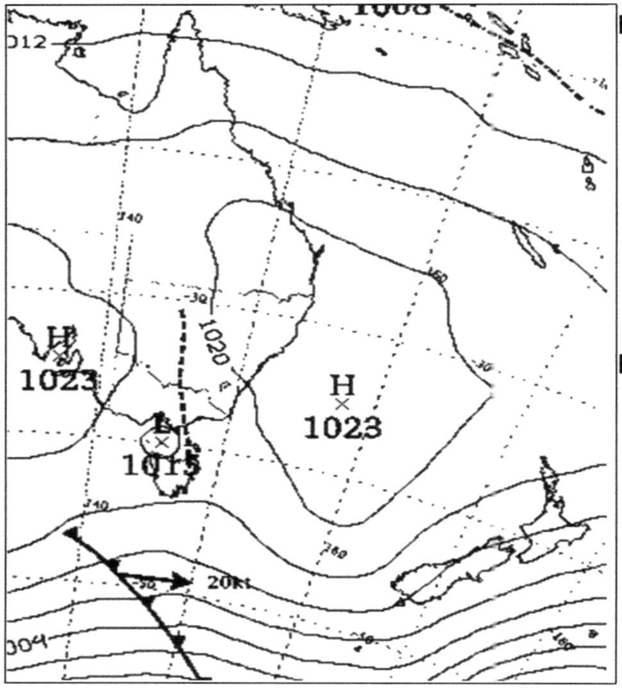

RAIN POTENTIAL

FROST/SNOW

APRIL **26** SATURDAY

The high pressure ridge over the northwest of New Zealand weakens as it slips across the North Island away to the east of the country, with a gentle southwest flow prevailing over the North Island. A stronger northwest to westerly flow however, prevails over the South Island ahead of another front associated with a depression traversing the Southern Ocean. Some light showers likely in the Hawkes Bay while mostly sunny, dry, and mild elsewhere. Chance of fog for a time in the east of the South Island and Auckland to the Waikato.

Auckland: Partly cloudy, brief showers, gentle breezes
Hamilton: Partly cloudy, mostly dry, calm air, chance of isolated showers
Wellington: Cloudy, warmer, dry, light breezes
Greymouth: Cloudy, fresh breezes, dry, mild day cool night
Christchurch: Sunny, warmer, dry, light breezes, chance of shallow fog patches
Dunedin: Partly cloudy, mild day cool night, dry, moderate shifting breezes

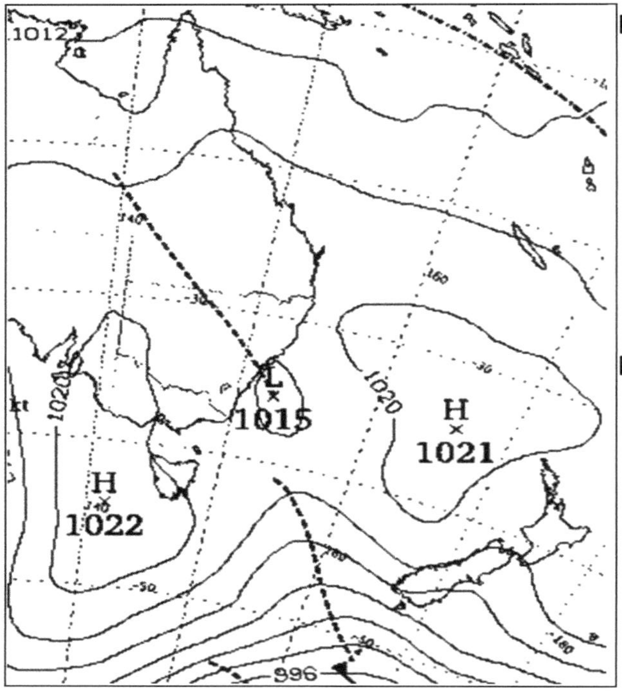

RAIN POTENTIAL

FROST/SNOW

APRIL 27 SUNDAY

While a couple of anticyclonic systems await in the Tasman Sea and to the north of the North Island, the prevailing weather will be dominated by a blustery wet front crossing up through the country from a depression in the Southern Ocean. Generally a mixed day with sunny spells and mostly overnight scattered rain activity in most districts. Some outbreaks of rain are likely to be heavy around the West Coast to Fiordland and around the North Island Central Plateau. A strong southwest flow prevails, lighter over the northern half of the North Island.

Auckland:	Partly cloudy, mild temperatures, occasional showers, windy
Hamilton:	Unsettled and breezy, mostly sunny, brief showers
Wellington:	Partly cloudy, dry, fresh to strong winds
Greymouth:	Changeable, partly cloudy, heavy overnight rain, blustery at times
Christchurch:	Cloudy, dry, fresh breezes easing later, chance of overnight fog
Dunedin:	Cloudy, brief passing showers, strong winds with occasional gusts

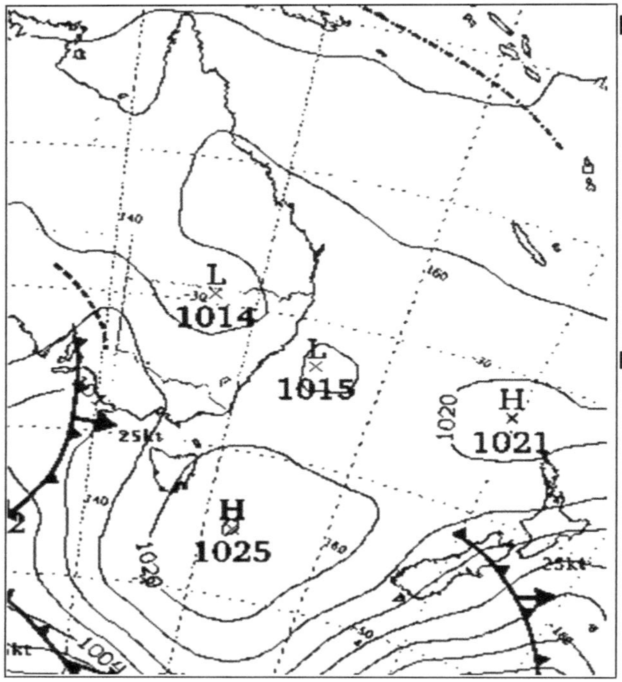

RAIN POTENTIAL

FROST/SNOW

APRIL 28 MONDAY

A large high behind the wet front moves onto New Zealand clearing rain activity to isolated overnight showers here and there, mostly in the South Island and western half of the North Island. Winds ease to a gentle to mostly calm west to southwest flow throughout. Some good sunny spells likely, especially in the North Island but cloudier in the South Island. Chance of frosts in the Otago district.

Auckland:	Mix of isolated showers and occasional sunny spells, fresh breezes
Hamilton:	Sunny, dry, gentle breezes
Wellington:	Cloudy, mostly dry, passing threats of rain, light breezes
Greymouth:	Mostly sunny and dry, calm atmosphere
Christchurch:	Partly cloudy, colder, dry, mostly calm air, possible fog patches
Dunedin:	Cloudy cooler, isolated showers, gentle breezes to mostly calm air

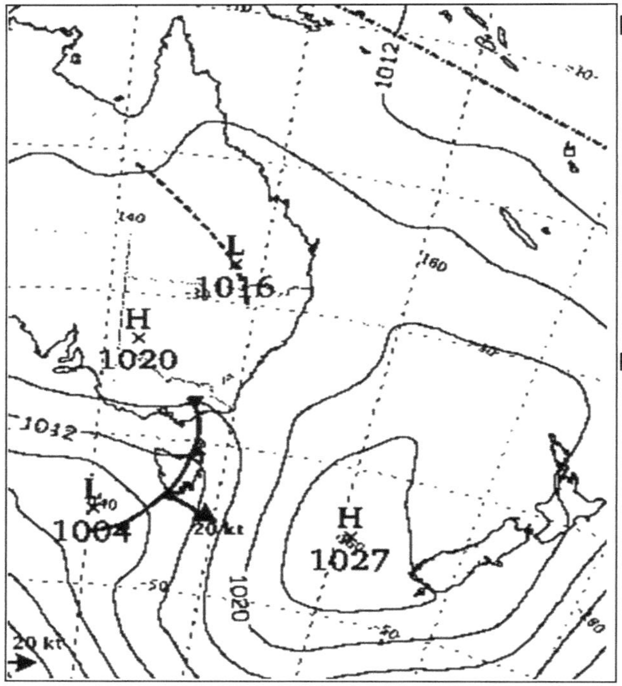

RAIN POTENTIAL

FROST/SNOW

APRIL 29 TUESDAY

A large anticyclone lingers over the country directing a very light westerly to mostly calm air over southern parts, backing southeasterly to the north. Mostly dry and sunny everywhere, except for some lingering showers around the East Cape. Cooler throughout the country with pockets of frost through the central North Island from the Waikato to Wellington and in the Otago region.

Auckland: Sunny, dry, mild day cool night, light shifting breezes
Hamilton: Sunny and dry, mild day cold night, mostly calm air
Wellington: Partly cloudy, isolated overnight showers, light shifting breezes
Greymouth: Sunny, dry, calm and pleasant air, cold overnight
Christchurch: Mostly sunny, dry, mostly calm air, cold overnight, possible fog
Dunedin: Sunny, dry, mostly calm air

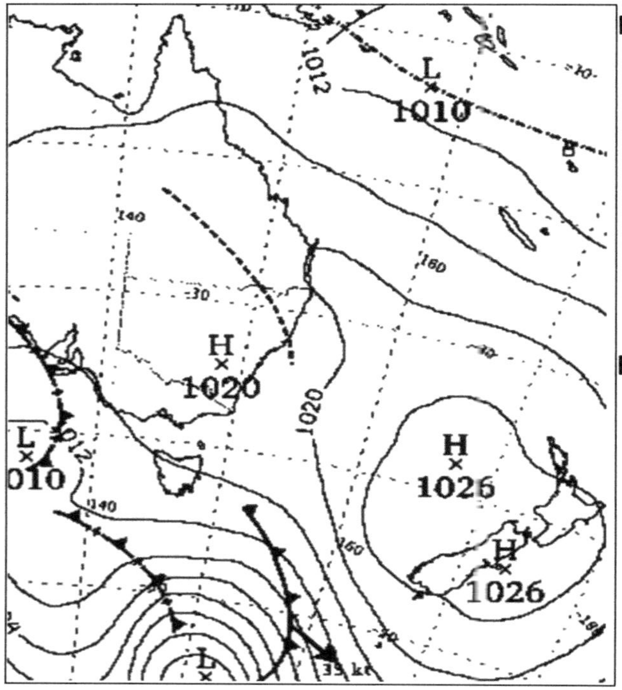

RAIN POTENTIAL

FROST/SNOW

APRIL 30 WEDNESDAY

Anticyclonic conditions continue to prevail over the country. Dry, mostly sunny, and a light westerly air throughout, backing easterly over the north of the country. Cool with frost pockets likely in the Waikato to the Central Plateau as well as in the Bay of Plenty, Wellington region and in Southland and Otago. Chance of fog in the Christchurch area.

Auckland: Sunny, dry, cooler, mostly calm air
Hamilton: Sunny, dry, mild day cold night, calm and serene air
Wellington: Partly cloudy and cool, dry, gentle shifting breezes
Greymouth: Cloudy, dry, cool overnight, light shifting breezes
Christchurch: Sunny, dry, gentle shifting breezes, chance of shallow fog patches
Dunedin: Cloudy, dry, increasing threats of rain, mostly calm and serene air

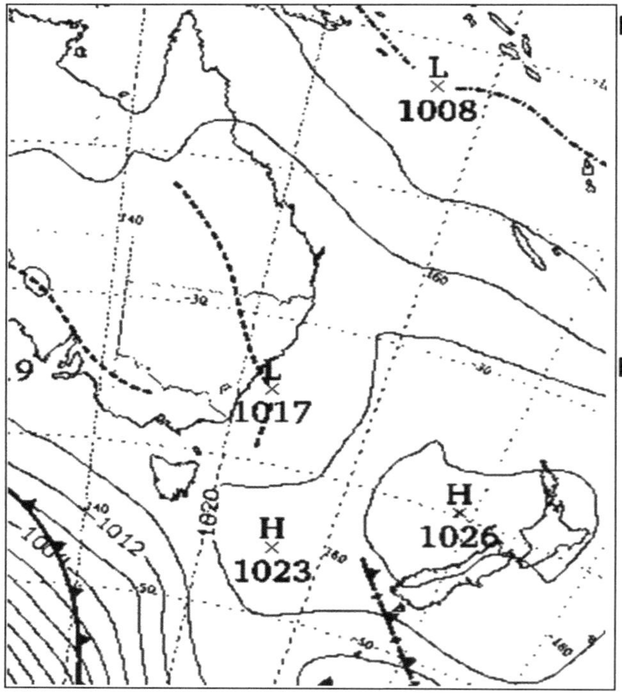

RAIN POTENTIAL

FROST/SNOW

MAY

PHASES OF THE MOON

1st	Northern Declination
4th	First Quarter
8th	Crossing Equator
11th	Apogee
12th	Full Moon
15th	Southern Declination
20th	Third Quarter
22nd	Crossing Equator
26th	Perigee #5
27th	New Moon
28th	Northern Declination

1	2	3	4	5	6	7	8	9	10	11	12	13	14	15	16	17	18	19	20	21	22	23	24	25	26	27	28	29	30	31
Thu	Fri	Sat	Sun	Mon	Tue	Wed	Thu	Fri	Sat	Sun	Mon	Tue	Wed	Thu	Fri	Sat	Sun	Mon	Tue	Wed	Thu	Fri	Sat	Sun	Mon	Tue	Wed	Thu	Fri	Sat
^			1Q				XhS			A	F			V					3Q		XhN				P5	N	^			

MONTHLY SUMMARY

2nd-5th Chance of fog in the northern half of the North Island including in the Waikato and Auckland regions as well as down through the east of the South Island including in Christchurch.
3rd Heavy rain likely in Fiordland.
4th-5th Possible fog patches in the Auckland district.
7th-8th Wet front brings pockets of heavy rain in the West Coast and over the top of the South Island.
8th-11th Wet front moves up through the North Island.
9th-10th Chance of fog in Christchurch.
10th Unusually cold overnight in the Ranfurly area, with frost pockets likely.

Over the next four weeks anticyclonic systems continue to be regularly situated to the east of the North Island maintaining more frequent northwesterlies than normal for this time of year, while low pressure systems will prevail more often over the south of Australia. Rainfall will continue to be lower than normal over the next four weeks, especially in the east as well as in parts of the north, while being above normal in the Nelson area. Some districts in the north and east of the North Island and along the Kaikoura coast may only see around 20% of their usual rainfall for this time of year, leading to severe soil moisture deficits in many eastern districts from Gisborne and into the Wairarapa, and throughout much of Central Otago. Other areas likely to see only around 50-75% of their usual rainfall will be parts of Northland, Auckland, Waikato to the Bay of Plenty, Coromandel, and Taupo regions, as well as in the Manawatu, Horowhenua, Kapiti, North Canterbury, and parts of inland Southland. In contrast, the Nelson district may see up to twice as much recorded rainfall than is typical for this time of year, mostly due to a heavy rain event expected at the start of June. The autumn season will continue to be milder than normal with more sunshine than is typical for this time of year in most districts, especially in the east of the country. Temperatures will reflect the drier and sunnier conditions over the next four weeks by being above average almost everywhere; some districts may be up to 1-2°C above average, while up to 2-2.5°C above average in parts of Marlborough, Nelson, Canterbury, and Otago.

11th Unusually warm day in Northland.
11th-13th Heavy rain in the West Coast, Nelson-Marlborough region and into parts of Canterbury.
12th Heavy rain in the Waikato. Possible misty patches in Christchurch.

13th-15th Fog possible in the Waikato to Auckland regions.
16th Heavy rain in the south of the South Island, particularly around Fiordland.
18th Possible shallow fog and haze in Christchurch.
19th Strong winds in the Wellington district.
19th-20th Further heavy rain likely in Fiordland.
23rd-26th Chance of fog patches in Christchurch.
26th Possible brief blustery spells in Christchurch.
28th-30th Chance of further patches of fog in the Christchurch region.
29th-31st Strong and unsettled winds in the Wellington region.
30th Heavy rain returns to Fiordland and as far north as Mt Cook.
31st Chance of fog in the Auckland region.

May 2025

Monday	Tuesday	Wednesday	Thursday	Friday	Saturday	Sunday
28	29	30 **Moon Sign**	**1** ^ **Gemini** Conjunct Jupiter Square Ascendant	**2** **Cancer** Square Venus Square Neptune Square Ascendant	**3** **Cancer** Square Mercury Square Ascendant	**4** **Leo** Trine Venus Conjunction Mars Trine Saturn Trine Neptune Opposition Pluto Square Ascendant
5 **1Q** **Leo** Square Sun Trine Mercury Square Ascendant	**6** **Leo** Square Uranus	**7** **Virgo** Trine Sun	**8** **XhS** **Virgo** Square Jupiter Opposition Saturn Trine Uranus Opposition Neptune	**9** **Libra** Opposition Venus Sextile Mars Opposition Neptune Trine Pluto	**10** **Libra** Trine Jupiter	**11** **A** **Scorpio** Opposition Mercury Square Pluto
12 **Scorpio** Opposition Sun Square Mars Trine Ascendant	**13** **FM** **Scorpio** Opposition Sun Trine Saturn Opposition Uranus Trine Neptune	**14** **Sagittarius** Trine Venus Trine Mars Trine Neptune	**15** **Sagittarius** Opposition Jupiter Square Ascendant	**16** **V** **Capricorn** Trine Mercury Square Saturn Square Neptune Square Ascendant	**17** **Capricorn** Trine Mercury Square Venus	**18** **Capricorn** Trine Sun Sextile Saturn Trine Uranus Sextile Neptune Conjunction Pluto
19 **Aquarius** Square Mercury Sextile Venus Opposition Mars	**20** **Aquarius** Square Sun Trine Jupiter Square Uranus	**21** **3Q** **Pisces**	**22** **Pisces** Sextile Mercury Square Jupiter Conjunct Saturn Sextile Uranus Conjunct Neptune	**23** **XhN** **Aries** Conjunct Venus Trine Mars	**24** **Aries** Conjunct Venus Sextile Jupiter	**25** **Taurus** Square Mars
26 **P5** **Taurus** Conjunct Mercury Sextile Saturn Conjunct Saturn Trine Pluto Square Ascendant	**27** **NM** **Gemini** Conjunct Sun Square Ascendant	**28** **Gemini** Conjunct Sun	**29** ^ **Gemini** Conjunct Jupiter Square Saturn Square Neptune Square Ascendant	**30** **Cancer** Square Ascendant	**31** **Cancer** Square Venus Trine Saturn Sextile Uranus trine Neptune Opposition Pluto Square Ascendant	1
2	3					

NORTH ISLAND RAINFALL

MAY 2025

d=mainly dry, n= fog, mist, dew or nonrecordable, l=light showers, p=passing showers, s=significant showers, r=rain, h=heavy falls

		Northland							Auckland					Bay Of Plenty				Waikato				Rot/Tau		Hawkes Bay			C Plateau			Taranaki			Manawatu			Lower North Island					
		CR	Kta	Ker	Rsl	Dar	Whg	Man	Wark	Hen	kH	Ak	Co	Tau	TeP	Wkt	Opo	Hml	TeA	Mata	Ot	Rt	Taup	Gis	Nap	Has	Taum	Chat	Waio	NP	Str	Wang	Pa	Dann	Lev	PN	Mast	UH	Wgt	all	Moon
1st	Thur	d	d	d	d	d	d	d	d	d	d	d	d	d	d	d	d	d	d	d	d	d	d	d	d	d	d	d	d	d	d	d	d	d	d	d	d	d	d	0	^
2nd	Fri	d	d	d	d	d	d	d	d	d	d	d	d	d	d	d	d	d	d	d	d	d	d	d	d	d	d	d	d	d	d	d	d	d	d	d	d	d	d	0	
3rd	Sat	d	d	d	n	d	d	d	d	d	d	d	h	s	d	d	d	d	d	l	d	d	d	n	d	d	d	d	d	d	d	d	d	d	d	d	d	d	d	43	
4th	Sun	d	d	l	d	d	l	d	d	d	d	d	r	l	l	l	l	d	d	l	d	d	d	d	d	d	d	d	d	d	d	d	d	d	d	d	d	d	d	20	
5th	Mon	d	r	d	d	s	d	d	d	d	d	d	d	d	d	d	d	d	d	d	d	d	d	d	d	d	d	d	d	d	d	d	d	d	d	d	d	d	d	15	1Q
6th	Tue	d	d	d	d	d	d	d	d	d	d	d	d	d	d	d	d	d	d	d	l	d	d	d	d	d	d	d	d	d	d	d	d	d	d	d	d	d	d	1	
7th	Wed	l	d	d	d	d	d	d	d	d	d	d	d	d	d	d	d	n	l	d	l	d	d	d	d	d	d	d	d	l	d	d	l	d	l	d	d	l	s	15	
8th	Thur	h	h	h	h	r	s	s	r	h	h	h	r	h	r	r	s	r	h	r	h	s	s	l	l	s	s	s	s	h	h	r	l	l	l	l	l	s	s	421	KhS
9th	Fri	l	l	l	l	s	d	n	s	r	s	s	s	r	s	h	h	l	l	s	s	l	r	r	s	s	s	s	r	s	l	s	l	s	l	l	r	s	r	236	
10th	Sat	l	l	d	l	s	d	d	l	l	l	l	h	s	h	l	l	n	l	d	l	s	n	l	d	d	l	d	d	l	l	l	l	d	l	d	d	d	d	94	
11th	Sun	s	r	s	l	s	l	l	l	r	h	r	r	h	h	r	h	h	h	h	h	h	h	s	s	s	h	h	s	h	h	s	r	l	s	s	l	s	s	664	A
12th	Mon	d	s	d	d	d	d	l	d	d	n	d	l	l	d	d	l	d	l	l	s	l	d	n	d	d	l	l	d	l	r	l	s	d	l	d	d	s	h	103	
13th	Tue	s	s	d	d	s	d	n	s	s	r	l	d	r	s	d	d	l	s	s	s	s	s	d	d	d	s	h	l	l	d	l	l	s	r	s	s	r	s	180	F
14th	Wed	d	d	d	d	d	d	d	d	l	d	d	d	d	d	d	d	d	l	d	d	d	d	d	d	d	d	d	d	d	d	l	d	d	d	d	l	d	d	5	
15th	Thur	d	d	d	d	d	d	d	d	d	d	d	d	d	d	d	d	d	d	d	d	d	d	d	l	l	d	d	d	d	d	d	d	d	d	d	d	d	d	5	
16th	Fri	d	d	d	d	d	d	d	d	d	d	d	d	d	d	d	d	d	d	d	d	d	d	d	d	d	d	d	d	d	d	d	l	d	d	d	d	d	d	1	V
17th	Sat	d	d	d	d	d	d	d	d	d	d	d	d	d	d	d	d	d	d	d	d	d	n	d	d	d	l	l	d	d	d	l	s	d	s	l	d	l	d	18	
18th	Sun	d	d	d	d	d	d	d	d	d	d	d	d	d	d	d	d	d	d	d	d	d	d	d	d	d	d	d	d	d	d	d	d	d	d	n	d	d	d	1	
19th	Mon	d	d	l	l	d	d	d	d	d	d	d	d	d	d	d	d	l	d	d	d	d	d	l	l	d	d	d	d	d	d	d	d	d	d	d	l	d	d	11	
20th	Tue	d	d	d	d	d	d	d	d	n	d	d	d	d	d	d	d	d	d	d	d	d	d	d	d	d	d	d	d	d	d	d	d	d	d	d	d	d	d	0	
21st	Wed	d	d	d	d	d	d	d	d	d	d	d	d	d	d	d	d	d	d	d	d	d	d	d	d	d	d	r	d	s	l	l	s	l	s	d	d	s	d	45	3Q
22nd	Thur	l	l	d	d	l	d	l	l	l	l	l	l	d	d	d	d	l	l	l	s	n	n	d	d	d	s	h	l	l	l	s	l	l	l	s	d	l	l	74	
23rd	Fri	l	d	d	d	d	d	d	d	l	l	d	d	d	d	d	d	d	d	d	d	d	d	d	d	d	n	d	d	d	d	d	d	d	d	d	l	d	d	4	KhN
24th	Sat	n	d	d	d	d	d	d	d	d	l	d	d	d	d	d	d	d	d	d	l	d	d	d	d	d	d	d	d	d	d	d	d	d	d	d	d	d	d	3	
25th	Sun	s	d	d	d	n	d	d	d	n	l	d	d	d	d	d	d	d	d	d	d	d	d	d	d	d	d	d	d	d	d	d	d	d	d	d	d	d	d	6	
26th	Mon	d	d	d	d	d	d	d	d	l	n	l	d	d	d	d	d	d	n	d	d	d	d	d	d	d	n	d	d	n	d	d	d	d	d	d	d	d	d	3	P5
27th	Tue	l	d	d	d	l	d	l	l	l	l	d	l	d	d	d	d	l	l	l	s	d	d	d	d	d	l	s	l	l	d	l	d	l	d	d	d	d	d	28	N
28th	Wed	d	d	d	d	d	d	d	d	l	l	l	d	d	d	d	d	d	d	d	d	d	d	d	d	d	d	d	l	d	d	d	d	d	d	d	d	d	d	4	
29th	Thur	d	d	d	d	d	d	d	d	d	d	d	d	d	d	d	d	d	d	d	d	d	d	d	d	d	d	d	d	d	d	d	d	d	d	d	d	d	d	0	^
30th	Fri	d	d	d	d	d	d	d	d	d	d	l	d	l	s	s	d	l	n	d	l	s	d	d	d	d	l	l	d	d	d	d	d	d	d	d	d	d	d	22	
31st	Sat	d	l	l	r	d	d	d	d	l	l	d	d	l	l	l	d	n	n	l	d	l	l	d	d	d	l	h	d	r	h	l	s	d	d	l	d	h	l	105	
estimate:		43	55	36	45	44	10	14	27	59	81	46	111	99	118	59	76	64	89	58	96	64	63	20	17	17	61	136	31	67	81	37	41	19	42	17	24	67	88	2124	
average:		88	140	140	147	122	215	173	131	100	91	112	105	122	142	125	102	120	100	109	138	137	120	125	87	87	137	274	101	150	186	77	99	93	104	87	102	133	120	4741	
trend:		drr	drr	drr	drr	drr	drr	drr	drr	drr	av	drr	av	drr	drr	drr	drr	drr	drr	drr	drr	drr	drr	drr	drr	drr	drr	drr	drr	drr	drr	drr	drr	drr	drr	drr	drr	drr	drr	drr	

SOUTH ISLAND RAIN

MAY 2025

d=mainly dry, n=fog, mist, dew or nonrecordable, l=light showers, p=passing showers, s=significant showers, r=rain, h=heavy falls

		West Coast				Top South Island				Canterbury				South Canterbury				Coastal Otago			Lakes			Inland Otago				Southland						
		West	Grey	Hok	BB	Ta	Mou	Nel	Ble	Dar	Ham	Kai	Ch	Rak	Meth	Ash	Tim	Oam	Dun	Bal	Twi	LP	LT	Crom	Qstn	Ran	Alex	TeA	Wtn	Gore	Inv	Puy Pt	all	Moon
1st	Thur	d	d	d	d	d	d	d	d	d	d	d	d	d	d	d	d	d	l	l	d	d	d	d	d	d	d	d	d	d	l	l	4	^
2nd	Fri	d	l	d	d	d	d	d	d	d	d	l	d	d	l	n	l	l	d	d	d	d	d	d	d	d	d	d	d	n	d	d	8	
3rd	Sat	d	l	l	d	d	d	d	d	d	d	d	d	d	n	n	d	d	d	d	d	d	d	d	d	d	d	d	d	d	d	d	3	
4th	Sun	d	d	d	d	d	d	d	d	d	d	d	d	d	d	d	l	l	s	s	d	d	d	l	s	l	l	h	r	r	s	r	94	
5th	Mon	d	d	d	d	d	d	d	d	d	d	d	d	d	d	d	d	l	d	d	d	d	d	s	s	l	l	s	l	d	d	l	32	1Q
6th	Tue	d	h	l	d	d	d	d	d	d	d	d	d	d	n	d	d	l	l	s	d	d	d	d	d	l	d	l	r	s	l	l	48	
7th	Wed	s	h	r	l	h	h	r	l	s	s	s	s	s	s	s	l	l	l	l	d	d	d	d	d	d	l	l	d	d	d	n	211	
8th	Thur	h	l	d	d	h	h	r	s	s	r	r	s	l	l	l	l	l	l	d	d	d	d	d	d	l	l	l	d	d	d	d	130	XhS
9th	Fri	l	d	d	d	l	s	s	s	s	s	s	s	l	l	l	d	d	l	l	d	d	d	d	d	d	d	d	d	d	d	l	48	
10th	Sat	d	d	d	d	d	d	d	d	d	d	d	d	d	d	d	d	d	d	d	d	d	d	d	d	d	d	d	d	d	d	d	1	
11th	Sun	r	r	h	r	h	h	r	r	s	s	s	s	r	s	s	s	r	n	d	d	d	s	l	l	d	l	s	d	d	d	d	226	A
12th	Mon	h	r	h	h	h	h	h	h	r	l	l	s	h	r	r	r	s	l	l	s	r	s	l	l	d	l	l	d	d	n	l	309	
13th	Tue	r	l	l	s	s	l	h	h	h	h	l	r	h	h	h	h	r	l	l	l	s	r	l	l	s	l	d	d	d	d	d	296	F
14th	Wed	l	l	d	d	d	d	d	l	d	d	d	d	d	d	l	d	d	d	d	s	h	d	d	d	d	d	d	d	d	d	d	35	
15th	Thur	d	l	s	h	d	d	d	d	d	d	d	d	d	d	d	d	d	d	d	d	d	d	d	s	d	d	s	d	d	l	s	52	
16th	Fri	r	s	r	r	d	d	d	d	d	d	d	d	d	d	d	d	d	d	d	l	d	d	n	l	d	d	l	l	d	l	r	59	V
17th	Sat	l	s	d	s	d	d	d	d	d	d	d	d	d	d	d	d	d	d	l	d	d	d	d	l	d	l	s	s	s	l	h	66	
18th	Sun	d	d	d	d	d	d	d	d	d	d	d	d	d	d	d	d	d	d	n	s	d	d	d	d	d	d	d	d	d	r	s	21	
19th	Mon	d	d	d	d	d	d	d	d	d	d	d	d	d	d	d	d	d	d	d	d	d	d	d	d	d	d	d	d	d	l	r	12	
20th	Tue	r	h	h	h	s	d	d	d	d	d	d	d	d	l	l	d	d	n	l	d	d	s	s	r	l	l	r	l	l	r	h	218	
21st	Wed	l	s	s	s	l	l	n	d	d	d	d	d	d	d	d	d	d	d	l	h	d	d	r	r	d	l	r	s	l	r	h	137	3Q
22nd	Thur	r	r	h	h	s	l	s	l	l	l	d	l	d	l	l	d	d	l	l	s	r	r	d	s	l	l	d	r	r	r	s	176	
23rd	Fri	d	d	d	d	d	d	d	d	d	d	d	d	d	d	d	d	d	d	d	r	d	d	d	d	d	d	d	d	d	l	s	21	XhN
24th	Sat	d	d	d	d	d	d	d	d	d	d	d	d	d	d	d	d	d	d	d	d	d	d	d	d	d	d	d	d	d	d	s	8	
25th	Sun	l	s	l	s	d	d	d	d	d	d	d	d	d	l	d	d	d	d	d	d	d	d	l	d	d	d	l	d	d	l	l	21	
26th	Mon	s	s	s	l	d	d	d	d	d	d	d	d	d	d	d	d	d	d	d	d	s	d	d	d	d	d	s	l	n	r	h	68	P5
27th	Tue	l	l	l	l	d	d	d	d	d	d	d	d	d	d	d	d	d	d	d	d	d	d	d	d	d	d	d	d	d	s	r	28	N
28th	Wed	l	l	n	d	d	d	d	d	d	d	d	d	d	d	d	d	d	d	d	d	d	d	d	d	d	d	d	d	d	l	s	12	
29th	Thur	d	d	d	d	d	d	d	d	d	d	d	d	d	d	d	d	l	d	d	d	d	d	d	d	d	d	d	d	n	l	l	6	^
30th	Fri	d	d	l	s	l	d	d	d	d	d	d	d	d	d	d	d	d	d	d	d	d	d	d	n	d	d	s	d	d	d	l	15	
31st	Sat	h	h	h	h	h	l	l	d	d	l	d	d	d	d	d	d	d	d	l	l	d	l	l	s	d	l	s	d	d	d	h	206	
estimate:		**149**	**187**	**221**	**206**	**187**	**112**	**98**	**80**	**63**	**58**	**35**	**43**	**70**	**70**	**54**	**52**	**43**	**15**	**23**	**62**	**56**	**42**	**33**	**60**	**12**	**17**	**109**	**48**	**41**	**76**	**242**	**2566**	
average:		162	198	242	357	184	136	80	72	66	99	84	67	69	71	63	42	42	80	66	55	73	53	38	65	33	22	105	96	89	107	204	3120	
trend:		drr	drr	drr	drr	av	drr	w tr	av	av	drr	drr	drr	av	av	av	av	av	drr	drr	av	drr	drr	av	av	drr	av	av	drr	drr	drr	w tr	drr	

NORTH ISLAND SUNSHINE ESTIMATES

MAY 2025

F=fine (8-12 hours of sunshine), pc= partly cloudy (4-7 hours), c=cloudy (1-3 hours), o=overcast (0 hours)

		Kta	Whg	Dar	Ak	Hml	Tau	Rt	Taum	NP	Wang	Taup	Gis	Nap	Wai	Dann	PN	Mast	Pa	UH	Wgt	ALL	Mocn
1st	Thur	pc	F	F	F	F	F	F	F	F	pc	F	c	pc	pc	pc	pc	pc	F	F	F	152	^
2nd	Fri	pc	pc	F	F	F	F	F	F	F	pc	F	c	pc	pc	c	pc	c	pc	c	pc	116	
3rd	Sat	F	c	F	F	c	o	o	pc	F	F	o	F	F	F	pc	F	pc	F	F	pc	131	
4th	Sun	pc	F	F	F	F	F	F	F	F	pc	F	F	F	F	pc	pc	pc	pc	pc	pc	154	
5th	Mon	pc	pc	pc	pc	pc	F	F	F	F	F	F	F	F	pc	c	F	c	c	c	pc	131	1Q
6th	Tue	F	F	F	F	pc	c	pc	F	pc	pc	pc	F	F	pc	F	pc	F	c	c	o	126	
7th	Wed	c	pc	pc	pc	pc	pc	pc	pc	c	pc	pc	F	F	pc	pc	pc	pc	c	c	c	95	
8th	Thur	pc	pc	pc	pc	pc	pc	pc	o	pc	o	pc	o	o	o	o	o	o	o	o	o	45	XhS
9th	Fri	pc	c	c	c	c	c	F	F	F	F	F	c	o	o	c	F	c	F	c	F	90	
10th	Sat	o	o	o	o	o	o	o	o	o	o	o	o	c	o	c	o	c	o	o	o	10	
11th	Sun	pc	pc	pc	pc	pc	pc	pc	pc	pc	pc	pc	pc	pc	F	pc	pc	pc	pc	c	pc	123	A
12th	Mon	pc	pc	pc	pc	c	c	c	c	c	c	c	pc	c	pc	pc	c	pc	pc	c	c	65	
13th	Tue	F	pc	F	F	pc	F	pc	pc	F	o	pc	c	o	o	pc	o	pc	pc	o	o	86	F
14th	Wed	pc	pc	pc	pc	pc	F	F	F	F	F	F	pc	o	o	pc	F	pc	F	F	F	141	
15th	Thur	pc	pc	pc	pc	pc	F	pc	c	F	c	pc	F	F	pc	pc	c	pc	pc	pc	pc	108	
16th	Fri	F	F	F	F	pc	F	F	pc	o	o	F	pc	pc	pc	pc	o	pc	o	o	o	111	V
17th	Sat	F	F	F	F	F	F	pc	c	pc	c	pc	F	pc	o	pc	c	pc	pc	pc	c	111	
18th	Sun	pc	F	F	F	F	F	F	F	F	F	F	pc	pc	pc	c	F	c	F	F	F	158	
19th	Mon	pc	pc	pc	pc	F	F	F	F	pc	F	F	pc	pc	pc	o	F	o	pc	pc	c	114	
20th	Tue	o	c	o	o	c	c	c	o	o	o	c	pc	c	o	pc	o	pc	o	o	o	29	
21st	Wed	pc	c	pc	pc	pc	pc	c	o	c	c	c	F	pc	pc	pc	c	pc	c	c	pc	81	3Q
22nd	Thur	pc	pc	pc	pc	F	F	F	pc	F	pc	F	pc	pc	pc	pc	pc	pc	F	pc	pc	128	
23rd	Fri	pc	c	pc	pc	pc	F	F	F	pc	o	F	F	F	F	pc	o	pc	pc	pc	pc	123	XhN
24th	Sat	c	c	c	c	c	c	pc	c	c	c	pc	pc	pc	pc	pc	c	pc	c	pc	o	67	
25th	Sun	o	o	c	c	pc	pc	c	pc	pc	pc	c	pc	pc	c	pc	pc	pc	pc	pc	pc	90	
26th	Mon	c	c	pc	pc	pc	pc	pc	o	pc	c	pc	F	F	pc	F	c	F	c	pc	pc	96	P5
27th	Tue	c	pc	pc	pc	pc	pc	F	c	F	c	F	F	pc	pc	F	c	F	pc	F	pc	123	N
28th	Wed	c	c	c	c	pc	F	F	F	F	c	F	F	F	F	pc	c	pc	F	F	pc	128	
29th	Thur	o	o	F	F	pc	pc	F	F	F	F	F	pc	F	F	o	F	o	F	F	F	128	^
30th	Fri	o	pc	c	c	o	o	o	o	o	o	o	o	pc	c	o	o	o	o	o	o	18	
31st	Sat	c	c	o	o	o	o	o	o	c	o	o	o	c	c	pc	o	pc	c	c	o	28	
est. hours:		152	143	166	166	166	189	178	162	181	123	178	175	170	147	140	123	140	147	128	134	3107	
av. hours:		149	140	125	142	131	164	148	92	144	124	129	148	153	123	116	105	120	130	129	128	2640	
trend:		av	av	more	more	more	more	more	more	more	av	more	more	more	more	more	more	more	more	av	av	more	

SOUTH ISLAND SUNSHINE ESTIMATES
MAY 2025

F=fine (8-12 hours of sunshine), pc= partly cloudy (4-7 hours), c=cloudy (1-3 hours), o=overcast (0 hours)

		Ta	Nel	Ble	Mou	West	Grey	Hok	Ham	Kai	Ch	Ash	Tim	Crom	Qstn	Bal	Dun	Wnt	Gore	Inv	Alex	ALL	Moon
1st	Thur	F	F	F	F	F	F	F	o	o	o	c	o	F	c	o	o	c	c	o	F	95	^
2nd	Fri	F	pc	o	pc	c	o	c	o	o	o	o	o	F	pc	o	o	c	c	pc	F	59	
3rd	Sat	c	pc	F	F	F	c	c	F	F	F	pc	pc	pc	pc	pc	pc	pc	pc	pc	pc	115	
4th	Sun	o	F	F	F	F	pc	pc	pc	pc	c	c	o	o	o	o	o	o	o	o	o	69	
5th	Mon	o	F	F	pc	pc	c	c	pc	pc	pc	pc	c	c	c	o	c	o	o	o	c	66	1Q
6th	Tue	c	o	c	c	o	o	o	o	o	o	o	o	c	o	o	o	o	o	o	c	10	
7th	Wed	o	o	o	o	o	c	c	o	o	o	o	c	o	o	o	o	o	o	o	o	8	
8th	Thur	pc	o	o	o	o	c	o	o	o	o	o	c	pc	pc	c	c	c	c	o	pc	37	XhS
9th	Fri	c	F	pc	c	c	F	F	pc	pc	pc	pc	pc	F	pc	F	F	F	F	F	F	137	
10th	Sat	o	o	o	o	o	c	c	pc	pc	c	c	c	pc	pc	c	c	pc	pc	pc	pc	54	
11th	Sun	F	pc	F	pc	pc	c	o	o	o	o	o	o	pc	o	o	o	o	o	o	pc	46	A
12th	Mon	pc	c	pc	c	o	o	o	o	o	o	o	o	o	o	o	o	o	o	o	o	15	
13th	Tue	pc	c	pc	o	o	F	F	c	c	c	c	o	pc	c	pc	pc	pc	pc	pc	pc	76	F
14th	Wed	F	F	F	F	pc	o	c	F	F	pc	F	pc	c	o	c	pc	c	c	pc	c	106	
15th	Thur	F	pc	F	pc	o	c	o	pc	pc	pc	pc	pc	c	c	c	pc	c	c	c	c	75	
16th	Fri	c	pc	pc	pc	pc	pc	pc	pc	pc	c	pc	pc	pc	pc	F	F	pc	pc	pc	pc	110	V
17th	Sat	c	pc	o	pc	c	pc	pc	pc	pc	pc	F	F	F	pc	F	F	pc	pc	F	F	119	
18th	Sun	pc	F	F	F	F	F	F	F	F	F	F	pc	F	pc	pc	F	F	F	F	F	166	
19th	Mon	F	c	c	c	pc	c	c	o	o	o	o	c	pc	c	o	o	o	o	o	pc	46	
20th	Tue	F	c	c	c	c	c	pc	pc	pc	pc	F	pc	pc	pc	pc	pc	c	c	pc	pc	98	
21st	Wed	pc	pc	pc	pc	o	c	o	c	c	c	pc	pc	pc	pc	c	c	c	c	c	pc	69	3Q
22nd	Thur	F	F	F	F	pc	F	F	F	F	F	F	F	F	pc	pc	pc	pc	pc	pc	F	150	
23rd	Fri	pc	F	F	F	F	pc	pc	F	F	F	F	F	F	pc	pc	F	pc	pc	pc	F	150	XhN
24th	Sat	F	c	c	c	pc	o	c	pc	pc	pc	c	pc	pc	pc	pc	c	c	c	o	pc	69	
25th	Sun	F	F	pc	F	c	c	pc	pc	pc	c	pc	pc	pc	pc	F	F	F	F	pc	pc	119	
26th	Mon	F	o	pc	c	o	o	o	F	F	F	F	pc	pc	pc	pc	c	c	c	c	pc	94	P5
27th	Tue	o	pc	F	pc	o	o	o	F	F	F	F	pc	F	pc	c	c	pc	pc	c	F	98	N
28th	Wed	F	F	F	F	F	pc	pc	F	F	pc	F	c	pc	pc	c	pc	pc	pc	pc	pc	119	
29th	Thur	pc	F	pc	c	F	o	o	pc	pc	pc	F	c	pc	pc	pc	F	pc	pc	pc	pc	106	^
30th	Fri	pc	o	c	pc	o	o	o	pc	pc	c	pc	pc	pc	c	pc	pc	pc	pc	F	pc	78	
31st	Sat	pc	c	c	F	c	pc	c	o	o	o	pc	o	F	pc	pc	pc	F	F	pc	F	90	
est. hours:		162	156	165	146	113	93	109	143	143	127	154	121	172	107	113	116	114	114	114	172	2651	
av. hours:		164	173	172	165	113	100	116	111	139	142	132	131	122	91	93	100	93	85	80	108	2430	
trend:		av	av	av	less	av	av	av	more	av	av	more	av	more	more	more	more	more	more	more	more	average	

MAY **1** THURSDAY

Anticyclonic conditions prevail over New Zealand bringing mostly dry, calm, and sunny conditions throughout the North Island, while a weak trough between the two high pressure centres crosses the South Island. This trough brings cloudy and mostly dry conditions throughout the South Island, except for the chance of some drizzle patches in Otago and fog in Canterbury. Chance of frosts in the Waikato to the Bay of Plenty and up into the Central Plateau of the North Island.

Auckland:	Sunny, dry, mild, mostly calm air, chance of overnight fog patches
Hamilton:	Calm and serene air, very sunny, dry, cold overnight, chance of fog
Wellington:	Sunny, dry, mild day cool night, light breezes to mostly calm air
Greymouth:	Sunny, dry, light breezes to mostly calm air
Christchurch:	Overcast, mild day cold night, dry, mostly calm air with chance of fog
Dunedin:	Calm and serene air, overcast, light showers

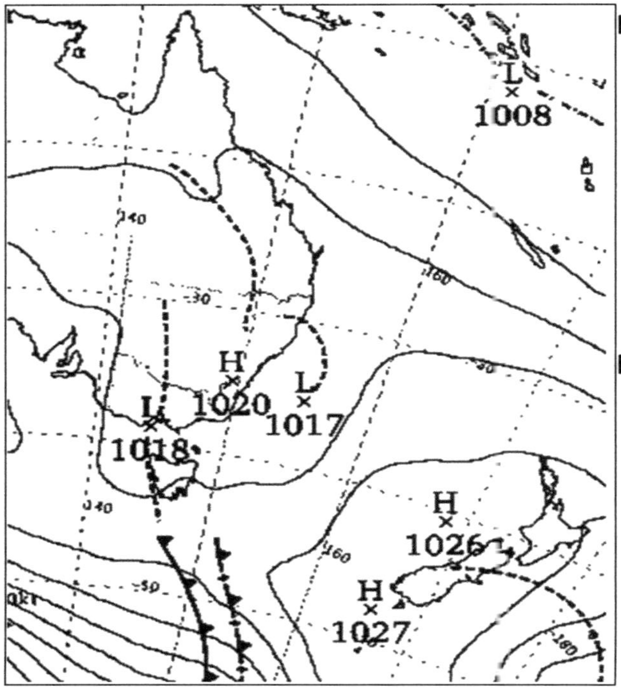

RAIN POTENTIAL

FROST/SNOW

MAY **2** FRIDAY

High pressure builds over New Zealand as it directs a mostly calm to light east to northeast flow over the country. Dry with milder temperatures almost everywhere, mostly sunny in the North Island and increased cloudy spells in the South Island. Chance of some isolated showers around South Canterbury and fog around Christchurch, as well as fog over the north of the North Island from Auckland to the Waikato.

Auckland:	Dry, mostly sunny, serene air, shallow fog likely
Hamilton:	Mostly sunny, dry, mild day cold night, calm atmosphere, shallow fog
Wellington:	Cloudy, calm air, dry, cooler
Greymouth:	Overcast, isolated showers, calm atmosphere
Christchurch:	Mostly overcast and dry, light breezes, overnight fog likely
Dunedin:	Overcast, cooler, mostly dry, odd threat of rain, calm atmosphere

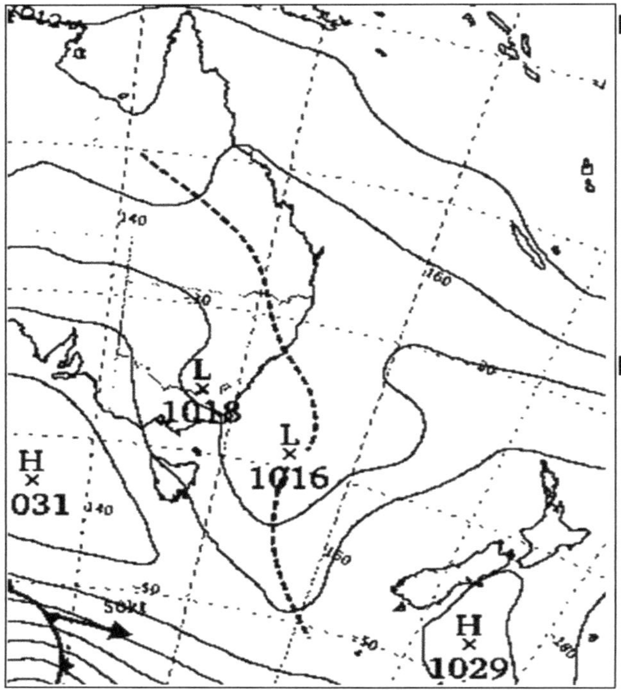

RAIN POTENTIAL

FROST/SNOW

MAY **3** SATURDAY

A large high pressure system continues to linger over New Zealand directing a mostly calm to light east to northerly air throughout. Chance of some isolated showers around the Coromandel to the Bay of Plenty, while fog possible Auckland to the Waikato over the next couple of days and in the east of the South Island. Dry and mostly sunny elsewhere. Cold overnight, particularly in the south of the country where frosts may be noted from Canterbury to Otago.

Auckland: Sunny, mostly dry, calm atmosphere, chance of overnight rain
Hamilton: Cloudy, warmer, dry, calm and serene air, chance of fog
Wellington: Partly cloudy, dry, calm atmosphere
Greymouth: Cloudy, isolated showers, mostly calm air
Christchurch: Mostly sunny, cool and dry, calm and serene air, possible fog patches
Dunedin: Partly cloudy, dry, calm and pleasant air

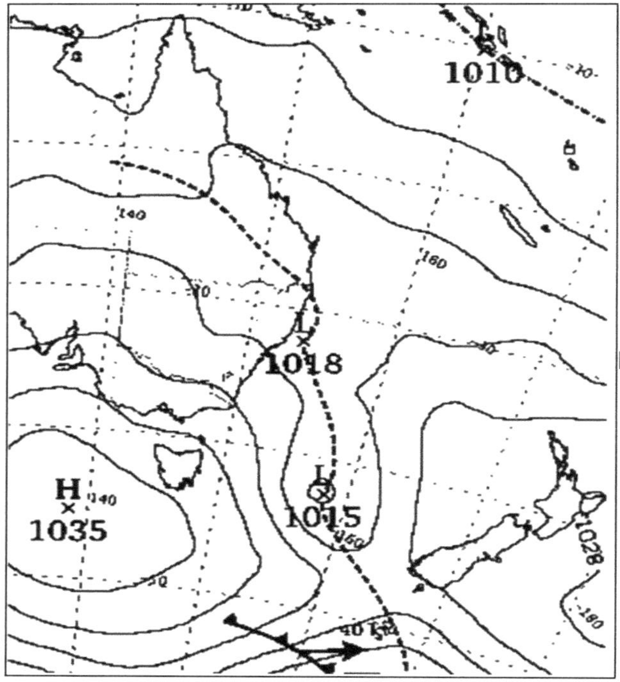

RAIN POTENTIAL

FROST/SNOW

MAY **4** SUNDAY

high pressure intensifies over the North Island with light shifting winds prevailing, mostly sunny skies and some isolated overnight showers around the Bay of Plenty, the rest of the North Island will be mostly dry. However, a trough behind the high and ahead of another high pressure ridge brings some rain through the south of the South Island with an easterly change. Colder temperatures in Otago may see wintry showers or snow in the higher ranges.

Auckland: Sunny, mostly dry, unsettled air, chance of overnight rain or fog
Hamilton: Sunny, cooler, dry, light shifting breezes, possible fog patches
Wellington: Warmer, dry, occasional cloudy spells, fluctuating breezes
Greymouth: Partly cloudy, dry, light breezes
Christchurch: Cloudy, dry, mild day cold night, gentle breezes
Dunedin: Warmer, overcast, light showers, gentle shifting breezes

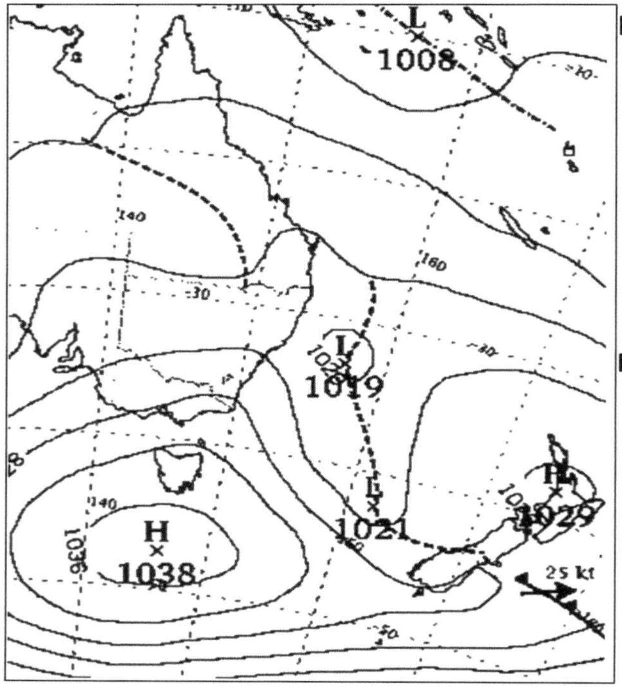

RAIN POTENTIAL

FROST/SNOW

MAY **5** MONDAY

High pressure continues to prevail over New Zealand directing a very light westerly to mostly calm air throughout the country with chance of fog overnight in Auckland to the Waikato. Mostly sunny in the North Island and central districts of the South Island, with increased cloudy spells elsewhere. A low pressure trough between two high pressure centres bring some light showers into Southland and Otago and chance of some isolated showers in Northland. Dry elsewhere with chance of frost pockets in the Waikato.

Auckland: Partly cloudy, dry, calm and serene air, fog likely
Hamilton: Mostly sunny, dry, warm day cold night, calm atmosphere, fog likely
Wellington: Partly cloudy, dry, calm atmosphere
Greymouth: Cloudy, calm air, dry, increasing threats of rain, warm day cool night
Christchurch: Partly cloudy, dry, calm and serene air, chance of shallow fog patches
Dunedin: Cloudy, cooler, dry, calm and pleasant air

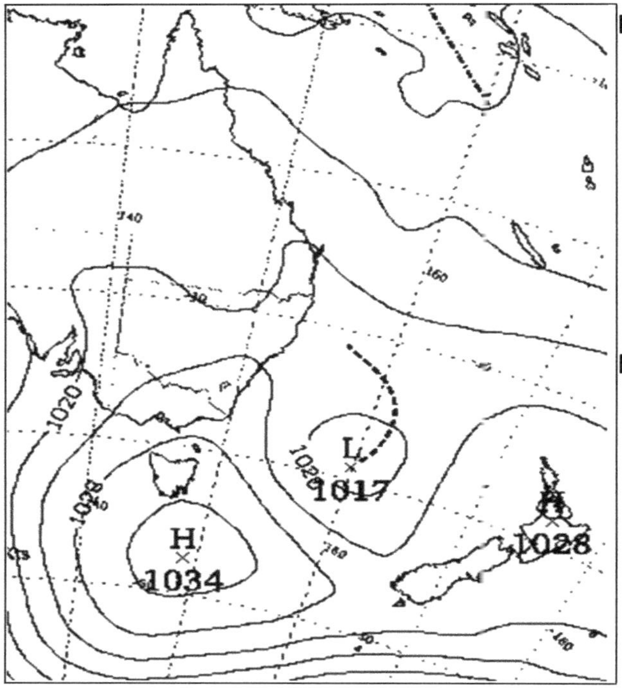

RAIN POTENTIAL

FROST/SNOW

MAY **6** TUESDAY

High pressure continues to dominate the country directing a light easterly air over the North Island, backing northwesterly over the South Island. A trough between the two high pressure centres spreads widespread dull and gloomy skies over the South Island, with some spotty showers in Southland and Otago. Sunnier and dry in the North Island, with chance of overnight fog patches continuing in the Auckland to Waikato regions.

Auckland: Gentle breezes to mostly calm air, sunny, dry
Hamilton: Partly cloudy and dry, warm day cool night, gentle breezes
Wellington: Overcast, shifting breezes, dry with increasing threats of rain
Greymouth: Overcast, heavy rain, light breezes
Christchurch: Overcast, warmer, dry with building threats of rain, mostly calm air
Dunedin: Overcast, spotty showers, gentle shifting breezes

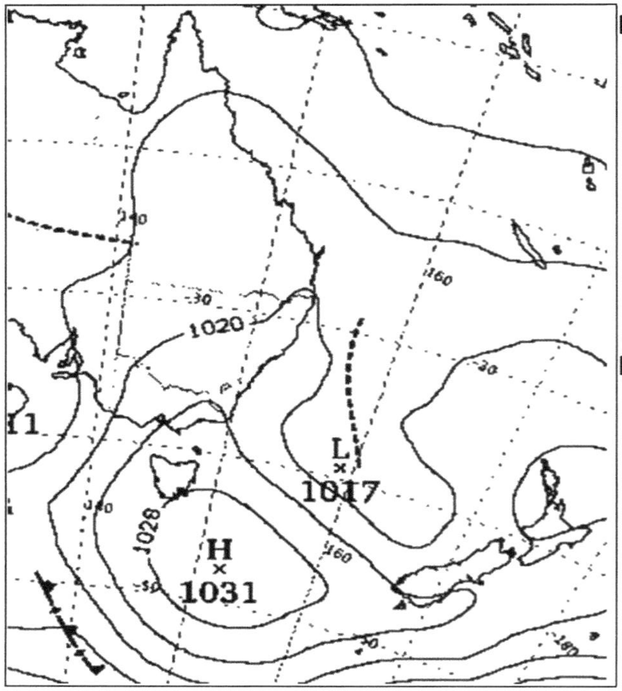

RAIN POTENTIAL

FROST/SNOW

MAY 7 WEDNESDAY

A low pressure trough continues to bring widespread cloud throughout the South Island and expands showers through the northern half of the South Island, while clearing in the south. Low pressure also moves onto the North Island with chance of isolated showers appearing overnight from Taranaki to Wellington, but remaining mostly dry and sunny elsewhere. Winds generally moderate in a westerly flow over the southern half of the South Island, backing northeasterly over the North Island.

Auckland: Gentle breezes, partly cloudy, dry with increasing threats of rain
Hamilton: Scattered cloud, brief showers with rain threatening, light breezes
Wellington: Cloudy, scattered showers, moderate breezes
Greymouth: Mostly cloudy, cooler, heavy rain, gentle breezes
Christchurch: Overcast, cooler, drizzle patches and showers, mostly calm air
Dunedin: Calm atmosphere, cloudy, light showers

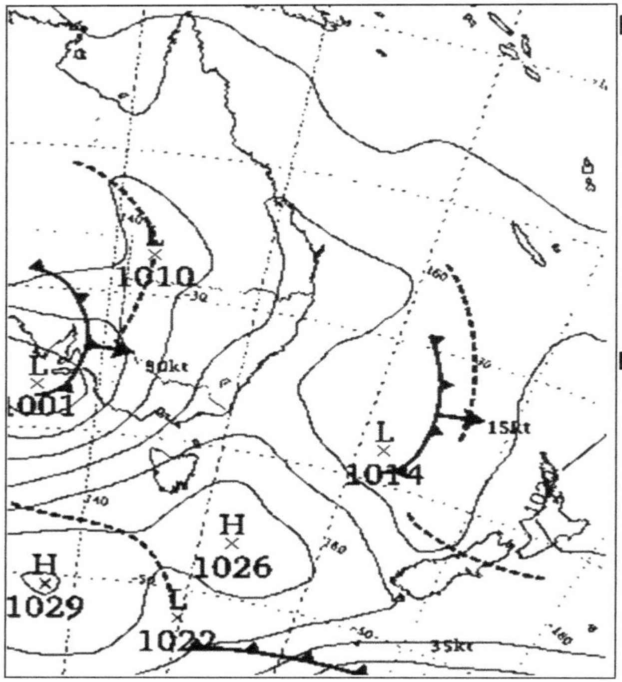

RAIN POTENTIAL

FROST/SNOW

MAY 8 THURSDAY

A weak depression stabilises over the country with widespread rain activity in most districts, more isolated in the south of the South Island, while heavy in the north of the South Island, and in parts of the North Island from Taranaki to Northland, as well as in the Bay of Plenty. Fresh westerly breezes over the North Island, backing easterly through the Cook Strait and over the South Island.

Auckland: Partly cloudy, rain, fluctuating breezes
Hamilton: Cloudy, rain, fluctuating breezes
Wellington: Overcast, scattered showers, fluctuating windy spells
Greymouth: Unsettled winds, cloudy, brief squalls, chance of thunderstorms
Christchurch: Fluctuating breezes, dull and gloomy, showers and drizzle patches
Dunedin: Fluctuating breezes, colder, cloudy, light showers

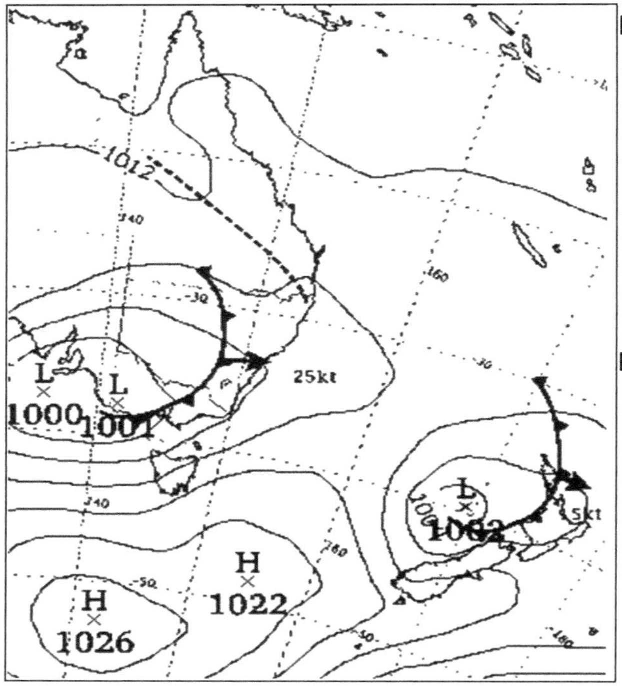

RAIN POTENTIAL

FROST/SNOW

MAY 9 FRIDAY

High pressure builds to a weak anticyclonic system centred over the country with moderate easterlies prevailing over the North Island, northerlies over the South Island. Mostly dry and sunny in the southern half of the South Island, with increased scattered cloud and scattered rain activity elsewhere. Chance of heavy falls at times around the East Cape, possible patches of fog in the Christchurch region, and scattered frosts in Southland and Otago.

Auckland:	Moderate breezes, cloudy, scattered showers
Hamilton:	Cloudy, light showers, gentle shifting breezes
Wellington:	Rain, occasional sunny outbreaks, cooler, fluctuating breezes
Greymouth:	Sunny, dry, mild temperatures, lighter shifting breezes
Christchurch:	Light shifting breezes, partly cloudy, morning showers likely
Dunedin:	Sunny, mild day cool night, brief morning showers, gentle breezes

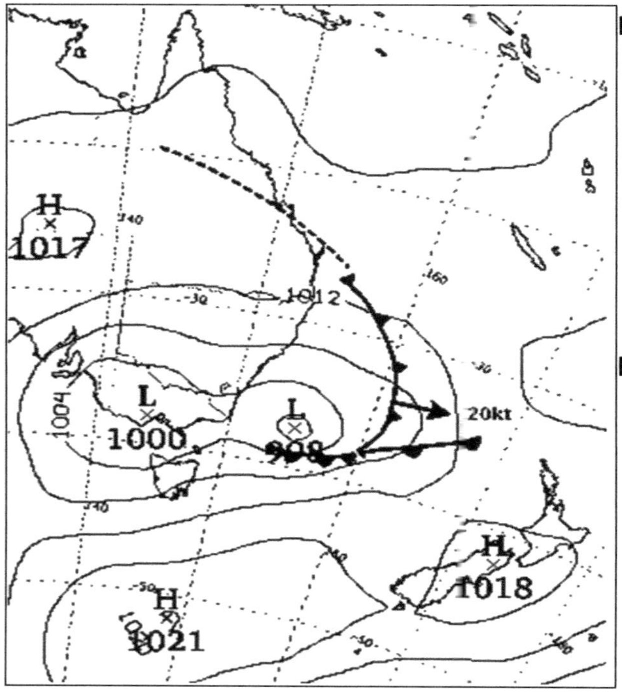

RAIN POTENTIAL

FROST/SNOW

MAY 10 SATURDAY

A complex of high and low pressure systems bring an unstable and strengthening north to northeasterly flow over the country. Dull and gloomy almost everywhere with shower activity easing as it contracts to the northern half of the North Island. Cold overnight temperatures in the Canterbury region to Southland may bring widespread frosts to many parts of the South Island.

Auckland:	Overcast, rain, moderate winds
Hamilton:	Overcast, isolated showers with rain threatening, light breezes
Wellington:	Overcast and cool, dry, passing threats of rain, moderate breezes
Greymouth:	Cloudy, dry, occasional threats of rain, moderate winds
Christchurch:	Cloudy, moderate breezes, dry, mild day cold night
Dunedin:	Cloudy, dry, moderate breezes

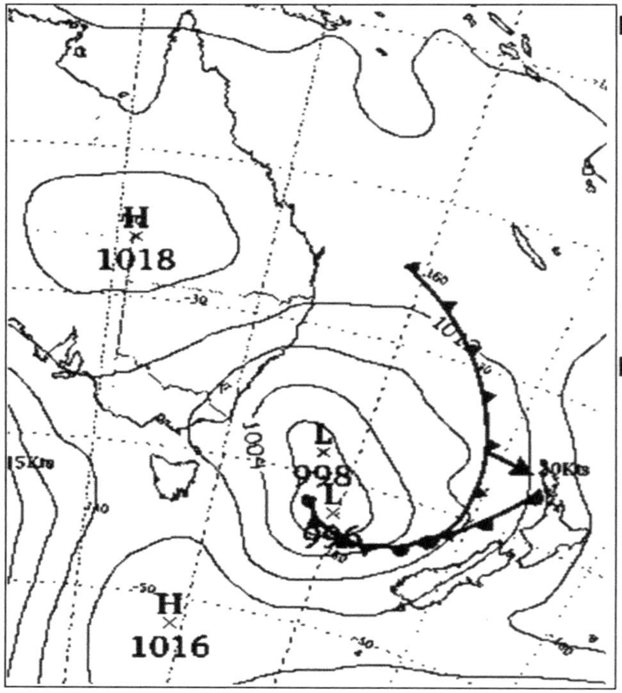

RAIN POTENTIAL

FROST/SNOW

MAY 11 SUNDAY

A deep trough between two low pressure centres crosses the country bringing widespread rain to many parts of the country, particularly heavy in central parts of the North Island, and west and north of the South Island, but more isolated to mostly dry conditions in Southland and Otago. Unstable and changeable winds in many districts, generally in a southeasterly flow over the South Island to northwesterlies over the North Island.

Auckland: Rain, some sunny outbreaks, occasional breezy spells
Hamilton: Heavy rain, some sunny spells, light breezes
Wellington: Changeable, mix of showers and sunny spells, moderate breezes
Greymouth: Cloudy, rain, windy spells
Christchurch: Overcast, rain, occasional breezy spells
Dunedin: Mostly calm air, overcast, isolated showers

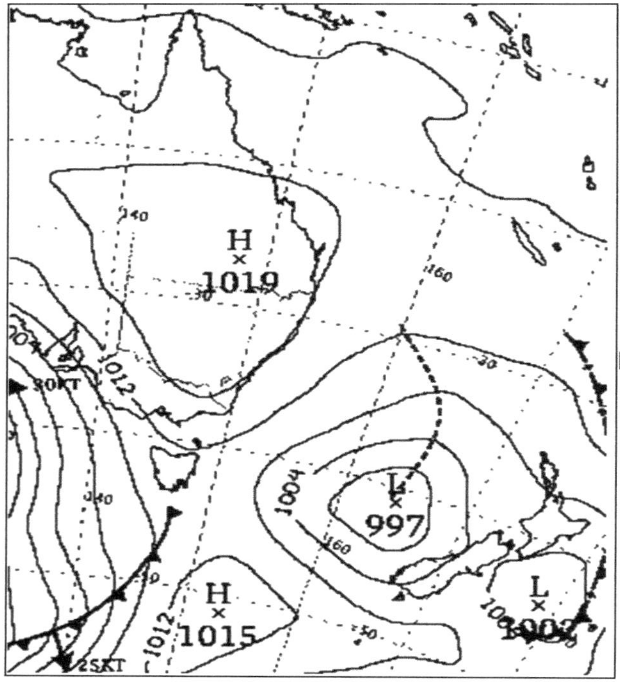

RAIN POTENTIAL

FROST/SNOW

MAY 12 MONDAY

A depression stalls over New Zealand bringing widespread stormy conditions throughout most of the South Island, while clearing in the North Island, with lingering showers through western parts from Auckland to Wellington. Warmer in the North Island and mainly dry in the north and east, and mostly dry in Southland. Unsettled westerly winds prevail over the North Island, backing easterly over the South Island.

Auckland: Cloudy, mostly dry, chance of morning showers, light breezes
Hamilton: Warmer, cloudy, dry, moderate breezes
Wellington: Cloudy, heavy rain, stormy, chance of localised pockets of flooding
Greymouth: Overcast, rain, fresh winds with occasional gusts
Christchurch: Stormy air, overcast, rain, unsettled winds for a time
Dunedin: Overcast, brief showers, moderate shifting breezes

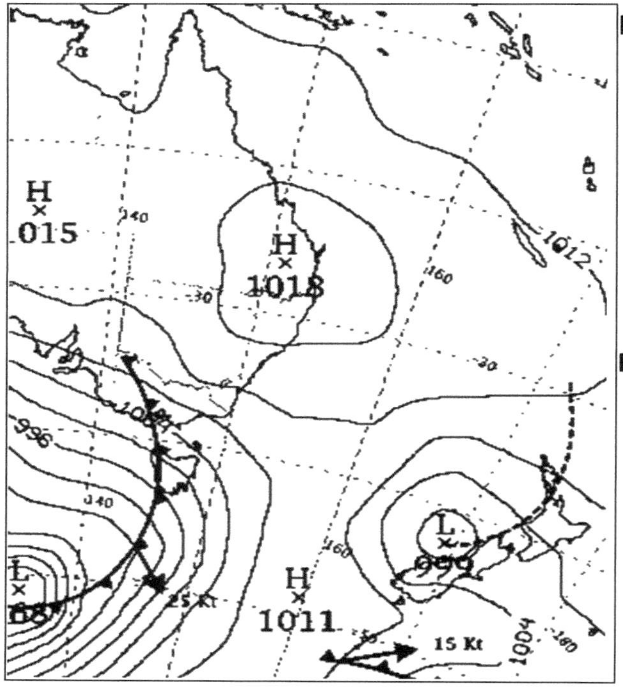

RAIN POTENTIAL

FROST/SNOW

MAY **13** TUESDAY

A low pressure trough between two weak high pressure ridges moves onto the country with winds easing and turning to a mostly light southwesterly flow. Widespread showers throughout the country, with shower activity also spreading into the east of the North Island but remaining mostly dry in Southland. Rain heavy at times in the West Coast to Canterbury, while more drizzly in the north of the North Island. Improving sunny spells.

Auckland:	Sunny day, overnight rain, gentle breezes
Hamilton:	Mostly sunny, brief passing showers, gentle breezes
Wellington:	Unsettled winds, mild temperatures, overcast, scattered showers
Greymouth:	Sunny, cooler, some light showers, winds easing to light
Christchurch:	Cloudy, rain, winds easing to light
Dunedin:	Mostly cloudy, isolated showers, cooler, light breezes

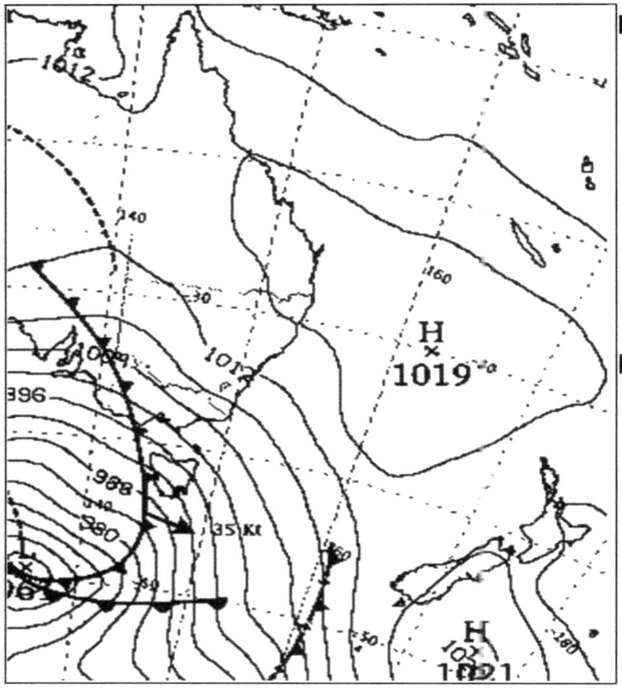

RAIN POTENTIAL

FROST/SNOW

MAY **14** WEDNESDAY

High pressure strengthens as it crosses New Zealand clearing rain activity in most parts of the country, increasing sunny spells and bringing colder overnight temperatures almost everywhere with frosts likely in many South Island districts. A very light to mostly calm easterly air prevails, fresher in the southern half of the South Island, and chance of fog in the Waikato to Auckland. Chance of brief showers in Otago, South Canterbury, northern parts of Taranaki to western fringes of the Waikato.

Auckland:	Partly cloudy, mostly dry, calm air, chance of showers or fog patches
Hamilton:	Calm and serene air, partly cloudy, dry, fog likely
Wellington:	Sunny, calm, cool, dry
Greymouth:	Overcast, brief showers, mild day cool night, windy at times
Christchurch:	Partly cloudy, dry, moderate breezes, cold overnight
Dunedin:	Cloudy, dry, light breezes freshening later in the day

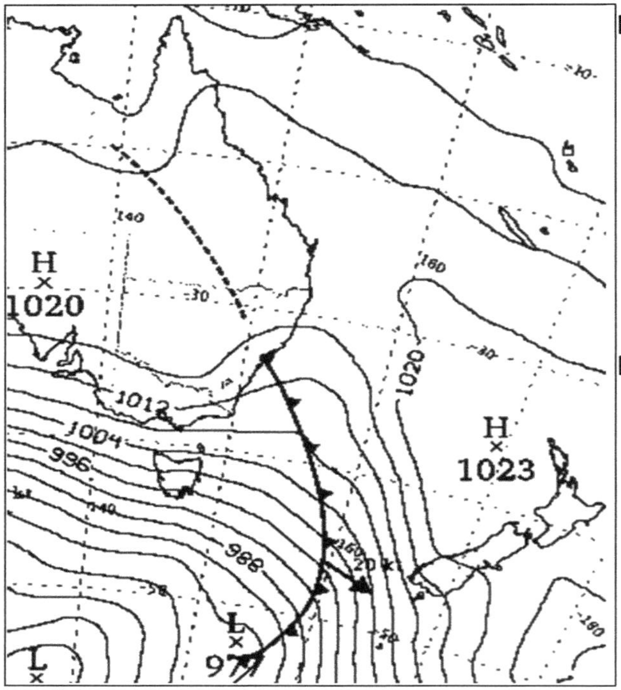

RAIN POTENTIAL

FROST/SNOW

MAY **15** THURSDAY

A deep depression south of New Zealand brushes through the southern half of the South Island bringing strong and gusty northwest winds and squally showers to the West Coast and Southland regions, while high pressure lingers over the North Island bringing mostly dry conditions and a light to mostly calm westerly air. Chance of fog patches in the Waikato to Auckland region. Mostly sunny over the north and east of the South Island and in most parts of the North Island.

Auckland: Partly cloudy, dry, mostly calm and serene air, fog possible
Hamilton: Partly cloudy, dry, mild day cool night, calm air, possible fog
Wellington: Partly cloudy, warmer, dry, fluctuating breezes
Greymouth: Unsettled and changeable winds, cloudy, cooler, showers
Christchurch: Partly cloudy, dry, unsettled and changeable winds
Dunedin: Cloudy, warmer, strong winds with occasional gusts, dry

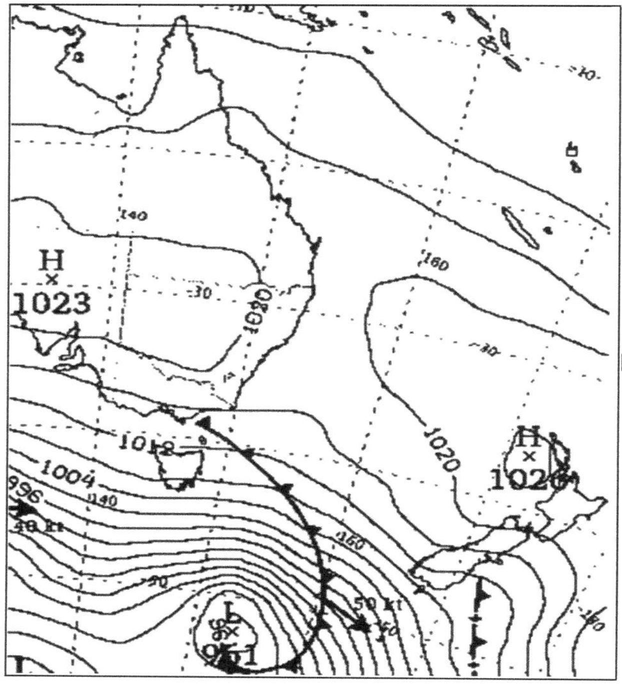

RAIN POTENTIAL

FROST/SNOW

MAY **16** FRIDAY

High pressure remains stalled over the North Island directing a gentle northwesterly flow, while an intense Southern Ocean depression extends a fresh northwest air over the South Island. A wet front crosses the South Island bringing scattered rain activity in western districts, while the east of the South Island and all of the North Island remain mostly sunny and dry.

Auckland: Sunny, dry, gentle breezes
Hamilton: Partly cloudy, dry, cooler, mostly calm air
Wellington: Overcast, dry with odd threat of rain, windy spells with possible gusts
Greymouth: Mostly cloudy, outbreaks of rain, moderate breezes
Christchurch: Warmer, cloudy, dry, light shifting breezes
Dunedin: Sunny, warmer, dry, winds easing to moderate

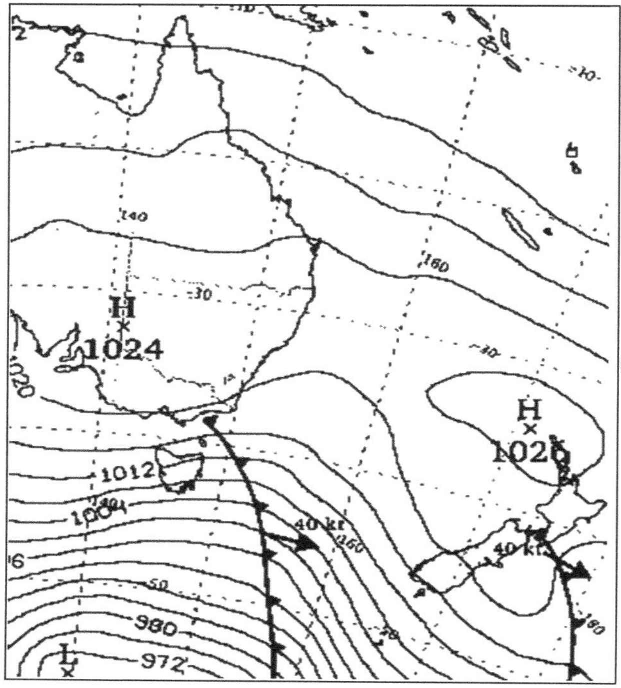

RAIN POTENTIAL

FROST/SNOW

MAY **17** SATURDAY

A blustery wet front crosses the South Island in a strong west to southwest flow, while high pressure remains stable over the North Island with a gentle westerly air prevailing. Mostly cloudy in central parts of the country from Taranaki to Marlborough/Nelson region, while mostly sunny elsewhere. Some after rain likely in the West Coast, Southland and coastal Otago districts, and possible isolated showers in the Taranaki to Manawatu regions of the North Island. Dry elsewhere.

Auckland:	Sunny and dry, gentle breezes, possible misty patches
Hamilton:	Mostly sunny and dry, mild day cool night, light breezes
Wellington:	Cloudy, warmer, dry, fresh breezes, strong at times
Greymouth:	Mix of sunny spells and rainy outbreaks, strong winds at times
Christchurch:	Partly cloudy, cooler, dry, occasional breezy spells
Dunedin:	Sunny, dry, cooler, fresh breezes

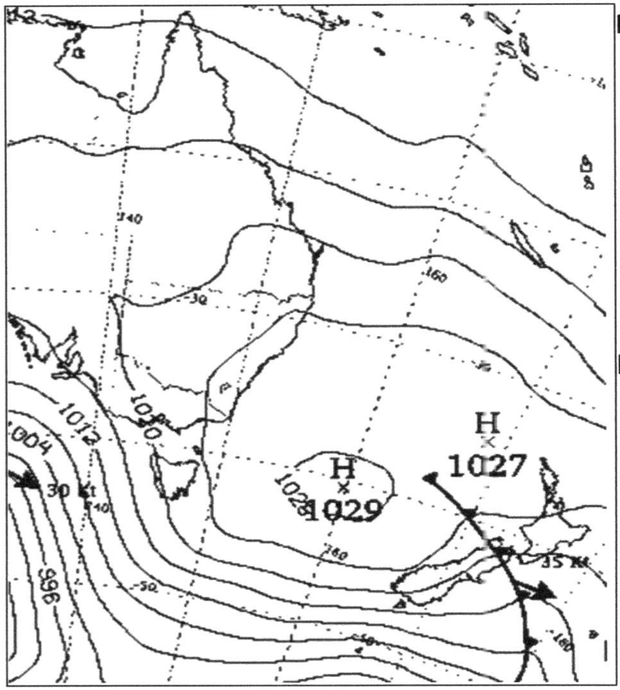

RAIN POTENTIAL

FROST/SNOW

MAY **18** SUNDAY

High pressure intensifies over New Zealand directing a light southeasterly flow over the North Island and north to westerly flow over the South Island. Mostly dry, warmer, and sunny throughout the country, except for some lingering showers in the south of the South Island but cold overnight with frosts likely in the south and east of the South Island, and chance of fog in Canterbury and sheltered parts of the North Island.

Auckland:	Gentle breezes and dry, sunny
Hamilton:	Sunny, dry, calm and pleasant air
Wellington:	Sunny and dry, mild day cool night, light shifting breezes
Greymouth:	Sunny, dry, mild day cool night, gentle shifting breezes
Christchurch:	Fluctuating breezes, dry, sunny, warm day cold night
Dunedin:	Sunny, dry, light breezes to mostly calm air, chance of misty patches

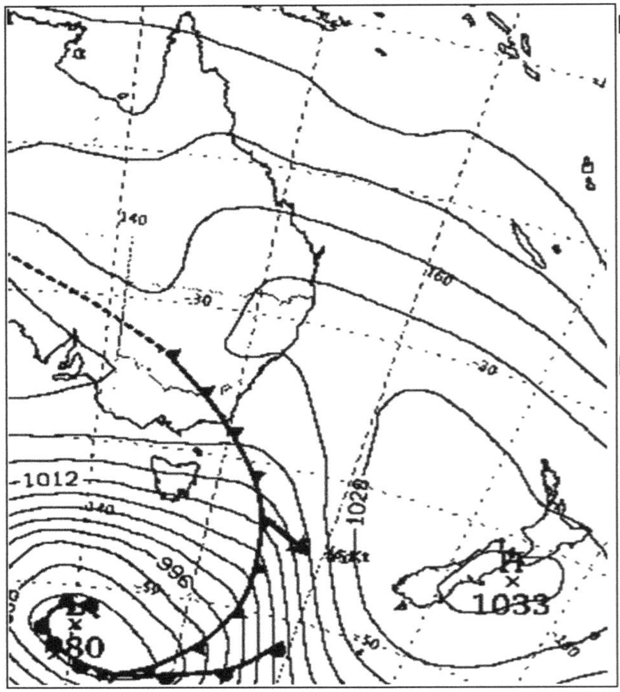

RAIN POTENTIAL

FROST/SNOW

MAY 19 MONDAY

High pressure departs to the east of New Zealand followed by a depression crossing to the south of the South Island. Winds generally light in a gentle westerly flow over the North Island, while cold, strong and blustery west to northwesterlies over the South Island and moving up through the Cook Strait. Widespread cloud over the South Island with some spotty showers around Fiordland and mostly dry elsewhere. Sunnier and mostly dry over the North Island, except for the chance of some isolated showers around the East Cape.

Auckland: Calm, dry and pleasant air, scattered cloudy spells
Hamilton: Sunny, mild day cold night, mostly dry, chance of evening showers
Wellington: Cloudy, dry, cooler, light breezes freshening later in the day
Greymouth: Cloudy, strengthening winds, mostly dry, increasing threats of rain
Christchurch: Blustery at times, overcast, cooler, dry with odd threat of rain
Dunedin: Overcast and blustery, dry, cooler, odd threat of rain

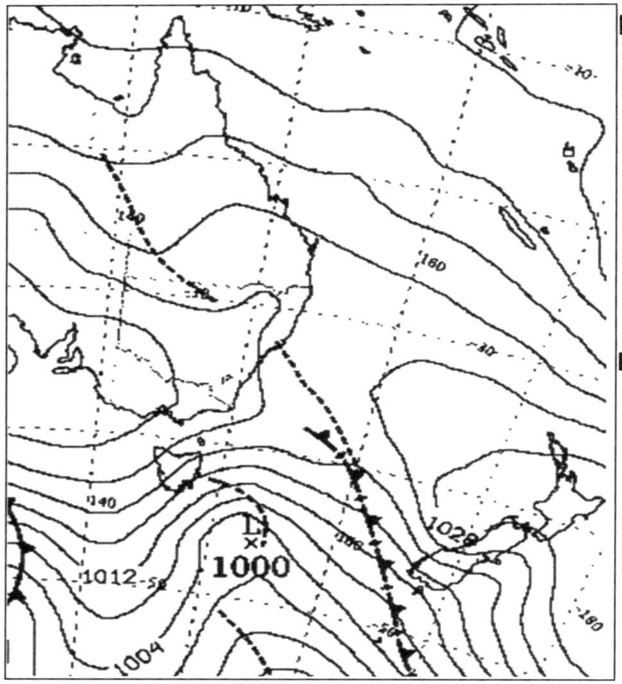

RAIN POTENTIAL

FROST/SNOW

MAY 20 TUESDAY

A wet front moves up through the country bringing a mix of sunny spells at first, and afternoon rain through the western half of the South Island, while remaining mostly dry but dull and gloomy with rain threatening at times over the North Island. Winds generally fresh and blustery at times, mostly in a west to southwest flow. Chance of frosts in the Waikato to the Bay of Plenty, while rain is likely to be heavy in Fiordland and the West Coast at times.

Auckland: Overcast, mostly dry with passing threats of rain, moderate breezes
Hamilton: Light breezes, cloudy, dry, cold overnight
Wellington: Unsettled winds, strong at times, overcast, dry with threats of rain
Greymouth: Cloudy, heavy rain, blustery winds easing later in the day
Christchurch: Partly cloudy, dry, fresh winds, slightly warmer
Dunedin: Partly cloudy, mostly dry, chance of evening showers

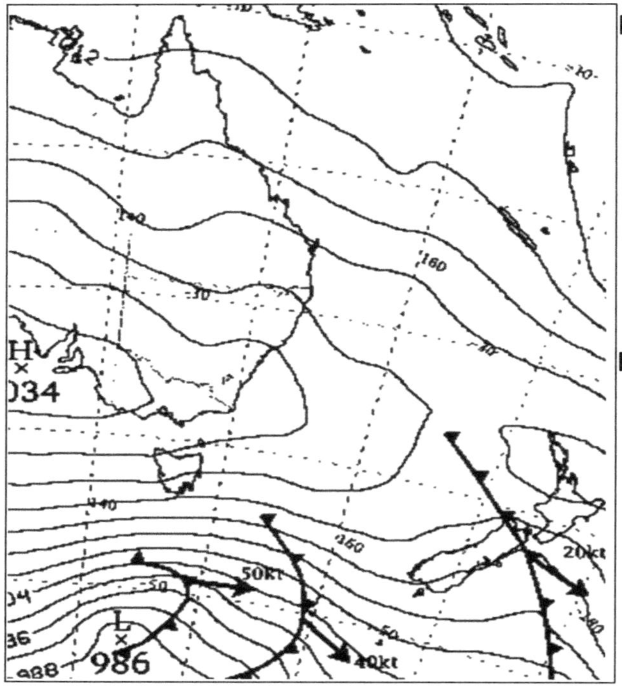

RAIN POTENTIAL

FROST/SNOW

MAY **21** WEDNESDAY

An unsettled and blustery west to southwest flow prevails across the country with the passage of a wet front associated with a deep depression crossing through the Southern Ocean. Mostly cloudy everywhere except for some sunny spells in the east and north of the North Island, and mostly dry Northland to Auckland and in the east of the North Island and Marlborough to Canterbury. Occasional showers elsewhere, and with heavy rain at times in Southland, Fiordland and the southern districts of the West Coast.

Auckland:	Partly cloudy, dry, windy spells at times
Hamilton:	Partly cloudy, dry, windy spells
Wellington:	Partly cloudy, dry, blustery winds, possible gales at times
Greymouth:	Cloudy, scattered showers, unsettled and blustery winds at times
Christchurch:	Strong winds, cloudy, mild day cool night, dry
Dunedin:	Cloudy, cooler, dry, blustery winds

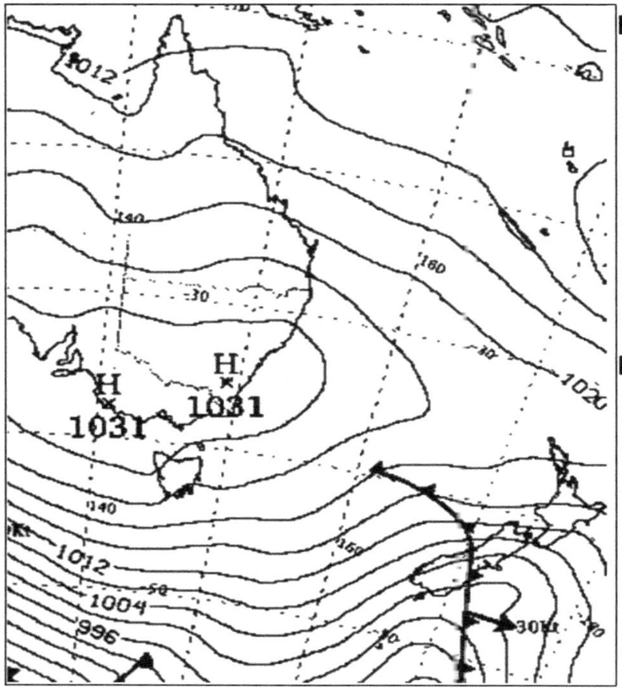

RAIN POTENTIAL

FROST/SNOW

MAY **22** THURSDAY

Low pressure prevails to the south of the country while a large anticyclonic system over the southeast of Australia ridges to the west of New Zealand. Winds ease to a moderate to light west to southwest flow throughout, fresher in the southeast of the South Island. Mostly dry in the east of the North Island with scattered overnight rain elsewhere, clearing to a mostly sunny day. Frosts likely in the south and east of the South Island.

Auckland:	Partly cloudy, light showers, fresh breezes
Hamilton:	Sunny, evening showers, light breezes to mostly calm air
Wellington:	Partly cloudy, isolated showers, mild day cool night, winds easing
Greymouth:	Winds easing, mostly sunny day, evening rain, heavy at times, colder
Christchurch:	Sunny, mostly dry, light breezes, chance of evening misty patches
Dunedin:	Partly cloudy, mostly dry, chance of evening showers

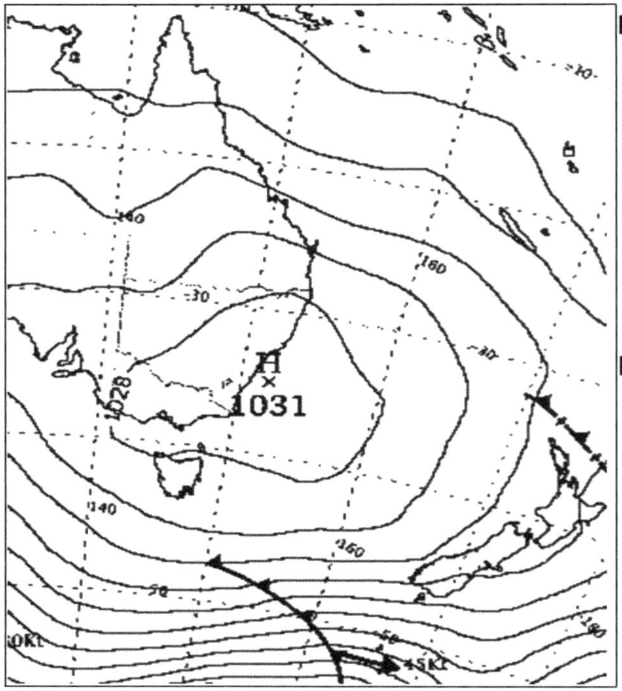

RAIN POTENTIAL

FROST/SNOW

MAY 23 FRIDAY

High pressure advances onto New Zealand with light southwesterly breezes prevailing throughout. Mostly dry, sunny, and cold everywhere except for some brief overnight showers in the south of the South Island and in the Waikato. Widespread frosts likely from the Waikato down through the Central Plateau to Wellington and east into the Bay of Plenty to the East Cape, as well as in Southland and Otago.

Auckland:	Mostly cloudy, cooler, isolated showers, gentle breezes
Hamilton:	Partly cloudy, dry, calm air, chance of frost pockets
Wellington:	Partly cloudy, dry, cooler, light breezes strengthening later in the day
Greymouth:	Partly cloudy, dry, cool, moderate breezes
Christchurch:	Sunny, dry, cold overnight, light breezes to mostly calm air
Dunedin:	Mostly sunny, warmer, dry, moderate breezes

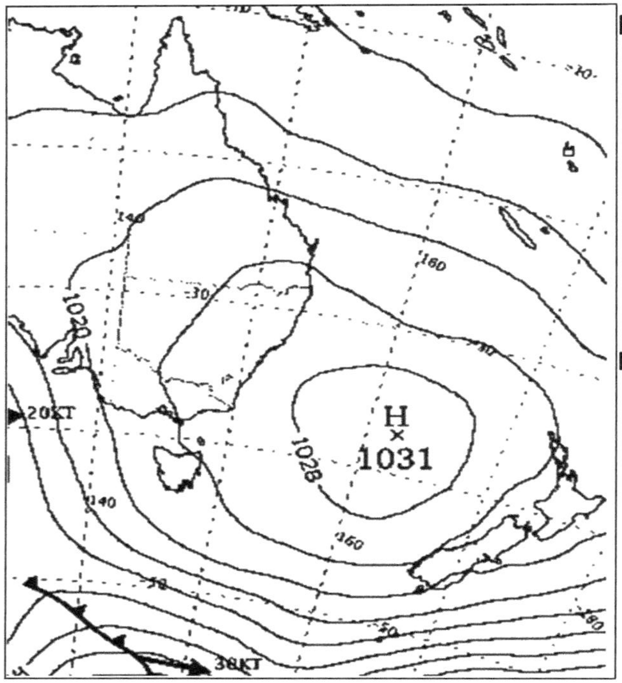

RAIN POTENTIAL

FROST/SNOW

MAY 24 SATURDAY

The anticyclonic system to the northwest of New Zealand weakens ahead of a Southern Ocean depression's wet front moving up into the south of the country. Cold southwest winds prevail, strengthening to strong and gusty particularly over the South Island, where winds are likely to reach gale force at times as far north as the Cook Strait. Some showers in Fiordland while mostly dry elsewhere with occasional threats of rain. Frosts likely in the Waikato to the Bay of Plenty and Central Plateau and in the east of the South Island.

Auckland:	Cloudy, isolated showers, strong breezes
Hamilton:	Cloudy, dry, possible frost pockets, freshening breezes at times
Wellington:	Overcast, dry, rain threatening, strong winds with occasional gusts
Greymouth:	Overcast, dry, strong winds with occasional gusts, static air
Christchurch:	Cloudy, strong winds with gusts, mild day cool night, dry
Dunedin:	Cloudy, cooler, dry, strong gusty winds

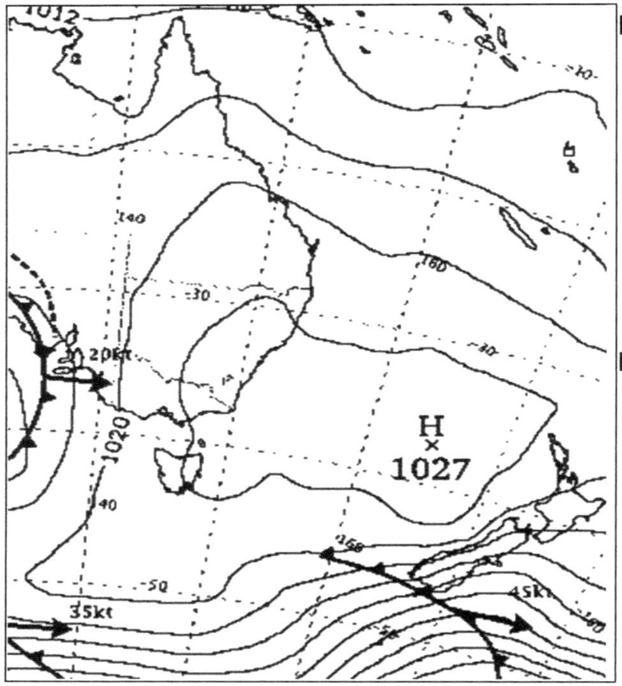

RAIN POTENTIAL

FROST/SNOW

MAY 25 SUNDAY

Anticyclonic over the North Island with a gentle westerly flow prevailing, while low pressure over the South Island directs a fresher westerly air. A trough between the two systems brings showers through the western half of the South Island, while remaining dry everywhere else. Cloudy in Northland, North Canterbury and northern parts of the West Coast, while mostly sunny elsewhere. Milder in most districts but frosts still likely in the south of the South Island.

Auckland: Cloudy, isolated showers, winds easing to light
Hamilton: Partly cloudy, dry, milder, mostly calm air
Wellington: Partly cloudy, dry, light breezes
Greymouth: Cloudy, scattered showers, windy at times
Christchurch: Cloudy, warmer, dry, gentle breezes, chance of overnight fog patches
Dunedin: Sunny, warmer, dry, unsettled breezes

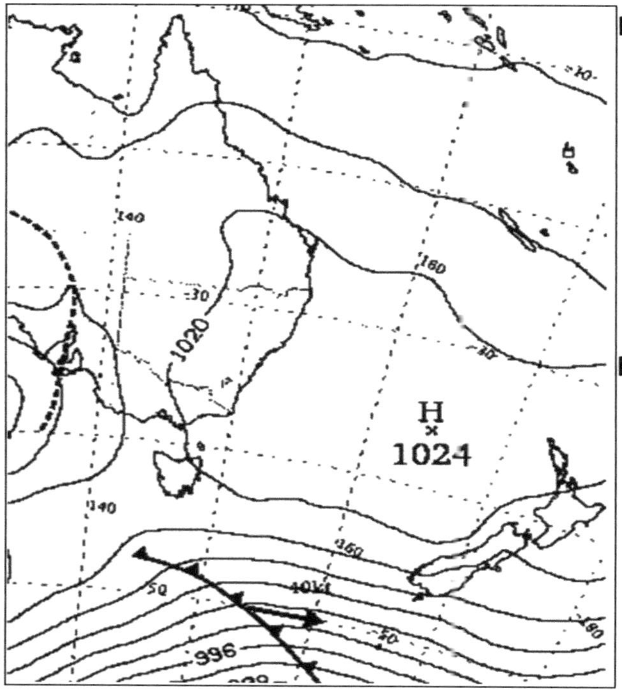

RAIN POTENTIAL

FROST/SNOW

MAY 26 MONDAY

High pressure contracts northwards maintaining dry, mild, and mostly sunny conditions over the North Island, with a westerly flow strengthening later in the day. A depression south of the country extends strong and blustery southwest winds up through the South Island and into the lower regions of the North Island, with winds up to gale force at times, including through the Cook Strait. A wet front associated with the depression brings overnight rain activity through the south and west of the South Island, mostly dry in the east.

Auckland: Cloudy, isolated showers, freshening winds
Hamilton: Partly cloudy, warmer, dry, light breezes strengthening later
Wellington: Partly cloudy, dry, strong winds, gales possible
Greymouth: Stormy air, overcast, squally showers
Christchurch: Sunny, dry, mild day cold night, strong winds, possible gales at times
Dunedin: Cloudy, dry, strong winds, chance of gale gusts

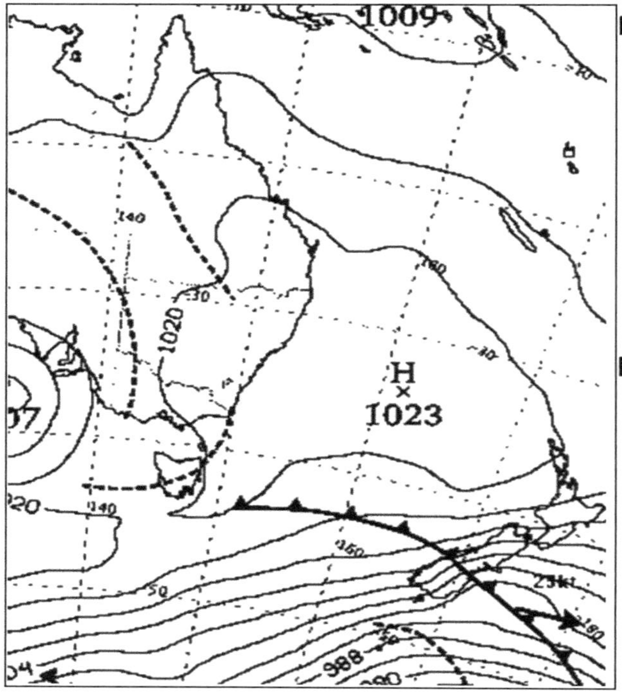

RAIN POTENTIAL

FROST/SNOW

MAY 27 TUESDAY

High pressure remains stalled in the northern Tasman Sea ridging over the northern half of the North Island, while a Southern Ocean depression extends a strong and blustery front up over the South Island and into the southern half of the North Island. Occasional showers from Southland to the Waikato, but remaining mostly dry and sunny in the far north of the country and eastern regions from the East Cape to Otago.

Auckland: Partly cloudy, spotty showers, fresh breezes
Hamilton: Mostly sunny, overnight light showers, moderate winds
Wellington: Partly cloudy, warmer, dry, strong blustery winds
Greymouth: Overcast and blustery air, passing squally showers
Christchurch: Sunny, warmer, dry, strong winds
Dunedin: Cloudy, dry, blustery

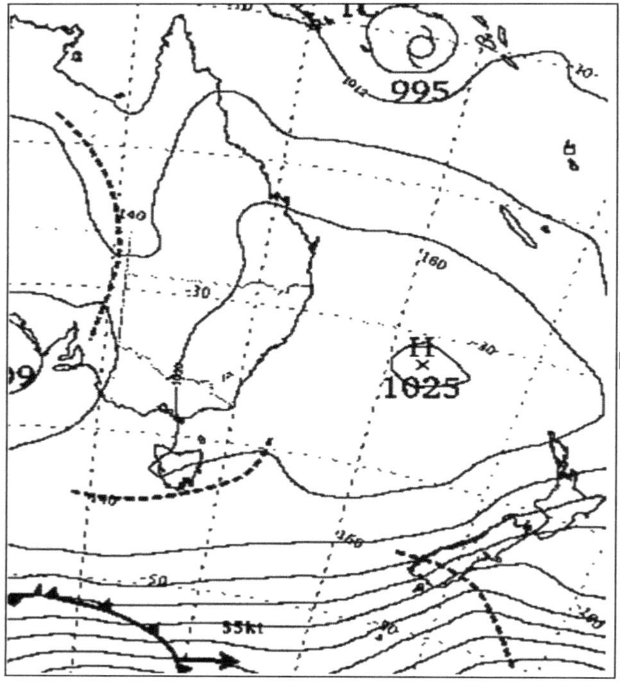

RAIN POTENTIAL

FROST/SNOW

MAY 28 WEDNESDAY

The strong southwest winds associated with the depression to the south of the country ease with showers contracting to the far southwest of the South Island, while the anticyclonic system in the Tasman Sea ridges further over the northern half of New Zealand with winds also easing to lighter over the North Island. Dry and cloudy in the far north of the North Island, while mostly sunny and dry elsewhere.

Auckland: Cloudy, isolated showers, winds easing to light
Hamilton: Partly cloudy, dry, mild temperatures, lighter breezes
Wellington: Partly cloudy and dry, mild day cool night, some windy spells
Greymouth: Mostly cloudy, brief showers, winds easing to moderate
Christchurch: Partly cloudy, dry, moderate breezes, cold overnight
Dunedin: Partly cloudy, dry, moderate breezes

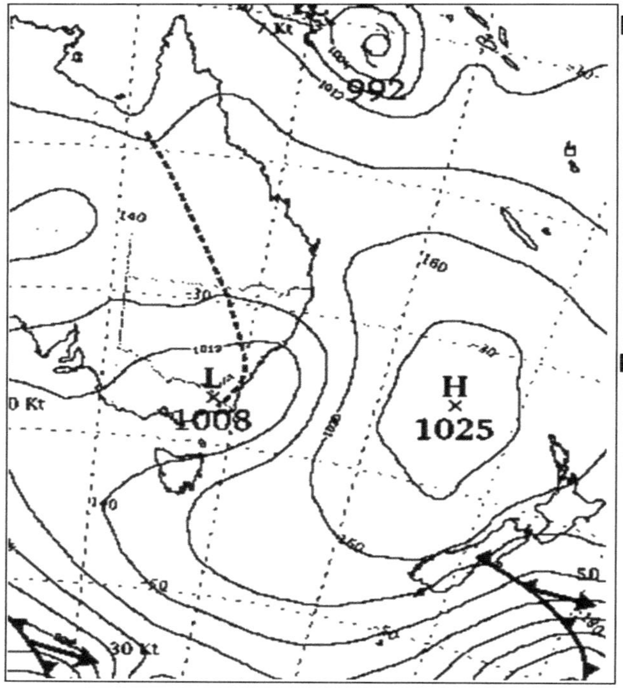

RAIN POTENTIAL

FROST/SNOW

MAY 29 THURSDAY

Anticyclonic over the country. Mostly sunny, dry, and colder everywhere except for the chance of some isolated showers in Fiordland, and misty patches in Canterbury and the Waikato to Auckland. Very light winds throughout. Frosts likely in the Waikato to central regions of the North Island and through Canterbury and Otago.

Auckland:	Sunny, dry, mild day cool night, mostly calm serene air
Hamilton:	Mostly cloudy, dry, cold overnight mild day, calm atmosphere
Wellington:	Sunny, dry, mild day cool night, mostly calm air
Greymouth:	Overcast, dry with occasional threats of rain, light breezes
Christchurch:	Partly cloudy, mostly calm air, dry, chance of overnight fog
Dunedin:	Sunny, dry, moderate breezes

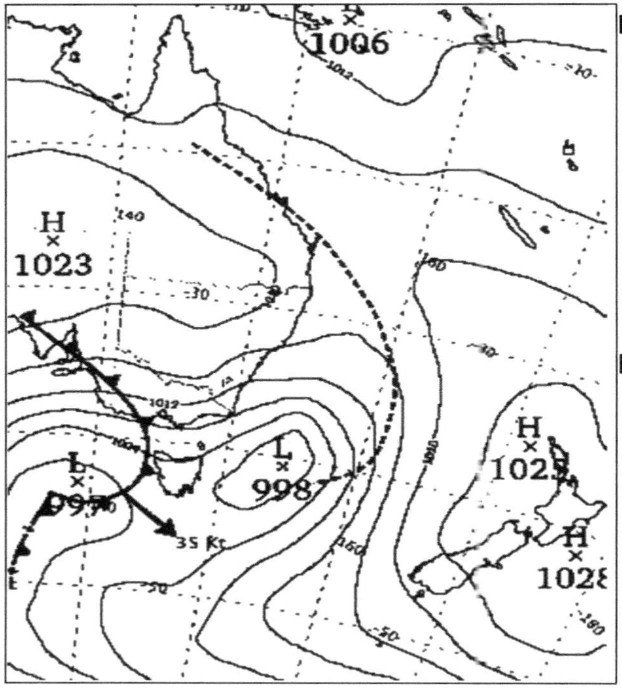

RAIN POTENTIAL

FROST/SNOW

MAY 30 FRIDAY

An intense depression crossing the southern Tasman Sea moves closer to New Zealand with strengthening northwest winds prevailing through the southern half of the North Island and down throughout the South Island, while high pressure remains over the northern half of the North Island, in a light northerly air. Cloudy almost everywhere, with some showers in the southwest of the South Island, and around the Bay of Plenty, and mostly dry elsewhere. Frosts extend in the south and east of the South Island.

Auckland:	Cloudy, isolated showers, light breezes
Hamilton:	Overcast, light showers, milder, moderate breezes
Wellington:	Overcast, strong blustery winds, dry with increasing threats of rain
Greymouth:	Overcast, dry, rain threatening, winds strengthening, possible gales
Christchurch:	Cloudy, cooler, dry, strong gusty winds
Dunedin:	Dry, mild, partly cloudy, strong winds with gusts

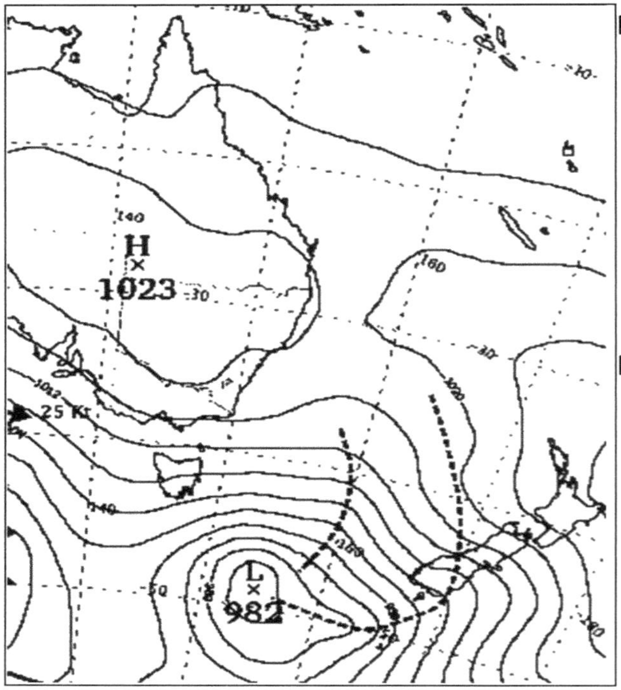

RAIN POTENTIAL

FROST/SNOW

MAY 31 SATURDAY

An intense depression south of New Zealand extends wet fronts up through the west and south of both islands, with heavy rain likely in Taranaki and all western districts of the South Island, with widespread cloud elsewhere and isolated showers in most other districts. Milder temperatures. Moderate northwest to westerly breezes prevail over the northern half of the North Island, while gusty westerly winds prevail elsewhere.

Auckland: Overcast, mild temperatures, isolated showers, some breezy spells
Hamilton: Overcast, isolated showers, windy spells
Wellington: Overcast, warmer, isolated showers, windy at times
Greymouth: Overnight heavy rain, strong gusty winds, brief sunny breakthroughs
Christchurch: Overcast, warmer, dry, strong winds, chance of gales
Dunedin: Mostly sunny, warmer, dry, strong gusty winds

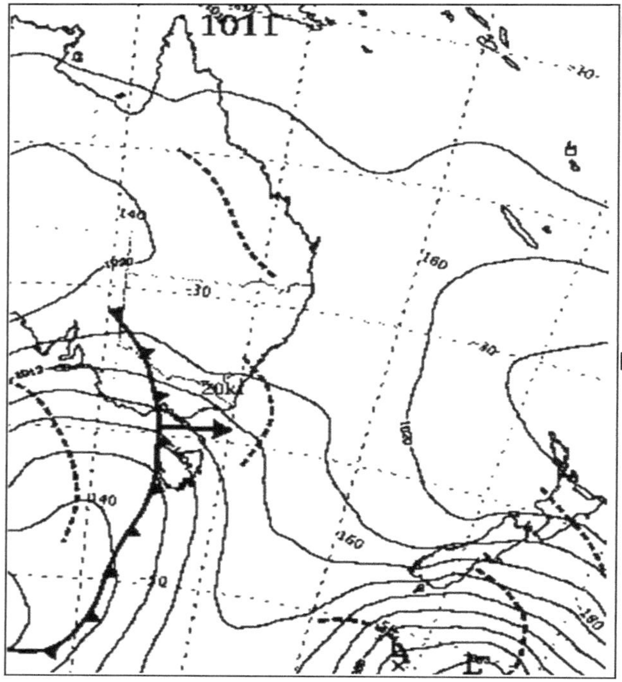

RAIN POTENTIAL

FROST/SNOW

JUNE

PHASES OF THE MOON

3rd	First Quarter
4th	Crossing Equator
7th	Apogee
11th	Full Moon
11th	Southern Declination
18th	Third Quarter
18th	Crossing Equator
23rd	Perigee #8
25th	New Moon
25th	Northern Declination

1 Sun	2 Mon	3 Tue	4 Wed	5 Thu	6 Fri	7 Sat	8 Sun	9 Mon	10 Tue	11 Wed	12 Thu	13 Fri	14 Sat	15 Sun	16 Mon	17 Tue	18 Wed	19 Thu	20 Fri	21 Sat	22 Sun	23 Mon	24 Tue	25 Wed	26 Thu	27 Fri	28 Sat	29 Sun	30 Mon
		1Q	XhS			A				F V							3Q XhN					P8		N ^					

MONTHLY SUMMARY

1st Chance of fog in the northern half of the North Island from the Waikato to Auckland.
1st-2nd Very heavy rain likely and possible pockets of localised flooding in the west of the country from Taranaki to the West Coast, including in the Nelson district, New Plymouth area, and along the Oakura River in Taranaki.
5th Chance of thunderstorms accompanying heavy rain in the Wellington region.
6th Very strong and blustery winds in Otago, with gale gusts around the peninsula.
7th-10th Chance of shallow fog patches in Christchurch.
10th Sub-zero frosts possible in the East Coast districts. Heavy rain likely in South Westland to Fiordland, including around the Franz Josef district to Mt Cook. Chance of fog in the Auckland and Waikato regions while winds will be very strong and blustery in the Wellington region.

Over the next four weeks the anticyclonic systems will pass more to the south of Australia while depressions will pass to the southeast of the South Island more often than is typical for this time of year. As such, a strong and cold southwesterly airflow will prevail through much of this outlook period, with particularly strong winds affecting the southern half of the country. During the next four weeks three major depressions are expected to develop near Australia before moving across New Zealand, with only the first week of this outlook period being predominately anticyclonic. This will make the next four weeks particularly wintry in nature, especially in the south of the country, where snowfalls to low levels can be expected in Southland and Otago and through the South Island high country passes. Despite this, in general, the northeast of the South Island will see below average rainfall, in contrast to the southern South Island districts which will see above average rainfall. Districts that may see only around three-quarters of their typical rainfall for this time of year will be parts of Northland, the Horowhenua, Kapiti to Wellington and including the Wairarapa, Nelson, Marlborough, Canterbury, and Fiordland. Meanwhile, the next four weeks are likely to bring up to twice as much rainfall than is typical for this time of year in parts of Southland and the Hawke's Bay, South Otago, and Westland. The next four weeks will also be sunnier than normal by around 20% more for this time of year in the Kapiti, Canterbury, and east Otago regions, with most other parts of the country likely to see around 10% more sunshine than normal. Temperatures will be around 0.5°C below average over this outlook period throughout most of the South Island with severe frosts in many districts, and with the Otago district up to 1°C cooler than average. In contrast, the north and east of the North Island, including Northland, Auckland, the Bay of Plenty, Gisborne, and Hawkes Bay regions can all anticipate being around half a degree warmer than normal. There is likely to be at least two major snowfall events during the next four weeks, affecting both the South Island as well as the North Island's Central Plateau.

11th Unusually mild day in the Rangiora area. Very heavy rain likely in the Arthurs Pass region. Light fog patches possible through the Waikato and into the Auckland region and in contrast, winds may reach gale force for a brief time in the Wellington region.
11th-19th Wet spell throughout the West Coast with occasional outbreaks of heavy rain, particularly around the 11th-12th.
12th-14th Light air in Canterbury may allow patches of fog in the Christchurch region.
17th-19th A strong southwesterly flow over the South Island brings snow to low levels in Southland and Otago, and including in the high country passes. Some snowfall likely to reach the Dunedin hill suburbs and in Queenstown. Chance that snowstorms may cause disruptions to power services. Winds may briefly reach gale force in the Christchurch region.
18th-21st A depression crosses the southeast of the country directing a strong and blustery southwest flow. Chance of disruptions to power services, particularly in some Southland districts.
21st Chance of misty patches in Christchurch.
23rd-2nd July Frequent severe frosts likely during this spell especially in inland Otago districts.
24th Chance of snowfall in the ranges around Wellington and on the Rimutaka Hills.
26th-30th Light air in the east of the South Island may see patches of fog in the Christchurch region.
27th-1st July Another depression crosses the southeast of the country.
27th Chance of scattered thunderstorms in the Wellington region.
30th Chance of snow in the Reefton area of the West Coast.
30th June-4th July A strong southwesterly returns over the South Island bringing further snow to low levels in Southland and Otago, and including in the high country passes. Snow also in the Central North Island, possibly causing the closure of the Desert Road for a time, as well as snow in the Reefton and Hanmer Forest areas. More settled conditions in the east of the country, however, widespread severe frosts are likely to prevail. Snow may reach sea level in some coastal parts of Southland and Otago.

June 2025

Monday	Tuesday	Wednesday	Thursday	Friday	Saturday	Sunday
26	27	28	29	30	31 Moon Sign	**1** Leo Sextile Sun Sextile Mercury Conjunct Mars Opposition Ascendant
2 Leo Trine Venus Conjunct Mars Sextile Jupiter Square Uranus	**3** 1Q Virgo Square Sun	**4** Virgo Square Mercury Trine Uranus	**5** XhS Libra Square Jupiter Opposition Saturn Trine Uranus Opposition Neptune Trine Neptune	**6** Libra Trine Sun	**7** A Libra Trine Mercury Opposition Venus Trine Jupiter Square Pluto	**8** Scorpio Square Pluto
9 Scorpio Square Mars Opposition Uranus	**10** Sagittarius Trine Saturn Opposition Uranus Trine Neptune Sextile Pluto	**11** FM Sagittarius Opposition Sun	**12** V Sagittarius Opposition Mercury Trine Venus Trine Mars Opposition Jupiter Square Saturn Square Neptune Square Ascendant	**13** Capricorn Opposition Mercury Trine Venus	**14** Capricorn Trine Uranus Conjunct Pluto	**15** Aquarius Square Venus Conjunct Pluto
16 Aquarius Trine Sun Opposition Mars Square Uranus	**17** Pisces Opposition Mars Trine Jupiter	**18** Pisces Trine Mercury	**19** 3Q XhN Aries Square Jupiter Conjunct Saturn Conjunct Neptune Sextile Pluto	**20** Aries Square Mercury	**21** Taurus Trine Mars Sextile Jupiter Square Pluto	**22** Taurus Conjunct Venus Square Ascendant
23 P8 Gemini Square Mars Sextile Saturn Conjunct Uranus Sextile Neptune Trine Pluto Square Ascendant	**24** Gemini Square Ascendant	**25** NM ^ Cancer	**26** Cancer Conjunct Sun Sextile Mars Conjunct Jupiter Square Saturn Square Neptune Square Ascendant	**27** Cancer Sextile Venus Square Ascendant	**28** Leo Conjunct Mercury Trine Saturn Saturn Neptune Opposition Pluto Square Ascendant	**29** Leo Square Venus
30 Virgo Conjunct Mars Sextile Jupiter Square Uranus	1					

NORTH ISLAND RAINFALL

JUNE 2025

d=mainly dry, n= fog, mist, dew or nonrecordable, l=light showers, p=passing showers, s=significant showers, r=rain, h=heavy falls

		Northland							Auckland					Bay Of Plenty				Waikato				Rot/Tau		Hawkes Bay			C Plateau			Taranaki			Manawatu			Lower North Island					
		CR	Kta	Ker	Rsl	Dar	Whg	Man	Wark	Hen	kH	Ak	Co	Tau	TeP	Wkt	Opo	Hml	TeA	Mata	Ot	Rt	Taup	Gis	Nap	Has	Taum	Chat	Waio	NP	Str	Wang	Pa	Dann	Lev	PN	Mast	UH	Wgt	all	Moon
1st	Sun	l	d	d	d	n	d	s	l	l	s	l	l	l	l	l	l	s	s	s	s	s	d	d	d	d	s	h	l	r	s	d	d	d	d	d	d	d	d	110	
2nd	Mon	l	l	l	n	d	d	l	n	l	l	d	n	l	s	s	l	d	d	d	d	s	n	d	d	d	l	s	d	h	h	l	d	d	d	d	d	l	l	85	
3rd	Tue	d	r	s	l	r	h	r	s	s	r	s	h	s	r	r	h	r	r	d	l	h	l	s	d	d	h	h	l	r	h	h	s	s	r	h	d	l	l	430	1Q
4th	Wed	d	d	d	l	d	d	d	d	l	l	d	d	d	d	d	d	n	l	d	l	d	l	d	d	d	l	l	n	d	d	d	d	d	d	d	d	d	d	13	
5th	Thur	d	l	l	d	d	d	d	d	l	l	d	d	d	d	d	d	d	d	d	l	d	d	d	d	d	n	d	l	l	d	d	d	d	d	l	d	d	d	9	XhS
6th	Fri	n	d	d	d	l	n	d	d	d	n	d	d	d	d	d	d	d	d	d	d	d	d	d	d	d	d	l	l	d	d	l	d	d	d	d	d	l	d	8	
7th	Sat	d	d	d	d	d	d	d	d	d	d	d	d	d	d	d	d	d	d	d	d	d	d	d	d	d	d	d	d	d	d	d	d	d	d	d	d	d	d	0	A
8th	Sun	d	d	d	d	d	d	d	d	d	d	d	d	d	d	d	d	d	d	d	d	d	d	d	d	d	d	d	d	d	d	d	d	d	d	d	d	d	d	0	
9th	Mon	d	d	d	d	d	d	d	d	d	d	d	d	d	d	d	d	d	d	d	d	d	d	d	d	d	d	d	d	d	d	d	d	d	d	d	d	d	d	0	
10th	Tue	d	d	d	d	d	d	d	d	d	d	d	d	d	d	d	d	d	d	d	d	d	d	d	d	d	n	d	d	d	d	d	d	d	l	d	d	d	d	1	
11th	Wed	d	d	d	d	d	d	d	d	d	d	d	d	d	d	d	d	d	n	d	d	d	d	d	d	d	d	d	d	d	d	d	d	d	d	d	d	d	d	1	F
12th	Thur	n	d	d	d	d	d	d	d	d	l	d	l	d	d	d	l	l	s	s	h	n	s	d	d	d	s	h	n	r	r	n	r	d	r	l	d	h	s	158	V
13th	Fri	d	l	s	d	s	s	l	l	s	s	l	r	s	s	d	l	r	r	d	s	r	s	d	d	d	s	h	s	s	n	h	d	l	l	l	l	n	d	183	
14th	Sat	n	l	d	d	s	d	d	l	s	r	r	l	l	d	d	d	l	l	l	l	d	d	l	d	d	n	d	d	l	s	d	d	d	d	d	d	d	d	53	
15th	Sun	d	l	d	s	l	d	l	s	l	s	l	l	d	s	d	d	l	l	d	l	d	d	d	d	d	d	n	l	n	d	l	d	d	l	d	d	d	d	40	
16th	Mon	l	l	d	d	s	n	l	s	r	r	s	s	l	l	d	d	s	s	d	s	l	l	d	d	d	s	h	s	r	l	l	d	l	d	s	d	d	d	135	
17th	Tue	n	r	d	n	s	l	s	d	n	l	l	d	d	d	d	d	l	l	d	l	d	l	d	d	d	s	r	l	l	l	l	d	d	l	s	d	d	d	61	
18th	Wed	s	l	l	d	l	l	l	l	l	s	n	d	l	d	d	d	l	l	s	l	l	d	d	d	d	s	h	s	l	l	r	d	l	l	s	d	d	d	80	
19th	Thur	d	d	d	d	l	d	d	n	l	l	n	l	d	d	d	d	l	l	d	l	d	l	d	d	d	l	h	r	l	n	s	d	l	d	l	l	d	d	56	3Q XhN
20th	Fri	h	h	h	h	s	r	s	s	l	l	l	l	n	d	d	d	l	l	d	l	d	d	d	d	d	d	d	d	d	d	d	d	d	d	d	d	d	d	137	
21st	Sat	r	s	r	h	s	l	s	r	s	s	s	h	h	h	h	r	r	r	s	h	s	r	h	h	h	h	h	h	h	h	h	d	r	s	r	s	l	n	555	
22nd	Sun	l	s	l	l	l	l	l	d	l	l	l	d	d	d	d	l	r	d	d	l	l	n	l	h	h	l	l	s	d	l	n	d	h	d	l	s	d	d	150	
23rd	Mon	d	n	d	d	d	d	d	d	l	s	l	d	d	l	s	d	d	d	d	d	d	l	l	n	l	d	l	l	d	d	d	d	l	d	n	l	d	l	25	P8
24th	Tue	s	l	r	r	l	s	l	d	l	d	d	r	s	l	d	l	d	d	d	n	d	d	d	l	n	d	d	d	d	d	d	d	l	d	d	l	d	l	69	
25th	Wed	s	d	l	r	d	r	l	d	d	d	d	h	d	d	d	d	l	d	l	n	l	d	h	s	s	d	l	d	d	d	d	d	d	d	d	d	l	l	110	N ^
26th	Thur	s	n	s	d	d	l	d	d	d	d	d	l	d	d	d	d	d	d	d	l	d	d	r	l	d	d	d	d	d	d	d	d	l	d	d	s	l	s	45	
27th	Fri	d	d	d	d	d	d	d	d	d	d	d	d	d	d	d	l	d	d	d	d	d	n	r	r	r	d	d	l	d	d	d	l	h	l	s	r	l	r	98	
28th	Sat	d	d	d	d	d	d	d	d	d	d	d	d	d	d	d	d	d	d	d	d	d	d	h	s	s	d	d	l	d	d	d	d	l	d	n	l	d	s	46	
29th	Sun	d	d	d	d	d	d	d	d	d	d	d	d	d	d	d	d	d	d	d	d	d	d	s	d	d	d	d	l	d	d	l	d	d	d	d	d	l	l	12	
30th	Mon	r	r	r	s	s	l	l	n	d	d	d	d	d	d	d	d	d	d	d	d	d	d	d	d	d	d	d	d	d	d	d	d	d	d	d	d	d	d	55	
estimate:		83	71	95	87	61	76	51	43	58	32	40	107	42	51	42	41	66	63	23	83	56	33	106	100	109	119	236	71	110	131	85	21	77	40	60	40	30	36	2724	
average:		122	150	185	167	146	170	196	173	148	135	125	126	140	188	130	134	117	119	137	153	132	122	97	85	77	144	246	93	152	188	85	106	95	109	97	100	137	122	5148	
trend:		drr	drr	drr	drr	drr	drr	drr	drr	drr	drr	drr	drr	drr	drr	drr	drr	drr	drr	drr	drr	drr	drr	av	wtr	wtr	drr	av	drr	drr	drr	av	drr	drr	drr	drr	drr	drr	drr	drr	

SOUTH ISLAND RAIN
JUNE 2025

d=mainly dry, n= fog, mist, dew or nonrecordable, l=light showers, p=passing showers, s=significant showers, r=rain, h=heavy falls

		West Coast				Top South Island				Canterbury				South Canterbury				Coastal Otago			Lakes			Inland Otago				Southland						
		West	Grey	Hok	BB	Ta	Mou	Nel	Ble	Dar	Ham	Kai	Ch	Rak	Meth	Ash	Tim	Oam	Dun	Bal	Twi	LP	LT	Crom	Qstn	Ran	Alex	TeA	Wtn	Gore	Inv	Puy Pt	all	Moon
1st	Sun	d	s	l	s	d	d	d	d	d	d	d	d	d	d	d	d	d	d	d	l	d	d	d	d	d	d	d	d	d	d	s	20	
2nd	Mon	h	h	h	h	h	h	h	r	l	s	l	d	d	d	l	n	n	l	l	d	r	s	l	l	d	d	l	l	s	l	s	422	
3rd	Tue	s	s	s	r	d	d	d	s	d	d	n	d	l	d	d	d	d	d	d	s	l	d	d	l	d	d	s	d	d	d	r	65	1Q
4th	Wed	h	h	r	h	l	l	l	n	d	d	d	d	d	s	d	d	d	d	d	r	d	d	d	d	d	d	l	d	d	l	h	110	
5th	Thur	r	r	h	s	d	d	d	d	d	d	d	d	d	d	d	d	d	l	l	d	d	d	l	l	d	d	s	s	s	s	r	92	XhS
6th	Fri	s	s	l	s	d	d	d	d	d	d	d	d	d	d	d	d	d	d	d	d	d	d	d	d	d	d	n	d	d	l	h	42	
7th	Sat	d	d	d	d	d	d	d	d	d	d	d	d	d	d	d	d	d	d	d	d	d	d	d	d	d	d	d	l	l	l	l	5	A
8th	Sun	d	d	d	d	d	d	d	d	d	d	d	d	d	d	d	d	d	d	d	d	d	d	d	d	d	d	d	d	d	d	d	0	
9th	Mon	d	d	d	l	d	d	d	d	d	d	d	d	d	d	d	d	d	d	d	d	d	d	d	d	d	d	d	d	d	d	l	2	
10th	Tue	d	d	n	l	d	d	d	d	d	d	d	d	d	d	d	d	d	d	d	d	d	d	d	d	d	d	s	l	d	l	r	23	
11th	Wed	l	s	h	h	n	d	d	d	d	d	d	d	d	d	d	d	d	l	l	d	d	l	h	h	n	l	l	d	l	d	d	156	F
12th	Thur	h	h	h	h	h	s	l	l	d	d	d	d	d	l	d	d	d	r	s	h	r	s	s	r	d	s	l	l	l	s	h	348	V
13th	Fri	l	r	s	l	l	l	d	d	r	s	h	d	r	h	d	l	l	s	l	r	h	l	s	n	s	l	d	s	l	s	l	152	
14th	Sat	l	l	d	d	r	s	l	d	d	d	d	d	d	d	s	d	d	d	d	d	l	d	d	d	d	d	d	d	d	l	s	39	
15th	Sun	s	r	s	h	d	d	d	d	d	d	d	d	d	d	l	d	d	l	s	d	d	l	d	d	d	d	s	l	s	h	r	100	
16th	Mon	r	r	h	h	l	l	d	d	d	d	d	d	d	d	d	d	d	d	l	d	d	d	d	d	d	d	l	l	s	r	r	108	
17th	Tue	r	s	s	r	d	d	d	d	d	d	d	d	d	d	d	d	d	l	l	n	d	d	d	d	d	n	l	l	s	s	h	73	
18th	Wed	h	h	r	r	l	l	d	d	l	d	l	l	d	l	l	l	l	l	h	d	d	d	l	l	d	l	r	h	h	h	r	197	
19th	Thur	h	s	s	l	d	l	d	d	d	d	d	d	d	d	d	d	d	s	s	d	d	d	d	d	l	d	l	r	r	r	s	93	3Q XhN
20th	Fri	d	d	d	d	d	d	d	d	l	d	l	d	d	l	l	d	d	s	l	d	d	d	d	d	d	d	d	s	r	l	l	39	
21st	Sat	d	d	d	d	d	d	d	d	d	d	l	d	d	d	d	d	d	d	d	d	d	d	d	d	d	d	d	d	d	d	l	3	
22nd	Sun	d	d	d	d	d	d	d	l	d	d	l	d	d	d	d	d	d	d	d	d	d	d	d	d	d	d	l	d	l	s	r	23	
23rd	Mon	d	d	d	d	d	d	d	d	d	d	d	d	d	d	d	d	d	d	d	d	d	d	d	d	d	d	d	d	l	l	l	6	P8
24th	Tue	d	d	l	d	d	d	d	d	d	d	d	d	d	d	d	d	d	d	d	d	l	d	d	d	d	d	l	s	l	l	s	20	
25th	Wed	d	d	d	d	d	d	d	d	l	d	l	d	s	l	d	d	d	d	d	l	d	s	d	d	d	d	d	d	d	d	d	15	N ^
26th	Thur	d	d	d	d	d	d	d	d	l	s	d	s	d	l	l	l	d	s	l	d	d	d	d	d	d	d	d	l	l	d	l	28	
27th	Fri	d	d	d	d	d	d	d	s	s	r	s	l	l	l	s	d	l	n	d	d	d	d	d	d	d	d	d	d	d	d	d	43	
28th	Sat	d	d	d	d	d	d	d	d	d	d	d	d	d	d	d	d	n	d	d	d	d	d	d	d	d	d	d	d	d	d	r	11	
29th	Sun	d	d	d	d	d	d	d	d	d	d	d	d	d	d	d	d	d	d	d	d	d	d	d	d	d	d	l	l	l	l	s	11	
30th	Mon	n	d	d	d	d	d	d	d	d	d	d	d	d	l	l	d	d	l	d	d	d	d	d	d	l	d	d	d	l	n	l	6	
estimate:		152	191	205	259	130	95	103	30	22	30	33	11	21	35	18	5	4	42	68	57	53	25	34	54	7	13	54	70	80	114	239	2253	
average:		187	203	225	315	193	130	87	62	71	106	71	65	67	68	60	40	40	80	65	48	67	54	32	55	26	20	102	78	80	97	190	2984	
trend:		drr	drr	drr	drr	drr	drr	wtr	drr	drr	drr	drr	drr	drr	drr	drr	drr	drr	drr	av	av	drr	drr	av	av	drr	av	drr	av	av	wtr	wtr	drr	

NORTH ISLAND SUNSHINE ESTIMATES

JUNE 2025

F=fine (8-12 hours of sunshine), pc= partly cloudy (4-7 hours), c=cloudy (1-3 hours), o=overcast (0 hours)

		Kta	Whg	Dar	Ak	Hml	Tau	Rt	Taum	NP	Wang	Taup	Gis	Nap	Wai	Dann	PN	Mast	Pa	UH	Wgt	ALL	Mocn
1st	Sun	pc	pc	pc	pc	o	pc	pc	pc	pc	F	pc	F	pc	F	c	F	c	F	F	pc	110	
2nd	Mon	c	c	o	o	o	o	o	o	c	o	o	c	c	o	pc	o	pc	o	o	o	24	
3rd	Tue	F	pc	F	F	pc	pc	pc	pc	F	pc	pc	c	pc	pc	pc	pc	pc	pc	pc	F	132	1Q
4th	Wed	pc	pc	pc	pc	pc	F	F	F	F	pc	F	F	F	pc	pc	pc	pc	F	F	F	143	
5th	Thur	pc	pc	c	c	pc	pc	c	pc	pc	pc	c	pc	pc	pc	pc	pc	pc	pc	pc	pc	96	XhS
6th	Fri	c	pc	c	c	F	F	F	F	pc	pc	F	F	F	F	pc	pc	pc	pc	pc	F	134	
7th	Sat	pc	pc	F	F	F	F	F	F	F	F	F	o	pc	pc	pc	F	pc	F	pc	F	149	A
8th	Sun	F	F	F	F	F	F	F	F	c	c	F	F	F	F	o	c	o	c	pc	o	120	
9th	Mon	F	pc	o	o	pc	F	F	F	o	pc	F	F	F	F	F	pc	F	pc	F	F	134	
10th	Tue	F	F	pc	pc	pc	pc	pc	o	o	o	pc	F	F	F	pc	o	pc	o	o	c	96	
11th	Wed	pc	F	F	F	pc	pc	pc	c	pc	pc	pc	F	F	pc	pc	pc	pc	o	o	o	106	F
12th	Thur	c	c	c	c	c	o	o	o	c	c	o	c	c	c	pc	c	pc	pc	pc	pc	48	V
13th	Fri	pc	pc	pc	pc	F	pc	F	F	pc	pc	F	pc	pc	o	pc	pc	pc	c	pc	c	111	
14th	Sat	pc	pc	pc	pc	c	pc	pc	pc	pc	c	pc	pc	pc	pc	o	c	o	pc	c	pc	85	
15th	Sun	o	c	c	c	o	c	c	o	c	c	c	c	pc	pc	pc	c	pc	o	o	o	46	
16th	Mon	pc	c	pc	pc	pc	pc	F	pc	pc	c	F	F	F	pc	pc	c	pc	F	F	F	125	
17th	Tue	pc	pc	pc	pc	c	pc	c	c	c	pc	c	pc	pc	pc	pc	pc	pc	pc	pc	F	91	
18th	Wed	pc	c	pc	pc	pc	pc	0	pc	pc	pc	0	pc	pc	0.5	pc	pc	pc	pc	pc	pc	87	
19th	Thur	pc	c	pc	pc	pc	c	c	c	pc	pc	c	F	pc	pc	pc	pc	pc	F	F	pc	96	3Q XhN
20th	Fri	o	o	o	o	o	o	o	o	o	o	o	o	o	o	o	o	o	o	o	o	1	
21st	Sat	c	c	pc	pc	o	c	o	c	c	o	o	c	o	o	o	o	o	o	o	o	19	
22nd	Sun	c	c	o	o	pc	F	pc	pc	F	o	pc	pc	pc	o	o	o	o	pc	c	c	63	
23rd	Mon	pc	pc	c	c	c	pc	o	pc	F	F	o	pc	pc	c	o	F	o	pc	pc	pc	79	P8
24th	Tue	pc	pc	c	c	c	o	o	pc	F	pc	o	o	c	o	o	pc	o	F	o	c	55	
25th	Wed	o	o	pc	pc	c	c	o	pc	F	pc	o	o	c	c	pc	pc	pc	pc	c	pc	64	N ^
26th	Thur	F	F	F	F	pc	F	F	F	F	pc	F	pc	c	c	c	pc	c	pc	c	c	120	
27th	Fri	F	F	c	c	pc	c	c	pc	F	pc	c	o	c	c	o	pc	o	pc	c	c	73	
28th	Sat	F	pc	F	F	F	F	F	F	pc	pc	F	o	c	c	c	pc	c	F	o	c	119	
29th	Sun	c	pc	pc	pc	pc	F	F	F	F	c	F	F	pc	F	pc	c	pc	pc	pc	F	133	
30th	Mon	o	o	o	o	o	o	o	o	o	o	o	o	o	o	o	o	o	o	o	o	4	
est. hours:		153	134	133	133	134	152	131	154	150	114	131	138	157	115	118	114	118	140	116	127	2662	
av. hours:		124	111	91	110	103	132	113	68	117	100	99	124	125	128	92	84	95	101	100	98	2115	
trend:		more	more	more	more	more	more	more	more	more	more	more	more	more	av	more	more	more	more	more	more	more	

SOUTH ISLAND SUNSHINE ESTIMATES
JUNE 2025

F=fine (8-12 hours of sunshine), pc= partly cloudy (4-7 hours), c=cloudy (1-3 hours), o=overcast (0 hours)

		Ta	Nel	Ble	Mou	West	Grey	Hok	Ham	Kai	Ch	Ash	Tim	Crom	Qstn	Bal	Dun	Wnt	Gore	Inv	Alex	ALL	Moon
1st	Sun	F	pc	F	pc	pc	pc	c	pc	pc	pc	c	o	c	c	o	o	c	c	pc	c	77	
2nd	Mon	pc	pc	pc	pc	c	c	pc	pc	pc	pc	pc	pc	pc	pc	pc	pc	pc	pc	pc	pc	89	
3rd	Tue	pc	pc	F	pc	o	o	o	c	c	c	pc	pc	pc	c	pc	pc	pc	pc	pc	pc	86	1Q
4th	Wed	pc	F	F	F	pc	pc	pc	pc	pc	F	pc	pc	c	o	o	o	c	c	o	c	93	
5th	Thur	F	pc	F	pc	c	pc	pc	F	F	F	pc	c	pc	c	o	c	c	c	o	pc	96	XhS
6th	Fri	F	F	F	pc	c	c	c	F	F	F	F	F	F	pc	o	pc	c	c	c	F	114	
7th	Sat	F	F	F	pc	pc	pc	F	F	F	F	pc	c	pc	pc	pc	pc	pc	pc	pc	pc	117	A
8th	Sun	o	F	F	F	pc	c	o	F	F	pc	F	pc	c	o	o	o	o	o	c	c	76	
9th	Mon	pc	F	F	F	o	o	o	F	F	F	F	F	pc	pc	pc	F	pc	pc	pc	pc	126	
10th	Tue	o	pc	F	pc	c	o	o	c	c	o	o	o	o	o	o	o	c	c	c	o	36	
11th	Wed	F	c	o	c	o	o	o	pc	pc	c	pc	c	pc	o	c	o	c	c	pc	pc	52	F
12th	Thur	pc	pc	F	pc	c	pc	c	pc	pc	pc	pc	c	o	o	c	o	o	o	o	o	56	V
13th	Fri	F	o	pc	o	pc	pc	F	pc	pc	pc	F	F	pc	c	pc	pc	pc	pc	pc	pc	112	
14th	Sat	pc	F	F	F	o	c	pc	F	F	F	F	F	pc	pc	pc	pc	pc	pc	pc	pc	126	
15th	Sun	pc	c	c	c	o	o	o	c	c	o	o	o	o	c	o	o	o	o	o	o	21	
16th	Mon	F	F	F	pc	c	c	c	F	F	F	F	pc	pc	c	pc	pc	pc	pc	o	pc	117	
17th	Tue	pc	pc	F	pc	o	c	c	pc	pc	pc	pc	pc	pc	c	o	pc	o	o	o	pc	68	
18th	Wed	c	pc	pc	c	o	c	o	pc	pc	pc	F	pc	pc	o	o	c	c	c	c	pc	67	
19th	Thur	F	F	F	pc	c	pc	pc	F	F	F	pc	pc	pc	c	pc	o	o	o	c	pc	97	3Q XhN
20th	Fri	F	c	o	c	o	c	pc	o	o	o	pc	c	pc	pc	o	pc	pc	pc	F	pc	75	
21st	Sat	F	pc	o	pc	o	F	F	F	F	pc	F	F	o	pc	pc	F	pc	pc	pc	o	108	
22nd	Sun	F	F	F	F	o	c	pc	pc	pc	F	c	c	pc	pc	pc	F	c	c	o	pc	95	
23rd	Mon	pc	F	F	F	c	c	c	pc	pc	c	pc	pc	o	o	pc	pc	c	c	pc	o	80	P8
24th	Tue	F	pc	c	pc	c	pc	F	pc	pc	pc	pc	pc	pc	c	c	c	pc	pc	pc	pc	90	
25th	Wed	F	F	pc	pc	c	pc	pc	c	c	o	c	o	c	c	o	o	o	o	o	c	61	N ^
26th	Thur	pc	pc	o	pc	pc	F	F	o	o	o	o	o	pc	pc	c	o	pc	pc	pc	pc	73	
27th	Fri	o	F	pc	pc	pc	F	F	pc	pc	pc	pc	pc	pc	pc	pc	F	pc	pc	pc	pc	118	
28th	Sat	o	pc	F	pc	c	pc	F	pc	pc	pc	F	pc	o	pc	o	pc	c	c	pc	o	87	
29th	Sun	pc	pc	F	pc	o	o	c	pc	pc	pc	pc	c	c	c	pc	c	c	c	pc	c	78	
30th	Mon	c	o	o	o	o	o	o	c	c	c	o	o	c	pc	pc	F	pc	pc	pc	c	53	
est. hours:		182	193	186	159	66	104	119	159	159	153	163	123	107	77	83	112	93	93	103	107	2542	
av. hours:		135	143	146	135	103	86	102	85	119	119	119	118	89	75	82	86	87	90	77	86	2082	
trend:		more	more	more	more	less	more	more	more	more	more	more	av	more	av	av	more	av	av	more	more	more	

JUNE **1** SUNDAY

A low pressure trough moves onto the west of the country as a high pressure system moves away to the east. A light north to northeasterly air prevails throughout, backing to strong and blustery west to northwest winds later in the day. Dry at first with fog likely in the northern half of the North Island, before rain activity moves onto the country, with isolated showers in the west of the South Island and showers in most parts of the North Island from Northland to the Manawatu. Mostly dry in the east of the country.

Auckland: Partly cloudy, occasional showers, light breezes, chance of fog
Hamilton: Overcast, showers, light breezes freshening later, chance of mist
Wellington: Partly cloudy, dry, gentle breezes strengthening later in the day
Greymouth: Mostly cloudy, scattered showers, moderate winds
Christchurch: Partly cloudy, dry, mostly calm air
Dunedin: Overcast, dry, warm day cool night, gentle breezes

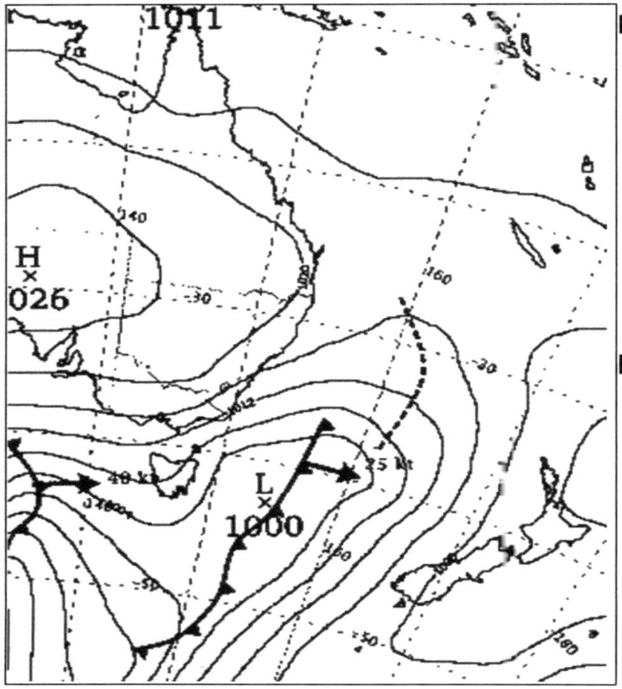

RAIN POTENTIAL

FROST/SNOW

JUNE **2** MONDAY

An intense depression centred to the west of Southland directs strong and gusty west to northwest winds throughout the country, with gales likely particularly around coastal districts in the west and through the Cook Strait. Wet fronts associated with the depression brings heavy rain in western districts from Taranaki to the West Coast, with chance of localised flooding including in northern parts of the South Island, while mostly dry in Northland and from the East Cape to Wellington, with showers elsewhere.

Auckland: Overcast, isolated showers, strong winds, chance of gale gusts
Hamilton: Overcast, warmer, mostly dry, threats of rain, windy with gusts
Wellington: Stormy air, overcast, brief squalls, gale winds likely
Greymouth: Stormy air, cloudy, heavy rain, strong winds with gusts
Christchurch: Partly cloudy, cooler, dry, strong gusty winds
Dunedin: Cloudy, colder, strong winds with gusts, passing squalls

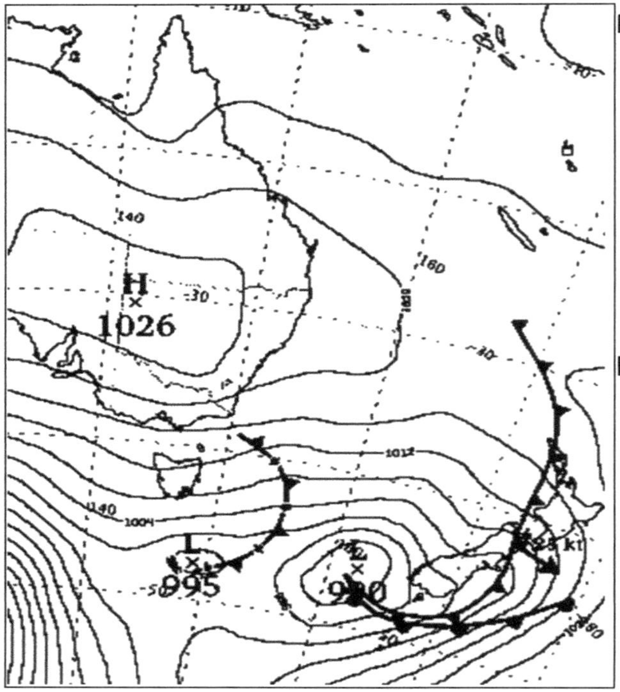

RAIN POTENTIAL

FROST/SNOW

JUNE **3** TUESDAY

The stormy conditions associated with the depression to the south of the South Island weakens ahead of a high pressure ridge to the north of the country. Westerly winds ease throughout, and rain eases to showers over the South Island, while intensifies to heavy rain in the west and north of the North island, and in the Bay of Plenty with outbreaks of rain in most other districts. Occasional sunny spells particularly in the North Island.

Auckland:	Overnight rain, sunny day, cooler, lighter breezes
Hamilton:	Moderate breezes, overnight rain, some sunny spells
Wellington:	Mostly sunny, winds easing, light showers
Greymouth:	Unsettled and windy, overcast, scattered showers
Christchurch:	Unsettled and breezy, cloudy, warmer, dry
Dunedin:	Partly cloudy, dry, moderate breezes

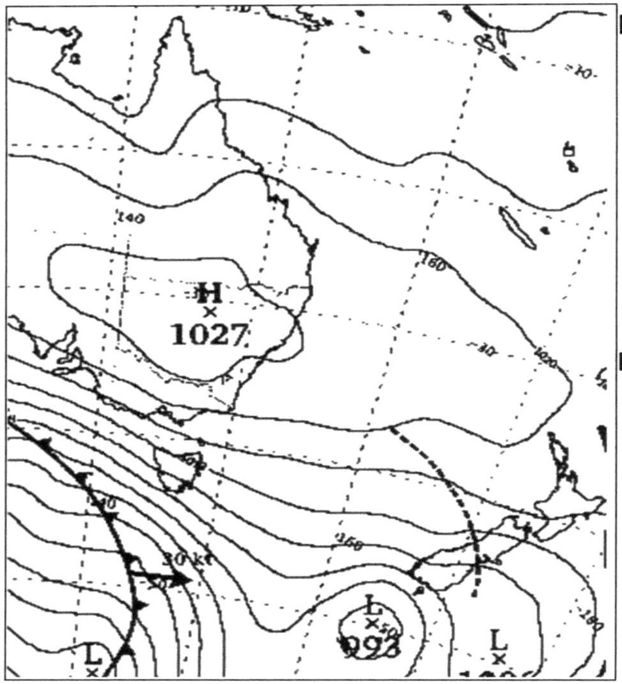

RAIN POTENTIAL

FROST/SNOW

JUNE **4** WEDNESDAY

A large anticyclonic system over southeast Australia extends a high pressure ridge over the North Island and a light east to southeast air, while another Southern Ocean depression brings strong northwest winds and stormy conditions over much of the South Island, including heavy rain in the western regions from Nelson to Fiordland. Mostly sunny in the east of the South Island and in the North Island. Some isolated showers Auckland to the Central Plateau, while dry elsewhere.

Auckland:	Partly cloudy, isolated showers, light breezes
Hamilton:	Partly cloudy, mostly dry, gentle breezes, chance of overnight mists
Wellington:	Sunny, breezy spells, dry
Greymouth:	Changeable, heavy rain, some sunny spells, cooler, windy spells
Christchurch:	Mostly sunny, dry, mild day cold night, light breezes
Dunedin:	Overcast, dry with some threats of rain, blustery at times

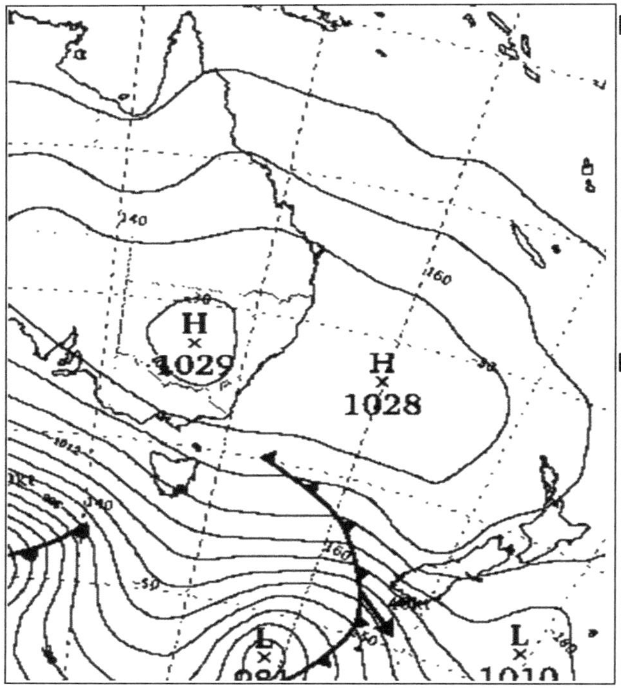

RAIN POTENTIAL

FROST/SNOW

JUNE 5 THURSDAY

High pressure ridges over the North Island directing a moderate southwest flow, with mostly dry, occasional sunny spells and mild temperatures throughout. In the South Island, however, a blustery wet front brings a strong, cold west to southwest flow everywhere, with winds up to gale gusts in the southern districts, and a wet front bringing squally showers through the west and south of the South Island, but mostly dry in Marlborough to Canterbury regions.

Auckland:	Mostly cloudy, spotty showers, fresh breezes with occasional gusts
Hamilton:	Partly cloudy, dry, fresh winds
Wellington:	Partly cloudy, mild temperatures, dry, windy spells
Greymouth:	Windy, partly cloudy, rain
Christchurch:	Mostly sunny, dry, fresh breezes at times
Dunedin:	Cloudy, colder, windy, light showers

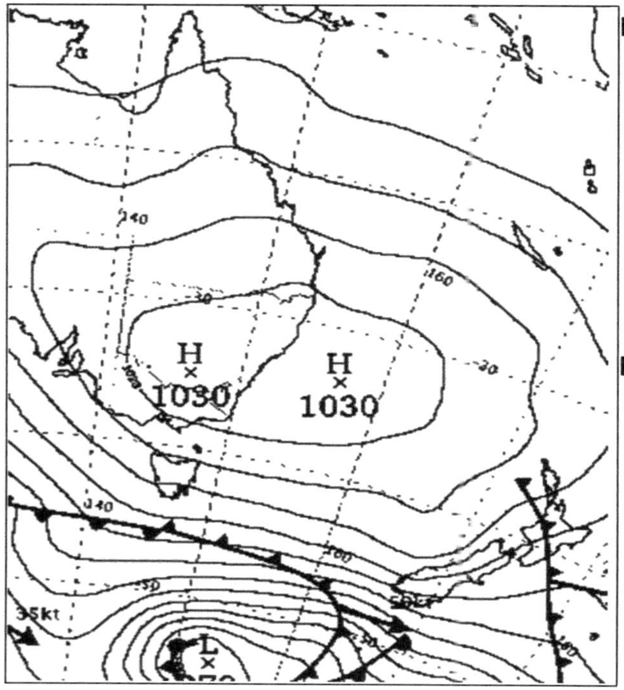

RAIN POTENTIAL

FROST/SNOW

JUNE 6 FRIDAY

An anticyclonic system stalls in the Tasman Sea while a Southern Ocean depression directs a very strong and blustery southwest flow throughout New Zealand. Wet fronts bring further rain into the west of the South Island, with mostly dry and sunnier conditions elsewhere. The colder air sees the likelihood of frosts in the south of the South Island, central districts of the North Island, as well as possibly around the East Cape of the North Island.

Auckland:	Mostly cloudy and dry, chance of isolated showers, wind gusts
Hamilton:	Unsettled and blustery winds, mostly sunny, dry
Wellington:	Sunny, dry, strong gusty winds, chance of gales at times
Greymouth:	Cloudy, rain, strong winds
Christchurch:	Sunny, dry, warmer day cool night, blustery winds
Dunedin:	Mostly cloudy, dry, milder, strong winds possibly to gale force

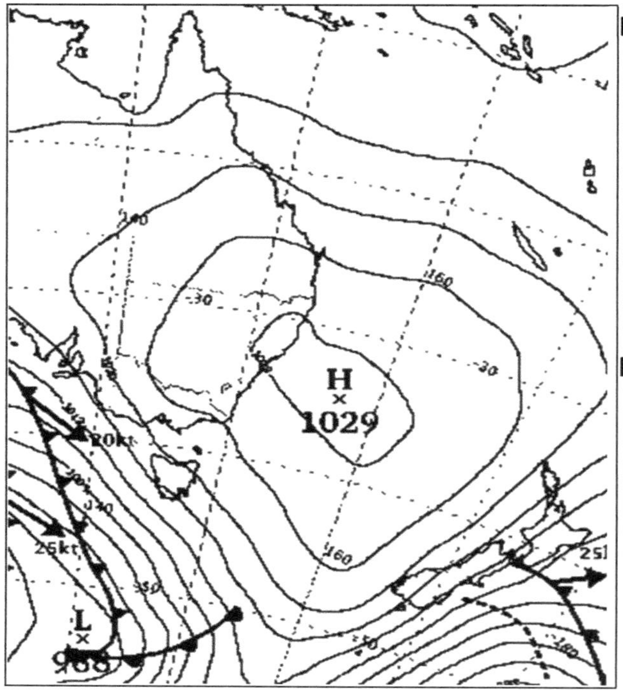

RAIN POTENTIAL

FROST/SNOW

JUNE **7** SATURDAY

An anticyclonic system moves onto New Zealand easing winds to a light northwest flow over the South Island, southerly over the North Island, and with the chance of fog patches in Canterbury districts. Rain activity clears to dry throughout the country, with good sunny hours in many parts, but cloudier in the south of the South Island. Colder everywhere with frosts in the south and east of the South Island and in central parts of the North Island including in the Waikato to the Bay of Plenty.

Auckland: Sunny, dry, light breezes
Hamilton: Sunny, gentle breezes, dry, mild day cold night with chance of frost
Wellington: Sunny, warmer, dry, winds easing to light
Greymouth: Winds easing to light, partly cloudy, dry
Christchurch: Sunny, cool, dry, mostly calm air, chance of overnight fog patches
Dunedin: Partly cloudy, colder, dry, winds easing to light

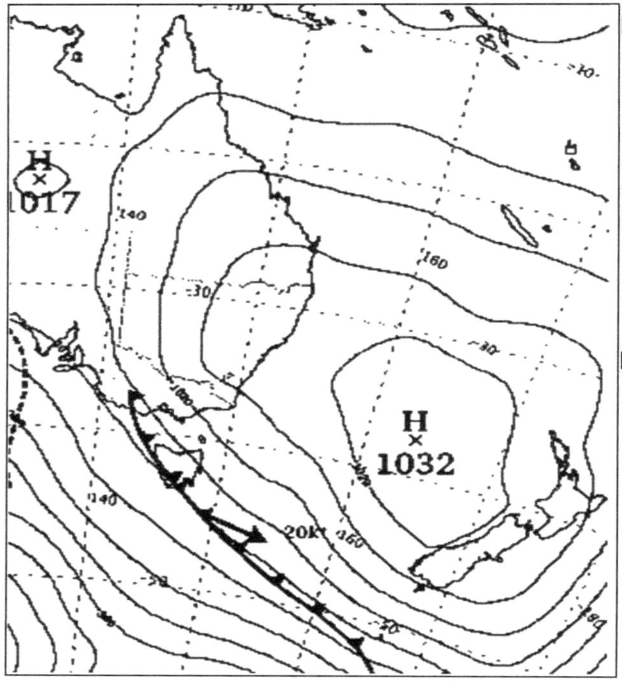

RAIN POTENTIAL

FROST/SNOW

JUNE **8** SUNDAY

High pressure builds over New Zealand directing a light westerly air throughout the country. Cold and dry everywhere, with frosts in central parts of the North Island from the Waikato to Wellington as well as around the Bay of Plenty, and in Southland and Otago. Mostly sunny in Canterbury, dull and gloomy in the south of the South Island, and with occasional cloudy spells elsewhere.

Auckland: Cooler, sunny dry, light shifting breezes to mostly calm air
Hamilton: Sunny, colder, dry, frost likely, calm air
Wellington: Cooler, overcast, dry with occasional threats of rain, shifting breezes
Greymouth: Cloudy, dry, moderate breezes
Christchurch: Mostly calm, dry, sunny spells, chance of frost and fog patches
Dunedin: Overcast, dry, moderate breezes

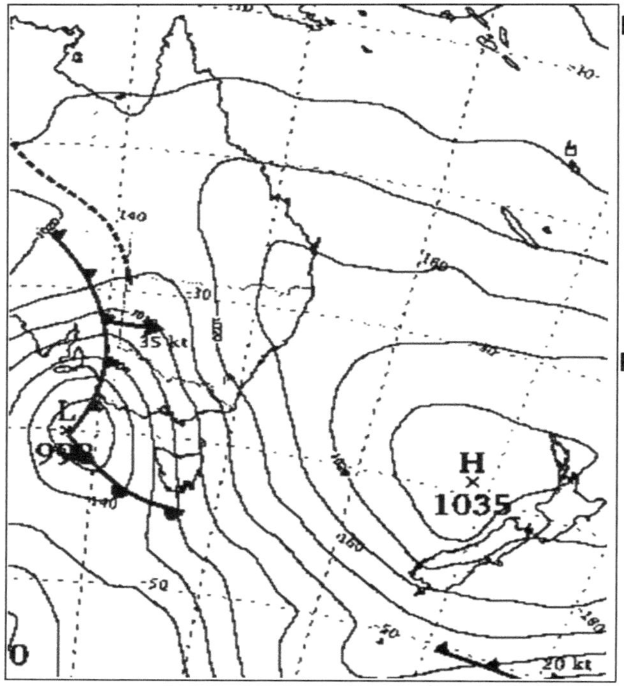

RAIN POTENTIAL

FROST/SNOW

JUNE 9 MONDAY

Anticyclonic throughout New Zealand with a light westerly flow prevailing, mostly calm and with fog patches likely in the east of the country including in Canterbury. Dry and sunny in most districts, with frosts south of Auckland to the Manawatu in the North Island, and in Otago.

Auckland: Overcast, dry with odd threat of rain, light breezes, chance of fog
Hamilton: Partly cloudy, dry, cold overnight, light shifting breezes, chance of fog
Wellington: Mostly sunny and dry, fluctuating breezes
Greymouth: Overcast, dry, light breezes
Christchurch: Calm serene air, warmer day cold night, sunny, dry, chance of fog
Dunedin: Sunny, dry, light breezes

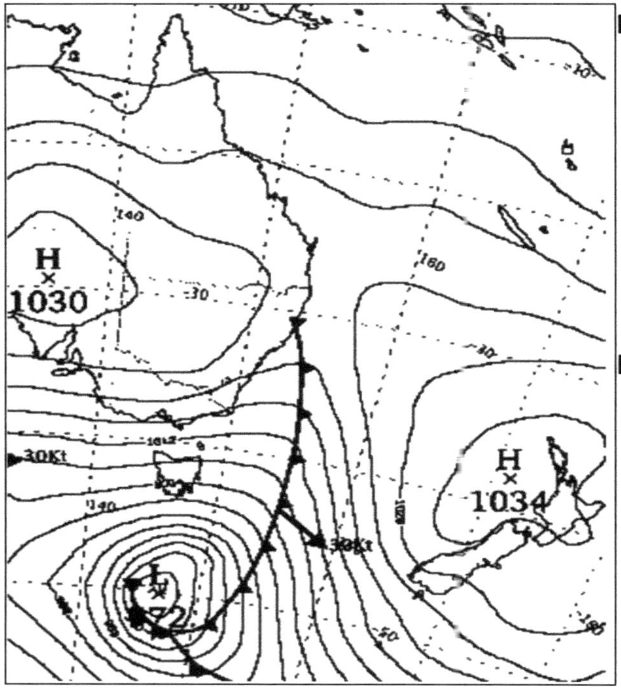

RAIN POTENTIAL

FROST/SNOW

JUNE 10 TUESDAY

High pressure continues to prevail over the northern half of the country under a light westerly air over the North Island, while fresher westerlies over the South Island. Dry and mostly sunny in the North Island with frosts from the Waikato to the Manawatu including in the Bay of Plenty. A depression to the south of New Zealand extends a wet front up through the South Island bringing some rain into Fiordland and Southland, while dry and very cloudy conditions prevail elsewhere. Chance of wintry showers in Fiordland and frosts in the ranges.

Auckland: Mostly sunny, dry, light shifting breezes, chance of fog
Hamilton: Partly cloudy, dry, mild day cold night, fluctuating breezes
Wellington: Cloudy, dry, blustery
Greymouth: Unsettled breezes, overcast, dry
Christchurch: Dry, dull and gloomy air, moderate breezes, cold overnight
Dunedin: Overcast, warmer, dry with odd threat of rain, fresh breezes

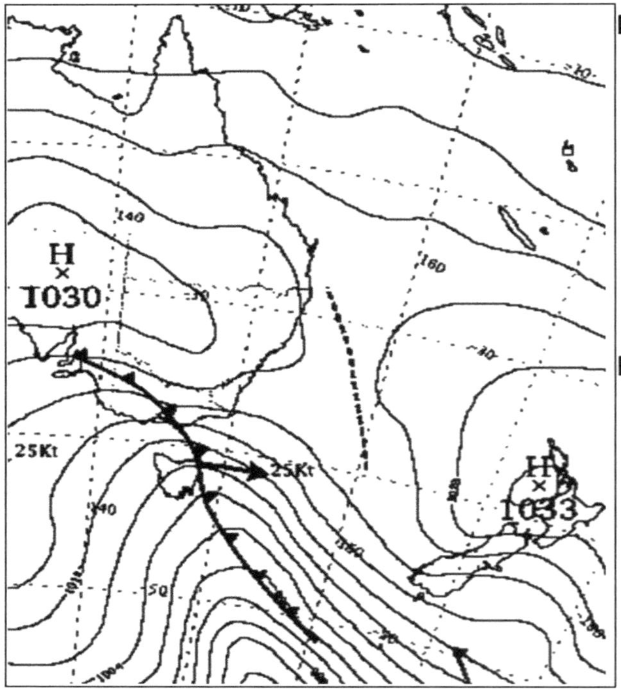

RAIN POTENTIAL

FROST/SNOW

JUNE 11 WEDNESDAY

Low pressure crosses the South Island while a high pressure system departs the North Island with a moderate northwest flow prevailing throughout, fresher through the Cook Strait with possible gales at times, and lighter elsewhere. Mostly sunny, milder, and dry throughout the North Island with chance of fog patches Auckland to the Waikato, while widespread cloud prevails over the South Island with outbreaks of rain, heavy at times, in the West Coast to the Southern Alps, and dry in the eastern and northern districts.

Auckland:	Mostly sunny, dry, cool overnight, gentle breezes, fog possible
Hamilton:	Dry, mild day cold night, fog patches likely, occasional sunny spells
Wellington:	Overcast, dry with increasing threats of rain, windy, chance of gales
Greymouth:	Overcast, scattered showers, moderate winds
Christchurch:	Cloudy, warmer, dry, moderate breezes
Dunedin:	Overcast, light showers, moderate breezes

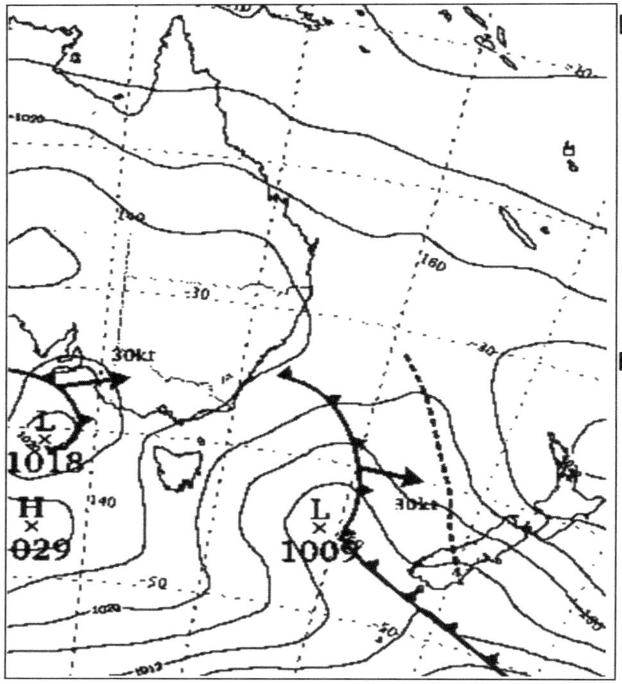

RAIN POTENTIAL

FROST/SNOW

JUNE 12 THURSDAY

Low pressure dominates New Zealand bringing a stormy air throughout the country with fresh to strong westerly winds prevailing. Mostly dry in Northland, the East Cape to the Hawkes Bay, and in Canterbury, with rain activity elsewhere, with heavy rain likely in the west and south of the South Island and around Taranaki to the Kapiti Coast.

Auckland:	Cloudy, isolated showers, fresh breezes at times
Hamilton:	Cloudy, milder, light showers, windy spells
Wellington:	Mix of showers and sunny spells, strong winds, possible gales
Greymouth:	Stormy air, mix of sunny spells and heavy rain, chance of flooding
Christchurch:	Partly cloudy, dry, light shifting breezes, possible fog patches
Dunedin:	Cooler, overcast, rain, fluctuating breezes

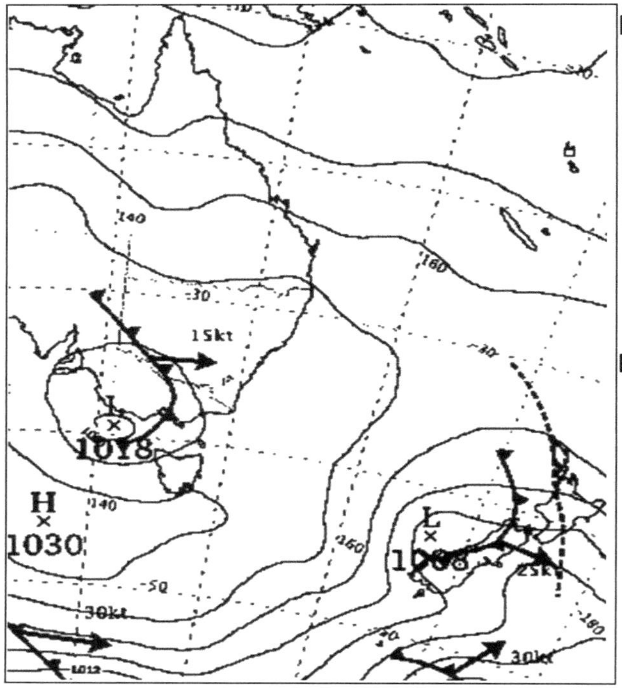

RAIN POTENTIAL

FROST/SNOW

JUNE **13** FRIDAY

Low pressure weakens over the country with winds easing to a light southwest flow throughout while a high pressure system ridges to the west of the country. A mixed day with sunny spells and outbreaks of rain everywhere, although mostly dry in the East Cape to the Hawkes Bay. Colder with frosts in the Waikato to the Bay of Plenty and Central Plateau as well as in the east and south of the South Island. Chance of fog in Canterbury, and light snow on the higher reaches.

Auckland:	Mix of showers and sunny spells, light breezes
Hamilton:	Changeable, mix of rain and sunny spells, light breezes
Wellington:	Cloudy, warmer day cooler night, dry, fluctuating breezes
Greymouth:	Mostly sunny, overnight rain, light breezes to mostly calm air
Christchurch:	Partly cloudy, dry, mild day cold night, serene air, fog possible
Dunedin:	Colder, mix of sunny spells and showers, mostly calm air

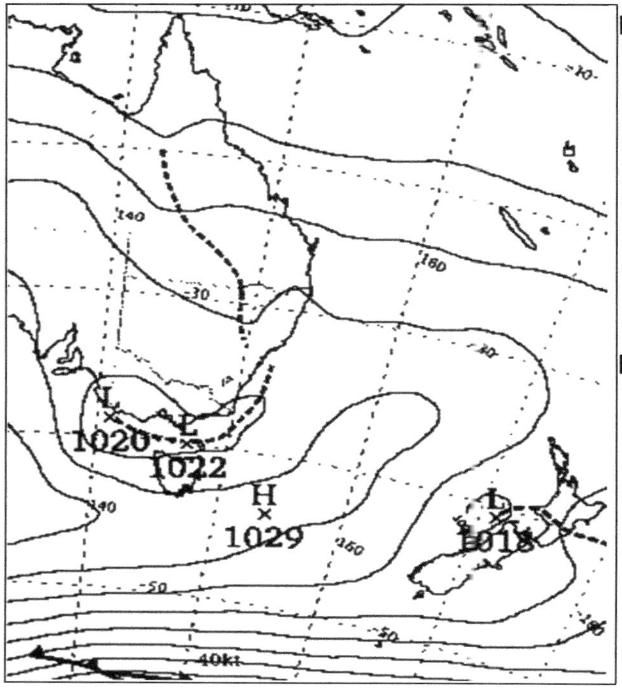

RAIN POTENTIAL

FROST/SNOW

JUNE **14** SATURDAY

High pressure ridges onto the North Island, while low pressure continues to sweep through the South Island, with a southwesterly flow prevailing, strong in the south, lighter over the North Island. Some lingering showers in the west of both islands, while mostly dry with good sunny spells in the east. Cold with frosts in the south and east of the South Island, and in the Wellington and Bay of Plenty regions of the North Island. Fog likely in Canterbury.

Auckland:	Rain, some sunny breakthroughs, cooler, gentle breezes
Hamilton:	Cloudy, light showers, mostly calm air
Wellington:	Dry and cooler, partly cloudy, strong breezes at times
Greymouth:	Cloudy, brief showers, cold overnight, breezy spells
Christchurch:	Mostly sunny, cold with frosts, dry, fog patches possible
Dunedin:	Partly cloudy, mostly dry, cool, fresh breezes at times

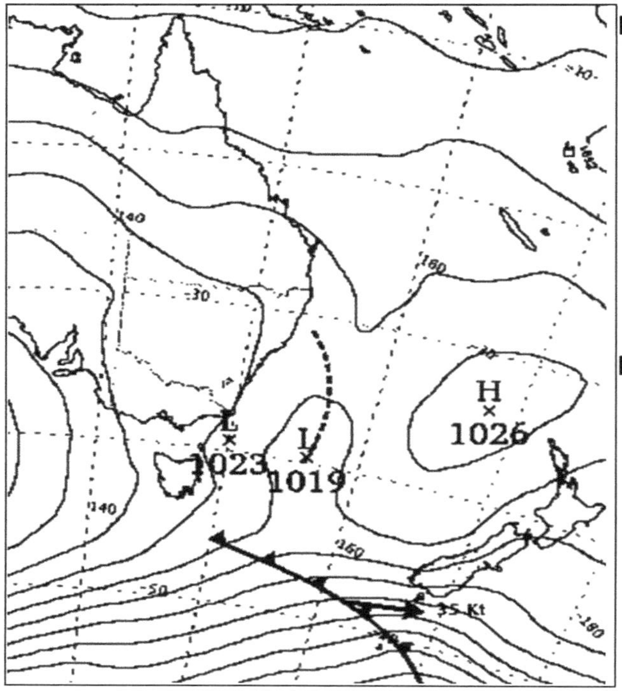

RAIN POTENTIAL

FROST/SNOW

JUNE 15 SUNDAY

A Southern Ocean depression spreads strong and blustery southwest winds up throughout the country, with gales likely in many western and central districts, including through the Cook Strait. Heavy rain in the west of the South Island, with wintry showers in Southland and Otago, and occasional showers in Northland to Auckland. Mostly dry and gloomy with occasional threats of rain elsewhere.

Auckland: Cloudy, scattered showers, windy spells
Hamilton: Overcast, cooler, isolated showers, fresh winds at times
Wellington: Overcast, dry, stormy air with strong winds and gales likely
Greymouth: Stormy, overcast, rain, colder, chance of gales and thunderstorms
Christchurch: Overcast, cool, dry, frosts, strong winds, chance of gales
Dunedin: Stormy air, overcast, isolated showers, chance of winds to gale force

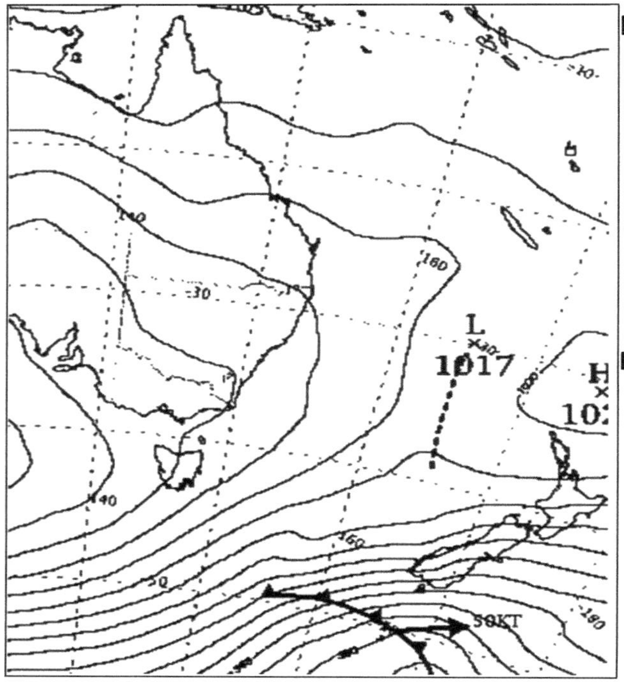

RAIN POTENTIAL

FROST/SNOW

JUNE 16 MONDAY

Low pressure continues to bring strong and blustery west to southwest winds throughout the country, with outbreaks of rain in all western parts of the country from Northland to Southland and Fiordland, while remaining mostly dry and sunny in eastern regions from the East Cape to Canterbury.

Auckland: Partly cloudy, scattered showers, some heavy, fresh to strong winds
Hamilton: Mix of sunny spells and showers, warmer, strengthening breezes
Wellington: Sunny, dry, strong gusty winds, chance of gales at times
Greymouth: Stormy static air, cloudy, rain, strong winds with chance of gale gusts
Christchurch: Unsettled and blustery winds, mostly sunny, warmer, dry
Dunedin: Partly cloudy, dry, very strong and unsettled winds, possible gales

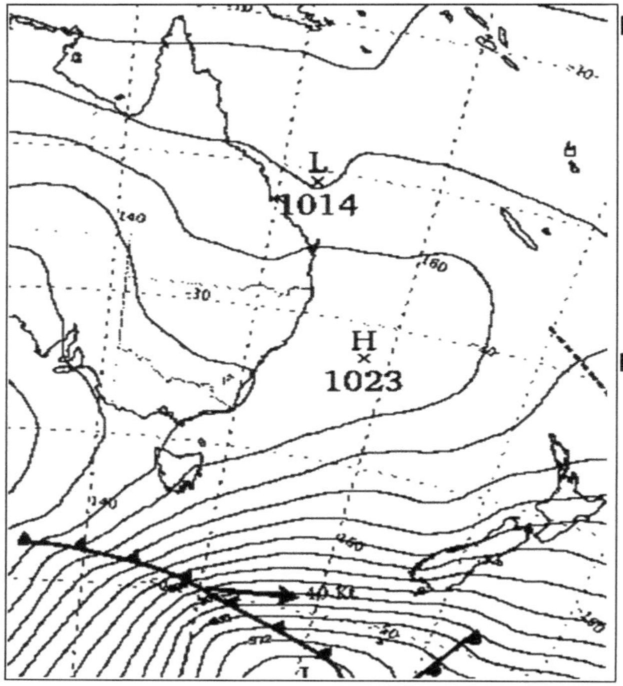

RAIN POTENTIAL

FROST/SNOW

JUNE **17** TUESDAY

An unstable southwesterly flow associated with a Southern Ocean depression flows across the country bringing stormy conditions with strong to gale force winds almost everywhere. Wintry showers prevail through the western half of the country, including the likelihood of snow to low levels in the south of the South Island. Mostly dry in the east of the country from the Bay of Plenty to Canterbury.

Auckland:	Partly cloudy, isolated showers, strong gusty winds
Hamilton:	Cloudy, light showers, strong blustery winds
Wellington:	Mostly sunny, warmer, blustery, winds possibly to gale force at times
Greymouth:	Cloudy, blustery, scattered showers, chance of gales
Christchurch:	Dry, partly cloudy, milder overnight, strong winds, gales possible
Dunedin:	Partly cloudy, light showers, strong winds, gales possible

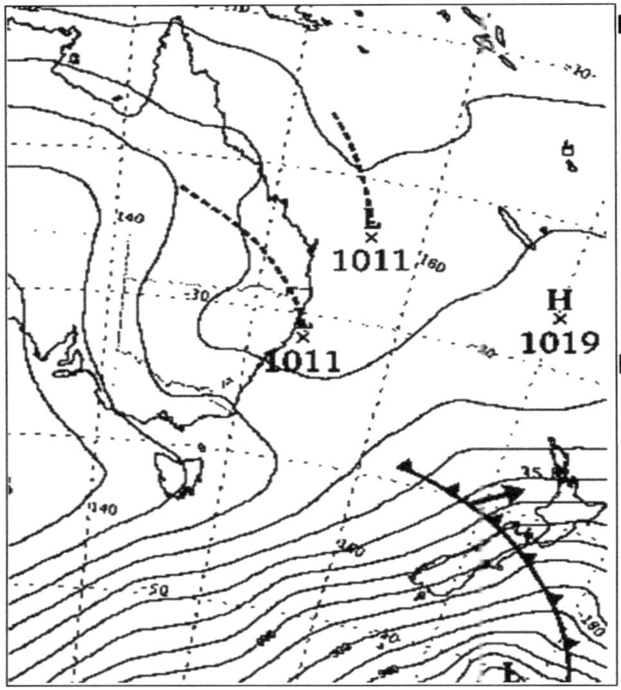

RAIN POTENTIAL

FROST/SNOW

JUNE **18** WEDNESDAY

Storm conditions continue throughout the country with a blustery west to southwest flow prevailing, up to gale force at times in most regions. Wet fronts bring heavy precipitation throughout the west and south of the South Island and lighter rain over most of the North Island, although likely to remain mostly dry in the east from the East Cape to Canterbury. Snow likely to low levels in the South Island. Occasional sunny outbreaks in most North Island regions.

Auckland:	Partly cloudy, gusty winds, squally showers, chance of gale gusts
Hamilton:	Unsettled and windy, partly cloudy, light showers
Wellington:	Partly cloudy, dry, strong winds, gales possible
Greymouth:	Cloudy, storm with heavy rain, colder, strong winds, gales likely
Christchurch:	Unsettled winds, gales possible, partly cloudy, isolated showers
Dunedin:	Cloudy, colder, squally showers, strong blustery winds

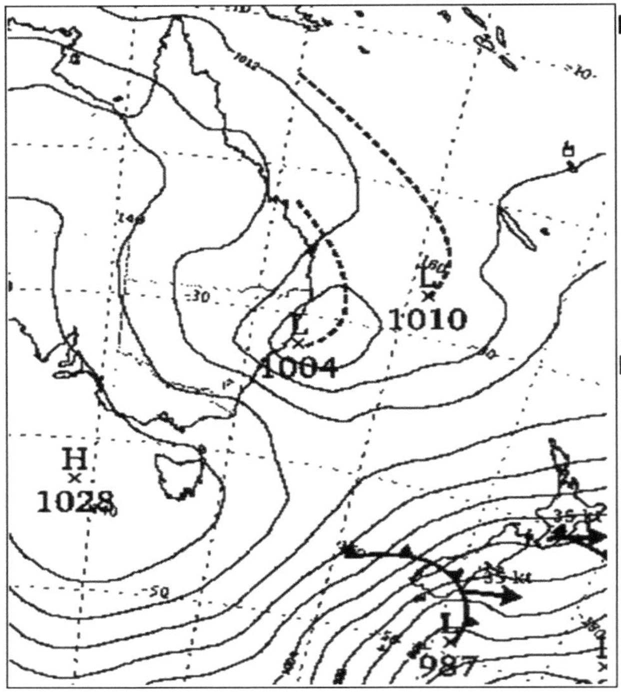

RAIN POTENTIAL

FROST/SNOW

JUNE 19 THURSDAY

The depression that brought stormy conditions over the past couple of days moves away to the east of the country while a high pressure ridge nears the west of the South Island and another low pressure trough approaches the north of the North Island. Southwest winds ease but still remain fresh throughout. A mixed day everywhere with some sunny spells in central New Zealand, cloudier to the north and south, and outbreaks of wintry precipitation in the west and south of the South Island and central parts of the North Island.

Auckland: Partly cloudy, light showers, winds easing to moderate
Hamilton: Cloudy, winds easing to moderate, light showers
Wellington: Partly cloudy, cooler, dry, winds easing
Greymouth: Changeable, some sunny spells, afternoon rain, colder, winds easing
Christchurch: Sunny, colder, dry, fresh breezes, frosts likely
Dunedin: Winds easing to moderate, overcast, cold, occasional showers

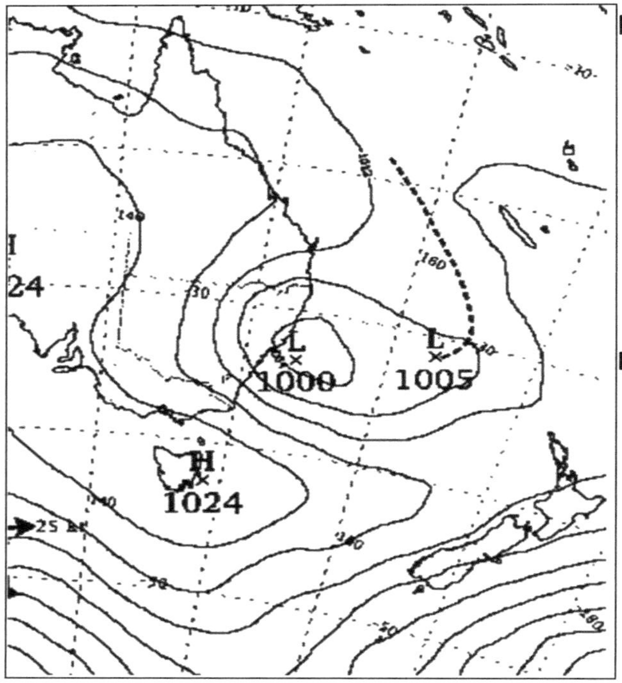

RAIN POTENTIAL

FROST/SNOW

JUNE 20 FRIDAY

A low pressure trough between two depressions crosses the country directing a westerly flow over the South Island, easterly over the North Island. Cold and dull and gloomy everywhere with lingering wintry showers in the south of the South Island. Rain moves into Northland with some heavy falls at times, and isolated showers down through Auckland and into the Waikato. Mostly dry with passing threats of rain elsewhere.

Auckland: Moderate breezes, overcast, passing showers
Hamilton: Overcast, isolated showers, cooler, moderate breezes
Wellington: Overcast, cooler, mostly dry, threats of rain, windy spells
Greymouth: Cloudy, dry, light breezes
Christchurch: Overcast, cool with frosts, light shifting breezes
Dunedin: Mix of showers and sunny spells, cold, light shifting breezes

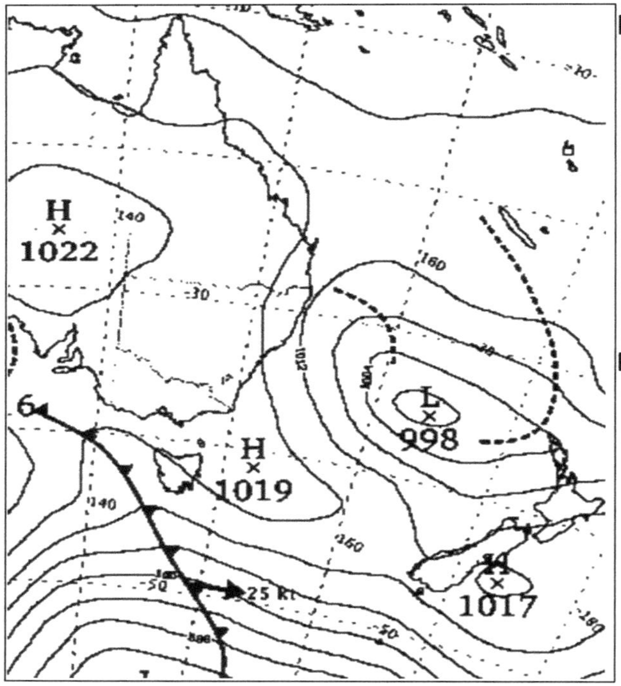

RAIN POTENTIAL

FROST/SNOW

JUNE 21 SATURDAY

A low pressure trough associated with a Southern Ocean depression directs a light air over the South Island, and mostly dry, sunny conditions with widespread frosts prevailing including possible fog patches in Canterbury. In contrast, a depression over the North Island brings unsettled easterly winds, widespread heavy rain in all North Island districts, and snow likely in the Central Plateau. Heavy rain may bring pockets of localised flooding in susceptible areas.

Auckland: Cloudy, pockets of rain, fluctuating breezes
Hamilton: Overcast, colder, rain, fluctuating winds
Wellington: Dull and gloomy, isolated showers, unsettled and fluctuating winds
Greymouth: Mostly sunny, cool, dry, mostly calm air, chance of frost pockets
Christchurch: Partly cloudy, cold with severe frosts, dry, calm air, chance of fog
Dunedin: Mostly sunny, dry, cold with overnight frosts, light breezes

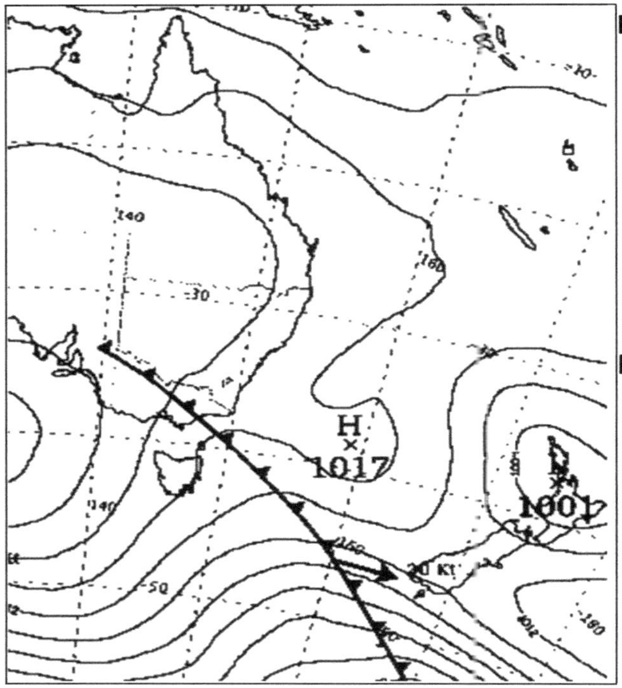

RAIN POTENTIAL

FROST/SNOW

JUNE 22 SUNDAY

A light east to southeasterly flow within a low pressure trough prevails over New Zealand. Widespread frosts throughout the South Island with wintry showers in Southland and in the higher reaches, and mostly dry with occasional sunny spells elsewhere. In the North Island, temperatures will be milder, with heavy rain easing to occasional showers in most districts overnight, but still heavy in the Hawkes Bay to the Wairarapa. Mostly sunny day in central regions of the North Island.

Auckland: Overcast, warmer, brief showers, winds easing to light
Hamilton: Rain, warmer, some sunny spells, light winds
Wellington: Cloudy, dry, light breezes
Greymouth: Cloudy, dry, mostly calm air, possible frost pockets
Christchurch: Light breezes, mostly sunny, cold with frosts, dry
Dunedin: Mostly sunny, dry, cold with frost, light shifting breezes

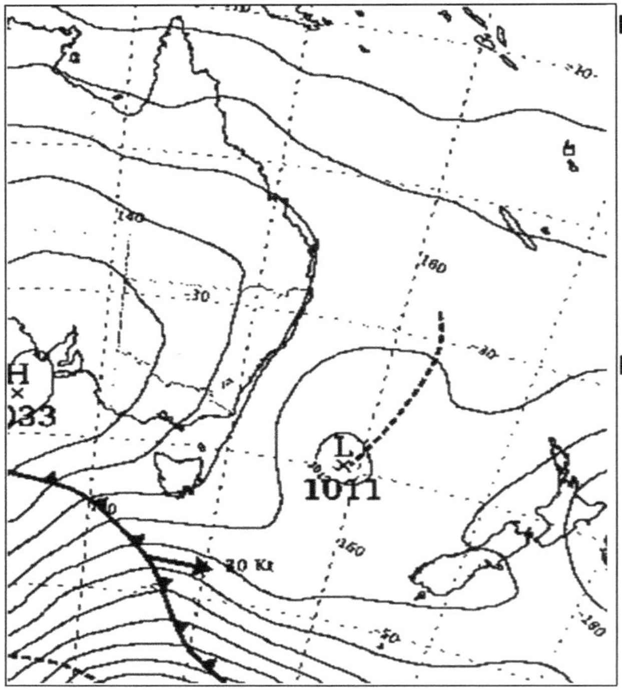

RAIN POTENTIAL

FROST/SNOW

JUNE 23 MONDAY

Low pressure continues over New Zealand directing a light northeasterly flow throughout, with chance of fog patches in many sheltered districts. Mostly cloudy and dry with widespread frosts in the South Island, although some isolated showers may move into Southland overnight. Mostly cloudy in the northern half of the North Island and in the Wairarapa, sunnier through central parts from Taranaki to the East Cape, and some isolated overnight showers likely in most regions except Northland.

Auckland: Cloudy, occasional showers, mostly calm air
Hamilton: Cloudy, dry, mild day cool night, mostly calm air
Wellington: Mostly calm air, mix of sunny spells and isolated showers
Greymouth: Cloudy, dry, cool, calm and serene air
Christchurch: Cloudy, dry, cool with severe overnight frosts, calm air, chance of fog
Dunedin: Partly cloudy, cold, dry, frosts, calm and serene air

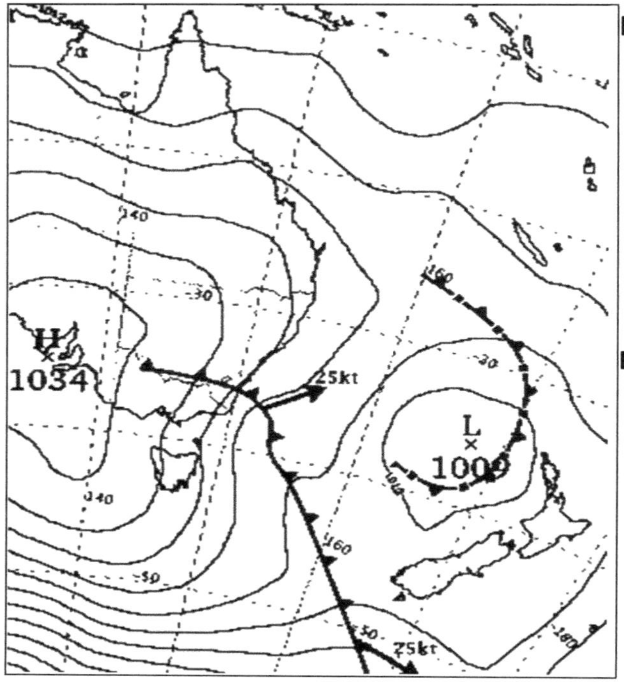

RAIN POTENTIAL

FROST/SNOW

JUNE 24 TUESDAY

A complex of low pressure prevails over the North Island in a light to moderate easterly flow, while high pressure ridges over the South Island, directing moderate westerly breezes at first, backing easterly later in the day. Mostly cloudy in the North Island spotty showers from Northland to the Bay of Plenty, while mostly dry in the western and southern regions. In the South Island, frosts will be widespread, with snow flurries and wintry overnight showers in southern regions, while mostly dry in the northern half.

Auckland: Cloudy, isolated pockets of rain, fluctuating breezes
Hamilton: Cloudy and dry, light breezes
Wellington: Cloudy, cool, isolated showers, chance of snow flurries on ranges
Greymouth: Partly cloudy, dry, light shifting breezes
Christchurch: Partly cloudy, dry, cold with frosts, light shifting breezes
Dunedin: Cloudy, dry, cold and frosty air, gentle breezes

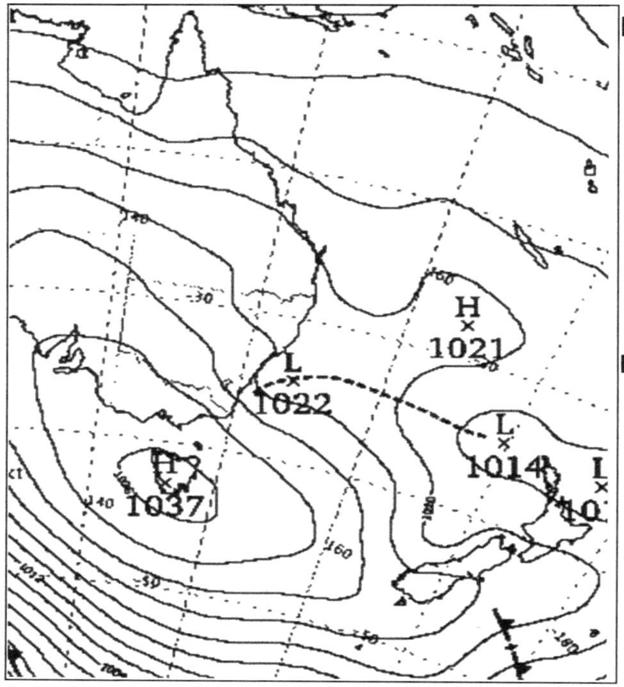

RAIN POTENTIAL

FROST/SNOW

JUNE **25** WEDNESDAY

A high pressure system crossing the southern Tasman Sea moves onto the country directing a light westerly flow over the South Island while turning southeasterly over the North Island. Widespread frosts continue over the South Island with some spotty wintry showers in Canterbury, while milder with outbreaks of rain in the east and far north of the North Island. Mainly dry elsewhere.

Auckland: Partly cloudy, pockets of heavy overnight rain, mostly dry day
Hamilton: Cloudy, cooler, isolated showers, gentle breezes
Wellington: Partly cloudy, overnight showers, chance of snow flurries in ranges
Greymouth: Slightly warmer, dry, partly cloudy, gentle shifting breezes
Christchurch: Light shifting breezes, overcast, dry, cool, fog possible
Dunedin: Overcast, dry with odd misty patch possible, mostly calm air

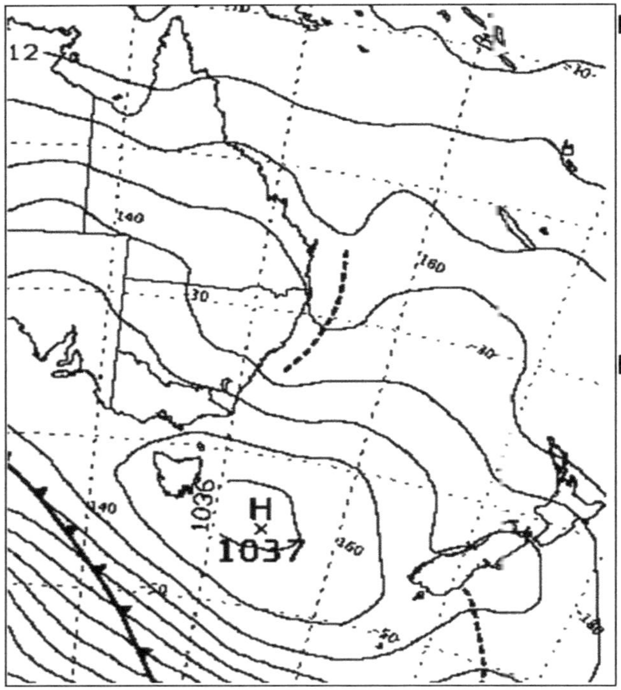

RAIN POTENTIAL

FROST/SNOW

JUNE **26** THURSDAY

A high pressure system crosses to the south of the country extending a light to moderate southeasterly flow throughout, while a low pressure trough lingers to the north of the North Island. Cold everywhere with frosts in many districts from the Waikato southwards. Some lingering overnight showers in the east of the country from the East Cape to Otago, while mostly dry in western regions. Occasional sunny spells in most parts of the country except remaining dull and gloomy from the Wairarapa to Canterbury.

Auckland: Sunny, cooler, mostly dry, fresh breezes
Hamilton: Partly cloudy, dry, mild day cold overnight, moderate breezes
Wellington: Cloudy, cooler, scattered showers, windy spells
Greymouth: Mostly sunny, dry, mild day cold night, light breezes
Christchurch: Overcast, scattered showers, gentle breezes to calm air, possible fog
Dunedin: Overcast, showers, mostly calm air

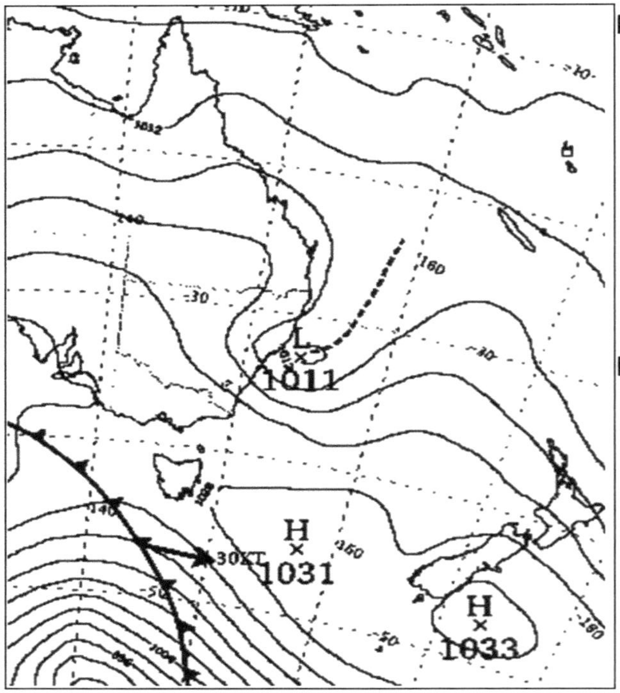

RAIN POTENTIAL

FROST/SNOW

JUNE **27** FRIDAY

Moderate east to southeasterly breezes continue to prevail over the country, more unsettled at times through the centre of New Zealand. Overnight rain likely in the east of the country from the East Cape to Wellington, with some heavy falls and possible thunderstorms at times, and showers into the northeast of the South Island and misty drizzles down into Canterbury. Mostly dry elsewhere. Severe frosts in many parts of the South Island and frosts in the Waikato to the Bay of Plenty and Central Plateau.

Auckland:	Cloudy, cooler, dry, fresh winds
Hamilton:	Cloudy, cooler, dry, chance of frost pockets, moderate breezes
Wellington:	Unsettled windy spells, cloudy, rain
Greymouth:	Mostly sunny, dry, slightly warmer, light shifting breezes
Christchurch:	Partly cloudy, light showers, cold, frosts likely and fog possible
Dunedin:	Mostly sunny, cold, dry, overnight showers likely, gentle breezes

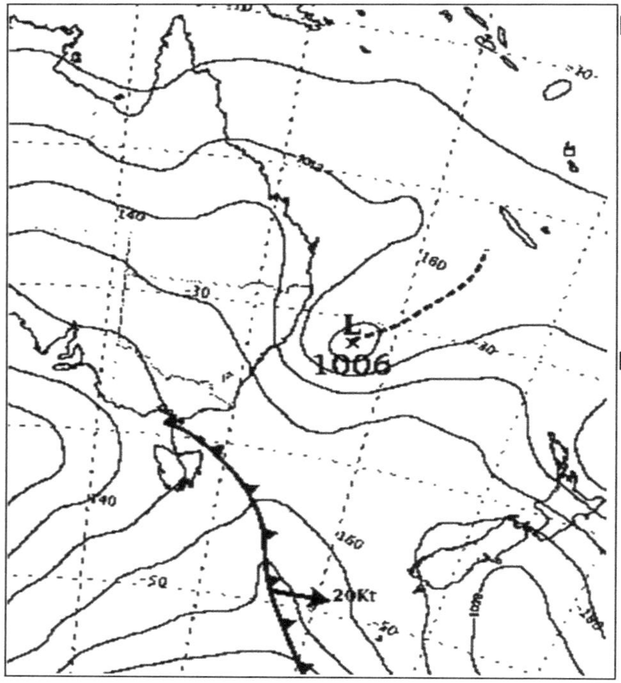

RAIN POTENTIAL

FROST/SNOW

JUNE **28** SATURDAY

A weak anticyclonic system moves onto New Zealand bringing a mostly calm and serene air throughout. Widespread frosts throughout the South Island and over much of the North Island except in northern and eastern regions. Some rain lingering overnight once more in eastern districts from the East Cape to Wellington, and in Fiordland to southern parts of the West Coast. Mostly dry and sunny elsewhere with fog patches possible.

Auckland:	Sunny, dry, mild day cold night, light breezes
Hamilton:	Sunny, dry, mild day cold night, pockets of frost likely, light breezes
Wellington:	Cloudy, cool, scattered showers, windy spells
Greymouth:	Cloudy, dry, colder, serene air, chance of frost in sheltered pockets
Christchurch:	Partly cloudy, cold, severe frosts, dry, calm air, fog likely
Dunedin:	Partly cloudy, dry, calm atmosphere, cold

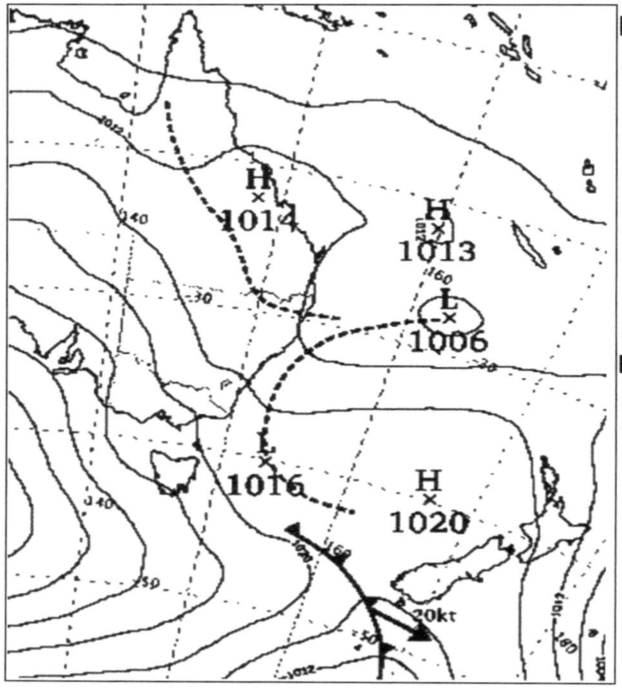

RAIN POTENTIAL

FROST/SNOW

JUNE 29 SUNDAY

An unstable northeasterly flow begins to develop over the country as a depression in the Tasman Sea nears. Generally, mostly dry, very cold and frosty everywhere except for some isolated snow flurries likely in Southland, spotty showers overnight on the East Cape, and milder overnight temperatures in Northland, Auckland, and the Hawkes Bay. Fog likely in sheltered districts, especially in Canterbury. Mostly cloudy in the South Island, while mostly sunny in the North Island.

Auckland: Partly cloudy, dry, light breezes freshening later in the day
Hamilton: Partly cloudy, dry, calm air, frosts
Wellington: Mostly sunny, light showers, moderate breezes, cold overnight
Greymouth: Overcast, colder, dry, light breezes
Christchurch: Partly cloudy, dry, cold with frosts, light breezes, chance of fog
Dunedin: Cloudy, dry, moderate breezes

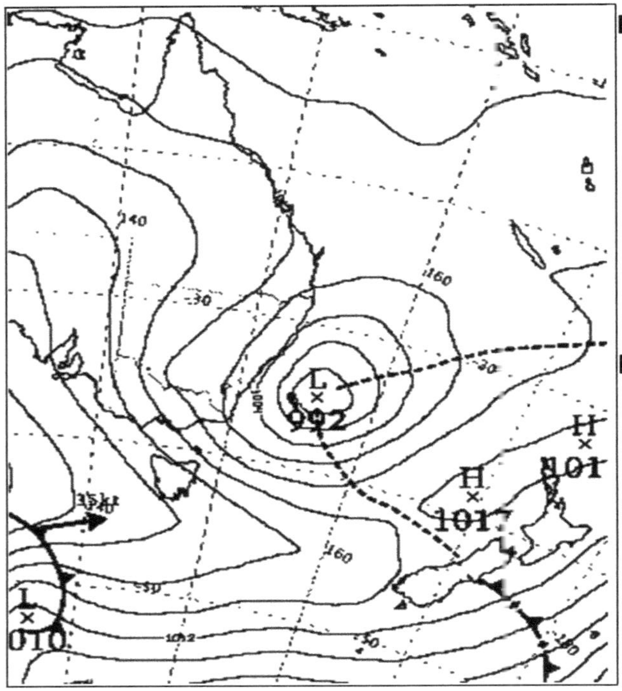

RAIN POTENTIAL

FROST/SNOW

JUNE 30 MONDAY

An intense depression that strengthened in the Tasman Sea nears New Zealand strengthening winds into a fresher northeasterly flow. Some showery activity mostly overnight in Northland, while mainly dry elsewhere. Very cold with severe frosts in many parts of the country, except in Northland, Auckland, the Hawkes Bay, and the north and west of the South Island. Cloud thickens everywhere with increasing threats of rain.

Auckland: Overcast, mostly dry with storm threatening, strengthening winds
Hamilton: Overcast, dry with threats of rain, frost likely, fresh breezes at times
Wellington: Overcast, cool, dry, moderate breezes
Greymouth: Dry, dull and gloomy, freshening breezes later in the day
Christchurch: Cloudy, cold with overnight frosts, dry, light breezes
Dunedin: Mostly sunny, isolated showers, cold with frost pockets, light breezes

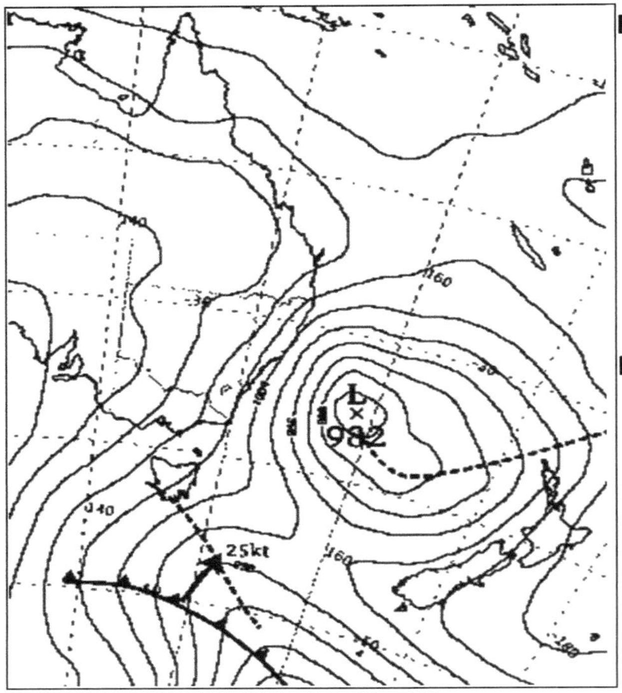

RAIN POTENTIAL

FROST/SNOW

JULY

PHASES OF THE MOON

1st	Crossing Equator
2nd	First Quarter
5th	Apogee
9th	Southern Declination
10th	Full Moon
16th	Crossing Equator
18th	Third Quarter
20th	Perigee #11
22nd	Northern Declination
24th	New Moon
29th	Crossing Equator

1	2	3	4	5	6	7	8	9	10	11	12	13	14	15	16	17	18	19	20	21	22	23	24	25	26	27	28	29	30	31
Tue	Wed	Thu	Fri	Sat	Sun	Mon	Tue	Wed	Thu	Fri	Sat	Sun	Mon	Tue	Wed	Thu	Fri	Sat	Sun	Mon	Tue	Wed	Thu	Fri	Sat	Sun	Mon	Tue	Wed	Thu
XhS	1Q			A				V	F						XhN		3Q		P11		^		N					XhS		

MONTHLY SUMMARY

23rd June-2nd July Frequent severe frosts likely during this spell especially in inland Otago districts.
27th June-1st July A depression crosses the southeast of the country in a blustery southwest flow.
30th June-4th July A strong southwesterly returns over the South Island bringing further snow to low levels in Southland and Otago, and including in the high country passes. Snow also in the Central North Island, possibly causing the closure of the Desert Road for a time, as well as snow in the Reefton and Hanmer Forest areas. More settled conditions in the east of the country, however, widespread severe frosts are likely to prevail. Snow may reach sea level in some coastal parts of Southland and Otago.
1st Scattered thunderstorms and blustery winds in the Wellington region.
1st-3rd A band of thunderstorms cross the northern half of the North Island bringing scattered thunderstorms through the Waikato and Auckland regions.
1st-6th A series of wet fronts cross the North Island.
2th Very severe frosts likely in many districts, particularly severe in the Tara Hills.
2nd-5th Widespread rain, often heavy, in the West Coast.
3rd Snow in Otago may reach sea-level and accumulate on the hills around Dunedin.
4th Very strong winds around the Otago Peninsula region. Chance of fog patches in the Wellington region.
5th Stormy air with strong gusty winds in the Wellington region and chance of snow in the surrounding hills.
6th Winds may reach gale force through the Cook Strait, possibly affecting ferry services for a time. Severe frosts in Otago and Southland.
7th-10th A depression crosses the southeast of the country.
8th Further severe overnight frosts likely in the south of the South Island.
9th-10th Heavy rain likely in the West Coast and into the Nelson-Marlborough region with chance of localised flooding, particularly in the Hokitika district to Takaka.
10th A wet front crossing the North Island brings heavy rain to most districts, particularly in the Kerikeri and Te Puke areas. Heavy rain also likely in the Lakes region of the South Island and around Coastal Otago.

Over the next four weeks, up to five depressions will form in the central Tasman Sea and track over the north of the North Island, or just above, with three of them

17th-1st Aug A period of severe frosts and freezing fogs with temperatures constantly below zero in most days in Central Otago during this spell. In contrast,

likely in the last week to ten days of July. In contrast, anticyclones will pass to the south of the country and over the southern half of the South Island. As such, north-easterly winds will prevail over most of the North Island and easterly over the South Island. This weather pattern will make the next four weeks a period of extremes and contrasts, with the chance of flooding events, tornadoes, and severe frosts. Overall, there will be above normal rain in the north and east of the North Island, as well as down the coastal fringes of South Canterbury and Otago and including the Hawke's Bay and Northland, where up to twice as much rainfall than is normal over this period. In contrast, the north and west of the South Island will be drier than average for this time of year, possibly only seeing around half their usual rainfall for this time of year. Storms bringing gales and heavy rain during this outlook period may see flooding events in Northland in the second week of this outlook period, the Hawke's Bay at the end of July, and Otago to south Canterbury in the last week of this outlook period. Sunshine hours will be around 20% more in the north Westland districts, as well as in parts of Canterbury, and Otago districts, while below average in eastern parts from Gisborne to Marlborough and around the average elsewhere. The unsettled conditions over the next four weeks will see average temperatures around 1°C below the norm in many parts of Otago, Southland and Canterbury. In contrast, Northland, the Coromandel, Waikato, and Gisborne districts are likely to be warmer by a similar variant. It will also be around average to half a degree warmer in other parts of the North Island, while average temperatures can be expected elsewhere. Other major weather events that may be noted during the next four weeks may be swarms of tornadoes, particularly in Taranaki, thunderstorms, spells of disruptive fog, gale winds, severe frosts, and snow.

11th Very mild temperatures around the East Coast, especially in the Gisborne area.
11th-12th Chance of fog in the north of the South Island possibly disrupting airport services at Nelson Airport to Wellington.
11th-17th Widespread rain over the North Island with regular heavy falls.
13th-14th Possible fog in the Christchurch area.
14th-16th A depression crosses the North Island in an unstable northwesterly flow bringing scattered thunderstorms in many places and the chance of tornados forming in the west of the North Island from Auckland to Taranaki as well as the chance of a possible tornado being seen in the Tauranga area.

temperatures are likely to be slightly milder than normal for this time of year over most of the North Island.

19th-22nd Depression crosses the north of the North Island. During this period, easterly winds prevail and may reach gale force at times, as well as bringing heavy rainfall in many parts of Northland including the chance of widespread flooding, and with the storm conditions bringing damage to infrastructures in some parts. This storm system may also bring severe winds, high rainfall, and localised damage to other districts from Auckland to the Coromandel.

22nd-31st Sub-zero temperatures prevail almost everywhere throughout the southern half of the South Island, including in Central Otago to inland Southland during this time. Chance of sea-ice being observed around some southern coasts, including in Fiordland and into Doubtful Sounds.

23rd Fog possible in Christchurch.

25th-28th Depression crosses the North Island in a fresh to strong northeasterly flow, tracking mostly easterly over the South Island. During this period, heavy rain may cause major flooding in some parts of Northland, and the Hawkes Bay, including near Hastings.

26th-28th Chance of fog in Christchurch.

28th Very cold in Central Otago with severe overnight frost minimas likely that may set new records.

30th-31st Another depression crosses the North Island. Chance of thunderstorms and heavy rain particularly in the Bay of Plenty, and Waikato to Auckland regions, and possible gale gusts in the Wellington region. Further south, chance of fog in the Christchurch region may develop into drizzle.

July 2025

Monday	Tuesday	Wednesday	Thursday	Friday	Saturday	Sunday
30 **Moon Sign**	**1** **Virgo** Conjunct Mars	**2** **XhS** **Virgo** Trine Venus Square Jupiter Opposition Saturn Trine Uranus Opposition Neptune Trine Pluto	**3** **1Q** **Libra** Square Sun Square Jupiter	**4** **Libra**	**5** **A** **Scorpio** Square Mercury Trine Jupiter Square Pluto	**6** **Scorpio** Trine Sun
7 **Scorpio** Opposition Venus Trine Saturn Opposition Uranus Trine Neptune Sextile Pluto	**8** **Sagittarius** Trine Mercury Square Mars	**9** **V** **Sagittarius** Square Ascendant	**10** **Capricorn** Trine Mars Opposition Jupiter Square Saturn Square Neptune Square Ascendant	**11** **FM** **Capricorn** Opposition Sun	**12** **Aquarius** Trine Venus Sextile Saturn Trine Uranus Sextile Neptune Conjunct Pluto	**13** **Aquarius** Opposition Mercury
14 **Pisces** Trine Jupiter Square Uranus	**15** **Pisces** Trine Sun Square Venus Opposition Mars	**16** **XhN** **Aries** Trine Sun Conjunct Saturn Sextile Uranus Conjunct Neptune Sextile Pluto	**17** **Aries** Trine Mercury Sextile Venus	**18** **3Q** **Aries** Square Sun Square Pluto	**19** **Taurus** Square Mercury Trine Mars	**20** **Taurus** Sextile Sun Sextile Saturn Conjunct Uranus Sextile Neptune Trine Pluto Square Ascendant
21 **P11** **Gemini** Sextile Mercury Conjunct Venus Square Mars Square Ascendant	**22** **^** **Gemini** Square Saturn Square Neptune Square Ascendant	**23** **Cancer** Conjunct Jupiter Square Ascendant	**24** **Cancer**	**25** **NM** **Cancer** Conjunct Sun Trine Saturn Sextile Uranus Trine Neptune Opposition Pluto Square Ascendant	**26** **Leo** Conjunct Mercury Square Ascendant	**27** **Leo** Sextile Venus Square Uranus
28 **Virgo** Sextile Jupiter	**29** **XhS** **Virgo** Square Venus Conjunct Mars	**30** **Libra** Sextile Sun Square Jupiter Opposition Saturn Trine Uranus Opposition Neptune Trine Pluto	**31** **Libra**	1	2	3
4	5					

NORTH ISLAND RAINFALL

JULY 2025

d=mainly dry, n= fog, mist, dew or nonrecordable, l=light showers, p=passing showers, s=significant showers, r=rain, h=heavy falls

		Northland							Auckland					Bay Of Plenty				Waikato				Rot/Tau		Hawkes Bay			C Plateau			Taranaki			Manawatu			Lower North Island					
		CR	Kta	Ker	Rsl	Dar	Whg	Man	Wark	Hen	MkH	Ak	Co	Tau	TeP	Wkt	Opo	Hml	TeA	Mata	Ot	Rt	Taup	Gis	Nap	Has	Taum	Chat	Waio	NP	Str	Wang	Pa	Dann	Lev	PN	Mast	UH	Wgt	all	Moon
1st	Tue	h	h	h	h	s	r	h	h	r	h	s	h	r	h	s	l	s	h	r	h	h	r	n	s	l	r	r	s	h	h	s	r	l	r	s	s	r	s	534	
2nd	Wed	l	s	l	n	s	l	l	s	l	l	l	l	l	l	l	d	s	r	r	h	l	l	d	d	d	r	h	s	s	s	s	s	r	s	s	l	r	s	213	XhS
3rd	Thur	l	s	s	l	s	s	s	s	r	r	s	h	l	l	d	l	r	r	l	r	s	r	d	d	l	h	h	h	r	r	s	l	r	s	s	l	l	d	273	1Q
4th	Fri	l	d	d	l	l	n	l	l	s	s	l	l	d	n	d	d	s	l	l	s	n	l	d	d	l	l	h	h	s	9	d	d	l	d	l	d	d	d	106	
5th	Sat	d	l	d	d	s	l	l	l	s	s	s	l	d	d	d	d	n	d	s	r	n	d	d	d	d	n	r	l	l	l	d	d	d	d	d	d	d	d	64	A
6th	Sun	l	l	l	l	s	l	l	l	s	s	l	d	d	d	d	d	n	l	d	l	d	l	l	l	s	d	s	l	l	l	l	d	s	d	l	s	s	l	76	
7th	Mon	n	d	d	d	d	d	d	d	d	d	d	d	d	d	d	d	d	d	d	d	d	d	d	d	d	d	d	d	d	d	d	d	d	d	d	l	l	d	3	
8th	Tue	d	d	d	d	d	d	d	d	d	d	d	d	d	d	d	d	d	d	d	d	d	d	d	d	d	d	l	d	d	d	d	d	d	d	d	d	d	d	1	
9th	Wed	l	l	l	l	l	l	d	d	d	d	d	d	d	d	d	d	d	d	d	d	d	d	d	d	d	d	d	d	d	d	d	d	d	d	d	d	d	d	9	V
10th	Thur	h	h	h	h	h	h	h	h	h	h	h	h	h	h	h	r	h	h	h	h	h	h	r	r	r	h	h	s	h	h	h	h	r	h	r	s	h	h	1034	
11th	Fri	h	h	s	h	l	l	l	l	s	r	s	r	r	h	h	h	r	r	s	r	r	r	n	d	d	r	h	l	s	h	s	r	l	r	r	l	r	s	480	F
12th	Sat	s	s	h	h	r	h	h	r	h	h	r	h	h	h	r	r	h	s	r	r	h	s	d	d	d	r	h	l	h	h	r	h	l	h	s	s	h	h	622	
13th	Sun	l	l	l	l	s	n	l	s	l	s	l	d	l	l	l	l	r	l	s	r	s	r	d	d	d	h	h	r	l	s	s	l	l	r	s	l	l	d	210	
14th	Mon	d	d	d	d	l	d	n	l	l	l	l	s	d	d	d	l	l	d	d	l	n	l	d	d	d	s	r	l	s	r	l	l	d	s	l	d	s	l	75	
15th	Tue	l	d	l	l	l	s	n	l	h	h	h	r	r	r	d	h	h	h	h	r	h	s	l	d	d	r	h	l	h	h	l	s	d	s	l	d	s	l	355	
16th	Wed	l	l	l	l	l	d	l	n	s	d	r	d	d	l	h	d	l	s	l	s	l	r	d	s	r	h	h	h	h	h	h	r	h	r	h	h	h	r	375	XhN
17th	Thur	l	s	r	r	h	r	h	h	h	h	h	h	h	h	r	r	h	h	d	r	r	l	h	h	h	s	l	l	d	l	n	d	r	d	n	r	d	l	532	
18th	Fri	d	s	h	s	l	r	l	s	s	s	l	d	d	d	d	d	d	d	d	d	d	d	s	l	l	n	l	d	d	d	d	d	d	d	d	l	d	s	112	3Q
19th	Sat	r	d	l	l	l	d	l	d	n	d	d	l	d	d	d	d	d	d	d	d	d	d	d	d	d	d	d	d	d	d	d	d	l	d	l	d	d	d	22	
20th	Sun	h	h	h	h	h	h	h	r	s	l	l	l	s	l	d	d	d	d	d	d	n	d	d	d	n	n	d	d	l	n	d	s	d	s	l	l	s	r	345	
21st	Mon	h	h	h	h	h	h	h	h	h	l	s	h	r	r	s	d	s	s	d	s	r	d	n	l	l	d	d	l	l	d	d	n	l	d	d	l	l	s	940	P11
22nd	Tue	l	d	s	h	l	r	r	l	d	d	d	r	d	d	d	d	d	d	d	d	d	d	s	r	h	d	l	s	d	d	d	n	s	d	d	s	l	s	142	^
23rd	Wed	d	d	d	d	d	d	d	d	d	d	d	d	d	d	d	d	d	d	d	d	d	d	s	s	s	d	d	l	d	d	d	d	s	d	d	s	l	r	41	
24th	Thur	s	l	h	h	d	r	s	d	d	d	d	d	d	d	d	d	d	d	d	d	d	d	l	l	l	d	d	d	d	d	d	d	l	d	d	d	d	d	76	
25th	Fri	r	l	r	h	l	r	s	d	d	d	d	l	d	d	d	d	d	d	d	d	d	d	l	s	d	d	d	d	d	d	d	d	d	d	d	d	d	n	74	N
26th	Sat	r	h	h	h	h	h	h	h	r	n	l	h	d	l	l	d	l	d	l	l	l	d	d	l	l	d	d	d	d	d	d	d	l	d	d	d	d	d	492	
27th	Sun	l	s	l	d	s	d	l	h	r	s	r	h	h	s	l	l	r	r	s	s	l	l	h	h	h	s	r	l	r	s	l	d	s	l	l	l	d	d	367	
28th	Mon	l	r	l	d	s	s	r	l	h	l	l	s	d	d	d	d	d	d	d	d	d	d	l	h	h	d	l	r	d	l	l	d	h	d	l	s	d	d	235	
29th	Tue	d	d	d	d	d	d	d	d	l	d	d	l	d	d	d	d	d	d	d	d	d	d	h	h	h	n	d	l	d	d	d	d	l	d	d	l	s	l	100	XhS
30th	Wed	s	s	s	l	d	l	l	d	d	d	l	s	h	h	h	r	s	s	h	s	h	r	d	r	h	l	l	l	r	s	s	l	s	l	l	l	d	d	309	
31st	Thur	l	s	l	s	l	l	l	d	d	d	d	d	d	l	h	s	d	d	d	d	r	l	r	r	r	l	l	d	r	s	l	l	l	n	s	l	l	l	132	
estimate:		211	303	470	427	290	448	360	284	277	202	149	443	211	315	180	117	153	164	154	213	192	102	135	239	241	182	357	128	190	294	94	106	144	124	87	92	145	127	8348	
average:		129	147	184	167	134	202	184	159	156	136	130	143	125	159	117	122	125	134	161	159	125	107	122	100	80	137	285	106	157	213	72	107	109	103	77	97	140	137	5247	
trend:		wtr	wtr	wtr	wtr	wtr	wtr	wtr	wtr	wtr	wtr	wtr	wtr	wtr	wtr	wtr	av	wtr	wtr	av	wtr	wtr	av	wtr	wtr	wtr	wtr	wtr	wtr	wtr	wtr	wtr	av	wtr	wtr	av	av	av	av	wtr	

SOUTH ISLAND RAIN

JULY 2025

d=mainly dry, n= fog, mist, dew or nonrecordable, l=light showers, p=passing showers, s=significant showers, r=rain, h=heavy falls

		West Coast				Top South Island				Canterbury				South Canterbury				Coastal Otago			Lakes			Inland Otago				Southland						
		West	Grey	Hok	BB	Ta	Mou	Nel	Ble	Dar	Ham	Kai	Ch	Rak	Meth	Ash	Tim	Oam	Dun	Bal	Twi	LP	LT	Crom	Qstn	Ran	Alex	TeA	Wtn	Gore	Inv	Puy Pt	all	Moon
1st	Tue	s	s	s	s	s	s	s	s	l	n	d	n	d	l	n	d	d	d	d	d	d	d	d	d	d	d	d	d	d	d	d	48	
2nd	Wed	h	h	h	h	s	r	r	l	r	s	n	s	r	s	r	s	s	s	d	d	d	r	l	l	d	d	d	s	s	s	r	247	XnS
3rd	Thur	r	h	h	h	s	r	s	l	d	s	d	l	d	l	n	d	d	s	h	h	d	d	s	r	l	s	r	h	h	h	s	280	1Q
4th	Fri	h	r	s	l	s	s	n	d	d	d	d	d	d	d	d	d	d	s	r	s	d	d	d	d	l	l	s	r	h	r	r	130	
5th	Sat	r	r	s	l	d	d	d	d	l	d	d	l	d	l	d	d	d	l	s	d	d	d	d	d	s	l	d	r	h	s	s	100	A
6th	Sun	d	d	d	d	d	l	d	s	l	l	d	d	d	l	d	d	d	s	l	d	d	d	d	d	d	d	d	l	n	d	l	23	
7th	Mon	d	d	d	d	d	d	d	d	d	d	d	d	d	d	d	d	d	d	d	d	d	d	d	d	d	d	l	l	d	d	d	3	
8th	Tue	d	d	d	d	d	d	d	d	d	d	d	d	d	d	d	d	d	d	d	d	d	d	d	d	d	d	d	d	d	d	d	0	
9th	Wed	r	r	r	r	l	l	d	d	s	h	s	s	d	r	l	l	d	d	d	r	d	s	l	s	d	d	l	d	d	d	d	134	/
10th	Thur	h	h	h	h	h	h	h	h	r	l	r	h	r	s	h	h	h	h	h	h	h	r	r	r	r	s	r	r	h	h	r	1139	
11th	Fri	s	l	s	h	r	s	s	s	d	l	d	d	d	l	n	d	d	d	s	d	s	l	s	s	d	l	h	r	r	r	h	167	F
12th	Sat	r	r	s	r	r	s	r	l	d	l	l	d	d	d	d	l	l	n	d	r	d	d	l	l	d	l	s	l	l	d	s	112	
13th	Sun	h	h	h	h	l	l	l	d	s	l	l	r	d	s	s	r	h	s	l	l	d	l	l	l	s	s	s	s	s	s	r	305	
14th	Mon	d	l	s	l	d	d	d	d	l	d	d	d	d	l	d	d	d	d	l	d	d	d	n	s	d	n	l	l	l	s	s	37	
15th	Tue	r	r	s	h	s	s	s	s	l	l	l	d	d	s	s	s	r	s	l	d	d	l	r	r	s	r	d	l	l	l	l	160	
16th	Wed	d	d	d	d	r	r	l	s	s	r	h	l	r	s	r	s	s	s	l	s	s	l	d	n	s	n	d	d	l	l	s	172	XhN
17th	Thur	d	d	d	d	d	d	d	d	l	s	r	d	d	s	s	n	l	d	d	l	d	d	d	d	l	d	d	d	d	d	d	34	
18th	Fri	d	d	d	d	d	d	d	d	d	s	d	d	d	d	d	d	d	d	d	d	d	d	d	d	d	d	d	d	d	d	d	8	ƆQ
19th	Sat	d	d	d	d	l	d	d	d	d	l	d	d	d	d	d	d	d	l	l	d	d	d	d	d	d	d	d	l	l	s	l	17	
20th	Sun	s	d	d	d	s	s	l	l	l	d	n	n	d	l	l	l	d	n	l	d	d	d	d	d	d	d	d	d	n	n	d	33	
21st	Mon	d	d	d	d	d	d	d	d	l	s	n	l	l	l	l	l	l	l	d	d	d	d	d	d	d	s	d	d	d	d	d	20	F11
22nd	Tue	d	d	d	d	d	d	d	d	l	r	s	s	d	s	s	d	l	d	d	d	d	d	d	d	d	d	d	d	d	d	d	34	^
23rd	Wed	d	d	d	d	d	d	d	d	d	l	d	d	l	d	d	d	d	d	d	d	d	d	d	d	d	d	d	d	d	d	d	4	
24th	Thur	d	d	d	d	d	d	d	d	d	d	d	d	d	d	d	d	d	d	d	d	d	d	d	d	d	d	d	d	d	l	d	1	
25th	Fri	d	d	d	d	d	d	d	d	d	d	d	d	d	d	d	n	d	l	d	d	d	d	d	d	d	d	d	d	l	d	d	2	N
26th	Sat	d	d	d	d	d	d	d	d	d	d	d	d	d	d	d	d	d	d	l	d	d	d	d	d	d	d	d	d	d	d	d	2	
27th	Sun	d	d	d	d	n	d	d	l	d	d	s	d	d	d	d	d	d	d	d	d	d	d	d	d	d	d	d	d	d	d	d	5	
28th	Mon	d	d	d	d	d	d	d	l	l	d	s	l	d	l	l	d	d	d	d	d	d	d	d	d	d	d	d	d	d	d	l	11	
29th	Tue	d	d	d	d	d	d	d	d	d	d	d	n	d	d	d	d	d	d	d	d	d	d	d	d	d	d	d	d	d	d	d	1	XhS
30th	Wed	d	d	d	d	r	s	l	l	l	d	s	l	d	d	d	d	d	d	d	d	d	d	d	d	d	d	n	d	d	d	d	35	
31st	Thur	d	d	d	d	l	l	l	s	d	d	l	d	d	d	d	d	d	d	d	d	d	d	d	d	d	d	d	d	d	l	s	16	
estimate:		234	302	291	237	218	172	79	53	57	84	98	76	44	64	73	51	82	61	83	130	88	42	40	46	42	41	73	103	107	85	112	3277	
average:		175	188	220	307	176	147	85	57	76	114	83	65	76	80	64	37	37	77	56	53	70	57	32	55	26	18	85	68	60	75	166	2895	
trend:		w tr	w tr	w tr	drr	w tr	w tr	av	av	drr	drr	w tr	w tr	drr	drr	av	w tr	w tr	drr	w tr	w tr	w tr	drr	av	av	w tr	w tr	drr	w tr	w tr	av	drr	w tr	

NORTH ISLAND SUNSHINE ESTIMATES
JULY 2025

F=fine (8-12 hours of sunshine), pc= partly cloudy (4-7 hours), c=cloudy (1-3 hours), o=overcast (0 hours)

		Kta	Whg	Dar	Ak	Hml	Tau	Rt	Taum	NP	Wang	Taup	Gis	Nap	Wai	Dann	PN	Mast	Pa	UH	Wgt	ALL	Moon
1st	Tue	c	c	c	c	c	c	o	o	c	o	o	c	pc	c	pc	o	pc	o	o	o	42	
2nd	Wed	pc	c	pc	pc	c	c	c	c	c	pc	c	pc	pc	c	pc	pc	pc	pc	pc	pc	82	XhS
3rd	Thur	pc	pc	pc	pc	pc	c	pc	c	pc	c	pc	pc	pc	c	pc	c	pc	F	F	F	99	1Q
4th	Fri	pc	pc	pc	pc	pc	c	pc	o	pc	c	pc	F	pc	pc	pc	c	pc	F	F	F	108	
5th	Sat	pc	c	c	c	pc	pc	pc	c	pc	o	pc	pc	c	o	o	o	o	c	c	o	57	A
6th	Sun	pc	pc	F	F	F	F	F	F	pc	F	F	pc	pc	pc	o	F	o	pc	c	pc	128	
7th	Mon	pc	pc	F	F	pc	pc	F	F	F	pc	F	o	c	c	o	pc	o	pc	pc	pc	108	
8th	Tue	o	o	o	o	o	o	o	o	o	o	o	o	o	o	o	o	o	o	o	o	2	
9th	Wed	o	o	o	o	o	o	o	o	o	o	o	o	o	o	o	o	o	o	o	o	0	V
10th	Thur	pc	c	o	o	o	o	o	o	o	c	o	o	o	o	c	c	c	o	o	o	16	
11th	Fri	c	o	o	o	o	o	o	o	o	o	o	c	c	c	c	o	c	o	o	o	13	F
12th	Sat	pc	pc	pc	pc	pc	c	o	o	c	c	o	c	c	c	c	c	c	o	o	c	51	
13th	Sun	F	pc	pc	pc	pc	pc	pc	pc	pc	pc	pc	F	F	F	pc	pc	pc	pc	pc	pc	118	
14th	Mon	pc	pc	pc	pc	o	pc	o	o	o	o	o	c	pc	pc	pc	o	pc	o	o	o	43	
15th	Tue	c	c	pc	pc	o	pc	c	o	c	c	c	F	pc	pc	c	c	c	o	c	o	58	
16th	Wed	c	c	o	o	o	o	o	o	pc	o	o	o	o	o	o	o	o	o	o	o	9	XhN
17th	Thur	pc	c	o	o	c	o	c	c	pc	F	c	o	c	pc	c	F	c	F	pc	pc	66	
18th	Fri	c	c	F	F	pc	F	pc	pc	c	F	pc	pc	F	F	pc	F	pc	F	F	F	133	3Q
19th	Sat	c	c	pc	pc	o	c	o	o	o	o	o	c	pc	c	o	o	o	o	o	o	24	
20th	Sun	o	o	o	o	o	o	o	o	o	o	o	o	o	o	o	o	o	o	o	o	0	
21st	Mon	c	o	o	o	o	o	o	o	o	o	o	o	o	o	o	o	o	c	c	o	5	P11
22nd	Tue	F	pc	F	F	pc	pc	c	F	F	c	c	o	o	c	o	c	o	pc	c	o	82	^
23rd	Wed	pc	F	F	F	F	F	F	F	F	c	F	c	c	c	c	c	c	F	c	pc	115	
24th	Thur	pc	c	pc	pc	pc	pc	F	F	F	F	F	pc	pc	pc	pc	F	pc	F	pc	pc	132	
25th	Fri	o	o	c	c	c	pc	pc	F	F	F	pc	o	c	o	c	F	c	F	pc	pc	85	N
26th	Sat	pc	c	o	o	o	o	o	o	o	o	o	o	o	o	o	o	o	c	pc	pc	24	
27th	Sun	pc	c	c	c	c	c	c	pc	c	o	c	pc	o	o	o	o	o	o	o	o	31	
28th	Mon	pc	c	F	F	pc	F	pc	pc	c	pc	pc	c	o	o	o	pc	o	pc	o	c	80	
29th	Tue	pc	pc	pc	pc	o	c	c	o	F	pc	c	o	o	o	o	pc	o	pc	c	o	55	XhS
30th	Wed	pc	pc	pc	pc	o	c	o	o	o	o	o	o	o	o	o	o	o	o	o	o	26	
31st	Thur	c	c	o	o	o	o	o	o	o	c	o	pc	pc	pc	c	c	c	c	c	pc	35	
est. hours:		120	97	115	115	87	105	91	86	97	87	91	85	92	73	61	87	61	109	76	89	1825	
av. hours:		142	134	110	140	120	152	129	90	133	120	120	130	138	125	102	103	107	122	120	117	2454	
trend:		less	less	av	less	less	less	less	av	less	less	less	less	less	less	less	less	less	less	less	less	less	

SOUTH ISLAND SUNSHINE ESTIMATES

JULY 2025

F=fine (8-12 hours of sunshine), pc= partly cloudy (4-7 hours), c=cloudy (1-3 hours), o=overcast (0 hours)

		Ta	Nel	Ble	Mou	West	Grey	Hok	Ham	Kai	Ch	Ash	Tim	Crom	Qstn	Bal	Dun	Wnt	Gore	Inv	Alex	ALL	Moon
1st	Tue	F	o	pc	c	o	o	o	c	c	c	o	o	o	o	o	o	o	o	o	o	23	
2nd	Wed	pc	pc	F	pc	o	o	o	pc	pc	pc	pc	c	c	o	o	c	c	c	o	c	53	XhS
3rd	Thur	F	c	F	o	o	c	c	pc	pc	pc	F	F	c	o	c	pc	c	c	c	c	70	1Q
4th	Fri	F	pc	F	pc	o	c	o	F	F	F	F	F	pc	pc	o	c	o	o	c	pc	88	
5th	Sat	o	pc	c	c	o	c	pc	c	c	pc	pc	pc	c	o	c	c	c	c	pc	c	56	A
6th	Sun	o	F	F	pc	c	F	F	F	F	F	F	F	pc	pc	c	c	c	c	c	pc	114	
7th	Mon	o	F	F	pc	pc	pc	F	F	F	F	F	pc	pc	pc	c	pc	pc	pc	pc	pc	128	
8th	Tue	o	o	o	o	o	o	o	o	o	o	o	o	o	c	c	pc	pc	pc	pc	o	22	
9th	Wed	c	o	o	o	o	o	o	o	o	o	o	o	o	o	o	o	o	o	o	o	2	V
10th	Thur	pc	o	c	c	o	o	o	c	c	c	pc	pc	o	c	pc	pc	c	c	pc	o	52	
11th	Fri	o	o	c	o	o	o	o	c	c	o	c	c	o	c	c	pc	c	c	c	o	23	F
12th	Sat	pc	o	c	o	o	o	o	pc	pc	c	c	o	o	o	o	o	o	o	o	o	21	
13th	Sun	c	pc	pc	pc	c	c	c	c	c	c	c	pc	pc	c	pc	pc	pc	pc	pc	pc	78	
14th	Mon	F	o	c	o	o	o	o	pc	pc	pc	pc	c	o	o	o	o	o	o	o	o	36	
15th	Tue	F	c	c	c	o	o	c	o	o	o	o	o	o	o	c	o	pc	pc	c	o	31	
16th	Wed	o	o	o	o	pc	pc	pc	o	o	o	o	o	c	o	c	o	pc	pc	pc	c	39	XhN
17th	Thur	c	F	pc	F	F	F	F	pc	pc	pc	pc	pc	pc	pc	pc	pc	F	F	F	pc	132	
18th	Fri	pc	F	F	pc	pc	F	F	F	F	pc	F	F	pc	pc	c	F	c	c	o	pc	124	3Q
19th	Sat	F	o	o	o	pc	F	F	c	c	c	pc	pc	pc	pc	o	c	c	c	o	pc	63	
20th	Sun	F	c	o	o	pc	pc	F	o	o	o	c	pc	pc	pc	o	c	pc	pc	pc	pc	82	
21st	Mon	F	pc	c	c	F	F	F	o	o	o	c	c	c	pc	o	o	c	c	o	c	70	P11
22nd	Tue	F	F	F	pc	F	F	F	o	o	o	pc	pc	o	pc	o	c	pc	pc	pc	o	92	^
23rd	Wed	pc	F	F	pc	F	F	F	F	F	F	F	F	c	pc	pc	F	c	c	pc	c	129	
24th	Thur	o	F	F	pc	pc	F	F	pc	pc	pc	pc	pc	o	o	c	c	pc	pc	c	o	98	
25th	Fri	F	F	pc	pc	F	F	F	pc	pc	F	c	c	o	pc	pc	pc	pc	pc	pc	o	118	N
26th	Sat	pc	pc	o	pc	pc	F	F	pc	pc	pc	pc	pc	o	pc	pc	F	pc	pc	pc	o	107	
27th	Sun	o	o	o	o	pc	c	pc	o	o	o	c	o	o	pc	pc	F	F	F	pc	o	54	
28th	Mon	pc	pc	c	pc	pc	F	F	c	c	c	o	o	c	pc	pc	pc	F	F	pc	c	94	
29th	Tue	F	pc	o	pc	pc	F	F	o	o	c	o	o	o	pc	pc	F	pc	pc	pc	o	81	XhS
30th	Wed	c	o	o	o	c	c	c	o	o	o	c	pc	o	pc	pc	F	F	F	pc	o	58	
31st	Thur	F	pc	F	pc	pc	pc	F	o	o	c	pc	pc	o	o	c	c	c	c	pc	o	69	
est. hours:		149	131	122	103	110	136	150	104	104	104	122	117	58	85	80	121	118	118	115	58	2206	
av. hours:		155	157	157	147	124	103	121	102	132	124	123	124	99	87	97	101	96	85	91	92	2317	
trend:		av	less	less	less	less	more	more	av	less	less	av	av	less	av	less	more	more	more	more	less	average	

JULY 1 TUESDAY

A cold, strong and unsettled westerly flow prevails across the country with low pressure bringing widespread dull and gloomy conditions to most districts, and rain throughout the North Island and showers down into the west of the South island. In contrast, much of Canterbury, Otago, and Southland will be mostly dry with increasing threats of rain. Snow likely in the Central Plateau and higher reaches of the southern North Island, as well as overnight in Southland and Otago to low levels.

Auckland:	Stormy, blustery winds, cloudy, heavy rain, chance of thunderstorms
Hamilton:	Cloudy, showers, chance of frosts, possible hail, unsettled breezes
Wellington:	Overcast, windy spells, scattered showers, chance of thunderstorms
Greymouth:	Fluctuating winds, overcast, cool, scattered showers
Christchurch:	Cloudy, cold, isolated showers, light breezes
Dunedin:	Overcast, dry with threats of rain, cold, frosts, strengthening winds

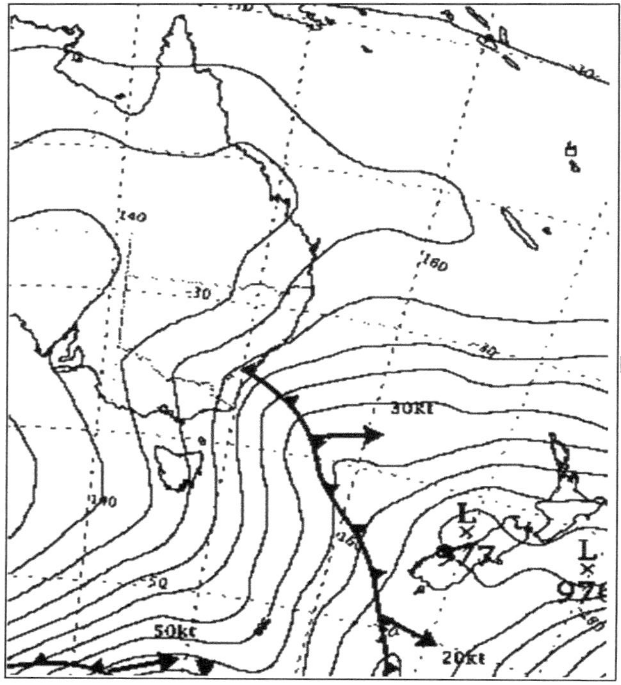

RAIN POTENTIAL

FROST/SNOW

JULY 2 WEDNESDAY

An intense Southern Ocean depression directs a stormy throughout New Zealand, with strong west to southwest winds prevailing, up to gale force in many places. Bands of thunderstorm likely to bring heavy rain at times in the west of both islands as well as from Auckland to the Waikato, with squally showers elsewhere. Wintry showers may bring snow to low levels in the South Island.

Auckland:	Cloudy, thunderstorms likely, showers, strong winds, possible gales
Hamilton:	Cloudy, showers, usty winds, chance of gales and thunderstorms
Wellington:	Overnight showers, sunny spells, chance of tl understorms and gales
Greymouth:	Storm, overcast, heavy rain, gusts and gale winds likely
Christchurch:	Changeable, sunny spells, showers, frosts, chance of hail and galess
Dunedin:	Cloudy, cold, wintry showers, blustery, chanc of gales and snow

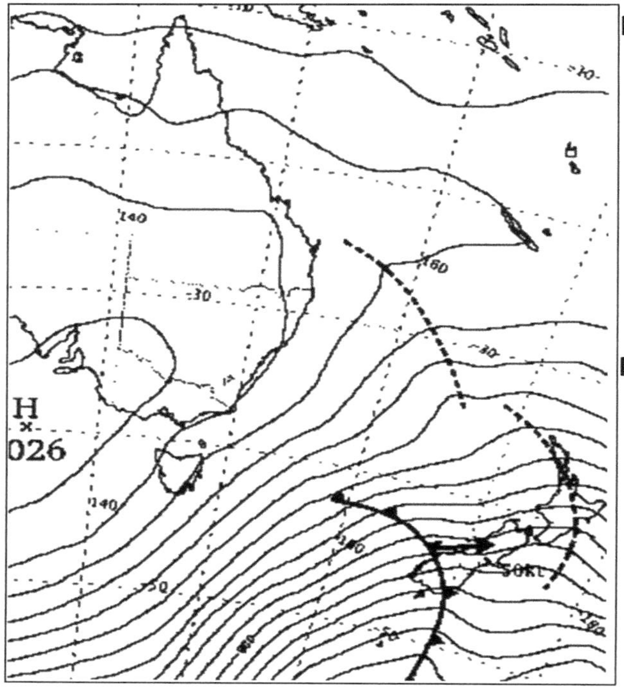

RAIN POTENTIAL

FROST/SNOW

JULY **3** THURSDAY

Storm conditions continue throughout the country with southwest gale winds prevailing everywhere, widespread rain and scattered thunderstorms, and snow likely to low levels in the South Island and on the Central Plateau.

Auckland: Partly cloudy, rain, gale winds likely, chance of thunderstorms
Hamilton: Partly cloudy, rain, gale winds likely, chance of thunderstorms
Wellington: Sunny, dry, strong and blustery winds to gale force at times
Greymouth: Cloudy, stormy winds, heavy rain, cold overnight
Christchurch: Partly cloudy, frosts, isolated showers, blustery, possible gales
Dunedin: Some sunny spells, wintry showers, gales and snow possible

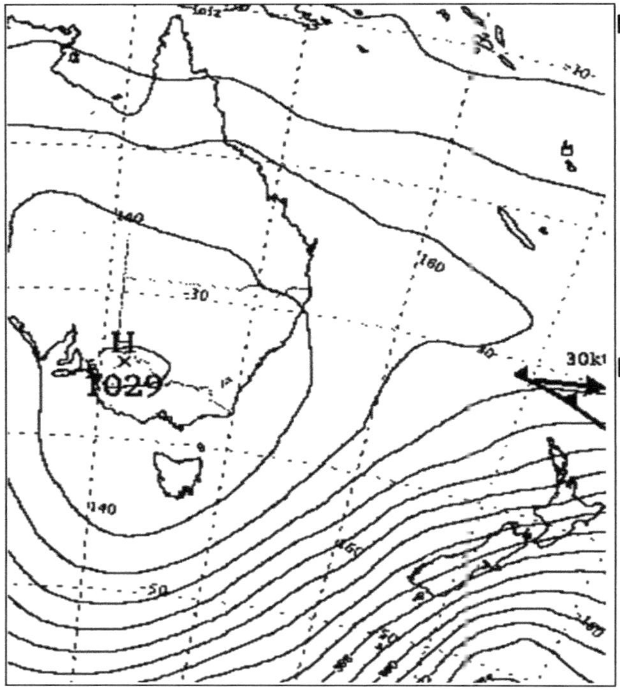

RAIN POTENTIAL

FROST/SNOW

JULY **4** FRIDAY

Storm conditions begin to ease across the country as a large high pressure system over Australia extends a ridge near the west of New Zealand. However, strong and stormy southwest winds continue to bring gales and outbreaks of rain and possible thunderstorms in most parts of the country, but with increasing sunny spells and drier conditions at times in the east from the Bay of Plenty to Canterbury. Further snow likely in Southland and the Central Plateau.

Auckland: Partly cloudy, showers, blustery, chance of thunderstorms and gales
Hamilton: Partly cloudy, occasional showers, strong and blustery winds
Wellington: Sunny, dry, strong winds to gale force likely
Greymouth: Strong blustery winds with gusts, cloudy, rain
Christchurch: Strong gusty winds, sunny, colder, dry, frosts
Dunedin: Cloudy, some wintry showers, strong stormy winds, chance of snow

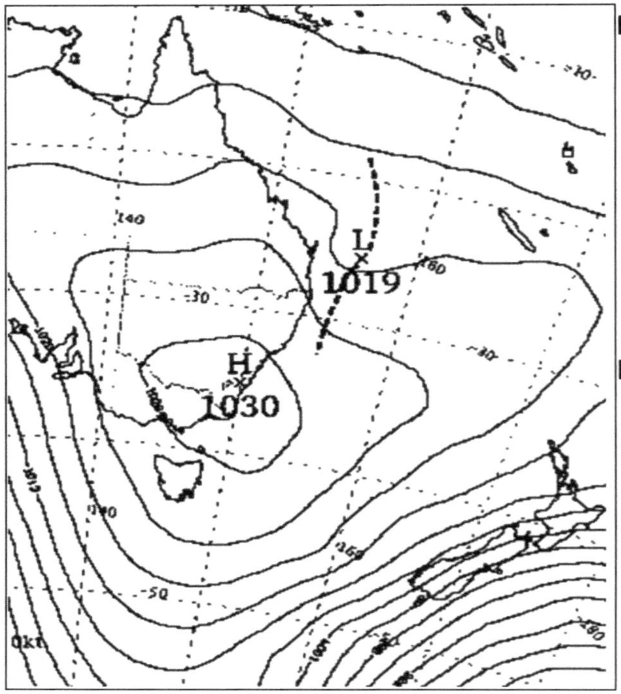

RAIN POTENTIAL

FROST/SNOW

JULY **5** SATURDAY

Very little change today as the high pressure ridge remains stable to the west of New Zealand while strong and blustery southwest winds associated with an intense depression now moved to the southeast of the country continues to bring gales and gusts throughout. Widespread cloud and occasional outbreaks of rain continues particularly in the west and south of the country, while mostly dry with good sunny spells in the east from the Bay of Plenty to Marlborough.

Auckland: Cloudy, showers, strong winds easing later in the day
Hamilton: Windy, isolated showers, some sunny spells
Wellington: Overcast, dry, threats of rain, strong gusty winds at times
Greymouth: Cloudy, rain, milder, strong winds
Christchurch: Partly cloudy, warmer, brief showers, unsettled winds with gusts
Dunedin: Cloudy, brief showers, strong winds easing later in the day

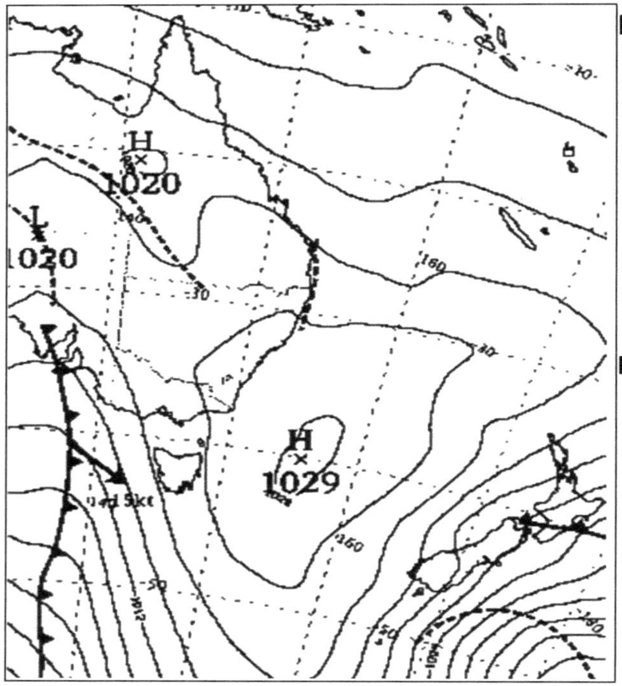

RAIN POTENTIAL

FROST/SNOW

JULY **6** SUNDAY

A high pressure ridge moves onto the country easing winds to light southwesterlies. Some lingering overnight wintry showers in most districts, clearing to a mostly dry and sunny day, particularly over the North Island and in the east of the South Island. Very cold with frosts in most districts, except in Northland.

Auckland: Overnight showers, sunny day, cooler, moderate winds
Hamilton: Sunny, light breezes, mostly dry, chance of frost and fog patches
Wellington: Brief showers, colder, some sunny spells, fresh breezes
Greymouth: Mostly sunny and dry, winds easing to light
Christchurch: Mostly sunny, colder, dry, winds easing to light
Dunedin: Cloudy, cold, wintry showers, winds easing to light, snow possible

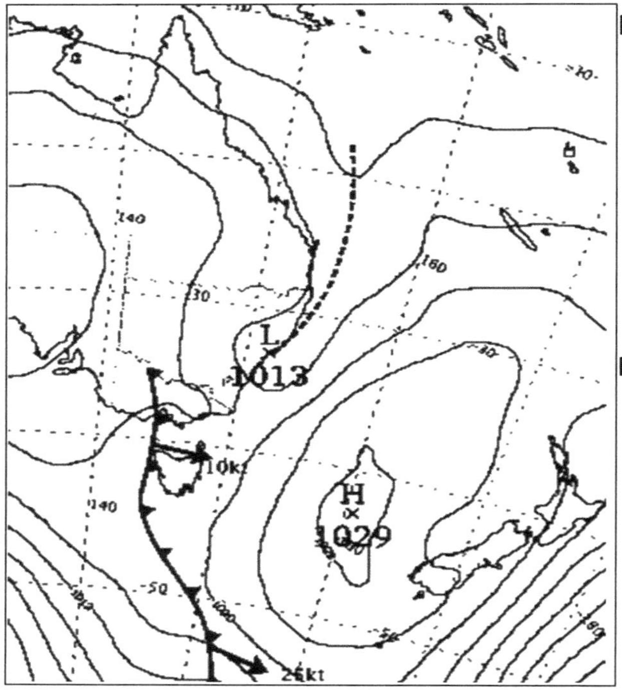

RAIN POTENTIAL

FROST/SNOW

JULY **7** MONDAY

Anticyclonic. Very cold with widespread frosts almost everywhere except in Northland. Mostly dry, sunny, and calm throughout.

Auckland: Sunny, dry, cold overnight light to mostly calm air
Hamilton: Partly cloudy, cold, frosts, dry, calm and serene air
Wellington: Mostly calm and cold air, partly cloudy, dry
Greymouth: Partly cloudy, dry, calm air, chance of frost and fog patches
Christchurch: Sunny, calm and cold air, frosts, dry
Dunedin: Partly cloudy, dry, cold, calm air

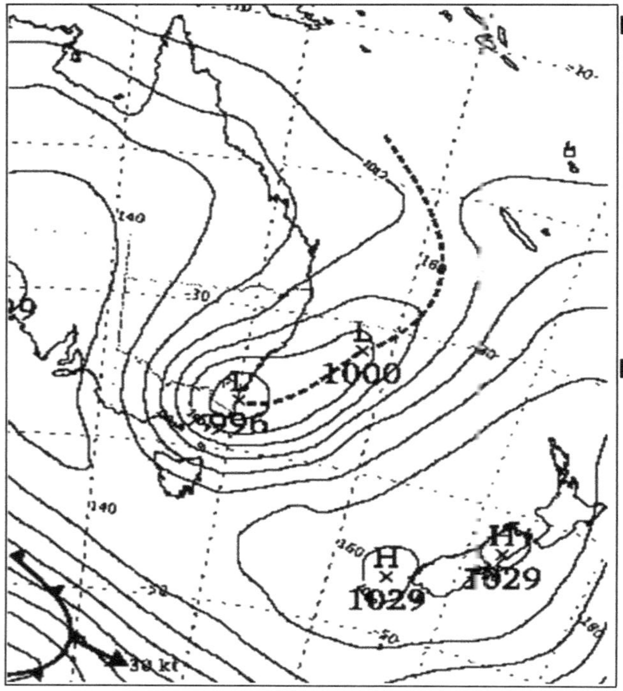

RAIN POTENTIAL

FROST/SNOW

JULY **8** TUESDAY

High pressure maintains a very cold and dry air throughout, with a moderate northeasterly change. An intense depression crossing the Tasman Sea sees cloud building in all districts, with increasing threats of rain developing.

Auckland: Gloomy air, dry with threats of rain, cold, moderate breezes
Hamilton: Overcast, dry, light breezes, frost pockets likely
Wellington: Overcast, dry, light breezes, cold overnight, odd threat of rain
Greymouth: Overcast, dry, threats of rain, moderate breezes, chance of frost
Christchurch: Overcast, cold with severe frosts, dry, light breezes
Dunedin: Partly cloudy, dry, cold with frosts, light breezes to mostly calm air

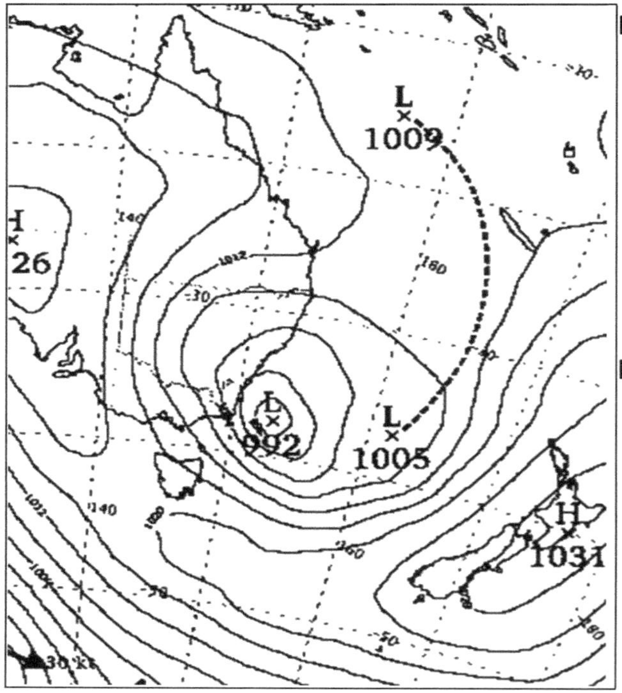

RAIN POTENTIAL

FROST/SNOW

JULY **9** WEDNESDAY

As the high pressure ridge moves away from the east of New Zealand, an intense depression moves onto the west of the country, with strengthening northeast winds prevailing throughout. Widespread cloud everywhere, with rain threatening but remaining mostly dry in the North Island, while rain, heavy at times, moves onto the South Island, with the chance of localised flooding possible in northern parts of the South Island, and further snowfall likely in the high country.

Auckland: Overcast, dry with rain threatening, freshening winds
Hamilton: Overcast, dry, threats of rain, light breezes freshening later in the day
Wellington: Overcast, dry with increasing threats of rain, windy spells
Greymouth: Stormy air, overcast, colder, rain
Christchurch: Overcast, cold, frosts, wintry showers, possible snow, windy at times
Dunedin: Overcast, cold, frosts, dry with rain threatening, blustery at times

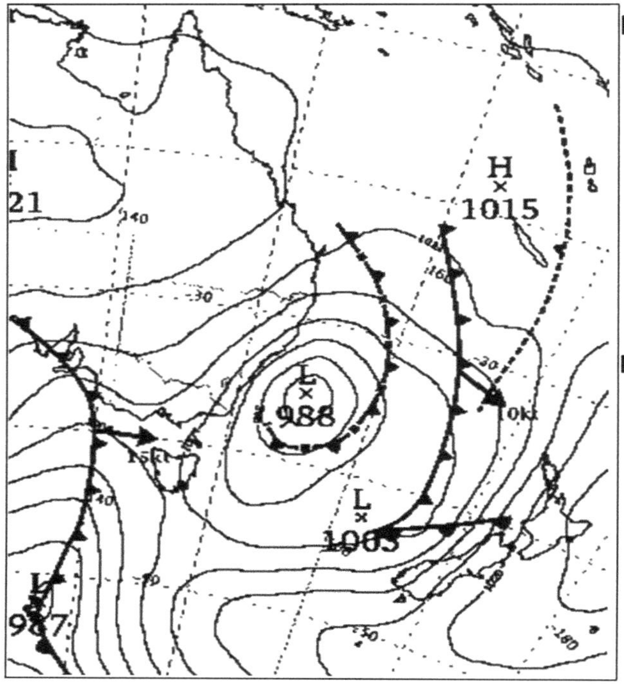

RAIN POTENTIAL

FROST/SNOW

JULY **10** THURSDAY

A complex of depressions bring unsettled and stormy conditions throughout New Zealand again, with a series of wet fronts bringing widespread cloud and heavy rain in most parts of the country. Heavy rain may see outbreaks of localised flooding in Northland, the Bay of Plenty, the South Island Lakes districts, Otago, and West Coast districts. Winds may reach gale force once more, particularly through the Cook Strait. Snow likely in the higher reaches.

Auckland: Stormy, warmer, mostly cloud, heavy rain, downpours, gales likely
Hamilton: Stormy air, gale winds likely, dull and gloomy air with heavy rain
Wellington: Overcast, milder, heavy rain, strong winds
Greymouth: Heavy rain, possible flooding, dull and gloomy air, unsettled winds
Christchurch: Cloudy, warmer, heavy rain, chance of sleet, unsettled winds
Dunedin: Partly cloudy, heavy rain, milder temperatures, fluctuating breezes

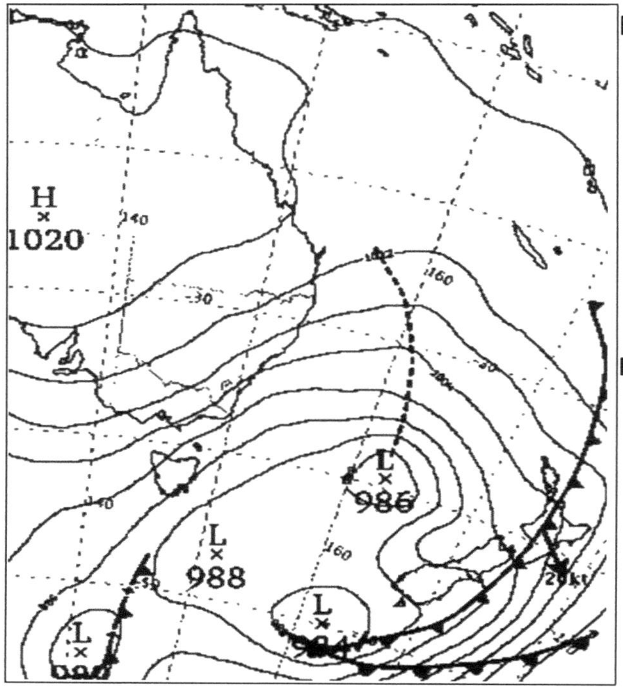

RAIN POTENTIAL

FROST/SNOW

JULY 11 FRIDAY

Wet fronts associated with a depression crossing the country continues to bring widespread cloud and outbreaks of rain throughout the country, with heavy falls still likely in a number of districts through central parts of the North Island and in the southwest of the South Island. Winds remain fresh to strong in a northwest flow. Further snow likely in in the south of the South Island and possibly in the Central Plateau.

Auckland: Overcast, outbreaks of rain, strong winds
Hamilton: Overcast, rain, fresh winds
Wellington: Overcast, scattered showers, fresh to strong winds
Greymouth: Overcast, passing showers, moderate breezes
Christchurch: Overcast, warmer, dry, fresh breezes
Dunedin: Cloudy, dry, warmer, moderate shifting breezes

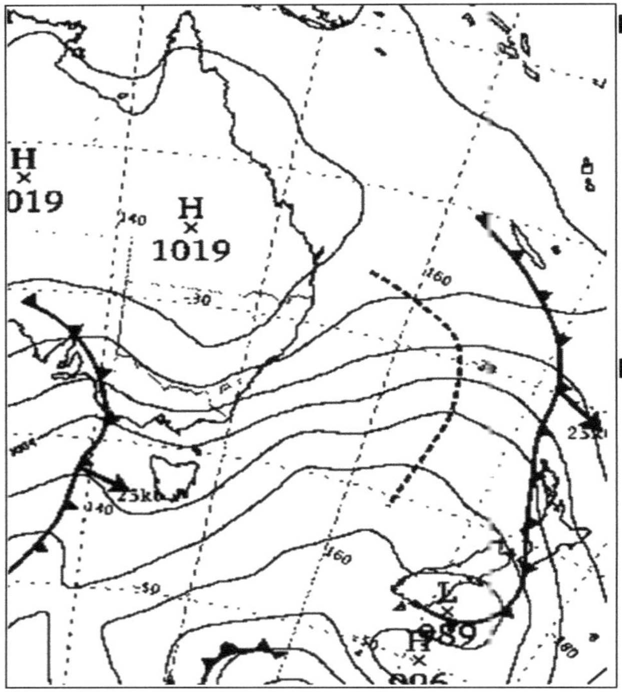

RAIN POTENTIAL

FROST/SNOW

JULY 12 SATURDAY

Low pressure continues to bring fresh to strong west to northwest winds over the North Island, lighter westerly flow over the South Island. A series of wet fronts associated with the depression brings further dull and gloomy conditions everywhere, and heavy rain in the northern half of the North Island, with outbreaks of occasional rain elsewhere. Milder in the North Island while frosty temperatures continue in the south of the South Island.

Auckland: Heavy rain, brief sunny spells, windy, chance of thunderstorms
Hamilton: Cloudy, heavy rain, fresh winds
Wellington: Cloudy, heavy rain, windy at times, static air
Greymouth: Overcast, rain, moderate winds
Christchurch: Cloudy, dry, mild, light breezes, chance of overnight fog
Dunedin: Overcast, isolated showers, light breezes

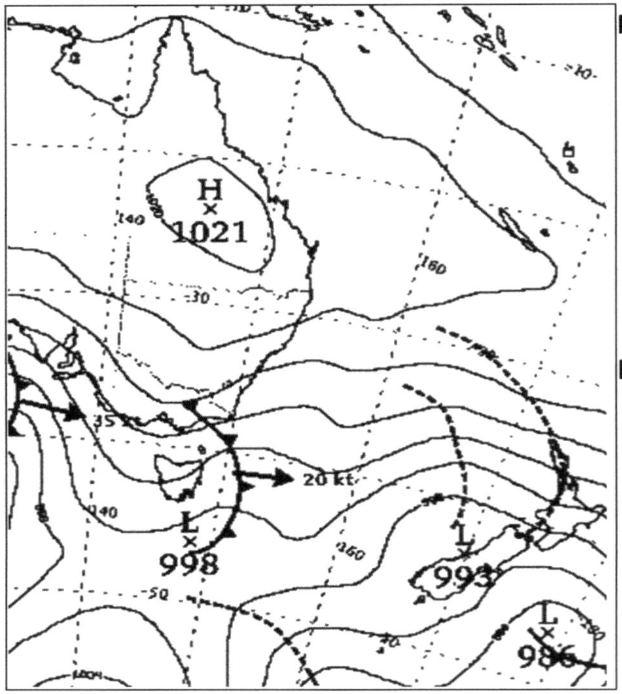

RAIN POTENTIAL

FROST/SNOW

JULY 13 SUNDAY

Low pressure maintains a fresh westerly flow over the country. A mixed day with good sunny spells over the North Island and outbreaks of afternoon rain, heavy at times especially in the Waikato, while lighter elsewhere. Widespread cloud and scattered wintry showers continue throughout the South Island, more drizzly and persistent in eastern parts, while heavier in the west.

Auckland: Changeable, mix of showers and sunny spells, breezy, static air
Hamilton: Changeable, partly cloudy, outbreaks of rain, light breezes
Wellington: Partly cloudy, dry, moderate to fresh breezes
Greymouth: Cloudy, heavy rain, moderate winds, static air
Christchurch: Cloudy, rain, light breezes
Dunedin: Changeable, colder, mix of sunny spells and showers, light breezes

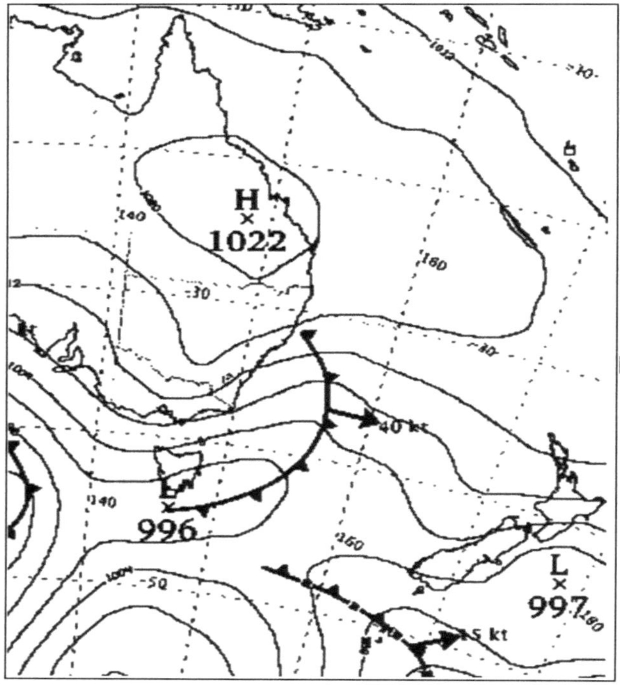

RAIN POTENTIAL

FROST/SNOW

JULY 14 MONDAY

Low pressure continues to dominate the country directing a fresh to strong and unsettled west to northwest flow over the North Island, backing southeasterly over the South Island. Cold and cloudy almost everywhere, with some lingering showers in the west and south of the South Island, and in the central west of the North Island from Taranaki to the Central Plateau where rain may be accompanied by passing thunderstorm activity.

Auckland: Mostly cloudy, showers, unsettled winds, chance of thunderstorms
Hamilton: Overcast, showers, mild day cold night, fluctuating breezes, static air
Wellington: Fluctuating breezes, overcast, brief showers
Greymouth: Overcast, some light showers, light shifting breezes
Christchurch: Partly cloudy, mostly dry, gentle breezes, chance of frost and fog
Dunedin: Overcast, dry, threats of rain, mostly calm air

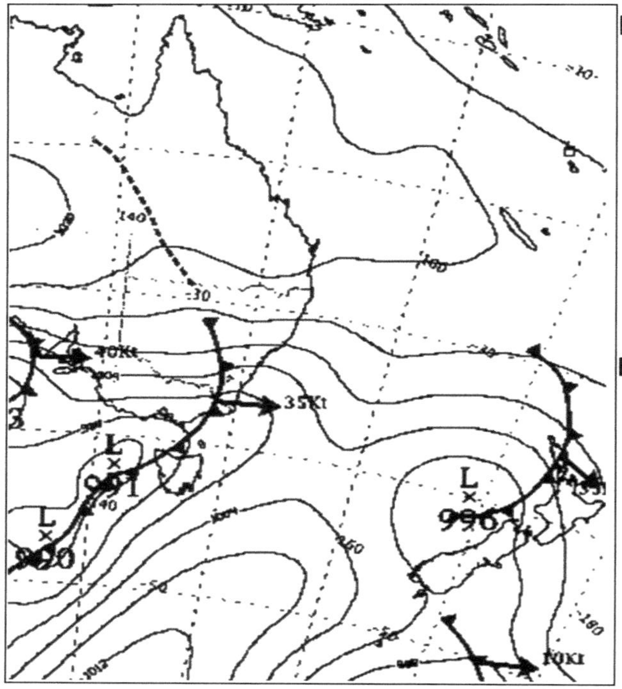

RAIN POTENTIAL

FROST/SNOW

JULY 15 TUESDAY

Low pressure continues to dominate the country with a fresh to moderate northwest flow prevailing over the North Island while colder southerlies prevail over the South Island. Widespread cloud and rain activity spreads throughout the country, occasionally accompanied by thunderstorm activity, particularly over the North Island. Rain likely to be very heavy from Auckland to the Bay of Plenty, and precipitation more wintry in the South Island.

Auckland: Heavy rain, possible thunderstorms, sunny spells, moderate breezes
Hamilton: Overcast, heavy rain, chance of thunderstorms, moderate breezes
Wellington: Overcast, brief showers, moderate shifting breezes
Greymouth: Dull and gloomy air, rain, light breezes freshening later in the day
Christchurch: Overcast, dry with odd threat of rain, light breezes
Dunedin: Overcast, colder, scattered showers, light fluctuating breezes

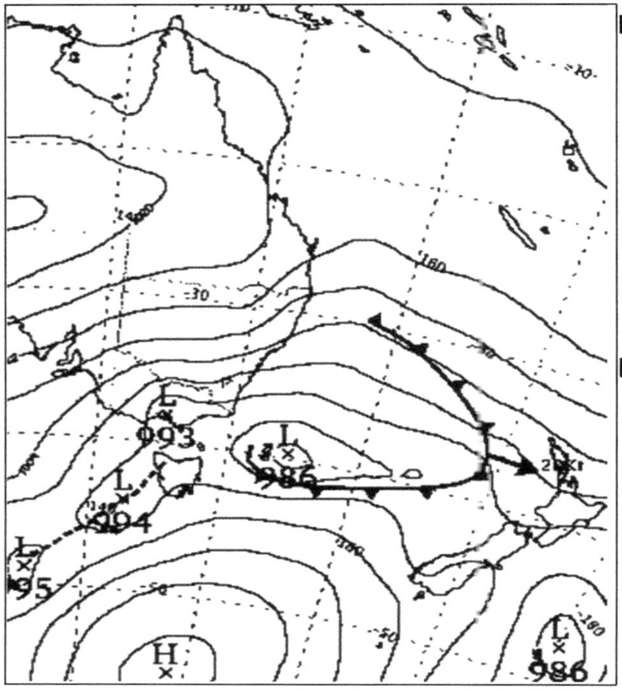

RAIN POTENTIAL

FROST/SNOW

JULY 16 WEDNESDAY

A depression continues to slowly cross the country, while a weak high pressure system to the south of the South Island extends a weak ridge into the far south. Dull and gloomy everywhere, with moderate easterly winds over the South Island, and northeasterlies over the North Island. Rain activity can be expected in all parts of the country, with very heavy rain and scattered thunderstorms particularly in the south and west of the North Island and through the Cook Strait, while lighter rain activity prevails elsewhere.

Auckland: Overcast, pockets of rain, fluctuating winds
Hamilton: Overcast, light showers, fluctuating breezes
Wellington: Overcast, rain, shifting windy spells
Greymouth: Partly cloudy, dry, fluctuating breezes
Christchurch: Cold, overcast, passing drizzle patches, moderate breezes
Dunedin: Overcast, colder, wintry showers, windy, chance of hail show

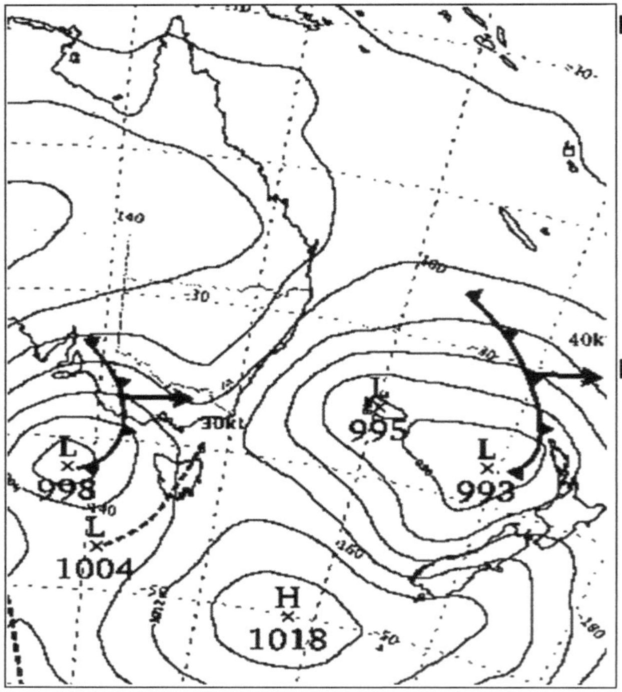

RAIN POTENTIAL

FROST/SNOW

JULY 17 THURSDAY

High pressure centres to the south of New Zealand extend a ridge up over the South Island, clearing clouds and rain activity and bringing very cold temperatures and scattered freezing fogs from Southland northward, and with only some isolated showers lingering around Canterbury to Marlborough. A couple of low pressure centres however, continue to bring widespread cloud and rain throughout the North Island, with very heavy falls in eastern parts from Northland to the Manawatu, with lighter showers to the west.

Auckland: Dull and gloomy air, heavy rain, cooler, moderate breezes
Hamilton: Cloudy, cool, heavy rain, moderate breezes
Wellington: Partly cloudy, light showers, colder, occasional breezy gusts
Greymouth: Mostly sunny, dry, moderate shifting breezes
Christchurch: Partly cloudy, cold with frosts, dry, moderate shifting breezes
Dunedin: Partly cloudy, dry, cold with frosts, light shifting breezes

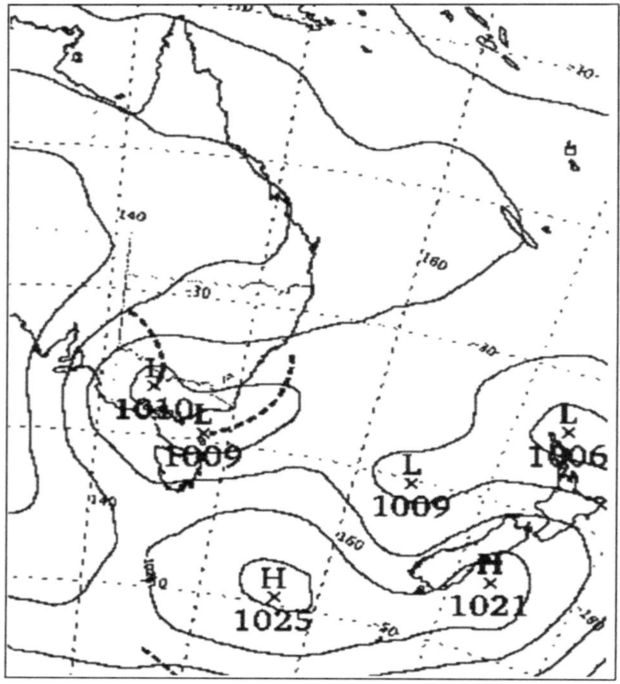

RAIN POTENTIAL

FROST/SNOW

JULY 18 FRIDAY

A weak low pressure trough sits to the north of New Zealand while a high pressure ridge lingers around the southwest of the country. Rain clears to mostly sunny almost everywhere, except in the far south of the South Island. Some lingering evening showers likely in the north of the South Island and in pockets over North Island, although mostly dry in Taranaki to the Wairarapa. Very cold southwest air prevails with widespread frosts south of Auckland, and freezing fogs likely especially in the South Island.

Auckland: Mostly sunny day, evening rain, light breezes
Hamilton: Partly cloudy, cooler, dry, mostly calm air
Wellington: Sunny, cold, frosts, scattered evening showers, mostly calm air
Greymouth: Mostly sunny, colder, dry, mostly calm air, frost pockets likely
Christchurch: Partly cloudy, dry, cold with severe frosts, calm and serene air
Dunedin: Sunny, dry, cold, severe frosts, mostly calm and serene atmosphere

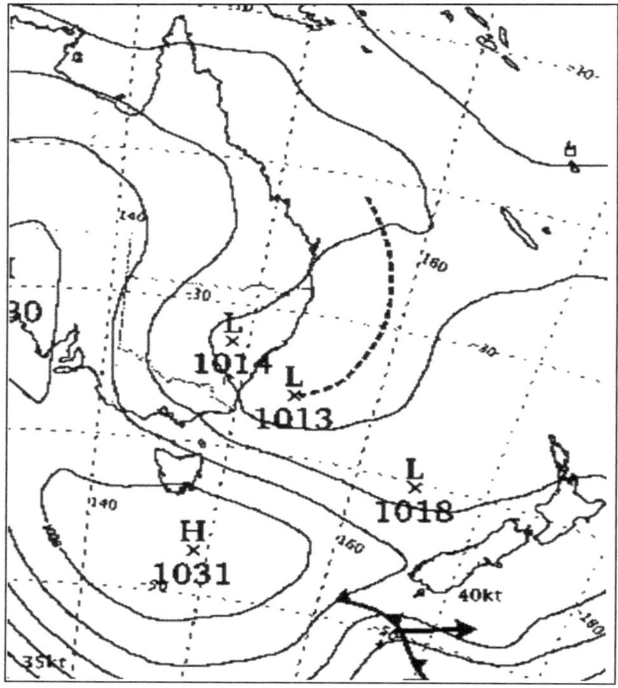

RAIN POTENTIAL

FROST/SNOW

JULY **19** SATURDAY

A high pressure system remains stalled over Tasmania, but maintains a very cold southwest air over the South Island, while a depression crossing the northern Tasman Sea brings a moderate northeasterly flow over the North Island. Mostly dry, cold, cloudy, and frosty in many districts south of the Waikato, with freezing fogs likely in Southland and Otago. Chance of some brief evening drizzle patches in the south and east of the South Island, and in Northland.

Auckland:	Partly cloudy, isolated showers, moderate breezes
Hamilton:	Dry, dull and gloomy air, light breezes
Wellington:	Overcast, dry, rain threatening, light breezes, chance of frost pockets
Greymouth:	Light breezes, sunny, dry, frosts
Christchurch:	Cloudy, light breezes, dry, cold with severe frosts
Dunedin:	Cloudy, light breezes, brief showers, frosts, chance of snow flurries

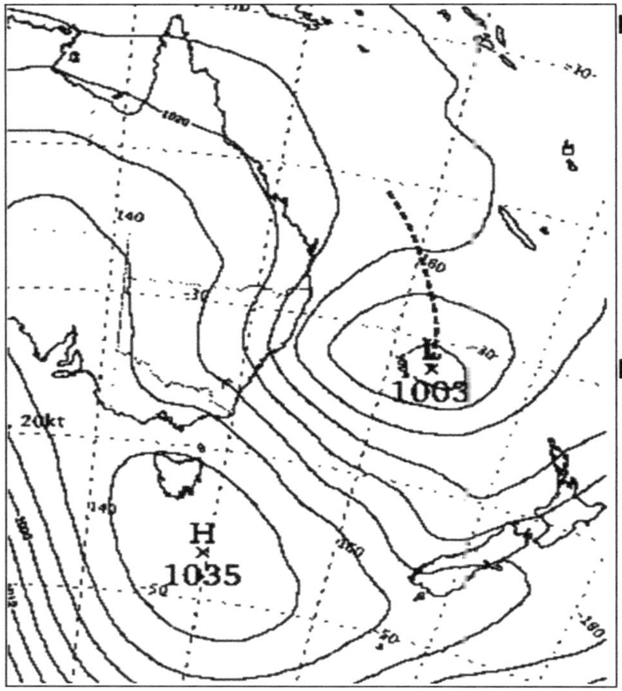

RAIN POTENTIAL

FROST/SNOW

JULY **20** SUNDAY

A large anticyclone remains stalled in the southern Tasman Sea while a depression north of New Zealand intensifies as it meets with the high pressure ridge. Strong and gusty northeast to easterly winds prevail over much of the country while a lighter westerly flow prevails over the far south of the South Island. Winds may reach gale force at times over the North Island. Storm conditions develop in Northland with heavy rain and possible flooding, while the south and central parts of the South Island remain dry and frosty. Occasional showers elsewhere.

Auckland:	Overcast, outbreaks of rain, strong winds
Hamilton:	Overcast, dry, strengthening winds
Wellington:	Overcast, cool, rain, freshening winds, chance of gale gusts at times
Greymouth:	Partly cloudy, dry, some frost pockets likely, fresh breezes at times
Christchurch:	Cloudy, frosts, chance of wintry showers or snow flurries
Dunedin:	Cloudy, cold, isolated showers, light breezes, chance of snow flurri

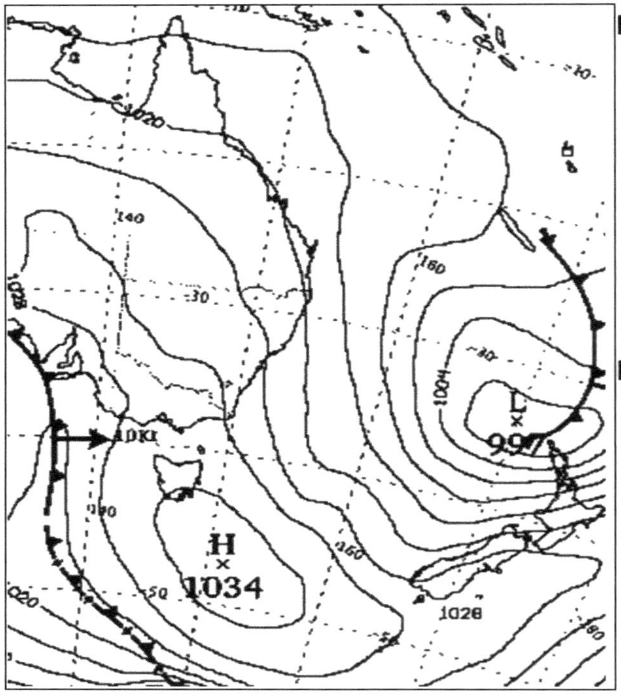

RAIN POTENTIAL

FROST/SNOW

JULY **21** MONDAY

A large anticyclone remains stalled in the southern Tasman Sea while a depression north of New Zealand intensifies as it meets with the high pressure ridge. Strong and gusty northeast to easterly winds prevail over much of the country while a lighter westerly flow prevails over the far south of the South Island. Winds may reach gale force at times over the North Island. Storm conditions develop in Northland with heavy rain and possible flooding, while the south and central parts of the South Island remain dry and frosty. Occasional showers elsewhere.

Auckland:	Overcast, outbreaks of rain, strong winds
Hamilton:	Overcast, dry, strengthening winds
Wellington:	Overcast, cool, rain, freshening winds, chance of gale gusts at times
Greymouth:	Partly cloudy, dry, some frost pockets likely, fresh breezes at times
Christchurch:	Cloudy, light air, frosts, chance of wintry showers or snow flurries
Dunedin:	Cloudy, cold, isolated showers, light breezes, chance of snow flurries

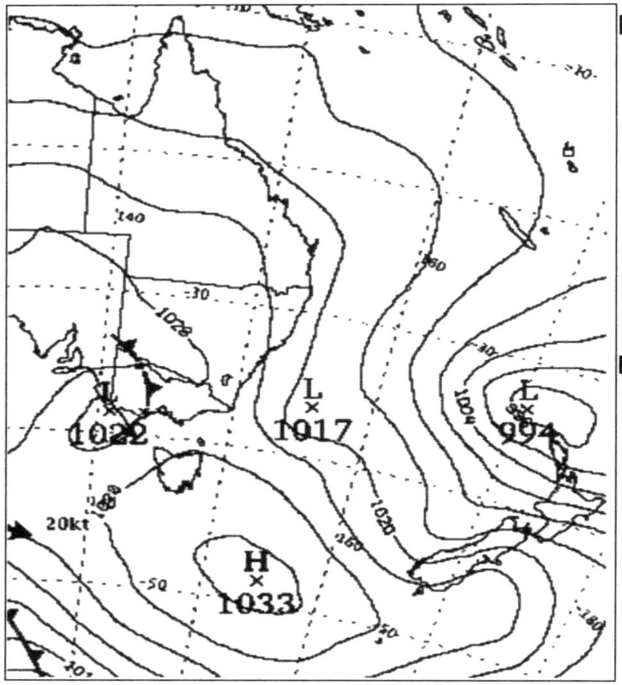

RAIN POTENTIAL

FROST/SNOW

JULY **22** TUESDAY

A high pressure system to the southwest of New Zealand extends its ridge further up the country as the depression over the North Island drifts away to the east. Winds remain fresh to strong in a northwesterly flow over the North Island, while lighter and backing southeasterly over the South Island. Overnight showers prevail in the east of the country from Canterbury to the East Cape, as well as in Northland to Auckland, while mostly dry elsewhere and with improving sunny spells. Very cold and frosty in the South Island with chance of freezing fog in places.

Auckland:	Mostly sunny day, pockets of evening rain, fresh breezes
Hamilton:	Partly cloudy, dry, windy spells
Wellington:	Overcast, cold, scattered showers, windy spells
Greymouth:	Dry, milder, sunny, gentle breezes
Christchurch:	Overcast, scattered showers, chance of fog and snow flurries
Dunedin:	Cloudy and cold, dry, light shifting breezes

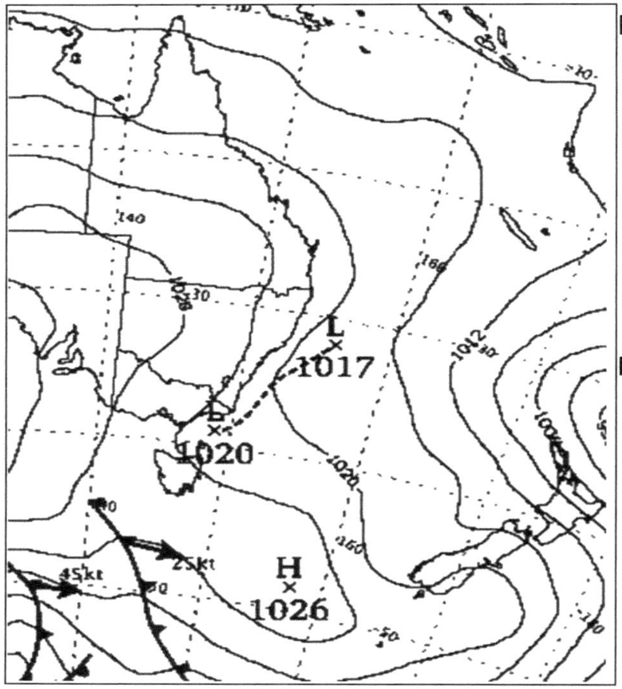

RAIN POTENTIAL

FROST/SNOW

JULY **23** WEDNESDAY

High pressure weakens over the country while a large low pressure trough lingers in the Tasman Sea. A very light southeasterly flow prevails over the North Island and gentle westerly over the South Island. Sunny and dry with widespread freezing frost and fog patches throughout the South Island. Some lingering overnight showers remain in the east of the North Island from the East Cape to Wellington, while mainly dry and mostly sunny elsewhere, cold and with frosts likely from Auckland to Wellington.

Auckland:	Sunny, dry, warm day cold night, light air
Hamilton:	Sunny, dry, cold overnight with frost likely, mostly calm air
Wellington:	Cold, rain, some sunny breakthroughs, moderate breezes
Greymouth:	Sunny and dry, mild day cold night, light shifting breezes
Christchurch:	Sunny, cold with frosts, dry, gentle shifting breezes, fog likely
Dunedin:	Sunny, dry, cold with frosts, light breezes

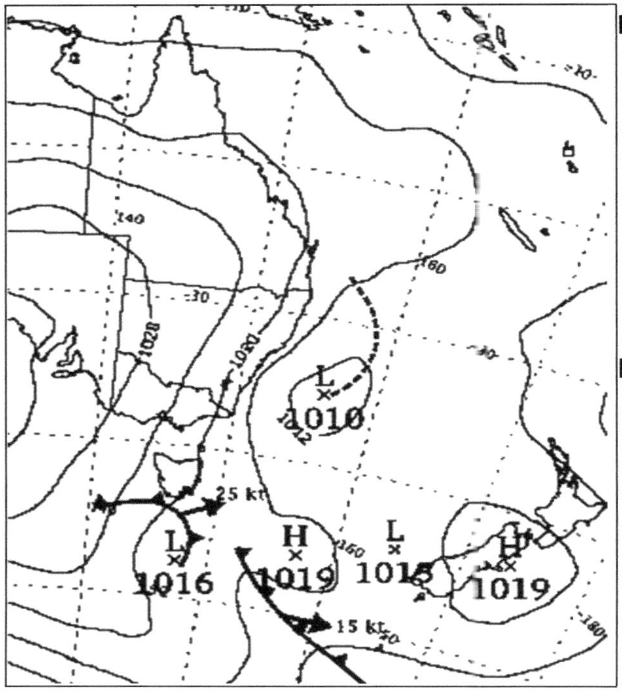

RAIN POTENTIAL

FROST/SNOW

JULY **24** THURSDAY

A complex of high pressure ridges and low pressure troughs direct a light easterly flow throughout the country. Milder in Northland with some spotty shower activity likely and milder around the East Cape to the Hawkes Bay again with the chance of isolated overnight showers. Mostly dry with widespread frosts elsewhere, cloudy in the south of the South Island and sunnier elsewhere.

Auckland:	Partly cloudy, dry, cooler, mostly calm air
Hamilton:	Calm, dry, frosts, some sunny spells
Wellington:	Partly cloudy, dry, light breezes
Greymouth:	Sunny and dry, mild day cold night, mostly calm air
Christchurch:	Partly cloudy, cold, frosts, mostly calm air, fog likely
Dunedin:	Cold with frosts, cloudy, dry, mostly calm and serene air

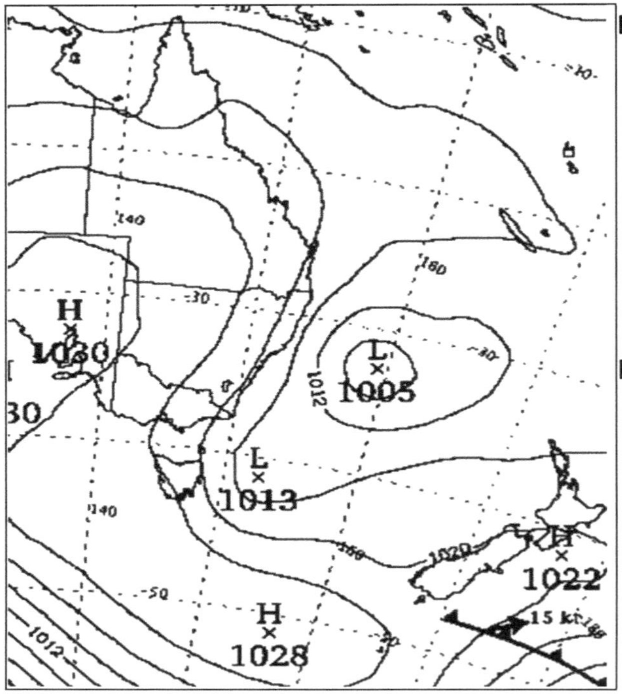

RAIN POTENTIAL

FROST/SNOW

JULY **25** FRIDAY

A large anticyclonic system directs a light easterly flow over the country, while a couple of weak depressions in the Tasman Sea bring a fresher easterly breeze over the northern half of the North Island and some showers into Northland and isolated showers around the Hawkes Bay. Generally, mostly dry, sunny, and very cold with frosts elsewhere, and possible freezing fog patches in the South Island.

Auckland: Cloudy, mostly dry, chance of overnight showers, moderate breezes
Hamilton: Cloudy, dry, gentle breezes
Wellington: Partly cloudy, isolated overnight showers, light breezes
Greymouth: Sunny, dry, gentle breezes, cold overnight
Christchurch: Sunny, dry, frosts, light breezes, fog possible
Dunedin: Partly cloudy, brief overnight showers, mostly calm air

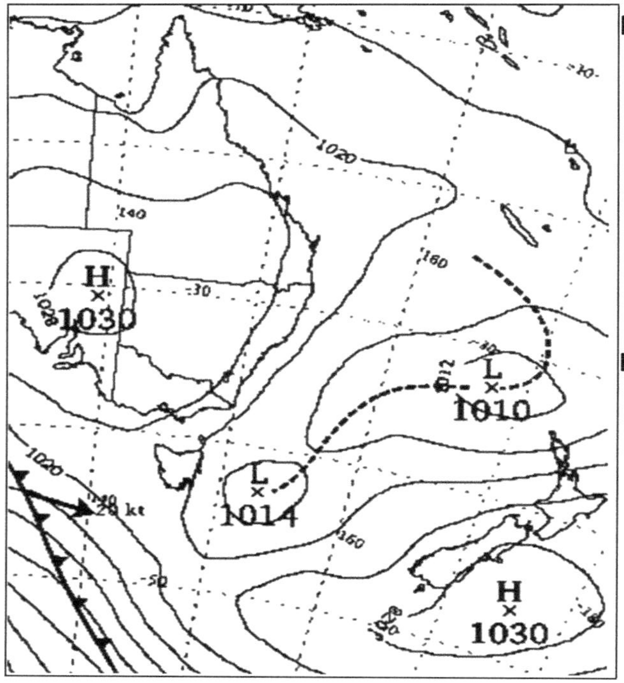

RAIN POTENTIAL

FROST/SNOW

JULY **26** SATURDAY

A large high departs to the east of New Zealand while a low pressure trough moves onto the country, with a moderate easterly flow prevailing throughout. Very dull and cloudy over the North Island with heavy rain in Northland and showers into the Bay of Plenty and around the East Cape to the Hawkes Bay. Mostly dry elsewhere. Mostly sunny in the South Island, but very cold with severe frosts in many districts and possible freezing fog here and there.

Auckland: Overcast, heavy rain with chance of localised flooding, fresh breezes
Hamilton: Overcast, isolated showers, gentle breezes
Wellington: Fresh breezy spells, partly cloudy, dry, mild day cold night
Greymouth: Dry, sunny, cold overnight, moderate breezes
Christchurch: Partly cloudy and dry, light breezes, frost and fog patches likely
Dunedin: Sunny, dry, light breezes, chance of frost and fog patches

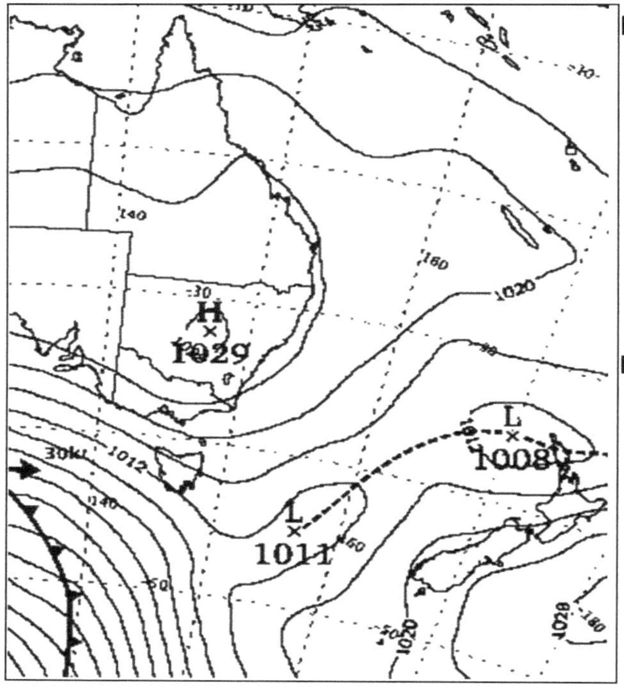

RAIN POTENTIAL

FROST/SNOW

JULY **27** SUNDAY

A weak depression moves onto the North Island bringing an unsettled easterly flow and widespread cloud and outbreaks of rain in most districts. In contrast, a high pressure ridge is maintained over the South Island, with moderate easterly breezes prevailing, mostly dry conditions and very cold with severe frosts. Cloud increasing over the northern half of the South Island with the chance of some isolated overnight showers.

Auckland: Mostly cloudy, heavy rain, possible flooding, fluctuating breezes
Hamilton: Cloudy, rain, light shifting breezes
Wellington: Overcast, dry, milder, fresh breezes
Greymouth: Cloudy, dry, mild day cold night, light shifting breezes
Christchurch: Overcast, dry with increasing threats of rain, frosts and chance of fog
Dunedin: Sunny and dry, gentle breezes, pockets of frost and fog likely

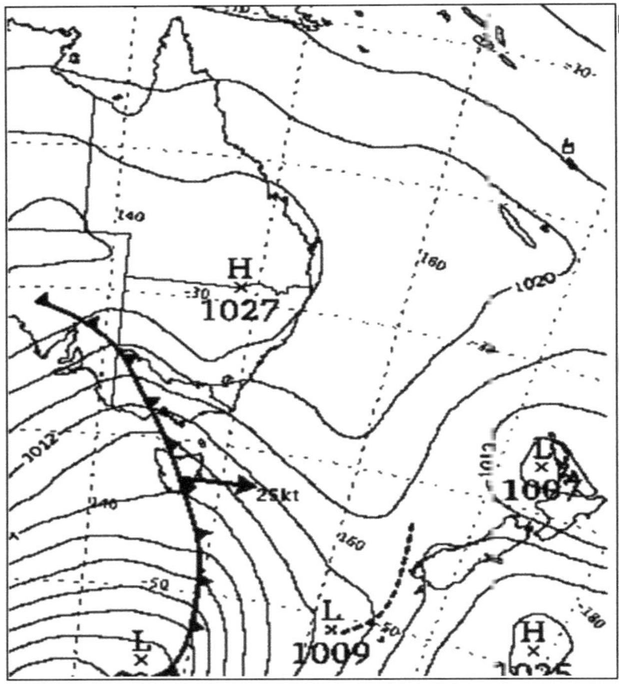

RAIN POTENTIAL

FROST/SNOW

JULY **28** MONDAY

Low pressure continues to bring scattered shower activity throughout most of the North Island, but sunnier and mostly dry around the Bay of Plenty, and a mostly easterly flow continues to prevail. High pressure continues to dominate the South Island, with a light easterly air also prevailing. Milder throughout the South Island and mostly dry except for the chance of some isolated showers around Marlborough to North Canterbury.

Auckland: Overnight rain, sunny and warmer day, moderate breezes
Hamilton: Partly cloudy, mostly dry, warmer day cool night, light breezes
Wellington: Cloudy, dry, windy at times
Greymouth: Sunny, warmer, dry, mostly calm air
Christchurch: Cloudy, cold, light wintry showers, gentle shifting breezes
Dunedin: Partly cloudy, dry, light shifting breezes

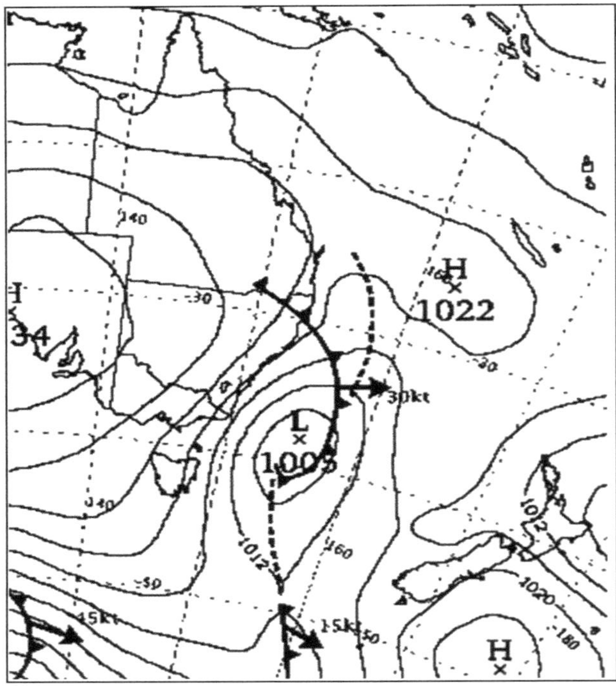

RAIN POTENTIAL

FROST/SNOW

JULY 29 TUESDAY

A depression in the north Tasman Sea extends a light northeasterly flow over the North Island while high pressure to the southeast of the country directs a gentle easterly flow over the South Island. A wet front within the low pressure trough brings some rain down through the east of the North Island from the East Cape to Wellington, while the rest of the country is mostly dry. Cold with frosts in the South Island and around the Bay of Plenty.

Auckland: Partly cloudy, isolated showers, light breezes to mostly calm air
Hamilton: Calm, dry, warmer, dull and gloomy with widespread cloud
Wellington: Overcast, light drizzly showers, gentle breezes
Greymouth: Mostly sunny, warm day cold night, light breezes
Christchurch: Mostly calm air, cloudy, isolated showers, chance of fog and flurries
Dunedin: Mostly sunny, milder, dry, light breezes

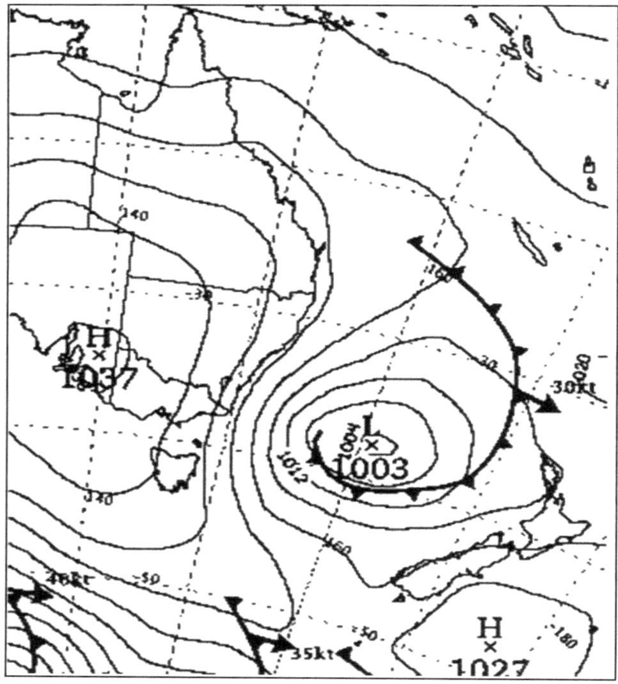

RAIN POTENTIAL

FROST/SNOW

JULY 30 WEDNESDAY

A depression in the north Tasman Sea remains almost stalled as it weakens while the high pressure ridge to the southeast of the country strengthens, with a light to moderate northeasterly flow prevailing throughout. Widespread cloud and rain spreads throughout the North Island and into the top of the South Island, while the central and southern districts of the South Island remain mostly dry and frosty.

Auckland: Cloudy, isolated overnight showers, gentle breezes
Hamilton: Overcast, scattered showers, light breezes
Wellington: Overcast, dry, moderate breezes
Greymouth: Cloudy and dry, moderate breezes
Christchurch: Light breezes, overcast, brief showers
Dunedin: Sunny, dry, gentle breezes

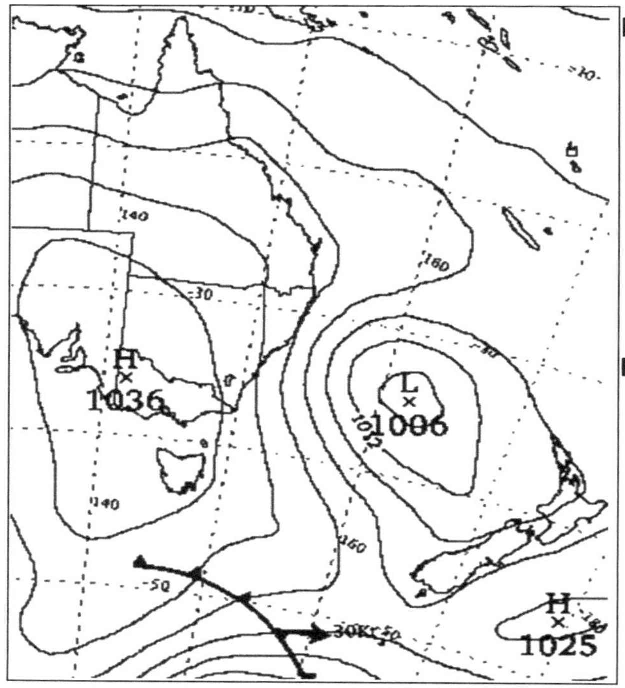

RAIN POTENTIAL

FROST/SNOW

JULY 31 THURSDAY

A weak depression slowly crosses the country bringing widespread cloud and scattered outbreaks of rain throughout the North Island and into the north of the South Island while a wet front brushes up through Fiordland bringing some showers into that region as well. Mostly dry in central and eastern parts of the South Island with severe frosts in the east and south. A very light northerly air prevails over the North Island, while an easterly backing southerly air covers the South Island.

Auckland: Overcast, dry, threats of rain, chance of thunderstorms and mists
Hamilton: Overcast, dry with chance of fog patches, calm atmosphere
Wellington: Calm and serene air, brief overnight showers, some sunny spells
Greymouth: Partly cloudy, dry, warmer, light breezes
Christchurch: Cloudy, dry, light shifting breezes
Dunedin: Cloudy, dry, light breezes, frost pockets likely

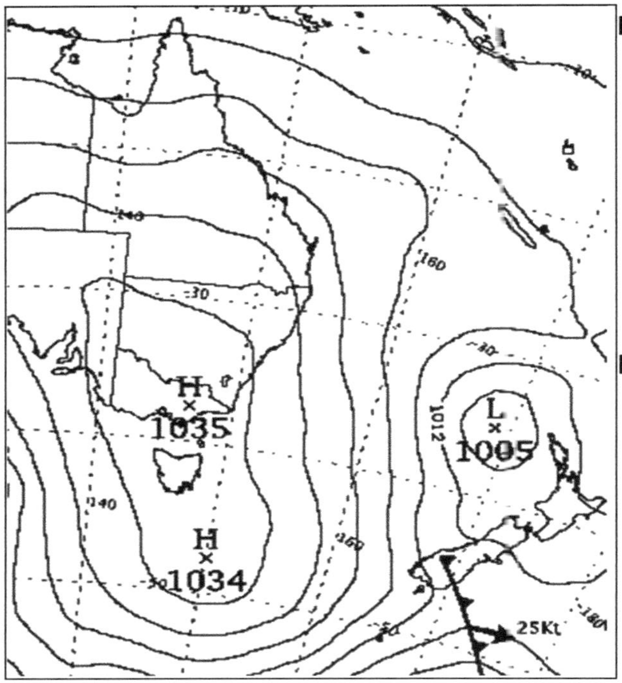

RAIN POTENTIAL

FROST/SNOW

AUGUST

PHASES OF THE MOON

1st	First Quarter
1st	Apogee
5th	Southern Declination
9th	Full Moon
12th	Crossing Equator
14th	Perigee #12
16th	Third Quarter
18th	Northern Declination
23rd	New Moon
25th	Crossing Equator
29th	Apogee
31st	First Quarter

1	2	3	4	5	6	7	8	9	10	11	12	13	14	15	16	17	18	19	20	21	22	23	24	25	26	27	28	29	30	31
Fri	Sat	Sun	Mon	Tue	Wed	Thu	Fri	Sat	Sun	Mon	Tue	Wed	Thu	Fri	Sat	Sun	Mon	Tue	Wed	Thu	Fri	Sat	Sun	Mon	Tue	Wed	Thu	Fri	Sat	Sun
1Q A				V				F			XhN		P12		3Q		^					N		XhS				A		1Q

MONTHLY SUMMARY

17th July-1st Aug A period of severe frosts, freezing fogs, with temperatures constantly below zero in most days in Central Otago during this spell. In contrast, temperatures are likely to be slightly milder than normal for this time of year over most of the North Island.
1st Wet front crosses the North Island bringing further widespread rain, heavy at times in the Waikato to the Bay of Plenty.
3rd Misty for a time in Christchurch.
5th Unusually mild day in Christchurch.
6th-7th Patches of fog possible in the Auckland region.
8th-10th Depression crosses the country in a fresh easterly flow bringing widespread rain throughout most of the country. Chance of heavy rain during this period that may cause localised flooding in some parts of south Canterbury and Otago and in some southern parts of Northland to the Coromandel and Tauranga. Light misty drizzles likely in the Christchurch area.
10th Chance of funnel formation winds developing in the Taranaki region, while patches of fog possible in the Auckland region.

Over the next four weeks, anticyclonic systems will move through the Tasman Sea before crossing through central New Zealand, while depressions will track further to the south of the country. As such, an unsettled southwesterly flow will prevail for much of the coming four weeks, especially over the east of the country, and around three depressions are regularly expected about every ten days. Rainfall will be close to normal or slightly above along the western coasts, while below average in the north and east. It will be unusually drier than normal in parts of the South Island, where parts of Marlborough, Banks Peninsula and inland parts of Otago may only see around a third of their usual amount of rainfall for this time of year, where soil moisture levels are likely to still continue to be below the norm. Eastern districts from the southern parts of the Wairarapa to Otago, as well as into the Nelson district may only see around half their usual amount of rainfall. Northland, the Waikato, Manawatu, Wanganui districts and the western parts of Southland can expect around 25% more rainfall than normal. Sunshine levels will be up to 20% more over this next four weeks compared to the norm for this time of year in Wellington, Nelson, and inland parts of south Canterbury with Marlborough, north Canterbury, and eastern parts of Otago around 10% sunnier. In contrast, the Manawatu can expect to be cloudier than usual and all other parts of the country likely to see sunshine hours typical for this time of year.

Temperatures will be around half a degree of the norm over most of the South Island for what is typical for this outlook period, while above average by up to a degree more than normal in the North Island and across the Cook Strait into central Marlborough and Nelson.

11th Chance of misty fog patches in the Auckland region.
13th-14th Chance of thunderstorms in the Waikato to Auckland regions, accompanying heavy rain at times.
15th Depression crosses the country.
15th-16th Chance of misty patches and shallow fog in Christchurch. Outbreaks of heavy rain likely in the West Coast, Nelson-Marlborough districts and into parts of Canterbury.
18th-19th Mixed spell in Christchurch with morning fog and strong blustery winds by the next day.
19th-22nd Very strong northwest winds, at first affecting Northland before moving down through central and southern parts of the country. Winds may reach gale force in places with the chance of causing some damage to infrastructure, particularly in the Otago region.
20th-22nd Strong northwest winds around the Wairarapa to Wellington may bring some disruptions to ferry services.
22nd Heavy rain in Southland including in the Milford Sounds area.
25th Severe frosts in the Mt Cook Village area.
26th-27th Depression crosses the country. Heavy rain in Northland with the chance of localised flooding in the Kerikeri to Warkworth areas.
30th Very cold in the Hanmer district with chance of sub-zero frosts.

August 2025

Monday	Tuesday	Wednesday	Thursday	Friday	Saturday	Sunday
28	29	30	31 **Moon Sign**	**1** **Scorpio** Trine Venus Square Pluto	**2** **1Q A** **Scorpio** Square Sun Square Mercury Trine Jupiter	**3** **Scorpio** Sextile Mars Trine Saturn Opposition Uranus Trine Neptune
4 **Sagittarius** Trine Sun Trine Mercury Trine Saturn Opposition Uranus Trine Neptune	**5** **Sagittarius** Trine Sun	**6** **V** **Capricorn** Opposition Venus Square Mars Square Saturn Square Neptune Square Ascendant	**7** **Capricorn** Opposition Jupiter	**8** **Capricorn** Opposition Mercury Trine Mars Sextile Saturn Trine Uranus Sextile Neptune Conjunct Pluto	**9** **FM** **Aquarius** Opposition Sun	**10** **Aquarius** Square Uranus
11 **Pisces** Trine Venus Trine Jupiter	**12** **XhN** **Pisces** Opposition Mars Conjunct Saturn Conjunct Neptune	**13** **Aries** Trine Mercury Square Venus Opposition Mars Square Jupiter	**14** **Aries** Trine Sun	**15** **P12** **Taurus** Square Mercury	**16** **3Q** **Taurus** Square Sun Conjunct Uranus Trine Pluto Square Ascendant	**17** **Gemini** Sextile Mercury Trine Mars Conjunct Uranus Trine Pluto Square Ascendant
18 **Gemini** Sextile Sun Square Ascendant	**19** **^** **Cancer** Square Mars Conjunct Jupiter Square Ascendant	**20** **Cancer** Conjunct Venus Conjunct Jupiter Square Ascendant	**21** **Leo** Conjunct Mercury Sextile Mars Trine Saturn Trine Neptune Opposition Pluto Square Ascendant	**22** **Leo**	**23** **NM** **Leo** Conjunct Sun Square Uranus	**24** **Virgo** Conjunct Sun Square Uranus
25 **Virgo** Sextile Jupiter	**26** **XhS** **Virgo** Sextile Venus Opposition Saturn Trine Uranus Opposition Neptune Trine Pluto	**27** **Libra** Conjunct Mars Square Jupiter	**28** **Libra** Square Pluto	**29** **Scorpio** Sextile Sun Square Venus Square Pluto	**30** **A** **Scorpio** Square Mercury Trine Jupiter	**31** **1Q** **Sagittarius** Square Sun Trine Venus Trine Saturn Opposition Uranus Trine Neptune Sextile Pluto
1	2					

NORTH ISLAND RAINFALL
AUGUST 2025

d=mainly dry, n= fog, mist, dew or nonrecordable, l=light showers, p=passing showers, s=significant showers, r=rain, h=heavy falls

		Northland							Auckland					Bay Of Plenty				Waikato				Rot/Tau		Hawkes Bay			C Plateau			Taranaki			Manawatu			Lower North Island					
		CR	Kta	Ker	Rsl	Dar	Whg	Man	Wark	Hen	MkH	Ak	Co	Tau	TeP	Wkt	Opo	Hml	TeA	Mata	Ot	Rt	Taup	Gis	Nap	Has	Taum	Chat	Waio	NP	Str	Wang	Pa	Dann	Lev	PN	Mast	UH	Wgt	all	Moon
1st	Fri	l	l	l	l	r	l	s	s	r	h	r	r	h	h	h	r	h	h	r	h	h	r	r	r	r	s	l	s	s	l	l	l	s	d	d	l	l	l	380	
2nd	Sat	d	d	d	d	d	d	l	l	l	n	d	d	d	l	d	d	d	d	d	d	n	d	l	d	d	d	d	l	d	l	d	l	d	d	d	l	l	l	17	1Q A
3rd	Sun	l	l	d	d	d	n	d	n	l	l	n	d	d	d	d	d	l	l	d	l	d	d	d	d	d	d	s	s	l	l	d	d	d	d	d	d	d	d	21	
4th	Mon	l	l	d	d	l	d	l	d	l	d	l	d	d	d	d	d	l	l	d	l	d	d	d	d	d	l	l	l	l	d	d	d	d	d	d	d	d	d	17	
5th	Tue	d	l	d	d	l	n	l	d	l	d	l	n	d	d	d	d	l	l	d	l	d	l	d	d	d	l	s	s	l	l	n	d	l	d	d	d	d	d	34	
6th	Wed	d	d	d	l	s	n	d	d	s	s	l	d	d	d	d	d	l	d	d	l	d	d	d	d	d	n	l	d	d	d	d	d	d	d	d	d	d	d	24	V
7th	Thur	l	d	d	d	d	d	d	d	l	d	d	d	d	d	d	d	l	l	d	s	d	d	d	d	d	s	r	d	l	d	d	d	d	l	d	d	d	d	30	
8th	Fri	h	h	h	h	h	h	h	h	h	r	r	d	l	s	l	d	l	s	h	d	l	n	d	d	d	d	l	d	d	d	d	n	d	d	d	d	l	d	424	
9th	Sat	n	s	l	s	s	l	s	h	l	h	h	h	h	h	h	h	h	h	h	h	h	h	r	r	h	h	h	r	h	h	r	r	r	r	s	h	h	h	931	F
10th	Sun	l	l	r	h	s	s	l	s	s	s	l	l	l	l	d	d	l	s	r	s	s	r	d	d	d	r	h	l	h	h	s	l	l	l	s	l	s	s	230	
11th	Mon	r	r	l	l	s	l	n	l	s	l	l	l	d	d	d	d	l	s	s	h	l	s	d	d	d	h	h	r	s	h	s	h	n	r	s	l	h	h	292	
12th	Tue	d	s	d	d	s	l	l	s	s	s	l	l	n	d	d	d	d	s	s	l	l	l	d	d	d	l	s	s	n	l	n	d	d	n	d	d	l	l	65	XhN
13th	Wed	l	s	l	l	r	l	s	s	l	r	l	l	l	l	l	n	s	s	s	s	l	l	d	l	l	s	r	r	l	l	s	l	s	s	l	d	d	d	142	
14th	Thur	s	d	l	l	d	l	d	l	r	h	r	s	l	l	d	d	s	s	s	l	n	d	d	d	d	l	l	l	l	l	d	l	s	s	r	d	d	d	115	
15th	Fri	s	l	l	l	s	l	l	s	s	h	l	h	h	h	h	r	h	h	h	h	h	h	l	l	l	h	h	r	h	h	r	s	r	r	r	s	s	r	584	P12
16th	Sat	l	l	d	d	l	d	l	l	l	l	n	d	d	d	d	d	l	l	l	l	n	l	d	d	d	s	r	l	s	r	s	h	s	s	s	d	r	s	126	3Q
17th	Sun	s	s	s	l	s	l	l	r	s	s	s	l	l	s	s	s	r	r	h	r	r	s	d	d	d	h	r	r	r	s	h	s	l	s	s	s	s	d	281	
18th	Mon	s	r	r	s	h	h	r	s	r	s	s	r	s	l	d	n	l	l	s	s	l	l	l	l	s	s	r	s	l	l	h	d	r	l	s	s	l	d	243	
19th	Tue	d	d	d	d	d	d	d	d	l	d	d	d	d	d	d	d	n	l	l	n	d	d	d	d	d	l	h	l	l	d	d	d	l	l	l	d	d	d	28	^
20th	Wed	d	d	d	d	l	d	d	l	n	d	d	d	l	n	l	l	l	s	l	l	d	d	d	d	d	n	s	d	l	l	d	d	d	n	d	d	d	d	28	
21st	Thur	d	d	d	d	s	d	l	d	d	d	d	d	d	d	d	d	n	d	d	l	d	n	d	d	d	s	h	s	s	r	r	n	l	d	s	d	d	d	75	
22nd	Fri	l	l	l	l	l	d	l	s	r	s	n	l	l	s	s	s	r	h	s	h	r	r	d	d	l	h	h	h	s	h	r	r	r	r	r	s	h	r	378	
23rd	Sat	n	l	d	l	l	l	l	d	r	d	s	s	s	s	s	s	l	s	h	l	s	l	l	l	s	l	s	s	l	l	d	l	s	d	l	r	n	d	128	N
24th	Sun	d	l	d	d	d	d	l	d	d	l	l	l	d	d	l	d	d	d	d	d	d	l	h	l	s	d	d	d	d	l	d	d	l	d	d	l	d	l	42	
25th	Mon	d	d	d	d	d	s	n	d	d	d	d	s	d	d	d	d	d	d	d	d	d	d	s	l	l	d	d	d	d	d	d	d	n	d	d	d	d	l	17	
26th	Tue	h	h	h	r	l	r	l	l	d	d	d	r	n	l	d	d	n	d	d	d	d	d	l	l	l	d	d	d	d	d	d	d	d	d	d	d	d	d	106	XhS
27th	Wed	s	h	h	h	h	h	h	h	h	h	h	h	h	h	h	r	r	r	h	r	h	h	l	l	d	r	s	l	h	h	s	h	l	r	s	s	h	h	989	
28th	Thur	s	d	l	d	d	l	d	l	l	s	l	d	d	s	d	s	s	s	s	s	l	s	l	l	l	r	h	l	s	r	s	l	s	l	s	l	l	l	145	
29th	Fri	l	l	l	l	l	l	s	s	s	r	l	s	s	r	s	s	h	r	s	r	l	r	d	l	s	r	h	h	h	r	s	l	r	s	r	d	l	d	283	
30th	Sat	d	d	d	d	n	d	d	d	l	l	l	d	d	d	d	d	d	d	d	d	d	d	l	l	l	l	d	d	d	l	l	l	n	l	d	l	l	s	27	A
31st	Sun	l	l	d	l	d	l	n	l	l	s	l	d	d	d	d	d	d	d	d	d	d	l	s	d	d	d	d	l	d	d	d	d	n	d	d	d	d	l	22	1Q
estimate:		102	153	170	153	160	185	195	210	195	224	147	224	161	196	139	106	230	199	235	225	178	133	63	47	60	239	387	149	189	256	139	108	99	105	119	75	151	116	6223	
average:		116	135	187	160	124	252	158	147	116	109	110	127	120	189	130	132	100	130	144	142	130	110	97	82	67	133	255	92	140	187	75	90	89	97	87	105	128	120	4912	
trend:		drr	wtr	drr	av	wtr	drr	wtr	wtr	wtr	wtr	wtr	wtr	wtr	av	av	drr	wtr	wtr	wtr	wtr	wtr	wtr	drr	drr	av	wtr	wtr	wtr	wtr	wtr	wtr	wtr	av	av	wtr	drr	wtr	av	wtr	

SOUTH ISLAND RAIN
AUGUST 2025

d=mainly dry, n= fog, mist, dew or nonrecordable, l=light showers, p=passing showers, s=significant showers, r=rain, h=heavy falls

		West Coast				Top South Island				Canterbury				South Canterbury				Coastal Otago			Lakes			Inland Otago				Southland						
		West	Grey	Hok	BB	Ta	Mou	Nel	Ble	Dar	Ham	Kai	Ch	Rak	Meth	Ash	Tim	Oam	Dun	Bal	Twi	LP	LT	Crom	Qstn	Ran	Alex	TeA	Wtn	Gore	Inv	Puy Pt	all	Moon
1st	Fri	d	d	d	d	l	l	l	d	d	l	d	l	d	n	d	d	d	l	l	d	d	d	d	d	d	d	l	s	l	l	l	27	
2nd	Sat	l	n	d	d	d	d	d	d	d	d	d	d	d	d	d	d	d	d	d	d	d	d	d	d	d	d	l	d	d	s	s	20	1Q A
3rd	Sun	r	r	r	r	d	d	d	d	d	d	d	d	d	l	n	d	d	l	l	d	d	d	d	d	d	d	s	s	s	s	s	95	
4th	Mon	r	r	s	h	d	d	d	d	d	d	d	d	d	d	d	d	d	d	d	d	d	d	d	d	d	d	s	l	d	d	s	67	
5th	Tue	s	s	s	r	l	l	n	d	d	d	d	d	d	d	d	d	d	n	d	d	d	d	l	r	d	d	r	d	l	l	r	73	
6th	Wed	n	s	r	s	d	d	d	d	d	d	d	d	d	d	d	d	d	d	d	l	d	d	l	d	d	d	l	d	d	d	s	30	V
7th	Thur	d	l	r	l	d	d	d	d	d	d	d	d	d	d	d	d	d	d	l	d	d	d	d	d	d	d	r	d	d	s	s	41	
8th	Fri	d	l	l	s	d	d	d	d	d	d	d	d	d	d	d	d	d	l	s	d	d	d	l	d	d	d	s	s	l	r	h	79	
9th	Sat	s	d	l	n	h	r	r	h	h	r	r	h	h	h	h	h	h	r	d	d	s	r	l	n	l	s	d	d	l	l	d	331	F
10th	Sun	h	s	s	d	s	l	l	l	d	d	d	n	l	l	s	s	h	h	h	s	l	s	s	l	r	r	n	r	s	r	n	313	
11th	Mon	l	s	d	d	r	h	r	s	n	d	d	d	d	d	d	d	d	d	d	l	d	d	d	d	d	d	d	d	d	d	s	68	
12th	Tue	l	s	d	d	d	d	d	l	d	d	d	d	d	d	d	d	d	d	d	d	d	d	d	d	d	d	d	d	d	l	s	21	XhN
13th	Wed	s	s	l	l	d	d	d	d	d	n	d	d	d	d	d	d	d	d	d	d	d	l	d	s	d	d	l	d	d	d	l	24	
14th	Thur	n	d	d	d	l	d	d	d	d	d	d	d	d	d	d	d	d	d	d	d	d	d	d	d	d	d	d	d	d	d	r	18	
15th	Fri	h	h	h	h	h	h	r	r	h	s	l	h	h	h	h	h	h	s	l	d	r	l	l	d	s	l	l	l	l	l	s	385	P12
16th	Sat	h	h	r	h	r	r	s	d	d	d	d	d	l	l	n	l	d	l	d	s	d	l	d	d	d	d	d	l	l	l	l	121	3Q
17th	Sun	r	r	s	l	d	l	l	d	d	d	d	d	d	l	d	l	d	l	l	d	d	d	d	d	l	d	d	l	l	s	l	57	
18th	Mon	d	d	d	d	d	l	d	d	d	d	d	d	d	d	d	d	d	l	d	d	d	d	d	d	d	d	d	l	n	l	l	6	
19th	Tue	l	r	r	s	d	d	d	d	d	d	d	d	d	d	d	d	d	d	d	d	d	d	d	d	d	d	s	d	d	d	r	50	^
20th	Wed	l	s	s	r	d	d	d	d	d	d	d	d	d	d	d	d	d	d	d	d	d	d	d	l	d	d	s	d	d	d	h	59	
21st	Thur	s	d	l	r	l	l	l	d	d	d	d	d	d	d	d	d	d	d	d	d	d	l	l	h	d	d	h	d	d	d	h	125	
22nd	Fri	h	h	h	h	h	s	l	d	l	l	s	n	l	l	s	l	d	l	l	h	h	h	s	h	l	s	h	h	r	h	h	380	
23rd	Sat	l	s	d	d	d	d	d	d	l	s	d	s	d	l	l	n	d	l	l	h	h	d	l	l	d	n	d	s	l	r	r	121	N
24th	Sun	d	d	l	d	s	s	l	d	l	s	s	l	d	s	l	d	n	d	l	l	d	l	d	d	d	d	d	l	d	l	s	50	
25th	Mon	d	d	d	d	d	d	d	d	d	l	d	d	d	l	d	d	d	d	d	l	d	d	d	d	d	d	d	d	d	d	d	5	
26th	Tue	d	d	d	d	d	d	d	d	d	d	d	d	d	d	d	d	d	d	d	d	d	d	d	d	d	d	d	d	d	d	d	1	XhS
27th	Wed	s	r	r	r	h	h	r	h	l	n	l	l	l	l	l	d	d	d	d	d	d	l	d	d	d	d	d	d	d	d	d	166	
28th	Thur	s	h	h	r	h	l	d	l	d	d	d	d	d	l	d	d	l	s	d	d	l	d	d	l	d	d	d	d	n	l	n	100	
29th	Fri	l	d	l	d	n	d	d	d	d	d	d	d	d	d	d	d	d	d	d	d	d	d	d	d	d	d	l	d	l	s	l	11	
30th	Sat	d	d	d	d	d	d	d	l	n	d	d	d	d	n	d	d	d	d	d	d	d	d	d	d	d	d	d	d	d	d	l	5	A
31st	Sun	d	d	d	d	d	d	d	d	d	d	d	d	d	d	d	d	d	d	d	d	d	d	d	d	d	d	d	d	d	d	s	8	1Q
estimate:		172	211	176	176	155	147	62	68	59	31	28	84	65	68	73	60	84	116	78	84	70	42	18	76	22	22	125	59	37	101	285	2853	
average:		182	208	225	313	230	156	80	67	85	123	84	52	80	87	60	35	35	75	46	45	79	60	33	62	26	15	85	65	62	60	˙52	2967	
trend:		av	av	drr	drr	drr	av	drr	av	drr	drr	drr	w tr	drr	drr	w tr	w tr	w tr	w tr	w tr	w tr	av	drr	drr	w tr	av	av	w tr	av	drr	w tr	w tr	drr	

NORTH ISLAND SUNSHINE ESTIMATES

AUGUST 2025

F=fine (8-12 hours of sunshine), pc= partly cloudy (4-7 hours), c=cloudy (1-3 hours), o=overcast (0 hours)

		Kta	Whg	Dar	Ak	Hml	Tau	Rt	Taum	NP	Wang	Taup	Gis	Nap	Wai	Dann	PN	Mast	Pa	UH	Wgt	ALL	Moon
1st	Fri	F	pc	pc	pc	pc	pc	pc	pc	F	pc	pc	c	pc	c	c	pc	c	pc	c	pc	95	
2nd	Sat	pc	pc	pc	pc	F	F	pc	pc	pc	F	pc	F	F	F	pc	F	pc	pc	pc	pc	132	1Q A
3rd	Sun	pc	pc	pc	pc	c	F	c	pc	pc	c	c	F	F	F	pc	c	pc	F	F	F	124	
4th	Mon	pc	pc	pc	pc	c	pc	c	c	pc	c	c	F	c	c	o	c	o	o	c	c	63	
5th	Tue	pc	pc	pc	pc	c	c	pc	c	pc	c	pc	F	F	F	pc	c	pc	pc	F	F	115	
6th	Wed	F	pc	pc	pc	pc	pc	pc	c	c	c	pc	F	pc	pc	c	c	c	pc	pc	F	100	V
7th	Thur	pc	pc	pc	pc	c	pc	pc	c	pc	pc	pc	F	F	F	pc	pc	pc	c	o	c	98	
8th	Fri	c	o	c	c	c	o	o	o	o	c	o	o	c	c	o	c	o	c	c	c	19	
9th	Sat	pc	pc	pc	pc	c	pc	pc	c	c	o	pc	c	pc	c	c	o	c	c	o	o	60	F
10th	Sun	pc	pc	pc	pc	c	F	pc	o	c	c	pc	pc	pc	F	c	c	c	c	c	c	78	
11th	Mon	pc	pc	pc	pc	o	c	c	c	pc	pc	c	c	c	pc	c	pc	c	o	o	c	65	
12th	Tue	pc	pc	pc	pc	pc	pc	pc	pc	c	c	pc	pc	F	pc	c	c	c	pc	pc	pc	96	XhN
13th	Wed	c	c	pc	pc	pc	pc	pc	pc	F	c	pc	c	pc	pc	c	c	c	pc	c	pc	88	
14th	Thur	o	c	o	o	c	o	o	o	o	o	o	o	o	o	o	o	o	o	o	o	6	
15th	Fri	pc	pc	pc	pc	F	F	c	pc	pc	pc	c	F	F	pc	pc	pc	pc	pc	pc	pc	110	P12
16th	Sat	pc	pc	pc	pc	c	pc	c	c	pc	c	c	pc	pc	pc	pc	c	pc	pc	c	pc	90	3Q
17th	Sun	pc	c	c	c	pc	pc	c	c	pc	c	c	pc	pc	pc	c	c	c	pc	pc	pc	75	
18th	Mon	pc	pc	F	F	pc	F	F	F	F	pc	F	F	F	F	pc	pc	pc	pc	pc	F	150	
19th	Tue	pc	pc	pc	pc	pc	pc	c	c	c	c	c	F	pc	pc	pc	c	pc	pc	pc	F	100	^
20th	Wed	c	pc	F	F	pc	F	F	pc	pc	pc	F	F	F	F	pc	pc	pc	c	c	c	121	
21st	Thur	pc	pc	pc	pc	pc	pc	c	o	o	c	c	c	pc	pc	o	c	o	o	o	o	52	
22nd	Fri	c	c	c	c	pc	pc	c	c	pc	pc	c	o	pc	c	o	pc	o	pc	c	pc	62	
23rd	Sat	F	pc	F	F	F	F	pc	F	F	pc	pc	F	pc	pc	pc	pc	pc	F	pc	c	137	N
24th	Sun	F	F	F	F	c	pc	c	pc	pc	F	c	c	c	pc	pc	F	pc	pc	pc	pc	108	
25th	Mon	o	o	pc	pc	pc	pc	F	pc	pc	pc	F	pc	pc	pc	pc	pc	pc	F	pc	F	110	
26th	Tue	o	o	o	o	o	o	o	o	o	pc	o	o	c	c	pc	pc	pc	pc	pc	pc	40	XhS
27th	Wed	pc	pc	pc	pc	pc	pc	c	c	pc	o	c	pc	c	c	c	o	c	o	c	o	63	
28th	Thur	o	o	o	o	c	o	o	o	o	c	o	o	c	pc	c	c	c	o	c	c	28	
29th	Fri	F	pc	F	F	F	F	F	F	F	c	F	pc	pc	pc	pc	c	pc	pc	c	pc	128	
30th	Sat	pc	pc	F	F	F	F	F	pc	F	pc	F	pc	pc	pc	pc	pc	pc	F	c	pc	123	A
31st	Sun	pc	pc	pc	pc	pc	F	F	F	pc	c	F	F	F	F	pc	c	pc	pc	pc	F	139	1Q
est. hours:		156	148	157	157	137	171	136	105	154	102	136	164	160	167	113	102	113	133	113	148	2772	
av. hours:		156	140	133	140	136	161	150	107	153	131	129	155	159	147	121	117	118	143	139	136	2771	
trend:		av	av	more	more	av	av	av	av	av	less	av	av	av	more	av	less	av	av	less	av	average	

SOUTH ISLAND SUNSHINE ESTIMATES
AUGUST 2025

F=fine (8-12 hours of sunshine), pc= partly cloudy (4-7 hours), c=cloudy (1-3 hours), o=overcast (0 hours)

		Ta	Nel	Ble	Mou	West	Grey	Hok	Ham	Kai	Ch	Ash	Tim	Crom	Qstn	Bal	Dun	Wnt	Gore	Inv	Alex	ALL	Moon
1st	Fri	pc	F	F	F	F	F	F	pc	pc	pc	F	F	c	c	c	pc	c	c	c	c	110	
2nd	Sat	F	c	c	c	o	o	c	F	F	pc	pc	o	pc	pc	pc	pc	c	c	c	pc	71	1Q A
3rd	Sun	F	F	F	F	o	o	o	pc	pc	pc	pc	c	pc	c	pc	pc	c	c	c	pc	87	
4th	Mon	c	pc	c	pc	o	o	o	F	F	F	F	F	F	c	pc	pc	pc	pc	pc	F	105	
5th	Tue	o	pc	F	pc	c	pc	pc	F	F	F	F	F	F	pc	c	pc	pc	pc	pc	F	120	
6th	Wed	o	pc	F	pc	c	o	o	pc	pc	F	pc	c	o	c	o	c	pc	pc	o	o	64	V
7th	Thur	o	pc	F	pc	c	o	o	pc	pc	c	c	o	o	o	o	o	o	o	o	o	38	
8th	Fri	F	o	o	o	o	o	o	o	o	o	o	o	o	o	o	o	o	o	o	o	10	
9th	Sat	pc	c	o	o	o	o	o	c	c	c	c	o	o	o	o	o	o	o	o	o	16	F
10th	Sun	F	o	o	o	o	pc	F	o	o	o	o	o	pc	pc	pc	F	o	o	c	pc	53	
11th	Mon	o	pc	c	pc	c	pc	pc	o	o	o	o	o	o	c	o	c	pc	pc	pc	o	52	
12th	Tue	pc	pc	pc	pc	c	o	o	F	F	pc	pc	c	o	c	o	o	o	o	c	o	60	XhN
13th	Wed	pc	F	pc	F	c	pc	pc	o	o	c	o	o	c	pc	o	c	c	c	o	c	57	
14th	Thur	pc	o	o	o	o	o	o	o	o	o	o	o	c	o	c	o	pc	pc	pc	c	28	
15th	Fri	F	pc	F	pc	c	c	pc	F	F	pc	pc	o	c	c	o	o	o	o	o	c	75	P12
16th	Sat	pc	pc	F	pc	c	c	c	pc	pc	pc	pc	c	pc	o	c	pc	pc	pc	pc	pc	97	3Q
17th	Sun	pc	pc	F	pc	F	F	F	pc	pc	pc	F	pc	F	pc	c	c	pc	pc	pc	F	123	
18th	Mon	c	F	F	pc	o	c	pc	F	F	pc	F	pc	pc	c	pc	pc	c	c	pc	pc	105	
19th	Tue	F	pc	F	pc	c	o	o	F	F	pc	pc	pc	F	c	pc	pc	pc	pc	pc	F	106	^
20th	Wed	pc	pc	pc	F	o	c	c	c	c	pc	F	pc	pc	pc	pc	pc	pc	pc	c	pc	85	
21st	Thur	F	o	c	c	o	o	o	c	c	c	c	pc	o	o	c	c	o	o	c	o	40	
22nd	Fri	F	F	pc	pc	c	c	pc	F	F	F	F	pc	pc	pc	c	pc	pc	pc	c	pc	110	
23rd	Sat	o	pc	pc	pc	c	pc	F	o	o	o	o	pc	pc	pc	pc	pc	pc	pc	pc	pc	83	N
24th	Sun	c	F	F	pc	F	F	F	pc	pc	pc	c	F	F	pc	c	F	F	F	pc	F	136	
25th	Mon	pc	pc	c	pc	F	F	F	pc	pc	pc	F	F	F	pc	pc	F	F	F	F	F	143	
26th	Tue	F	c	pc	c	pc	o	o	pc	pc	pc	pc	pc	o	o	c	pc	o	o	c	o	65	XhS
27th	Wed	pc	pc	pc	o	c	o	o	o	o	o	c	c	o	o	o	o	c	c	c	o	26	
28th	Thur	F	c	c	c	c	c	c	c	c	c	c	c	c	o	o	o	o	o	c	c	39	
29th	Fri	o	F	pc	F	c	pc	pc	pc	pc	pc	F	pc	pc	c	pc	pc	pc	pc	pc	pc	117	
30th	Sat	F	F	F	F	pc	F	F	F	F	F	F	F	F	pc	F	F	F	F	F	F	178	A
31st	Sun	F	F	F	F	c	pc	pc	F	F	F	F	pc	pc	pc	c	c	c	c	c	pc	122	1Q
est. hours:		166	170	171	151	81	103	116	150	150	142	152	127	123	85	87	115	103	103	99	123	2519	
av. hours:		166	169	181	161	124	124	137	115	145	148	142	144	147	121	116	113	122	122	119	141	2757	
trend:		av	av	av	av	less	less	less	more	av	av	av	less	less	less	less	av	less	less	less	less	average	

AUGUST 1 FRIDAY

A large anticyclonic system centred over the southeast of Australia extends a high pressure ridge over the west of New Zealand, directing a fresh southwest flow over the South Island, backing southeasterly over the North Island. A wet trough ahead of the high brings widespread rain throughout the North Island, with heavy falls likely in the Waikato to the Bay of Plenty. Some isolated showers also likely in the north and the south of the South Island, with mostly dry and frosty conditions elsewhere.

Auckland: Partly cloudy, outbreaks of rain, some heavy falls, unsettled winds
Hamilton: Partly cloudy, heavy rain, possible thunderstorms, windy at times
Wellington: Partly cloudy, some showers, fresh winds with occasional gusts
Greymouth: Sunny, cool and dry, light breezes
Christchurch: Partly cloudy, brief showers, cold, fresh breezes
Dunedin: Partly cloudy, light showers, chance of snow on the hills

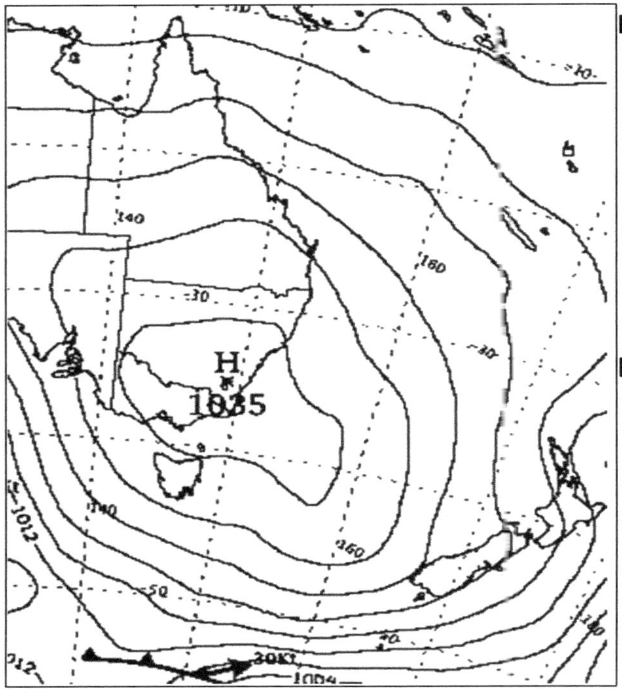

RAIN POTENTIAL

FROST/SNOW

AUGUST 2 SATURDAY

A very cold west to southwest low crosses the country directed from an intense depression south of the South Island, while a large high slowly drifts into the Tasman Sea off the Australian coast. Rain clearing to isolated showers mostly in the Waikato and east of the North Island, and in the southwest of the South Island. Mostly dry elsewhere with good sunny spells. Very cold throughout with widespread and severe frosts in many districts including as far north as Auckland.

Auckland: Unsettled winds, partly cloudy, isolated showers, chance of hail
Hamilton: Sunny, dry, moderate breezes, overnight frost
Wellington: Partly cloudy, colder, brief showers, blustery at times
Greymouth: Overcast, isolated showers, frost pockets, blustery at times
Christchurch: Partly cloudy, dry, cool with severe frosts, windy
Dunedin: Partly cloudy, slightly warmer, dry, strong blustery winds

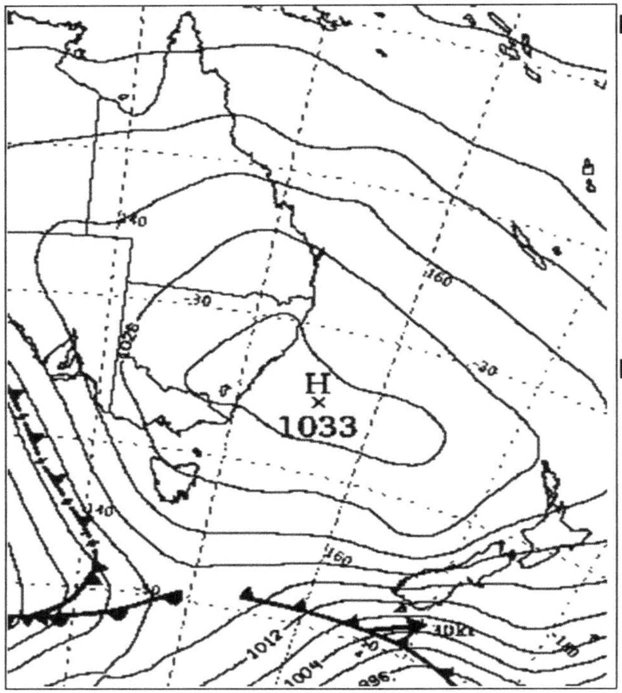

RAIN POTENTIAL

FROST/SNOW

AUGUST 3 SUNDAY

High pressure crosses to the north of New Zealand bringing a fresh westerly flow and good sunny spells throughout the North Island with only some lingering showers likely around the Central Plateau to Taranaki. In the South Island, however, a stronger northwest flow prevails over the southern half of the island, lighter over the northern half, and a wet front brings rain throughout the west and south of the South Island.

Auckland: Partly cloudy, isolated showers, moderate winds
Hamilton: Cloudy, light showers, gentle breezes, chance of frost pockets
Wellington: Sunny, dry, slightly warmer, moderate winds
Greymouth: Overcast, colder, rain, winds easing to moderate
Christchurch: Partly cloudy, warmer day, frosts, dry, chance of misty patches
Dunedin: Windy, warmer, mix of sunny spells and showers

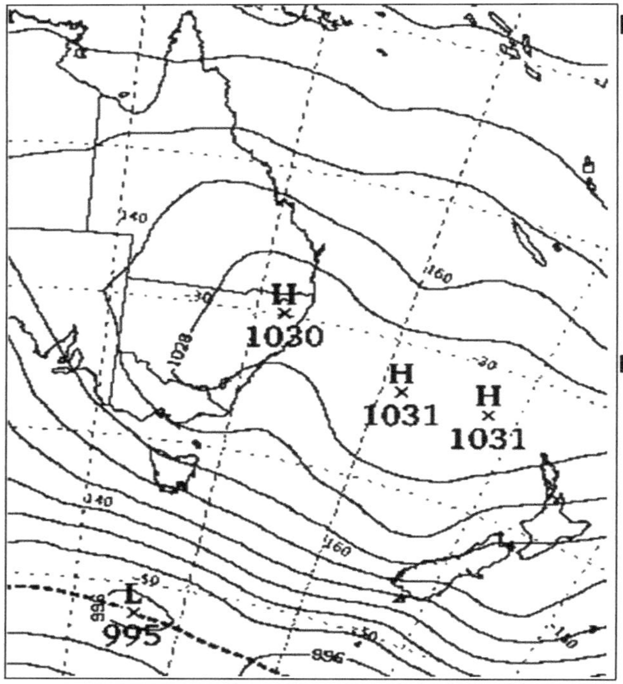

RAIN POTENTIAL

FROST/SNOW

AUGUST 4 MONDAY

A weakening high pressure system continues to slowly cross the north of New Zealand bringing a lighter westerly flow over the north of the North Island, and fresher westerlies over the southern half, with increased cloud in all districts, although remaining mostly dry. Low pressure continues to bring strong westerly winds over the South Island, with outbreaks of rain in the west of the island, but remaining mostly dry in eastern regions.

Auckland: Partly cloudy, isolated showers, fresh breezes at times
Hamilton: Cloudy, warmer, isolated showers, fresh breezes
Wellington: Cloudy, warmer, dry, strong breezes
Greymouth: Overcast, slightly warmer, rain, strong gusty winds
Christchurch: Sunny, cool, dry, strong winds at times
Dunedin: Partly cloudy, dry, blustery winds, chance of gales

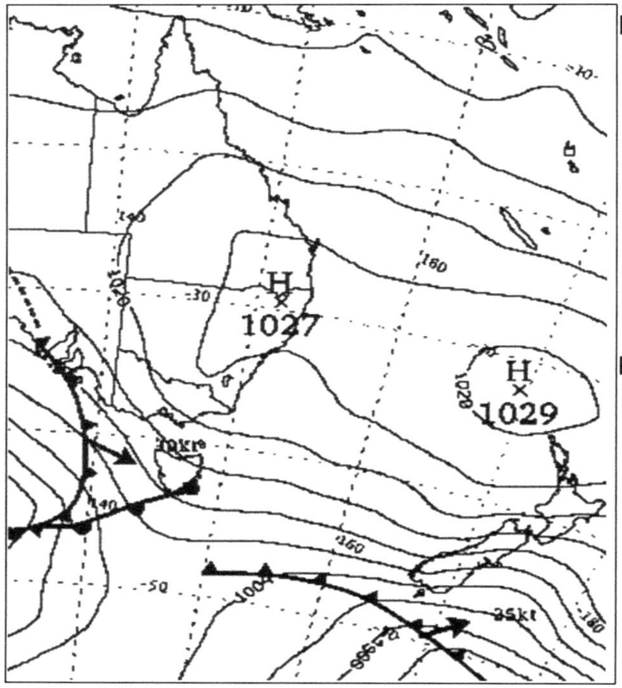

RAIN POTENTIAL

FROST/SNOW

AUGUST 5 TUESDAY

A large anticyclonic system north of the North Island continues to bring a moderate westerly flow, mostly dry, and occasional sunny spells throughout, except for the chance of some isolated morning showers likely in the central districts. A stronger west to northwest flow covers the South Island with a mix of sunny spells and showers in the west and south of the South Island, while sunnier and mostly dry through eastern regions. Milder temperatures throughout.

Auckland: Partly cloudy, light showers, warmer, light breezes
Hamilton: Cloudy, light showers, moderate breezes
Wellington: Sunny, dry, warmer, some breezy spells
Greymouth: Changeable, mix of showers and sunny spells, moderate winds
Christchurch: Sunny, warmer, dry, light breezes
Dunedin: Partly cloudy, slightly warmer, isolated showers, moderate breezes

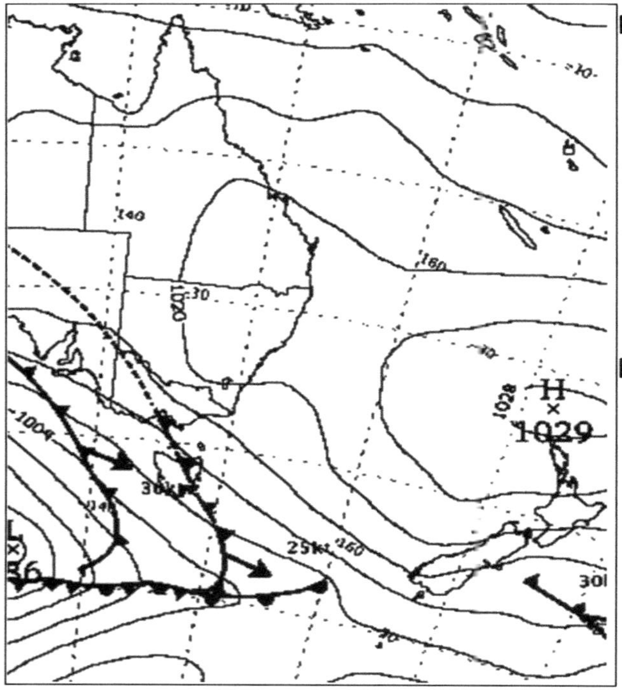

RAIN POTENTIAL

FROST/SNOW

AUGUST 6 WEDNESDAY

Anticyclonic over most of the country, with some isolated showers likely in North and to Auckland, and lingering around the West Coast for a brief time. Mostly sunny in the Bay of Plenty to the Hawkes Bay, and Marlborough to Canterbury regions, while mostly cloudy elsewhere. Westerly winds prevail, lighter over the North Island with chance of fog patches Auckland to the Waikato, while occasionally fresh over the South Island.

Auckland: Partly cloudy, showery pockets, light breezes
Hamilton: Light breezes to mostly calm air, cloudy, isolated showers
Wellington: Sunny, dry, moderate breezes
Greymouth: Warmer, overcast, scattered showers, fresh breezes at times
Christchurch: Partly cloudy, warmer day cold night, light shifting breezes
Dunedin: Cloudy, dry, light breezes

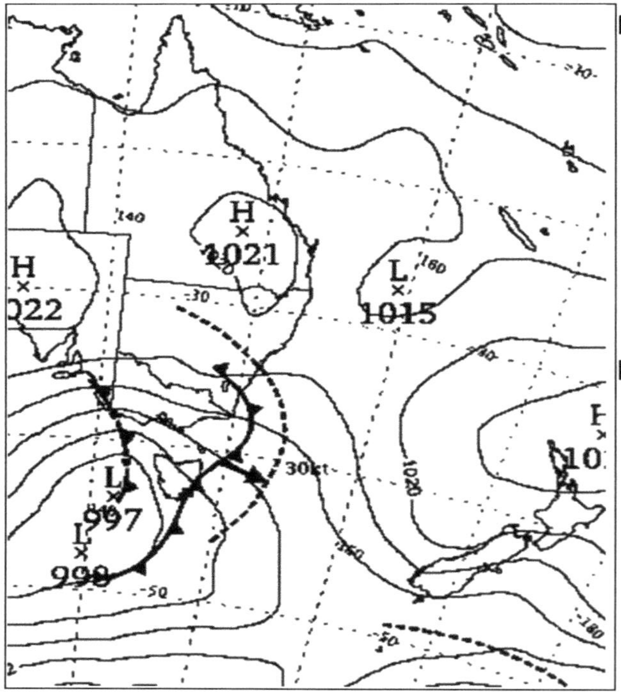

RAIN POTENTIAL

FROST/SNOW

AUGUST 7 THURSDAY

Low pressure advances closer to the western coasts of New Zealand behind the high departing to the east. Mostly dry and sunny in the north and east of the North Island, and mostly dry with increasing threats of rain through the east of the South Island. Widespread cloud prevails elsewhere with some spotty shower activity now and then. Chance of fog in Auckland to the Waikato overnight.

Auckland: Partly cloudy, dry, warmer day cooler night, fog patches possible
Hamilton: Mostly calm air, cloudy, spotty showers, chance of misty patches
Wellington: Cloudy, cooler, dry, moderate shifting breezes
Greymouth: Overcast, some brief showers, cooler, moderate breezes
Christchurch: Cloudy, dry, chance of frost, light breezes to mostly calm air
Dunedin: Overcast, dry with increasing threats of rain, gentle breezes

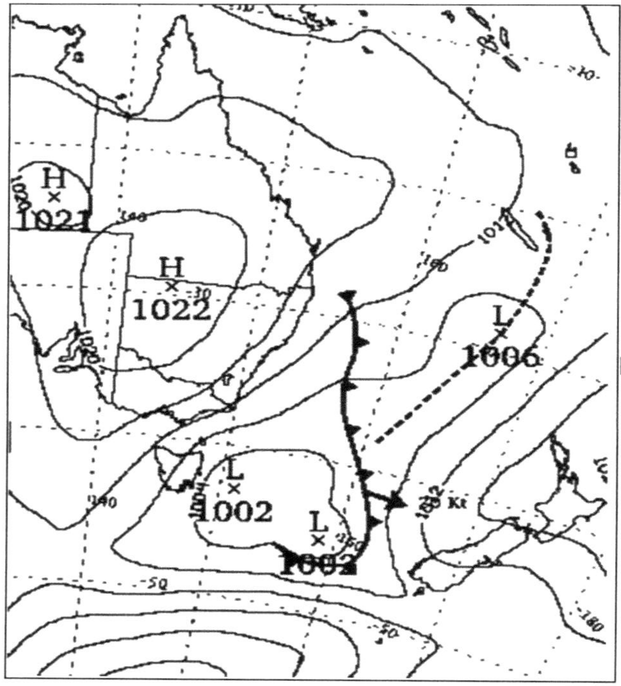

RAIN POTENTIAL

FROST/SNOW

AUGUST 8 FRIDAY

A deep depression moves onto the country bringing strengthening southwest winds throughout. Widespread cloud everywhere, with heavy rain developing in Northland to Auckland with showers into the Waikato, and rain in Fiordland and Southland. Mostly dry with increasing threats of rain elsewhere.

Auckland: Stormy, blustery, heavy rain, chance of flooding and thunderstorms
Hamilton: Cloudy, light showers, blustery winds, chance of thunderstorms
Wellington: Cloudy, dry with increasing threats of rain, strong blustery winds
Greymouth: Overcast, cool, light showers, windy with strong gusts
Christchurch: Overcast, cooler, dry, threats of rain building, strong winds, static air
Dunedin: Cooler, overcast, light showers, strong and blustery winds

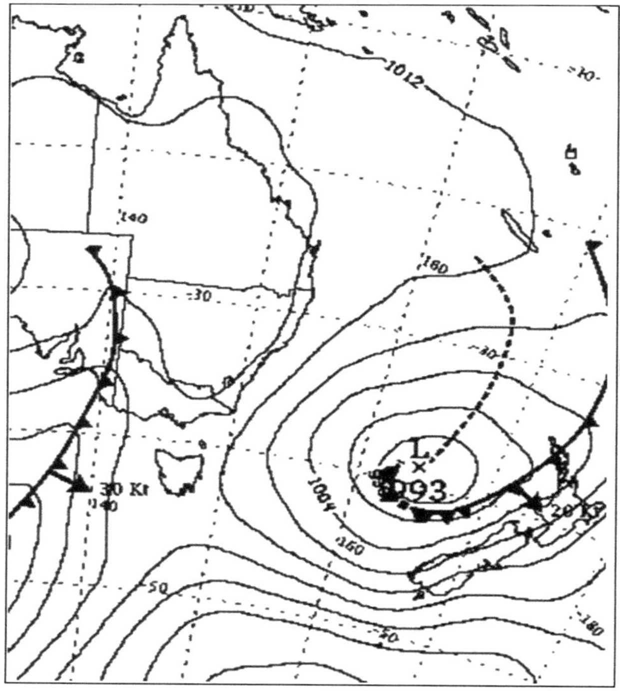

RAIN POTENTIAL

FROST/SNOW

AUGUST **9** SATURDAY

Stormy conditions continue throughout the country with an intense depression to the west of the country bringing strong northwest winds over the North Island backing easterly over the South Island. Widespread cloud and heavy rain almost everywhere, with chance of localised flooding in parts of Northland to the Coromandel and Bay of Plenty. Mostly dry for a time in the southwest of the South Island.

Auckland:	Stormy air, heavy rain, flooding possible, possible thunderstorms
Hamilton:	Cloudy, heavy rain with chance of localised flooding, winds easing
Wellington:	Stormy air, heavy rain, winds easing, chance of thunderstorms
Greymouth:	Overcast, mostly dry with passing threats of rain, unsettled winds
Christchurch:	Cloudy, heavy rain, wind gusts, chance of thunderstorms
Dunedin:	Static air, overcast, colder, rain, stormy winds

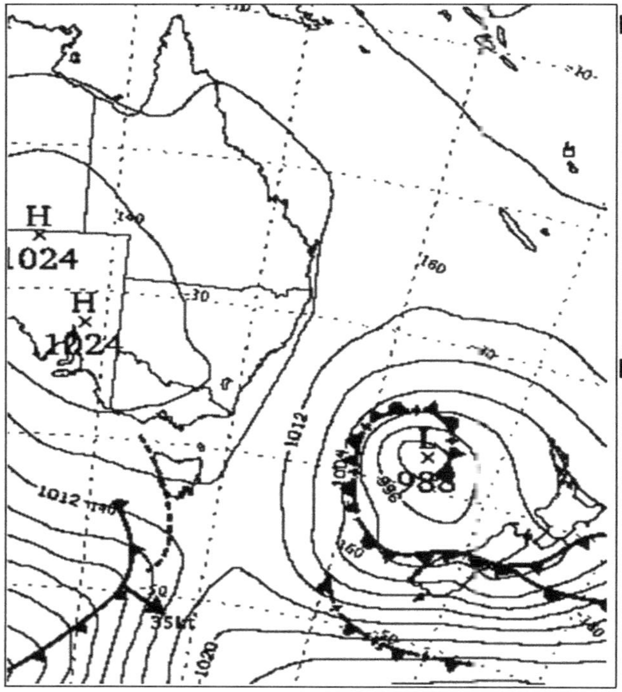

RAIN POTENTIAL

FROST/SNOW

AUGUST **10** SUNDAY

The depression over New Zealand decays with a wet trough between two low pressure centres forming through central parts. Northwest winds ease over the North Island, while unsettled easterly breezes prevail over the South Island, with unsettled and fluctuating breezes at times. Scattered rain and possible thunderstorms prevail throughout the country, with some brief sunshine breakthroughs developing. Chance of snow in the south of the South Island and in the higher reaches.

Auckland:	Partly cloudy, scattered showers, fresh breezes
Hamilton:	Fluctuating winds, cloudy, scattered showers
Wellington:	Cloudy, fluctuating breezes, squally showers at times
Greymouth:	Mostly calm air, mix of sunny spells and showers, cooler
Christchurch:	Overcast, isolated showers, mostly calm air, chance of misty patches
Dunedin:	Stormy air, heavy rain, chance of flooding, sunny spells, winds easing

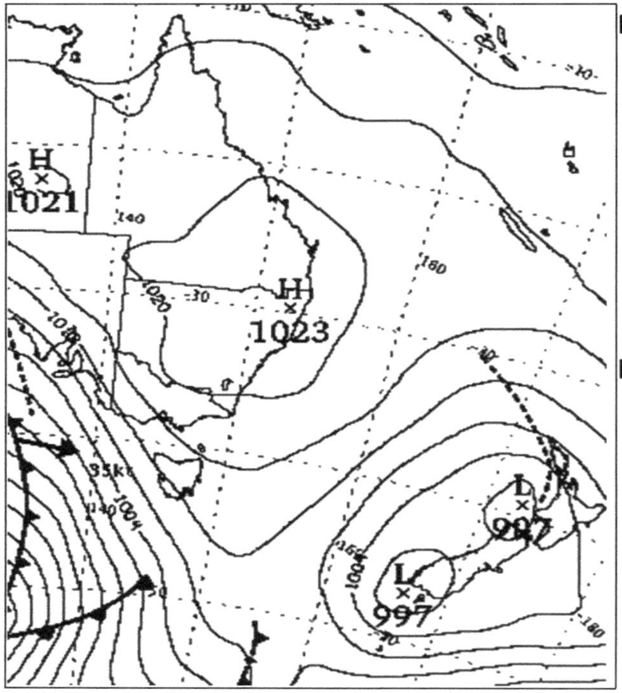

RAIN POTENTIAL

FROST/SNOW

AUGUST **11** MONDAY

A depression weakens as it lingers over New Zealand directing a lighter easterly flow over the North Island, backing northwesterly over the South Island. Shower activity persists over much of the North island, except in the East Cape to the Hawkes Bay, and outbreaks of heavy rain likely around Taranaki. Rain activity eases across the South Island and contracts to mostly around the Cook Strait regions of Nelson and Marlborough, with some isolated showers lingering in Fiordland. Dry elsewhere.

Auckland: Moderate breezes, mix of sunny spells and showers
Hamilton: Mild day cool night, overcast, brief drizzly showers, gentle breezes
Wellington: Cloudy, heavy rain, shifting breezy spells
Greymouth: Warmer, partly cloudy, occasional showers, light breezes
Christchurch: Overcast, light breezes, dry, chance of misty patches
Dunedin: Cloudy, fluctuating breezes, dry

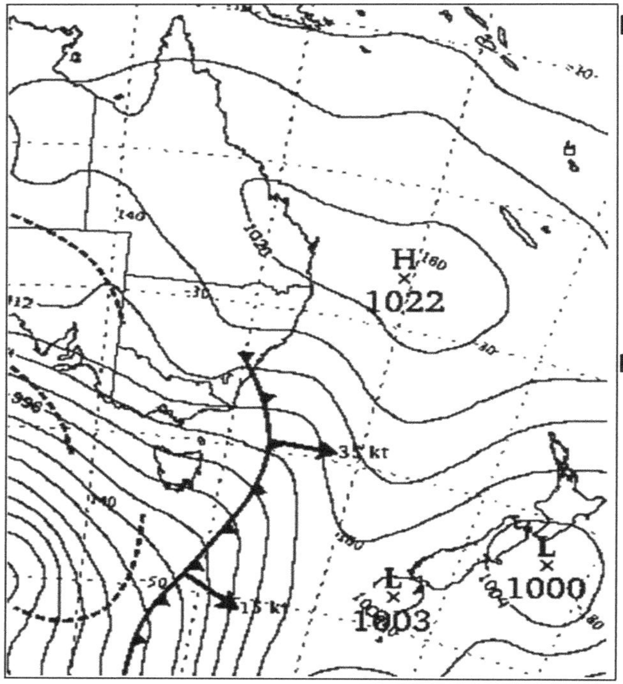

RAIN POTENTIAL

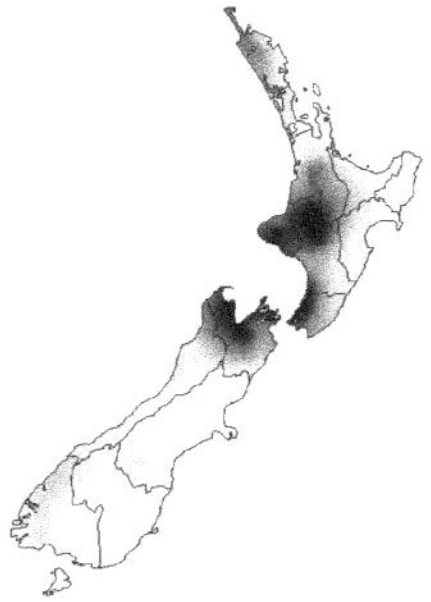

FROST/SNOW

AUGUST **12** TUESDAY

A low pressure trough flows up through the country in a mostly light southwesterly flow over the South Island, more westerly over the North Island, with high pressure ridging to the north of the country. A mix of sunny spells and showers prevail over the western half of the North Island, while mostly dry in the east from the Bay of Plenty to Wellington. Widespread cloud and some light showers in the south and west of the South Island, while sunnier, colder with frosts, and dry down through the eastern half of the South Island.

Auckland: Changeable, partly cloudy, occasional showers, fresh breezes
Hamilton: Cloudy, cooler, dry, moderate breezes
Wellington: Partly cloudy, isolated showers, fresh breezes
Greymouth: Overcast, scattered showers, light breezes
Christchurch: Partly cloudy, cold, dry, gentle breezes
Dunedin: Overcast, dry, light breezes to mostly calm air

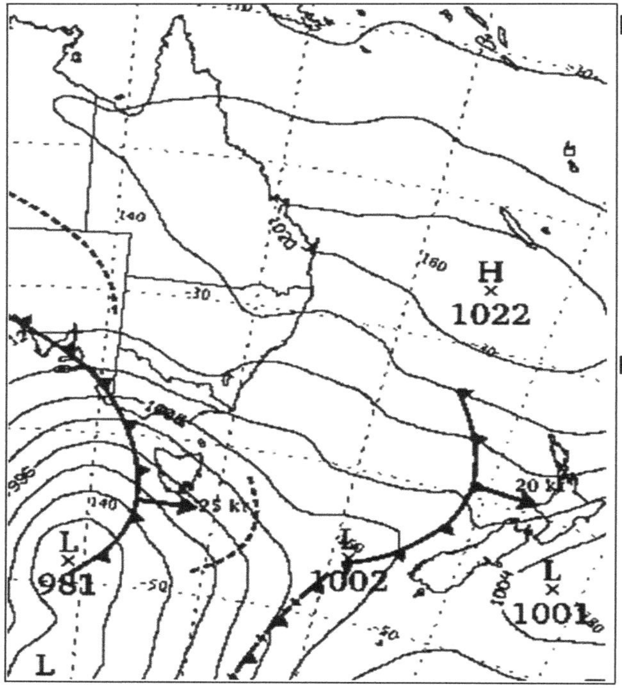

RAIN POTENTIAL

FROST/SNOW

AUGUST 13 WEDNESDAY

A high pressure system ridges to the north of New Zealand while low pressure prevails in a light to moderate southwesterly flow throughout. The passage of a thundery wet front brings another mixed day over the North Island with some sunny spells and outbreaks of rain in most districts. Widespread cloud prevails over the South Island with spotty showers in the western and southern regions, and mostly dry with threats of rain elsewhere.

Auckland: Partly cloudy, scattered showers, moderate breezes
Hamilton: Light breezes, mix of showers and sunny spells
Wellington: Cloudy, dry, light breezes
Greymouth: Cloudy, cool, occasional showers, light breezes with odd breezy spell
Christchurch: Cloudy, dry with passing threats of rain, light air
Dunedin: Cloudy, cooler, dry, light breezes

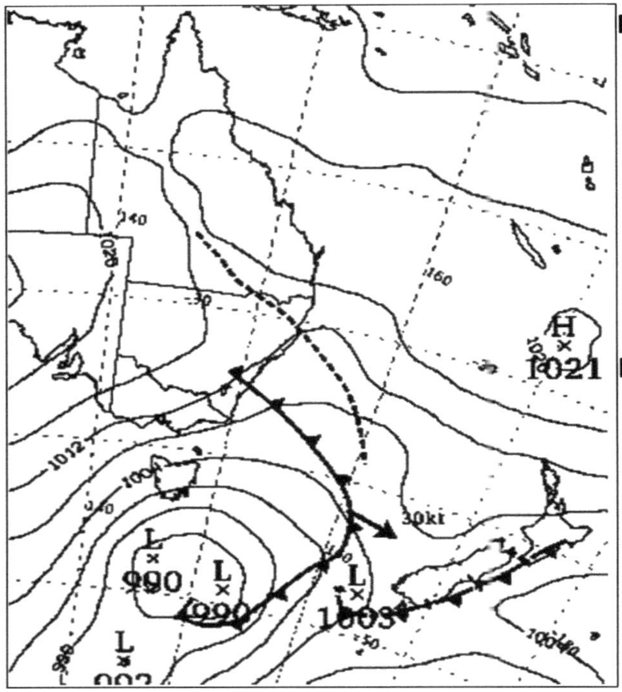

RAIN POTENTIAL

FROST/SNOW

AUGUST 14 THURSDAY

A complex of low pressure centres moves onto New Zealand bringing a stormy band of rain activity and strong winds throughout; northwesterly over the North Island, southeasterly over the South Island. Widespread cloud everywhere, with occasional showers in most North Island districts. Heavier rain at times in Fiordland, while mostly dry with passing threats of rain expected elsewhere south of Wellington.

Auckland: Overcast, blustery, rain, possible thunderstorms
Hamilton: Cloudy, showers, fresh breezes, possible thunderstorms
Wellington: Overcast, slightly cooler, dry with odd threat of rain, windy at times
Greymouth: Overcast, dry with rain threatening, freshening winds
Christchurch: Overcast, dry, rain threatening at times, freshening winds
Dunedin: Overcast, dry, moderate breezes

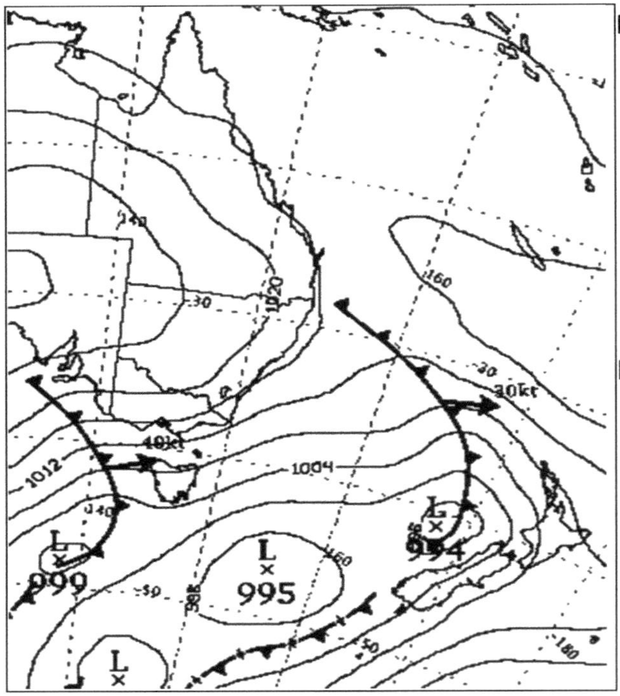

RAIN POTENTIAL

FROST/SNOW

AUGUST 15 FRIDAY

Strong and unsettled westerly winds prevail over New Zealand as a complex of low pressure centres bring stormy conditions throughout the country. Widespread overnight heavy rain in many districts with occasional sunny spells during the day, particularly over the northern half of the South Island and throughout the North Island. Snow likely in the higher reaches and southern parts of the South Island.

Auckland: Partly cloudy, spells of heavy rain, breezy, chance of thunderstorms
Hamilton: Heavy rain, windy, some sunny spells, chance of thunderstorms
Wellington: Wet and windy, some sunny spells
Greymouth: Cloudy, cool, heavy rain, blustery winds
Christchurch: Sunny spells, heavy downpours, chance of thunderstorms and hail
Dunedin: Overcast, cold, showers, strong winds, chance of snow

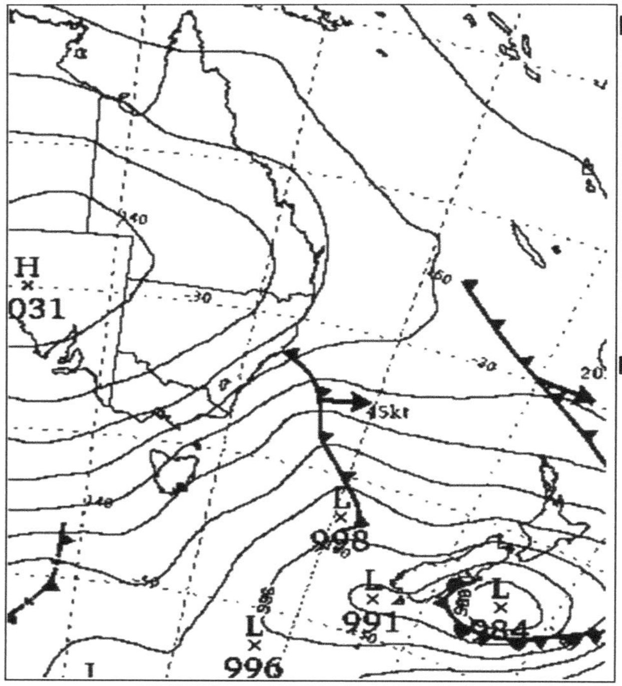

RAIN POTENTIAL

FROST/SNOW

AUGUST 16 SATURDAY

Low pressure depressions continue to direct very strong and unsettled westerly winds across the North Island and southwesterly over the South Island. A very cold and stormy wet front brings heavy rain into the west of both islands, while eastern regions, Northland and Southland to Otago clear to mostly dry with occasional sunny spells. Chance of snow overnight in the South Island passes and Central Plateau.

Auckland: Partly cloudy, brief showers, fresh windy spells
Hamilton: Cloudy, spotty showers, fresh breezes
Wellington: Changeable, mix of squally showers and sunny spells, gusty winds
Greymouth: Unsettled and windy, widespread cloud, heavy rain
Christchurch: Partly cloudy, unsettled winds, mostly dry
Dunedin: Partly cloudy, brief evening showers, fluctuating winds

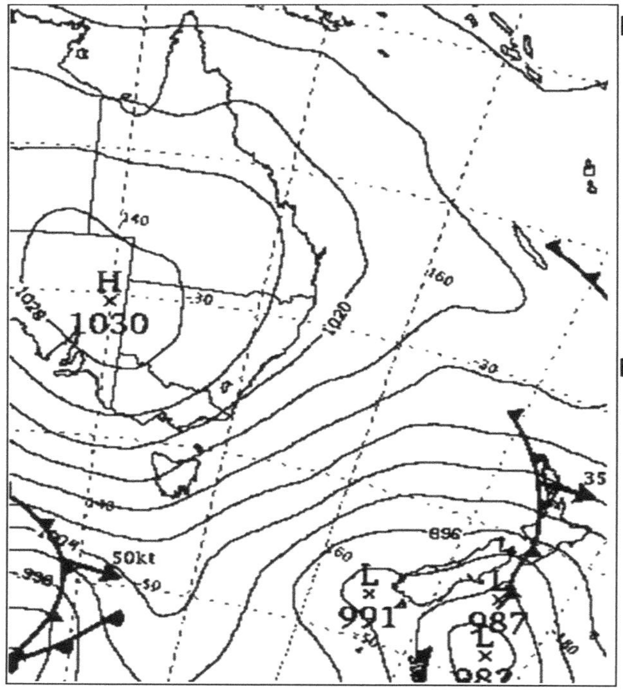

RAIN POTENTIAL

FROST/SNOW

AUGUST **17** SUNDAY

A depression to the south of New Zealand continues to direct fresh to strong westerly winds over the Northland and an easing southwesterly flow over the South Island. Outbreaks of heavy rain over most of the North Island and into the northwest of the South Island, with some isolated overnight showers in Southland and Otago; mostly dry in Canterbury and around the East Cape to the Hawkes Bay. Very frosty with good sunny spells in the South Island and mostly cloudy in the North Island.

Auckland: Cloudy, scattered showers, fresh breezes at times
Hamilton: Mostly cloudy, cooler, rain, moderate breezes
Wellington: Partly cloudy, dry, fresh breezes
Greymouth: Changeable, sunny day, evening rain, moderate breezes
Christchurch: Partly cloudy, colder, mostly dry, chance of frost and fog
Dunedin: Cloudy, milder, light showers, moderate breezes

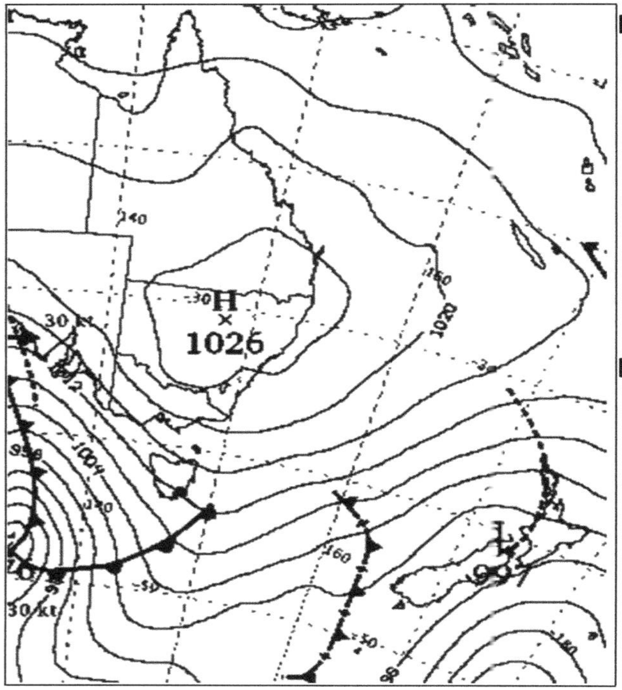

RAIN POTENTIAL

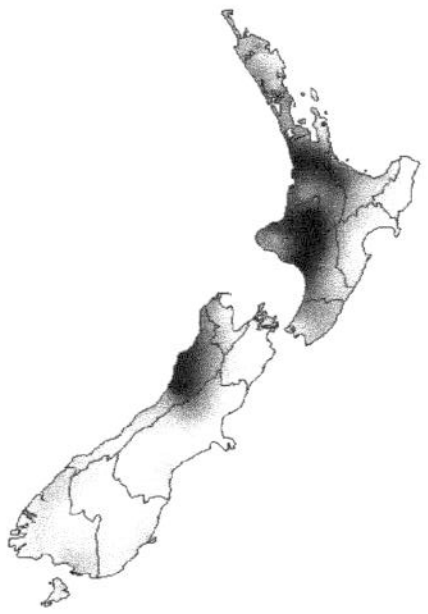

FROST/SNOW

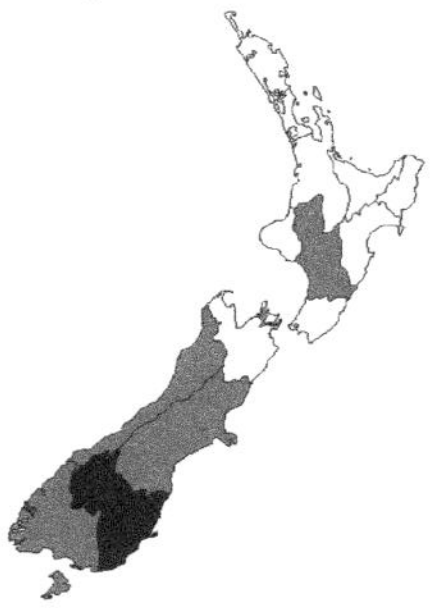

AUGUST **18** MONDAY

Low pressure prevails over the country in a lighter west to southwest flow, while a large high in the Coral Sea extends a ridge towards New Zealand. Mostly dry, cold and frosty in the South Island, with good sunshine in Canterbury to Otago, and mostly cloudy elsewhere. In the North Island. Widespread overnight rain persists in most districts, with sunny spells during the day. Very cold from the Waikato to Wellington, with chance of snow in the Central Plateau.

Auckland: Sunny day, evening rain, winds easing to moderate
Hamilton: Cooler, partly cloudy, evening showers, chance of frost pockets
Wellington: Partly cloudy, slightly cooler, dry, moderate winds
Greymouth: Cloudy, cool, dry, fresh breezes
Christchurch: Partly cloudy, cold, dry, light breezes, possible mists and frost
Dunedin: Partly cloudy, isolated showers, moderate breezes

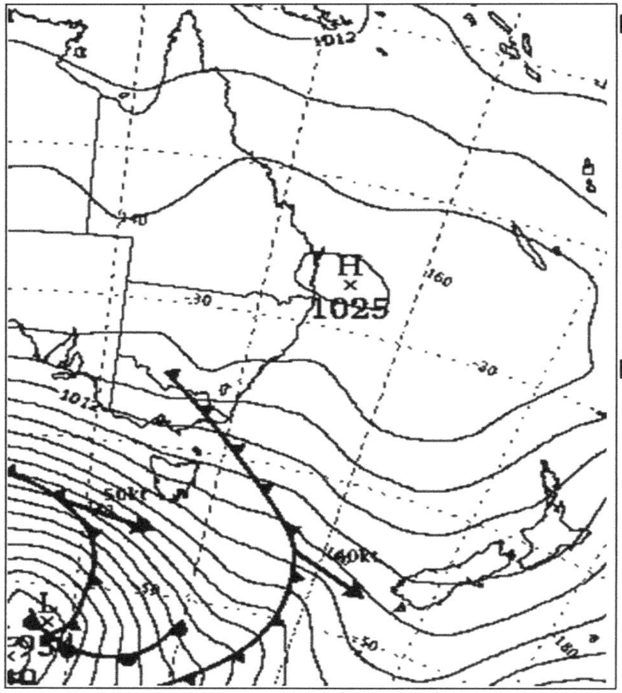

RAIN POTENTIAL

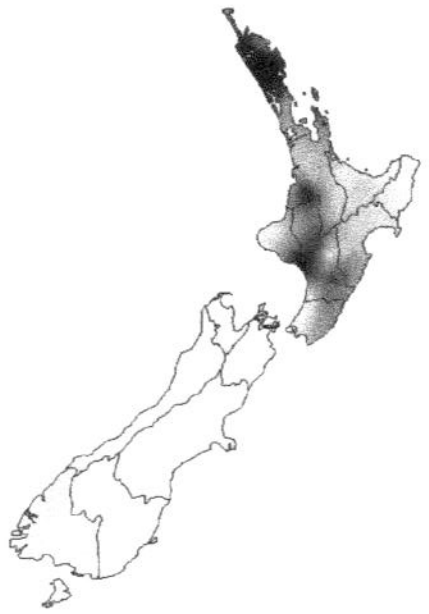

FROST/SNOW

AUGUST 19 TUESDAY

A high pressure ridge to the north of New Zealand clears rain over most of the North Island, except for some lingering showers in central parts from the Waikato to the Manawatu with other districts mostly dry and sunny. In the South Island, a moderate west to northwest flow prevails ahead of an intense depression approaching the south of the country, with showers throughout the West Coast to Fiordland, while mostly dry, sunny, and with frosts elsewhere.

Auckland: Partly cloudy, mostly dry, chance of evening showers, fresh breezes
Hamilton: Partly cloudy, isolated showers, moderate breezes
Wellington: Freshening breezes, dry, sunny
Greymouth: Overcast, rain, fresh winds with occasional gusts
Christchurch: Partly cloudy, dry, overnight frosts, moderate breezes
Dunedin: Partly cloudy, dry, mild breezes freshening later in the day

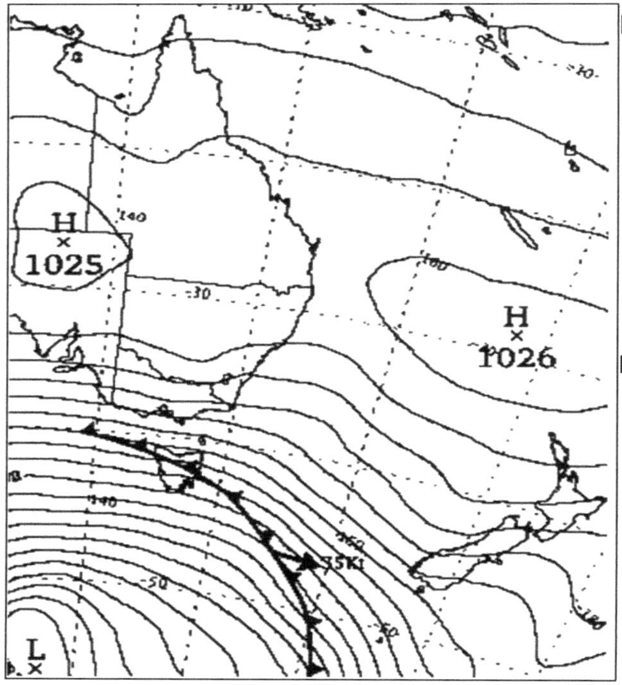

RAIN POTENTIAL

FROST/SNOW

AUGUST 20 WEDNESDAY

An intense depression in the Southern Ocean directs strong to gale force west to northwest winds across the country, which may bring damage to local infrastructures in a number of regions. Heavy rain at times in the west of the South Island and isolated showers in the western half of the North Island. Mostly dry through the east of the country. Milder temperatures everywhere.

Auckland: Partly cloudy, mostly dry, warmer, fresh breezes
Hamilton: Partly cloudy, light showers, warmer, blustery at times
Wellington: Cloudy, warmer, dry, strong winds with gusts
Greymouth: Mostly cloudy, some showers, strong winds, possibly gales
Christchurch: Partly cloudy, strong and blustery winds, dry, warmer
Dunedin: Strong and gusty winds, cloudy, warmer, dry, possible gales

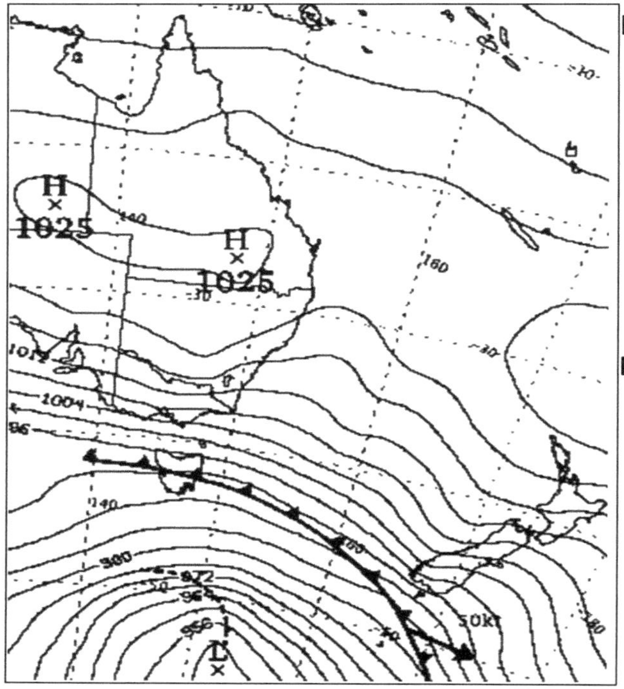

RAIN POTENTIAL

FROST/SNOW

AUGUST 21 THURSDAY

An intense Southern Ocean depression continues to direct stormy and damaging west to northwest winds throughout the country. Mostly cloudy but with milder temperatures everywhere. Heavy rain likely in the west of the North Island from Taranaki to the Manawatu, and in the South Island West Coast to Fiordland and into the western half of Southland and Otago. Mostly dry with increasing threats of rain in eastern regions.

Auckland: Partly cloudy, dry, warmer day, strong winds with occasional gusts
Hamilton: Unsettled and blustery winds, partly cloudy, isolated showers
Wellington: Overcast, gusty winds, gales possible, mostly dry, threats of rain
Greymouth: Overcast, dry, rain threatening, strong gusty winds, possible gales
Christchurch: Strong and unsettled winds, cloudy, dry
Dunedin: Cloudy, blustery winds, warmer, dry

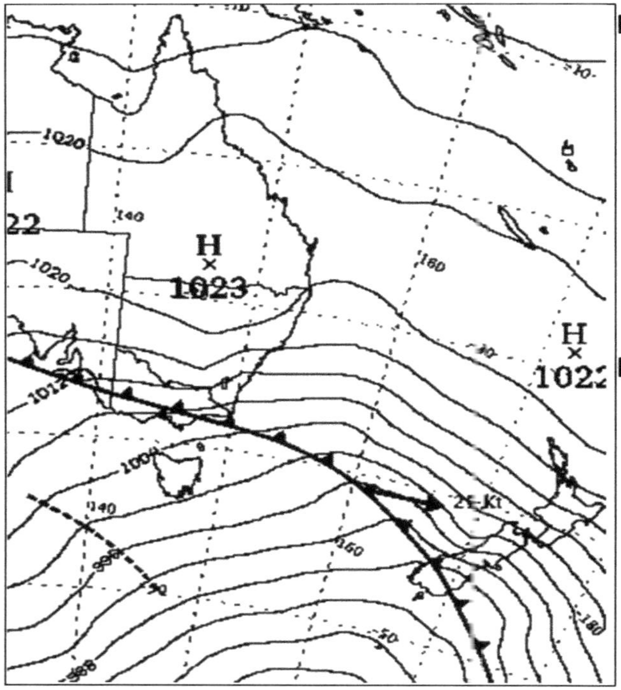

RAIN POTENTIAL

FROST/SNOW

AUGUST 22 FRIDAY

A stormy wet front crosses the country in a strong flow with winds turning southwesterly overnight. Widespread stormy conditions almost everywhere overnight, with heavy falls likely in most parts of the country except in Northland and the eastern fringes of both islands. Some sunny spells during the day. Chance of snow in the central North Island and in the south and east of the South Island.

Auckland: Cloudy, warmer, scattered showers, strong winds, occasional gusts
Hamilton: Windy, partly cloudy, overnight rain
Wellington: Overnight rain, some sunny spells, strong winds with gusts
Greymouth: Cloudy, heavy rain, winds easing
Christchurch: Partly cloudy, isolated showers, fresh breezes at times
Dunedin: Mild day cold night, mix of sunny spells and isolated showers, windy

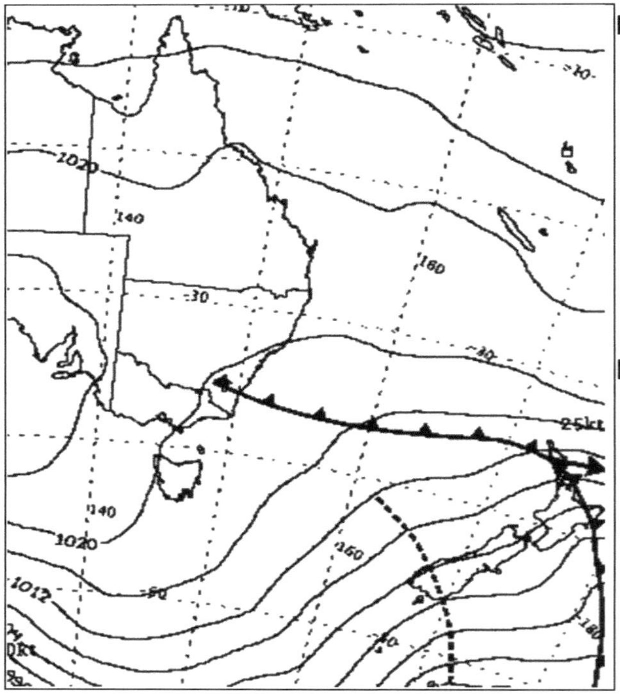

RAIN POTENTIAL

FROST/SNOW

AUGUST 23 SATURDAY

An anticyclonic system in the north of the Tasman Sea extends a high pressure ridge onto the north of New Zealand establishing a light to mostly calm southwesterly air over the country as the depression of the past few days moves away to the east. Overnight rain eases to showers throughout the country, with cold and frosty temperatures prevailing and chance of snow in the higher reaches of both Islands. Cloudy throughout the northern half of the South Island, while mostly sunny elsewhere.

Auckland: Overnight showers, sunny, cooler day, mostly calm air
Hamilton: Mostly sunny, overnight showers, mostly calm air
Wellington: Cloudy, cooler, dry, light breezes
Greymouth: Cooler, partly cloudy, overnight showers, gentle breezes
Christchurch: Overcast, colder, occasional showers, gentle shifting breezes
Dunedin: Partly cloudy, light showers, mostly calm air, chance of snow flurries

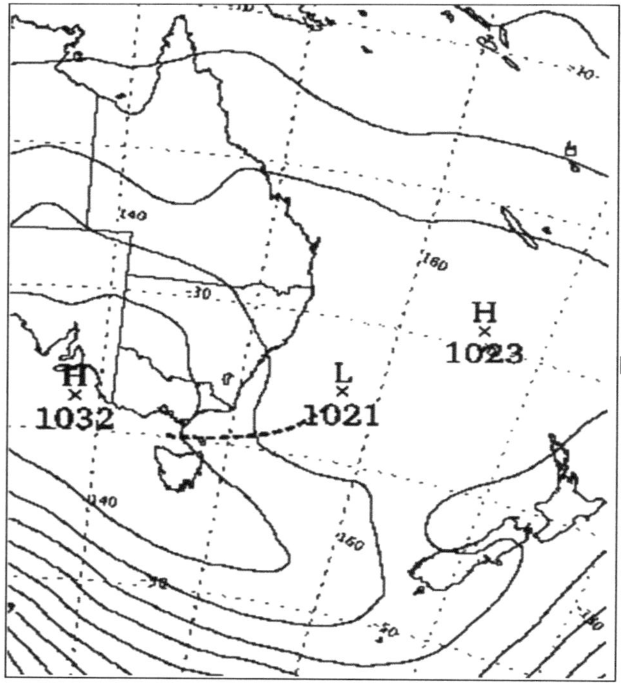

RAIN POTENTIAL

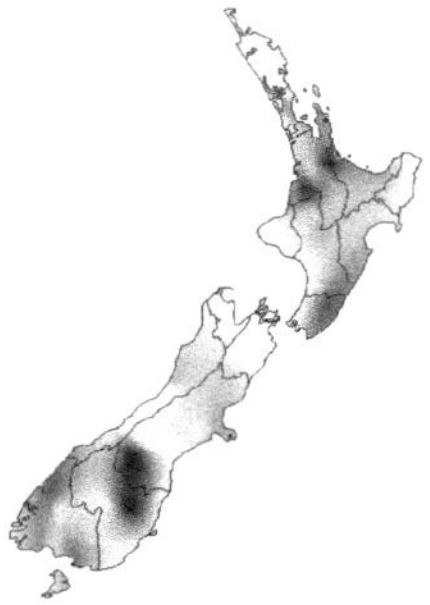

FROST/SNOW

AUGUST 24 SUNDAY

High pressure moves onto the south of the country while a low pressure trough drifts past the north of New Zealand, with a gentle westerly air prevailing over the South Island, turning easterly over the North Island. Mostly dry in Northland and down through the central regions of the North Island, as well as in Otago, with isolated overnight showers elsewhere. Cold with widespread frosts throughout the South Island and in the Waikato. Mostly sunny day.

Auckland: Sunny, cooler, mostly dry day, overnight showers, gentle breezes
Hamilton: Cloudy, dry, cooler, gentle breezes
Wellington: Partly cloudy, colder, light showers, light shifting breezes
Greymouth: Light shifting breezes, sunny, dry
Christchurch: Partly cloudy, light showers, frosts, chance of hail or snow flurries
Dunedin: Mostly sunny and dry, colder, frost likely, calm air

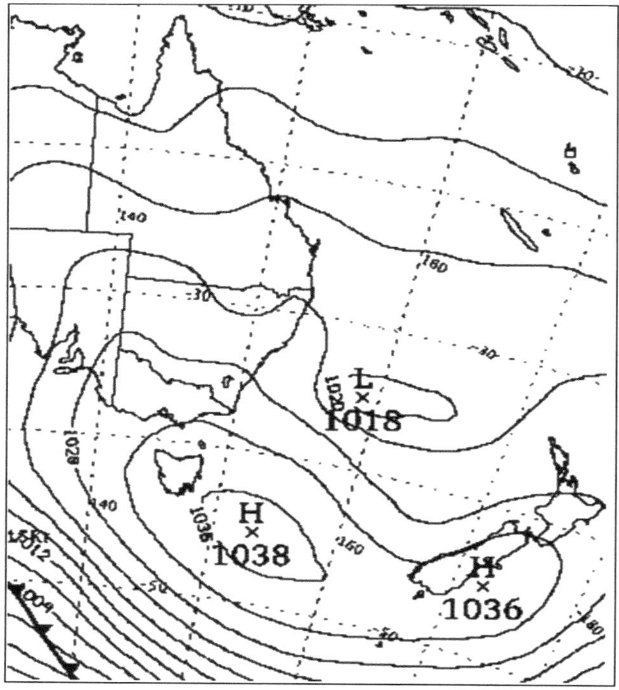

RAIN POTENTIAL

FROST/SNOW

AUGUST **25** MONDAY

A strong high pressure system drifts over New Zealand bring a light to moderate northeasterly flow throughout, while a depression enters the northern Tasman Sea to the west of the North Island. Widespread frosts in the South Island and through the central North Island from the Waikato to the Central Plateau. Mostly dry and sunny everywhere except for some isolated overnight showers lingering around the East Cape to the Hawkes Bay and increasing cloud developing in Northland.

Auckland: Mostly cloudy, dry, chance of isolated showers, moderate breezes
Hamilton: Partly cloudy, dry, cold over night, gentle breezes
Wellington: Sunny, cold, isolated overnight showers, windy at times
Greymouth: Sunny, dry, cool, fresh breezes
Christchurch: Partly cloudy, dry, cold with severe frosts, light shifting breezes
Dunedin: Sunny, dry, cold with frost and fog patches likely

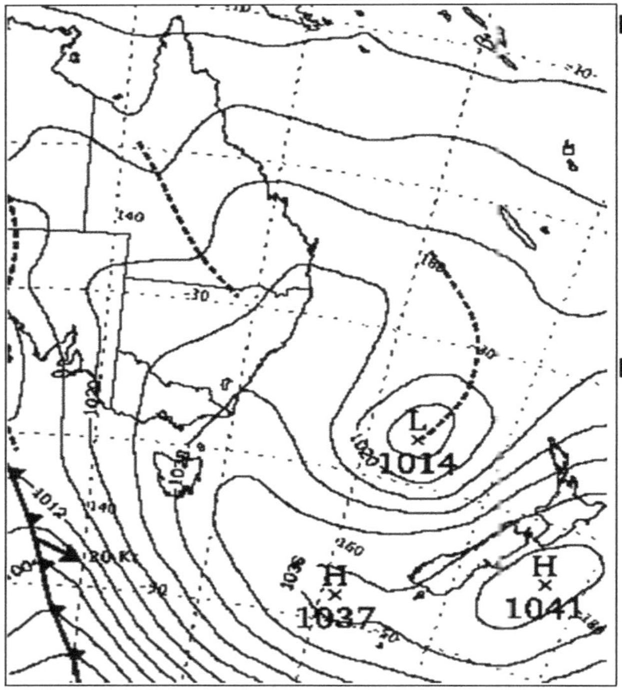

RAIN POTENTIAL

FROST/SNOW

AUGUST **26** TUESDAY

A depression to the west of New Zealand moves closer to New Zealand behind a large high moving away to the east, with freshening northeasterly winds developing. Widespread cloud throughout the North Island and in the southern half of the South Island with rain moving into Northland. Mostly dry elsewhere and sunny in the northern half of the South Island. Cold with frosts over most of the South Island and into the south and west of the North Island.

Auckland: Overcast, mostly dry and static air, strong winds, gales possible
Hamilton: Overcast, milder, mostly dry, threats of rain, moderate winds
Wellington: Unsettled winds, partly cloudy, dry with rain threatening
Greymouth: Overcast, mostly dry, mild day cold night, fresh breezes
Christchurch: Cold and dry, some sunny spells, frosts, strong winds at times
Dunedin: Partly cloudy, dry, cold, moderate breezes

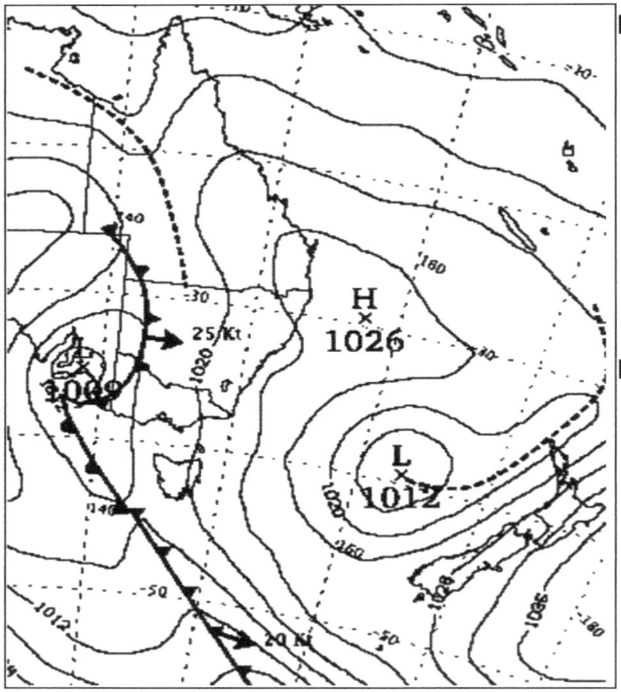

RAIN POTENTIAL

FROST/SNOW

AUGUST 27 WEDNESDAY

A weak depression crosses the country bringing westerly change over the North Island, while maintaining a northeast to easterly flow over the South Island. Widespread cloud throughout with heavy rain, including the chance of localised flooding, in many northern, central and western regions of the North Island as well as in the top of the South Island, with rain down through the West Coast and isolated showers into Canterbury and East Cape to the Hawkes Bay. Mostly dry in Southland and Otago.

Auckland:	Overnight rain, sunny spells, chance of thunderstorms and flooding
Hamilton:	Cloudy, rain, winds easing to light
Wellington:	Overcast, heavy rain, moderate winds
Greymouth:	Overcast, rain, fluctuating winds
Christchurch:	Overcast, warmer, brief showers, fluctuating breezes
Dunedin:	Overcast, dry with rain threatening, mostly calm air, chance of frost

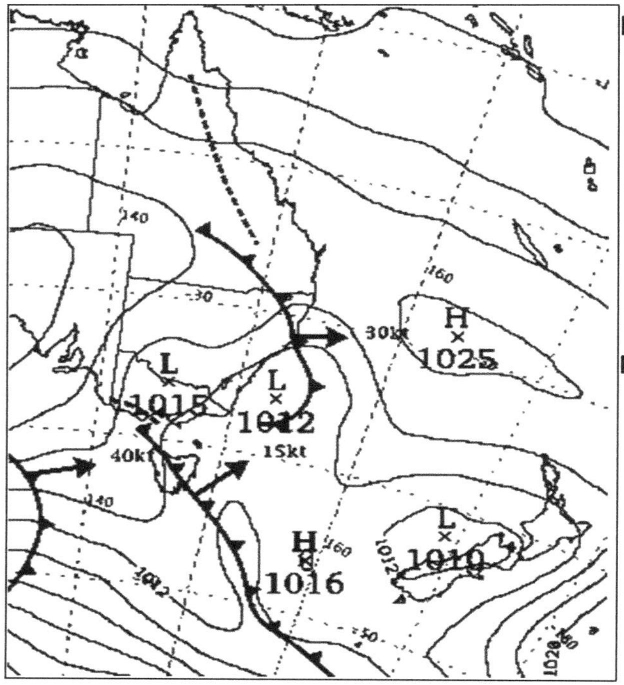

RAIN POTENTIAL

FROST/SNOW

AUGUST 28 THURSDAY

A complex of low pressure centres cross the country with a wet front continuing to bring widespread cloud and rain activity throughout the North Island, and into the western half of the South Island, where some heavy rain can still be anticipated. Northwest breezes prevail over the North Island while a light to mostly calm air stabilises over the South Island for a time.

Auckland:	Overcast, warmer, brief showers, moderate breezes
Hamilton:	Moderate breezes, warmer, cloudy, scattered showers
Wellington:	Cloudy, isolated showers, warmer, fluctuating breezes
Greymouth:	Cloudy, cooler, heavy rain, light shifting breezes
Christchurch:	Cloudy, mostly dry, odd threat of rain, colder, light shifting breezes
Dunedin:	Overcast, showers, light breezes, chance of overnight light snow

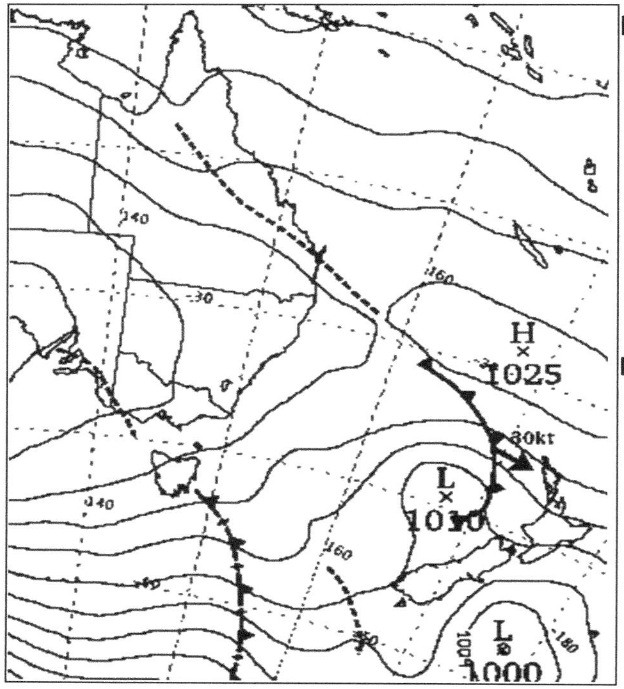

RAIN POTENTIAL

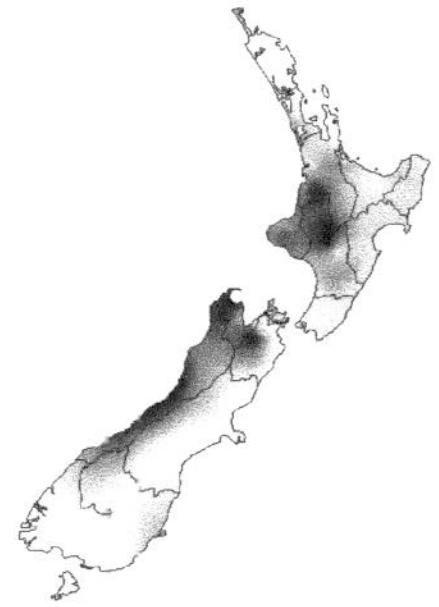

FROST/SNOW

AUGUST **29** FRIDAY

High pressure that has been building to the north and west of New Zealand ridges onto the country directing a moderate southwesterly flow throughout. Heavy rain overnight in the North Island clears to a mostly dry and sunny day, while a mostly sunny and dry day with cold overnight temperatures and frosts prevail over the South Island.

Auckland:	Overnight showers, sunny day, moderate breezes
Hamilton:	Partly cloudy, outbreaks of heavy rain mostly overnight, light breezes
Wellington:	Dry, mild, occasional sunny spells, fresh breezes
Greymouth:	Partly cloudy, dry, warmer day cooler night, gentle breezes
Christchurch:	Partly cloudy, dry, cold with frosts, light breezes
Dunedin:	Partly cloudy, dry, moderate breezes

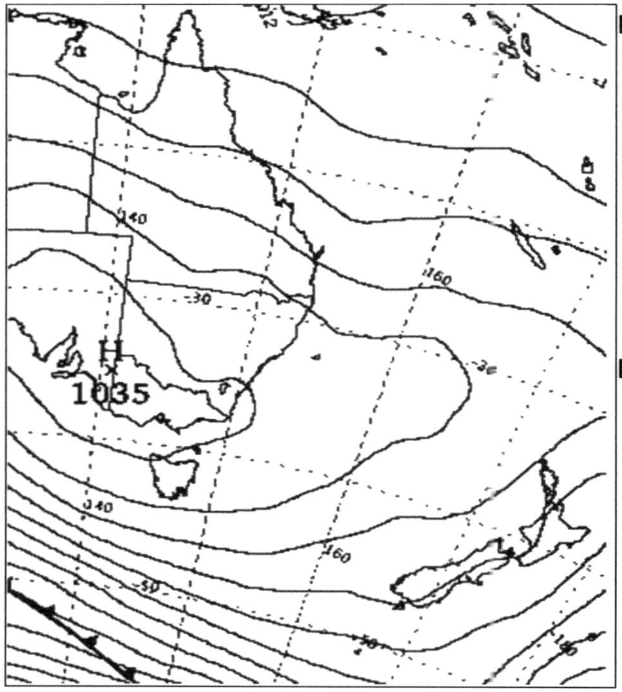

RAIN POTENTIAL

FROST/SNOW

AUGUST **30** SATURDAY

A large anticyclonic system strengthens as it crosses the Tasman Sea, directing a very light west backing southerly air over the country. Mostly dry and sunny conditions throughout, except for the chance of some isolated showers around the western fringes of the North Island and possible patches of fog in sheltered districts in both islands. Severe frosts likely in the east and south of the South Island.

Auckland:	Light breezes to mostly calm air, partly cloudy, spotty showers
Hamilton:	Mostly sunny, dry, warmer day cold night, light breezes
Wellington:	Partly cloudy, some showers, moderate breezes
Greymouth:	Light shifting breezes, sunny, dry, mild day cold night
Christchurch:	Sunny, slightly warmer, dry, gentle shifting breezes
Dunedin:	Sunny, dry, gentle breezes, chance of frost pockets in sheltered parts

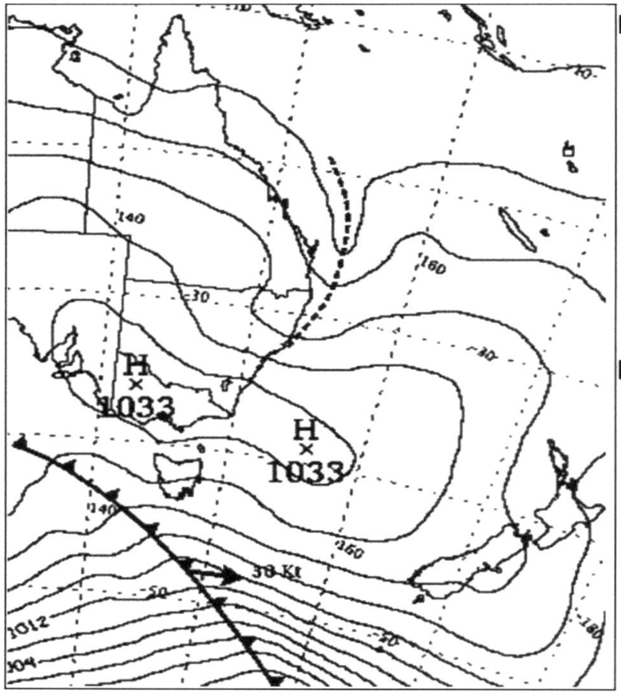

RAIN POTENTIAL

FROST/SNOW

AUGUST 31 SUNDAY

High pressure prevails over the North Island in a gentle to mostly calm southwest flow, with mostly sunny, dry, and frosty conditions everywhere, except for the chance of some isolated showers around the East Cape. Very cold, mostly sunny and dry throughout the South Island, with widespread frosts, some which may be severe in Otago, Southland, as well as sharp frosts in the Bay of Plenty. A blustery wet front advances onto the southwest of the country as another Southern Ocean depression crosses, bringing increased cloud and some showers into Southland, and strong and gusty southwest winds developing.

Auckland: Partly cloudy, occasional showers, light breezes
Hamilton: Partly cloudy, dry, mostly calm air, chance of overnight frost pockets
Wellington: Sunny, morning showers, cooler, fresh shifting breezes
Greymouth: Partly cloudy, dry, strengthening winds, chance of gale gusts
Christchurch: Sunny, dry, mild day cold with frosts overnight, moderate breezes
Dunedin: Unsettled blustery winds at times, cloudy, dry, cool

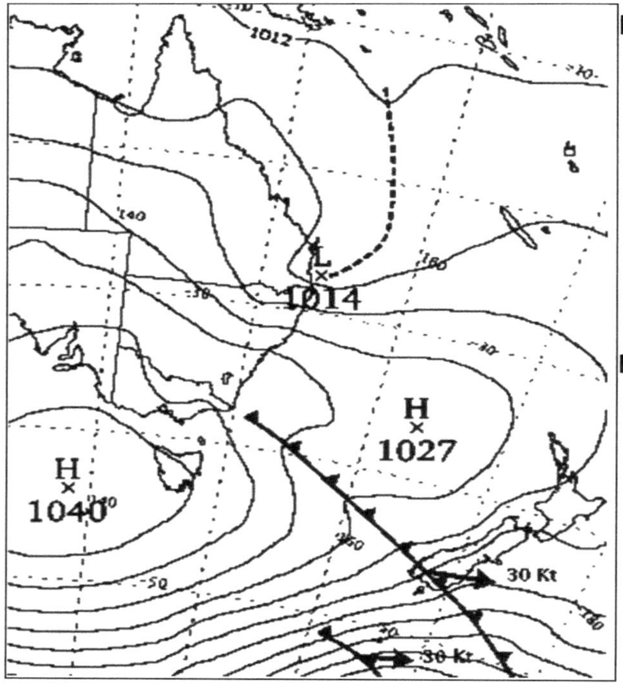

RAIN POTENTIAL

FROST/SNOW

SEPTEMBER

PHASES OF THE MOON

1st	Southern Declination
7th	Full Moon
8th	Crossing Equator
10th	Perigee #9
14th	Third Quarter
14th	Northern Declination
22nd	New Moon
22nd	Crossing Equator
26th	Apogee
29th	Southern Declination
30th	First Quarter

1	2	3	4	5	6	7	8	9	10	11	12	13	14	15	16	17	18	19	20	21	22	23	24	25	26	27	28	29	30
Mon	Tue	Wed	Thu	Fri	Sat	Sun	Mon	Tue	Wed	Thu	Fri	Sat	Sun	Mon	Tue	Wed	Thu	Fri	Sat	Sun	Mon	Tue	Wed	Thu	Fri	Sat	Sun	Mon	Tue
V						F	XhN		P9				3Q ^								N XhS				A			V	1Q

MONTHLY SUMMARY

3rd-4th Fog patches possible in Christchurch.
6th Depression crosses the country.
6th-7th Chance of further fog in Christchurch.
9th-10th Shallow fog patches possible in the Canterbury and Christchurch areas.
9th-14th Wet spell in the West Coast.
10th Unusually warm day in the Dunedin area

Anticyclonic systems coming from the south and east will be the predominant weather systems over the next four weeks, bringing lighter, mostly easterly winds, and milder conditions throughout most of the country. Rainfall will be fairly typical to slightly above in parts of Northland, along the east of the North Island including in the Gisborne region, and in parts of Otago and Fiordland, but only around half the usual amount of rainfall is expected to Taranaki, Wanganui, the Manawatu, Westland, Marlborough, and Banks Peninsula districts. All other regions can expect around 75% of their usual amount of recorded rainfall compared to their average for this outlook period. Soil moisture levels will still remain below average in eastern parts of Marlborough and in north Canterbury. The next four weeks will be sunnier than average in the Manawatu, Kapiti, Westland, Dunedin, and inland districts of the South Island, cloudier than average in Northland, and around the norm elsewhere. It will be warmer than normal for this time of year in the north of the North Island and west of the South Island, with most other parts of the country likely to recorded temperatures close to the seasonal average for this time of year. Some snow can still be expected during this outlook period, including to low levels in the east of the South Island in the first week ahead.

11th Occasional wind gusts in the Christchurch area.
11th-13th Strong blustery winds in the Wellington region.
14th-15th Chance of snow in the east of the South Island, possibly reaching low-lying parts of Canterbury, including around Ashburton, Methven, Twizel, Tekapo, Darfield and Geraldine areas and snow flurries may be observed in Christchurch. The Lindis, Crown Range and other Passes may be closed for a time. Cold conditions may cause stock losses, particularly of new-born lambs. Further north, strong northwest winds may reach gale force for a time around the Wairarapa on the eastern coast north of Wellington.
16th-17th A cold southerly flow over the southern half of the country brings chance of severe frosts in sheltered parts from Fairlie to Hanmer, as well as around Mt Cook.

17th-18th Chance of shallow patches of fog in the Christchurch area.
20th-23rd Chance of further fog in the Christchurch area.
21st Chance of severe frosts in the Mt Cook region.
24th Heavy rain likely in the West Coast.
25th Strong blustery winds for a time in the Christchurch area.
25th-27th Chance of drizzly fog patches over the North Island, including in the Auckland to Waikato districts and in Wellington.
29th-30th Heavy rain in Northland likely, with chance of brief localised flooding in the Kaitaia area.
30th Very warm day in the Wanganui district.

September 2025

Monday	Tuesday	Wednesday	Thursday	Friday	Saturday	Sunday
1 **Sagittarius** Square Sun Trine Venus Sextile Mars	**2** **V** **Sagittarius** Trine Mercury Square Saturn Square Neptune Square Ascendant	**3** **Capricorn** Trine Sun Opposition Jupiter	**4** **Capricorn** Square Mars Opposition Jupiter	**5** **Aquarius** Opposition Venus Trine Uranus Conjunct Pluto	**6** **Aquarius** Opposition Venus Trine Mars	**7** **Pisces** Opposition Mercury Square Uranus
8 **FM** **Pisces** Opposition Sun Trine Jupiter	**9** **XhN** **Aries** Conjunct Saturn Sextile Uranus Conjunct Neptune Sextile Pluto	**10** **Aries** Trine Venus Opposition Mars Square Jupiter	**11** **P9** **Taurus** Square Pluto	**12** **Taurus** Trine Sun Trine Mercury Square Venus Sextile Jupiter Square Ascendant	**13** **Gemini** Conjunct Uranus Trine Pluto Square Ascendant	**14** **3Q** **Gemini** Square Sun Square Mercury Trine Mars Square Ascendant
15 **^** **Cancer** Square Saturn Square Neptune Square Ascendant	**16** **Cancer** Conjunct Jupiter Square Ascendant	**17** **Leo** Square Mars Trine Saturn Sextile Uranus Trine Neptune Opposition Pluto Square Ascendant	**18** **Leo** Square Ascendant	**19** **Leo** Conjunct Venus Sextile Mars Square Uranus	**20** **Virgo**	**21** **Virgo** Conjunct Sun Opposition Saturn Trine Uranus Opposition Neptune Trine Pluto
22 **NM XhS** **Virgo**	**23** **Libra** Conjunct Sun Conjunct Mercury Trine Uranus Opposition Neptune Trine Pluto	**24** **Libra** Square Jupiter	**25** **Scorpio** Sextile Venus Conjunct Mars Square Pluto	**26** **A** **Scorpio** Trine Jupiter	**27** **Scorpio** Trine Jupiter Trine Saturn Opposition Uranus Trine Neptune	**28** ***DLS Starts*** **Sagittarius** Square Venus
29 **V** **Sagittarius** Square Saturn Square Ascendant	**30** **1Q** **Capricorn** Square Sun Sextile Mars Square Neptune Square Ascendant	1 **Moon Sign**	2	3	4	5
6	7					

NORTH ISLAND RAINFALL
SEPTEMBER 2025

d=mainly dry, n= fog, mist, dew or nonrecordable, l=light showers, p=passing showers, s=significant showers, r=rain, h=heavy falls

		Northland							Auckland					Bay Of Plenty				Waikato				Rot/Tau		Hawkes Bay			C Plateau			Taranaki			Manawatu			Lower North Island					
		CR	Kta	Ker	Rsl	Dar	Whg	Man	Wark	Hen	MkH	Ak	Co	Tau	TeP	Wkt	Opo	Hml	TeA	Mata	Ot	Rt	Taup	Gis	Nap	Has	Taum	Chat	Waio	NP	Str	Wang	Pa	Dann	Lev	PN	Mast	UH	Wgt	all	Moon
1st	Mon	l	d	d	d	d	d	n	d	l		l	d	d	d	d	d	l	d	d	l	d	d	d	d	d	n	d	n	l	d	d	d	d	l	l	n	d	d	16	
2nd	Tue	l	s	d	d	s	d	n	n	d	[illegible]	d	d	d	d	d	d	d	d	d	d	d	d	l	d	d	d	d	d	d	d	d	d	l	d	d	d	d	d	19	V
3rd	Wed	d	d	d	d	d	d	d	d	d		d	d	d	d	d	d	d	d	d	d	d	d	d	d	d	d	d	n	l	d	d	d	d	d	d	d	d	d	4	
4th	Thur	d	d	d	d	d	d	d	d	d		n	d	d	d	d	d	d	d	d	l	d	d	d	d	d	n	n	d	l	d	d	d	d	d	d	d	d	d	4	
5th	Fri	l	d	d	d	d	d	l	d	d	[illegible]	d	d	d	d	d	d	d	d	d	d	d	d	d	d	d	d	l	d	n	d	d	d	d	l	l	d	d	d	8	
6th	Sat	s	s	h	r	s	h	h	s	l		l	s	l	s	s	l	l	l	l	d	l	d	l	d	d	d	d	d	d	d	d	d	d	l	d	d	d	d	141	
7th	Sun	s	l	h	r	s	h	r	h	s	[illegible]	s	h	r	s	r	r	h	h	d	h	r	l	h	h	h	r	r	r	l	s	s	d	r	l	s	r	l	d	462	
8th	Mon	d	d	d	d	l	l	s	s	l		s	s	d	d	l	l	l	l	d	l	d	d	s	l	n	l	s	s	l	n	d	d	l	l	s	s	l	d	72	F
9th	Tue	d	d	d	d	d	d	d	d	d	[illegible]	d	d	d	d	d	d	l	l	r	l	d	d	d	d	d	l	h	s	n	l	d	l	s	l	s	l	l	d	51	XhN
10th	Wed	d	d	d	d	d	d	d	d	s		l	l	d	d	d	d	s	l	d	l	l	l	d	d	l	s	r	s	l	l	l	d	n	d	l	l	d	d	53	
11th	Thur	d	d	d	d	d	d	d	d	d	[illegible]	d	d	d	d	d	d	d	d	d	d	d	d	d	d	d	l	d	d	d	d	d	d	d	d	d	d	d	d	1	P9
12th	Fri	d	d	d	d	l	d	l	l	s	[illegible]	l	d	d	l	d	l	l	l	d	s	l	l	d	d	d	r	h	s	s	h	l	d	d	l	l	d	l	d	94	
13th	Sat	d	d	d	d	d	d	d	d	l	[illegible]	d	l	d	d	d	d	d	n	d	l	d	d	d	d	d	l	l	n	l	d	d	d	d	l	d	d	d	d	8	
14th	Sun	d	d	d	d	d	d	d	d	d		d	d	l	d	d	d	l	l	d	l	n	n	d	d	d	l	s	d	l	l	d	d	d	s	n	d	s	d	28	3Q
15th	Mon	l	l	d	d	r	l	s	s	r	[illegible]	r	s	r	h	r	h	h	h	l	h	r	s	h	s	s	r	h	h	l	h	s	s	r	l	l	r	s	s	399	[illegible]
16th	Tue	l	l	r	s	l	r	r	l	s	[illegible]	s	d	r	r	s	s	s	s	s	s	s	d	r	s	s	n	d	d	d	d	d	d	s	d	d	l	l	s	154	
17th	Wed	l	d	l	l	d	l	s	d	d	[illegible]	d	r	d	d	d	d	d	d	d	d	d	d	d	d	d	d	d	d	d	d	d	d	d	d	d	d	d	d	22	
18th	Thur	l	d	l	l	d	r	r	s	d	[illegible]	d	l	d	d	d	d	d	d	d	d	d	d	d	d	d	n	d	d	d	d	d	d	d	d	d	d	d	d	39	
19th	Fri	h	h	h	h	l	r	s	l	l	[illegible]	d	d	d	d	d	d	d	d	d	d	d	d	d	d	d	d	d	d	l	l	d	d	d	d	d	d	d	d	139	
20th	Sat	r	l	h	h	s	h	h	h	s		s	r	s	s	s	s	s	s	s	s	s	d	h	h	h	l	l	l	d	d	d	d	s	d	d	l	d	l	335	
21st	Sun	l	l	d	d	n	d	d	d	n	[illegible]	d	d	d	l	n	d	d	d	d	n	d	d	r	s	s	d	d	l	d	d	l	d	s	d	d	l	d	d	46	
22nd	Mon	d	d	d	d	l	l	d	d	d		l	d	d	d	d	d	d	l	n	l	d	d	d	d	d	d	d	d	d	d	d	d	d	d	d	d	d	d	10	N KhS
23rd	Tue	d	d	d	d	d	d	d	d	d	c	d	d	d	d	d	d	d	d	d	d	d	d	d	d	d	d	d	d	d	d	d	d	d	d	d	d	d	d	0	
24th	Wed	l	l	d	d	d	d	d	d	d	c	d	l	l	l	l	l	d	d	d	d	d	d	d	d	d	d	l	d	d	s	d	d	d	d	d	d	d	d	17	
25th	Thur	n	l	d	d	d	l	l	d	d	c	d	d	d	d	d	n	l	l	d	l	d	d	d	d	d	l	h	d	s	s	d	r	l	h	l	l	l	l	91	
26th	Fri	l	d	s	s	d	s	r	h	s	[illegible]	s	s	s	r	l	l	s	s	d	r	s	l	d	d	d	l	l	l	d	l	d	d	d	d	d	d	d	d	133	▲
27th	Sat	d	d	d	d	d	s	s	h	d	c	s	d	l	h	r	s	l	l	r	h	s	h	l	d	d	l	s	l	d	d	d	d	l	d	l	d	d	d	164	
28th	Sun	d	d	d	d	d	d	d	d	n	[illegible]	l	d	d	d	d	d	l	l	d	l	d	s	d	d	d	s	s	l	l	s	l	d	l	l	l	n	d	d	35	
29th	Mon	d	d	d	d	d	d	d	d	d	[illegible]	d	d	l	l	l	d	d	d	d	l	l	l	l	s	s	l	l	s	l	d	l	d	l	d	d	l	d	l	41	▼
30th	Tue	h	h	h	h	l	r	s	l	l	c	d	d	d	l	d	d	d	d	d	d	l	l	d	d	l	d	d	d	d	d	d	d	n	d	d	l	d	d	293	1Q
estimate:		179	132	174	101	47	145	119	140	55	[illegible]	55	88	54	82	65	59	65	78	35	95	53	45	121	68	92	73	150	64	32	83	16	18	55	45	32	40	22	16	2878	
average:		94	127	179	122	99	132	93	137	114	[illegible]	92	104	95	162	92	114	90	104	130	123	122	97	72	55	50	126	266	90	125	172	62	81	82	87	72	77	113	97	4154	
trend:		wtr	av	av	drr	drr	wtr	wtr	av	drr	[illegible]	drr	drr	drr	drr	drr	drr	drr	drr	drr	drr	drr	drr	wtr	wtr	wtr	drr	drr	drr	drr	drr	drr	drr	drr	drr	drr	drr	drr	drr	drr	

SOUTH ISLAND RAIN

SEPTEMBER 2025

d=mainly dry, n= fog, mist, dew or nonrecordable, l=light showers, p=passing showers, s=significant showers, r=rain, h=heavy falls

		West Coast				Top South Island				Canterbury				South Canterbury				Coastal Otago			Lakes			Inland Otago				Southland						
		West	Grey	Hok	BB	Ta	Mou	Nel	Ble	Dar	Ham	Kai	Ch	Rak	Meth	Ash	Tim	Oam	Dun	Bal	Twi	LP	LT	Crom	Qstn	Ran	Alex	TeA	Wtn	Gore	Inv	Puy Pt	all	Moon
1st	Mon	l	l	l	d	d	d	d	d	d	d	d	d	d	d	d	d	d	l	l	d	d	d	d	l	d	d	d	s	s	s	s	32	
2nd	Tue	d	d	d	d	d	d	d	d	d	d	d	d	d	n	d	d	d	d	d	d	d	d	d	d	d	d	d	l	s	s	s	15	V
3rd	Wed	l	l	l	d	d	d	d	d	d	d	d	d	d	d	d	d	d	d	d	d	d	d	d	d	d	d	d	d	d	s	l	12	
4th	Thur	d	l	d	d	d	d	d	d	d	d	d	d	d	d	n	d	d	d	d	d	d	d	d	d	d	d	d	d	d	d	s	10	
5th	Fri	d	l	l	l	d	d	d	d	d	d	d	d	d	d	d	d	d	d	d	d	d	d	d	d	d	d	s	d	d	l	r	26	
6th	Sat	d	d	d	d	d	d	d	d	d	d	d	d	d	d	d	d	d	d	d	d	d	d	d	d	d	d	d	d	d	d	l	1	
7th	Sun	d	l	s	s	d	d	d	d	d	d	d	d	d	d	d	d	d	d	d	d	d	d	d	d	d	d	d	d	d	n	r	27	
8th	Mon	s	d	d	d	d	d	d	l	d	d	d	d	d	d	d	d	d	l	d	d	d	d	d	l	d	d	l	l	l	l	s	20	F
9th	Tue	h	h	h	s	l	l	n	d	d	d	d	d	d	d	d	d	l	l	s	d	d	l	n	l	d	d	s	s	s	s	r	124	XhN
10th	Wed	r	r	r	s	l	l	l	d	l	d	l	l	d	d	d	l	d	l	s	d	d	n	d	d	d	d	l	r	r	s	l	82	
11th	Thur	s	l	l	r	d	d	d	d	d	d	d	d	d	d	d	d	d	d	l	s	d	d	d	l	d	d	d	d	d	l	l	33	P9
12th	Fri	r	r	h	r	s	l	l	d	d	d	d	d	d	d	d	d	d	d	d	d	r	l	l	r	d	l	s	l	l	l	r	114	
13th	Sat	l	l	s	l	d	d	d	d	d	d	d	d	d	d	d	d	d	d	d	r	d	d	d	d	d	d	d	d	d	d	h	49	
14th	Sun	r	r	h	h	l	n	d	d	l	d	d	d	h	r	r	r	r	h	r	d	h	r	s	h	r	l	h	h	h	h	h	383	3Q
15th	Mon	s	l	d	d	n	l	l	s	r	h	s	h	d	r	r	s	l	l	l	h	s	s	l	l	s	s	l	l	l	l	l	164	^
16th	Tue	d	d	d	d	d	d	d	d	d	l	l	s	d	d	d	d	d	d	d	l	d	d	d	d	d	d	d	d	d	d	d	11	
17th	Wed	d	d	d	d	d	d	d	d	d	s	d	d	d	d	d	d	d	d	d	d	d	d	d	d	d	d	d	d	d	l	s	9	
18th	Thur	d	d	d	d	d	d	d	d	d	d	d	d	d	d	d	d	d	d	d	d	d	d	d	d	d	d	d	d	d	d	r	11	
19th	Fri	d	n	d	d	d	d	d	d	d	d	d	d	d	d	d	d	d	d	d	d	d	d	d	d	d	d	l	l	l	s	s	18	
20th	Sat	d	d	d	d	d	d	d	l	d	l	l	d	d	d	d	d	d	d	d	d	d	d	d	d	d	d	d	l	d	l	r	16	
21st	Sun	d	d	d	d	l	l	l	d	d	d	d	d	d	d	d	d	d	d	d	d	d	d	d	d	d	d	d	d	l	l	s	13	
22nd	Mon	l	d	d	d	d	d	d	d	d	d	d	d	d	d	d	d	d	d	d	d	d	d	d	d	d	d	d	d	d	d	l	4	N XhS
23rd	Tue	d	d	d	d	d	d	d	d	d	d	d	d	d	d	d	d	d	d	d	d	d	d	d	d	d	d	d	d	d	d	d	1	
24th	Wed	h	h	h	h	h	r	r	s	d	n	d	d	d	d	d	d	d	d	d	d	r	l	d	d	d	d	l	l	l	l	s	223	
25th	Thur	n	l	l	s	d	s	s	s	s	d	d	l	h	r	l	d	l	d	d	s	d	d	l	d	s	d	d	d	l	d	d	80	
26th	Fri	r	r	s	h	r	l	l	d	l	d	d	d	d	l	d	d	d	d	d	d	s	s	d	d	d	d	l	d	d	l	s	98	A
27th	Sat	l	s	s	n	l	l	l	d	d	d	d	d	d	l	d	d	d	l	d	s	d	d	d	d	d	d	d	s	n	s	s	44	
28th	Sun	d	d	d	d	l	d	d	l	d	d	d	d	d	d	d	d	d	d	d	l	d	d	d	d	d	d	d	l	l	l	s	10	
29th	Mon	d	d	d	d	d	d	d	d	d	d	d	d	d	d	d	d	d	d	d	d	d	d	d	d	d	d	d	d	d	d	d	0	V
30th	Tue	d	d	d	d	d	d	d	d	d	d	l	d	d	d	d	d	d	d	d	d	d	d	d	d	d	d	d	d	d	d	d	1	1Q
estimate:		143	116	119	143	56	26	29	16	23	29	13	27	40	40	23	18	15	29	27	58	54	30	12	42	23	9	51	77	59	79	206	1631	
average:		165	183	225	403	200	113	85	65	60	97	55	45	72	74	55	45	45	67	52	50	85	52	29	65	27	20	93	64	57	77	174	2899	
trend:		drr	drr	drr	drr	drr	drr	drr	drr	drr	drr	drr	drr	drr	drr	drr	drr	drr	drr	drr	av	drr	drr	drr	drr	av	drr	drr	wtr	av	av	wtr	drr	

NORTH ISLAND SUNSHINE ESTIMATES
SEPTEMBER 2025

F=fine (8-12 hours of sunshine), pc= partly cloudy (4-7 hours), c=cloudy (1-3 hours), o=overcast (0 hours)

		Kta	Whg	Dar	Ak	Hm	Tau	Rt	Taum	NP	Wang	Taup	Gis	Nap	Wai	Dann	PN	Mast	Pa	UH	Wgt	ALL	Moon
1st	Mon	pc	pc	pc	pc	c	c	pc	pc	pc	c	pc	c	o	c	c	c	c	pc	pc	pc	71	
2nd	Tue	pc	F	F	F	F	F	F	F	F	pc	F	F	F	F	F	pc	F	F	F	F	176	V
3rd	Wed	F	c	pc	pc	pc	F	F	F	F	c	F	F	F	F	F	c	F	F	F	F	146	
4th	Thur	F	pc	c	c	pc	pc	F	F	pc	c	F	F	F	F	F	c	F	o	c	F	125	
5th	Fri	o	c	pc	pc	F	F	F	F	pc	pc	F	F	F	F	F	pc	F	pc	pc	F	134	
6th	Sat	pc	o	o	o	o	o	o	o	o	pc	o	o	c	o	c	pc	c	pc	pc	pc	31	
7th	Sun	c	c	c	c	c	pc	c	pc	F	c	c	o	pc	pc	c	c	c	c	c	c	64	
8th	Mon	F	pc	pc	pc	F	F	pc	pc	F	c	pc	F	F	F	c	c	c	pc	pc	F	129	F
9th	Tue	pc	F	F	F	pc	F	pc	pc	pc	pc	pc	F	F	pc	F	pc	F	pc	pc	pc	146	XhN
10th	Wed	F	pc	pc	pc	pc	F	F	F	F	pc	F	F	F	F	F	pc	F	F	F	F	169	
11th	Thur	F	pc	pc	pc	pc	pc	pc	pc	c	c	pc	F	F	F	F	c	F	pc	pc	F	125	P9
12th	Fri	c	c	pc	pc	F	pc	pc	pc	pc	c	pc	F	F	F	pc	c	pc	F	pc	pc	121	
13th	Sat	pc	o	o	o	pc	pc	pc	pc	pc	c	pc	F	pc	F	pc	c	pc	c	pc	pc	86	
14th	Sun	pc	pc	c	c	c	c	c	o	o	o	c	F	F	pc	o	o	o	o	o	o	48	3Q
15th	Mon	o	o	o	o	c	o	o	c	F	pc	o	o	F	pc	pc	pc	pc	pc	pc	pc	66	^
16th	Tue	pc	pc	F	F	F	F	F	F	F	F	F	pc	o	c	F	F	F	F	pc	F	161	
17th	Wed	pc	pc	pc	pc	F	F	F	F	F	F	F	F	pc	F	F	F	F	F	F	F	179	
18th	Thur	c	o	c	c	c	c	c	c	c	c	c	F	F	F	F	c	F	c	pc	F	88	
19th	Fri	o	o	o	o	o	pc	o	c	pc	pc	o	o	F	c	c	pc	c	F	pc	pc	65	
20th	Sat	c	c	pc	pc	pc	c	c	c	F	o	c	o	o	o	o	o	o	pc	c	F	55	
21st	Sun	F	pc	pc	pc	F	F	F	pc	F	F	F	F	o	F	F	F	F	F	F	pc	170	
22nd	Mon	F	F	F	F	F	F	F	F	o	o	F	F	F	F	F	o	F	c	pc	F	151	N XhS
23rd	Tue	pc	pc	pc	pc	pc	F	pc	c	c	pc	pc	F	F	F	F	pc	F	c	pc	F	122	
24th	Wed	c	c	c	c	c	c	c	o	o	c	c	pc	F	pc	pc	c	pc	o	o	c	52	
25th	Thur	F	F	F	F	pc	F	pc	c	pc	pc	pc	F	pc	pc	pc	pc	pc	F	F	pc	133	
26th	Fri	F	pc	c	c	c	o	o	o	c	pc	o	pc	F	pc	pc	pc	pc	pc	o	o	69	A
27th	Sat	F	F	F	F	c	F	F	pc	pc	pc	F	F	pc	F	pc	pc	pc	pc	pc	pc	132	
28th	Sun	pc	pc	c	c	o	pc	c	c	c	c	c	c	pc	o	o	c	o	o	o	c	50	
29th	Mon	c	c	pc	pc	c	c	c	pc	pc	pc	c	pc	o	pc	pc	pc	pc	F	pc	pc	86	V
30th	Tue	pc	o	o	o	c	c	pc	pc	F	F	pc	c	c	c	pc	F	pc	F	pc	F	101	1Q
est. hours:		161	125	137	137	142	188	160	149	163	124	160	206	193	197	176	124	176	176	152	204	3250	
av. hours:		167	153	153	149	144	164	149	109	160	143	140	173	173	172	129	120	140	150	153	156	2997	
trend:		av	less	less	av	av	more	av	more	av	less	more	more	more	more	more	av	more	more	av	more	average	

SOUTH ISLAND SUNSHINE ESTIMATES
SEPTEMBER 2025

F=fine (8-12 hours of sunshine), pc= partly cloudy (4-7 hours), c=cloudy (1-3 hours), o=overcast (0 hours)

		Ta	Nel	Ble	Mou	West	Grey	Hok	Ham	Kai	Ch	Ash	Tim	Crom	Qstn	Bal	Dun	Wnt	Gore	Inv	Alex	ALL	Moon
1st	Mon	F	pc	pc	c	F	F	F	F	F	F	F	F	pc	pc	pc	pc	pc	pc	pc	pc	142	
2nd	Tue	F	F	F	F	pc	F	F	F	F	F	F	pc	F	pc	pc	c	pc	pc	c	F	150	V
3rd	Wed	pc	F	F	F	pc	pc	F	F	F	F	F	F	F	pc	F	F	F	F	F	F	170	
4th	Thur	pc	F	F	F	c	pc	pc	F	F	F	F	pc	F	pc	pc	F	pc	pc	F	F	153	
5th	Fri	F	F	F	F	c	F	pc	F	F	F	F	F	F	pc	F	F	F	F	F	F	166	
6th	Sat	pc	pc	F	pc	pc	F	F	F	F	pc	F	F	F	pc	pc	F	c	c	c	F	141	
7th	Sun	F	F	F	pc	c	pc	pc	c	c	c	c	c	pc	pc	pc	c	c	c	pc	pc	94	
8th	Mon	pc	F	F	F	o	o	o	F	F	F	c	pc	pc	c	pc	o	c	c	o	pc	86	F
9th	Tue	F	F	F	F	c	c	pc	c	c	c	c	pc	pc	c	pc	c	o	o	c	pc	86	XhN
10th	Wed	F	F	F	F	c	c	o	F	F	pc	pc	pc	F	F	pc	F	F	F	F	F	140	
11th	Thur	pc	F	F	F	c	c	o	pc	pc	pc	F	pc	pc	pc	pc	pc	pc	pc	pc	pc	102	P9
12th	Fri	F	F	F	F	pc	o	o	F	F	F	F	F	F	pc	F	F	pc	pc	pc	F	149	
13th	Sat	F	F	F	F	pc	c	o	o	o	o	pc	pc	F	pc	pc	F	pc	pc	pc	F	100	
14th	Sun	F	c	c	pc	o	c	c	o	o	o	c	o	pc	c	pc	pc	pc	pc	F	pc	61	3Q
15th	Mon	F	F	F	F	F	F	F	c	c	F	F	pc	F	pc	pc	c	c	c	c	F	131	^
16th	Tue	F	F	F	F	F	F	F	F	F	F	F	F	F	F	pc	F	F	F	pc	F	187	
17th	Wed	F	F	F	F	pc	o	o	F	F	F	F	F	F	F	pc	F	F	F	F	F	170	
18th	Thur	F	F	F	F	pc	pc	F	F	F	F	F	F	pc	pc	pc	c	c	c	c	pc	139	
19th	Fri	F	F	pc	F	F	F	F	pc	pc	F	o	pc	F	F	pc	pc	pc	pc	F	F	154	
20th	Sat	c	F	c	F	F	F	F	F	F	pc	pc	F	F	pc	pc	pc	pc	pc	pc	F	136	
21st	Sun	pc	F	F	F	pc	pc	F	F	F	F	F	F	F	pc	pc	pc	pc	pc	pc	F	151	
22nd	Mon	F	F	F	F	pc	pc	F	F	F	F	F	F	F	c	pc	pc	pc	pc	c	F	146	N XhS
23rd	Tue	pc	c	pc	pc	o	o	o	pc	pc	c	pc	pc	c	c	c	pc	o	o	c	c	58	
24th	Wed	pc	pc	pc	pc	c	c	c	o	o	c	pc	pc	pc	c	o	c	o	o	pc	pc	55	
25th	Thur	pc	pc	F	F	c	c	pc	F	F	F	F	c	F	F	F	F	F	F	F	F	154	
26th	Fri	c	pc	c	pc	c	c	c	F	F	F	F	F	F	pc	pc	pc	pc	pc	F	F	117	A
27th	Sat	F	c	F	pc	pc	c	pc	F	F	F	F	F	F	F	pc	F	pc	pc	c	F	144	
28th	Sun	F	pc	c	pc	pc	F	F	o	o	o	c	o	F	pc	pc	pc	pc	pc	pc	F	101	
29th	Mon	pc	pc	pc	c	F	F	F	c	c	o	pc	F	F	F	F	F	F	F	F	F	153	V
30th	Tue	o	F	c	F	F	F	F	F	F	pc	F	F	F	F	F	F	F	F	F	F	178	1Q
est. hours:		221	242	236	226	141	161	178	204	204	195	223	198	236	174	176	185	157	157	166	236	3915	
av. hours:		173	186	186	184	142	132	142	144	168	165	147	153	175	158	135	128	136	140	134	163	3091	
trend:		more	more	more	more	av	more	more	more	more	more	more	more	more	av	more	more	more	more	more	more	more	

SEPTEMBER 1 MONDAY

An anticyclonic system drifting over Tasmania extends a high pressure ridge onto the west of New Zealand directing a moderate to light southwest flow throughout. Mostly cloudy and dry over the North Island with the chance of some misty patches in sheltered parts, and pockets of frosts in the Bay of Plenty to the Manawatu. Mostly sunny over the South Island with some brief morning showers in Southland, Otago, and more isolated up the West Coast, mainly dry elsewhere, and frosts in southern and eastern districts.

Auckland: Partly cloudy, mostly dry after morning showers, moderate breezes
Hamilton: Cloudy, brief showers, moderate breezes
Wellington: Partly cloudy, dry, slightly warmer, freshening winds
Greymouth: Sunny day, brief morning showers, moderate breezes
Christchurch: Sunny, light and pleasant breezes, mild day cold night, dry
Dunedin: Light breezes, occasional sunny spells, brief morning showers

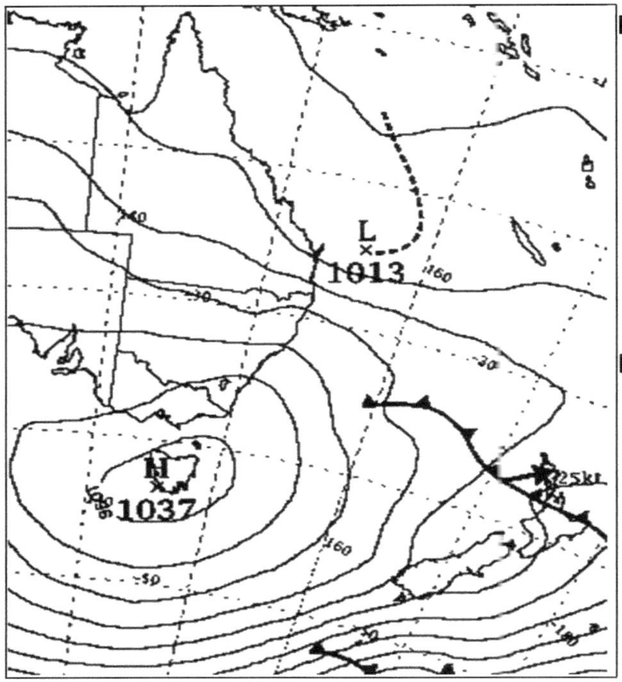

RAIN POTENTIAL

FROST/SNOW

SEPTEMBER 2 TUESDAY

A large high centred off the Australian southeast coast edges across the Tasman Sea and directs a light southwest breeze to mostly calm air over the North Island and moderate southwest breezes over the South Island. Cool and with frosts from the Waikato to Wellington and Canterbury to Southland, and good sunny spells in most districts. Chance of some brief morning showers in Southland, around the East Cape, and in Northland. Dry elsewhere.

Auckland: Sunny, mostly dry, calm atmosphere, chance of misty patches
Hamilton: Sunny, dry, mild day cold night, gentle breezes
Wellington: Sunny, dry, fresh breezes, mild day cool night
Greymouth: Sunny, dry, fresh breezes for a time, mild day cold night
Christchurch: Sunny, dry, light breezes, frost and fog patches possible
Dunedin: Cloudy, dry, moderate breezes

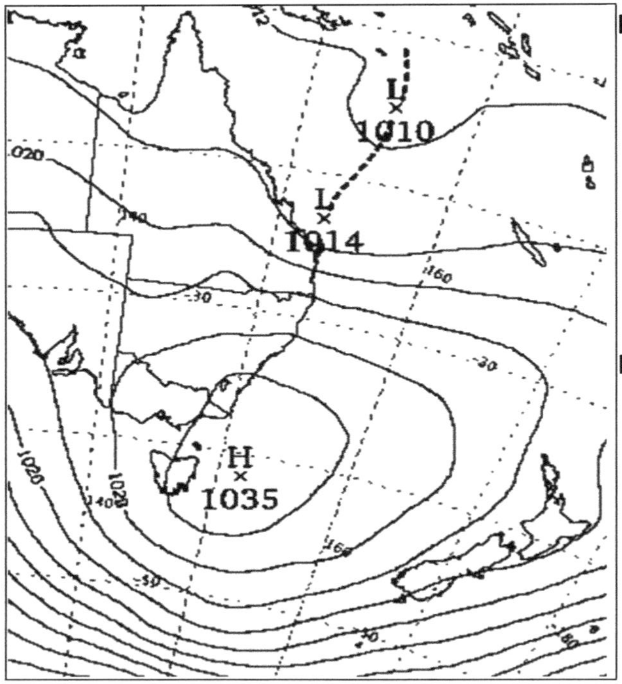

RAIN POTENTIAL

FROST/SNOW

SEPTEMBER **3** WEDNESDAY

A large anticyclonic system continues to cross the Tasman Sea bringing a mostly dry and sunny day throughout the country with widespread frosts from the Waikato to the Manawatu and Canterbury to Southland and only the chance of some isolated showers around the Southland coastal fringes and Fiordland. A moderate west to southwest flow prevails over the South Island with chance of fog patches in Canterbury, and lighter southwest flow to mostly calm air over the North Island.

Auckland: Cloudy, mostly dry and calm air, possible brief drizzle patches
Hamilton: Partly cloudy, dry, chance of frost pockets, mostly calm air
Wellington: Sunny, warmer, dry, light air
Greymouth: Partly cloudy, light showers, moderate winds
Christchurch: Sunny, dry, light air, possible fog
Dunedin: Sunny, dry, cold overnight, light breezes

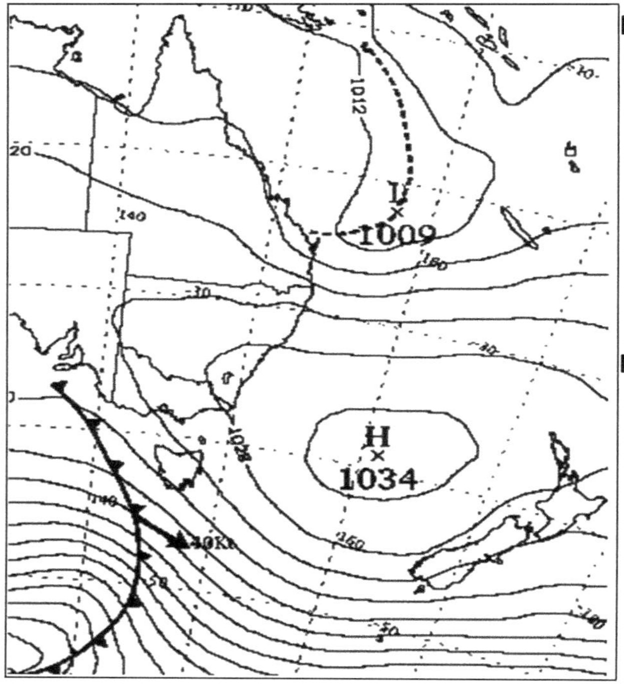

RAIN POTENTIAL

FROST/SNOW

SEPTEMBER **4** THURSDAY

A high pressure system weakens as it crosses the country in a moderate westerly flow over the North Island, while an approaching wet front strengthens the westerly flow over the South Island. Mostly sunny, dry, and cold throughout with frosts once more from the Waikato to the Manawatu, as well as almost throughout the South Island. An approaching wet front associated with another Southern Ocean depression brushes some showers into Fiordland for a brief time.

Auckland: Cloudy, mostly dry except for isolated drizzle patches, light breezes
Hamilton: Partly cloudy, dry, mostly calm air
Wellington: Unsettled windy spells, sunny, dry
Greymouth: Changeable, partly cloudy, isolated showers, fluctuating breezes
Christchurch: Sunny, dry, frost patches likely, moderate breezes
Dunedin: Sunny, warmer day, cold night, dry, fresh winds

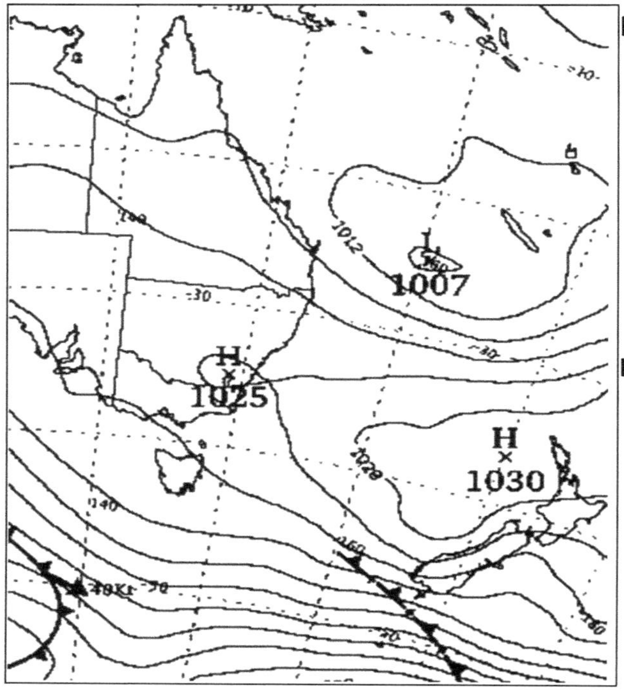

RAIN POTENTIAL

FROST/SNOW

SEPTEMBER **5** FRIDAY

High pressure lingers over the North Island with a westerly air prevailing, but backing easterly over the far north, while low pressure to the south of the country brings a moderate westerly flow over the South Island. Milder in most districts, but still some frosts likely from the Waikato to the Manawatu, and in the West Coast, Southland, and Otago regions. Mostly sunny and dry everywhere except for some showery spells possible in Southland and Fiordland.

Auckland: Partly cloudy, dry, light shifting breezes
Hamilton: Mostly sunny, dry, mild day cold night, gentle breezes
Wellington: Sunny, dry, moderate breezes
Greymouth: Partly cloudy, brief showers, light breezes
Christchurch: Light breezes, warmer day overnight frosts, sunny, dry, chance of fog
Dunedin: Partly cloudy, dry, warmer, light breezes

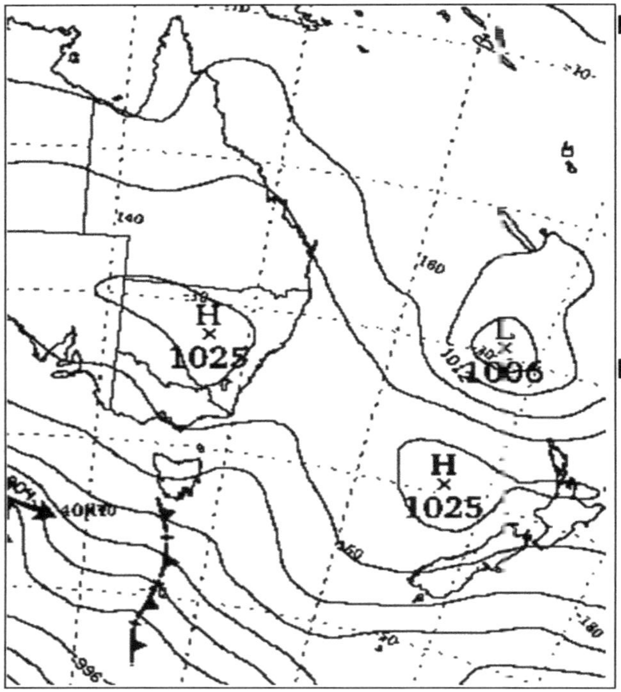

RAIN POTENTIAL

FROST/SNOW

SEPTEMBER **6** SATURDAY

A depression moves down onto the north of New Zealand as the high departs, bringing strengthening and blustery easterly winds and dull and gloomy conditions throughout the North Island with milder temperatures and rain spreading from Northland down to the Bay of Plenty as the day progresses. A trough between two high pressure systems however, crosses the South Island as the high departs to the east, with a gentle westerly air prevailing and chance of fog in the east. Mostly sunny and dry throughout the South Island with frosts in many southern and eastern districts.

Auckland: Overcast, light showers, fluctuating windy spells
Hamilton: Overcast, light showers, slightly warmer, fluctuating breezes
Wellington: Partly cloudy, dry, mild day cool night, fluctuating breezes
Greymouth: Sunny, dry, mostly calm air
Christchurch: Partly cloudy, mild day frosts overnight, dry, calm and pleasant air
Dunedin: Sunny, dry, moderate breezes

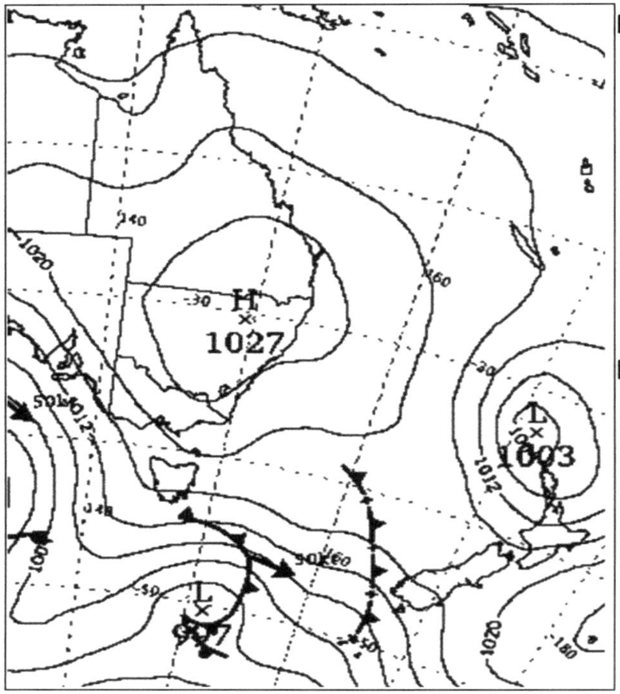

RAIN POTENTIAL

FROST/SNOW

SEPTEMBER **7** SUNDAY

A strong and stormy southwest flow moves up across the country bringing a mix of brief sunny spells and outbreaks of rain throughout the North Island, with heavy rain likely in eastern regions. In contrast, the departing depression may only bring some rain into the West Coast and Fiordland, while the rest of the South Island will be mostly dry with frosts and occasional sunny spells.

Auckland: Cloudy, warmer, rain with heavy falls, strong winds, possible gales
Hamilton: Unsettled breezy spells, cloudy, mild, heavy rain, static air possible
Wellington: Cloudy, dry, unsettled winds
Greymouth: Milder, mix of sunny spells and showers, fluctuating winds
Christchurch: Unsettled breezes, cloudy, dry
Dunedin: Cloudy, dry, slightly cooler, light breezes, occasional blustery spells

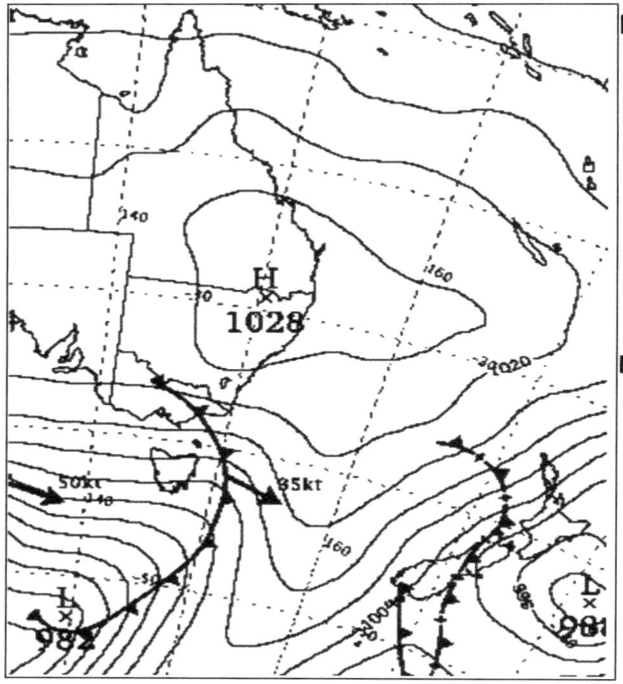

RAIN POTENTIAL

FROST/SNOW

SEPTEMBER **8** MONDAY

A strong and unstable westerly flow prevails over New Zealand as a Southern Ocean depression crosses to the south of the country, while a large high pressure system ridges to the north of New Zealand. A mixed day of sunny spells and occasional showers throughout most of the country, except for mostly dry and sunny with frosts in Canterbury. Chance of wintry showers and flurries in the higher reaches of the South Island.

Auckland: Unsettled and blustery air, mix of sunny spells and showers
Hamilton: Sunny, isolated showers, freshening winds
Wellington: Dry, sunny, fresh to strong winds
Greymouth: Overcast, dry with rain threatening, strengthening winds
Christchurch: Sunny, cold, windy spells, chance of frost in sheltered parts
Dunedin: Overcast, blustery at times, scattered showers

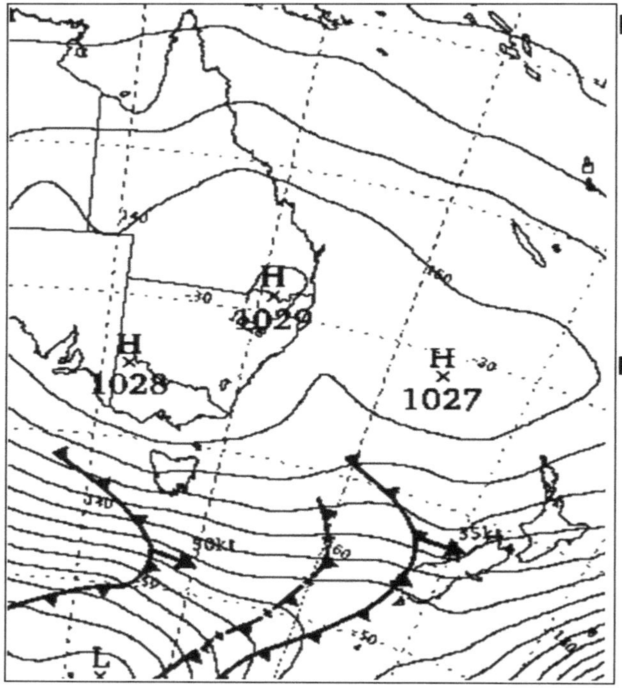

RAIN POTENTIAL

FROST/SNOW

SEPTEMBER **9** TUESDAY

Low pressure brings a very strong and blustery southwest flow across the country while high pressure remains almost stalled to the north of New Zealand. A passing wet front brings increased cloud throughout the South Island with rain in most districts except in Marlborough to Canterbury, which will be mostly dry and milder with passing threats of rain. Sunnier over the North Island with some evening showers in central regions from the Waikato to the Manawatu, while mostly dry in Northland and the east from the Bay of Plenty to the Hawkes Bay.

Auckland:	Sunny, dry, strong gusty winds, possible gale gusts
Hamilton:	Partly cloudy, isolated showers, strong gusty winds
Wellington:	Partly cloudy, dry, strong gusty winds, gales possible
Greymouth:	Stormy air, cloudy, heavy rain, chance of thunderstorms
Christchurch:	Cloudy, dry, strong winds at times, frost patches possible
Dunedin:	Cloudy, light squally showers, blustery winds

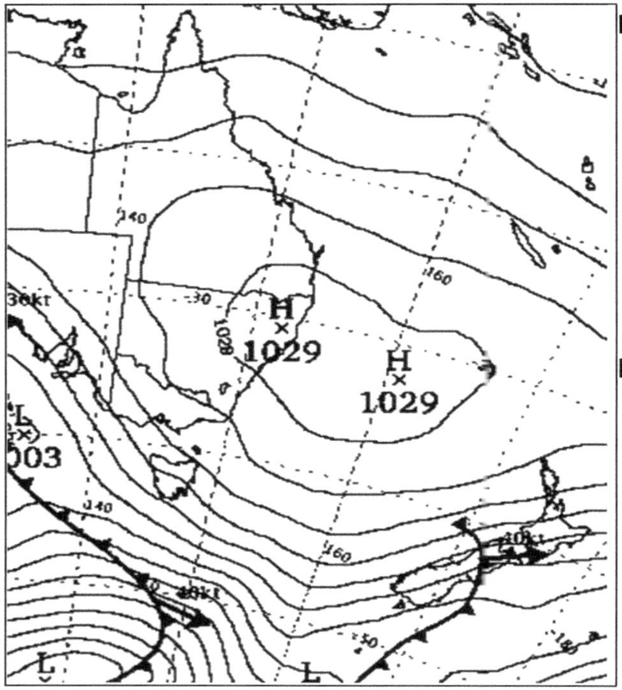

RAIN POTENTIAL

FROST/SNOW

SEPTEMBER **10** WEDNESDAY

A strengthening high pressure system moves onto the north of New Zealand bringing a mix of sunny spells and afternoon showers and lighter southwesterly winds over the North Island. Low pressure however, continues to bring evening showers in most regions in the South Island, although remaining mostly dry in Canterbury to North Otago, and strong westerly winds prevailing. Frosts still likely in the east and south of the South Island and in the south of the North Island to the Bay of Plenty.

Auckland:	Partly cloudy, occasional showers, winds easing to moderate
Hamilton:	Partly cloudy, scattered showers, breezy at times
Wellington:	Sunny, dry, mild day cool night, unsettled and changeable winds
Greymouth:	Static air, cloudy, rain, blustery
Christchurch:	Partly cloudy, isolated showers, windy spells, chance of hail and frost
Dunedin:	Sunny, warmer day cooler night, isolated showers, unsettled winds

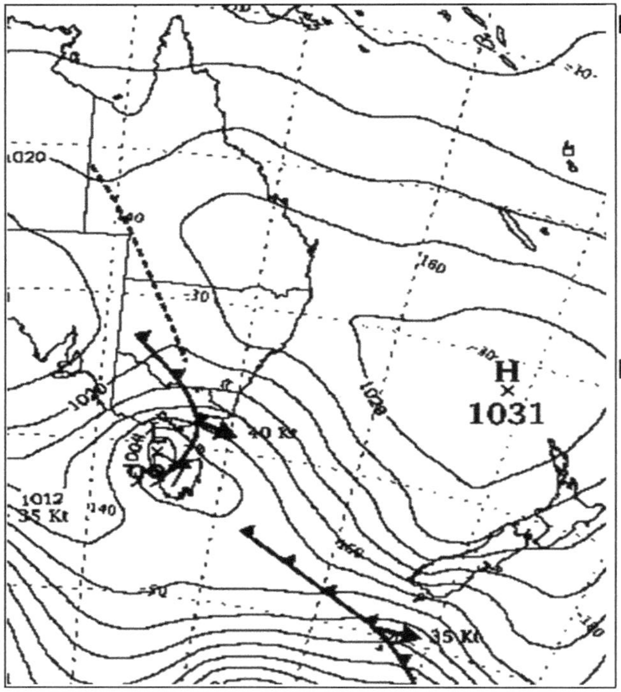

RAIN POTENTIAL

FROST/SNOW

SEPTEMBER **11** THURSDAY

High pressure maintains a mostly dry influence over the North Island with moderate northwest to westerly breezes prevailing, plenty of sunny spells, and milder temperatures in most districts. In contrast, an unstable and blustery northwest flow associated with a depression off the southwest coasts of the South Island extends a wet front into the west of the South Island with scattered showers in Westland to the Alps, while remaining mostly dry elsewhere with milder temperatures and good sunny spells.

Auckland:	Partly cloudy, dry, fresh breezes
Hamilton:	Partly cloudy, dry, light breezes
Wellington:	Strong and unsettled winds, sunny, dry
Greymouth:	Cloudy, brief showers, strong and unsettled winds
Christchurch:	Partly cloudy, dry, blustery winds with chance of gale gusts at times
Dunedin:	Partly cloudy, dry, warmer, strong blustery winds

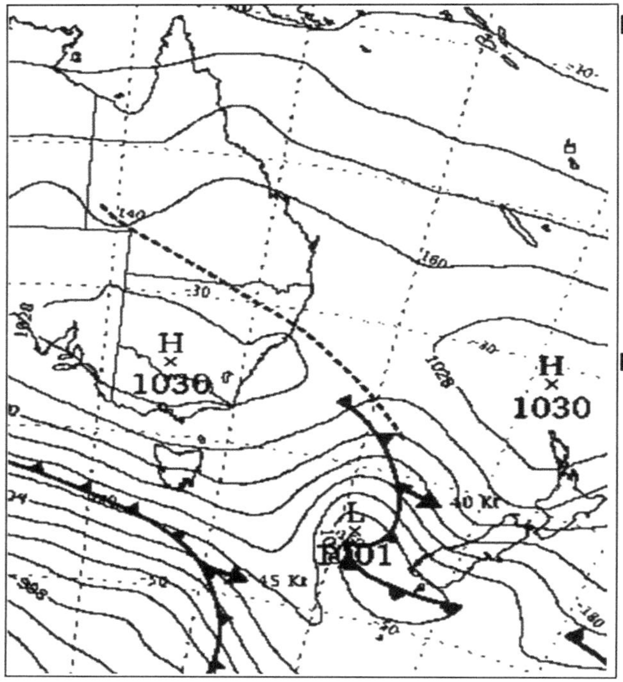

RAIN POTENTIAL

FROST/SNOW

SEPTEMBER **12** FRIDAY

Low pressure to the south of New Zealand pushes a wet front up over the country bringing scattered rain into most districts as the day progresses except remaining mostly dry in eastern coastal fringes from the East Cape to Canterbury. Mostly sunny everywhere. Winds strong and blustery in west to southwest flow, with winds possibly reaching gale force through the Cook Strait and western parts of the South Island.

Auckland:	Mix of sunny spells and showers, strong winds, chance ofgale gusts
Hamilton:	Mostly sunny, light showers, fresh breezes
Wellington:	Partly cloudy, warmer, dry, fresh to strong winds
Greymouth:	Overcast, rain, strong winds
Christchurch:	Sunny, warmer, dry, strong winds, possible gales
Dunedin:	Sunny, dry, warm day cool night, strong winds with occasional gusts

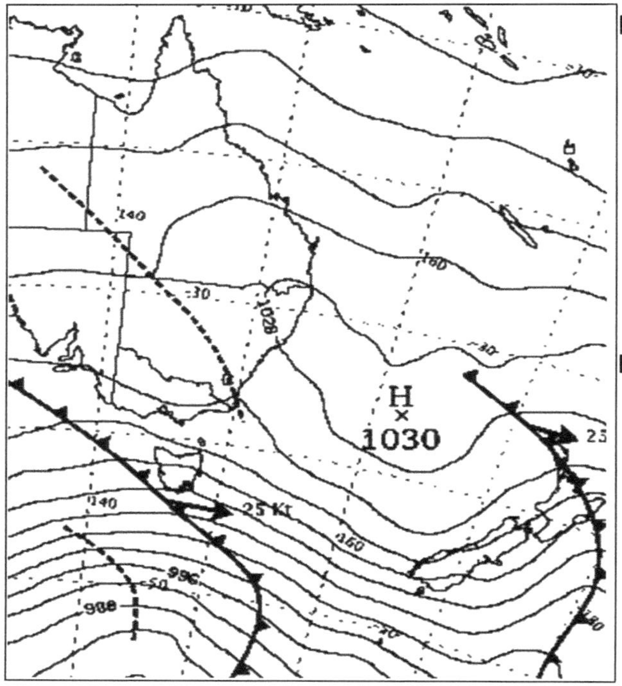

RAIN POTENTIAL

FROST/SNOW

SEPTEMBER **13** SATURDAY

High pressure centred to the north of New Zealand directs a light west to southwest flow over the North Island and dry, warmer, and mostly sunny day throughout, cloudier over Northland and chance of some overnight frost pockets in the Waikato to the Central Plateau. In contrast, a blustery northwest flow associated with a Southern Ocean depression crosses the South Island bringing milder temperatures throughout and some heavy falls of rain into the southwest regions. Mostly dry elsewhere.

Auckland:	Overcast, isolated showers winds easing to light
Hamilton:	Partly cloudy, mostly dry, threats of rain, light breezes
Wellington:	Partly cloudy, dry, slightly cooler, windy
Greymouth:	Cloudy, cooler, light showers, strong winds
Christchurch:	Dull and gloomy air, cooler dry, windy at times
Dunedin:	Partly cloudy, dry, slightly cooler, blustery at times

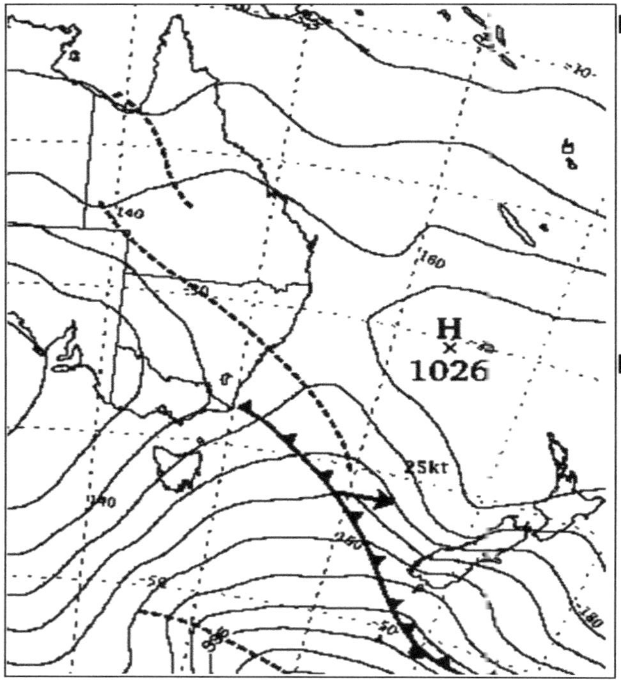

RAIN POTENTIAL

FROST/SNOW

SEPTEMBER **14** SUNDAY

A wet front associated with a Southern Ocean depression sweeps up through the country in a stormy southerly flow over the South island, backing westerly over the North Island. Widespread cloud throughout the country, with heavy rain over much of the South Island and showers spreading into the central parts of the North Island. Mostly dry elsewhere. Wintry showers may bring a return of snow southern parts of the South Island including in Canterbury, and winds may reach gale force over the central North Island.

Auckland:	Cloudy, cooler, mostly dry, threats of rain, fresh breezes at times
Hamilton:	Cloudy, moderate breezes, isolated showers
Wellington:	Overcast, strong winds, dry with increasing threats of rain
Greymouth:	Cloudy, rain, winds easing to moderate
Christchurch:	Overcast, dry, threats of rain, milder, light breezes, chance of flurries
Dunedin:	Cloudy, warmer day cold night, heavy rain, moderate breezes

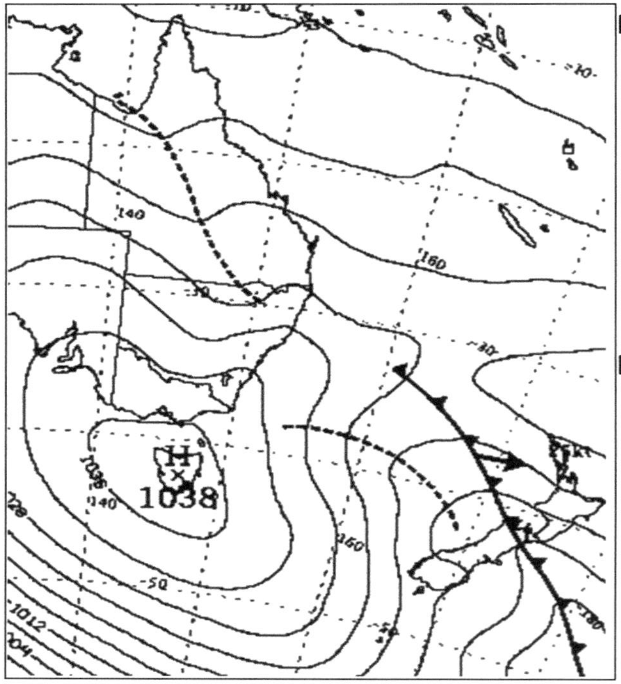

RAIN POTENTIAL

FROST/SNOW

SEPTEMBER 15 MONDAY

A stormy wet front continue to spread up through the country followed by a high pressure ridge associated with a large high crossing through the southern Tasman Sea. Widespread, mostly overnight rain, throughout the country, with heavy rain and storm conditions moving up over the North Island, while easing in the South Island. Snow still likely in the south and east of the South Island as well as in central parts of the North Island. Winds ease and change into an easterly flow later in the day. Widespread cloud in the northern half of the North Island, while increased sunny spells elsewhere.

Auckland: Overcast, drizzle patches and outbreaks of rain, warmer, light air
Hamilton: Cloudy, heavy rain, light breezes to mostly calm air
Wellington: Changeable, partly cloudy, scattered showers, shifting windy spells
Greymouth: Moderate winds, sunny, mostly dry, brief evening showers possible
Christchurch: Colder, sunny, evening rain, chance of snow, fresh breezes
Dunedin: Cloudy, spotty showers, colder, light shifting breezes

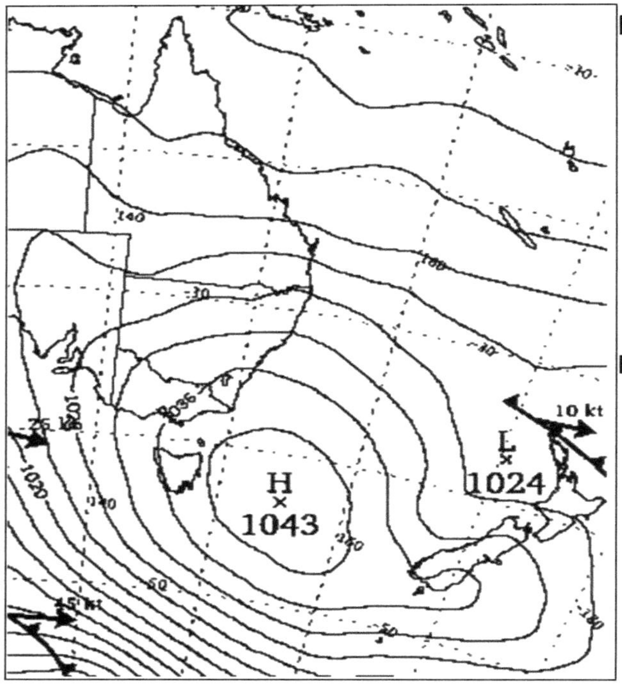

RAIN POTENTIAL

FROST/SNOW

SEPTEMBER 16 TUESDAY

A large anticyclonic system moves onto the country with a lighter westerly air prevailing over the South Island, backing easterly over the North Island. Very cold with frosts in the Waikato to the Central Plateau and throughout most of the South Island. Rain clears to mostly dry with good sunny spells breaking through in the South Island, while overnight showers continue over the eastern and northern regions of the North Island. Mostly dry elsewhere.

Auckland: Sunny, evening showers, colder, moderate breezes
Hamilton: Mostly sunny, occasional showers, colder, mostly calm air
Wellington: Sunny, cooler, occasional showers, moderate breezes
Greymouth: Sunny, dry, calm air, cold overnight
Christchurch: Sunny, showers, severe frosts, possible fog, chance of light snow
Dunedin: Sunny, dry, colder, chance of frosts, mostly calm air

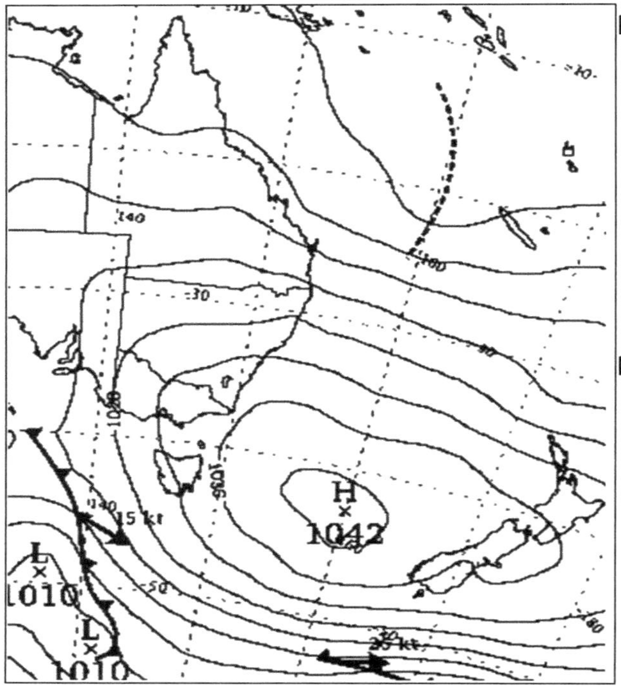

RAIN POTENTIAL

FROST/SNOW

SEPTEMBER **17** WEDNESDAY

A gentle westerly flow and very cold anticyclonic air prevails over the country with severe frosts and scattered fogs in many parts of the country except in Northland and around the Hawkes Bay. Very sunny and mostly dry throughout, with only some isolated showers likely in North Canterbury and in the north of the North Island.

Auckland:	Partly cloudy, mostly dry except for isolated pockets of rain, calm air
Hamilton:	Sunny, dry, calm and cool air with overnight frost likely
Wellington:	Dry, sunny, mostly calm air, chance of frosts in sheltered parts
Greymouth:	Dull and gloomy air, dry, cold overnight, mostly calm air
Christchurch:	Sunny, dry, gentle breezes, frost and fog likely
Dunedin:	Sunny, cool, dry, gentle breezes

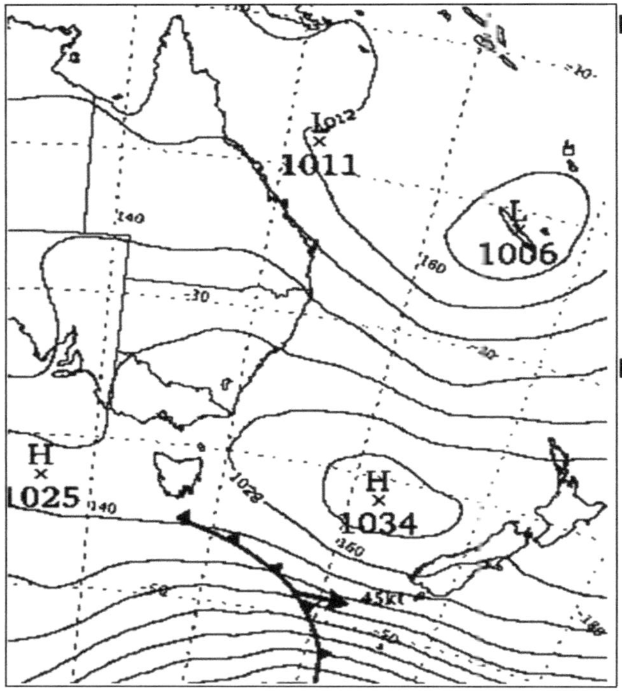

RAIN POTENTIAL

FROST/SNOW

SEPTEMBER **18** THURSDAY

A high pressure ridge maintains a very light westerly air throughout most of the country except for freshening breezes in the southeast of the South Island from a wet front brushing up through Fiordland for a time, bringing some showers into the region. The far north of the North Island may also see some showers and increasing cloud as a depression to the north of the country advances nearer to New Zealand. Mostly dry an elsewhere, with good sunny spells from the East Cape to Canterbury. Frosts in many districts from the Waikato southwards.

Auckland:	Cloudy, calm, spotty showers
Hamilton:	Cloudy, dry, slightly warmer, calm and serene air
Wellington:	Sunny, dry, slightly warmer, light breezes
Greymouth:	Partly cloudy, dry, mild day cold night, unsettled breezes for a time
Christchurch:	Sunny, dry, frosts, light breezes, fog possible
Dunedin:	Cloudy, dry, moderate breezes

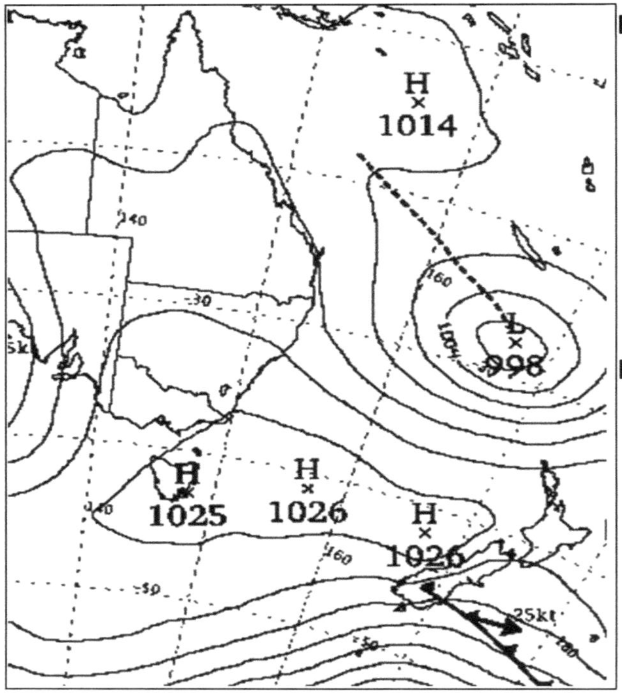

RAIN POTENTIAL

FROST/SNOW

SEPTEMBER 19 FRIDAY

A deep depression north of New Zealand moves closer to the North Island directing very strong and stormy northwest winds across the northern half of the North Island and some heavy rain into Northland. In contrast, an anticyclonic system to the west of New Zealand ridges over the south of the country with very a very light southerly air prevailing, while a trough between two highs brings some isolated showers in Southland and Fiordland. Dry elsewhere. Very cold with widespread frosts over the South Island, while milder over the North Island.

Auckland: Overcast, isolated showers, strong winds
Hamilton: Overcast, dry with increasing threats of rain, strengthening winds
Wellington: Partly cloudy and dry, moderate breezes
Greymouth: Sunny, mostly dry, cool overnight with chance of misty patches
Christchurch: Partly cloudy, warmer, chance of frost patches, dry, gentle breezes
Dunedin: Cloudy, slightly warmer, dry, calm atmosphere

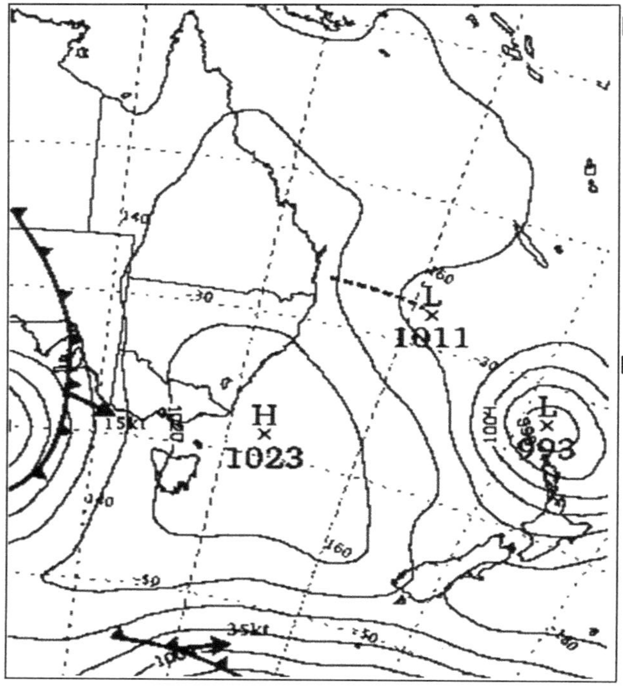

RAIN POTENTIAL

FROST/SNOW

SEPTEMBER 20 SATURDAY

A depression brings further widespread cloud and rain throughout the north and east of the North Island as well as another low pressure trough bringing some spotty showers overnight into Fiordland and Southland. Mostly dry and sunny in the west of the North Island and over the remaining South Island regions. Cold with severe frosts in the South Island, with chance of fog in the east of the South Island, while a strong and blustery south to southeast flow prevails over the North Island.

Auckland: Changeable, mix of sunny spells and rain, blustery, possible gales
Hamilton: Partly cloudy, scattered showers, windy at times
Wellington: Unsettled winds, sunny, brief overnight showers
Greymouth: Sunny, dry, mild day cool night, gentle shifting breezes
Christchurch: Partly cloudy, colder, dry, gentle shifting breezes, possible fog
Dunedin: Partly cloudy, cooler, dry, gentle shifting breezes

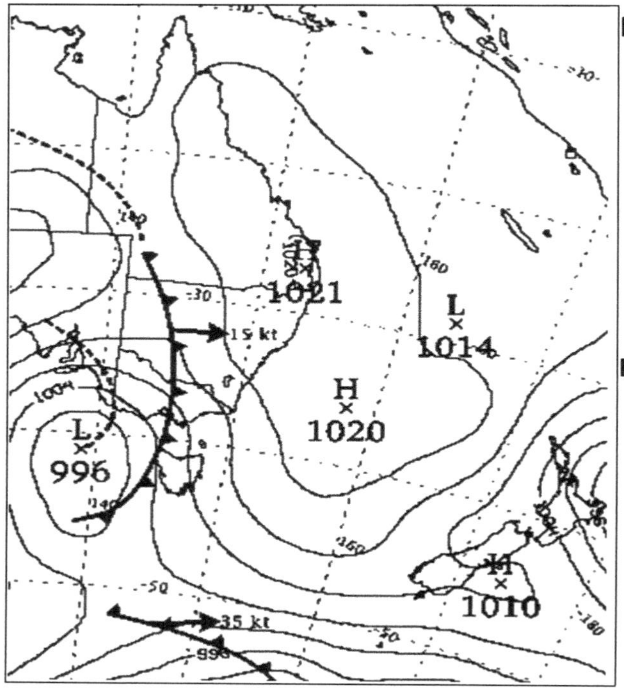

RAIN POTENTIAL

FROST/SNOW

SEPTEMBER 21 SUNDAY

A weak anticyclonic system ridges over the northwest of the country, directing a gentle southwest flow, while low pressure prevails in a fresher westerly flow over the southern parts of the South Island. Some lingering showers prevail overnight in the east of the North Island from the East Cape to Wellington, isolated showers around the northern fringes of the South Island and in Fiordland. Mostly dry with good sunny spells throughout the day. Frosts likely in the south and east of the South Island and in sheltered parts around the Bay of Plenty.

Auckland:	Partly cloudy, mostly dry, winds easing, chance of isolated showers
Hamilton:	Sunny, dry, warmer day cooler night, calm and pleasant air
Wellington:	Moderate breezes, dry, scattered cloudy spells, mild day cold night
Greymouth:	Partly cloudy, light breezes to mostly calm air, dry
Christchurch:	Calm and serene air, sunny, dry, frost and fog likely
Dunedin:	Partly cloudy and dry, light breezes

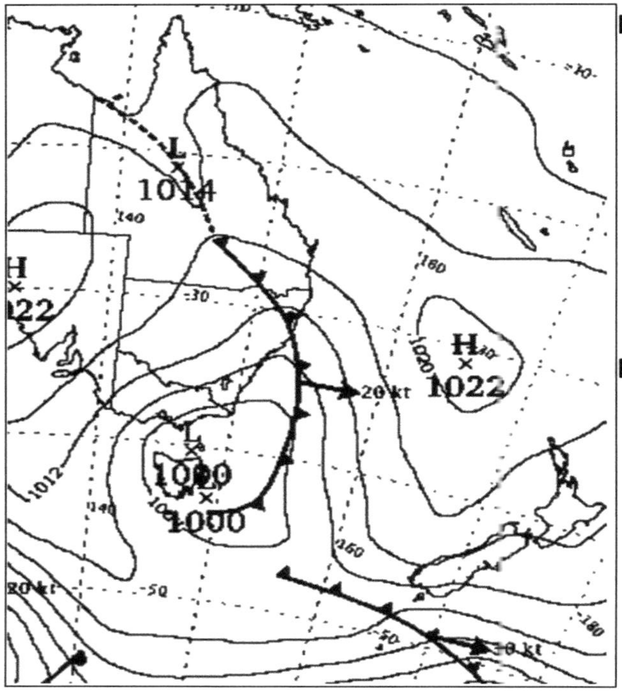

RAIN POTENTIAL

FROST/SNOW

SEPTEMBER 22 MONDAY

High pressure prevails over New Zealand directing a light westerly flow throughout. Mostly dry everywhere with good sunny spells, except for the chance of some isolated showers around Fiordland. Frosts likely in the Waikato to the Central Plateau and East Cape in the North Island and in the most parts of the South Island, except milder in Canterbury where fog patches may be observed.

Auckland:	Partly cloudy, spotty showers, gentle shifting breezes
Hamilton:	Sunny, chance of frost, dry, gentle breezes to mostly calm air
Wellington:	Sunny, calm, dry and pleasant air
Greymouth:	Partly cloudy, dry, mostly calm air
Christchurch:	Sunny, dry, mostly calm air, some frost likely, chance of fog
Dunedin:	Warmer, dry, mostly cloudy, calm air

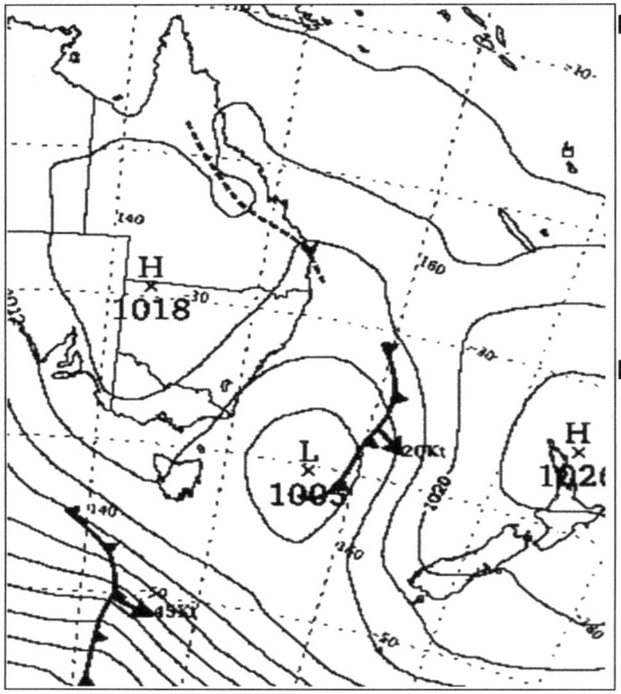

RAIN POTENTIAL

FROST/SNOW

SEPTEMBER 23 TUESDAY

High pressure prevails over the north of New Zealand directing a light northerly air over the North Island, while a fresher north to northwesterly flow prevails over the northern half of the South Island and fresh northwesterly breezes associated with low pressure to the south flows through the southern half of the South Island. Mostly dry throughout, with good sunny spells in the north and east of the North Island and mostly cloudy and dull elsewhere. Some frosts likely in the Waikato, Bay of Plenty, Southland and Otago regions.

Auckland: Dry and mild, occasional sunny spells, light breezes
Hamilton: Cloudy, dry, light shifting breezes to mostly calm and pleasant air
Wellington: Mostly sunny, dry, gentle shifting breezes, warmer
Greymouth: Overcast, dry with increasing of threats of rain, unsettled winds
Christchurch: Cloudy, dry, moderate breezes, chance of frost in sheltered parts
Dunedin: Unsettled breezy spells, dry, frequent cloudy spells

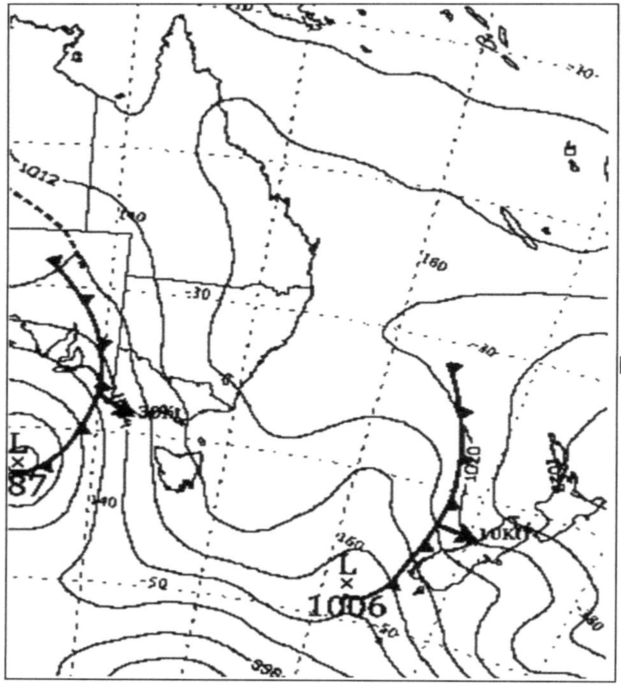

RAIN POTENTIAL

FROST/SNOW

SEPTEMBER 24 WEDNESDAY

The high pressure ridge contracts up over the North Island with a moderate northwesterly flow prevailing, mostly cloudy and dry conditions throughout, except for the chance of some isolated showers briefly in Taranaki. A wet trough associated with the low pressure south of the country however, brings outbreaks of rain mostly overnight throughout the west and north of the South Island, w th heavy falls likely in the northwest districts, but mostly dry in Canterbury, and isolated showers in Southland and Otago. Milder temperatures everywhere.

Auckland: Cloudy, mostly dry, chance of overnight showers, moderate breezes
Hamilton: Cloudy, dry, pleasant air, light breezes
Wellington: Cloudy, dry, breezy at times
Greymouth: Cloudy, warmer, heavy rain, moderate breezes
Christchurch: Warmer, cloudy, dry, gentle breezes
Dunedin: Cloudy, dry, gentle shifting breezes

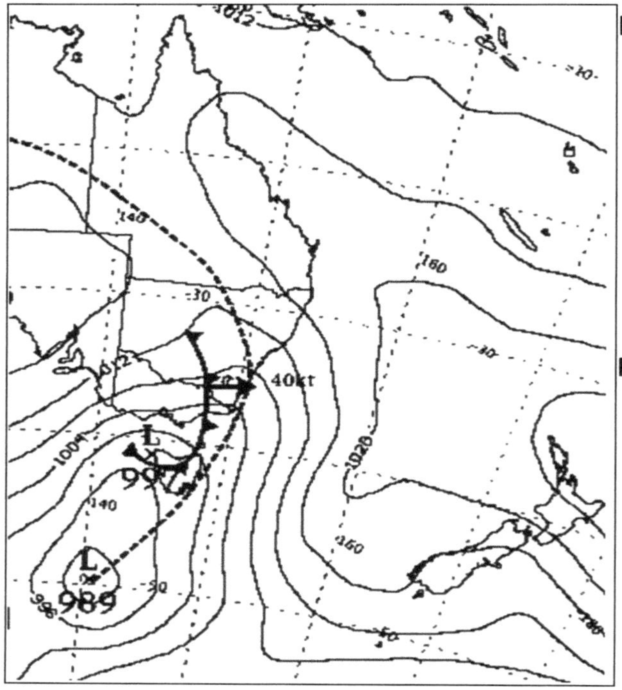

RAIN POTENTIAL

FROST/SNOW

SEPTEMBER **25** THURSDAY

A weak high pressure centre maintains a mostly sunny, calm, and dry day throughout most of the country. However, overnight showers can be expected in the southern half of the North Island and most South Island districts, while Northland to the East Cape, Southland, most of Otago, and North Westland will be mostly dry with the odd passing threat of rain.

Auckland:	Sunny, warmer, dry, calm and serene air
Hamilton:	Partly cloudy, isolated showers, serene air, chance of fog patches
Wellington:	Partly cloudy, light showers, calm air
Greymouth:	Cloudy, isolated showers, moderate breezes
Christchurch:	Sunny, overnight showers, freshening breezes later in the day
Dunedin:	Sunny, cooler, dry, light breezes

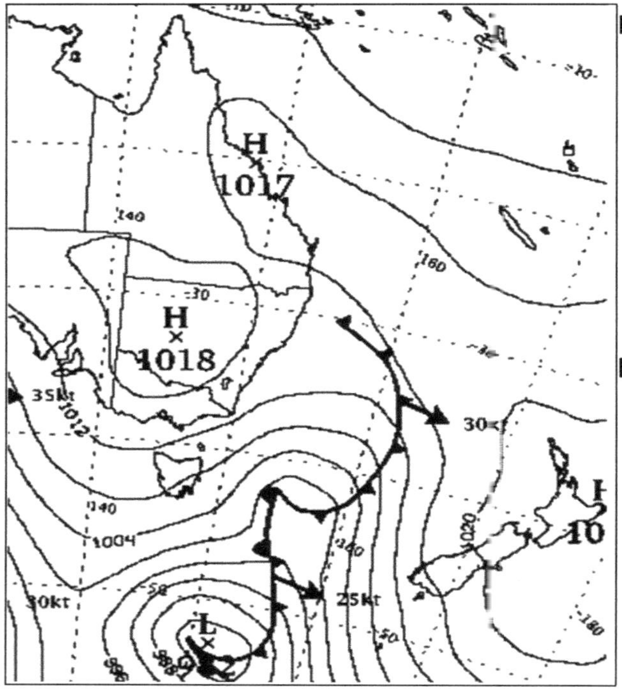

RAIN POTENTIAL

FROST/SNOW

SEPTEMBER **26** FRIDAY

A wet front associated with a Southern Ocean depression passes up through the country in a strengthening westerly flow, bringing scattered rain activity throughout the western half of the South Island, with only the eastern coastal areas likely to remain dry and sunny. A low pressure trough ahead of a weak anticyclonic system to the north of New Zealand also brings widespread cloud and showers down through the North Island from Northland to the Central Plateau, but remaining mostly dry from the East Cape to Wellington.

Auckland:	Cloudy and wet, warmer, light breezes to mostly calm air
Hamilton:	Cloudy, warmer, light showers, mostly calm air
Wellington:	Overcast, dry, mild temperatures, chance of gale winds at times
Greymouth:	Cloudy, rain, windy spells
Christchurch:	Sunny, colder, dry, fresh breezes
Dunedin:	Partly cloudy, dry, gentle breezes

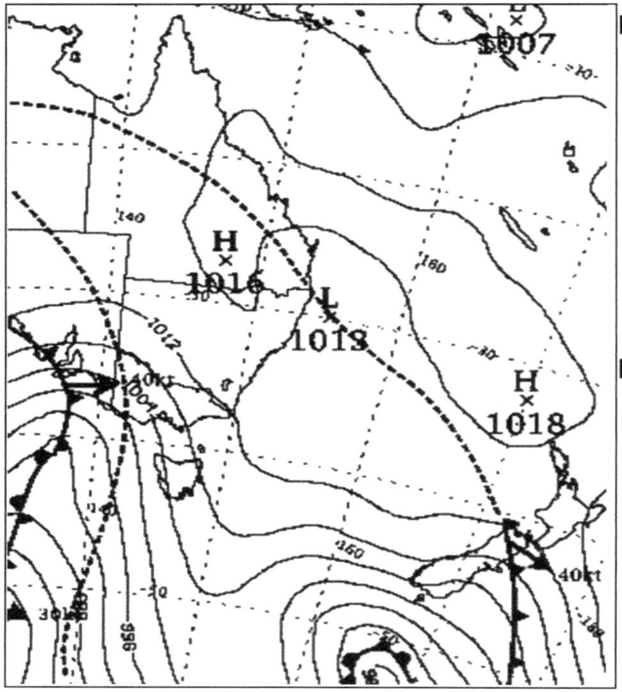

RAIN POTENTIAL

FROST/SNOW

SEPTEMBER **27** SATURDAY

A complex of weak high pressure centres and low pressure centres descend onto New Zealand with a very light northeasterly flow predominately prevailing. Mostly sunny and dry throughout the day, but with overnight drizzly showers in most parts of the country from Northland to the Manawatu, and isolated showers in most parts of the South Island except in Marlborough to central Canterbury.

Auckland: Sunny, pockets of heavy rain, calm air, chance of fog overnight
Hamilton: Cloudy, calm, light showers, patches of fog possible
Wellington: Partly cloudy, dry, light breezes
Greymouth: Cloudy, occasional showers, gentle breezes
Christchurch: Sunny, warmer day cold night, dry, gentle breezes
Dunedin: Sunny, overnight showers, warmer day colder night, gentle breezes

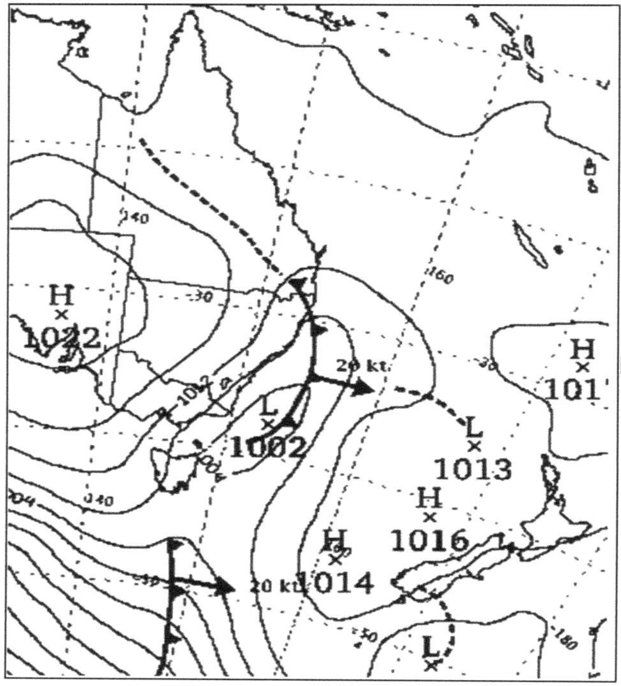

RAIN POTENTIAL

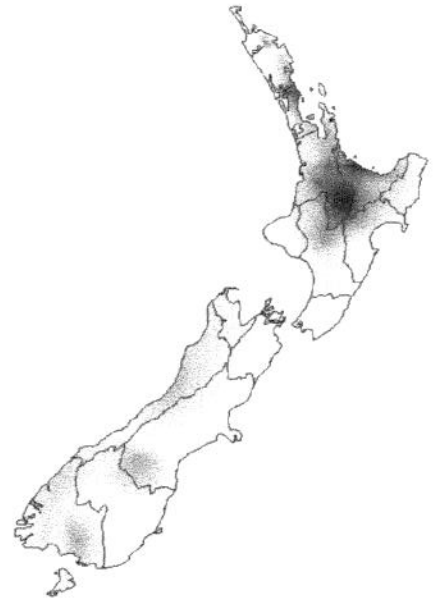

FROST/SNOW

SEPTEMBER **28** SUNDAY

A weak high pressure centre crosses the South Island while a low pressure centre crosses the North Island in a light but changeable air throughout. Mostly cloudy and dry throughout the North Island with overnight drizzly showers in central parts, while mostly sunny and dry in the west of the South Island and cloudy and dry in the east, after the passing of some isolated overnight showers mostly in Fiordland.

Auckland: Cloudy, isolated showers, mostly calm to light shifting breezes
Hamilton: Overcast, isolated showers, gentle shifting breezes
Wellington: Cloudy, dry, gentle shifting breezes, chance of misty patches
Greymouth: Calm and serene air, sunny, dry, cool
Christchurch: Overcast, dry, mostly calm and pleasant air
Dunedin: Partly cloudy, colder, dry, calm and pleasant air

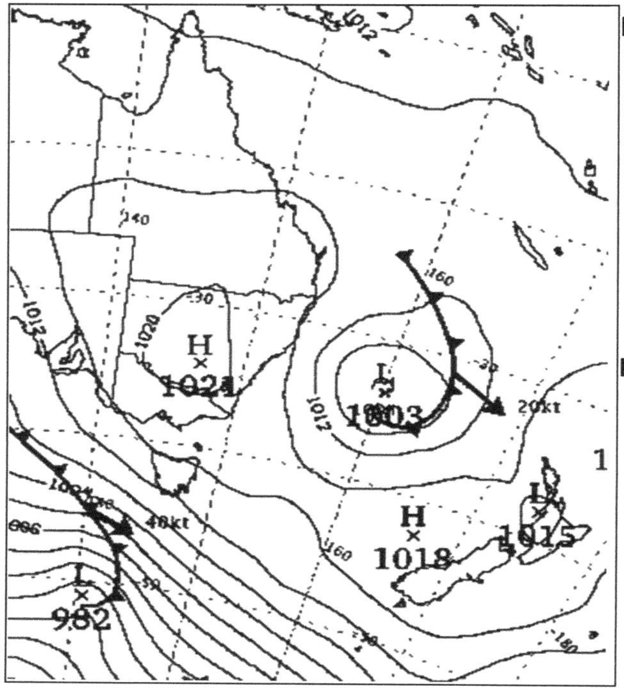

RAIN POTENTIAL

FROST/SNOW

SEPTEMBER **29** MONDAY

A weak depression directs a moderate northeasterly flow and mostly cloudy conditions over the North Island with some showers from the Waikato to the Manawatu as well as around the East Coast. In contrast, a weak high pressure system directs a mostly calm air over the South Island, mostly sunny and dry conditions, with the return of frosts in the southern regions.

Auckland: Partly cloudy, mostly dry, chance of overnight showers, light breezes
Hamilton: Cloudy, dry, cool overnight, mostly calm air
Wellington: Calm and serene air, cooler, partly cloudy, brief showers
Greymouth: Calm and serene air, sunny, dry
Christchurch: Overcast, colder, calm air, dry, fog likely
Dunedin: Sunny, dry, calm and cool atmosphere

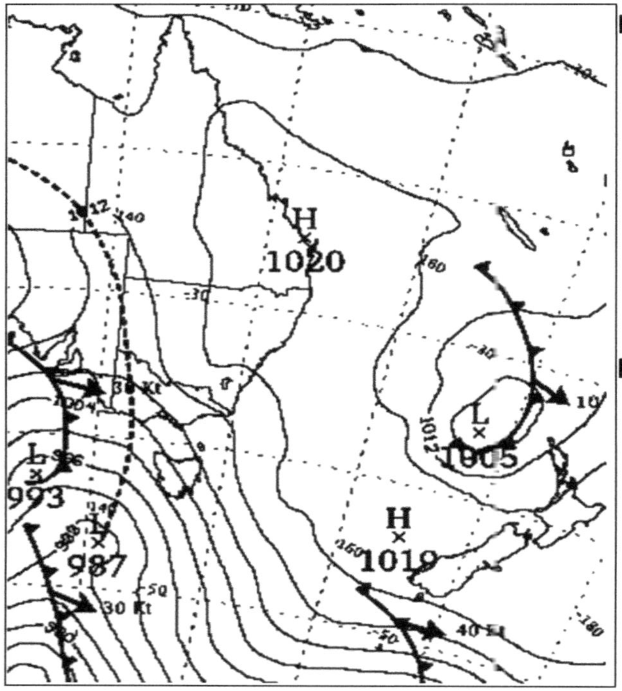

RAIN POTENTIAL

FROST/SNOW

SEPTEMBER **30** TUESDAY

Strong and blustery easterly winds prevail over the northern half of the North Island as a weak depression passes to the north of the country, bringing some heavy rain for a time into Northland districts, while the remaining North Island regions will be mostly dry, sunnier in the southwest and cloudier in the east. Anticyclonic conditions prevail over the South Island in a light northwesterly flow, with dry and mostly sunny conditions everywhere and frosts from Canterbury to Southland.

Auckland: Overcast, isolated showers, fluctuating breezes
Hamilton: Gentle shifting breezes, cloudy, warmer, dry
Wellington: Light fluctuating breezes, sunny, dry
Greymouth: Light shifting breezes, sunny, dry
Christchurch: Partly cloudy, slightly warmer, dry, gentle breezes, chance of fog
Dunedin: Sunny, slightly warmer, dry, light breezes to mostly calm air

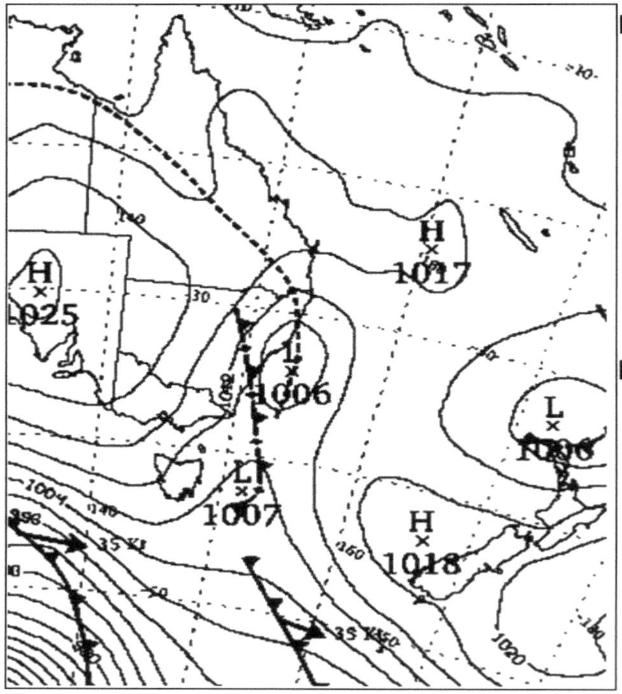

RAIN POTENTIAL

FROST/SNOW

OCTOBER

PHASES OF THE MOON

6th	Crossing Equator
7th	Full Moon
8th	Perigee #6
12th	Northern Declination
13th	Third Quarter
19th	Crossing Equator
21st	New Moon
24th	Apogee
26th	Southern Declination
29th	First Quarter

1	2	3	4	5	6	7	8	9	10	11	12	13	14	15	16	17	18	19	20	21	22	23	24	25	26	27	28	29	30	31
Wed	Thu	Fri	Sat	Sun	Mon	Tue	Wed	Thu	Fri	Sat	Sun	Mon	Tue	Wed	Thu	Fri	Sat	Sun	Mon	Tue	Wed	Thu	Fri	Sat	Sun	Mon	Tue	Wed	Thu	Fri
					XhN	F	P6				^	3Q						XhS		N			A		V			1Q		

MONTHLY SUMMARY

1st-3rd Possible fog patches in Christchurch.
3rd-4th Chance of fog in the Wellington region
4th A thunderstorm-embedded trough crosses Auckland and into the Waikato bringing hail and lightning strikes and the chance of interruption to some power supplies.
8th Chance of hail showers in the lower North Island, including in the Wellington region.
9th Very high rainfall likely in the Fiordland district, including at Milford Sound.
10th Heavy rain in the Bruce Bay to Hokitika area.

Over the next four weeks, the anticyclonic influence of the previous four weeks will be replaced by more frequent depressions tracking to the south of the country. As such, blustery westerly winds will often prevail, frequently to gale force at times when accompanying stormy and colder than typical conditions. There may be up to 7 days or more during this outlook period when winds may reach gale force, and up to a couple of weeks with particularly strong blustery winds prevailing. Rainfall will be up to twice as much as the norm for this time of year in parts of Westland, in the Hanmer district, and parts of north Canterbury, as well as around 25% wetter in many other districts, particularly in the South Island. Only in the far north of Northland, parts of the Bay of Plenty, and Hawke's Bay will rainfall be around 25% below average, while the remainder of the country will be fairly close to the norm for this time of year. While soil moisture levels will recover to good levels in Fiordland, Gisborne, and the Wairarapa, soil moisture levels in Marlborough and Central Otago will be below average for this time of year. It will be sunnier than normal in the north and east of the country, including in Northland, Gisborne, northern parts of Canterbury, in Otago and Southland, with sunshine hours generally typical for this time of year elsewhere. It will be cooler than average over the next four weeks, particularly over much of the South Island, where parts of north Westland and South Island inland districts may be up to 1.5°C or more below the average of this outlook period, with remaining South Island districts to the lower North Island districts cooler by around 0.5°C overall. In contrast, it will be around half a degree warmer or more in Northland, the Coromandel, Gisborne, and Hawke's Bay districts. Overall, the country as a whole will average around half a degree below the norm for this outlook period. Snowfall events are still likely to occur in the South Island over the next four weeks, particularly over the second half of October, and even on the North Island's Desert Road at the end of October.

11th Very strong and blustery northwest winds likely in the south and east of the country, particularly in the Wellington region.
11th-12th Chance of fog in Christchurch.
11th-13th Scattered thunderstorms likely to pass through the northern half of the North Island, particularly through Auckland and into the Waikato districts. Strong blustery winds with possible gale gusts in the Wellington area.
11th-15th Depression crossing the North Island brings widespread rain activity throughout, except for mostly dry in the East Coast-Hawkes Bay region.
11th-18th Period of widespread heavy rain in the West Coast to Fiordland and into the far northwest of the South Island.
12th-13th Chance of hail or light snow and strong wind gusts in Christchurch.
12th-17th Chance of snowfall in many high country areas, including in the Mt Cook district, Arthurs Pass and Tekapo districts, as well as in the North Island Central Plateau region around the Desert Road.
13th Chance of thunderstorms in the North Auckland area, including into the Rodney district with possible lightning strikes causing some localised damage.
14th Unusually cold with sub-zero frosts likely in the Waiouru district. Brief respite from the heavy rain likely in the West Coast.
14th-18th Strong west to northwest winds to gale force at times may be disruptive in the east and south of the South island. Strong winds may cause disruptions at times in the Wairarapa, Wellington, Mt Cook, and Southland districts.
17th Stormy air in the Wellington region with possible gale gusts, thunderstorms and heavy rain that may bring some surface flooding including in the Hutt Valley to Petone area.
18th Heavy rain in North Canterbury, possibly causing the Waiau River to overflow.
19th Unstable winds in the Dargaville area may develop funnel formations. Chance of misty patches in Christchurch.
20th Heavy rain in the Arthurs Pass area likely.
20th-21st Chance of heavy rain in the top of the South Island and into Canterbury as well as snowfall in many high country areas of the South Island, including the Arthurs Pass area.
20th-28th Spell of scattered rain throughout most of the North Island.
22nd Chance of heavy rain in central North Island districts that may bring some brief surface flooding.
23rd Very strong and blustery northwest winds likely in the east and south of both islands, particularly in the Wellington area.

24th-25th Chance of snowfall in many high country areas of the South Island. More snowfall likely in the Arthurs Pass area again. Blustery winds persist in the Wellington region.

24th-27th Further heavy rain likely in the west of the South Island from Fiordland to Bruce Bay.

26th-27th Very strong and blustery northwest winds likely in the south and east of the country, particularly in the Wairarapa.

27th Heavy rain in the central North Island likely, particularly in the Ohakune area.

27th-28th Chance of snowfall in the Southern Alps and southern Canterbury high country including around Tekapo and Arthurs Pass again, with possible flurries as far to the east as Fairlie. Snow may also fall later in the North Island Central Plateau around the Desert Road. Chance of heavy rain also in the central North Island, particularly in the Ohakune region.

30th-31st Very strong and blustery northwest winds likely in the Wairarapa to Wellington regions and in the southern parts of the South Island.

October 2025

Monday	Tuesday	Wednesday	Thursday	Friday	Saturday	Sunday
29	30 **Moon Sign**	**1** **Capricorn** Square Mercury Trine Venus Opposition Jupiter	**2** **Capricorn** Square Mars Sextile Saturn Trine Uranus Conjunct Pluto	**3** **Aquarius** Trine Sun	**4** **Aquarius** Trine Mercury Square Uranus	**5** **Pisces** Opposition Venus Trine Mars
6 **XhN** **Pisces** Opposition Venus Trine Jupiter Conjunct Saturn Conjunct Neptune	**7** **FM** **Aries** Opposition Sun	**8** **Aries** Opposition Mercury Square Jupiter Square Pluto	**9** **P6** **Taurus** Opposition Mars	**10** **Taurus** Trine Venus Sextile Saturn Conjunct Uranus Sextile Neptune Trine Pluto Square Ascendant	**11** **Gemini** Trine Sun Square Ascendant	**12** **^** **Gemini** Square Venus Square Saturn Square Uranus Square Neptune Square Ascendant
13 **Cancer** Square Sun Trine Mercury Trine Mars Square Ascendant	**14** **3Q** **Cancer** Sextile Venus Conjunct Jupiter Trine Saturn Sextile Uranus Trine Neptune Opposition Pluto Square Ascendant	**15** **Leo** Square Mercury Square Mars Square Ascendant	**16** **Leo** Sextile Sun Square Uranus	**17** **Virgo**	**18** **Virgo** Sextile Mars Sextile Jupiter Opposition Saturn	**19** **XhS** **Libra** Conjunct Venus Opposition Saturn Trine Uranus Opposition Neptune Trine Pluto
20 **Libra** Conjunction Venus	**21** **Libra** Conjunct Venus	**22** **NM** **Libra** Conjunct Sun Square Jupiter Square Pluto	**23** **Scorpio**	**24** **A** **Scorpio** Conjunct Mercury Conjunct Mars Trine Jupiter Trine Saturn Opposition Uranus Trine Neptune	**25** **Sagittarius** Opposition Uranus Trine Neptune	**26** **Sagittarius** Sextile Venus
27 **V** **Sagittarius** Sextile Sun Square Saturn Square Neptune Square Ascendant	**28** **Capricorn** Square Venus	**29** **Capricorn** Sextile Mars Opposition Jupiter Sextile Saturn Trine Uranus Conjunct Pluto	**30** **1Q** **Aquarius** Conjunct Pluto	**31** **Aquarius** Trine Venus Square Mars	1	2
3	4					

NORTH ISLAND RAINFALL

OCTOBER 2025

d=mainly dry, n= fog, mist, dew or nonrecordable, l=light showers, p=passing showers, s=significant showers, r=rain, h=heavy falls

		Northland							Auckland					Bay Of Plenty				Waikato				Rot/Tau		Hawkes Bay			C Plateau			Taranaki			Manawatu			Lower No-th Island					
		CR	Kta	Ker	Rsl	Dar	Whg	Man	Wark	Hen	MkH	Ak	Co	Tau	TeP	Wkt	Opo	Hml	TeA	Mata	Ot	Rt	Taup	Gis	Nap	Has	Taum	Chat	Waio	NP	Str	Wang	Pa	Dann	Lev	PN	Mast	UH	Wgt	all	Moon
1st	Wed	s	l	r	r	h	h	h	s	l	d	d	d	d	d	d	d	d	d	d	d	d	d	d	l	d	d	d	d	d	d	d	d	d	d	d	n	d	d	142	
2nd	Thur	d	l	l	d	d	l	l	l	d	d	d	r	l	n	d	n	d	d	d	d	d	l	l	d	d	d	d	d	d	d	d	d	d	d	d	d	d	d	27	
3rd	Fri	l	r	l	l	s	s	d	d	d	d	d	l	d	l	l	s	d	s	d	d	s	l	s	d	d	l	d	d	l	l	d	d	d	d	d	d	d	d	62	
4th	Sat	d	r	l	l	s	s	s	r	r	h	r	d	l	s	s	s	s	s	d	h	r	s	d	s	l	h	h	h	s	r	h	s	r	r	h	h	r	s	385	
5th	Sun	l	r	l	l	s	l	d	s	s	l	s	d	d	d	l	h	s	l	h	d	h	r	s	s	r	s	s	s	s	r	s	s	r	s	s	r	r	r	246	
6th	Mon	l	l	d	d	l	l	s	l	n	d	l	h	s	s	n	s	s	l	d	l	l	l	r	l	l	d	d	l	d	d	l	l	s	r	s	h	s	s	145	XhN
7th	Tue	d	d	d	d	d	n	l	d	d	d	d	d	d	d	d	d	d	d	d	d	d	d	l	d	d	d	d	d	d	d	d	d	l	d	d	l	s	s	15	F
8th	Wed	d	d	d	d	d	d	d	d	d	d	d	d	d	d	d	d	d	d	d	d	d	d	l	l	d	d	d	d	d	d	d	n	n	d	d	d	l	l	7	
9th	Thur	d	d	d	d	d	d	d	d	d	d	d	d	d	d	d	d	d	d	d	d	d	d	d	d	d	d	d	d	d	d	d	d	d	d	d	d	d	d	0	P6
10th	Fri	d	l	d	l	d	l	l	d	d	d	d	l	d	n	s	d	d	d	d	d	n	d	d	d	d	d	d	d	l	l	d	d	d	d	d	l	d	n	18	
11th	Sat	l	s	r	r	r	s	r	r	r	s	s	r	s	r	s	s	l	s	s	l	s	l	d	d	d	r	r	n	l	r	l	s	l	r	s	s	r	l	259	
12th	Sun	l	d	r	s	s	s	s	r	h	s	r	h	s	s	l	s	h	r	r	h	s	s	d	s	s	h	h	h	r	h	s	l	s	l	s	d	r	d	417	^
13th	Mon	l	d	s	l	r	s	s	h	h	r	s	h	r	r	l	l	r	r	r	h	s	s	d	d	d	r	h	r	r	r	l	l	s	l	l	s	r	l	319	
14th	Tue	d	d	l	d	n	l	s	n	h	l	l	s	l	n	d	l	l	l	l	l	n	l	d	d	l	l	r	l	l	d	d	l	d	d	d	l	l	l	71	3Q
15th	Wed	l	d	l	n	s	l	l	l	s	s	l	s	l	l	l	l	s	s	l	r	l	l	d	d	d	h	r	s	r	h	s	r	s	s	s	l	h	s	217	
16th	Thur	d	d	d	d	d	d	d	n	d	d	d	d	d	d	d	d	d	d	d	l	d	d	d	d	d	l	s	d	d	l	d	d	l	n	d	d	l	d	14	
17th	Fri	d	d	d	d	d	d	d	d	d	d	d	d	d	d	d	d	d	d	d	s	d	d	d	d	d	d	l	d	d	d	d	d	d	d	d	d	n	n	8	
18th	Sat	d	d	d	n	d	d	d	d	d	d	d	d	d	d	d	d	d	d	d	d	d	d	d	d	d	d	l	d	d	s	d	h	d	l	d	l	h	h	109	
19th	Sun	d	d	n	l	d	d	d	d	d	d	d	d	d	d	d	d	n	l	s	s	l	n	d	d	d	s	h	l	l	h	r	s	s	s	s	n	h	s	134	XhS
20th	Mon	d	d	l	l	d	d	d	d	l	l	d	d	l	s	l	l	l	l	l	l	s	l	l	d	d	s	s	l	s	s	l	h	d	l	d	h	h	h	164	
21st	Tue	r	d	s	l	s	l	s	s	s	s	s	h	s	h	r	r	r	h	r	h	r	s	h	l	l	h	h	s	h	h	s	r	l	r	s	r	h	h	433	
22nd	Wed	d	d	l	d	l	d	l	s	r	s	s	l	l	l	l	l	s	r	s	h	s	s	n	d	d	h	h	s	l	r	s	h	r	s	s	d	r	l	261	N
23rd	Thur	d	d	d	d	d	d	d	d	n	l	n	l	d	d	d	d	l	l	n	l	n	d	d	d	d	l	h	r	l	d	l	d	l	l	d	d	l	l	55	
24th	Fri	l	d	h	r	r	s	r	r	h	r	r	h	h	h	r	h	s	r	r	h	h	s	l	l	l	h	h	r	r	h	s	h	s	h	s	s	r	s	468	A
25th	Sat	s	d	d	d	s	d	l	s	l	l	l	l	d	l	d	l	s	s	s	s	l	l	d	d	d	s	r	s	s	s	s	d	r	d	l	d	s	d	111	
26th	Sun	l	d	l	d	l	l	l	s	l	l	s	r	l	l	l	l	l	s	l	s	l	s	l	l	l	r	h	r	s	r	s	l	s	s	s	l	s	s	175	
27th	Mon	d	d	d	d	d	d	d	d	d	d	d	d	d	d	d	d	d	l	d	n	d	d	d	d	d	l	h	l	d	h	r	n	d	s	r	s	s	l	85	V
28th	Tue	s	d	l	s	r	r	r	r	h	r	r	r	r	r	r	r	s	s	d	r	r	s	l	r	r	h	h	h	r	h	h	s	h	s	r	r	l	r	459	
29th	Wed	s	d	l	d	s	s	l	r	s	h	l	d	r	r	l	r	s	r	d	h	r	l	h	r	r	d	l	s	d	l	s	d	s	d	n	s	d	l	250	
30th	Thur	d	d	d	d	d	d	d	d	d	d	d	d	d	s	d	d	d	d	d	d	d	d	l	d	d	d	d	d	d	d	d	d	d	d	d	d	d	d	10	1Q
31st	Fri	d	d	d	d	d	d	d	l	l	d	d	d	d	d	d	d	l	l	d	l	d	l	d	n	d	l	r	l	s	l	s	l	d	l	l	d	l	d	41	
estimate:		52	45	84	62	123	113	115	119	183	123	89	211	87	137	77	120	108	135	114	185	125	73	111	45	51	234	389	199	116	254	134	164	131	109	97	155	269	173	5108	
average:		64	110	136	99	96	172	116	118	88	79	92	91	115	134	100	97	97	105	114	127	125	105	65	55	47	140	260	90	137	214	80	102	92	101	95	77	126	102	4163	
trend:		drr	drr	drr	drr	wtr	drr	av	av	wtr	wtr	av	wtr	drr	av	drr	wtr	wtr	wtr	av	wtr	av	drr	wtr	av	av	wtr	wtr	wtr	drr	wtr	wtr	wtr	wtr	av	av	wtr	wtr	wtr	wtr	

SOUTH ISLAND RAIN

OCTOBER 2025

d=mainly dry, n= fog, mist, dew or nonrecordable, l=light showers, p=passing showers, s=significant showers, r=rain, h=heavy falls

		West Coast				Top South Island				Canterbury				South Canterbury				Coastal Otago			Lakes			Inland Otago				Southland						
		West	Grey	Hok	BB	Ta	Mou	Nel	Ble	Dar	Ham	Kai	Ch	Rak	Meth	Ash	Tim	Oam	Dun	Bal	Twi	LP	LT	Crom	Qstn	Ran	Alex	TeA	Wtn	Gore	Inv	Puy Pt	all	Moon
1st	Wed	d	d	d	d	d	d	d	d	d	d	d	d	d	d	d	d	d	d	d	d	d	d	d	d	d	d	d	d	d	d	d	0	
2nd	Thur	d	d	d	l	d	d	d	d	d	d	d	d	d	d	d	d	d	d	d	d	d	d	d	d	d	d	d	l	d	l	l	4	
3rd	Fri	r	r	r	h	s	l	l	d	l	d	d	d	d	s	s	s	s	h	d	d	d	r	s	h	h	s	r	s	s	l	r	243	
4th	Sat	s	r	h	h	h	h	r	l	r	r	l	s	r	r	r	r	l	l	d	r	h	h	d	l	l	l	d	d	l	d	d	305	
5th	Sun	s	d	d	d	s	l	s	s	l	l	s	l	d	l	l	l	d	d	d	r	d	d	d	d	d	d	d	d	d	l	d	61	
6th	Mon	d	d	d	d	r	l	l	d	s	s	l	l	s	s	l	d	n	s	n	d	d	d	d	d	d	d	d	l	d	n	d	50	XhN
7th	Tue	l	d	d	d	s	d	d	d	s	d	l	l	d	l	l	d	d	d	d	d	d	d	d	d	d	d	d	d	d	d	d	16	F
8th	Wed	d	d	d	d	d	d	d	d	d	d	d	d	d	d	d	d	d	d	d	d	d	d	d	d	d	d	d	d	d	d	d	0	
9th	Thur	d	d	l	d	l	d	d	d	d	d	d	d	d	d	d	d	d	d	d	d	d	d	d	d	d	d	d	d	d	d	d	2	P6
10th	Fri	r	s	h	h	h	l	l	d	d	d	d	d	d	d	d	d	d	d	n	d	h	s	s	h	d	l	h	l	l	l	h	252	
11th	Sat	h	h	h	h	h	h	r	r	d	d	d	d	d	d	d	d	d	l	d	h	h	r	l	s	d	n	r	s	l	s	r	301	
12th	Sun	h	h	h	h	h	r	s	s	d	l	d	d	d	d	s	d	l	l	d	h	s	l	s	s	d	l	r	d	d	n	h	228	^
13th	Mon	r	h	h	s	s	s	s	l	s	s	l	r	s	s	r	s	r	h	s	l	l	l	s	r	s	s	r	r	r	r	r	272	
14th	Tue	d	d	d	d	n	d	l	d	d	l	n	s	d	d	d	d	d	n	d	s	d	d	d	l	d	d	l	d	d	l	s	25	3Q
15th	Wed	h	h	h	h	h	r	s	r	d	h	d	d	d	l	d	l	d	d	d	d	h	r	l	r	l	d	s	l	l	s	h	323	
16th	Thur	s	h	h	h	r	l	s	l	s	h	d	d	s	s	d	n	d	d	s	h	r	r	l	h	d	d	h	r	s	r	h	284	
17th	Fri	h	h	h	h	s	n	d	d	l	r	l	l	l	l	l	d	d	l	l	h	h	s	s	s	l	l	h	s	l	r	h	259	
18th	Sat	h	h	h	l	h	h	r	s	r	h	r	r	l	s	d	d	d	d	l	h	d	d	d	d	d	d	d	l	l	l	l	284	
19th	Sun	r	s	s	l	d	d	n	n	s	r	l	s	l	l	d	d	d	d	l	d	d	d	d	d	d	d	n	d	l	l	s	61	XI-S
20th	Mon	s	d	d	d	h	h	h	h	n	r	r	d	d	d	d	d	d	d	d	d	d	n	d	d	d	d	d	d	d	d	n	127	
21st	Tue	s	r	s	n	h	h	h	h	h	h	h	h	h	l	h	r	r	s	d	d	s	l	l	d	l	l	d	d	d	d	d	382	
22nd	Wed	h	l	d	l	s	s	l	c	s	s	d	l	l	l	l	l	r	h	r	l	d	s	n	d	l	d	d	d	n	d	d	99	M
23rd	Thur	s	l	l	d	d	d	d	c	d	d	d	d	l	d	n	d	d	d	d	d	d	d	d	d	d	d	d	d	d	d	d	10	
24th	Fri	h	h	h	h	h	h	r	s	d	d	d	d	d	d	l	d	d	s	r	d	d	d	s	r	s	l	h	h	h	h	h	433	A
25th	Sat	s	s	h	h	s	r	n	c	s	r	s	l	d	d	s	d	d	s	s	l	d	d	d	l	l	d	d	l	s	l	l	108	
26th	Sun	h	h	h	r	s	s	l	c	l	d	l	d	d	d	l	d	d	d	l	d	d	d	d	l	n	d	l	l	l	l	s	124	
27th	Mon	r	h	h	h	h	s	l	s	l	r	d	d	l	l	l	s	d	r	s	d	h	r	r	h	l	r	h	s	s	s	h	302	V
28th	Tue	l	s	l	h	d	s	l	r	r	s	s	l	r	r	h	l	l	s	s	h	d	l	l	s	s	d	s	s	l	d	l	181	
29th	Wed	r	d	d	d	l	d	d	c	l	l	l	l	s	l	l	n	d	d	d	d	d	l	d	d	d	d	d	d	d	d	l	33	
30th	Thur	n	l	l	s	d	d	d	c	d	d	d	d	d	d	d	d	d	d	d	d	d	d	l	r	d	l	r	l	d	l	h	62	1Q
31st	Fri	h	h	h	h	l	n	n	c	d	d	d	d	d	d	d	d	d	l	s	d	r	l	d	l	d	d	r	l	l	r	s	152	
estimate:		382	342	423	401	421	257	141	110	111	211	78	75	83	69	93	56	48	102	65	180	217	113	56	160	45	36	194	97	79	103	236	4981	
average:		202	240	272	461	247	119	87	73	69	107	63	45	88	91	59	52	52	75	66	60	89	53	39	75	40	30	118	76	83	87	173	3388	
trend:		w tr	w tr	w tr	drr	w tr	w tr	w tr	w tr	w tr	w tr	w tr	w tr	av	drr	w tr	av	av	w tr	av	w tr	w tr	w tr	w tr	w tr	av	av	w tr	w tr	av	w tr	w tr	w tr	

NORTH ISLAND SUNSHINE ESTIMATES
OCTOBER 2025

F=fine (8-12 hours of sunshine), pc= partly cloudy (4-7 hours), c=cloudy (1-3 hours), o=overcast (0 hours)

		Kta	Whg	Dar	Ak	Hml	Tau	Rt	Taum	NP	Wang	Taup	Gis	Nap	Wai	Dann	PN	Mast	Pa	UH	Wgt	ALL	Moon
1st	Wed	pc	c	F	F	c	c	pc	F	F	F	pc	c	pc	pc	F	F	F	F	F	F	136	
2nd	Thur	c	c	pc	pc	o	c	o	o	F	pc	o	o	pc	c	pc	pc	pc	pc	pc	F	83	
3rd	Fri	pc	c	c	c	c	c	c	o	c	pc	c	F	c	pc	c	pc	c	pc	pc	c	68	
4th	Sat	F	F	c	c	c	pc	pc	c	pc	c	pc	c	F	pc	o	c	o	pc	o	o	73	
5th	Sun	pc	pc	pc	pc	o	c	c	pc	F	o	c	c	pc	o	o	o	o	o	o	o	55	
6th	Mon	F	F	F	F	F	pc	F	F	pc	F	F	c	o	pc	o	F	o	F	c	pc	133	XhN
7th	Tue	F	F	F	F	F	F	F	F	F	pc	F	c	pc	pc	F	pc	F	F	c	pc	167	F
8th	Wed	F	F	F	F	pc	F	F	F	F	F	F	c	pc	pc	F	F	F	F	F	F	193	
9th	Thur	pc	pc	F	F	o	c	o	pc	F	F	o	F	pc	F	F	F	F	F	pc	F	135	P6
10th	Fri	c	c	c	c	o	c	c	o	c	pc	c	pc	F	pc	o	pc	o	o	o	o	51	
11th	Sat	c	pc	c	c	pc	c	pc	c	o	c	pc	pc	pc	pc	pc	c	pc	o	o	c	62	
12th	Sun	pc	F	pc	pc	pc	pc	pc	c	pc	pc	pc	F	F	F	pc	pc	pc	pc	pc	F	124	^
13th	Mon	F	F	F	F	F	F	F	pc	pc	pc	F	pc	F	pc	pc	pc	pc	F	pc	pc	141	
14th	Tue	F	F	F	F	F	F	pc	F	pc	pc	pc	F	pc	pc	pc	pc	pc	c	c	c	140	3Q
15th	Wed	F	F	F	F	F	F	F	pc	F	F	F	F	F	F	F	F	F	F	F	F	194	
16th	Thur	F	F	F	F	F	F	pc	pc	F	F	pc	F	F	F	F	F	F	F	F	F	192	
17th	Fri	F	F	pc	pc	F	F	F	F	c	pc	F	c	F	F	pc	pc	pc	o	o	c	131	
18th	Sat	pc	F	pc	pc	pc	pc	pc	o	o	o	pc	pc	F	pc	pc	o	pc	o	o	o	77	
19th	Sun	c	o	c	c	c	c	o	c	c	pc	o	c	c	pc	pc	pc	pc	F	F	F	69	XhS
20th	Mon	o	o	o	o	o	o	o	o	o	o	o	c	o	o	o	o	o	o	o	o	7	
21st	Tue	F	pc	pc	pc	pc	pc	pc	pc	pc	c	pc	pc	pc	pc	o	c	o	pc	c	o	91	
22nd	Wed	F	pc	pc	pc	F	pc	c	c	c	o	c	pc	F	pc	pc	o	pc	o	o	c	85	N
23rd	Thur	c	pc	c	c	pc	F	pc	o	pc	c	pc	F	F	F	F	c	F	c	pc	F	108	
24th	Fri	F	F	F	F	pc	F	F	pc	F	pc	F	F	F	F	F	pc	F	F	F	F	179	A
25th	Sat	pc	F	pc	pc	pc	pc	pc	c	c	c	pc	F	pc	pc	pc	c	pc	c	c	pc	99	
26th	Sun	F	F	pc	pc	pc	pc	c	pc	c	c	c	F	F	F	pc	c	pc	pc	pc	pc	112	
27th	Mon	pc	pc	c	c	F	pc	F	c	o	o	F	F	c	c	F	o	F	F	pc	pc	108	V
28th	Tue	F	pc	pc	pc	pc	pc	pc	c	c	o	pc	c	c	c	o	o	o	c	o	o	54	
29th	Wed	F	F	F	F	F	F	F	F	F	pc	F	c	c	pc	pc	pc	pc	F	F	F	179	
30th	Thur	F	pc	pc	pc	F	F	pc	c	c	o	pc	F	F	F	pc	o	pc	c	c	F	109	1Q
31st	Fri	F	pc	pc	pc	F	F	F	F	pc	F	F	F	F	F	F	F	F	F	F	F	205	
est. hours:		220	214	191	191	174	202	178	144	162	141	178	206	211	206	162	141	162	171	136	170	3559	
av. hours:		183	175	165	180	175	201	186	146	192	182	183	211	207	194	167	147	185	178	186	193	3636	
trend:		more	more	more	av	av	av	av	av	less	less	av	av	av	av	av	av	less	av	less	less	average	

SOUTH ISLAND SUNSHINE ESTIMATES
OCTOBER 2025

F=fine (8-12 hours of sunshine), pc= partly cloudy (4-7 hours), c=cloudy (1-3 hours), o=overcast (0 hours)

		Ta	Nel	Ble	Mou	West	Grey	Hok	Ham	Kai	Ch	Ash	Tim	Crom	Qstn	Bal	Dun	Wnt	Gore	Inv	Alex	ALL	Moon
1st	Wed	o	F	F	F	pc	F	F	pc	pc	pc	c	c	pc	c	c	pc	o	o	pc	pc	98	
2nd	Thur	o	pc	F	pc	c	pc	c	pc	pc	F	F	o	o	c	o	o	o	o	o	o	66	
3rd	Fri	o	o	c	o	c	c	o	o	o	o	o	o	o	o	o	o	o	o	c	o	10	
4th	Sat	F	c	o	o	c	pc	pc	o	o	c	o	c	c	c	o	c	c	c	c	c	43	
5th	Sun	pc	o	o	o	F	F	F	o	o	pc	pc	c	pc	c	pc	o	c	c	c	pc	66	
6th	Mon	o	pc	pc	c	F	F	F	o	o	o	c	pc	F	pc	pc	pc	pc	pc	pc	F	108	XhN
7th	Tue	o	F	F	F	F	F	F	F	F	F	F	F	F	F	F	F	F	F	F	F	197	F
8th	Wed	c	F	F	F	pc	F	F	F	F	F	F	F	F	pc	pc	F	pc	pc	F	F	165	
9th	Thur	pc	c	pc	c	c	pc	c	pc	pc	pc	pc	c	c	c	o	o	c	c	c	c	65	P6
10th	Fri	pc	o	o	o	o	o	o	pc	pc	pc	pc	c	pc	pc	pc	c	pc	pc	pc	pc	68	
11th	Sat	c	o	c	o	o	o	c	c	c	c	c	c	pc	pc	pc	pc	pc	pc	pc	pc	69	
12th	Sun	F	pc	F	pc	o	c	c	pc	pc	pc	pc	pc	pc	c	c	pc	o	o	o	pc	86	^
13th	Mon	F	F	pc	F	F	F	F	pc	pc	pc	F	F	F	F	F	pc	F	F	F	F	178	
14th	Tue	c	pc	pc	pc	c	o	o	c	c	pc	F	pc	c	o	pc	pc	pc	pc	pc	c	75	3Q
15th	Wed	F	F	F	F	c	pc	pc	F	F	F	F	F	pc	c	pc	pc	F	F	pc	pc	159	
16th	Thur	pc	F	F	F	F	F	F	F	F	F	F	F	pc	pc	pc	pc	pc	pc	F	pc	178	
17th	Fri	o	o	pc	pc	o	o	o	pc	pc	pc	F	F	F	F	c	pc	c	c	pc	F	97	
18th	Sat	c	pc	pc	pc	pc	pc	pc	c	c	pc	pc	pc	F	F	F	F	F	F	F	F	133	
19th	Sun	F	pc	F	pc	pc	pc	F	pc	pc	pc	pc	pc	F	F	F	F	F	F	F	F	162	XhS
20th	Mon	pc	o	o	o	o	o	o	o	o	o	c	c	pc	F	F	F	F	F	F	pc	80	
21st	Tue	pc	c	c	c	pc	pc	c	o	o	o	o	o	o	o	o	o	o	o	o	o	23	
22nd	Wed	pc	pc	c	F	o	c	c	o	o	o	o	pc	pc	pc	o	o	c	c	F	pc	67	N
23rd	Thur	pc	pc	F	pc	c	o	o	pc	pc	pc	pc	o	pc	c	c	pc	o	o	o	pc	77	
24th	Fri	F	F	F	pc	o	c	c	o	o	o	c	c	pc	pc	pc	c	F	F	F	pc	93	A
25th	Sat	c	F	F	pc	o	c	c	pc	pc	c	pc	c	c	c	o	c	c	c	c	c	69	
26th	Sun	F	pc	F	pc	o	o	o	F	F	F	F	F	pc	o	c	pc	pc	pc	pc	pc	121	
27th	Mon	F	F	F	F	pc	pc	F	F	F	F	F	pc	c	o	pc	c	c	c	c	c	127	V
28th	Tue	F	c	pc	c	pc	pc	pc	pc	pc	pc	pc	F	F	F	pc	F	F	F	F	F	144	
29th	Wed	F	F	F	F	pc	F	F	F	F	F	F	F	pc	pc	c	F	c	c	o	pc	171	
30th	Thur	F	F	F	F	c	pc	pc	F	F	F	F	F	F	pc	F	pc	pc	pc	pc	F	163	1Q
31st	Fri	F	F	F	F	c	c	pc	F	F	F	F	F	F	F	pc	F	c	c	c	F	157	
est. hours:		170	191	214	185	117	147	149	171	171	185	190	172	185	142	134	151	144	144	165	185	3312	
av. hours:		201	212	225	211	159	144	158	173	200	198	171	177	214	189	163	146	161	152	155	189	3598	
trend:		less	av	av	less	less	av	av	av	less	av	more	av	less	less	less	av	less	av	av	av	average	

OCTOBER **1** WEDNESDAY

An anticyclonic system departs New Zealand to the east in a gentle easterly flow over the North Island, backing northwesterly over the South Island, while a weak depression in the northern Tasman Sea extends a low pressure trough onto the north of the country bringing some heavy rain into Northland, while dry elsewhere. Good sunny spells in most districts except for mostly dull and gloomy conditions in Southland, Otago, and around the East Cape. Particularly milder around Taranaki, while some frosts still likely in the east and south of the South Island.

Auckland: Sunny, mostly dry, gentle breezes, chance of brief morning showers
Hamilton: Cloudy, dry, calm and pleasant air
Wellington: Sunny, dry, warmer day cooler night, calm air
Greymouth: Calm and pleasant air, warmer, mostly sunny and dry
Christchurch: Partly cloudy, slightly warmer, dry, calm air, fog likely
Dunedin: Partly cloudy, dry and calm air

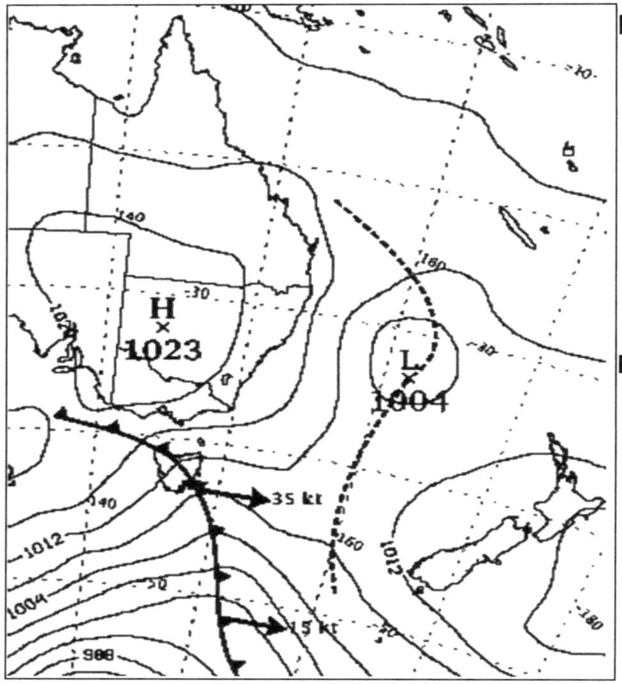

RAIN POTENTIAL

FROST/SNOW

OCTOBER **2** THURSDAY

A large high over the southeast of Australia extends a high pressure ridge over the Tasman Sea while a low pressure trough to crosses the country in a light to moderate northerly flow. Cloud develops throughout the country with increasing threats of rain as a wet front nears the south of the South Island and a weak depression to the north of New Zealand passes by. Mostly dry everywhere except for some isolated showers likely in the far north and east of the North Island.

Auckland: Partly cloudy, some pockets of showers, calm atmosphere
Hamilton: Overcast, dry, mild day cool night, gentle breezes
Wellington: Sunny, warmer, dry, gentle breezes
Greymouth: Cloudy, dry, light breezes
Christchurch: Mostly sunny, dry, gentle breezes, chance of misty patches
Dunedin: Overcast, colder day milder night, dry, rain threatening, light breezes

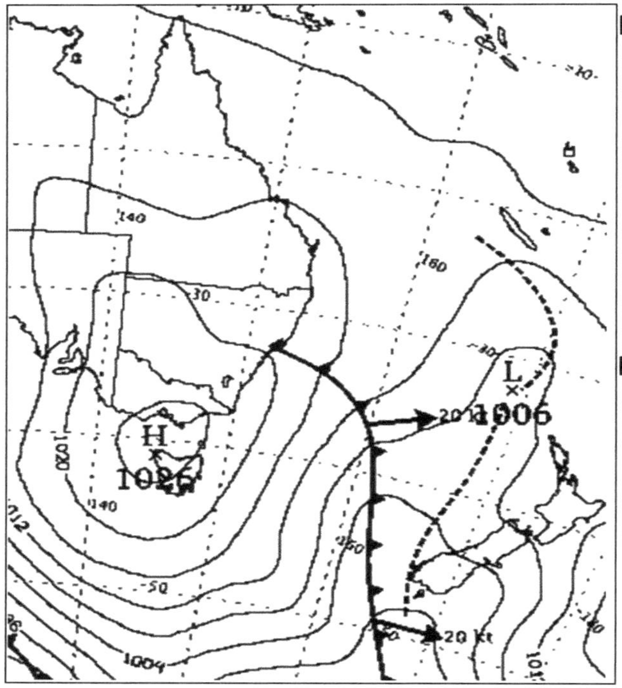

RAIN POTENTIAL

FROST/SNOW

OCTOBER 3 FRIDAY

A depression that was north of New Zealand yesterday slips down through the Tasman Sea directing a northerly flow over the country, widespread cloud and rain activity throughout the South Island. Cloudy with light showers developing over the northern half of the North Island, and remaining cloudy and dry with increasing threats of rain over the lower half of the North island.

Auckland:	Cloudy, mostly dry with increasing threats of rain, light breezes
Hamilton:	Cloudy, warmer, dry, light shifting breezes
Wellington:	Cloudy, dry, light shifting breezes
Greymouth:	Cloudy, rain, fluctuating breezes
Christchurch:	Overcast, dry, gentle shifting breezes
Dunedin:	Overcast, fluctuating winds, heavy rain, chance of thunderstorms

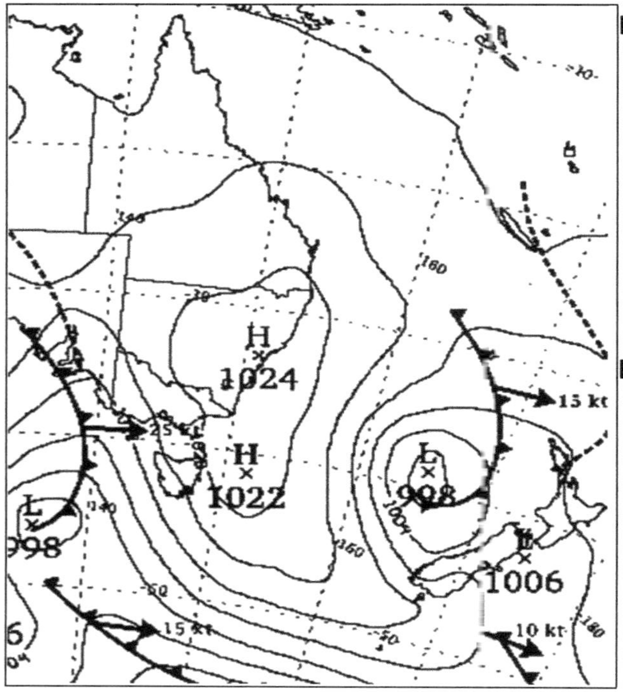

RAIN POTENTIAL

FROST/SNOW

OCTOBER 4 SATURDAY

A depression to the west of the North Island brings stormy conditions over much of the country, with unsettled winds throughout, northwesterly over the North Island and southeasterlies over the South Island. The wet front brings heavy rain through central New Zealand occasionally accompanying thunderstorms and chance of hail showers. In contrast, a weak high pressure centre crosses Southland bringing mostly dry but cloudy conditions to the area.

Auckland:	Cloudy, thunderstorms likely, rain, heavy falls, fluctuating breezes
Hamilton:	Cloudy, light showers, fluctuating breezes, chance of thunderstorms
Wellington:	Overcast, scattered showers, unsettled and changeable winds
Greymouth:	Partly cloudy, rain, fluctuating and shifting winds
Christchurch:	Cloudy, scattered showers, colder, fluctuating breezes
Dunedin:	Cloudy, isolated showers, colder, unsettled and changeable winds

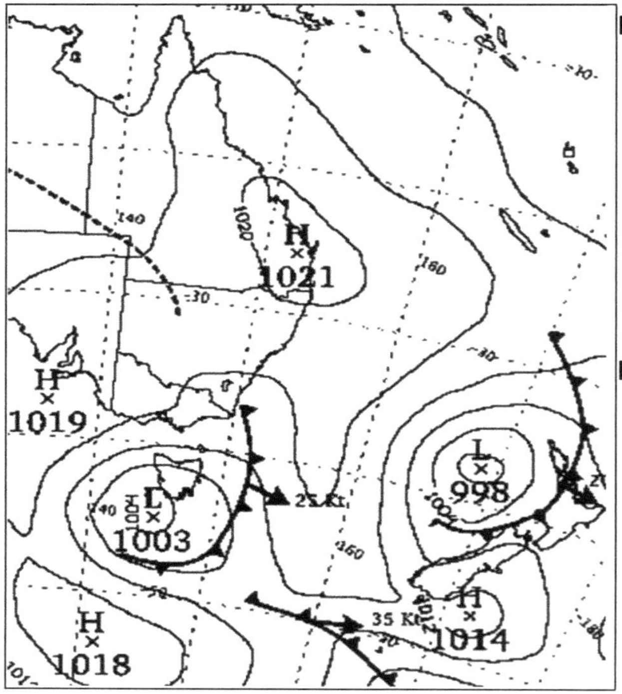

RAIN POTENTIAL

FROST/SNOW

OCTOBER **5** SUNDAY

A depression crosses the North Island with changeable and unsettled winds throughout, widespread cloud and outbreaks of rain, possibly accompanying thunderstorms. A low pressure trough also covers the South Island directing a fresh south to southeasterly flow and bringing widespread cloud and showers to most districts, although remaining mostly dry in Southland and rain clearing to spotty showers through Canterbury, Otago, and the West Coast.

Auckland: Partly cloudy, scattered showers, shifting breezy spells
Hamilton: Overcast, scattered showers, fluctuating breezes
Wellington: Overcast, colder, rain, fluctuating breezes
Greymouth: Sunny, warmer, dry, gentle shifting breezes
Christchurch: Partly cloudy, light showers, gentle breezes
Dunedin: Overcast, mostly dry, cool, moderate breezes

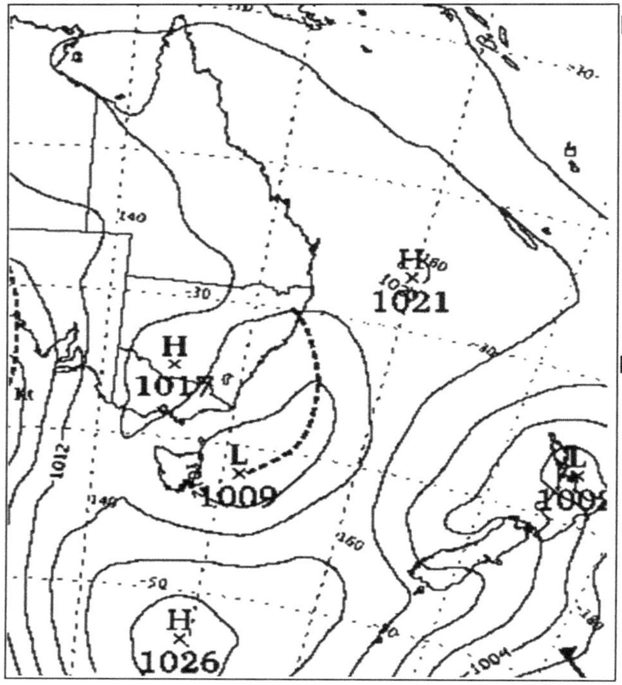

RAIN POTENTIAL

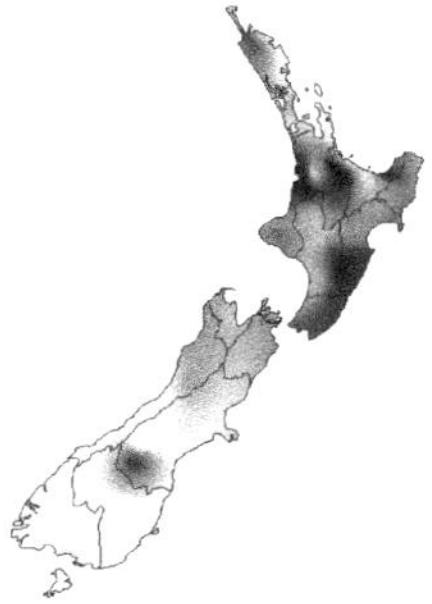

FROST/SNOW

OCTOBER **6** MONDAY

A low pressure trough brings a northerly flow over the North Island while a high pressure system to the south of the South Island directs an easterly flow and colder temperatures over the island. A mix of sunny spells and showers prevail over the North Island and into the northern half of the South Island, while a mostly dry, cold, and sunny day prevails in the southern half of the South Island.

Auckland: Cooler, sunny, pockets of showers, some heavy, calm air, static air
Hamilton: Partly cloudy, cooler, scattered showers, mostly calm air
Wellington: Cloudy, cool, occasional showers, fluctuating breezes
Greymouth: Sunny, warmer day cooler night, dry, light shifting breezes
Christchurch: Overcast, brief showers, moderate breezes
Dunedin: Cloudy, cool, occasional showers, fluctuating breezes

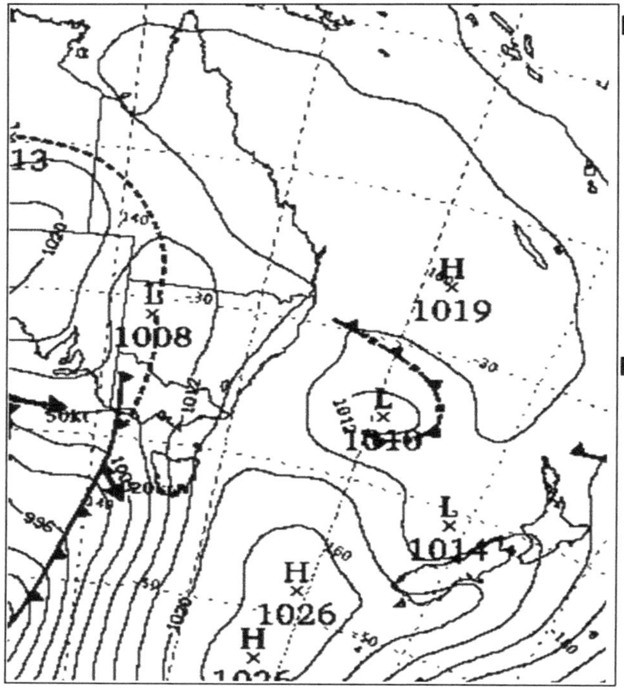

RAIN POTENTIAL

FROST/SNOW

OCTOBER **7** TUESDAY

An anticyclonic system sits over the country directing a light south to southeast flow over the North Island and a north to northwest flow over the South Island. Mostly sunny and colder throughout the country, with chance of frosts from the Waikato to Wellington and in the south and east of the South Island. Isolated showers likely over the northern half of the South Island to Wellington, and mostly dry elsewhere.

Auckland:	Sunny, dry, gentle breezes
Hamilton:	Sunny, dry, mild day cold right, gentle shifting breezes
Wellington:	Cooler, changeable, mix of sunny spells and showers, gentle breezes
Greymouth:	Sunny, slightly cooler, dry, calm and pleasant air
Christchurch:	Sunny, light shifting breezes, isolated showers, colder
Dunedin:	Sunny, dry, gentle shifting breezes

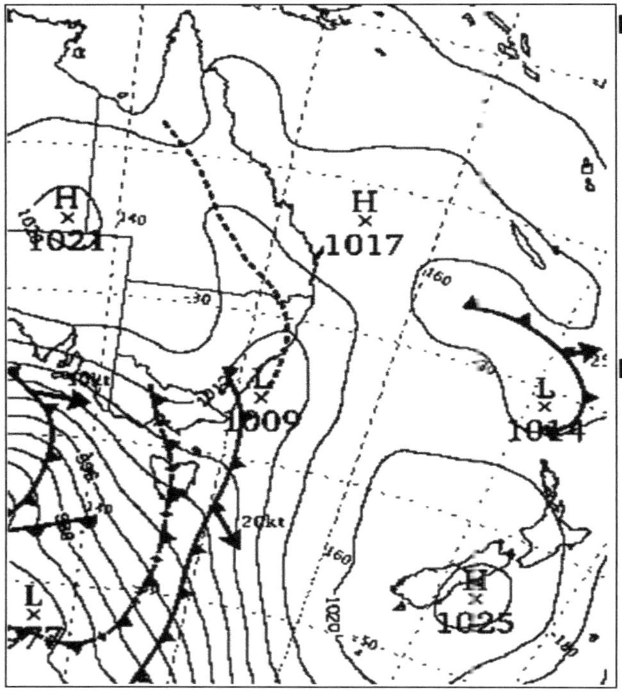

RAIN POTENTIAL

FROST/SNOW

OCTOBER **8** WEDNESDAY

Anticyclonic conditions continue to prevail over the country in a light easterly flow, backing northwesterly over the South Island. Overnight frosts in many districts from the Waikato to Wellington and in Canterbury to Southland. Mostly dry with occasional cloudy spells throughout.

Auckland:	Sunny, dry, gentle breezes to mostly calm air
Hamilton:	Partly cloudy, dry, chance of frost pockets, calm and serene air
Wellington:	Calm and sunny air, light showers, cold overnight
Greymouth:	Sunny, dry, gentle breezes freshening later in the day
Christchurch:	Sunny, dry, cold overnight, light breezes
Dunedin:	Moderate shifting breezes, sunny, warmer, dry

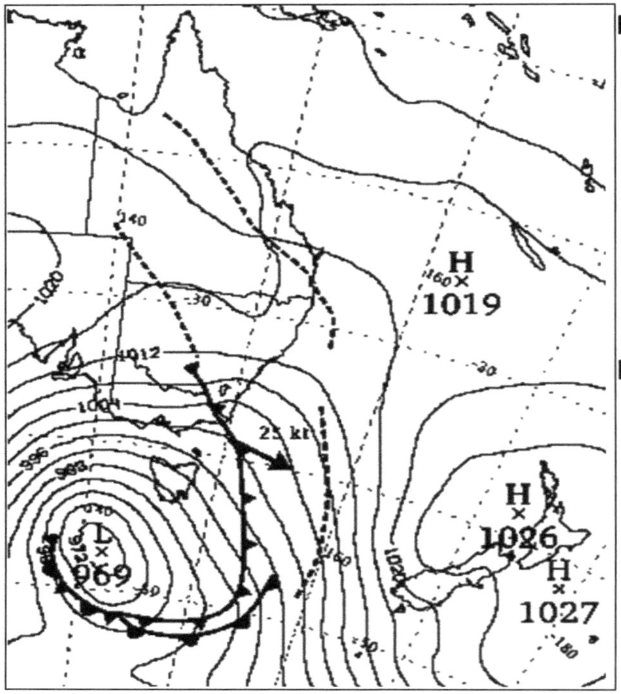

RAIN POTENTIAL

FROST/SNOW

OCTOBER **9** THURSDAY

An unsettled and blustery low pressure front moves onto the south of the country behind the high moving away to the east. Strong north to northwest winds prevail over the South Island, possibly reaching gale force at times, while more moderate northeast breezes over Northland backing northwesterly elsewhere over the North Island. Mostly dry and milder throughout, with only Southland to Otago and around the Bay of Plenty likely to see some frost pockets. Cloud increasing almost everywhere with increasing threats of rain, particularly in the southern half of the South Island.

Auckland: Sunny, dry, moderate breezes
Hamilton: Overcast, warmer, dry, freshening breezes
Wellington: Sunny, warmer, dry, windy spells
Greymouth: Partly cloudy, dry, increasing threats of rain, strengthening winds
Christchurch: Partly cloudy, warmer day, chance of frost pockets, dry, blustery
Dunedin: Overcast, dry, strong gusty winds

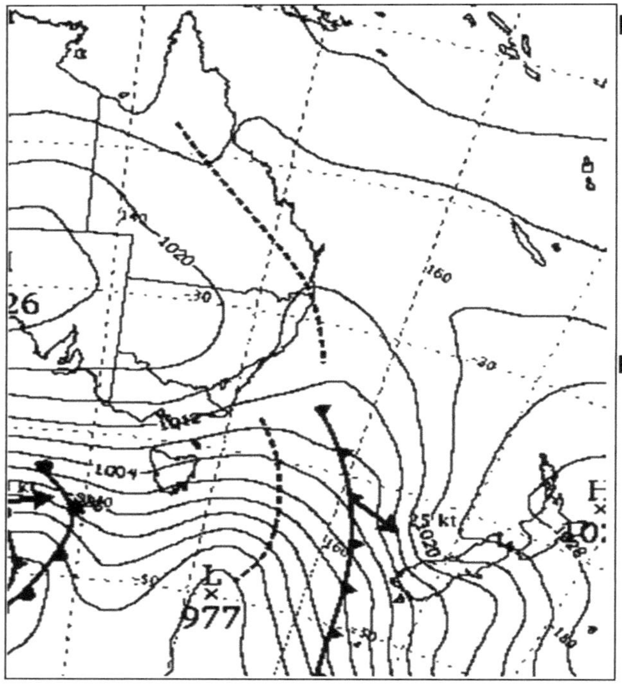

RAIN POTENTIAL

FROST/SNOW

OCTOBER **10** FRIDAY

A blustery wet front crosses the South Island bringing strong and stormy west to northwest winds throughout and heavy rain into the western half of the South Island with showers into the east. Milder temperatures everywhere, and mostly dry and cloudier over the North Island with increasing threats of rain.

Auckland: Cloudy, mostly dry, increasing threats of rain, fresh winds
Hamilton: Overcast, dry with increasing threats of rain, strong winds at times
Wellington: Overcast, isolated showers, strong winds, possible gales at times
Greymouth: Stormy air, overcast, squally showers
Christchurch: Unsettled and blustery winds, partly cloudy, warmer, dry
Dunedin: Cloudy, warmer, dry, strong winds with occasional gusts

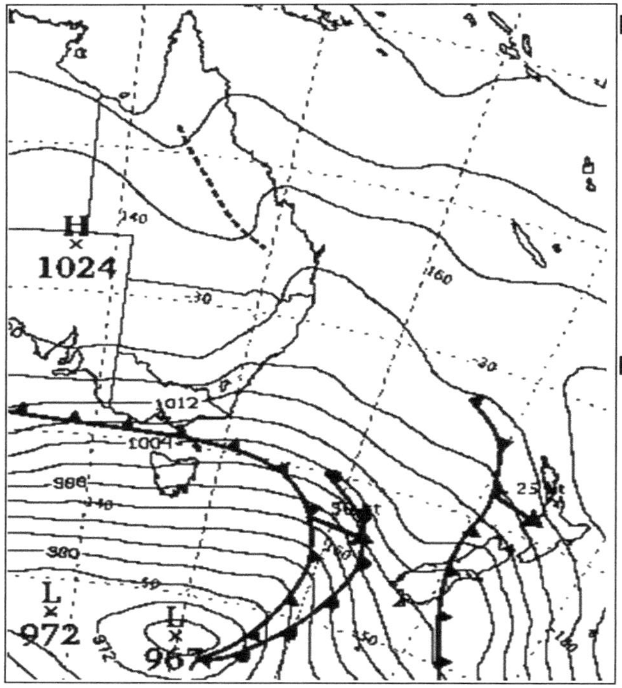

RAIN POTENTIAL

FROST/SNOW

OCTOBER **11** SATURDAY

A disturbed wet front crosses the country in a blustery and stormy west to northwest flow, bringing very strong winds and scattered thunderstorms, possible hail showers, and rain into many districts, although mostly dry in the eastern coastal fringes from the East Cape to the Hawkes Bay and Canterbury. Heavy rain at times likely in Northland and in the west and north of the South Island

Auckland: Cloudy, thunderstorms likely, rain, strong gusty winds
Hamilton: Unsettled blustery air, partly cloudy, brief squally showers, static air
Wellington: Cloudy, strong gusty winds, passing squalls, chance of gale winds
Greymouth: Overcast, cooler, heavy rain, strong winds with gusts
Christchurch: Dry, cloudy, mild day cool night, strong gusty winds
Dunedin: Partly cloudy, light showers, windy

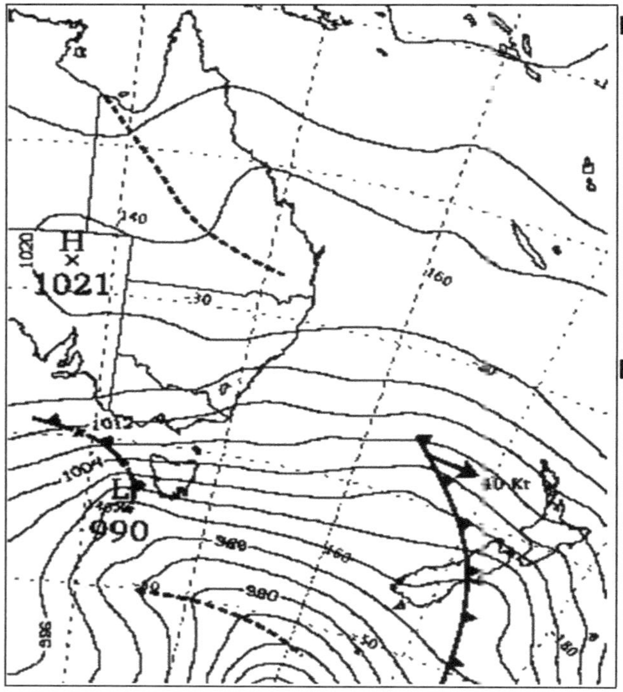

RAIN POTENTIAL

FROST/SNOW

OCTOBER **12** SUNDAY

Storm conditions continue to prevail across the country, particularly in the west and central parts of New Zealand, while coastal fringes in Canterbury and around the East Cape may remain mostly dry. Very strong and stormy west to northwest winds prevail everywhere, and heavy falls likely in the west of both islands. Scattered thunderstorms likely in many districts, with hail showers at times. Snow may be observed in the South Island high country as well as in the North Island Central Plateau.

Auckland: Stormy air, gusty winds, heavy rain, brief sunny breakthroughs
Hamilton: Stormy, chance of thunderstorms, heavy rain, sunny spells, blustery
Wellington: Mostly sunny and dry, strong winds, gales possible
Greymouth: Cloudy, cooler, heavy rain, strong winds possibly to gale force
Christchurch: Partly cloudy, dry, cooler, strong fluctuating winds
Dunedin: Cloudy, isolated showers, unsettled breezy spells

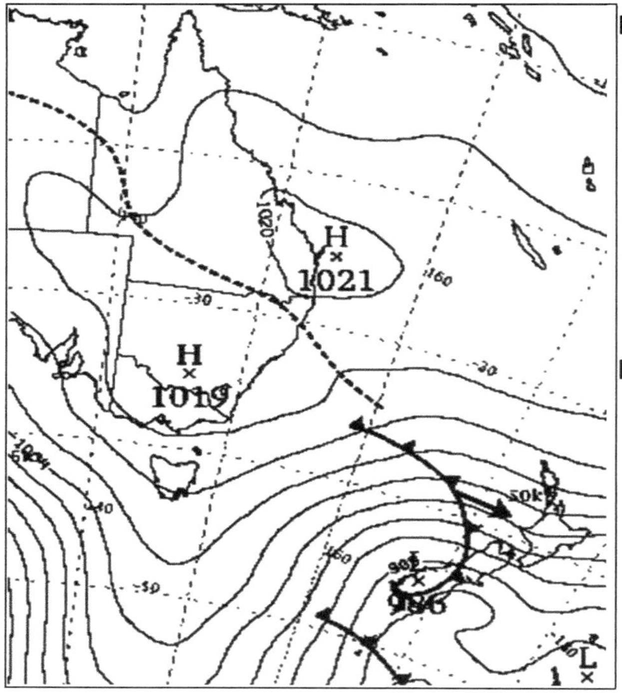

RAIN POTENTIAL

FROST/SNOW

OCTOBER 13 MONDAY

A large high to the northwest of New Zealand begins to push the low pressure away from New Zealand with a unsettled southwest change. Winds ease in the west, but remain stormy in the east. Scattered thunderstorms and widespread rain, mostly overnight likely, with heavy falls over the northern half of the North Island and in the southwest of the South Island. Chance of snow in the Central Plateau and south of the South Island, and hail accompanying thunderstorm activity. Rain clearing to a mostly dry and sunny day.

Auckland: Windy, mostly sunny, heavy evening rain, possible thunderstorms
Hamilton: Mostly sunny, rain, windy, chance of evening thunderstorms
Wellington: Partly cloudy, strong blustery winds, occasional squally showers
Greymouth: Changeable, sunny day, overnight heavy rain, cooler, winds easing
Christchurch: Partly cloudy, rain, chance of hail or light snow, freshening breezes
Dunedin: Changeable, occasional sunny spells, heavy rain, light breezes

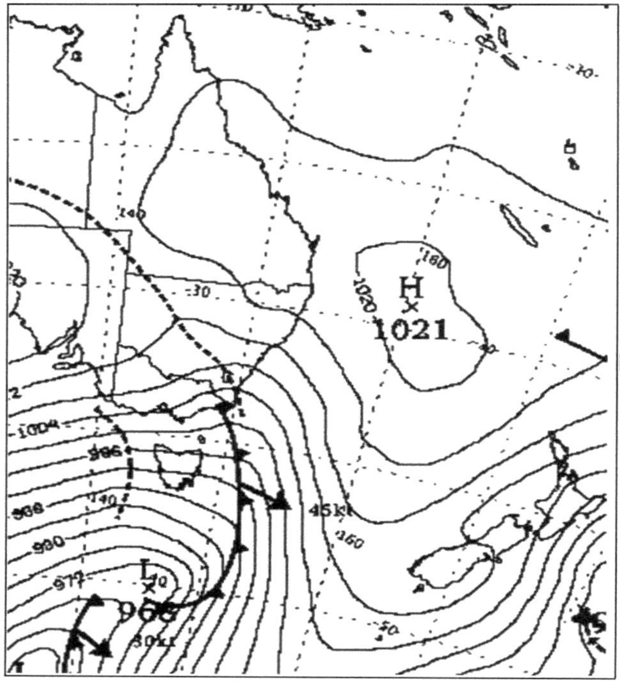

RAIN POTENTIAL

FROST/SNOW

OCTOBER 14 TUESDAY

Despite a high pressure influence over the north of New Zealand, another stormy wet front approaches the southwest of the country. Mostly dry, colder, and sunny over the North Island before showers develop later in the day in most parts except the east from the East Cape to the Hawkes Bay. Very cold and cloudier over the South Island with isolated outbreaks of showers here and there, but with strong gusty winds and gales possible especially through the Cook Strait. Chance of further light snow in the higher reaches of both islands.

Auckland: Sunny, blustery, cooler, showers, possible evening thunderstorms
Hamilton: Sunny, chance of overnight frosts, light showers, windy at times
Wellington: Cloudy, colder, isolated showers, strong blustery winds
Greymouth: Overcast, mostly dry, mild day cold night, strong and gusty winds
Christchurch: Blustery winds, mostly cloudy, some showers, possible snow flurries
Dunedin: Unsettled winds, partly cloudy, cooler, isolated showers

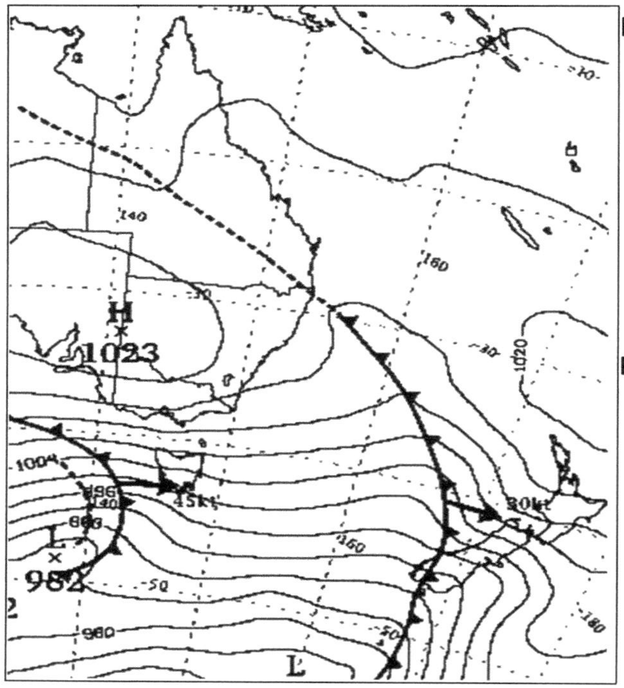

RAIN POTENTIAL

FROST/SNOW

OCTOBER 15 WEDNESDAY

Another very unstable wet front crossing the Tasman Sea moves onto New Zealand, while a high pressure ridge lingers over the far north of the North Island. After a mostly sunny day, evening storms arrive once more throughout the South Island and the western half of the North Island, with heavy rain in many western areas from Fiordland to Taranaki, lighter showers elsewhere, and mostly dry around the East Cape and Christchurch. Winds up to gale force at times, and cold overnight temperatures may see snow again on the southern ranges, and Central Plateau.

Auckland: Very sunny, warmer, dry day overnight showers, blustery at times
Hamilton: Changeable, sunny, some showers, fresh to strong winds
Wellington: Sunny day, evening rain, strong winds to gale force at times likely
Greymouth: Cloudy, heavy rain, strong blustery winds, chance of gales
Christchurch: Sunny, dry, very strong winds at times, mild day cold night
Dunedin: Partly cloudy, dry, warmer, strong blustery winds, chance of gales

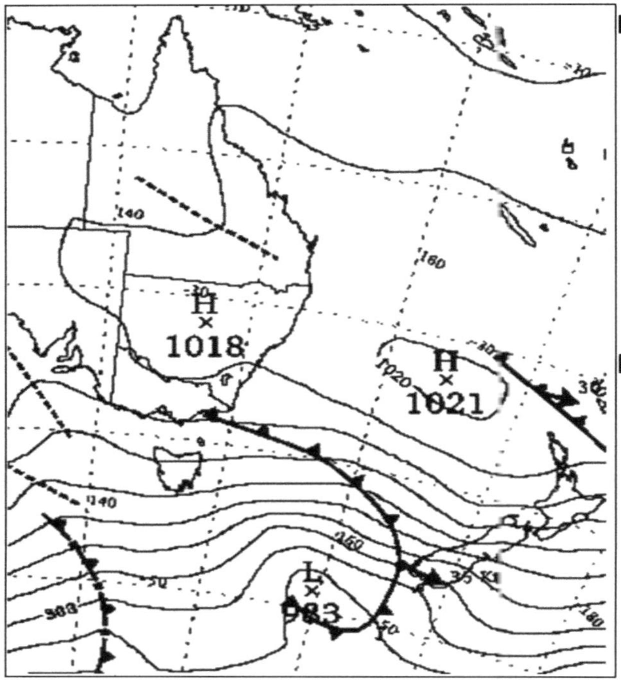

RAIN POTENTIAL

FROST/SNOW

OCTOBER 16 THURSDAY

The high pressure ridge to the north of the country moves down over the North Island bringing a lighter westerly flow and a mostly dry and partly cloudy day throughout, with only some isolated lingering showers likely around Taranaki. In contrast, low pressure brings a strong west to northwest flow and another wet front with widespread heavy rain throughout the South Island overnight and the chance of further snowfalls in the higher reaches, after a mostly sunny and dry day.

Auckland: Sunny, dry, strong breezes, warm day cool night
Hamilton: Sunny, dry, windy at times
Wellington: Sunny, warmer, dry, blustery at times
Greymouth: Changeable, sunny day, heavy overnight rain, windy spells
Christchurch: Sunny, dry, warmer, fresh and shifting winds
Dunedin: Partly cloudy, dry, shifting breezy spells

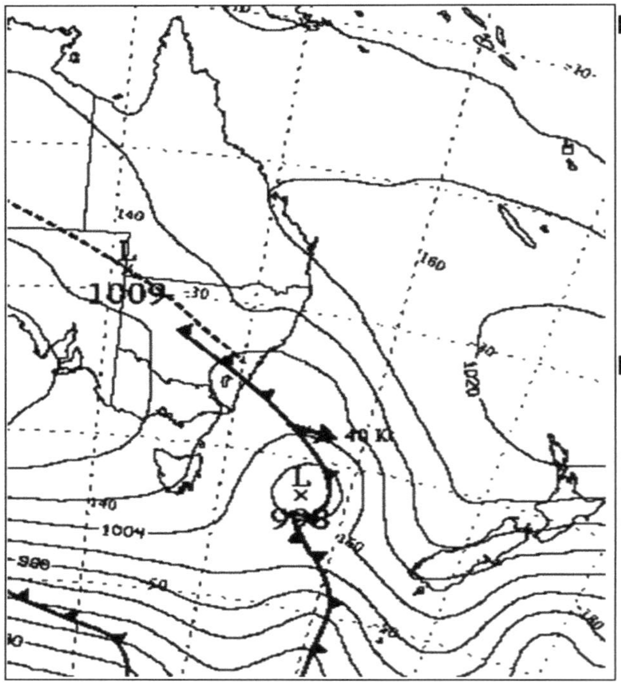

RAIN POTENTIAL

FROST/SNOW

OCTOBER 17 FRIDAY

While a high pressure ridge brings a mostly dry, sunny and pleasant day with moderate northwesterly breezes across the North Island, the South Island again sees ongoing storm activity as a depression continues to cross to the south of the country and directing a stormy wet front up through the regions in a moderate southwest flow. Widespread overnight rain can be expected throughout the South Island with heavy falls in western regions and lighter to isolated showers to the east. Although temperatures may become milder over the South Island at this time, some snow may still be observed in the higher reaches.

Auckland: Partly cloudy, dry, warmer, fresh breezes
Hamilton: Sunny, dry, unsettled breezy spells
Wellington: Cloudy, isolated showers, warmer, stormy air
Greymouth: Overcast, fluctuating winds, heavy rain
Christchurch: Partly cloudy, light showers, strong fluctuating winds
Dunedin: Partly cloudy, cooler, light showers, shifting breezy spells

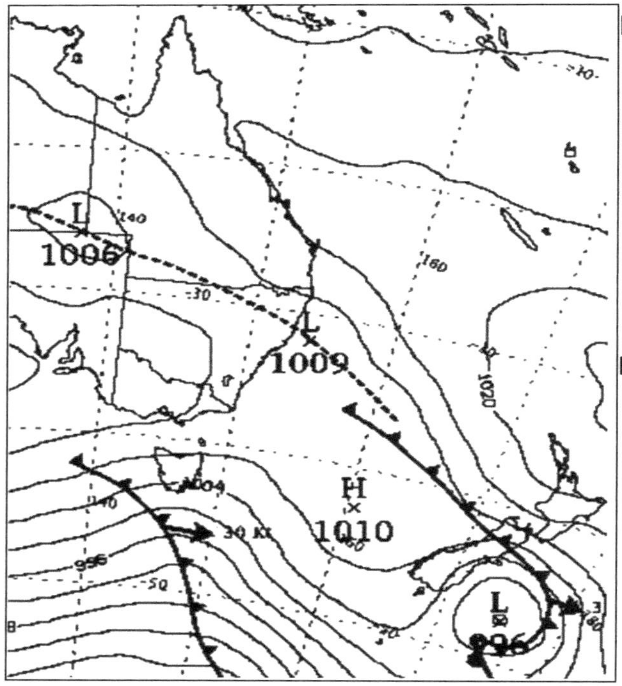

RAIN POTENTIAL

FROST/SNOW

OCTOBER 18 SATURDAY

The high pressure ridge maintains a mostly dry and sunny day over the northern half of the North Island. However, increased cloud and threats of rain move into the southern half of the North Island as an unstable wet front moves up through the South Island and across the Cook Strait, where winds may reach gale force at times. Very strong winds, overnight thunderstorms and heavy rain likely over the northern half of the South Island and into the lower North Island likely, while rain clears to occasional showers in the southern half of the South Island and increasing sunny spells.

Auckland: Unsettled and blustery winds, mostly cloudy, dry, warmer
Hamilton: Partly cloudy, warmer, dry, fresh to strong breezes
Wellington: Overcast, heavy rain, strong winds, chance of thunderstorms and gales
Greymouth: Mix of heavy rain and sunny spells, chance flooding, unsettled winds
Christchurch: Partly cloudy, warmer, rain, unsettled winds
Dunedin: Sunny, dry, light breezes, chance of misty patches

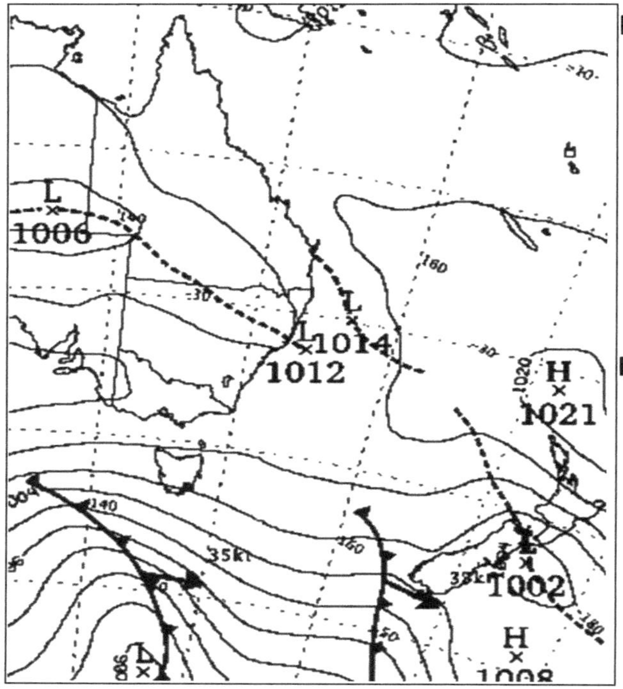

RAIN POTENTIAL

FROST/SNOW

OCTOBER 19 SUNDAY

The depression that has bought severe winds over much of the southern half of the country ease as the wet front departs. However, low pressure prevails everywhere, in a lighter north to northwest flow. Overnight rain can be expected from Auckland to Christchurch and down the west of the South Island, while mostly dry in Otago, South Canterbury, and from the East Cape to the Hawkes Bay. Widespread cloud over the northern half of the North Island, while sunnier to the south.

Auckland: Cloudy, warmer, dry, mostly calm air, chance of overnight drizzle
Hamilton: Cloudy, some drizzly patches, mostly calm air
Wellington: Changeable, sunny, overnight rain, possible thunderstorms
Greymouth: Changeable, mix of sunny spells and showers, light air
Christchurch: Colder, changeable, mix of sunny spells and showers, mostly calm air
Dunedin: Dry, sunny, mild day cold night, mostly calm air

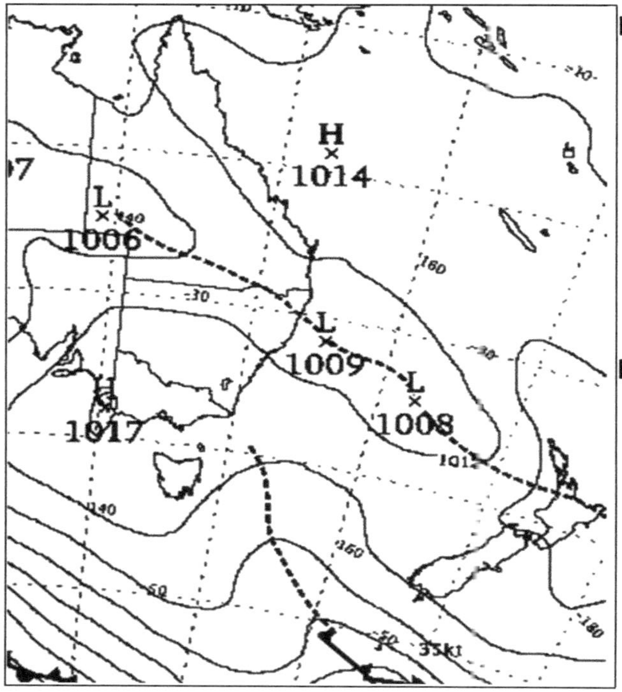

RAIN POTENTIAL

FROST/SNOW

OCTOBER 20 MONDAY

A complex of low pressure over the country and weak high pressure ridges to the north and southeast of New Zealand brings a moderate northeasterly flow over much of the country. Heavy rain likely through central New Zealand from North Canterbury to the Wairarapa with showers extending northwards into the Waikato and Auckland. Mostly dry in Northland and in the southern half of the South Island. Widespread cloud prevails from Northland to central Canterbury with sunnier spells in the far south of the South Island.

Auckland: Overcast, moderate breezes, isolated showers, cooler
Hamilton: Overcast, mostly calm air, some light showers
Wellington: Overcast, heavy rain, fresh shifting breezes
Greymouth: Overcast, dry, moderate shifting breezes
Christchurch: Overcast, dry, freshening breezes
Dunedin: Sunny, dry, gentle breezes to mostly calm air

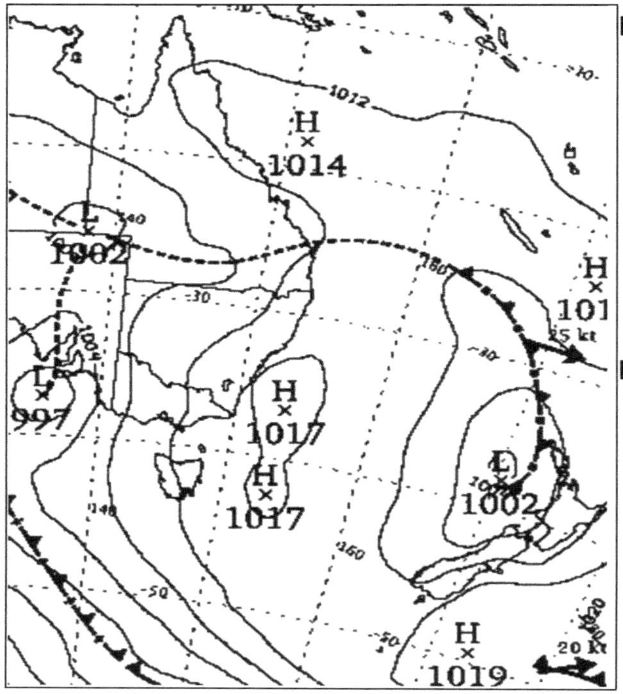

RAIN POTENTIAL

FROST/SNOW

OCTOBER 21 TUESDAY

Low pressure centres consolidate over the country directing westerly winds over the North Island and easterlies over the South Island. Widespread overnight rain throughout most of the country is likely except in the far southwest around Southland, with some good sunny spells during the day in the northern half of the North Island while dull and gloomy elsewhere. Cold temperatures in the South Island may see further brief snowfalls in the South Island high country. Heavy rain around Taranaki, the top of the South Island and East Cape to the Bay of Plenty may cause some localised surface flooding.

Auckland: Partly cloudy, overnight rain, light shifting breezes
Hamilton: Changeable, mix of sunny spells and rain, moderate breezes
Wellington: Overcast, cooler, heavy rain, unsettled and changeable winds
Greymouth: Overnight rain, partly cloudy day, fluctuating and shifting winds
Christchurch: Unsettled winds, overcast, heavy rain
Dunedin: Overcast, blustery at times, scattered showers, slightly warmer

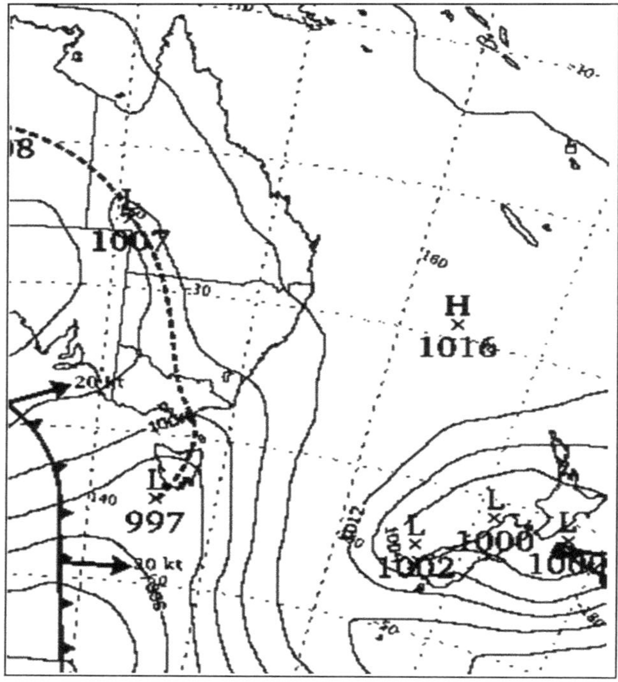

RAIN POTENTIAL

FROST/SNOW

OCTOBER 22 WEDNESDAY

Low pressure crosses to the south of the country while a weak high pressure ridge crosses close to the north of New Zealand, with a fresh westerly flow prevailing over the North Island and a milder westerly flow over the South Island. Further outbreaks of rain in most parts of the country, except in the far southwest of the South Island and from the East Cape to the Hawkes Bay. Mostly sunny in the north and east of the North Island while cloudier elsewhere.

Auckland: Mix of showers and sunny spells, windy at times
Hamilton: Changeable, sunny, overnight rain, fresh breezes
Wellington: Cloudy, cool, brief showers, windy spells
Greymouth: Cloudy, warmer, light showers overnight, moderate breezes
Christchurch: Overcast, light showers, light breezes
Dunedin: Overcast, colder, heavy rain, light breezes

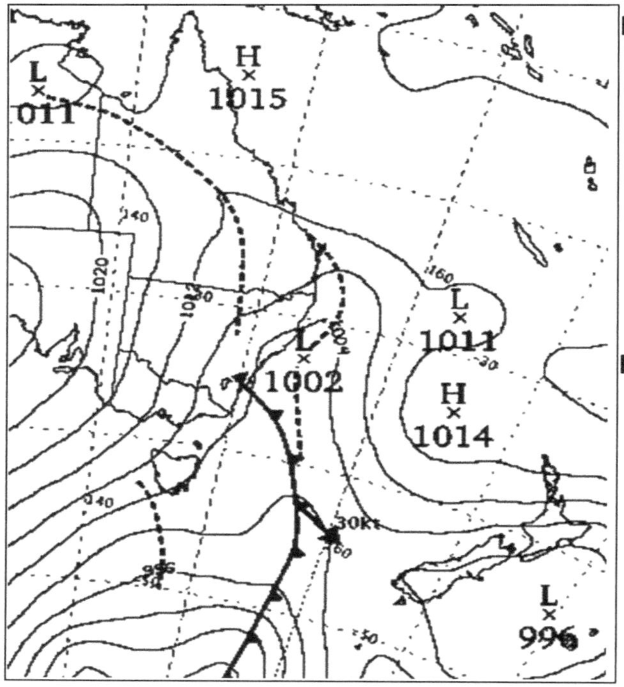

RAIN POTENTIAL

FROST/SNOW

OCTOBER 23 THURSDAY

A depression to the west of New Zealand directs a strong and blustery north to northwest flow throughout the country. Mostly cloudy and dull in the southern half of the South Island with occasional sunny breakthroughs elsewhere. Some light showers around the northwest of the South Island and into the west of the North Island from the Waikato to the Manawatu. Mainly dry in the east of the country, through the southern half of the South Island and north of Auckland.

Auckland: Cloudy, isolated showers, fresh breezy spells
Hamilton: Partly cloudy, cooler, light showers, unsettled and fluctuating winds
Wellington: Breezy, isolated showers, occasional sunny spells, warmer
Greymouth: Overcast, cooler, shifting breezy spells, isolated showers
Christchurch: Partly cloudy, dry, mild temperatures, windy spells
Dunedin: Partly cloudy, dry, gentle shifting breezes

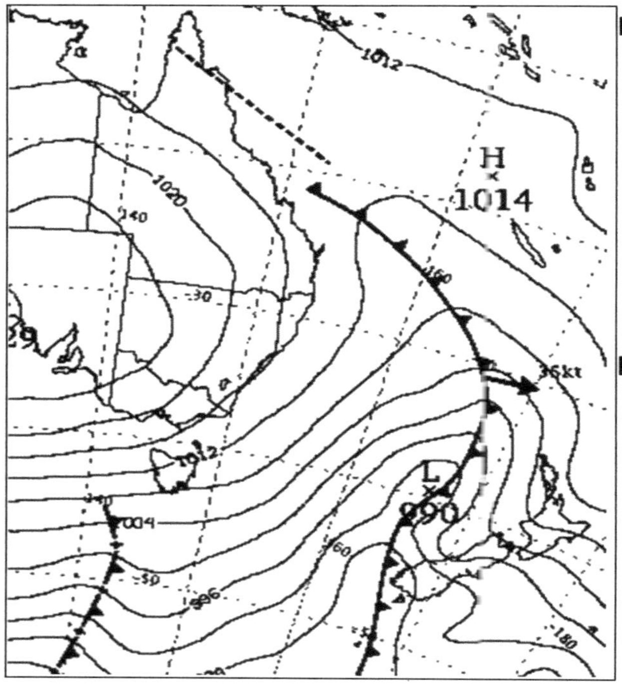

RAIN POTENTIAL

FROST/SNOW

OCTOBER 24 FRIDAY

A depression intensifies as it crosses New Zealand bringing stormy west to southwest winds throughout the country and stormy conditions almost everywhere. Widespread rain throughout most of New Zealand although mostly dry with increasing threats of rain in Canterbury. Heavy rain likely in the Bay of Plenty to Taranaki, and down throughout the west and south of the South Island. Gale winds likely through the Cook Strait and chance of overnight snow in the South Island high country again.

Auckland: Sunny, overnight heavy rain in places, blustery winds, possible gales
Hamilton: Mix of sunny spells and showers, warmer, strong winds with gusts
Wellington: Sunny day, overnight showers, strong stormy winds
Greymouth: Heavy rain, widespread cloudy, strong stormy winds, gales likely
Christchurch: Overcast, dry, warmer, strong gusty winds to near gales likely
Dunedin: Cloudy, strong winds with chance of gale gusts, squally showers

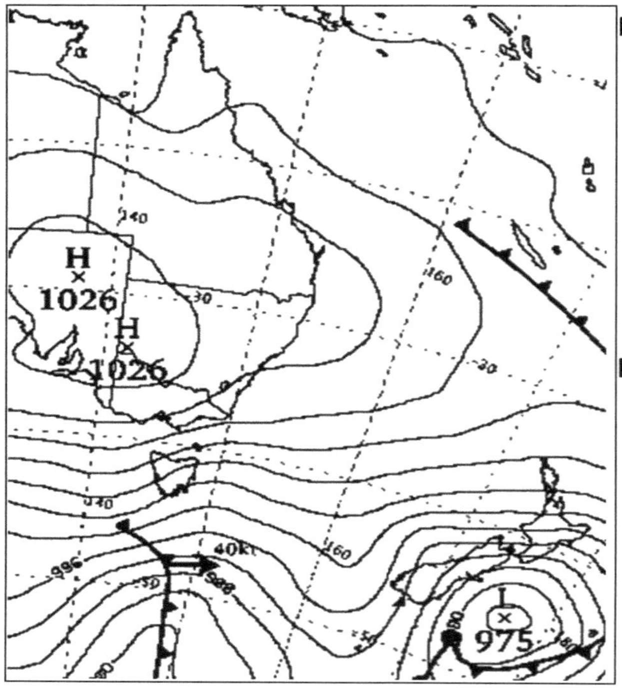

RAIN POTENTIAL

FROST/SNOW

OCTOBER **25** SATURDAY

A very strong and blustery west to southwest flow prevails over New Zealand with the passage of a depression crossing to the south of the country, with winds up to gale force at times particularly through the Cook Strait and up through the North Island. Widespread cloud throughout the South Island and scattered outbreaks of showers in most districts, heaviest around the West Coast and more isolated in the southeast. Mostly dry and sunny in the Bay of Plenty to the Hawkes Bay, while cloudier with scattered showers elsewhere over the North Island.

Auckland: Stormy air, strong winds, squalls, brief sunny spells, cooler
Hamilton: Slightly cooler, some sunny spells, overnight showers, gusty winds
Wellington: Partly cloudy, dry, strong winds, gales possible
Greymouth: Cloudy, cooler, blustery, squally showers
Christchurch: Cloudy, colder, brief showers, strong winds, chance of hail
Dunedin: Cloudy, colder, scattered showers, winds easing to moderate

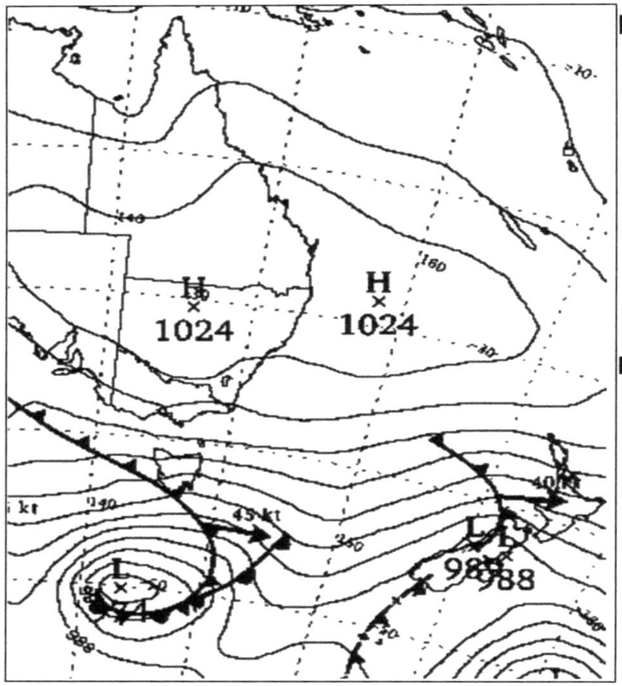

RAIN POTENTIAL

FROST/SNOW

OCTOBER **26** SUNDAY

A deep depression south of New Zealand continues to bring a mix of sunny spells and overnight rain throughout the North Island and in most regions of the South Island except for mostly dry and sunny in Canterbury, and cloudier with threats of rain in Otago. Winds very strong throughout the country in a west to northwesterly flow, with winds up to gale force at times including through the Cook Strait and up into the lower North Island. Chance of overnight snowfalls still possible in the South Island high country.

Auckland: Partly cloudy, scattered showers, strong winds
Hamilton: Partly cloudy, light showers, strong winds with occasional gusts
Wellington: Mix of sunny spells and showers, strong winds, gales possible
Greymouth: Overcast, heavy rain, strong blustery winds
Christchurch: Sunny, dry, mild day cold night, strong and unsettled winds
Dunedin: Partly cloudy, dry, unsettled winds

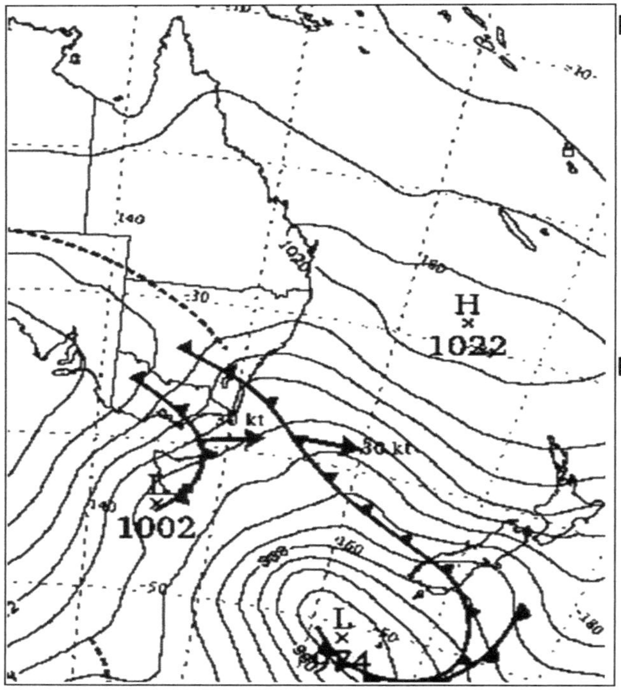

RAIN POTENTIAL

FROST/SNOW

OCTOBER **27** MONDAY

A wet front crosses up through the country bringing strong south to southwest winds over the South Island backing west to northwest over the North Island, with winds up to gale force at times. Widespread overnight heavy rain throughout the South Island and into the south and west of the North Island, while mostly dry in the north and east of the North Island. Despite very heavy rain at times in the western half of the South Island, some good sunny spells can be expected in most regions of the country for a time.

Auckland:	Blustery winds, cloudy, mostly dry with increasing threats of rain
Hamilton:	Mostly sunny, dry, strong winds possibly to gale force at times
Wellington:	Partly cloudy, isolated showers, strong gusty winds, chance of gales
Greymouth:	Chance of thunderstorms, heavy rain, some sunny spells, blustery
Christchurch:	Sunny, warmer, dry, strong winds to gale force likely
Dunedin:	Cloudy, warmer, rain, strong gusty winds, chance of gales at times

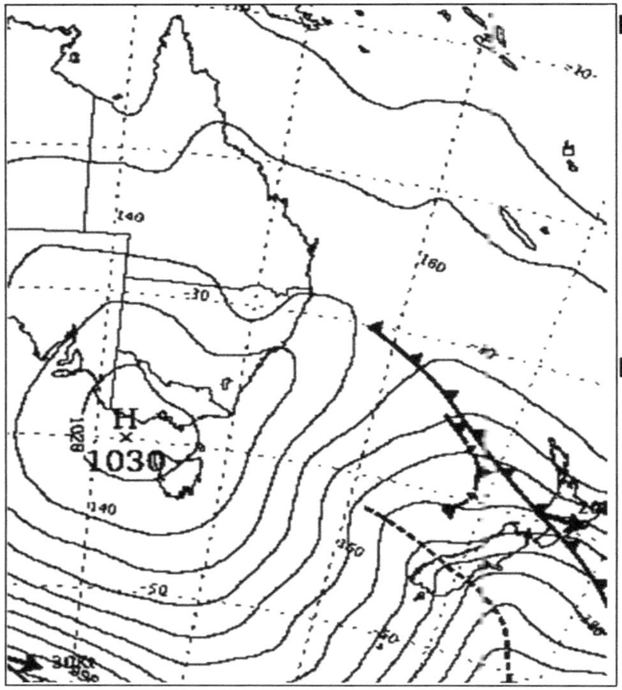

RAIN POTENTIAL

FROST/SNOW

OCTOBER **28** TUESDAY

A large high pressure system drifts into the Tasman Sea and ridges to the west of New Zealand while low pressure prevails throughout the country. Fluctuating winds prevail, westerly turning southeasterly over the South Island, and southeasterly turning westerly over the North Island. Dull and gloomy throughout the North Island with widespread and often heavy rain in many regions, especially from Taranaki to Wellington, while a mix of sunny spells and showers prevail throughout the South Island. Colder everywhere with chance of snow in the southern ranges as well as in the North Island Central Plateau.

Auckland:	Partly cloudy, warmer, rain, fresh winds easing
Hamilton:	Cloudy, scattered showers, mild day cool night, shifting breezy spells
Wellington:	Overcast, rain, warmer dry colder night, strong changeable winds
Greymouth:	Changeable, mix of showers and sunny spells, moderate breezes
Christchurch:	Partly cloudy, light showers, cold overnight, winds easing to light
Dunedin:	Sunny, colder, scattered showers, winds easing to light

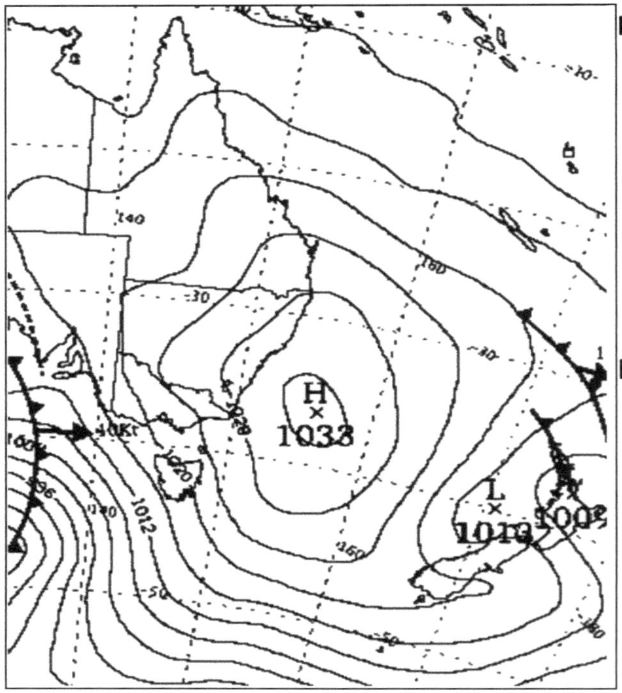

RAIN POTENTIAL

FROST/SNOW

OCTOBER **29** WEDNESDAY

A large high pressure system in the Tasman Sea ridges onto New Zealand easing winds throughout; northwesterlies over the South Island backing southeasterlies over the North Island. Rain activity clears to mostly dry and cloudy in the southern half of the South Island, while a mix of brief sunny spells and widespread shower activity prevails elsewhere, including heavy rain likely in the east of the North Island. Chance of overnight snow still in the Central Plateau.

Auckland:	Sunny day, overnight rain, cooler, light breezes
Hamilton:	Sunny and mild day, cold with overnight rain, moderate breezes
Wellington:	Sunny, colder, brief showers, winds easing
Greymouth:	Sunny, dry, mild day cold night, mostly calm and serene air
Christchurch:	Sunny, chance of frost, isolated wintry showers, chance of hail
Dunedin:	Sunny, dry, colder, light breezes

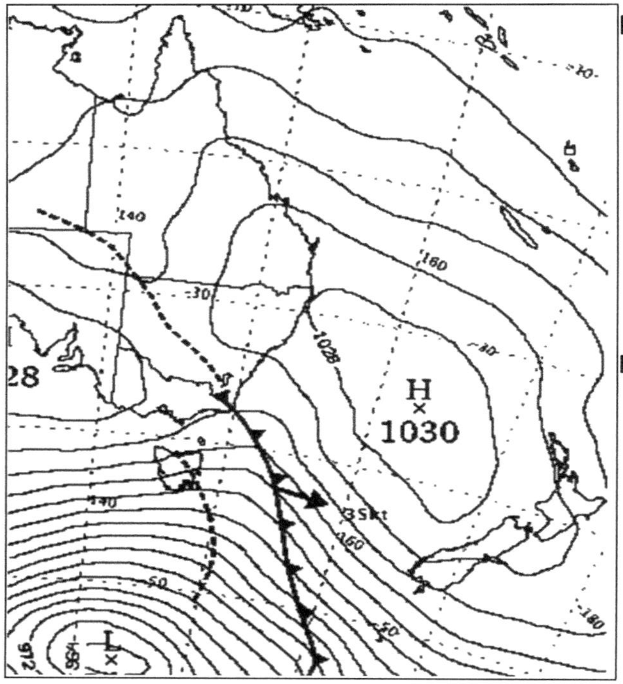

RAIN POTENTIAL

FROST/SNOW

OCTOBER **30** THURSDAY

An intense depression crossing to the south of New Zealand pushes a wet front once more up over the South Island with the high pressure ridge contracting northward ahead of the front. Heavy rain for a time in Southland and southern West Coast districts, while clearing to mostly dry and sunny elsewhere. Strengthening westerly winds prevail throughout, possibly up to gale force in the southwest of the country and through the Cook Strait. Chance of late frosts in the central North Island to the Bay of Plenty, as well as in the east of the South Island.

Auckland:	Cloudy, cool, dry, light breezes strengthening later in the day
Hamilton:	Partly cloudy, dry, windy spells, very cold, chance of frost pockets
Wellington:	Sunny, dry, strong gusty winds, chance of gales at times
Greymouth:	Partly cloudy, isolated showers, moderate winds strengthening later
Christchurch:	Sunny, dry, cool, strong blustery winds possibly to gale force at times
Dunedin:	Partly cloudy, dry, warmer, strong blustery winds

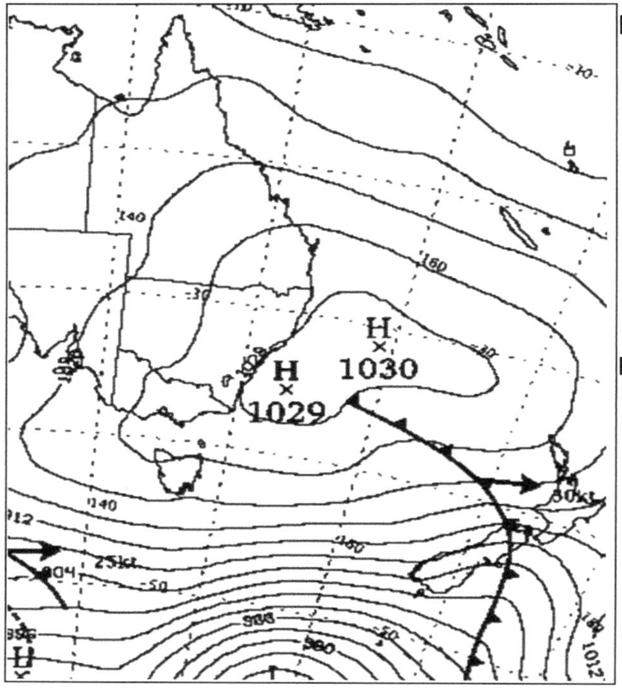

RAIN POTENTIAL

FROST/SNOW

OCTOBER **31** FRIDAY

A large high pressure system remains almost stalled to the northeast of New Zealand while a blustery low pressure system crosses to the south, with a strong southwest flow prevailing throughout the country. A changeable day with a mix of sunny spells and outbreaks of rain in most parts of the country, although slightly cloudier and mostly dry in the east of the country from the East Cape to Canterbury and in Northland. Chance of heavy rain for a time in the West Coast, and showers more isolated in the west of the North Island.

Auckland:	Partly cloudy, mostly dry, chance of morning showers, strong winds
Hamilton:	Sunny, morning showers, strong winds, possibly gales
Wellington:	Very sunny, dry, warmer day cool night, strong winds
Greymouth:	Cloudy, heavy rain, strong and blustery near gale winds
Christchurch:	Sunny, warmer, dry, strong blustery winds
Dunedin:	Sunny, warmer, brief morning showers, strong gusty winds

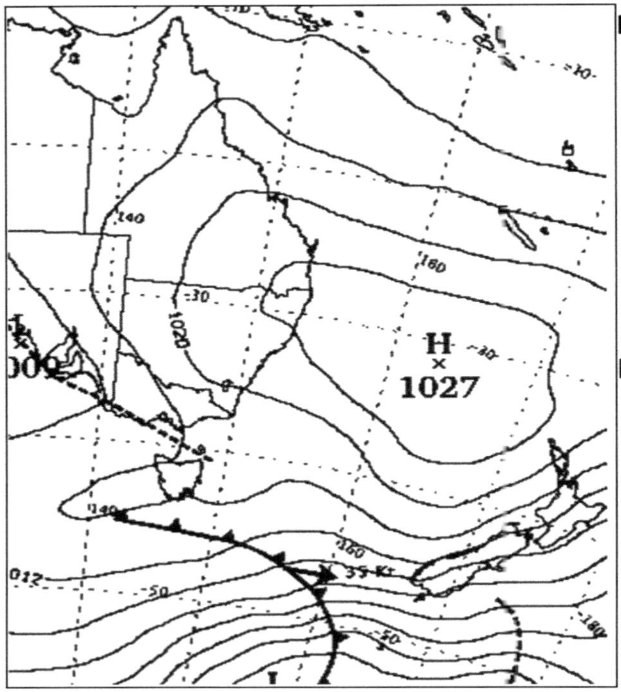

RAIN POTENTIAL

FROST/SNOW

NOVEMBER

PHASES OF THE MOON

2nd	Crossing Equator
5th	Perigee #1
6th	Full Moon
8th	Northern Declination
12th	Third Quarter
15th	Crossing Equator
20th	New Moon
20th	Apogee
22nd	Southern Declination
28th	First Quarter
29th	Crossing Equator

1	2	3	4	5	6	7	8	9	10	11	12	13	14	15	16	17	18	19	20	21	22	23	24	25	26	27	28	29	30
Sat	Sun	Mon	Tue	Wed	Thu	Fri	Sat	Sun	Mon	Tue	Wed	Thu	Fri	Sat	Sun	Mon	Tue	Wed	Thu	Fri	Sat	Sun	Mon	Tue	Wed	Thu	Fri	Sat	Sun
	XhN			P1	F		^				3Q			XhS					N A		V						1Q	XhN	

MONTHLY SUMMARY

2nd-3rd strong west to northwest winds to gale force at times may be disruptive in the east of south of both islands, from the East Cape and Manawatu, down through Kaikoura and Canterbury and into Otago and Southland, possibly disrupting power and damaging local infrastructure in the Bluff and Invercargill region. Strong winds may also cause problems in the lower North Island for a time, possibly bringing a brief closure to the Rimutaka Hill highway.
7th Chance of misty patches in the Wellington region, later turning to heavy rain.
8th-9th Possible patches of fog in Christchurch.
9th Very warm day in Canterbury with unusually high maximum temperatures likely in southern parts, particularly around Timaru. Heavy rain likely in Fiordland.

Over the next four weeks, a strong anticyclonic influence will return once more, with high pressure ridges moving from the Tasman Sea and across the country and bringing a mostly dry, southerly flow more frequent than is typical for this time of year. The second half of this outlook period will be particularly very dry and warm, with few districts seeing any rainfall. Overall, the next four weeks will bring very low rainfall amounts in parts of the South Island, including in the Nelson, Marlborough, and central Otago districts where many districts may see only between 50-75% of their usual rainfall. Similarly, the east of the North Island can also expect only around half their usual rainfall for this time of year, with severe soil moisture deficits affecting the east of the country from the Bay of Plenty to the Hawke's Bay and include the Taupo district, down through Wanganui and Wellington, Nelson, Marlborough and Canterbury and into Otago. In contrast to this, the Horowhenua may be around 25% wetter than average, while the Waikato, Auckland, and Northland all likely to see rainfall amounts that are close to average for this time of year. Sunshine levels will be below normal in the Far North, while above normal over much of the South Island and in the south and west of the North Island from Taranaki and Gisborne regions to Wellington. Temperatures during this outlook period will average close the norm overall for this time of year in most parts of the country, with cooler than average temperatures by around half a degree less than average over the first couple of weeks, and warmer than normal temperatures over, by a similar variant above the norm, in the second fortnight ahead.

13th Chance of shallow fog in the Christchurch area.
14th-17th Wet spell over most of the North Island with high rainfall likely in the coastal areas of the Waikato and in the Ruapehu district and hail showers possible in the lower North Island including in Wellington.
16th Cold overnight in Central Otago with the chance of sub-zero temperatures bringing pockets of frost.
17th-19th Chance of thunderstorms and hail showers in the Wellington region.
21st-22nd Chance of fog in Christchurch.
22nd Heavy rain in Fiordland likely.
23rd-24th Stormy air in the Wellington region with very strong and blustery southerly winds, possibly to gale force at times, and chance of scattered thunderstorms.
27th Light air in the Wellington to Christchurch regions with chance of fog patches
30th-6th December Unusually hot spell in the north and east of both islands.

November 2025

Monday	Tuesday	Wednesday	Thursday	Friday	Saturday	Sunday
27	28	29	30	31 **Moon Sign**	**1** **Pisces** Trine Sun Square Mercury Square Uranus	**2** **Pisces** Trine Jupiter Conjunct Saturn
3 **XhN** **Aries** Trine Mercury Trine Mars Conjunct Saturn Conjunct Neptune Sextile Pluto	**4** **Aries** Opposition Venus Square Jupiter	**5** **Taurus** Opposition Sun Square Pluto	**6** **FM P1** **Taurus** Opposition Sun	**7** **Gemini** Opposition Mercury Opposition Mars Conjunct Uranus Trine Pluto Square Ascendant	**8** **^** **Gemini** Square Saturn Square Ascendant	**9** **Cancer** Trine Venus Square Ascendant
10 **Cancer** Trine Sun Conjunct Jupiter Trine Saturn Trine Neptune Square Ascendant	**11** **Leo** Trine Mercury Square Venus Trine Mars Opposition Pluto Square Ascendant	**12** **3Q** **Leo** Square Sun	**13** **Virgo** Square Mercury Square Mars Square Uranus	**14** **Virgo** Opposition Saturn	**15** **XhS** **Virgo** Opposition Saturn Trine Uranus Opposition Neptune Trine Pluto	**16** **Libra**
17 **Libra** Square Jupiter	**18** **Scorpio** Conjunct Venus Square Pluto	**19** **Scorpio**	**20** **NM A** **Scorpio** Conjunct Sun Conjunct Venus Trine Jupiter Trine Saturn	**21** **Sagittarius** Conjunct Sun Conjunct Mercury Trine Jupiter Trine Saturn Opposition Uranus Trine Neptune Sextile Pluto	**22** **Sagittarius** Conjunct Mars	**23** **V** **Sagittarius** Square Saturn Square Neptune Square Ascendant
24 **Capricorn** Square Ascendant	**25** **Capricorn** Sextile Mercury Sextile Venus Opposition Jupiter	**26** **Aquarius** Sextile Sun Trine Uranus Conjunct Pluto	**27** **Aquarius** Square Mercury Sextile Mars	**28** **1Q** **Aquarius** Square Sun Square Venus Square Uranus	**29** **Pisces** Trine Mercury Square Mars	**30** **XhN** **Pisces** Trine Mercury trine Venus Trine Jupiter Conjunct Saturn Sextile Uranus Conjunct Neptune
1	2					

NORTH ISLAND RAINFALL
NOVEMBER 2025

d=mainly dry, n= fog, mist, dew or nonrecordable, l=light showers, p=passing showers, s=significant showers, r=rain, h=heavy falls

		Northland							Auckland					Bay Of Plenty				Waikato			
		CR	Kta	Ker	Rsl	Dar	Whg	Man	Wark	Hen	MH	Ak	Co	Tau	TeP	Wkt	Opo	Hml	TeA	Mata	Ot
1st	Sat	d	d	d	d	d	d	d	d	d	c	d	d	d	d	d	d	d	d	d	d
2nd	Sun	d	d	d	d	d	d	d	d	d	l	d	d	d	d	d	d	l	l	d	l
3rd	Mon	d	d	d	d	d	d	l	l	l	l	n	s	d	d	d	d	l	l	d	l
4th	Tue	d	d	d	d	l	l	l	l	r	l	l	l	d	d	d	d	d	d	d	d
5th	Wed	d	d	d	d	d	d	d	d	d	c	d	d	d	d	d	d	d	d	d	d
6th	Thur	d	d	d	d	d	d	d	d	d	c	d	d	d	d	d	d	d	d	d	d
7th	Fri	d	n	l	l	l	r	s	l	h	s	s	r	d	d	d	d	l	s	d	r
8th	Sat	l	d	d	l	d	n	d	d	l	l	d	d	d	d	d	d	d	d	d	d
9th	Sun	d	d	d	d	d	d	d	d	d	c	d	d	d	d	d	d	d	d	d	d
10th	Mon	d	d	d	d	d	d	d	d	d	c	d	d	d	d	d	d	d	d	d	d
11th	Tue	d	d	d	d	d	d	d	d	d	c	d	d	d	d	d	d	d	d	d	d
12th	Wed	d	d	d	d	d	d	d	d	l	l	d	d	l	n	l	d	l	d	d	l
13th	Thur	d	d	d	d	d	d	d	s	s	l	s	l	l	l	l	l	d	l	d	d
14th	Fri	d	d	d	d	d	d	n	d	d	c	d	d	d	d	d	d	d	d	d	l
15th	Sat	d	l	n	d	l	l	s	s	h	h	r	s	s	r	s	s	h	h	h	h
16th	Sun	s	h	h	h	r	s	s	r	r	l	s	l	d	d	d	d	d	d	d	l
17th	Mon	r	h	h	h	h	h	h	r	r	s	s	s	l	n	n	d	d	d	d	d
18th	Tue	l	l	r	r	s	s	s	s	l	c	n	l	d	d	n	d	d	d	d	d
19th	Wed	d	d	d	d	d	d	d	d	d	c	d	d	d	d	d	d	d	d	d	d
20th	Thur	d	d	d	d	d	d	d	d	d	c	d	d	d	d	d	d	d	d	d	d
21st	Fri	d	d	d	d	d	d	d	d	d	c	d	s	d	d	d	d	d	d	d	d
22nd	Sat	l	d	l	d	d	d	d	d	d	c	d	d	d	d	d	d	d	d	d	r
23rd	Sun	l	l	l	l	d	d	n	d	d	l	l	l	d	d	d	d	n	l	d	d
24th	Mon	l	s	d	d	s	d	s	s	s	s	s	s	l	l	n	d	s	r	s	r
25th	Tue	d	l	d	d	l	l	s	l	l	s	l	s	d	d	d	d	l	l	d	l
26th	Wed	d	d	d	d	d	d	d	d	d	l	d	d	n	d	d	d	l	s	d	l
27th	Thur	n	d	d	d	d	d	n	s	l	l	d	l	s	s	d	d	l	l	d	r
28th	Fri	n	d	d	d	d	d	d	d	d	n	d	d	d	d	d	d	d	d	s	d
29th	Sat	d	d	d	d	d	d	d	d	d	d	d	d	d	d	d	d	d	d	d	d
30th	Sun	d	d	d	d	d	d	d	d	d	d	d	d	d	d	d	d	d	d	d	d
estimate:		32	61	59	53	65	61	57	65	91	7[illegible]	47	64	21	19	8	7	57	86	34	102
average:		61	87	128	89	78	105	99	93	82	9[illegible]	80	85	82	131	85	98	95	103	103	117
trend:		drr	drr	drr	drr	drr	drr	drr	drr	av	dr[illegible]	drr	drr	drr	drr	drr	drr	drr	drr	drr	drr

		Rot/Tau		Hawkes Bay			C Plateau			Taranaki			Manawatu			Lower North Island					
		Rt	Taup	Gis	Nap	Has	Taum	Chat	Waio	NP	Str	Wang	Pa	Dann	Lev	PN	Mast	UH	Wgt	all	Moon
1st	Sat	d	d	d	d	d	d	s	l	l	d	d	d	l	l	l	d	d	d	12	
2nd	Sun	d	d	d	d	d	l	l	d	l	d	d	d	d	d	d	n	d	d	15	
3rd	Mon	n	d	d	d	d	d	s	d	d	d	d	d	d	d	d	d	d	d	23	XI-N
4th	Tue	d	d	d	d	d	d	d	d	d	d	d	d	d	d	d	l	d	d	23	
5th	Wed	d	d	d	d	d	d	d	d	d	d	d	d	d	d	d	d	d	d	0	
6th	Thur	d	d	l	l	l	d	d	l	d	d	d	d	s	l	l	l	d	d	17	F [illegible]1
7th	Fri	d	d	l	d	d	d	d	n	d	d	d	d	d	d	d	d	d	n	88	
8th	Sat	d	d	d	d	d	d	d	d	d	d	d	d	d	d	d	d	d	d	7	′
9th	Sun	d	d	d	d	d	d	d	d	d	d	d	d	d	d	d	d	d	d	0	
10th	Mon	d	d	d	d	d	d	d	d	d	d	d	d	d	d	d	d	d	d	0	
11th	Tue	d	d	d	d	d	d	d	d	l	n	d	l	d	9	d	d	s	d	21	
12th	Wed	d	l	l	d	d	s	r	d	s	l	l	l	d	l	d	d	n	n	38	3Q
13th	Thur	n	d	d	d	d	d	d	d	d	d	d	d	d	d	d	d	d	d	30	
14th	Fri	d	d	d	d	d	d	r	d	s	s	d	s	n	s	l	d	r	l	57	
15th	Sat	r	s	l	n	d	h	h	h	h	r	r	r	r	h	h	l	r	s	642	XI-S
16th	Sun	d	l	d	d	d	l	d	l	d	d	d	d	d	d	d	d	d	d	131	
17th	Mon	d	d	l	s	s	d	d	d	d	d	d	d	d	d	d	d	d	d	218	
18th	Tue	d	d	n	d	l	d	d	d	d	d	d	d	d	d	d	d	d	d	64	
19th	Wed	d	d	l	l	d	d	l	d	d	d	d	d	d	d	d	d	d	d	5	
20th	Thur	d	d	d	d	d	d	d	d	d	d	d	d	d	d	d	d	d	d	0	N A
21st	Fri	d	d	l	l	d	d	d	d	d	d	d	d	d	d	d	d	d	d	12	
22nd	Sat	d	d	l	s	l	d	n	l	d	d	d	d	d	d	d	d	d	d	27	
23rd	Sun	d	d	d	d	d	d	d	d	d	d	d	d	d	d	d	d	d	d	17	V
24th	Mon	l	s	d	l	l	s	h	h	s	r	h	h	r	r	s	r	h	s	288	
25th	Tue	d	d	l	d	l	l	r	l	d	l	d	l	d	d	d	l	l	d	58	
26th	Wed	d	l	d	d	d	l	r	l	r	l	l	l	r	r	r	s	l	s	91	
27th	Thur	r	l	l	d	s	l	l	d	d	d	l	d	l	d	d	d	d	d	62	
28th	Fri	d	d	d	d	d	d	d	d	d	d	d	d	d	d	d	d	d	d	8	1Q
29th	Sat	d	d	d	d	d	d	d	d	d	d	d	d	d	d	d	d	d	d	0	
30th	Sun	d	d	d	d	d	d	d	d	d	d	d	d	d	d	d	d	d	d	0	XI-N
estimate:		23	16	16	19	21	63	207	52	59	30	34	40	37	101	58	26	66	16	1951	
average:		105	95	55	60	52	129	247	82	120	163	67	84	85	87	82	65	111	87	3667	
trend:		drr	drr	drr	drr	drr	drr	drr	drr	drr	drr	drr	drr	drr	wtr	drr	drr	drr	drr	drr	

		SOUTH ISLAND RAIN NOVEMBER 2025 d=mainly dry, n= fog, mist, dew or nonrecordable, l=light showers, p=passing showers, s=significant showers, r=rain, h=heavy falls																																	
		West Coast				Top South Island				Canterbury				South Canterbury				Coastal Otago			Lakes			Inland Otago				Southland						
		West	Grey	Hok	BB	Ta	Mou	Nel	Ble	Dar	Ham	Kai	Ch	Rak	Meth	Ash	Tim	Oam	Dun	Bal	Twi	LP	LT	Crom	Qstn	Ran	Alex	TeA	Wtn	Gore	Inv	Puy Pt	all	Moon
1st	Sat	r	l	l	r	d	d	d	d	d	d	l	d	d	d	d	d	l	l	l	h	d	d	d	l	l	d	s	l	s	s	r	90	
2nd	Sun	s	r	h	r	l	l	l	d	d	l	d	d	d	d	d	d	d	l	s	d	d	l	d	l	n	d	s	l	s	s	l	105	
3rd	Mon	l	l	s	r	n	l	d	d	d	d	d	d	d	d	d	d	d	d	d	d	s	d	d	d	d	d	s	l	l	l	h	57	XhN
4th	Tue	l	s	s	s	l	d	d	d	n	l	d	d	d	d	l	l	l	s	r	s	d	l	s	r	l	s	r	r	h	h	r	142	
5th	Wed	l	l	l	l	d	d	d	d	d	d	d	d	d	d	d	d	d	d	d	r	d	d	d	d	d	d	d	d	d	l	r	32	
6th	Thur	l	d	d	d	d	d	d	d	s	s	s	l	l	s	l	l	d	l	l	d	d	d	d	d	d	d	d	d	l	d	d	45	F P1
7th	Fri	r	d	d	d	d	l	l	d	d	d	l	d	d	d	d	d	d	d	d	d	d	d	d	d	d	d	d	d	d	d	d	18	
8th	Sat	d	l	d	d	d	d	d	d	d	d	d	d	d	d	d	d	d	d	d	d	d	d	d	d	d	d	d	d	d	d	d	1	^
9th	Sun	d	l	l	l	d	d	d	d	d	d	d	d	d	d	d	d	d	d	d	d	d	d	d	d	d	d	l	d	d	d	d	4	
10th	Mon	d	d	d	s	d	d	d	d	d	d	d	d	d	d	d	d	d	l	l	d	d	d	d	d	d	d	r	s	d	s	h	69	
11th	Tue	9	r	r	h	d	d	d	d	d	d	d	d	d	d	d	d	d	d	l	d	d	d	d	l	d	l	s	s	s	r	h	153	
12th	Wed	d	d	d	d	d	d	d	d	n	d	d	d	d	n	s	d	l	d	d	l	d	d	d	d	l	d	d	d	d	d	d	10	3Q
13th	Thur	d	d	d	d	d	d	d	d	d	d	d	d	d	d	d	d	d	d	d	d	d	d	d	d	d	d	d	d	d	d	l	2	
14th	Fri	s	s	s	d	r	d	d	n	d	d	d	d	d	d	d	d	d	d	l	d	d	d	d	d	l	d	d	d	d	n	s	42	
15th	Sat	s	s	s	l	l	l	l	l	l	l	l	d	l	l	l	s	l	l	n	d	d	d	d	d	d	d	d	s	s	s	s	80	XhS
16th	Sun	d	l	l	d	d	d	d	d	l	d	d	l	d	d	l	n	d	l	s	d	d	d	d	d	d	d	d	s	r	l	l	35	
17th	Mon	d	d	l	d	d	d	d	d	d	d	d	d	d	d	d	d	d	l	d	d	d	d	d	d	d	d	d	l	l	s	s	15	
18th	Tue	d	l	l	d	l	d	n	d	d	d	d	d	d	d	d	l	l	d	d	d	d	d	d	d	d	d	d	d	n	d	l	7	
19th	Wed	d	d	d	d	d	d	d	d	d	d	d	d	d	d	d	l	d	d	d	d	d	d	d	d	d	d	d		d	d	l	3	
20th	Thur	d	d	d	d	d	d	d	d	d	d	d	l	d	d	d	d	d	d	d	d	d	d	d	d	d	d	d	l	n	d	l	3	N A
21st	Fri	d	d	d	d	d	d	d	l	d	d	l	d	d	d	d	d	d	n	d	d	d	d	d	d	d	d	d	l	l	d	l	8	
22nd	Sat	d	l	d	d	d	d	d	d	d	d	l	d	d	d	d	d	d	d	d	d	d	d	d	d	l	d	d	d	d	d	l	8	
23rd	Sun	r	h	h	h	d	d	d	d	d	d	d	d	d	d	d	d	d	s	l	d	h	d	d	l	d	d	r	s	s	r	h	161	V
24th	Mon	r	r	s	s	l	n	l	l	h	h	r	h	r	h	h	r	h	s	s	l	d	r	r	l	s	s	l	r	s	s	l	279	
25th	Tue	d	l	d	d	s	d	l	d	l	l	l	l	d	l	d	l	d	n	l	s	d	d	d	d	d	d	d	d	l	l	l	31	
26th	Wed	r	s	s	l	d	d	d	l	d	l	s	d	d	d	l	l	d	d	d	d	d	d	d	d	l	s	d	d	l	l	l	49	
27th	Thur	d	d	d	d	d	d	d	l	d	d	d	d	d	l	d	d	d	d	s	d	d	d	d	d	d	d	d	d	d	d	l	6	
28th	Fri	d	d	d	d	d	d	d	d	d	d	d	d	d	d	d	d	d	d	d	d	d	d	d	d	d	d	d	d	d	d	d	0	1Q
29th	Sat	d	d	d	d	d	d	d	d	d	d	d	d	d	d	d	d	d	d	d	d	d	d	d	d	d	d	d	d	d	l	d	1	
30th	Sun	d	d	n	d	d	d	d	d	d	d	d	d	d	d	d	d	d	d	d	d	d	d	d	d	d	d	d	d	d	d	d	0	XhN
estimate:		**104**	**102**	**103**	**143**	**22**	**8**	**8**	**9**	**27**	**37**	**33**	**27**	**20**	**31**	**41**	**27**	**25**	**32**	**36**	**53**	**24**	**13**	**14**	**19**	**18**	**17**	**58**	**62**	**77**	**81**	**187**	**1456**	
average:		175	210	260	398	194	97	75	50	59	96	61	47	69	72	63	47	47	80	52	51	68	46	34	62	39	28	76	69	94	85	163	2967	
trend:		drr	drr	drr	drr	drr	drr	drr	drr	drr	drr	drr	drr	drr	drr	drr	drr	drr	drr	drr	**av**	drr	drr	drr	drr	drr	drr	drr	**av**	drr	**av**	wtr	drr	

NORTH ISLAND SUNSHINE ESTIMATES
NOVEMBER 2025

F=fine (8-12 hours of sunshine), pc= partly cloudy (4-7 hours), c=cloudy (1-3 hours), o=overcast (0 hours)

		Kta	Whg	Dar	Ak	Hml	Tau	Rt	Taum	NP	Wang	Taup	Gis	Nap	Wai	Dann	PN	Mast	Pa	UH	Wgt	ALL	Moon
1st	Sat	pc	o	c	c	pc	F	pc	pc	F	pc	pc	F	F	F	c	pc	c	F	pc	F	123	
2nd	Sun	F	pc	pc	pc	F	F	F	c	c	o	F	F	F	F	pc	o	pc	pc	pc	F	136	
3rd	Mon	F	pc	o	o	F	F	F	F	F	c	F	F	F	F	F	c	F	F	pc	F	172	XhN
4th	Tue	F	F	F	F	F	F	F	F	F	pc	F	F	F	F	F	pc	F	F	F	F	224	
5th	Wed	F	pc	pc	pc	F	F	F	F	pc	o	F	F	F	F	F	o	F	F	F	F	184	
6th	Thur	pc	pc	F	F	pc	pc	F	pc	F	pc	F	c	o	pc	pc	pc	pc	F	c	pc	123	F P1
7th	Fri	c	pc	c	c	F	F	F	F	F	F	F	pc	pc	pc	pc	F	pc	F	pc	F	147	
8th	Sat	pc	F	F	F	F	F	F	F	F	F	F	F	F	F	F	F	F	F	F	F	230	^
9th	Sun	c	F	F	F	F	F	F	F	F	F	F	F	F	F	F	F	F	F	F	F	220	
10th	Mon	F	F	F	F	F	F	F	F	F	F	F	F	F	F	F	F	F	F	pc	F	223	
11th	Tue	F	F	pc	pc	o	c	o	o	c	o	o	F	c	o	c	o	c	o	o	o	50	
12th	Wed	pc	o	o	o	pc	F	pc	F	pc	F	pc	pc	F	F	F	F	F	F	pc	F	131	3Q
13th	Thur	pc	pc	pc	pc	c	pc	c	pc	o	pc	c	F	F	F	pc	pc	pc	pc	pc	pc	113	
14th	Fri	o	o	o	o	o	o	o	o	o	o	o	o	F	c	c	o	c	o	o	o	18	
15th	Sat	o	o	o	o	c	c	c	pc	pc	pc	c	o	pc	F	F	pc	F	F	F	pc	94	XhS
16th	Sun	o	o	o	o	o	o	o	o	c	F	o	o	o	o	pc	F	pc	F	pc	F	52	
17th	Mon	c	o	c	c	pc	c	c	F	F	pc	c	o	F	F	F	pc	F	F	F	F	122	
18th	Tue	F	pc	pc	pc	F	F	F	pc	F	pc	F	c	F	pc	pc	pc	pc	pc	pc	F	136	
19th	Wed	c	pc	F	F	F	F	F	F	F	F	F	F	F	F	pc	F	pc	F	F	F	207	
20th	Thur	pc	F	F	F	F	F	F	F	F	c	F	pc	F	pc	o	c	o	pc	c	c	148	N A
21st	Fri	pc	F	F	F	pc	F	F	F	F	F	F	pc	F	pc	F	F	F	F	pc	F	170	
22nd	Sat	c	pc	pc	pc	F	F	pc	F	F	F	pc	F	F	F	F	F	F	F	F	F	178	
23rd	Sun	F	F	F	F	F	F	F	pc	pc	c	F	F	F	pc	pc	c	pc	o	c	c	133	V
24th	Mon	pc	pc	F	F	F	F	F	pc	pc	pc	F	pc	pc	c	pc	pc	pc	F	pc	F	145	
25th	Tue	pc	F	F	F	F	F	F	pc	pc	o	F	F	F	pc	pc	o	pc	o	o	pc	134	
26th	Wed	o	c	o	o	c	o	o	o	c	pc	o	pc	o	o	pc	pc	pc	pc	pc	pc	45	
27th	Thur	pc	pc	F	F	F	F	F	F	F	F	F	F	F	pc	F	F	F	F	F	F	208	
28th	Fri	pc	F	pc	pc	pc	F	F	F	F	F	F	F	F	F	F	F	F	F	F	F	216	1Q
29th	Sat	pc	F	F	F	F	F	F	F	F	F	F	F	F	F	F	F	F	F	F	F	221	
30th	Sun	pc	pc	F	F	F	F	F	F	F	F	F	F	F	F	F	F	F	F	F	F	215	XhN
est. hours:		165	177	186	186	225	270	244	227	258	207	244	255	274	248	212	207	212	253	196	269	4515	
av. hours:		190	192	182	187	197	215	201	164	204	205	194	215	214	185	177	172	199	198	204	210	3905	
trend:		less	av	av	av	more	more	more	more	more	av	more	more	more	more	more	more	av	more	av	more	more	

SOUTH ISLAND SUNSHINE ESTIMATES
NOVEMBER 2025

F=fine (8-12 hours of sunshine), pc= partly cloudy (4-7 hours), c=cloudy (1-3 hours), o=overcast (0 hours)

		Ta	Nel	Ble	Mou	West	Grey	Hok	Ham	Kai	Ch	Ash	Tim	Crom	Qstn	Bal	Dun	Wnt	Gore	Inv	Alex	ALL	Moon
1st	Sat	F	F	F	F	F	F	F	F	F	F	F	F	F	F	F	F	pc	pc	pc	F	193	
2nd	Sun	F	F	F	F	pc	F	F	F	F	F	F	F	F	F	F	F	pc	pc	pc	F	198	
3rd	Mon	o	F	F	F	o	o	o	F	F	F	F	pc	pc	o	pc	pc	pc	pc	c	pc	111	XhN
4th	Tue	c	F	F	F	c	F	F	F	F	F	F	F	F	F	F	F	F	F	F	F	207	
5th	Wed	F	c	F	c	c	pc	F	pc	pc	pc	pc	F	F	F	pc	c	pc	pc	pc	F	128	
6th	Thur	F	pc	pc	pc	F	F	F	pc	pc	F	pc	F	F	F	pc	pc	pc	pc	pc	F	158	F P1
7th	Fri	F	F	F	pc	pc	c	pc	F	F	F	F	F	F	F	F	F	F	F	F	F	215	
8th	Sat	pc	F	F	F	pc	F	F	F	F	F	F	F	F	F	F	F	F	F	F	F	231	^
9th	Sun	F	F	F	F	pc	pc	c	F	F	F	F	pc	c	pc	c	pc	c	c	c	c	137	
10th	Mon	c	F	F	F	pc	o	o	pc	pc	pc	pc	c	pc	pc	o	o	o	o	c	pc	88	
11th	Tue	pc	pc	o	pc	F	F	F	pc	pc	pc	pc	pc	F	F	F	F	F	F	F	F	162	
12th	Wed	F	F	F	F	F	F	F	c	c	pc	pc	F	F	F	F	F	F	F	F	F	196	3Q
13th	Thur	F	pc	F	pc	pc	pc	pc	F	F	F	F	pc	pc	c	c	c	c	c	c	pc	114	
14th	Fri	F	pc	pc	pc	c	o	o	F	F	F	F	pc	pc	F	c	c	c	c	pc	pc	111	
15th	Sat	F	F	F	F	pc	F	F	pc	pc	pc	pc	F	F	F	pc	F	pc	pc	pc	F	166	XhS
16th	Sun	F	F	c	F	F	F	F	pc	pc	pc	pc	pc	F	F	pc	pc	pc	pc	F	F	156	
17th	Mon	pc	F	F	F	pc	F	F	F	F	F	F	F	F	F	F	pc	F	F	F	F	187	
18th	Tue	F	F	pc	F	c	pc	pc	pc	pc	pc	pc	pc	F	F	F	F	F	F	F	F	163	
19th	Wed	F	F	F	F	F	F	pc	F	F	F	F	F	F	F	c	pc	F	F	c	F	206	
20th	Thur	pc	pc	o	c	F	F	pc	o	o	o	c	pc	F	F	pc	c	c	c	F	F	92	N A
21st	Fri	F	F	F	pc	pc	pc	pc	F	F	F	F	F	F	F	c	c	c	c	c	F	160	
22nd	Sat	pc	F	F	F	c	c	pc	F	F	F	F	F	c	pc	o	pc	o	o	o	c	121	
23rd	Sun	c	pc	pc	pc	c	c	pc	pc	pc	pc	c	pc	pc	pc	c	c	pc	pc	pc	pc	83	V
24th	Mon	F	F	pc	F	F	F	F	c	c	pc	pc	pc	pc	pc	F	pc	F	F	pc	pc	147	
25th	Tue	F	pc	F	F	c	o	c	pc	pc	pc	c	pc	pc	F	o	o	c	c	pc	pc	82	
26th	Wed	pc	pc	c	pc	pc	F	F	c	c	pc	c	c	pc	pc	c	c	c	c	pc	pc	86	
27th	Thur	F	F	F	F	F	F	F	F	F	F	F	F	F	F	F	F	F	F	pc	F	244	
28th	Fri	F	F	F	F	F	F	F	F	F	F	F	pc	F	F	o	c	c	c	c	F	176	1Q
29th	Sat	F	F	F	F	F	pc	c	F	F	F	F	F	F	F	pc	F	F	F	F	F	198	
30th	Sun	F	F	F	F	pc	pc	pc	F	F	F	F	F	F	F	F	F	F	F	F	F	225	XhN
est. hours:		272	294	280	283	195	198	208	246	246	264	262	247	276	267	169	187	185	185	201	276	4739	
av. hours:		228	225	230	224	173	161	182	168	202	215	181	185	229	208	170	161	177	168	176	211	3874	
trend:		more	more	more	more	more	more	more	more	more	more	more	more	more	more	av	more	av	av	more	more	more	

NOVEMBER 1 SATURDAY

A large high projects a high pressure ridge over the north of New Zealand and directs a moderate west to southwest flow over the North Island, while low pressure to the southeast of the country brings a fresh southwest flow over the South Island. Mostly cloudy in the far north of New Zealand and over the top half of the South Island while mostly sunny elsewhere. Outbreaks of showers in the west and south of the South Island and isolated showers in central parts of the North Island. Mainly dry elsewhere.

Auckland:	Cloudy, dry, mild temperatures, fresh breezes at times
Hamilton:	Partly cloudy, dry, mild temperatures, fresh breezes
Wellington:	Dry, warmer, sunny, fresh winds
Greymouth:	Fresh breezes, sunny day, brief morning showers
Christchurch:	Sunny, dry, fresh breezes
Dunedin:	Sunny, breezy, mild temperatures, brief morning showers

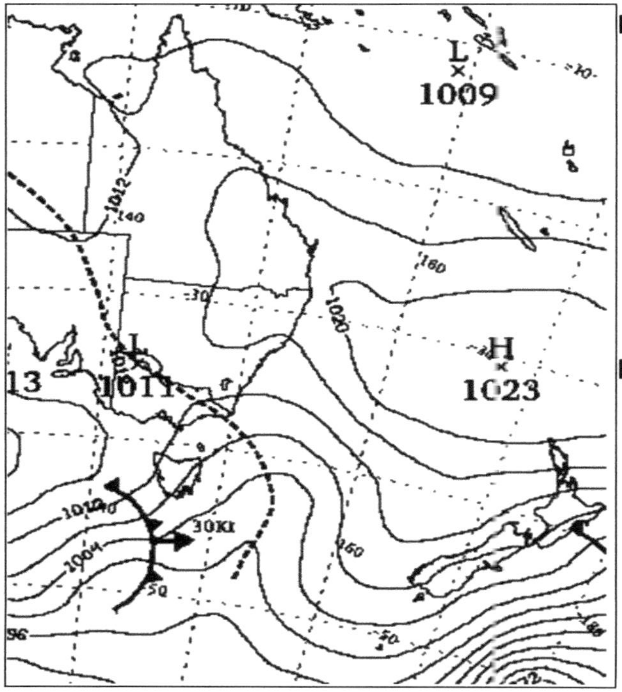

RAIN POTENTIAL

FROST/SNOW

NOVEMBER 2 SUNDAY

A Southern Ocean depression nears the southeast of the South Island directing very strong to storm force northwest winds throughout the South Island, more westerly over the North Island, which still lingers under a milder high pressure ridge. Mostly dry in the North Island except for some showers in the central regions to Taranaki. In contrast, a mix of sunny spells and squally showers prevail over the west and south of the South Island, with some heavy falls likely in the West Coast. The very strong prevailing west to northwest winds may see some damage to infrastructure in places.

Auckland:	Partly cloudy, mostly dry, strong gusty winds, possible gales
Hamilton:	Sunny, warmer, brief showers, strong winds, possible gales at times
Wellington:	Sunny, dry, cooler, strong winds, chance of storm force for a time
Greymouth:	Morning rain clearing to sunny day, strong winds, gales possible
Christchurch:	Stormy winds, very sunny and dry, warmer
Dunedin:	Sunny, cooler, light showers, strong winds to gale force likely

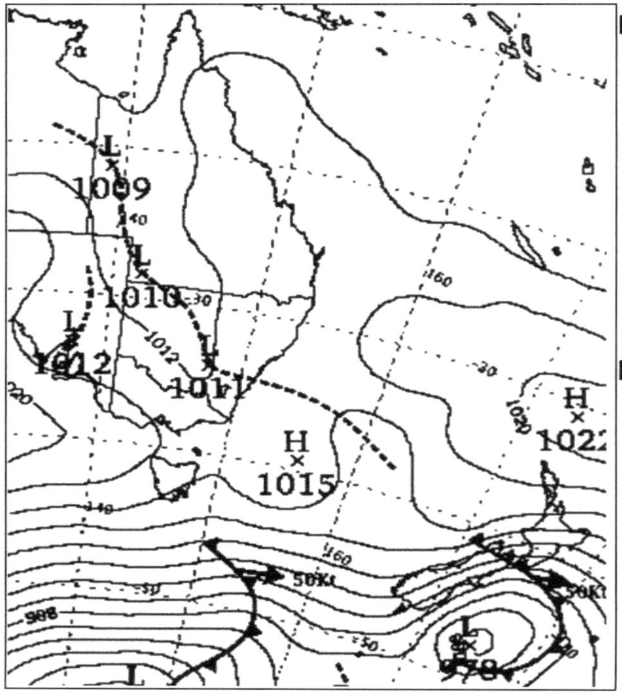

RAIN POTENTIAL

FROST/SNOW

NOVEMBER 3 MONDAY

A strong and gusty westerly flow prevails throughout the country, with winds likely to reach storm force at times over the South Island, gales through the Cook Strait and into the lower half of the North Island, while a lighter high pressure ridge extends a moderate westerly flow over the northern half of the North Island. Mostly dry and cloudy throughout the North Island except for some isolated showers through central parts, and mostly dry in Marlborough to Canterbury regions, with outbreaks of rain elsewhere.

Auckland: Unsettled breezes easing, overcast, drizzle patches and showers
Hamilton: Very sunny, brief morning showers, strong winds easing for a time
Wellington: Sunny, warmer, dry, strong winds to near gale force at times
Greymouth: Overcast, brief squalls, strong winds to gale force likely
Christchurch: Sunny, dry, warm day cool night, strong blustery winds
Dunedin: Partly cloudy, dry, strong winds to gale force at times

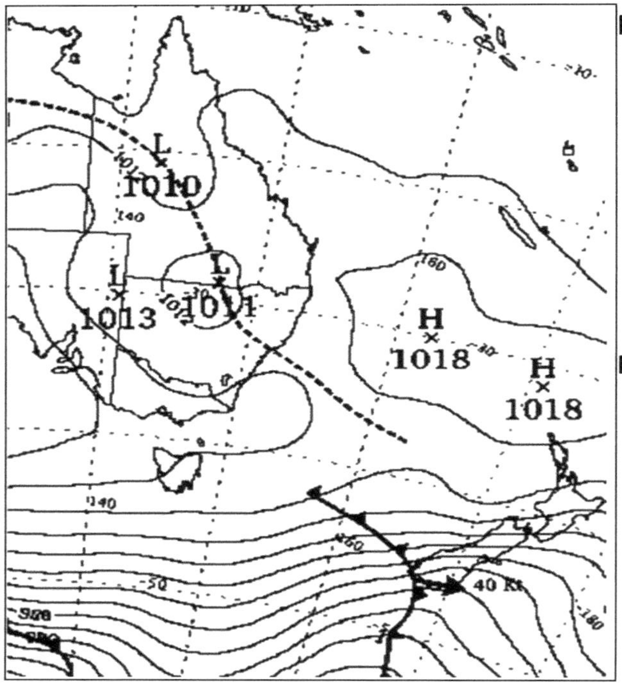

RAIN POTENTIAL

FROST/SNOW

NOVEMBER 4 TUESDAY

The high pressure ridge extends a mostly dry with regular cloudy spells over the North Island with light to moderate southwest winds prevailing. In contrast, strong and gusty westerly winds persist over the South Island with outbreaks of rain in most districts, heavy at times in Southland and Otago and more sporadic elsewhere.

Auckland: Sunny, mostly dry day overnight showers, moderate breezes
Hamilton: Sunny, dry, light breezes
Wellington: Very sunny, cooler, dry, moderate winds
Greymouth: Mix of sunny spells and showers, cooler, strong and unsettled winds
Christchurch: Very sunny, dry, mild day cold night, chance of frost patches, blustery
Dunedin: Very sunny day, overnight showers, blustery, chance of gales

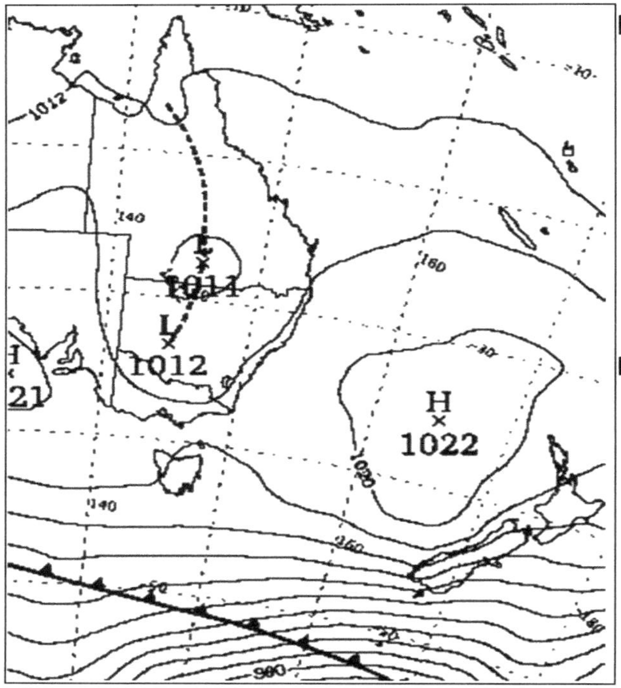

RAIN POTENTIAL

FROST/SNOW

NOVEMBER 5 WEDNESDAY

A couple of anticyclonic centres build in the Tasman Sea, extending a high pressure ridge along the west of the country and directing a strong southwesterly flow throughout. Increasing sunny spells everywhere, with some showers in Fiordland to Otago, and mostly dry elsewhere.

Auckland: Mostly cloudy and dry, fresh winds with occasional gusts
Hamilton: Sunny, cooler, dry, fresh breezes, blustery at times
Wellington: Sunny and dry, strong winds
Greymouth: Cloudy, warmer, brief showers, strong winds
Christchurch: Partly cloudy, milder, dry, strong gusty winds at times
Dunedin: Cloudy, warmer, dry, strong winds with occasional gusts

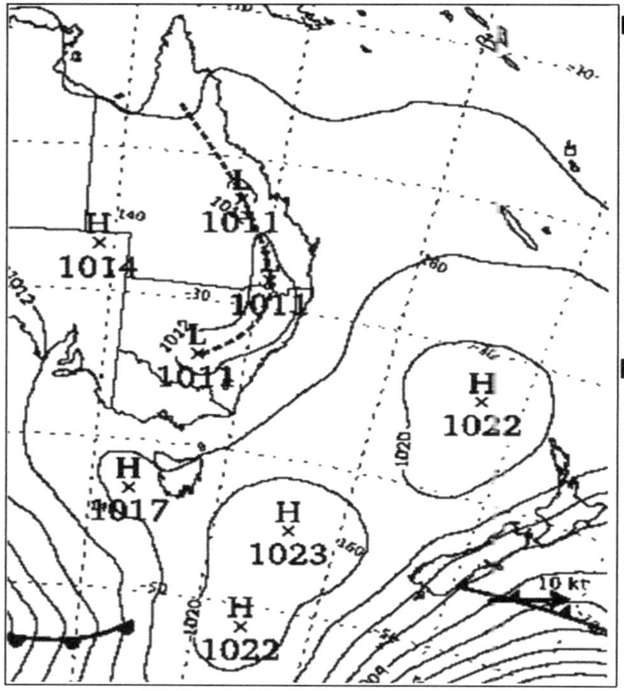

RAIN POTENTIAL

FROST/SNOW

NOVEMBER 6 THURSDAY

A large anticyclonic system crosses to the south of the country while a weak low pressure trough forms over central New Zealand. Light to moderate southwest breezes prevail at first, backing southeasterly later in the day. A mixed day through the east of the country from Otago to the East Cape with outbreaks of light showers between sunny spells. Mostly dry and sunny in the north and west of the North Island, and in South Canterbury, Southland, and southern parts of the West Coast to Fiordland.

Auckland: Sunny, dry, light breezes to mostly calm air
Hamilton: Partly cloudy, dry, gentle breezes
Wellington: Partly cloudy, dry, slightly warmer day cooler night, lighter breezes
Greymouth: Very bright and sunny, dry, mild day cool night, calm and pleasant air
Christchurch: Sunny, warmer, brief showers, moderate breezes
Dunedin: Partly cloudy, cooler, light showers, light shifting breezes

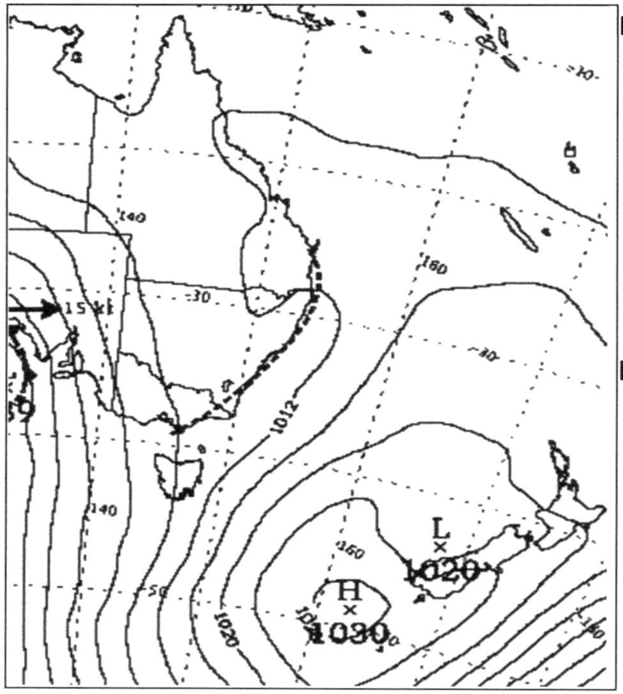

RAIN POTENTIAL

FROST/SNOW

NOVEMBER **7** FRIDAY

A large anticyclonic system crosses the country directing a light easterly flow over the North Island, backing westerly over the South Island. Cloudy with some showers Northland to the Waikato, and in the Nelson to North Westland regions, while mostly dry with occasional sunny spells elsewhere. Colder with frosts possible in the south and east of the South Island, Wellington to Taranaki and around the Bay of Plenty. Chance of fog in sheltered districts, possibly turning to drizzle in the Wellington region.

Auckland: Gentle breezes, cloudy, scattered showers and pockets of heavy rain
Hamilton: Very sunny, mostly calm air, chance of brief showers
Wellington: Very sunny, colder, isolated showers, gentle shifting breezes
Greymouth: Cloudy, warmer, dry, gentle shifting breezes
Christchurch: Colder, sunny, dry, light shifting breezes, chance of fog patches
Dunedin: Very sunny, cool and dry, mostly calm air

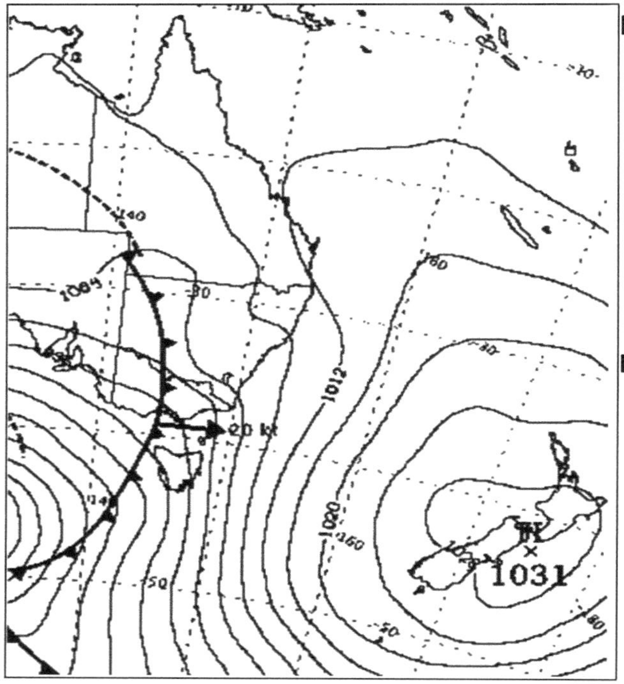

RAIN POTENTIAL

FROST/SNOW

NOVEMBER **8** SATURDAY

A large anticyclonic system slowly crosses the country in a gentle easterly air over the North Island, backing north to northwesterly over the South Island. Mostly dry and sunny throughout New Zealand with possible patches of fog in sheltered districts overnight, and risk of frosts from the Waikato to Wellington, in eastern parts of the North Island from the Bay of Plenty to the East Cape, and in Southland and Otago.

Auckland: Sunny, mostly dry except for odd isolated shower, calm air
Hamilton: Very sunny, dry, mild day cool night, mostly calm air
Wellington: Dry, cool, sunny, light shifting breezes
Greymouth: Sunny, colder, isolated showers, gentle breezes
Christchurch: Very sunny, dry, light breezes, chance of frost and fog patches
Dunedin: Very bright and sunny, dry, slightly warmer, light air

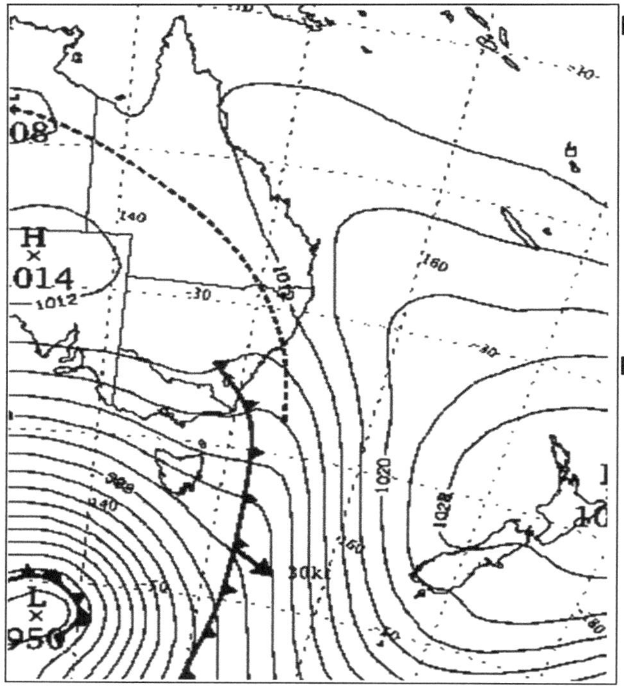

RAIN POTENTIAL

FROST/SNOW

NOVEMBER 9 SUNDAY

High pressure remains stalled over the country with light shifting breezes prevailing over the North Island, and fresher westerly breezes over the South Island and milder temperatures throughout. Sunny and dry almost everywhere except for a cold front crossing the southwest of the South Island bringing increased cloud and the odd threat of rain developing into Southland, Fiordland, and Otago.

Auckland: Sunny, warmer day cooler night, dry, light shifting breezes
Hamilton: Sunny, dry, warmer day cool night, light breezes to mostly calm air
Wellington: Sunny, slightly warmer, dry, moderate breezes
Greymouth: Partly cloudy, warmer, light showers, fresh breezes at times
Christchurch: Bright and sunny, slightly warmer, dry, moderate breezes
Dunedin: Partly cloudy, warmer, dry, moderate breezes

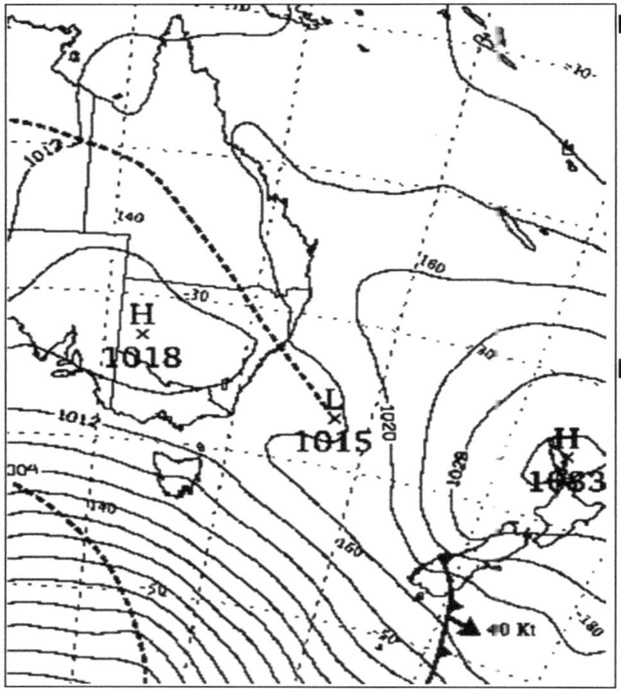

RAIN POTENTIAL

FROST/SNOW

NOVEMBER 10 MONDAY

A large high remains stalled over the North Island bringing light gentle breezes, increased scattered cloud and dry conditions throughout, with the chance of pockets of frost in the Waikato to the Bay of Plenty and Central Plateau. In contrast, a wet front moves onto the south of the South Island bringing rain into Fiordland and showers into Southland and Otago, in a moderate and warming westerly flow. Dry and mostly sunny elsewhere.

Auckland: Very sunny and dry, gentle breezes to mostly calm air
Hamilton: Sunny, dry, mild day cool night, gentle breezes to mostly calm air
Wellington: Sunny, dry, slightly warmer, moderate winds
Greymouth: Overcast, dry with passing threats of rain, unsettled breezes
Christchurch: Partly cloudy, dry, warm day cool night, gentle breezes
Dunedin: Overcast, very warm and humid air, some showers, gentle breezes

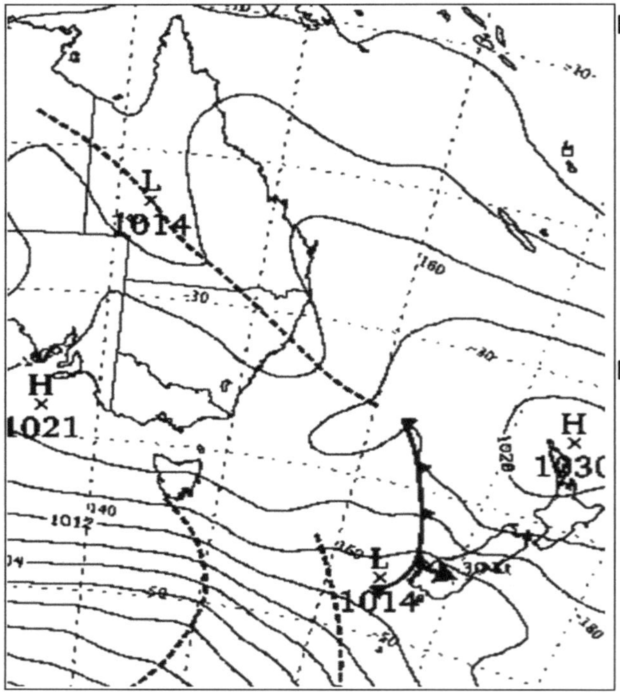

RAIN POTENTIAL

FROST/SNOW

NOVEMBER **11** TUESDAY

High pressure continues to bring a mostly cloudy but dry day over the North Island in a light to moderate northwest flow. Meanwhile, a low pressure trough between two high pressure centres crosses the South Island, with light westerly breezes prevailing, and a mostly dry and sunny day followed by evening rain in the west and south of the South Island. Chance of some showers brushing the south of the North Island later in the day.

Auckland: Partly cloudy, dry, moderate breezes
Hamilton: Overcast, dry with odd threat of rain, light breezes
Wellington: Overcast, dry, odd threat of rain, moderate winds
Greymouth: Mostly sunny, evening rain, light breezes
Christchurch: Partly cloudy, warm, dry, gentle breezes
Dunedin: Sunny, cooler, dry, gentle breezes

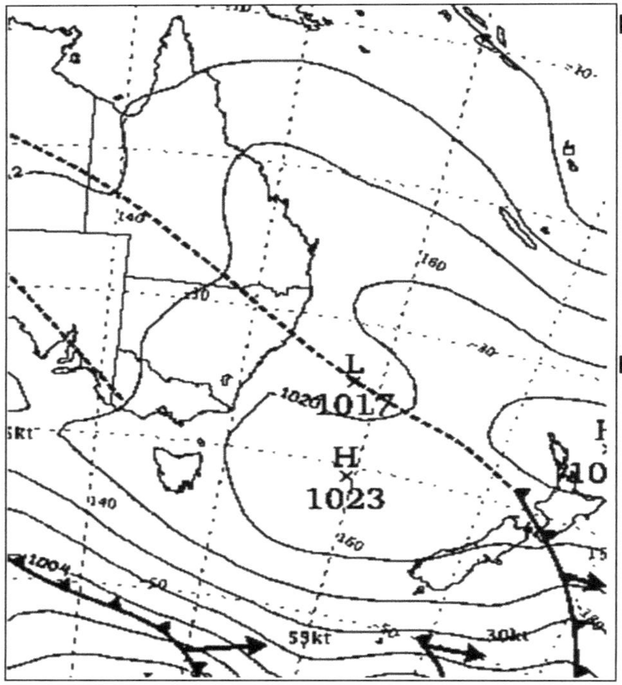

RAIN POTENTIAL

FROST/SNOW

NOVEMBER **12** WEDNESDAY

High pressure slowly crosses New Zealand in a mostly calm to light westerly air throughout. Mostly cloudy in Northland, mostly sunny elsewhere. Some evening showers likely around the Central Plateau to Taranaki and in South Canterbury, and dry elsewhere.

Auckland: Overcast, warmer, isolated showers, gentle breezes
Hamilton: Partly cloudy, isolated showers, mostly serene air
Wellington: Sunny, isolated evening showers, slightly cooler, mostly calm air
Greymouth: Mostly calm and pleasant air, very sunny, dry
Christchurch: Partly cloudy, dry, calm and serene air, chance of shallow fog patches
Dunedin: Sunny, dry, slightly cooler, calm and serene air

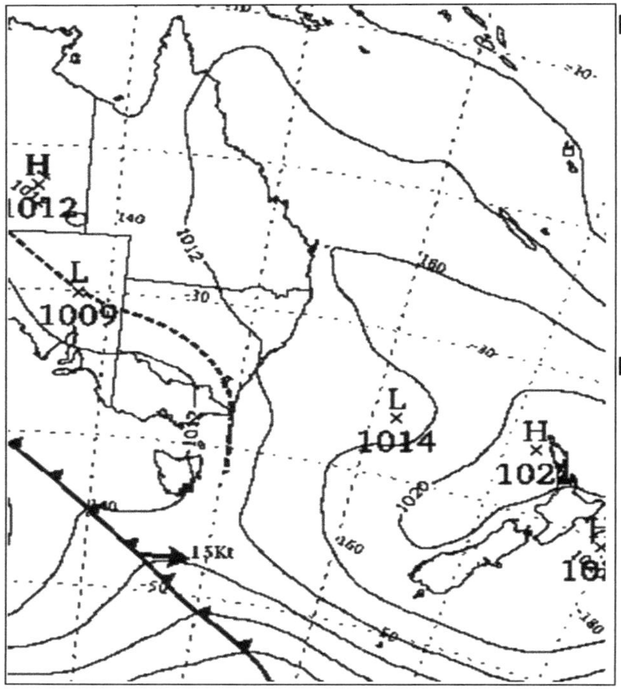

RAIN POTENTIAL

FROST/SNOW

NOVEMBER 13 THURSDAY

Mostly dry, calm, and sunny throughout the country as a large high continues to linger over New Zealand directing a very light northwesterly flow throughout. Cloudier in the south of the South Island and in the Auckland to Waikato regions, where some isolated evening showers may be observed. Chance of shallow fog in sheltered parts, particularly in Canterbury.

Auckland: Partly cloudy, light showers mostly calm air
Hamilton: Cloudy, warmer, dry, light air
Wellington: Gentle shifting breezes, partly cloudy, dry, warmer
Greymouth: Light and cool air, partly cloudy, dry
Christchurch: Calm and cool air, sunny, dry, chance of shallow fog patches
Dunedin: Cloudy, moderate breezes, cooler, dry

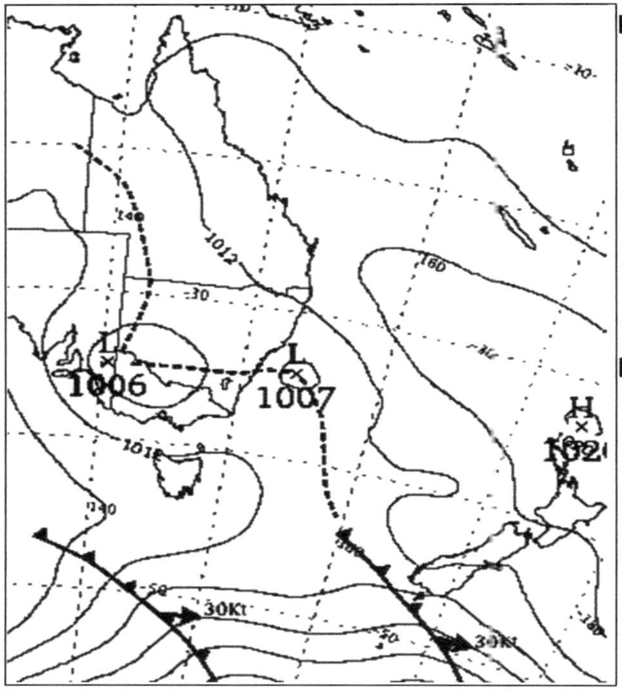

RAIN POTENTIAL

FROST/SNOW

NOVEMBER 14 FRIDAY

A depression moves up onto the South Island as the high departs to the northeast of the North Island, with a freshening southwest change over the South Island and moderate northwesterlies over the North Island. Mostly cloudy and dull over the North Island with some showers from Taranaki to Wellington, while mostly sunny with passing evening showers from the Nelson-Marlborough region to the West Coast and Fiordland. Dry elsewhere.

Auckland: Overcast, mostly dry, increasing threats of rain, fresh breezes
Hamilton: Overcast, mostly dry, increasing threats of rain, moderate breezes
Wellington: Overcast, brief showers, moderate shifting breezes
Greymouth: Overcast, rain, fluctuating breezes
Christchurch: Sunny, slightly warmer, dry, gentle shifting breezes
Dunedin: Cloudy, slightly warmer, dry, light shifting breezes

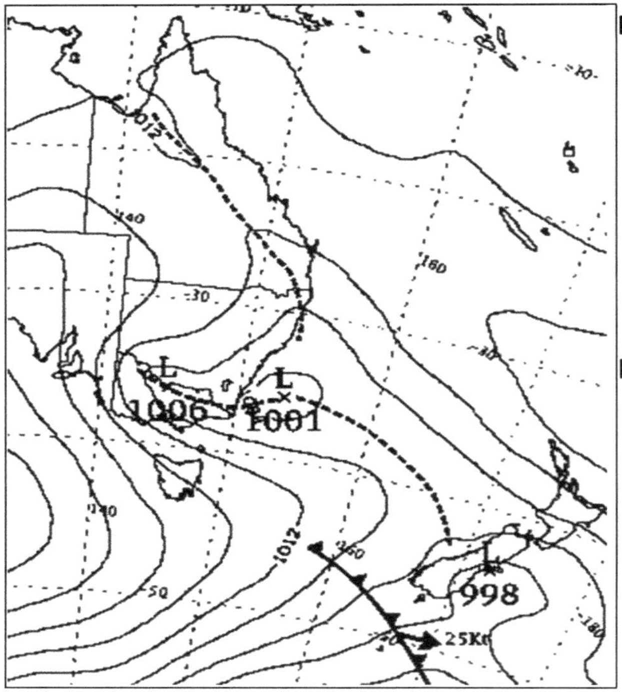

RAIN POTENTIAL

FROST/SNOW

NOVEMBER **15** SATURDAY

A strong southwest flow moves up over New Zealand with a wet front bringing scattered rain activity into most districts as the day progresses, with heavy falls at times around the Waikato to Taranaki and into the Manawatu and lighter scattered rain activity elsewhere, and chance of thunderstorms and hail showers particularly over the southern half of the North Island.

Auckland:	Overcast, rain, heavy downpours, fresh winds with occasional gusts
Hamilton:	Unsettled and blustery air, cloudy, heavy rain
Wellington:	Partly cloudy, occasional showers, fresh winds
Greymouth:	Sunny and mostly dry day, overnight rain, fresh to strong breezes
Christchurch:	Partly cloudy, dry, warmer day cooler night, fresh breezes at times
Dunedin:	Breezy, sunny, brief evening showers, colder

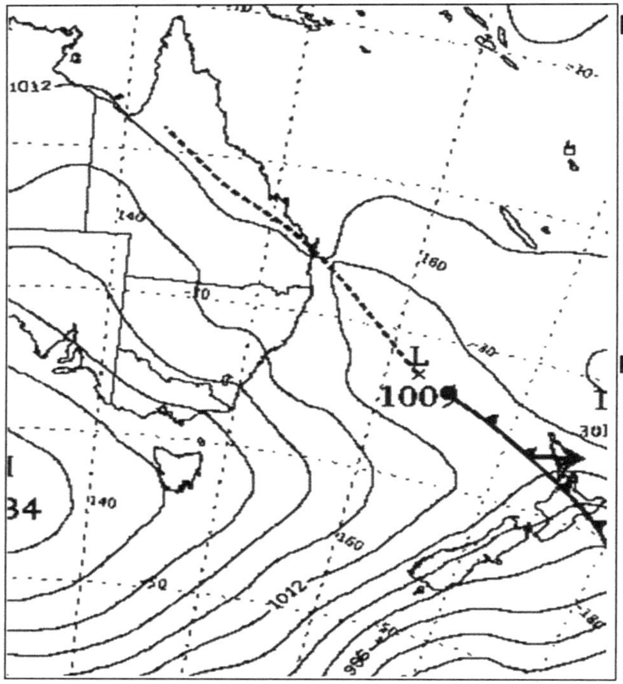

RAIN POTENTIAL

FROST/SNOW

NOVEMBER **16** SUNDAY

A cold change overnight may bring some pockets of frost into the south of the South Island as a weak depression directs a light to moderate westerly flow over the South Island, backing southeasterly over the North Island. Rain intensifies in Northland while clearing elsewhere to overnight isolated showers mostly in the Waikato and Southland to Otago, with a mostly dry day prevailing. A high pressure ridge nears the west of the country.

Auckland:	Overcast, rain, light breezes
Hamilton:	Overcast, dry with lingering threats of rain, gentle breezes
Wellington:	Sunny, dry, gentle shifting breezes
Greymouth:	Sunny, cooler, isolated showers, gentle breezes
Christchurch:	Partly cloudy, colder, passing showers, light breezes
Dunedin:	Partly cloudy, colder, isolated showers, moderate breezes

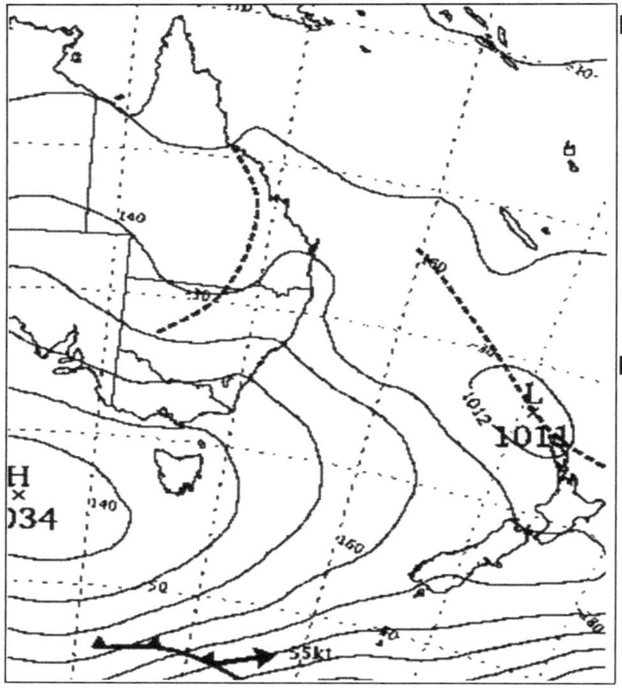

RAIN POTENTIAL

FROST/SNOW

NOVEMBER 17 MONDAY

A large anticyclonic system crossing the Australian Bight extends a high pressure ridge near the west coast of New Zealand directing a light to moderate southwest flow over the South Island and very cold overnight temperatures with chance of frosts from Southland to Canterbury. In contrast, a low pressure trough brings a calm but thundery air over the North Island with heavy rain in Northland, and showers into the Bay of Plenty to the Hawkes Bay, while mostly dry in the west and south of the North Island.

Auckland: Colder, cloudy, rain, light breezes
Hamilton: Partly cloudy, cooler, dry, calm and pleasant air
Wellington: Very sunny, dry, cooler, moderate shifting breezes
Greymouth: Sunny, dry, slightly warmer, moderate breezes
Christchurch: Sunny, dry, gentle shifting breezes
Dunedin: Shifting breezy spells, partly cloudy, isolated showers

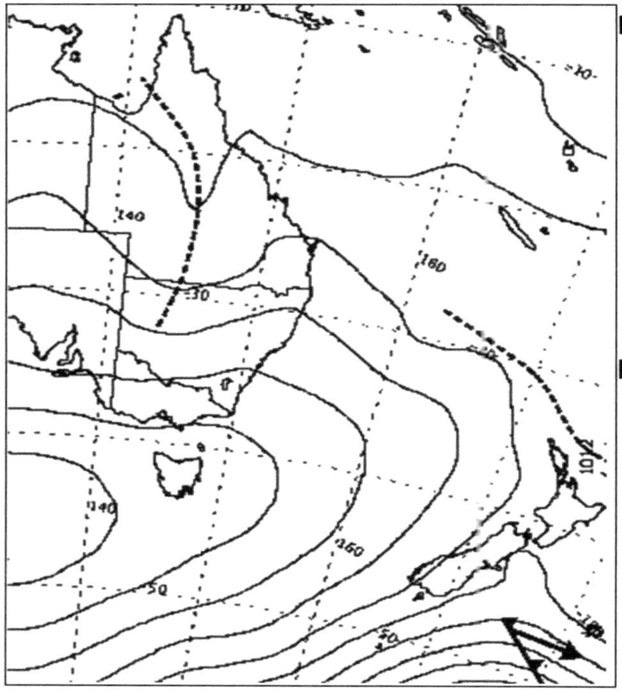

RAIN POTENTIAL

FROST/SNOW

NOVEMBER 18 TUESDAY

A large high moving over Tasmania extends a ridge along the western coasts of New Zealand, and directs a moderate southwesterly flow throughout. Mostly dry and sunny everywhere, with the chance of isolated frosts in the Bay of Plenty and Otago, and possible lingering showers overnight in Northland.

Auckland: Partly cloudy, warmer, light showers, gentle breezes
Hamilton: Mostly sunny, dry, warm day cool night, gentle breezes
Wellington: Sunny, dry, mild day cool night, light breezes
Greymouth: Partly cloudy, isolated showers, moderate breezes
Christchurch: Partly cloudy, dry, milder, light breezes freshening later in the day
Dunedin: Sunny, slightly warmer, dry, some windy spells, chance of gale gusts

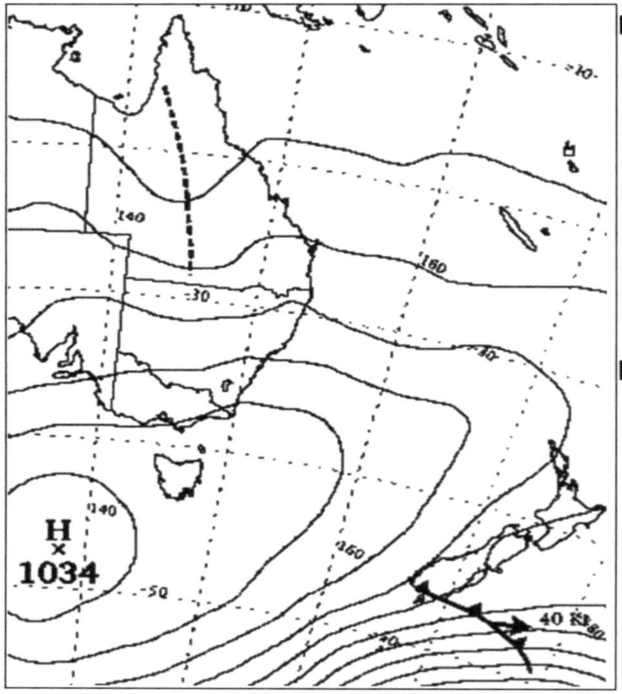

RAIN POTENTIAL

FROST/SNOW

NOVEMBER **19** WEDNESDAY

A large anticyclonic system crossing the Tasman Sea ridges over New Zealand in a gentle southwest flow throughout. Mostly sunny and dry everywhere with cold overnight temperatures in the south of the South Island, milder elsewhere.

Auckland: Mostly calm and pleasant air, sunny, dry
Hamilton: Dry, cooler, sunny, calm and pleasant air
Wellington: Sunny, dry, gentle shifting breezes
Greymouth: Mostly sunny, dry, mild day cold night, light breezes
Christchurch: Very sunny, dry, cooler, gentle breezes
Dunedin: Partly cloudy, dry, moderate breezes

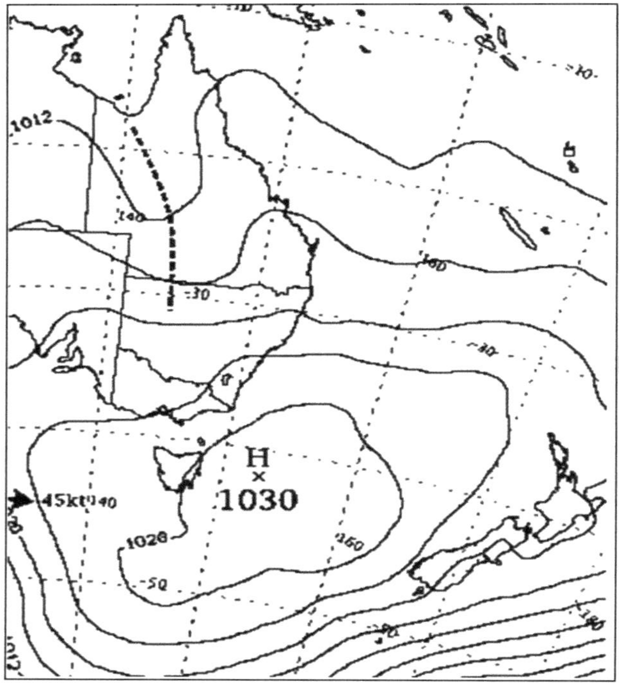

RAIN POTENTIAL

FROST/SNOW

NOVEMBER **20** THURSDAY

A large high covers the Tasman Sea and brings a mostly calm southwesterly air throughout the country. Dry and mostly sunny everywhere, except for increased cloudy spells in the Wellington region and down through the east of the South Island into Marlborough and Canterbury.

Auckland: Very sunny, dry, warm day cool night, pleasant serene atmosphere
Hamilton: Warmer, sunny, dry, calm and serene air
Wellington: Cloudy, dry, calm and pleasant air
Greymouth: Sunny, dry, slightly warmer, gentle breezes to mostly calm air
Christchurch: Calm atmosphere, cloudy, light overnight showers, chance of fog
Dunedin: Cloudy, dry, mostly calm air

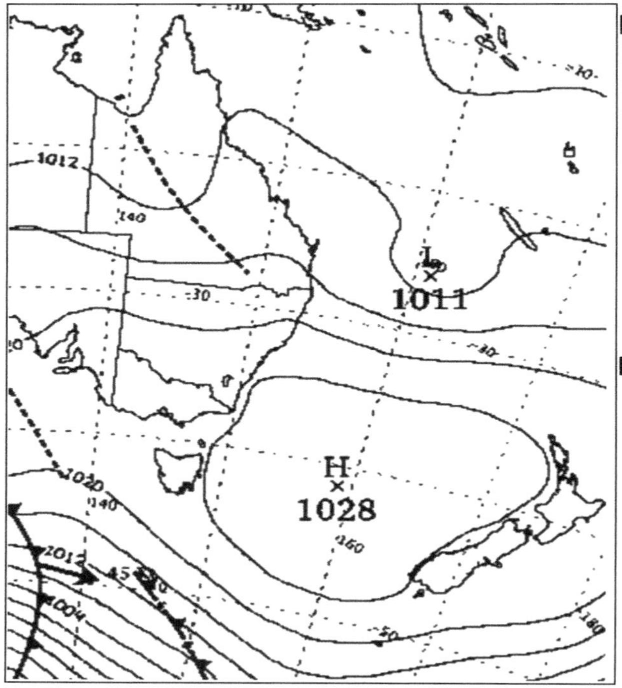

RAIN POTENTIAL

FROST/SNOW

NOVEMBER 21 FRIDAY

A large high covers New Zealand. Dry and mostly sunny everywhere with cold temperatures overnight in the east of the South Island. Chance of some isolated overnight showers briefly in Southland and around the East Coast to the Hawkes Bay. Possible fog patches in sheltered districts.

Auckland: Sunny, mostly dry, chance of overnight showers, calm atmosphere
Hamilton: Cloudy, dry, slightly cooler, calm atmosphere
Wellington: Sunny and dry, calm atmosphere
Greymouth: Dry, occasional cloudy spells, calm and pleasant air
Christchurch: Sunny, mostly dry, calm atmosphere, chance of misty patches
Dunedin: Cloudy, mostly dry, chance of overnight showers, calm air, mists

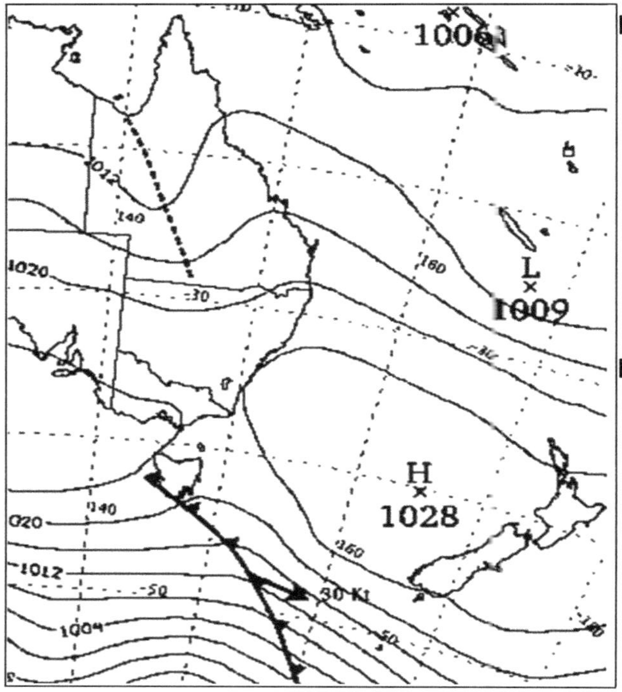

RAIN POTENTIAL

FROST/SNOW

NOVEMBER 22 SATURDAY

A wet front nears the south of the South Island with strengthening westerly winds building over the South Island, while mostly calm conditions prevail over the North Island as the high departs to the east. Some isolated overnight showers likely in the east of the North Island from the East Cape to the Wairarapa, with mostly dry elsewhere. Cloudy with increasing threats of rain in the south of the South Island ahead of an advancing front, sunnier elsewhere.

Auckland: Partly cloudy, dry, warmer, light breezes to mostly calm air
Hamilton: Very sunny, warmer, dry, calm atmosphere
Wellington: Very sunny, warmer, dry, gentle shifting breezes
Greymouth: Cloudy, light showers, moderate breezes at times
Christchurch: Sunny, dry, moderate breezes
Dunedin: Partly cloudy, dry, winds strengthening to fresh at times

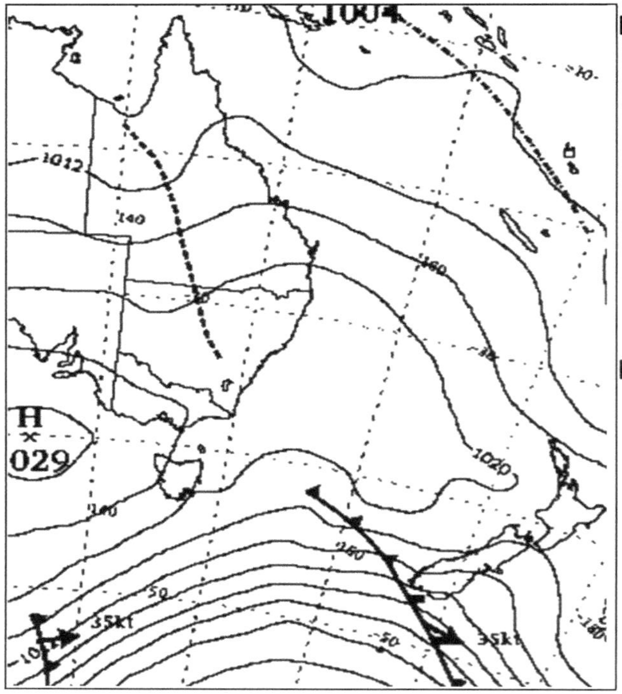

RAIN POTENTIAL

FROST/SNOW

NOVEMBER **23** SUNDAY

Low pressure to the south of New Zealand directs a blustery wet front up through the South Island in a stronger southwest to westerly flow, and bringing heavy rain in most districts except in eastern parts from Marlborough to Canterbury. In contrast, a lighter west to southwest flow and mostly dry and sunny conditions prevail over the North Island with increasing threats of rain later in the day.

Auckland: Sunny, mostly dry, some brief overnight showers likely, light breezes
Hamilton: Sunny, overnight showers, mild day cool night, gentle breezes
Wellington: Cloudy, dry, fresh breezes, static air
Greymouth: Cloudy, heavy rain, windy at times
Christchurch: Partly cloudy, warmer, dry, shifting breezy spells
Dunedin: Cloudy, warmer, scattered showers, moderate breezes

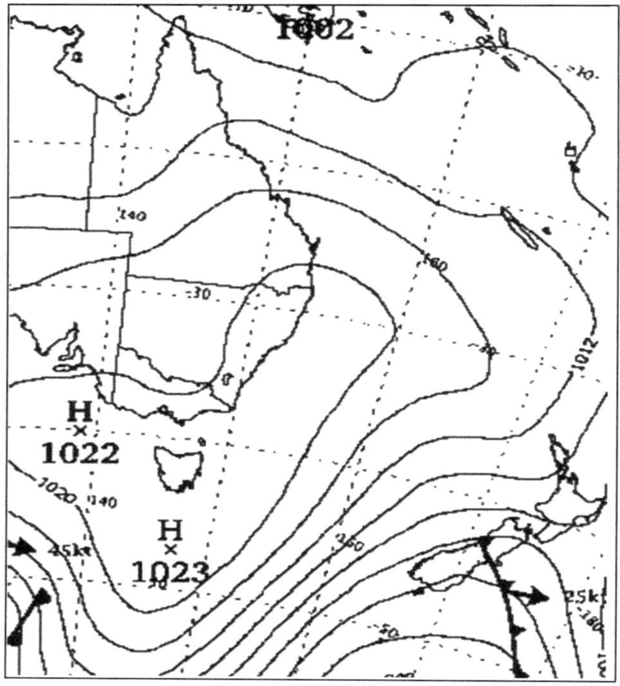

RAIN POTENTIAL

FROST/SNOW

NOVEMBER **24** MONDAY

A stormy air sweeps the country in a strong south to southwest flow throughout. A band of thunderstorms cross the country overnight, accompanied by blustery winds to gale force at times. Rain likely to be heavy in many parts of the lower North Island and into the top half of the South Island, with lighter rain activity elsewhere. Good sunny spells during the day.

Auckland: Sunny, overnight showers, unsettled windy spells
Hamilton: Changeable, overnight rain, sunny day, strong breezes
Wellington: Sunny day, overnight showers, chance of thunderstorms, windy
Greymouth: Changeable, overnight rain, sunny day, moderate breezes
Christchurch: Partly cloudy, heavy overnight rain, strong winds with gusts at times
Dunedin: Fresh breezes, partly cloudy, overnight rain, colder

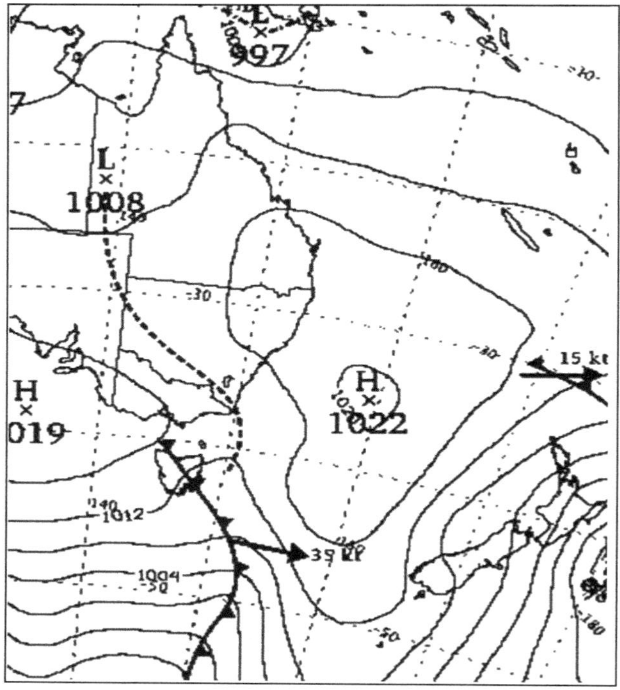

RAIN POTENTIAL

FROST/SNOW

NOVEMBER **25** TUESDAY

A weak front between two high pressure systems crosses up the country in a light west to southwest flow. A mixed day of sunny spells and showers in most parts of the country, with more cloud in Otago and central South Island regions, and sunnier in central parts of the North Island and Fiordland. Colder overnight temperatures in most districts, with chance of frosts in parts of the Waikato to the Bay of Plenty and down into Wellington, and in Otago.

Auckland: Very sunny, cooler, light overnight showers, mostly calm air
Hamilton: Sunny, light showers and cold overnight, gentle breezes
Wellington: Cloudy, cooler, dry, gentle breezes
Greymouth: Overcast, light showers, warmer, fluctuating breezes
Christchurch: Cloudy, colder, light showers, mostly calm air
Dunedin: Overcast, isolated showers, calm air, chance of misty patches

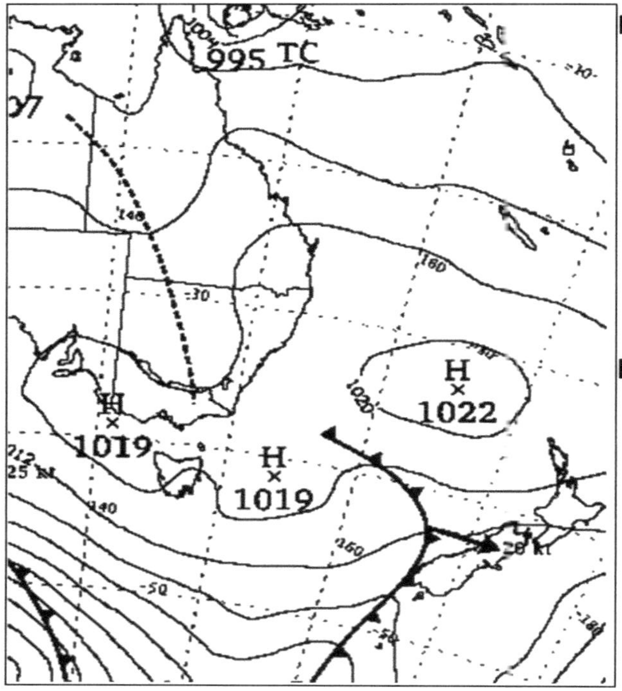

RAIN POTENTIAL

FROST/SNOW

NOVEMBER **26** WEDNESDAY

A complex of high and low pressure centres prevail over New Zealand with a gentle west to southwest air prevailing throughout. Increasing cloud in most districts, and particularly dull and gloomy in the North Island. Mostly dry in Northland, the Bay of Plenty, East Cape to the Hawkes Bay, and South Canterbury to Southland, with showers elsewhere and possible brief isolated overnight showers in Otago.

Auckland: Overcast, mostly dry, chance of overnight drizzle, light breezes
Hamilton: Cloudy, brief showers, mostly calm air
Wellington: Changeable, mix of sunny spells and showers, mostly calm air
Greymouth: Changeable, mix of sunny spells and showers, cooler, light breezes
Christchurch: Partly cloudy, dry, calm atmosphere
Dunedin: Cloudy, cool, dry, calm and serene air

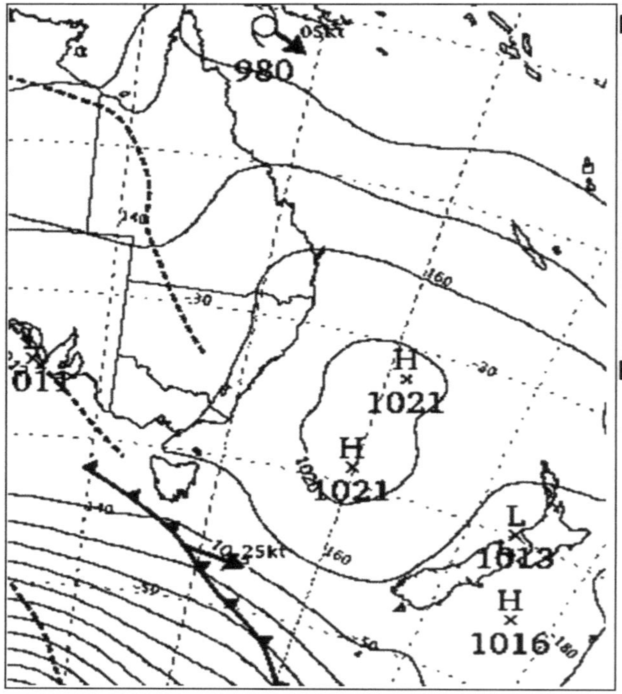

RAIN POTENTIAL

FROST/SNOW

NOVEMBER **27** THURSDAY

A couple of high pressure systems cross the country bringing a gentle southwest flow over the South Island, backing southeasterly over the North Island. Mostly dry and sunny day throughout with the chance of some lingering overnight showers over the North Island from Auckland to the Wairarapa, and in coastal Otago.

Auckland: Calm atmosphere, light overnight showers, sunny day
Hamilton: Light breezes, sunny day, morning showers, warmer day cooler night
Wellington: Bright and sunny, dry, mild day cold night, gentle shifting breezes
Greymouth: Very sunny, dry, mild day cool night, gentle shifting breezes
Christchurch: Very sunny, dry, mild day cold night, gentle breezes
Dunedin: Mostly calm air, very sunny, dry, warmer day cooler night

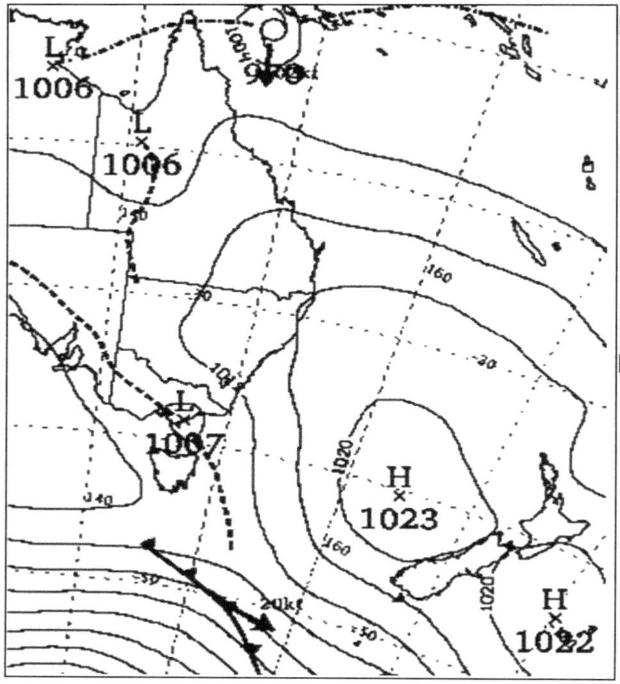

RAIN POTENTIAL

FROST/SNOW

NOVEMBER **28** FRIDAY

A couple of anticyclonic centres remain almost stalled over the country with a mostly calm air prevailing throughout, except for some slight west to southwest breezes around the eastern coasts of the South Island. Good sunny spells and dry almost everywhere, except for increased cloud in Southland and Otago.

Auckland: Dry, scattered cloudy spells, warmer, calm and pleasant air
Hamilton: Partly cloudy, warmer, dry, calm and pleasant air
Wellington: Sunny, dry, calm and pleasant air
Greymouth: Very sunny, dry and pleasant air, gentle breezes to mostly calm
Christchurch: Sunny, warmer, dry, gentle shifting breezes
Dunedin: Cloudy, warmer, dry, gentle breezes

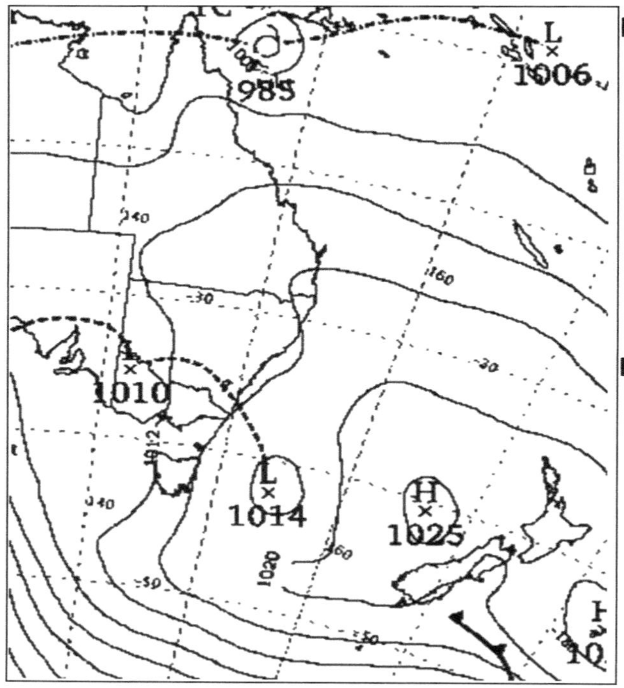

RAIN POTENTIAL

FROST/SNOW

NOVEMBER **29** SATURDAY

High pressure continues to bring a partly cloudy and dry day throughout the country in a gentle, to mostly calm air throughout.

Auckland:	Gentle shifting breezes, sunny, dry
Hamilton:	Sunny, dry, gentle breezes to mostly calm air
Wellington:	Moderate shifting breezes, dry, warmer, bright and sunny
Greymouth:	Cloudy, warmer, dry, mostly calm and pleasant air
Christchurch:	Mostly sunny, dry, slightly warmer, gentle breezes, chance of fog
Dunedin:	Sunny, warmer, dry, light breezes to mostly calm air

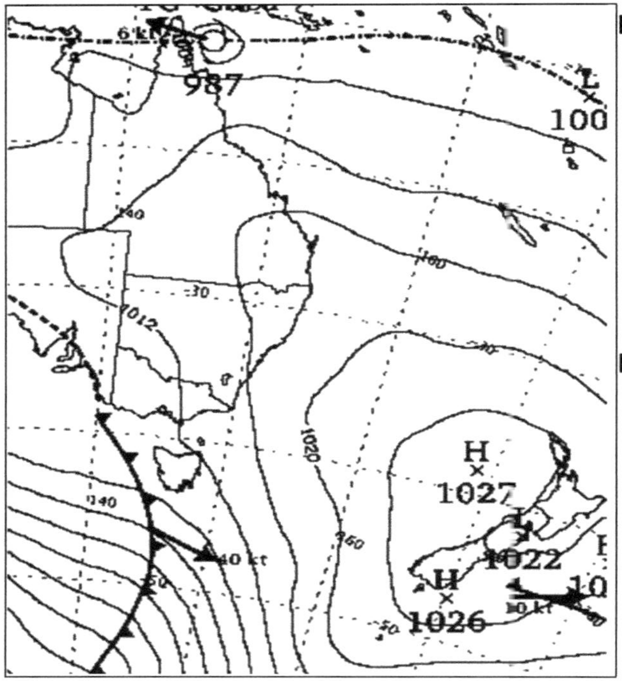

RAIN POTENTIAL

FROST/SNOW

NOVEMBER **30** SUNDAY

Anticyclonic conditions continue. Mostly sunny, warmer, and dry everywhere with a gentle westerly to mostly calm air prevailing.

Auckland:	Sunny, dry, pleasant and serene air
Hamilton:	Sunny, warmer, dry, calm and pleasant air
Wellington:	Sunny, warmer, dry, gentle shifting breezes
Greymouth:	Partly cloudy, dry, gentle shifting breezes
Christchurch:	Very bright and sunny, dry, gentle shifting breezes, possible fog
Dunedin:	Very bright and sunny, cooler, dry, gentle breezes to mostly calm air

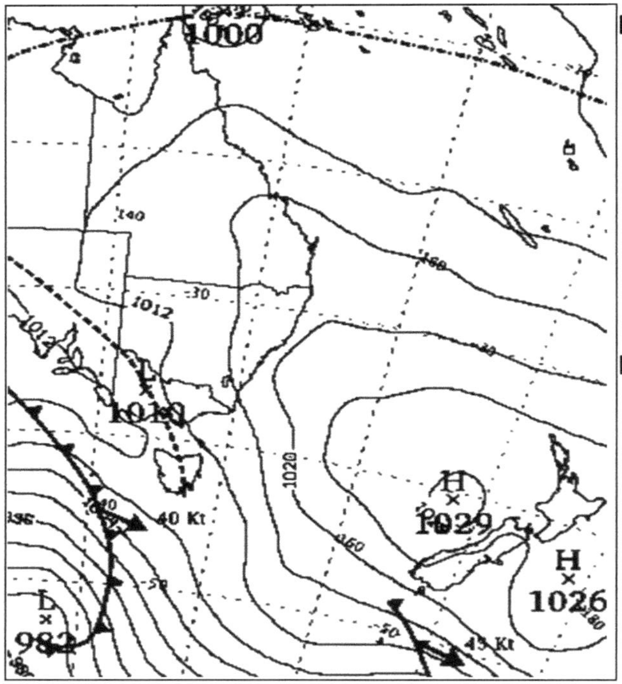

RAIN POTENTIAL

FROST/SNOW

DECEMBER

PHASES OF THE MOON

4th	Full Moon
4th	Perigee #2
5th	Northern Declination
11th	Third Quarter
12th	Crossing Equator
17th	Apogee
19th	Southern Declination
20th	New Moon
27th	First Quarter
27th	Crossing Equator

1	2	3	4	5	6	7	8	9	10	11	12	13	14	15	16
Mon	Tue	Wed	Thu	Fri	Sat	Sun	Mon	Tue	Wed	Thu	Fri	Sat	Sun	Mon	Tue
			F P2	^						3Q	XhS				

17	18	19	20	21	22	23	24	25	26	27	28	29	30	31
Wed	Thu	Fri	Sat	Sun	Mon	Tue	Wed	Thu	Fri	Sat	Sun	Mon	Tue	Wed
A		V	N							1Q XhN				

MONTHLY SUMMARY

30th Nov-6th December Unusually hot spell in the north and east of both islands.
1st-3rd Chance of light fog in the Christchurch region.
5th Very hot day in Blenheim. Strong blustery winds for a time in the Christchurch area.

Over the next four weeks frequent anticyclonic systems pass east of the South Island, while troughs of low pressure form in the northern Tasman Sea before crossing over the north of the country, which indicates the stabilising of a La Nina weather pattern. As such, a dry north-easterly flow will prevail more often than is typical for this time of year. This north-easterly flow will bring warmer than normal temperatures for this time of year with many districts in the north and west, as well as in Southland and Otago temperatures up to 0.5-1°C above the average, with the remaining parts of the country closer to the norm. Rainfall during this outlook period will be above normal in Northland and below normal in the Bay of Plenty, down through the southwest of the North Island and in inland parts of the South Island. It will be very wet at the start of this outlook period in Northland to the Bay of Plenty, with some districts possibly seeing up to twice as much rainfall as is typical for this time of year by the end of this four week period. The Bay of Plenty, south Canterbury, north Taranaki, Manawatu, Buller, coastal parts of Kaikoura and inland parts of Southland and Otago may only see around half their usual rainfall by the end of this outlook period. As such, soil moisture deficits will continue to be severe in Marlborough, Canterbury, and most of Otago, as well as increasing in Auckland to the Bay of Plenty region, in Taranaki, the Wairarapa, Manawatu, and Nelson districts. It will be sunnier than average in the south of the country, with many parts through Southland and Otago likely to record up to 20% more sun-hours than normal. In contrast, the northern half of the North Island will see below average sunshine, as well as being cloudier than average in the Gisborne, Wellington, Marlborough, and Nelson regions.

11th-20th Period of widespread and fairly persistent rain across the northern half of the North Island, particularly in Northland, where some outbreaks of heavy rain at times are likely.
18th Chance of overnight ground frosts around parts of the Hanmer Forest region, while unusually high maximum temperatures possible in the Wanaka district. Possible fog in Wellington. Windy with occasional gusts at times in Christchurch.

19th-27th Regular outbreaks of heavy rain in the West Coast of the South Island during this period. Chance of flooding for a time in the Franz Josef district.
21st Very hot day in Canterbury, particularly in the Christchurch area.
28th-29th Heavy rain at times in the top of the South Island districts including the chance of brief localised flooding in the Blenheim to Motueka region, including in Nelson.
29th-30th Wet front crosses the North Island bringing outbreaks of heavy rain in most districts, with lighter rain in the Bay of Plenty. Chance of brief surface flooding in the Central Plateau.

JANUARY 2026
2nd-9th Regular outbreaks of heavy rain at times in the Fiordland region.
9th Very strong and blustery northwest winds likely around the southwest coastal fringes of the South Island. Very hot day in the Culverden area.

December 2025

Monday	Tuesday	Wednesday	Thursday	Friday	Saturday	Sunday
1 **Aries** Trine Sun Trine Mars	**2** **Aries** Trine Mars Square Jupiter Square Pluto	**3** **Taurus**	**4** **Taurus** Opposition Mercury Opposition Venus Sextile Saturn Conjunction Uranus Sextile Neptune Trine Pluto Square Ascendant	**5** **FM P2 Gemini** Opposition Sun Square Ascendant	**6** **^ Gemini** Opposition Mars Square Saturn Square Neptune Square Ascendant	**7** **Cancer** Square Ascendant
8 **Leo** Trine Mercury Conjunct Jupiter Trine Saturn Sextile Uranus Trine Neptune Opposition Pluto Square Ascendant	**9** **Leo** Trine Sun Trine Venus Square Ascendant	**10** **Leo** Square Mercury Trine Mars Square Uranus	**11** **Virgo** Square Venus	**12** **3Q XhS Virgo** Square Sun Square Mars Sextile Jupiter Opposition Saturn Trine Uranus Opposition Neptune Trine Pluto	**13** **Libra** Trine Pluto	**14** **Libra** Sextile Sun Square Jupiter
15 **Scorpio** Sextile Mars Square Pluto	**16** **Scorpio**	**17** **A Scorpio** Trine Jupiter Trine Saturn Opposition Uranus Trine Neptune	**18** **Sagittarius** Conjunct Mercury	**19** **Sagittarius** Conjunct Mercury	**20** **NM V Capricorn** Conjunct Sun Conjunct Venus Square Saturn Square Ascendant	**21** **Capricorn** Conjunct Sun Conjunct Mars Square Neptune Square Ascendant
22 **Capricorn** Opposition Jupiter	**23** **Capricorn** Opposition Jupiter Trine Uranus Sextile Neptune Conjunct Pluto	**24** **Aquarius**	**25** **Aquarius** Square Uranus	**26** **Pisces** Sextile Mars	**27** **XhN Pisces** Square Mercury Trine Jupiter Conjunct Saturn Conjunct Neptune	**28** **1Q Aries** Square Sun Square Venus Square Mars Conjunct Neptune
29 **Aries** Trine Mercury Square Jupiter	**30** **Taurus** Trine Sun Trine Venus Trine Mars Square Pluto	**31** **Taurus** Sextile Jupiter Conjunct Uranus Square Ascendant	1 **Moon Sign**	2	3	4
5	6					

NORTH ISLAND RAINFALL
DECEMBER 2025

d=mainly dry, n= fog, mist, dew or nonrecordable, l=light showers, p=passing showers, s=significant showers, r=rain, h=heavy falls

		Northland							Auckland					Bay Of Plenty				Waikato				Rot/Tau		Hawkes Bay			C Plateau			Taranaki			Manawatu			Lower North Island					
		CR	Kta	Ker	Rsl	Dar	Whg	Man	Wark	Hen	MkH	Ak	Co	Tau	TeP	Wkt	Opo	Hml	TeA	Mata	Ot	Rt	Taup	Gis	Nap	Has	Taum	Chat	Waio	NP	Str	Wang	Pa	Dann	Lev	PN	Mast	UH	Wgt	all	Moon
1st	Mon	d	d	d	d	d	d	d	d	d	d	d	d	d	d	d	d	d	d	d	d	d	l	d	d	d	d	d	d	d	d	d	d	d	d	d	d	d	d	1	
2nd	Tue	d	d	d	d	d	d	d	d	d	d	d	d	d	d	d	d	d	d	d	d	d	d	d	d	d	d	d	d	d	d	d	d	d	d	d	d	d	d	0	
3rd	Wed	d	d	d	d	d	d	d	d	d	d	d	d	d	d	d	d	d	d	d	d	d	d	d	d	d	d	d	d	d	d	d	d	d	d	d	d	d	d	0	
4th	Thur	h	h	h	h	s	h	r	h	s	l	s	s	l	l	l	l	l	l	l	d	l	l	d	d	d	d	d	d	d	d	d	d	d	d	d	d	d	d	200	
5th	Fri	l	d	l	s	l	s	l	s	d	d	d	n	d	d	n	n	d	d	d	d	l	d	d	d	d	d	d	d	d	d	d	d	d	n	d	d	d	d	27	F ³2
6th	Sat	d	d	d	d	d	d	d	d	d	d	d	d	d	d	d	d	d	d	d	d	d	d	d	d	d	d	d	d	d	d	d	d	d	d	d	d	d	d	0	
7th	Sun	d	d	d	d	d	d	d	d	d	d	d	d	d	d	d	d	d	d	d	d	d	d	d	d	d	d	d	d	d	d	d	d	d	d	d	d	d	d	0	
8th	Mon	d	d	d	d	d	d	d	d	d	d	d	d	d	d	d	d	d	d	d	d	d	d	d	d	l	d	d	d	d	d	d	d	d	d	d	d	d	d	1	
9th	Tue	d	h	s	s	s	n	d	d	l	l	l	d	d	d	d	d	d	d	d	d	d	d	l	d	d	d	d	d	d	d	d	d	d	d	d	d	d	d	43	
10th	Wed	d	l	l	l	l	s	l	s	r	d	n	d	d	d	d	d	l	l	l	l	d	d	d	d	d	d	d	d	d	d	d	d	d	d	d	d	d	d	35	
11th	Thur	d	d	d	d	d	d	d	l	n	n	d	d	h	s	d	d	d	d	d	s	d	d	n	d	d	s	l	d	d	d	d	d	l	d	d	d	d	d	40	
12th	Fri	n	l	l	n	l	l	l	l	l	r	l	d	d	d	d	l	d	d	d	d	d	d	d	d	d	d	d	s	d	d	d	d	d	d	d	d	d	d	32	3Q KhS
13th	Sat	d	d	l	d	d	l	l	l	s	l	d	d	d	d	d	d	d	d	d	d	d	d	d	d	d	d	d	d	d	d	d	d	d	d	d	d	d	d	13	
14th	Sun	d	d	d	d	d	d	d	d	d	c	d	d	d	d	d	d	d	d	d	d	d	d	d	d	d	d	d	d	d	d	d	d	d	d	d	d	d	d	0	
15th	Mon	s	s	l	l	l	l	n	n	l	c	d	d	d	d	d	d	d	d	d	d	d	d	d	d	n	d	d	d	d	d	d	d	d	d	d	d	d	d	18	
16th	Tue	h	h	h	h	h	r	h	h	h	r	h	r	r	h	r	r	s	r	h	r	r	s	l	l	s	d	h	h	h	h	r	d	l	l	l	d	d	d	643	
17th	Wed	s	s	h	h	r	h	h	h	h	r	s	h	s	r	r	h	s	s	l	h	s	h	h	r	h	h	h	h	s	h	s	d	s	d	d	d	d	d	532	A
18th	Thur	r	d	h	h	d	h	n	d	d	c	d	d	d	d	d	d	d	d	d	l	d	l	h	r	s	l	d	d	d	d	d	d	d	d	l	d	d	d	118	
19th	Fri	h	r	h	h	s	h	s	s	l	r	d	h	s	r	l	l	s	l	d	r	s	l	d	s	l	l	s	l	l	s	n	d	l	d	l	n	d	d	241	
20th	Sat	h	r	h	h	s	h	r	h	s	l	s	r	l	l	l	d	l	l	d	l	s	s	d	l	s	s	l	l	l	l	s	r	l	h	l	l	r	l	301	N V
21st	Sun	s	s	h	s	s	r	l	s	l	c	d	s	l	l	d	d	l	l	d	l	s	l	d	d	d	d	d	d	d	d	d	d	d	d	d	d	d	d	81	
22nd	Mon	l	l	d	d	d	l	d	d	d	c	d	d	d	d	d	d	d	d	d	d	l	d	d	d	d	d	d	d	d	d	d	d	l	d	d	d	d	d	9	
23rd	Tue	n	d	d	d	d	l	d	l	d	c	d	d	d	d	d	d	d	d	d	d	d	d	d	d	d	d	d	d	d	d	d	d	d	d	d	d	n	d	4	
24th	Wed	s	d	d	d	d	d	d	n	d	c	d	d	d	d	d	l	d	d	d	d	d	l	d	d	d	l	l	d	d	l	d	d	d	d	d	d	d	d	11	
25th	Thur	d	d	d	d	d	d	d	d	d	c	d	d	d	d	d	r	d	d	d	d	l	l	d	l	s	l	d	d	d	n	d	d	d	d	d	d	d	d	18	
26th	Fri	d	d	d	d	d	d	d	n	d	l	d	d	d	d	d	d	d	d	d	d	l	d	d	d	d	d	d	d	d	l	d	d	d	d	d	d	d	d	4	
27th	Sat	d	n	d	d	d	d	d	n	d	c	l	l	l	d	d	n	d	d	d	d	d	d	d	d	d	d	n	d	d	d	d	d	d	d	d	d	l	d	7	XaN
28th	Sun	l	s	s	d	l	d	d	d	d	c	d	d	d	d	d	s	n	d	d	l	n	d	d	d	d	d	s	d	l	l	d	d	d	d	d	d	n	d	29	1Q
29th	Mon	r	h	h	h	h	r	r	h	h	h	h	d	r	r	n	r	h	h	h	h	h	h	s	h	h	h	h	h	s	r	h	r	h	h	h	h	h	h	834	
30th	Tue	d	d	d	d	d	d	d	n	d	c	d	d	n	s	n	s	l	l	d	s	s	h	h	h	h	h	h	h	l	r	h	r	h	r	r	h	s	s	427	
31st	Wed	d	d	d	d	d	d	d	d	l	c	d	h	r	s	d	l	d	d	d	r	r	l	l	l	s	d	d	d	d	d	d	d	l	d	d	l	d	d	83	
estimate:		170	160	251	192	128	199	132	153	122	85	66	101	78	97	30	102	43	51	60	105	75	81	108	91	121	119	169	115	41	86	68	31	89	59	44	49	44	32	3748	
average:		61	80	108	89	82	135	103	117	106	85	77	84	85	141	85	111	82	106	113	124	95	77	50	57	52	122	244	88	110	150	72	84	105	102	82	72	108	87	3732	
trend:		wtr	wtr	wtr	wtr	wtr	wtr	wtr	wtr	wtr	av	drr	wtr	av	drr	drr	av	drr	drr	drr	drr	drr	av	wtr	wtr	wtr	av	drr	wtr	drr	drr	av	drr	drr	drr	drr	drr	drr	drr	av	

SOUTH ISLAND RAIN

DECEMBER 2025

d=mainly dry, n= fog, mist, dew or nonrecordable, l=light showers, p=passing showers, s=significant showers, r=rain, h=heavy falls

		West Coast				Top South Island				Canterbury				South Canterbury				Coastal Otago			Lakes			Inland Otago				Southland						
		West	Grey	Hok	BB	Ta	Mou	Nel	Ble	Dar	Ham	Kai	Ch	Rak	Meth	Ash	Tim	Oam	Dun	Bal	Twi	LP	LT	Crom	Qstn	Ran	Alex	TeA	Wtn	Gore	Inv	Puy Pt	all	Moon
1st	Mon	d	d	d	d	d	d	d	d	d	d	d	d	d	d	d	d	d	d	d	d	d	d	d	d	d	d	d	d	d	d	d	0	
2nd	Tue	d	d	d	d	d	d	d	d	d	d	d	d	d	d	d	d	d	d	d	d	d	d	d	d	d	d	d	d	d	d	r	13	
3rd	Wed	d	d	d	d	d	d	d	d	d	d	d	d	d	d	d	d	d	d	d	d	d	d	d	d	d	d	d	d	d	d	r	12	
4th	Thur	l	s	r	s	d	d	d	d	d	d	d	d	d	d	d	d	d	d	l	d	d	d	d	d	d	d	n	l	s	s	s	55	
5th	Fri	s	s	l	d	d	d	d	d	d	d	d	d	d	d	d	d	d	d	d	d	d	d	d	d	d	d	d	d	d	d	r	24	F P2
6th	Sat	d	d	d	d	d	d	d	d	d	d	d	d	d	d	d	n	d	d	d	d	d	d	d	d	d	d	l	d	l	s	l	10	^
7th	Sun	d	d	d	d	d	d	d	d	d	d	d	d	d	n	d	d	d	d	d	d	d	d	d	d	d	d	d	d	d	h	r	26	
8th	Mon	d	d	n	d	d	d	l	d	s	l	l	s	s	s	l	s	s	l	l	l	l	d	l	d	d	d	d	d	n	s	l	56	
9th	Tue	d	d	d	d	d	d	d	d	d	d	d	d	d	d	d	d	d	d	d	d	d	d	d	d	d	d	d	d	d	d	s	8	
10th	Wed	n	d	d	n	d	d	d	d	l	s	l	d	l	s	l	r	l	l	d	d	d	l	d	l	l	l	l	d	l	l	s	59	
11th	Thur	d	n	l	d	n	l	d	d	d	d	d	d	d	l	d	l	d	d	d	d	d	d	d	d	d	d	d	d	d	d	n	6	
12th	Fri	d	d	d	d	d	d	d	d	r	d	d	l	l	s	n	d	l	d	d	d	d	d	d	d	l	d	d	d	d	d	l	24	3Q XhS
13th	Sat	d	d	d	d	d	d	d	d	l	d	d	d	l	n	l	d	d	d	d	d	d	d	d	d	d	d	d	d	d	d	d	5	
14th	Sun	d	d	d	d	d	d	d	d	d	l	d	d	d	l	n	d	d	d	d	d	d	d	d	d	d	d	d	d	d	d	d	2	
15th	Mon	d	d	d	d	d	d	d	d	d	d	d	d	d	d	d	d	d	d	d	d	d	d	d	d	d	d	d	d	d	d	d	0	
16th	Tue	d	d	d	d	d	d	d	d	n	d	d	d	d	l	d	d	d	l	d	d	d	d	d	d	d	d	l	l	d	l	l	11	
17th	Wed	d	d	d	d	d	d	d	d	d	l	n	d	l	l	l	d	d	d	d	d	d	d	d	d	d	d	d	d	d	d	d	7	A
18th	Thur	d	d	d	d	d	d	d	d	d	d	d	d	d	d	d	d	d	d	d	d	d	d	d	d	d	d	d	d	d	d	d	0	
19th	Fri	d	d	d	d	d	d	d	d	d	d	d	d	d	d	d	d	d	d	d	d	d	d	d	d	d	d	d	d	d	d	d	0	
20th	Sat	s	r	h	r	h	r	r	s	l	l	n	l	d	l	d	d	d	d	d	d	d	d	d	d	d	d	d	d	d	d	l	114	N V
21st	Sun	s	r	h	h	s	l	l	d	l	d	d	n	d	l	l	n	l	d	n	d	d	d	d	d	d	d	l	d	l	n	h	175	
22nd	Mon	h	h	h	h	s	l	d	d	l	d	d	l	l	l	d	d	d	s	s	r	d	d	d	l	d	d	r	s	s	h	h	244	
23rd	Tue	h	h	h	h	l	d	l	d	s	n	d	s	d	l	l	l	l	s	s	d	d	d	s	s	h	s	s	r	s	s	s	388	
24th	Wed	l	h	h	h	d	d	d	d	s	r	d	s	d	s	r	s	l	l	s	d	l	l	l	l	s	s	s	l	l	l	l	199	
25th	Thur	l	d	l	s	d	d	d	d	d	d	d	n	d	l	n	l	s	l	d	l	d	d	d	d	d	d	l	l	l	s	r	38	
26th	Fri	s	h	h	r	d	d	d	d	d	d	d	d	s	d	d	d	d	d	d	d	d	d	d	d	d	d	d	l	d	n	l	75	
27th	Sat	s	r	r	h	n	d	d	d	d	d	d	d	d	d	d	d	d	d	r	d	d	d	l	l	d	l	l	l	l	l	h	103	XhN
28th	Sun	h	r	s	r	h	s	h	l	r	h	r	s	s	s	s	r	h	h	s	r	l	l	s	l	h	s	l	l	l	n	l	312	1Q
29th	Mon	d	d	d	d	h	h	h	h	l	s	r	l	l	n	l	d	d	d	d	s	d	d	d	d	d	d	d	d	d	l	d	251	
30th	Tue	d	d	d	d	l	d	l	l	r	d	d	s	d	r	r	s	d	n	d	d	s	s	d	d	n	d	d	l	l	d	d	75	
31st	Wed	s	l	d	d	d	d	n	d	n	l	l	n	d	l	l	d	d	l	d	s	d	d	d	d	d	d	d	n	n	d	d	23	
estimate:		115	175	287	356	108	85	97	56	60	53	27	46	31	61	45	50	43	42	44	43	9	11	15	19	61	19	34	33	31	72	183	2314	
average:		202	223	250	445	157	94	75	55	66	78	63	55	80	95	67	72	73	87	69	77	75	50	49	60	54	33	100	86	95	87	184	3256	
trend:		drr	drr	wtr	drr	drr	av	wtr	av	av	drr	drr	av	drr	drr	drr	drr	drr	drr	drr	drr	drr	drr	drr	drr	av	drr	drr	drr	drr	drr	av	drr	

NORTH ISLAND SUNSHINE ESTIMATES
DECEMBER 2025

F=fine (8-12 hours of sunshine), pc= partly cloudy (4-7 hours), c=cloudy (1-3 hours), o=overcast (0 hours)

		Kta	Whg	Dar	Ak	Hml	Tau	Rt	Taum	NP	Wang	Taup	Gis	Nap	Wai	Dann	PN	Mast	Pa	UH	Wgt	ALL	Moon
1st	Mon	F	F	F	F	F	F	F	F	pc	pc	F	F	F	F	F	pc	F	pc	F	pc	199	
2nd	Tue	pc	pc	F	F	F	F	F	F	pc	F	F	F	F	F	F	F	F	pc	pc	pc	191	
3rd	Wed	c	o	pc	pc	pc	F	F	F	F	pc	F	F	F	F	F	pc	F	F	pc	pc	166	
4th	Thur	pc	o	c	c	c	o	pc	c	pc	c	pc	o	c	pc	F	c	F	c	F	F	84	
5th	Fri	F	pc	F	F	F	F	F	F	F	F	F	F	F	F	F	F	F	F	F	F	246	F P2
6th	Sat	F	pc	F	F	F	F	F	F	F	F	F	F	F	F	F	F	F	F	F	F	246	^
7th	Sun	F	c	pc	pc	F	F	F	F	F	F	F	F	F	F	F	F	F	F	F	F	215	
8th	Mon	F	F	c	c	F	F	F	F	F	F	F	F	F	F	F	F	F	F	c	F	188	
9th	Tue	c	c	c	c	F	F	F	F	F	F	F	F	F	F	F	F	F	F	F	F	192	
10th	Wed	pc	pc	pc	pc	F	F	F	F	F	F	F	F	F	F	pc	F	pc	F	c	F	147	
11th	Thur	c	pc	F	F	pc	F	pc	pc	F	F	pc	F	F	F	pc	F	pc	F	F	F	193	
12th	Fri	pc	pc	pc	pc	pc	pc	pc	F	F	F	pc	c	F	pc	pc	F	pc	F	F	pc	150	3Q XhS
13th	Sat	F	pc	F	F	F	F	F	c	F	pc	F	F	F	F	F	pc	F	pc	pc	F	174	
14th	Sun	F	pc	F	F	pc	F	pc	F	F	pc	pc	F	F	c	c	pc	c	F	c	pc	124	
15th	Mon	o	o	o	o	o	c	o	o	o	pc	o	pc	c	pc	pc	pc	pc	F	F	F	75	
16th	Tue	c	o	o	o	o	o	o	o	o	o	o	o	o	o	o	o	o	c	o	o	4	
17th	Wed	c	o	c	c	o	o	o	o	o	c	o	o	o	o	o	c	o	F	o	pc	31	A
18th	Thur	pc	o	F	F	c	pc	pc	pc	F	pc	pc	c	c	o	o	pc	o	pc	c	o	71	
19th	Fri	o	o	o	o	o	o	o	c	o	c	o	o	o	c	pc	c	pc	c	o	pc	25	
20th	Sat	o	o	o	o	o	o	o	o	o	c	o	pc	c	pc	o	c	o	o	o	o	19	N V
21st	Sun	o	o	o	o	o	o	o	o	o	c	o	F	pc	pc	o	c	o	pc	c	c	38	
22nd	Mon	c	pc	pc	pc	o	c	o	c	c	c	o	c	F	F	o	c	o	o	o	c	44	
23rd	Tue	pc	c	pc	pc	F	pc	c	c	F	pc	c	F	F	pc	F	pc	F	c	o	c	118	
24th	Wed	pc	pc	F	F	F	F	pc	pc	pc	pc	pc	c	o	o	o	pc	o	pc	o	o	92	
25th	Thur	pc	o	o	o	o	pc	c	pc	F	pc	c	pc	F	pc	pc	pc	pc	c	c	pc	84	
26th	Fri	pc	o	pc	pc	c	pc	c	o	pc	F	c	F	F	F	F	F	F	F	pc	F	129	
27th	Sat	c	pc	pc	pc	o	c	c	c	pc	pc	c	F	F	F	pc	pc	pc	pc	o	pc	103	XhN
28th	Sun	o	c	c	c	o	o	c	o	o	o	c	pc	pc	pc	o	o	o	o	o	o	22	1Q
29th	Mon	c	c	c	c	c	o	o	o	o	o	o	o	o	o	o	o	o	o	o	o	15	
30th	Tue	F	pc	F	F	pc	pc	c	pc	F	pc	c	o	pc	c	pc	pc	pc	pc	o	pc	114	
31st	Wed	F	c	o	o	pc	pc	pc	F	F	F	pc	pc	pc	pc	F	F	F	F	F	F	151	
est. hours:		160	106	156	156	169	198	173	174	214	207	173	216	221	209	173	207	173	219	153	193	3648	
av. hours:		218	208	202	224	220	232	218	185	229	219	211	234	231	213	192	181	216	215	221	226	4295	
trend:		less	less	less	less	less	less	less	av	av	av	less	av	av	av	av	more	less	av	less	less	less	

SOUTH ISLAND SUNSHINE ESTIMATES
DECEMBER 2025

F=fine (8-12 hours of sunshine), pc= partly cloudy (4-7 hours), c=cloudy (1-3 hours), o=overcast (0 hours)

		Ta	Nel	Ble	Mou	West	Grey	Hok	Ham	Kai	Ch	Ash	Tim	Crom	Qstn	Bal	Dun	Wnt	Gore	Inv	Alex	ALL	Moon
1st	Mon	F	F	F	F	F	F	pc	F	F	F	F	F	F	F	F	F	F	F	F	F	244	
2nd	Tue	F	F	F	F	pc	F	F	F	F	F	F	F	c	c	pc	c	c	c	c	c	161	
3rd	Wed	F	F	F	F	pc	c	c	F	F	F	F	pc	c	pc	c	pc	o	o	o	c	133	
4th	Thur	F	F	F	F	F	pc	F	F	F	F	F	pc	F	F	F	pc	F	F	pc	F	200	
5th	Fri	F	F	F	F	F	pc	c	F	F	F	F	F	F	F	F	F	F	F	F	F	195	F P2
6th	Sat	F	F	F	F	F	F	F	F	F	c	F	F	F	F	pc	F	F	F	F	F	218	^
7th	Sun	o	F	F	F	F	F	F	c	c	F	o	o	F	F	pc	F	pc	pc	F	F	145	
8th	Mon	F	F	F	F	F	F	F	F	F	F	F	F	F	F	F	F	F	F	pc	F	225	
9th	Tue	pc	F	F	F	F	pc	pc	F	F	c	pc	pc	pc	F	pc	pc	pc	pc	F	pc	154	
10th	Wed	F	o	pc	c	pc	F	F	c	c	pc	o	o	F	F	pc	F	pc	pc	F	F	134	
11th	Thur	c	F	F	F	c	pc	F	F	F	pc	pc	c	F	F	pc	c	pc	pc	pc	F	143	
12th	Fri	F	F	pc	F	pc	pc	F	o	o	o	o	c	F	F	pc	F	F	F	pc	F	134	3Q XhS
13th	Sat	F	F	F	F	pc	pc	pc	pc	pc	F	pc	c	F	F	F	F	F	F	F	F	186	
14th	Sun	F	o	F	pc	pc	c	pc	F	F	F	pc	pc	F	F	F	F	F	F	F	F	178	
15th	Mon	F	pc	F	pc	c	c	c	F	F	F	F	pc	F	F	F	F	pc	pc	F	F	170	
16th	Tue	o	F	pc	F	F	F	F	o	o	c	o	pc	F	F	F	pc	pc	pc	F	F	151	
17th	Wed	o	F	F	F	F	F	F	F	F	F	F	F	F	F	F	F	F	F	F	F	242	A
18th	Thur	o	F	pc	F	F	F	F	F	F	F	F	F	F	F	F	F	F	F	F	F	238	
19th	Fri	o	o	c	o	pc	pc	pc	F	F	F	F	pc	F	F	F	F	F	F	F	F	156	
20th	Sat	o	o	o	o	o	o	o	c	c	pc	pc	pc	c	c	o	o	o	o	c	c	31	N V
21st	Sun	o	o	o	o	c	c	o	pc	pc	F	F	pc	o	o	o	o	o	o	o	o	38	
22nd	Mon	pc	c	c	o	o	o	o	o	o	o	c	c	pc	pc	c	pc	o	o	o	pc	40	
23rd	Tue	pc	pc	pc	c	o	o	o	o	o	o	o	o	o	o	F	o	o	o	o	o	26	
24th	Wed	pc	o	c	o	o	o	o	o	o	o	o	o	c	o	o	o	o	o	o	c	17	
25th	Thur	o	pc	F	pc	pc	pc	pc	F	F	F	F	pc	F	pc	pc	F	F	F	F	F	161	
26th	Fri	o	c	F	c	c	o	o	F	F	pc	pc	o	F	c	pc	pc	c	c	pc	F	83	
27th	Sat	F	pc	F	F	o	o	o	pc	pc	pc	pc	c	o	o	c	o	o	o	c	o	65	XhN
28th	Sun	c	o	o	o	o	pc	pc	o	o	o	o	o	F	F	pc	F	F	F	F	F	94	1Q
29th	Mon	pc	o	o	c	pc	F	F	c	c	pc	pc	pc	F	pc	c	c	pc	pc	F	F	114	
30th	Tue	F	F	c	F	F	F	F	o	o	c	o	pc	F	F	pc	pc	pc	pc	pc	F	138	
31st	Wed	o	c	c	o	pc	pc	F	F	F	F	F	c	F	F	pc	F	F	F	F	F	157	
est. hours:		189	220	242	226	181	193	216	216	216	214	215	165	283	268	198	214	197	197	237	283	4368	
av. hours:		220	246	247	242	191	174	194	178	209	221	195	182	249	227	176	167	175	170	186	219	4068	
trend:		less	less	av	av	av	more	more	more	av	av	av	av	more	more	more	more	more	more	more	more	average	

DECEMBER **1** MONDAY

A large high pressure system lingers over New Zealand bringing a gentle westerly flow to mostly calm air throughout, with the chance of fog patches overnight particularly around Canterbury. Partly cloudy and dry everywhere, with higher temperatures establishing, particularly in the northern half of the North Island and Canterbury to Otago regions.

Auckland: Calm atmosphere, sunny and dry, warmer
Hamilton: Calm and pleasant air, sunny, warm, dry
Wellington: Gentle breezes, warmer, dry, passing cloudy spells
Greymouth: Sunny, dry, mild day cool night, gentle breezes
Christchurch: Sunny, dry, gentle breezes, chance of overnight fog patches
Dunedin: Very bright and sunny, dry, warmer, gentle breezes

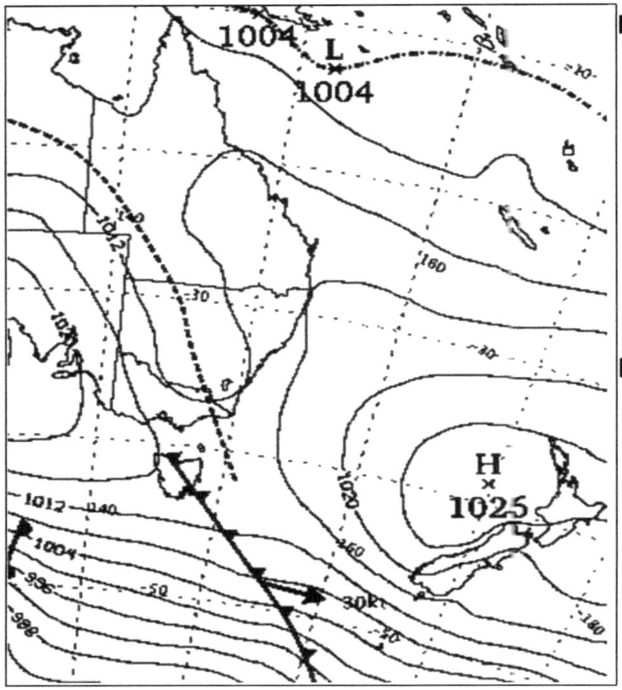

RAIN POTENTIAL

FROST/SNOW

DECEMBER **2** TUESDAY

A high pressure system remains stalled over New Zealand directing a gentle northwest flow over the south of the South Island and southwesterly elsewhere, with the chance of fog patches overnight still likely particularly around Canterbury. Partly cloudy and dry almost everywhere except for increased cloud in the south of the South Island, and some showers possible around Fiordland. High maximum temperatures likely in the East Cape to the Waikato, and Canterbury to Otago.

Auckland: Sunny, dry, warmer, pleasant light breezes
Hamilton: Sunny, dry, warm day cool night, gentle breezes
Wellington: Partly cloudy, cooler, gentle breezes
Greymouth: Sunny, dry, light breezes
Christchurch: Sunny, warmer, dry, gentle breezes, chance of shallow fog
Dunedin: Cloudy, dry, hot day mild night, gentle breezes

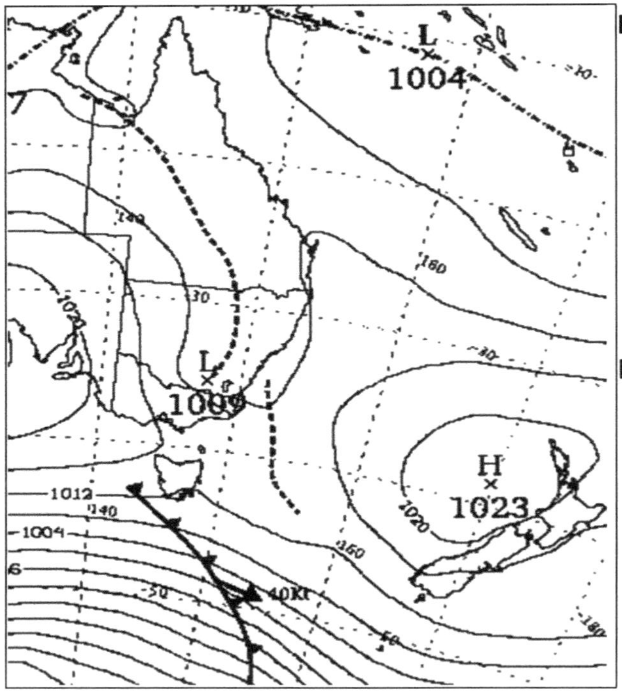

RAIN POTENTIAL

FROST/SNOW

DECEMBER **3** WEDNESDAY

A stalled high pressure system over New Zealand begins to weaken ahead of a low pressure wet front nearing the southwest coasts of the South Island. Mostly dry, warm, and sunny with a gentle westerly air throughout the North Island, with high maximum temperatures likely around the East Cape. In contrast, while mostly sunny and dry conditions also prevail in the South Island, a freshening westerly flow will extend a wet front into Fiordland and some outbreaks of showers. Still very hot in Canterbury.

Auckland: Partly cloudy, dry, light breezes to mostly calm air
Hamilton: Partly cloudy, dry, slightly cooler, gentle breezes
Wellington: Partly cloudy, dry, light shifting breezes
Greymouth: Cloudy, dry, warmer, moderate breezes
Christchurch: Very sunny, warmer, dry, gentle breezes, possible fog
Dunedin: Freshening breezes, dry, cooler, scattered cloudy spells

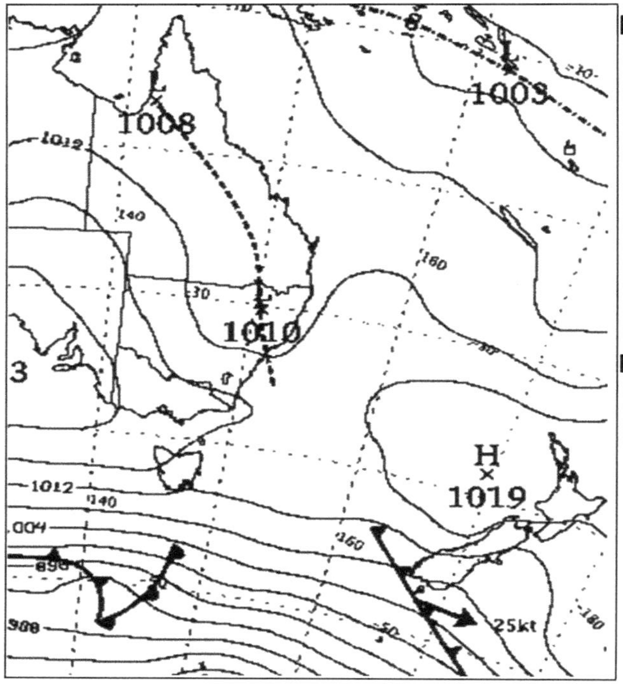

RAIN POTENTIAL

FROST/SNOW

DECEMBER **4** THURSDAY

A low pressure trough moves onto New Zealand ahead of a high pressure ridge building in the Tasman Sea. A light southerly air, widespread cloud build over the North Island with rain at times from Northland to the Bay of Plenty, while mostly dry south of the Central Plateau. A stronger and more unsettled west to southwest flow crosses the South Island, with good sunny spells throughout, and some evening showers in the west and southern regions, while dry in the north and east. Maximum temperatures still high in the East Cape and Canterbury.

Auckland: Cloudy, warmer, showers, moderate breezes
Hamilton: Cloudy, light breezes, warmer, brief showers
Wellington: Sunny, dry, fresh breezes with occasional gusts likely
Greymouth: Partly cloudy, scattered showers, unsettled breezes at times
Christchurch: Hot day cool night, sunny, dry, unsettled breezes
Dunedin: Dry, partly cloudy, warmer, fresh breezes

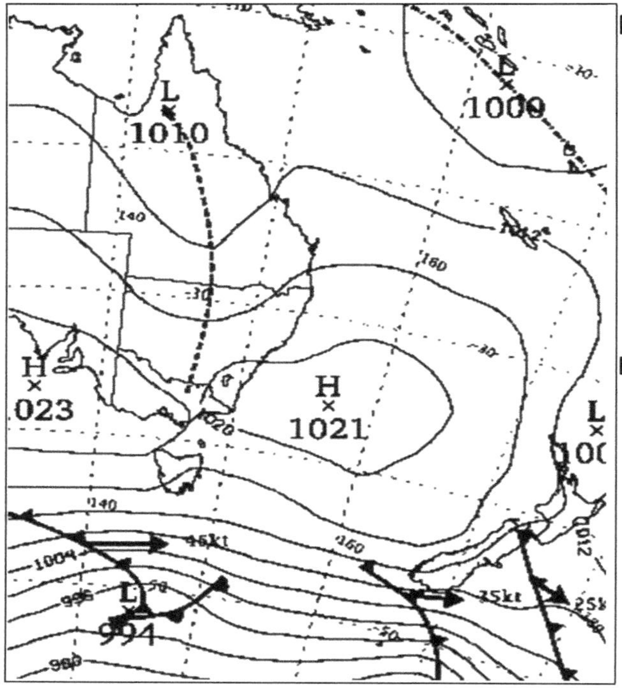

RAIN POTENTIAL

FROST/SNOW

DECEMBER **5** FRIDAY

The high pressure ridge remains almost stalled in the Tasman Sea while a low pressure system crossing through the Southern Ocean brings unsettled conditions throughout the South Island in a strong and blustery southwest flow and with outbreaks of evening rain likely in some western and southern parts, but remaining mostly dry, very warm, and sunny in the east. A lighter southwest flow, partly cloudy and mostly dry conditions prevail over the North Island, except for the chance of some isolated showers in Northland.

Auckland: Very sunny, mostly dry, chance of morning showers, fresh breezes
Hamilton: Very sunny and dry, light breezes
Wellington: Sunny, warmer, dry, shifting breezes
Greymouth: Cloudy, occasional showers, fresh breezes
Christchurch: Sunny, slightly cooler, dry, moderate breezes
Dunedin: Unsettled winds, dry, sunny

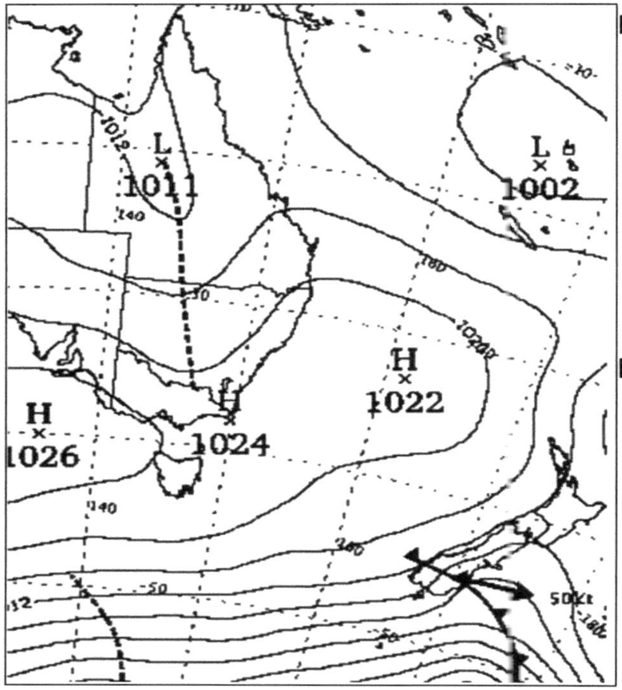

RAIN POTENTIAL

FROST/SNOW

DECEMBER **6** SATURDAY

A low pressure trough crosses the country bringing moderate northerly breezes throughout backing northeasterly for a time over the South Island. Mostly dry, sunny, and very hot throughout both islands except for the chance of some isolated showers around Southland.

Auckland: Sunny, dry, warm and pleasant air, light breezes
Hamilton: Very sunny, dry, warmer day cooler night, calm and pleasant air
Wellington: Very sunny, dry, warmer day cooler night, moderate shifting breezes
Greymouth: Very sunny, dry, moderate breezes
Christchurch: Cloudy, very hot day cool night, dry, fluctuating breezes
Dunedin: Sunny, dry, warmer day cooler night, occasional breezy spells

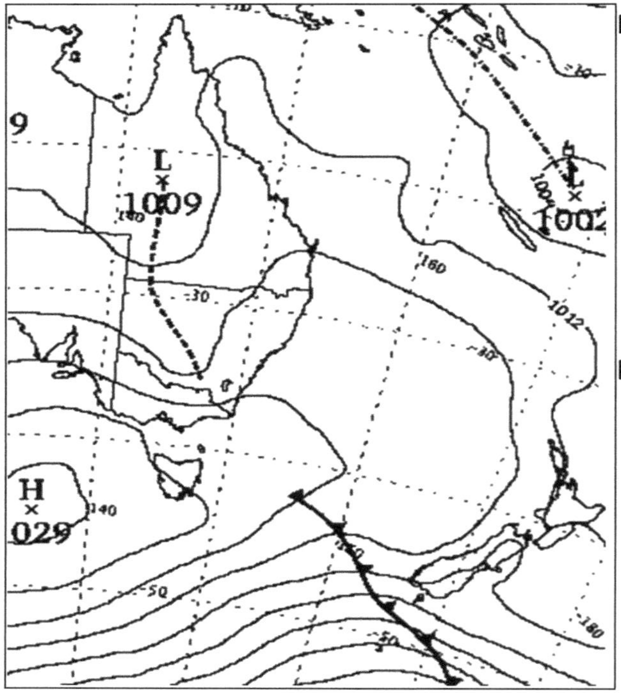

RAIN POTENTIAL

FROST/SNOW

DECEMBER **7** SUNDAY

A large anticyclonic system extends its high pressure ridge onto the country bringing a south to southwest change throughout and some showers into Fiordland and Southland. Mainly dry elsewhere, with long sunny spells and high temperatures over much of the North Island, but some cloudier pockets and cooler over the South Island.

Auckland: Partly cloudy, dry, gentle breezes
Hamilton: Sunny, dry, milder, gentle breezes to mostly calm air
Wellington: Sunny, dry, gentle breezes
Greymouth: Sunny, dry, light to moderate breezes
Christchurch: Light breezes to mostly calm air, sunny, dry, cooler
Dunedin: Mostly sunny, cooler, dry, gentle breezes

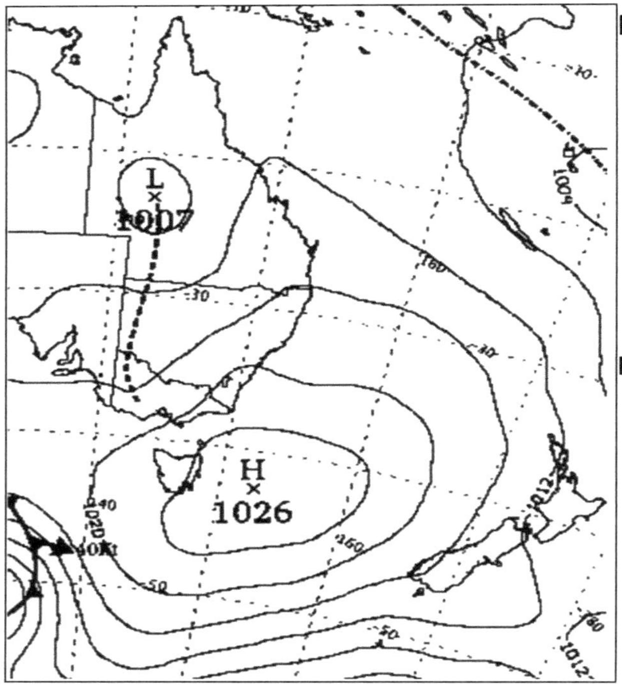

RAIN POTENTIAL

FROST/SNOW

DECEMBER **8** MONDAY

A large high pressure system drifts over New Zealand directing a light northwest flow to mostly calm air across the South Island, and moderate scutheasterlies over the North Island. Mostly sunny, warm, and dry throughout the North Island. In contrast, a mixed day of sunny spells and scattered evening showers prevail over the South Island, with cold overnight temperatures that may see some frost pockets in sheltered areas in Otago and Southland.

Auckland: Cloudy, dry, light pleasant breezes, cooler
Hamilton: Sunny, dry, slightly cooler, gentle breezes
Wellington: Very sunny, cooler, dry, moderate breezes
Greymouth: Very sunny and dry, slightly cooler, calm and serene air
Christchurch: Sunny, evening showers, slightly cooler, calm air
Dunedin: Sunny, cooler, brief showers, calm atmosphere

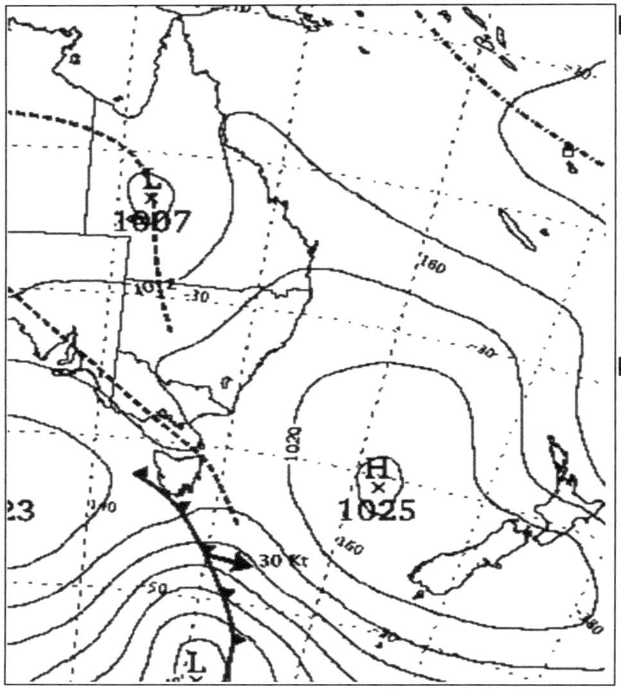

RAIN POTENTIAL

FROST/SNOW

DECEMBER **9** TUESDAY

A low pressure trough between two highs and associated with a Southern Ocean depression flows up through the country in a moderate southwest flow over the South Island backing northwesterly over the North Island. Mostly cloudy over the North Island with some showers Northland to Auckland, and mostly sunny in the South Island with some showers around Fiordland. Dry elsewhere.

Auckland: Cloudy, spotty showers, mostly calm air
Hamilton: Sunny, dry, warm day cool night, calm and pleasant air
Wellington: Sunny, dry, mild day cool night, occasional breezy spells
Greymouth: Partly cloudy, occasional breezy spells, dry, warmer
Christchurch: Cloudy, dry, gentle breezes
Dunedin: Shifting breezy spells, partly cloudy, dry

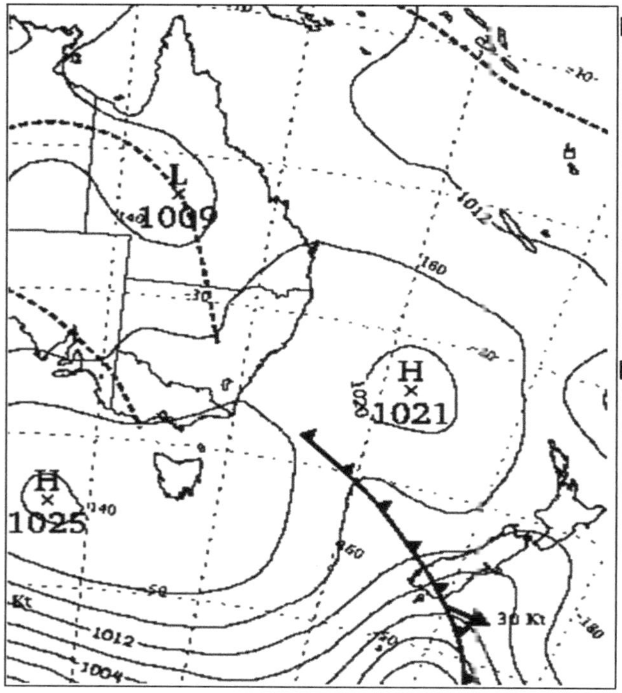

RAIN POTENTIAL

FROST/SNOW

DECEMBER **10** WEDNESDAY

A complex of low pressure centres cross the country bringing light but changeable breezes throughout the country, mostly northeasterly over the South Island, southeasterly over the North Island. Mostly sunny and dry day throughout with evening showers in Northland to Auckland, and in patches throughout the South Island. Warmer everywhere.

Auckland: Changeable, some showers, brief sunny spells, gentle breezes
Hamilton: Very sunny, isolated evening showers, calm atmosphere
Wellington: Mostly sunny, dry, gentle breezes
Greymouth: Sunny, dry, fluctuating breezes
Christchurch: Partly cloudy, dry, warmer, fluctuating breezes
Dunedin: Mostly sunny, warmer day cooler night, light showers, light breezes

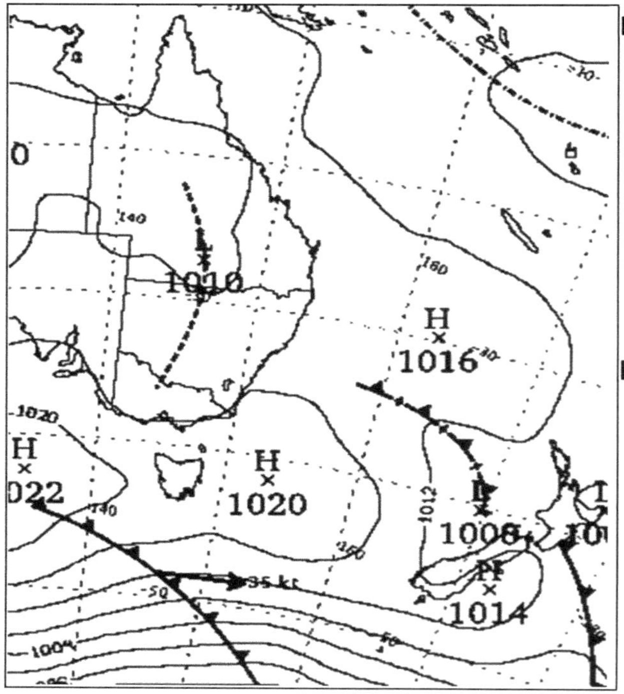

RAIN POTENTIAL

FROST/SNOW

DECEMBER **11** THURSDAY

Low pressure continues to linger over New Zealand in a gentle to mostly calm air throughout. A mostly sunny and dry day everywhere with some spotty evening showers likely in the North Island from Auckland to the Central Plateau, more frequent around the Bay of Plenty.

Auckland: Sunny, isolated showers, mostly calm air
Hamilton: Partly cloudy, dry, gentle breezes
Wellington: Very sunny, dry, cooler, mostly calm air
Greymouth: Partly cloudy, mostly dry, gentle breezes, chance of misty patches
Christchurch: Partly cloudy, cooler, dry, gentle breezes
Dunedin: Cloudy, colder, dry, moderate breezes

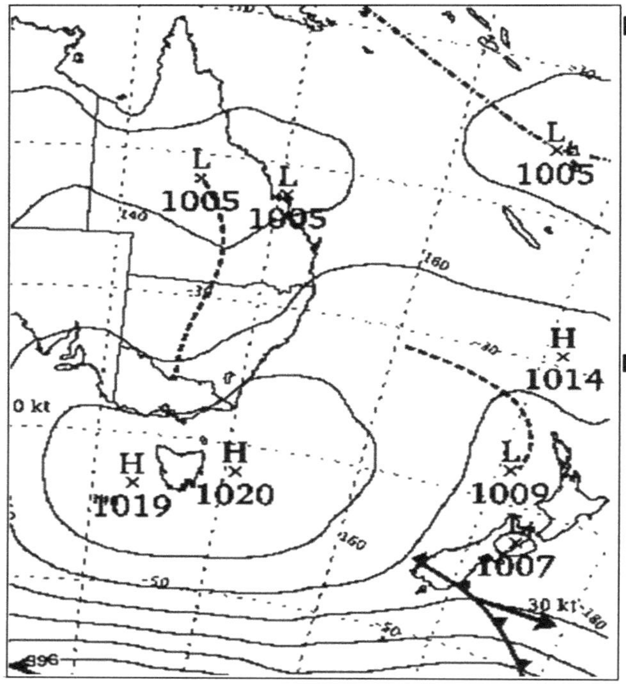

RAIN POTENTIAL

FROST/SNOW

DECEMBER **12** FRIDAY

A weak high pressure system moves onto New Zealand with a light west to southwest flow to mostly calm air prevailing. Partly cloudy and mostly dry throughout the country except for the chance of some brief evening showers in Northland to Auckland and around central Canterbury. Warmer.

Auckland: Cloudy and calm air, warmer, scattered showers
Hamilton: Calm and serene air, dry, scattered cloudy spells, cool overnight
Wellington: Partly cloudy, warmer, dry, calm atmosphere
Greymouth: Cloudy, calm and serene air, dry
Christchurch: Calm and serene air, overcast, brief showers, slightly warmer
Dunedin: Sunny, warmer, dry, calm and serene air, chance of misty patches

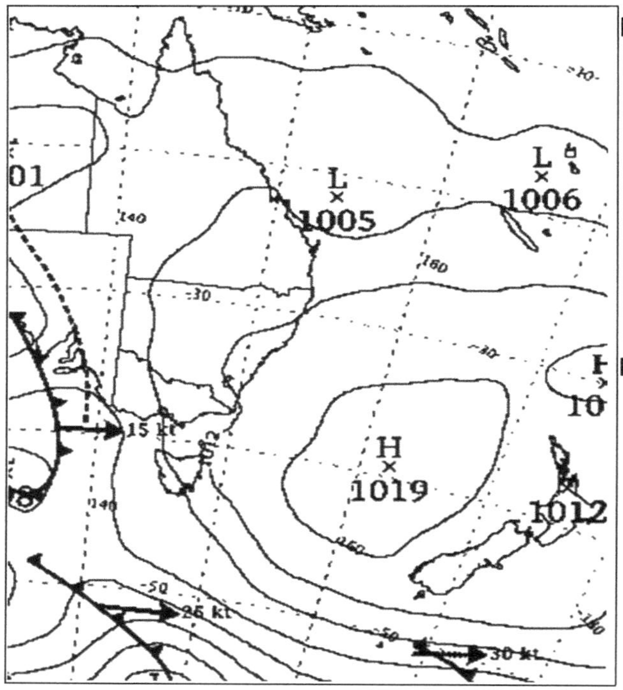

RAIN POTENTIAL

FROST/SNOW

DECEMBER **13** SATURDAY

High pressure builds over New Zealand bringing a light southwest flow to mostly calm air throughout. Mostly sunny and dry everywhere, and very warm in the North Island. Chance of some isolated showers briefly along the Canterbury coastal fringes and around Northland.

Auckland: Sunny, warm, spotty showers, gentle breezes to mostly calm air
Hamilton: Sunny, warmer, dry, gentle breezes to mostly calm
Wellington: Sunny, cooler, dry, light shifting breezes
Greymouth: Partly cloudy, dry, light breezes to mostly calm and pleasant air
Christchurch: Mostly sunny, cooler, dry, calm and pleasant air
Dunedin: Sunny, dry, calm and pleasant air, chance of misty patches

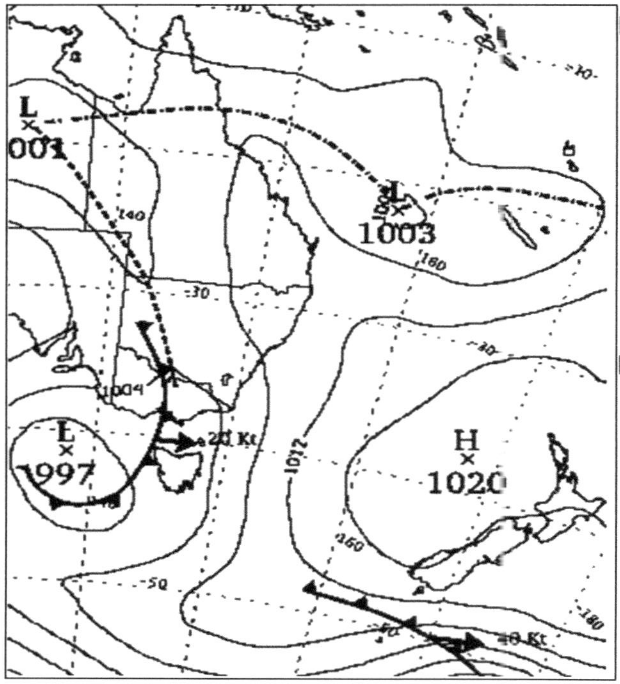

RAIN POTENTIAL

FROST/SNOW

DECEMBER **14** SUNDAY

Anticyclonic. Warmer, dry, and mostly sunny throughout the country with a mostly calm air throughout except for some light north to northwesterly breezes at times over the South Island.

Auckland: Sunny, dry, calm air
Hamilton: Calm and pleasant air, partly cloudy, warm day cool night, dry
Wellington: Cloudy, dry, calm and pleasant air
Greymouth: Cloudy, dry, light air
Christchurch: Calm atmosphere, sunny, dry, warmer
Dunedin: Very sunny, warmer, dry, gentle breezes

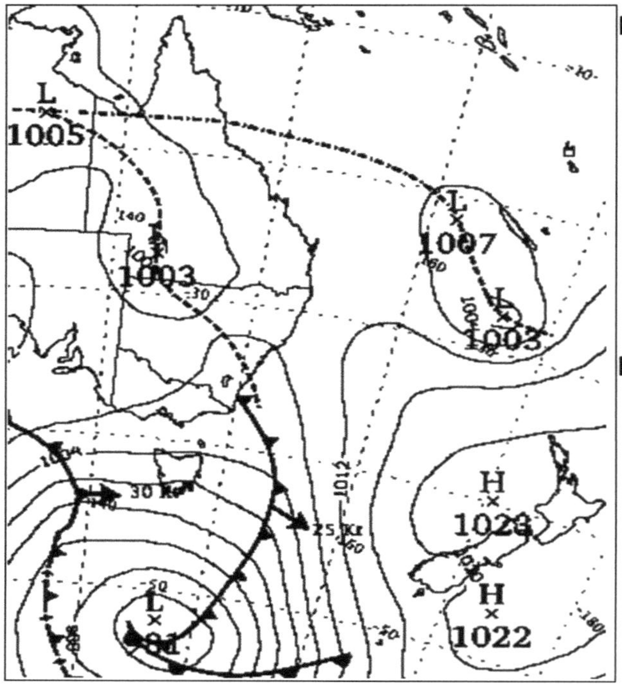

RAIN POTENTIAL

FROST/SNOW

DECEMBER **15** MONDAY

A weak depression to the north of New Zealand directs a northeasterly flow over the North Island and brings increasing cloud and threats of rain throughout the North Island. In contrast, a high pressure ridge still lingers over the South Island in a fresher west to northwest flow, and mostly dry and sunny conditions continuing. Warmer everywhere.

Auckland: Overcast, warmer, mostly dry, increasing threats of rain
Hamilton: Overcast, dry, some threats of rain, mostly calm air
Wellington: Sunny, dry, moderate breezes
Greymouth: Cloudy, freshening breezes, mild temperatures, dry
Christchurch: Sunny, dry, slightly cooler, gentle shifting breezes
Dunedin: Sunny, dry, unsettled breezy spells

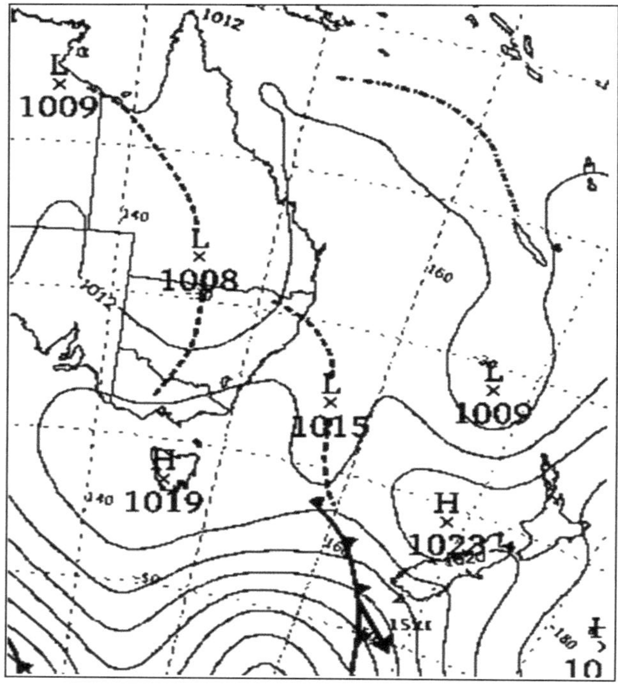

RAIN POTENTIAL

FROST/SNOW

DECEMBER **16** TUESDAY

A large high in the south Tasman Sea directs an east to southeast flow over the South Island with increasing cloudy spells, particularly in Marlborough to Canterbury, and while mostly dry, some isolated overnight showers are possible in Southland and Otago. In contrast, low pressure maintains very cloudy conditions throughout the North Island with heavy rain from Northland to the Central Plateau, and showers into the Manawatu and threatening rain into Wellington. High maximum temperatures everywhere, especially in Canterbury and Otago.

Auckland: Overcast, heavy rain, fluctuating breezes
Hamilton: Light breezes, overcast, scattered showers
Wellington: Overcast, dry, warmer, moderate breezes
Greymouth: Very sunny, warmer, dry, moderate breezes
Christchurch: Cloudy, very warm, dry, gentle shifting breezes
Dunedin: Partly cloudy, brief showers, moderate breezes

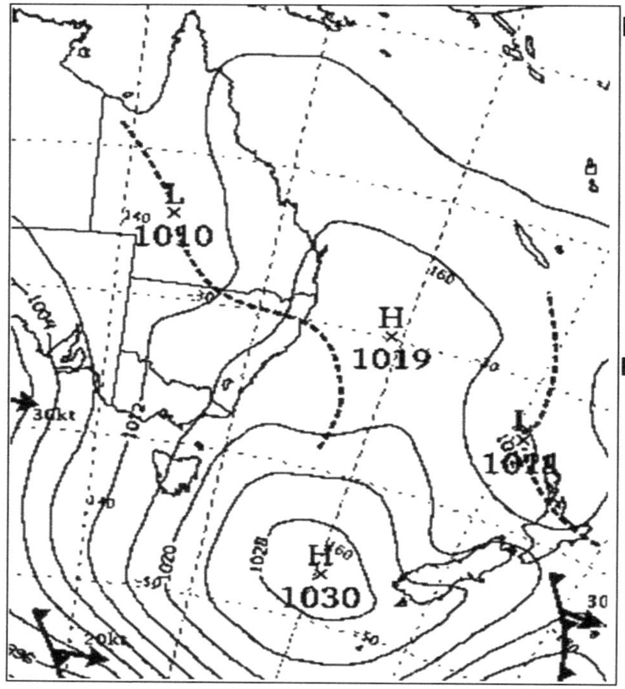

RAIN POTENTIAL

FROST/SNOW

DECEMBER **17** WEDNESDAY

New Zealand sits between a weak depression to the north of the country and an anticyclonic system to the southeast, with fresh to strong easterly winds prevailing over the North Island and lighter east to northeasterlies over the South Island. The low pressure system brings widespread cloud and heavy rain over much of the North Island, while the high pressure to the south brings a mostly dry day with good sunny spells over most of the South Island except for the chance of some isolated showers possible around Marlborough.

Auckland: Cloudy, heavy rain, fresh breezes at times
Hamilton: Overcast, scattered showers, warmer, moderate breezes
Wellington: Partly cloudy, cooler, dry, fresh breezes at times
Greymouth: Very bright and sunny, hot day cool night, dry, calm atmosphere
Christchurch: Very sunny, dry, cooler, gentle breezes
Dunedin: Sunny, cooler, dry, gentle breezes to mostly calm air

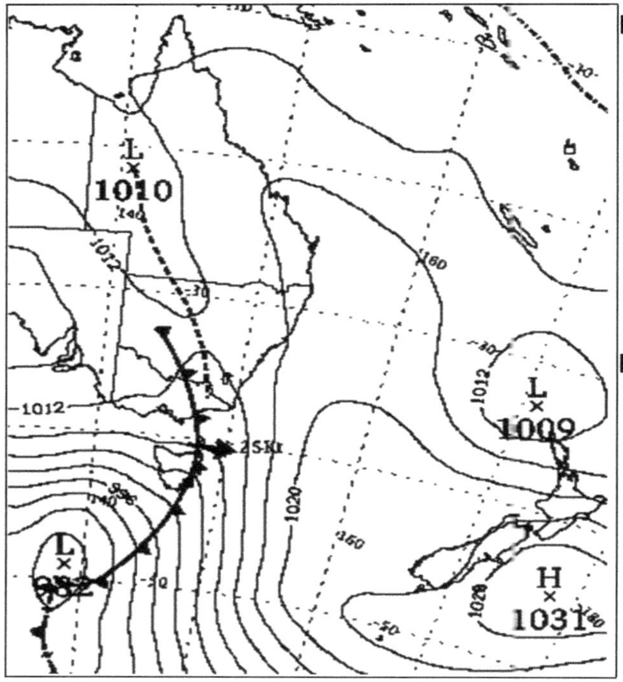

RAIN POTENTIAL

FROST/SNOW

DECEMBER **18** THURSDAY

Low pressure to the north of New Zealand continues to bring widespread cloudy spells and outbreaks of rain over the North Island from Northland to the Hawkes Bay accompanied by moderate to fresh east to northeast breezes. In contrast, high pressure prevails over the South Island with a moderate east to northeast flow over the northern half of the island while another high pressure system to the west generates a northwest flow over the southern regions. Mostly sunny and dry throughout the South Island with chance of frost patches in Canterbury.

Auckland: Sunny, mostly dry, possible brief drizzle patches, moderate breezes
Hamilton: Cloudy, cooler, dry, moderate breezes
Wellington: Overcast and dry, light breezes, chance of fog patches overnight
Greymouth: Very sunny, dry, slightly cooler, fluctuating breezes
Christchurch: Bright and sunny, dry, mild day cool night, moderate breezes
Dunedin: Very sunny and dry, light breezes

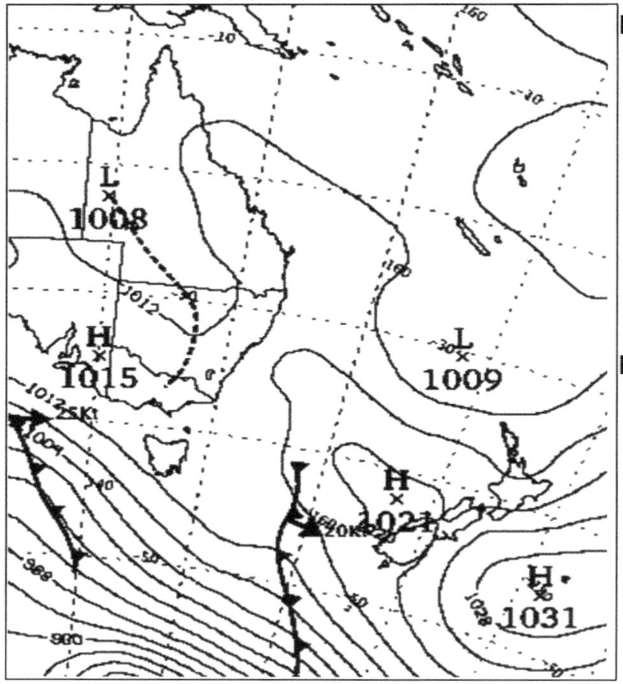

RAIN POTENTIAL

FROST/SNOW

DECEMBER **19** FRIDAY

A high pressure system establishes a firm influence over the country and directs a east to northeast flow throughout with a hot day everywhere. Cloudy with widespread showers persist over the North Island, while in contrast, a sunny and dry day persists over the South Island.

Auckland: Overcast, warmer and humid air, pockets of rain, fresh breezes
Hamilton: Overcast, warmer, occasional showers, fresh breezes
Wellington: Cloudy, warmer, dry, moderate breezes
Greymouth: Partly cloudy, dry with developing threats of rain, light breezes
Christchurch: Sunny, warmer, dry, mostly calm
Dunedin: Sunny, warmer, dry, light breezes to mostly calm air

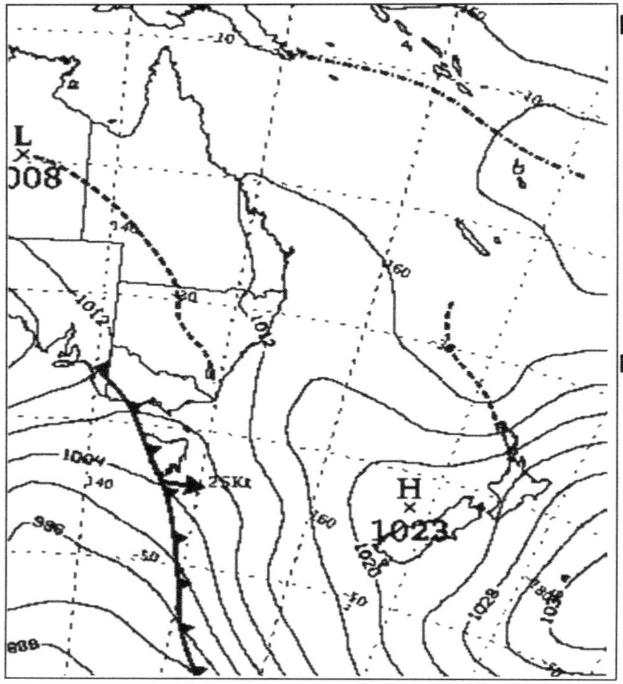

RAIN POTENTIAL

FROST/SNOW

DECEMBER **20** SATURDAY

An intense Southern Ocean depression advances onto the southwest of the country as the high pressure departs to the east with fresh to strong north to northeast breezes prevailing throughout. Dull and gloomy with high temperatures everywhere, and showers throughout the North Island and down through the north and western regions of the South Island. Mostly dry and very hot in Canterbury to Southland.

Auckland: Overcast, rain, fresh breezes
Hamilton: Overcast, light showers and drizzle patches, fresh breezes
Wellington: Overcast, overnight showers, hot and humid day, some breezy spells
Greymouth: Overcast, rain, moderate winds
Christchurch: Partly cloudy, light showers, moderate breezes
Dunedin: Overcast, dry, light breezes

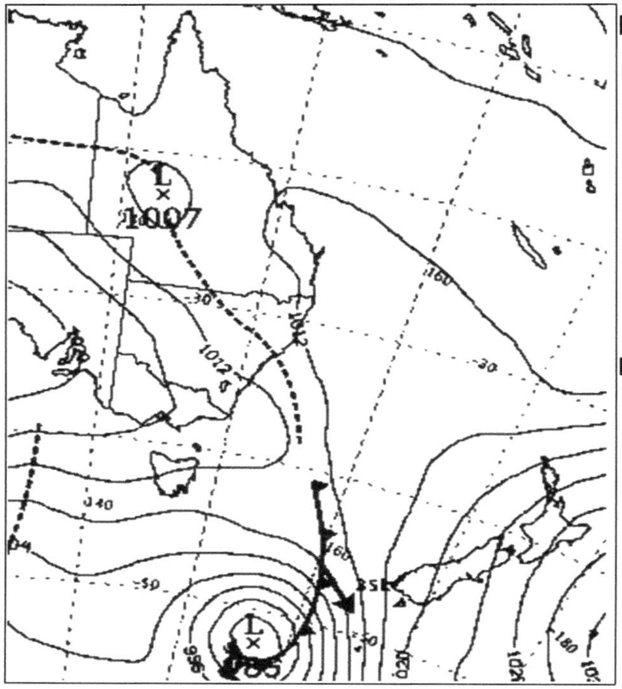

RAIN POTENTIAL

FROST/SNOW

DECEMBER **21** SUNDAY

A complex of low pressure centres linger over New Zealand directing an unstable airflow throughout, lighter easterlies at times over the North Island. Mostly cloudy, hot and humid everywhere with outbreaks of rain in Northland and along the West Coast and Fiordland regions, and showers from Auckland to the Central Plateau, in the Nelson-Marlborough region and around the foothills of Canterbury. Mostly dry in the east and south of both Islands.

Auckland:	Overcast, fluctuating breezes, pockets of showers
Hamilton:	Overcast, isolated showers, moderate to light breezes
Wellington:	Cloudy, dry, cooler, fluctuating breezes
Greymouth:	Cloudy, rain, fluctuating breezes
Christchurch:	Very warm, sunny, isolated showers, fluctuating breezes
Dunedin:	Overcast, warmer, dry, fluctuating breezes

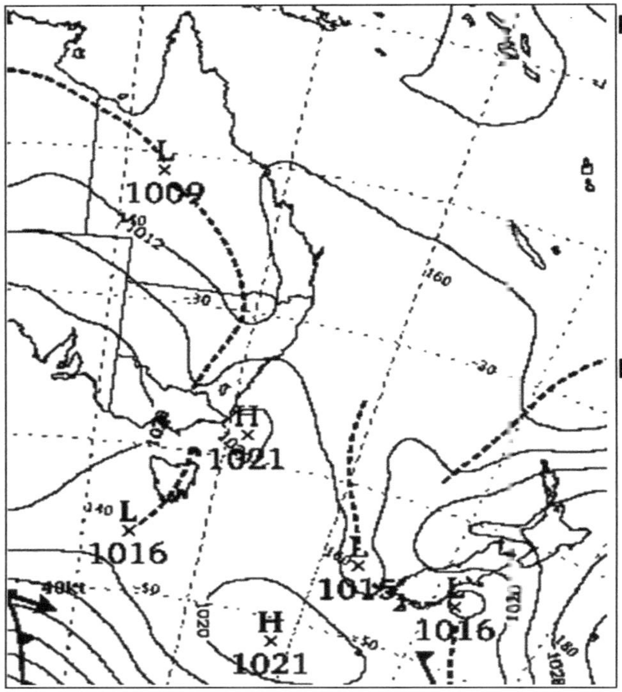

RAIN POTENTIAL

FROST/SNOW

DECEMBER **22** MONDAY

Low pressure brings a moderate westerly flow over the South Island with widespread cloud and outbreaks of rain in most districts, and very high temperatures into Canterbury. Widespread cloud, mostly dry and a light northeasterly air prevails over the North Island, hot around the East Cape and Taranaki to the King Country.

Auckland:	Cloudy, mostly dry with passing threats of rain, fresh breezes at times
Hamilton:	Overcast, mostly dry, some threats of rain, moderate breezes
Wellington:	Cloudy, dry, light breezes
Greymouth:	Overcast, heavy rain, moderate breezes
Christchurch:	Overcast, very hot and humid air, brief showers, light breezes
Dunedin:	Partly cloudy, passing showers, moderate breezes

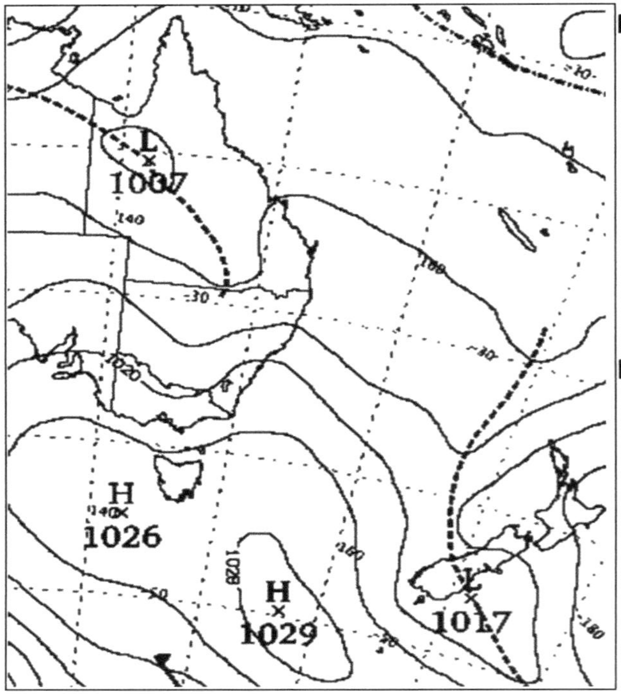

RAIN POTENTIAL

FROST/SNOW

DECEMBER **23** TUESDAY

A large high pressure system crosses to the south of the country while a weak low pressure trough flows to the north, with east to northeast breezes prevailing over the South Island. Mostly dry and sunny with high temperatures throughout the North Island under a light northeasterly flow. In contrast, widespread cloud and outbreaks of rain dominate the South Island, with heavy falls likely in the West Coast that may see some pockets of localised flooding.

Auckland: Partly cloudy, mostly dry, chance of overnight showers, light breezes
Hamilton: Sunny, dry, warmer day cooler night, gentle breezes
Wellington: Warm, dry, cloudy, gentle breezes
Greymouth: Overcast, heavy rain, unsettled breezes, chance of thunderstorms
Christchurch: Overcast, hot, humid, scattered showers, fluctuating breezes
Dunedin: Overcast, scattered showers, moderate breezes

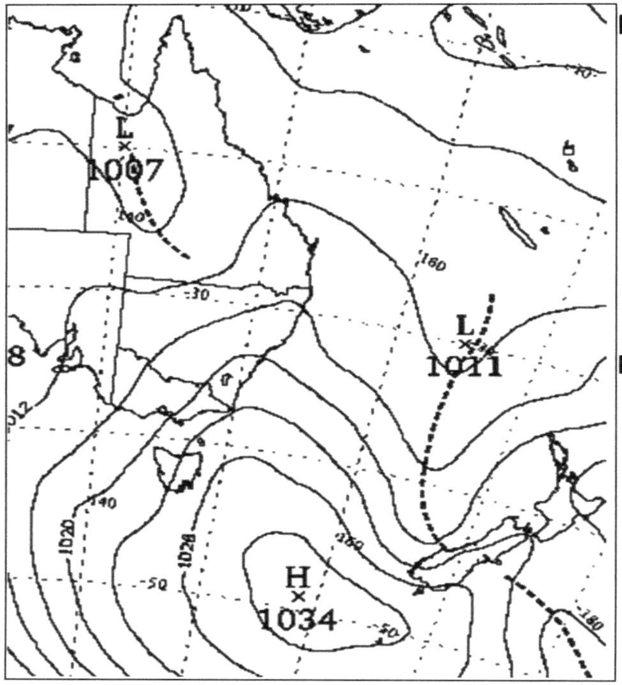

RAIN POTENTIAL

FROST/SNOW

DECEMBER **24** WEDNESDAY

A low pressure trough moves down through the country as the high pressure ridge moves away to the east. Mostly dry, hot, and sunny throughout the North Island under a gentle to mostly calm east to northeast flow. Windier northeasterlies prevail over the South Island bringing cooler temperatures, widespread cloud and outbreaks of showers into all South Island regions, with heavier outbreaks of rain in the West Coast and more isolated around Nelson-Marlborough.

Auckland: Sunny, warmer, dry, light breezes
Hamilton: Sunny, dry, hot, gentle breezes
Wellington: Overcast, dry, light shifting breezes
Greymouth: Overcast, heavy rain, static air, fresh winds
Christchurch: Overcast, colder, scattered showers, fresh breezes at times
Dunedin: Overcast, isolated showers, cooler, breezy at times

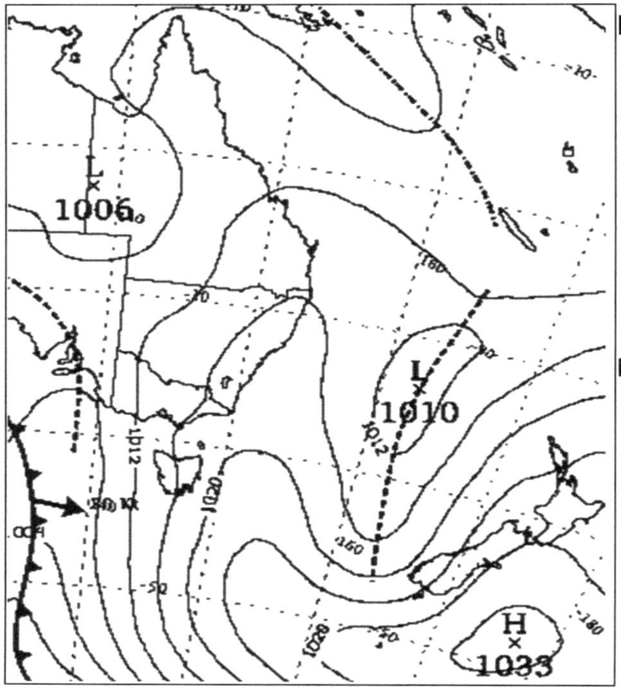

RAIN POTENTIAL

FROST/SNOW

DECEMBER **25** THURSDAY

A low pressure trough directs a light to moderate north to northeasterly flow over the country with very warm temperatures over the North Island and slightly milder over the South Island. Mostly cloudy in Northland and mostly sunny elsewhere. Some isolated showers likely in the Hawkes Bay, and occasional showers in the south of the South Island. Mostly dry elsewhere.

Auckland: Overcast, dry with passing threats of rain, moderate breezes
Hamilton: Hot, dry, overcast, dry with odd threat of rain, light breezes
Wellington: Partly cloudy, dry, cooler, gentle breezes
Greymouth: Partly cloudy, mostly dry, passing threats of rain, light breezes
Christchurch: Sunny, isolated showers, moderate shifting breezes
Dunedin: Sunny, warmer, brief overnight showers, light breezes

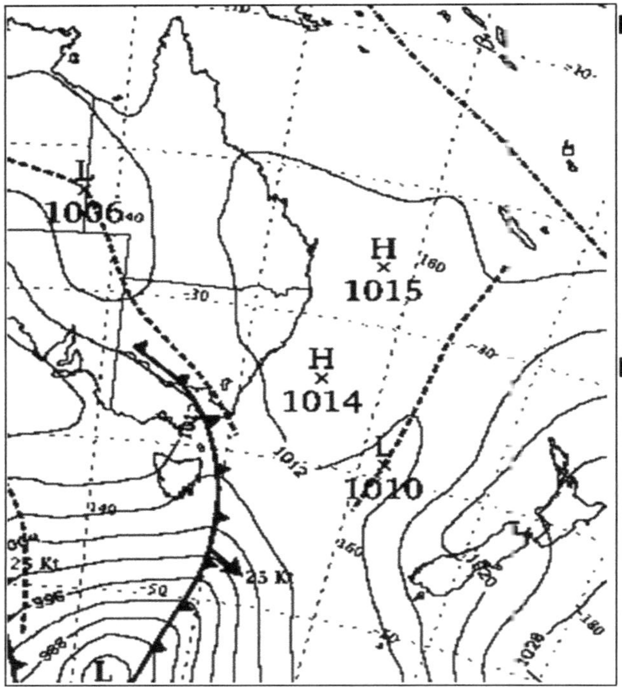

RAIN POTENTIAL

FROST/SNOW

DECEMBER **26** FRIDAY

A light northerly to mostly calm air crosses the North Island while a westerly change associated with a low pressure trough covers the South Island. Warmer throughout the country, with partly cloudy and mostly dry conditions throughout the North Island In the South Island, widespread cloud and some showers prevail over the West Coast and mostly dry with occasional cloudy spells prevail elsewhere.

Auckland: Partly cloudy, mostly dry with brief threats of rain, gentle breezes
Hamilton: Dry, cloudy, cooler, calm air
Wellington: Sunny, warm and dry, gentle breezes
Greymouth: Overcast, heavy rain, strong breezes
Christchurch: Partly cloudy, dry, moderate shifting breezes
Dunedin: Partly cloudy, dry, warmer, moderate breezes

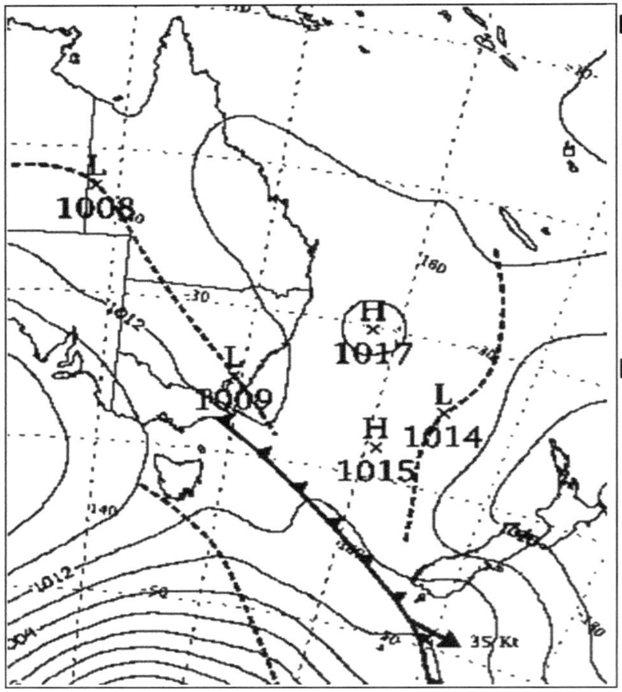

RAIN POTENTIAL

FROST/SNOW

DECEMBER 27 SATURDAY

Low pressure prevails over the country with light northeasterlies over the North Island and fresh westerlies over the South Island. Very hot throughout the country, with widespread cloud over the South Island and rain into the western and southern regions, while remaining mostly dry Marlborough to Canterbury. Mostly dry throughout the North Island, with widespread cloud in the Waikato to the Central Plateau and sunnier elsewhere.

Auckland: Partly cloudy, spotty showers, gentle breezes
Hamilton: Overcast, dry, light breezes
Wellington: Partly cloudy, warm, dry, moderate breezes
Greymouth: Overcast, rain, fresh breezes, chance of gale gusts at times
Christchurch: Partly cloudy, dry, very hot, breezy spells
Dunedin: Very hot day cool night, cloudy, dry, rain threatening, breezy spells

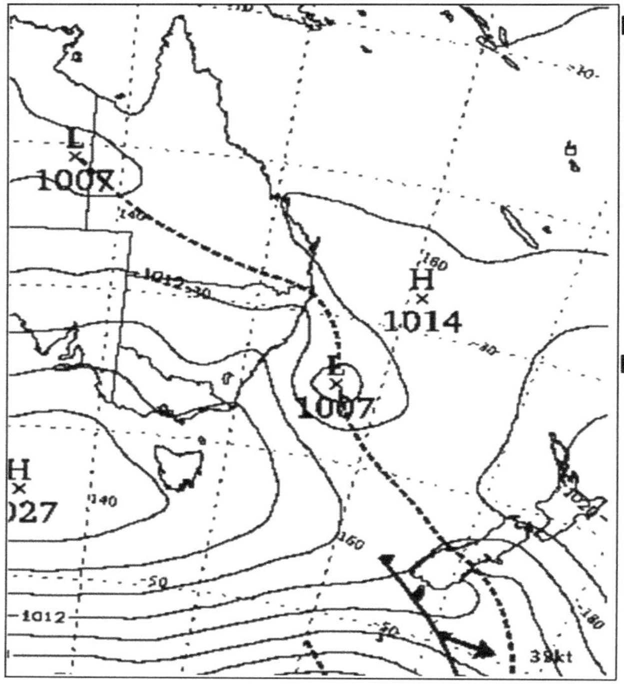

RAIN POTENTIAL

FROST/SNOW

DECEMBER 28 SUNDAY

A weak depression crosses the South Island directing a moderate westerly air and bringing widespread cloud and rain throughout, with heavy falls at times in northern parts of the South Island as well as along the coastal fringes of Otago. In contrast, a lighter northeasterly flow prevails over the North Island, with dull and gloomy conditions throughout and some isolated showers in Northland and around Taranaki to the King Country. Mostly dry with increasing threats of rain elsewhere. Very hot everywhere.

Auckland: Cloudy, humid, dry with increasing threats of rain, light breezes
Hamilton: Hot and humid, overcast, isolated showers, mostly calm air
Wellington: Overcast, dry, threats of rain, unsettled winds at times, static air
Greymouth: Changeable, overnight rain, fresh breezes, static air
Christchurch: Overcast, very hot, humid, showers, chance of thunderstorms
Dunedin: Overnight thunderstorms and heavy rain, sunny day, cooler, breezy

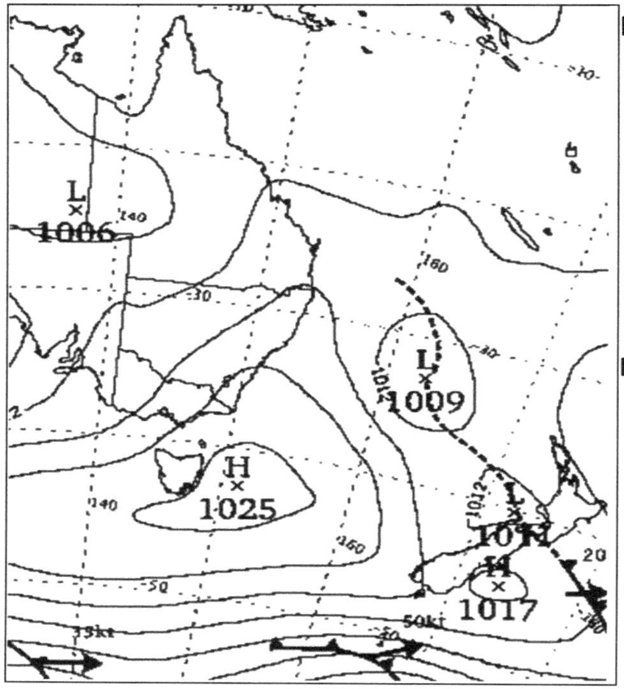

RAIN POTENTIAL

FROST/SNOW

DECEMBER 29 MONDAY

Low pressure brings a southwest flow and a cooler change across the South Island while a high pressure ridge moves onto the country. Mostly easterly winds prevail over the North island, backing westerly over Northland and with high temperatures persisting. Rain contracts to the north of the South Island and spreads throughout the North Island, with heavy falls in many regions that may bring some localised flooding.

Auckland: Cloudy, humid, heavy rain, fluctuating breezes
Hamilton: Cloudy, slightly cooler, heavy rain, shifting breezy spells, humid
Wellington: Stormy air, overcast, heavy rain, cooler, chance of thunderstorms
Greymouth: Very sunny, dry, mild temperatures, mostly calm air
Christchurch: Partly cloudy, colder, brief showers, calm atmosphere
Dunedin: Cloudy, mostly dry and calm air

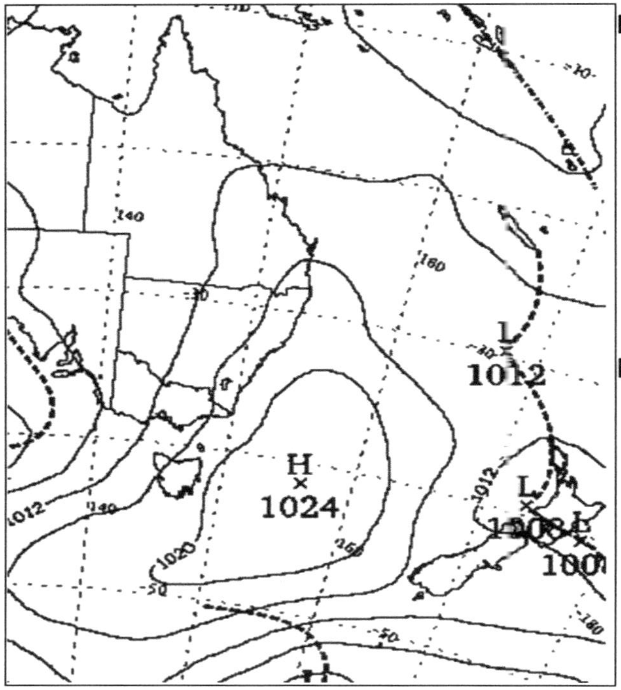

RAIN POTENTIAL

FROST/SNOW

DECEMBER 30 TUESDAY

High pressure ridges to the west of New Zealand directing a gentle southwesterly flow and cooler temperatures up through the country. A wet front brings outbreaks of rain into central regions of the South Island, as well as heavy rain persisting in the lower half of the North Island from the Central Plateau to Wellington, with the chance of localised flooding in some places. Mostly dry and sunny in Northland to the north Waikato, and in the west and south of the South Island.

Auckland: Sunny and mostly dry, warm, humid and mostly calm air
Hamilton: Partly cloudy, light showers, mostly calm air
Wellington: Changeable, rain, some sunny outbreaks, light breezes
Greymouth: Sunny, dry, gentle breezes
Christchurch: Cloudy, scattered showers, moderate breezes
Dunedin: Mostly cloudy, gentle breezes, isolated morning showers

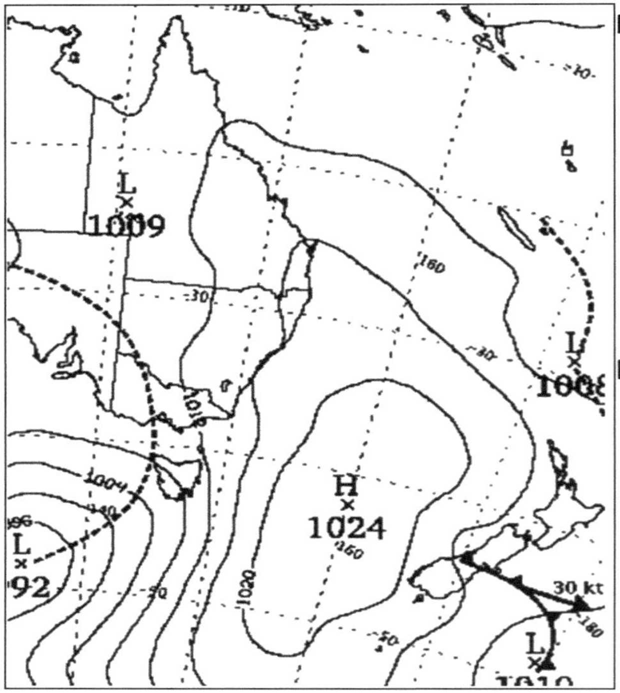

RAIN POTENTIAL

FROST/SNOW

DECEMBER 31 WEDNESDAY

A large high pressure system moves over the country with a trough between two high pressure centres forming. Light, changeable breezes prevail, generally southwesterly in the west of the country and southeasterly over the North Island, backing northeasterly over the South Island eastern parts. A changeable day with a mix of sunny spells and showers throughout the eastern half of the North Island, in the north of the South Island and in patches in Canterbury. Mostly dry and sunny in the west of the North Island and south of the South Island.

Auckland: Overcast, cooler, some pockets of rain, gentle breezes
Hamilton: Partly cloudy, dry, gentle breezes to mostly calm air
Wellington: Very bright and sunny, dry, gentle breezes
Greymouth: Calm and serene air, partly cloudy, brief morning showers
Christchurch: Sunny, mostly dry, chance of morning showers, gentle breezes
Dunedin: Sunny, cooler, light showers, gentle breezes

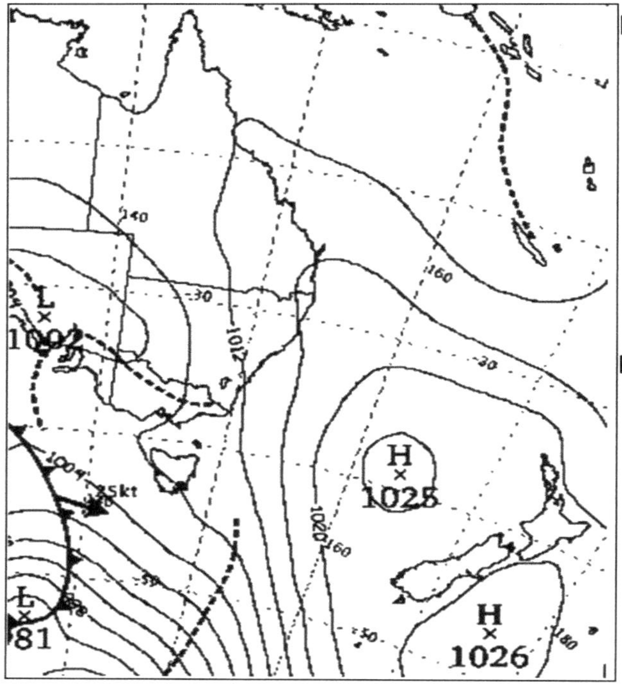

RAIN POTENTIAL

FROST/SNOW

APPENDICES

QUICK GLANCE RAIN AMOUNTS

NORTH ISLAND 2025

	Northland							Auckland					Bay Of Plenty				Waikato			
	CR	Kta	Ker	Rsl	Dar	Whg	Man	Wrk	Hen	[illegible]kH	Ak	Co	Tau	TeP	Wkt	Opo	Hml	TeA	Mata	Ot
Jan	av	av	wtr	wtr	drr	av	drr	drr	drr	av	drr	wtr	wtr	wtr	av	av	drr	wtr	drr	drr
Feb	wtr	wtr	wtr	wtr	wtr	drr	drr	drr	drr	drr	drr	drr	wtr	drr	drr	drr	wtr	av	drr	drr
Mar	drr	drr	drr	drr	drr	drr	drr	drr	av	drr	av	av	drr	drr	drr	drr	av	drr	drr	av
Apr	drr	av	wtr	wtr	wtr	wtr	wtr	wtr	av	drr	drr	wtr	av	wtr	drr	drr	drr	drr	drr	drr
May	drr	drr	drr	drr	drr	drr	drr	drr	drr	av	drr	av	drr	drr	drr	drr	drr	drr	drr	drr
Jun	drr	drr	drr	drr	drr	drr	drr	drr	drr	drr	drr	drr	drr	drr	drr	drr	drr	drr	drr	drr
Jul	wtr	wtr	wtr	wtr	wtr	wtr	wtr	wtr	wtr	wtr	wtr	wtr	wtr	wtr	wtr	av	wtr	wtr	av	wtr
Aug	drr	wtr	drr	av	wtr	drr	wtr	wtr	wtr	wtr	wtr	wtr	wtr	av	av	drr	wtr	wtr	wtr	wtr
Sep	wtr	av	av	drr	drr	wtr	wtr	av	drr	drr	drr	drr	drr	drr	drr	drr	drr	drr	drr	drr
Oct	drr	drr	drr	drr	wtr	drr	av	av	wtr	wtr	av	wtr	drr	av	drr	wtr	wtr	wtr	av	wtr
Nov	drr	drr	drr	drr	drr	drr	drr	drr	av	drr	drr	drr	drr	drr	drr	drr	drr	drr	drr	drr
Dec	wtr	wtr	wtr	wtr	wtr	wtr	wtr	wtr	wtr	av	drr	wtr	av	drr	drr	av	drr	drr	drr	drr
Year	wtr	av	wtr	wtr	wtr	drr	drr	drr	av	av	drr	wtr	drr	drr	drr	drr	av	av	drr	av

	Rot/Tau		Hawkes Bay			C Plateau			Taranaki			Manawatu			Lower North Island			
	Rt	Taup	Gis	Nap	Has	Taum	Chat	Waio	NP	Str	Wang	Pa	Dann	Lev	PN	Mast	UH	Wgt
Jan	wtr	av	drr	drr	av	drr	drr	drr	drr	drr	wtr	drr	drr	drr	drr	drr	av	av
Feb	drr	drr	drr	drr	av	drr	drr	drr	drr	drr	drr	av	drr	av	drr	drr	drr	drr
Mar	drr	wtr	drr	drr	av	wtr	wtr	wtr	wtr	drr	av	drr	drr	wtr	av	drr	drr	drr
Apr	wtr	drr	drr	drr	drr	drr	drr	drr	drr	drr	drr	drr	drr	drr	drr	drr	drr	drr
May	drr	drr	drr	drr	drr	drr	drr	drr	drr	drr	drr	drr	drr	drr	drr	drr	drr	drr
Jun	drr	drr	av	wtr	wtr	drr	av	drr	drr	drr	av	drr	drr	drr	drr	drr	drr	drr
Jul	wtr	av	wtr	wtr	wtr	wtr	wtr	wtr	wtr	wtr	wtr	av	wtr	wtr	av	av	av	av
Aug	wtr	wtr	drr	drr	av	wtr	wtr	wtr	wtr	wtr	wtr	wtr	av	av	wtr	drr	wtr	av
Sep	drr	drr	wtr	wtr	wtr	drr	drr	drr	drr	drr	drr	drr	drr	drr	drr	drr	drr	drr
Oct	av	drr	wtr	av	av	wtr	wtr	wtr	drr	wtr	wtr	wtr	wtr	av	av	wtr	wtr	wtr
Nov	drr	drr	drr	drr	drr	drr	drr	drr	drr	drr	drr	drr	drr	wtr	drr	drr	drr	drr
Dec	drr	av	wtr	wtr	wtr	av	drr	wtr	drr	drr	av	drr	drr	drr	drr	drr	drr	drr
Year	drr	drr	drr	drr	wtr	drr	drr	av	drr	drr	av	drr	drr	drr	drr	drr	drr	drr

SOUTH ISLAND 2025

	West Coast				Top South Island				Canterbury				South Canterbury			
	West	Grey	Hok	BB	Ta	Mou	Nel	Ble	Dar	Ham	Kai	Ch	Rak	Meth	Ash	Tim
Jan	av	av	drr	drr	wtr	av	wtr	drr	wtr	av	drr	av	av	wtr	wtr	av
Feb	wtr	drr	drr	drr	drr	drr	drr	drr	drr	drr	drr	drr	drr	drr	drr	drr
Mar	wtr	wtr	wtr	drr	drr	drr	drr	drr	drr	drr	drr	drr	drr	drr	av	wtr
Apr	drr	drr	drr	drr	drr	drr	drr	drr	drr	drr	drr	drr	drr	drr	drr	drr
May	drr	drr	drr	drr	av	drr	wtr	av	av	drr	drr	drr	av	av	av	av
Jun	drr	drr	drr	drr	drr	drr	wtr	drr	drr	drr	drr	drr	drr	drr	drr	drr
Jul	wtr	wtr	wtr	drr	wtr	wtr	av	av	drr	drr	wtr	wtr	drr	drr	av	wtr
Aug	av	av	drr	drr	drr	av	drr	av	drr	drr	drr	wtr	drr	drr	wtr	wtr
Sep	drr	drr	drr	drr	drr	drr	drr	drr	drr	drr	drr	drr	drr	drr	drr	drr
Oct	wtr	wtr	wtr	drr	wtr	wtr	wtr	wtr	wtr	wtr	wtr	wtr	av	drr	wtr	av
Nov	drr	drr	drr	drr	drr	drr	drr	drr	drr	drr	drr	drr	drr	drr	drr	drr
Dec	drr	drr	wtr	drr	drr	av	wtr	av	av	drr	drr	av	drr	drr	drr	drr
Year	av	drr	drr	drr	drr	drr	drr	drr	drr	drr	drr	drr	drr	drr	av	av

	Coastal Otago			Lakes			Inland Otago				Southland				
	Oam	Dun	Bal	Twi	LP	LT	Crom	Qstn	Ran	Alex	TeA	Wtn	Gore	Inv	Puy P
Jan	av	drr	wtr	wtr	av	drr	drr	drr	wtr	drr	drr	drr	av	drr	drr
Feb	drr	drr	drr	av	drr	drr	drr	drr	drr	drr	av	drr	drr	drr	wtr
Mar	av	drr	drr	wtr	drr	drr	drr	drr	av	av	av	wtr	wtr	wtr	wtr
Apr	av	drr	drr	drr	drr	drr	drr	drr	drr	drr	drr	drr	drr	drr	av
May	av	drr	drr	av	drr	drr	av	av	drr	av	av	drr	drr	drr	wtr
Jun	drr	drr	av	av	drr	drr	av	av	drr	av	drr	av	av	wtr	wtr
Jul	wtr	drr	wtr	wtr	wtr	drr	av	av	wtr	wtr	drr	wtr	wtr	av	drr
Aug	wtr	wtr	wtr	wtr	av	drr	drr	wtr	av	av	wtr	av	drr	wtr	wtr
Sep	drr	drr	drr	av	drr	drr	drr	drr	av	drr	drr	wtr	av	av	wtr
Oct	av	wtr	av	wtr	wtr	wtr	wtr	wtr	av	av	wtr	wtr	av	wtr	wtr
Nov	drr	drr	drr	av	drr	drr	drr	drr	drr	drr	drr	av	drr	av	wtr
Dec	drr	drr	drr	drr	drr	drr	drr	drr	av	drr	drr	drr	drr	drr	av
Year	av	drr	av	wtr	drr	drr	drr	drr	av	av	drr	av	drr	av	wtr

RAIN SUMMARY FOR REGIONS

RAINFALL % FOR REGIONS AS COMPARED TO AVERAGE

The following are estimated variations from district averages (average 100%). For example, for Whangarei, January's potential is for only 88% of their average rainfall and Warkworth only around 63% of the norm, i.e. a drier than normal month. In contrast, Russell can expect around 36% more rainfall than normal (136%), i.e. a wetter than average month.

JANUARY 2025

Northland							Greater Auckland					Bay Of Plenty				Gisborne / Hawkes Bay		
CR	Kta	Ker	Rsl	Dar	Whg	Man	Wark	Hen	MkH	Ak	Co	Tau	TeP	Wkt	Opo	Gis	Nap	Has
101%	104%	111%	136%	76%	88%	84%	63%	50%	111%	61%	161%	119%	153%	102%	110%	71%	52%	90%

Waikato				Thermal / Central Plateau				Taranaki / King Country				Lower North Island						
Hml	TeA	Mata	Ot	Rt	Taup	Chat	Waio	Taum	NP	Str	Wang	Pa	Dann	Lev	PN	Mast	UH	Wgt
73%	139%	70%	63%	117%	100%	57%	67%	62%	86%	57%	120%	52%	40%	85%	45%	46%	91%	89%

West Coast				Top South Island				Canterbury				South Canterbury				Coastal Otago		
West	Grey	Hok	BB	Ta	Mou	Nel	Ble	Dar	Ham	Kai	Ch	Rak	Meth	Ash	Tim	Oam	Dun	Bal
102%	99%	87%	61%	111%	87%	141%	61%	121%	99%	63%	87%	95%	120%	216%	100%	93%	81%	123%

S Hydrolakes		Inland Otago						Southland			
LP	LT	Twi	Alex	TeA	Crom	Qstn	Ran	Wtn	Gore	Inv	Puy Pt
97%	77%	120%	76%	47%	56%	63%	162%	74%	102%	80%	58%

North Is Overall	South Is Overall	New Zealand Overall
86%	89%	87%

FEBRUARY 2025

Northland							Greater Auckland					Bay Of Plenty				Gisborne / Hawkes Bay		
CR	Kta	Ker	Rsl	Dar	Whg	Man	Wark	Hen	MkH	Ak	Co	Tau	TeP	Wkt	Opo	Gis	Nap	Has
236%	204%	154%	189%	192%	40%	33%	29%	46%	71%	72%	53%	116%	60%	16%	34%	54%	50%	97%

Waikato				Thermal / Central Plateau				Taranaki / King Country				Lower North Island						
Hml	TeA	Mata	Ot	Rt	Taup	Chat	Waio	Taum	NP	Str	Wang	Pa	Dann	Lev	PN	Mast	UH	Wgt
115%	99%	79%	84%	35%	69%	72%	70%	64%	40%	53%	70%	112%	38%	113%	53%	45%	74%	50%

West Coast				Top South Island				Canterbury				South Canterbury				Coastal Otago		
West	Grey	Hok	BB	Ta	Mou	Nel	Ble	Dar	Ham	Kai	Ch	Rak	Meth	Ash	Tim	Oam	Dun	Bal
134%	59%	51%	45%	37%	44%	45%	78%	28%	69%	79%	36%	25%	48%	47%	27%	15%	53%	67%

S Hydrolakes		Inland Otago						Southland			
LP	LT	Twi	Alex	TeA	Crom	Qstn	Ran	Wtn	Gore	Inv	Puy Pt
53%	33%	121%	40%	92%	40%	40%	35%	75%	51%	85%	119%

North Is Overall	South Is Overall	New Zealand Overall
77%	60%	70%

MARCH 2025

Northland							Greater Auckland					Bay Of Plenty				Gisborne / Hawkes Bay		
CR	Kta	Ker	Rsl	Dar	Whg	Man	Wark	Hen	MkH	Ak	Co	Tau	TeP	Wkt	Opo	Gis	Nap	Has
16%	23%	50%	25%	70%	31%	25%	32%	107%	72%	102%	91%	75%	34%	61%	27%	21%	31%	97%

Waikato				Thermal / Central Plateau				Taranaki / King Country				Lower North Island						
Hml	TeA	Mata	Ot	Rt	Taup	Chat	Waio	Taum	NP	Str	Wang	Pa	Dann	Lev	PN	Mast	UH	Wgt
107%	86%	43%	110%	53%	123%	151%	146%	137%	114%	52%	106%	52%	78%	129%	116%	40%	68%	36%

West Coast				Top South Island				Canterbury				South Canterbury				Coastal Otago		
West	Grey	Hok	BB	Ta	Mou	Nel	Ble	Dar	Ham	Kai	Ch	Rak	Meth	Ash	Tim	Oam	Dun	Bal
117%	129%	112%	71%	27%	19%	42%	26%	57%	27%	23%	76%	54%	52%	104%	146%	98%	39%	81%

S Hydrolakes		Inland Otago						Southland			
LP	LT	Twi	Alex	TeA	Crom	Qstn	Ran	Wtn	Gore	Inv	Puy Pt
72%	56%	131%	106%	111%	46%	75%	78%	143%	126%	164%	128%

North Is Overall	South Is Overall	New Zealand Overall
69%	84%	76%

APRIL 2025

Northland							Greater Auckland					Bay Of Plenty				Gisborne / Hawkes Bay		
CR	Kta	Ker	Rsl	Dar	Whg	Man	Wark	Hen	MkH	Ak	Co	Tau	TeP	Wkt	Opo	Gis	Nap	Has
70%	100%	270%	336%	140%	212%	209%	202%	108%	85%	79%	255%	104%	124%	74%	60%	52%	45%	40%

Waikato				Thermal / Central Plateau				Taranaki / King Country				Lower North Island						
Hml	TeA	Mata	Ot	Rt	Taup	Chat	Waio	Taum	NP	Str	Wang	Pa	Dann	Lev	PN	Mast	UH	Wgt
54%	48%	60%	61%	121%	52%	43%	58%	23%	27%	21%	65%	20%	37%	63%	37%	49%	36%	41%

West Coast				Top South Island				Canterbury				South Canterbury				Coastal Otago		
West	Grey	Hok	BB	Ta	Mou	Nel	Ble	Dar	Ham	Kai	Ch	Rak	Meth	Ash	Tim	Oam	Dun	Bal
57%	49%	42%	33%	20%	30%	21%	42%	73%	48%	42%	72%	36%	52%	53%	47%	99%	38%	51%

S Hydrolakes		Inland Otago						Southland			
LP	LT	Twi	Alex	TeA	Crom	Qstn	Ran	Wtn	Gore	Inv	Puy Pt
33%	39%	38%	19%	30%	21%	39%	43%	60%	80%	59%	100%

North Is Overall	South Is Overall	New Zealand Overall
99%	48%	77%

MAY 2025

Northland							Greater Auckland					Bay Of Plenty				Gisborne / Hawkes Bay		
CR	Kta	Ker	Rsl	Dar	Whg	Man	Wark	Hen	MkH	Ak	Co	Tau	TeP	Wkt	Opo	Gis	Nap	Has
49%	39%	26%	30%	36%	5%	8%	21%	59%	89%	41%	106%	81%	83%	48%	74%	16%	20%	19%

Waikato				Thermal / Central Plateau				Taranaki / King Country				Lower North Island						
Hml	TeA	Mata	Ot	Rt	Taup	Chat	Waio	Taum	NP	Str	Wang	Pa	Dann	Lev	PN	Mast	UH	Wgt
53%	89%	53%	70%	47%	53%	50%	30%	45%	44%	44%	48%	42%	21%	41%	20%	24%	50%	74%

West Coast				Top South Island				Canterbury				South Canterbury				Coastal Otago		
West	Grey	Hok	BB	Ta	Mou	Nel	Ble	Dar	Ham	Kai	Ch	Rak	Meth	Ash	Tim	Oam	Dun	Ba
92%	95%	91%	58%	102%	82%	123%	111%	95%	59%	42%	64%	101%	99%	86%	125%	101%	19%	35%

S Hydrolakes		Inland Otago						Southland			
LP	LT	Twi	Alex	TeA	Crom	Qstn	Ran	Wtn	Gore	Inv	Puy Pt
77%	80%	112%	76%	103%	87%	93%	37%	50%	47%	71%	119%

North Is Overall	South Is Overall	New Zealand Overall
45%	82%	60%

JUNE 2025

Northland							Greater Auckland					Bay Of Plenty				Gisborne / Hawkes Bay		
CR	Kta	Ker	Rsl	Dar	Whg	Man	Wark	Hen	MkH	Ak	Co	Tau	TeP	Wkt	Opo	Gis	Nap	Has
68%	47%	51%	52%	42%	45%	26%	25%	39%	61%	32%	85%	30%	27%	32%	31%	109%	117%	141%

Waikato				Thermal / Central Plateau				Taranaki / King Country				Lower North Island						
Hml	TeA	Mata	Ot	Rt	Taup	Chat	Waio	Taum	NP	Str	Wang	Pa	Dann	Lev	PN	Mast	UH	Wgt
57%	53%	17%	55%	42%	27%	96%	76%	83%	72%	70%	100%	20%	81%	37%	62%	40%	22%	29%

West Coast				Top South Island				Canterbury				South Canterbury				Coastal Otago		
West	Grey	Hok	BB	Ta	Mou	Nel	Ble	Dar	Ham	Kai	Ch	Rak	Meth	Ash	Tim	Oam	Dun	Bal
81%	94%	91%	82%	67%	73%	119%	49%	31%	28%	46%	16%	31%	51%	30%	12%	10%	53%	105%

S Hydrolakes		Inland Otago						Southland			
LP	LT	Twi	Alex	TeA	Crom	Qstn	Ran	Wtn	Gore	Inv	Puy Pt
79%	46%	118%	67%	53%	106%	99%	28%	90%	100%	118%	126%

North Is Overall	South Is Overall	New Zealand Overall
53%	75%	61%

JULY 2025

Northland							Greater Auckland					Bay Of Plenty				Gisborne / Hawkes Bay		
CR	Kta	Ker	Rsl	Dar	Whg	Man	Wark	Hen	MkH	Ak	Co	Tau	TeP	Wkt	Opo	Gis	Nap	Has
164%	206%	255%	256%	216%	222%	195%	178%	177%	149%	115%	310%	169%	198%	154%	96%	111%	239%	301%

Waikato				Thermal / Central Plateau				Taranaki / King Country				Lower North Island						
Hml	TeA	Mata	Ot	Rt	Taup	Chat	Waio	Taum	NP	Str	Wang	Pa	Dann	Lev	PN	Mast	UH	Wgt
122%	122%	95%	134%	153%	95%	125%	121%	133%	121%	138%	131%	99%	132%	121%	113%	95%	103%	92%

West Coast				Top South Island				Canterbury				South Canterbury				Coastal Otago		
West	Grey	Hok	BB	Ta	Mou	Nel	Ble	Dar	Ham	Kai	Ch	Rak	Meth	Ash	Tim	Oam	Dun	Bal
134%	160%	132%	77%	124%	117%	93%	94%	76%	74%	118%	117%	58%	80%	113%	138%	221%	79%	148%

S Hydrolakes		Inland Otago						Southland			
LP	LT	Twi	Alex	TeA	Crom	Qstn	Ran	Wtn	Gore	Inv	Puy Pt
126%	74%	245%	230%	85%	126%	84%	162%	152%	178%	113%	67%

North Is Overall	South Is Overall	New Zealand Overall
159%	113%	143%

AUGUST 2025

Northland							Greater Auckland					Bay Of Plenty				Gisborne / Hawkes Bay		
CR	Kta	Ker	Rsl	Dar	Whg	Man	Wark	Hen	MkH	Ak	Co	Tau	TeP	Wkt	Opo	Gis	Nap	Has
88%	114%	91%	96%	129%	74%	124%	143%	168%	205%	133%	176%	134%	104%	107%	81%	65%	58%	89%

Waikato				Thermal / Central Plateau				Taranaki / King Country				Lower North Island						
Hml	TeA	Mata	Ot	Rt	Taup	Chat	Waio	Taum	NP	Str	Wang	Pa	Dann	Lev	PN	Mast	UH	Wgt
230%	153%	163%	159%	137%	121%	152%	162%	179%	135%	137%	185%	120%	111%	108%	137%	71%	118%	97%

West Coast				Top South Island				Canterbury				South Canterbury				Coastal Otago		
West	Grey	Hok	BB	Ta	Mou	Nel	Ble	Dar	Ham	Kai	Ch	Rak	Meth	Ash	Tim	Oam	Dun	Bal
94%	102%	78%	56%	67%	94%	78%	101%	70%	26%	34%	162%	81%	78%	121%	171%	239%	154%	169%

S Hydrolakes		Inland Otago						Southland			
LP	LT	Twi	Alex	TeA	Crom	Qstn	Ran	Wtn	Gore	Inv	Puy Pt
89%	70%	187%	145%	147%	55%	123%	84%	91%	59%	168%	188%

North Is Overall	South Is Overall	New Zealand Overall
127%	96%	115%

SEPTEMBER 2025

Northland							Greater Auckland					Bay Of Plenty				Gisborne / Hawkes Bay		
CR	Kta	Ker	Rsl	Dar	Whg	Man	Wark	Hen	MkH	Ak	Co	Tau	TeP	Wkt	Opo	Gis	Nap	Has
191%	104%	97%	83%	47%	110%	128%	102%	48%	83%	59%	84%	57%	51%	71%	51%	168%	124%	184%

Waikato				Thermal / Central Plateau				Taranaki / King Country				Lower North Island						
Hml	TeA	Mata	Ot	Rt	Taup	Chat	Waio	Taum	NP	Str	Wang	Pa	Dann	Lev	PN	Mast	UH	Wgt
72%	75%	27%	77%	44%	46%	56%	71%	58%	25%	48%	26%	23%	67%	51%	44%	51%	19%	16%

West Coast				Top South Island				Canterbury				South Canterbury				Coastal Otago		
West	Grey	Hok	BB	Ta	Mou	Nel	Ble	Dar	Ham	Kai	Ch	Rak	Meth	Ash	Tim	Oam	Dun	Bal
87%	64%	53%	35%	28%	23%	34%	25%	38%	30%	23%	60%	55%	54%	41%	40%	34%	43%	51%

S Hydrolakes		Inland Otago						Southland			
LP	LT	Twi	Alex	TeA	Crom	Qstn	Ran	Wtn	Gore	Inv	Puy Pt
64%	58%	116%	43%	54%	41%	64%	84%	121%	103%	103%	118%

North Is Overall	South Is Overall	New Zealand Overall
69%	56%	64%

OCTOBER 2025

Northland							Greater Auckland					Bay Of Plenty				Gisborne / Hawkes Bay		
CR	Kta	Ker	Rsl	Dar	Whg	Man	Wark	Hen	MkH	Ak	Co	Tau	TeP	Wkt	Opo	Gis	Nap	Has
82%	41%	62%	63%	128%	66%	99%	101%	208%	156%	96%	231%	75%	102%	77%	123%	170%	82%	109%

Waikato				Thermal / Central Plateau				Taranaki / King Country				Lower North Island						
Hml	TeA	Mata	Ot	Rt	Taup	Chat	Waio	Taum	NP	Str	Wang	Pa	Dann	Lev	PN	Mast	UH	Wgt
111%	128%	100%	146%	100%	69%	150%	221%	167%	85%	119%	167%	160%	142%	108%	102%	201%	213%	169%

West Coast				Top South Island				Canterbury				South Canterbury				Coastal Otago		
West	Grey	Hok	BB	Ta	Mou	Nel	Ble	Dar	Ham	Kai	Ch	Rak	Meth	Ash	Tim	Oam	Dun	Bal
189%	143%	155%	87%	170%	216%	161%	157%	161%	197%	123%	166%	94%	76%	158%	107%	92%	137%	99%

S Hydrolakes		Inland Otago						Southland			
LP	LT	Twi	Alex	TeA	Crom	Qstn	Ran	Wtn	Gore	Inv	Puy Pt
244%	212%	301%	119%	164%	143%	213%	114%	127%	96%	119%	136%

North Is Overall	South Is Overall	New Zealand Overall
123%	147%	298%

NOVEMBER 2025

Northland							Greater Auckland					Bay Of Plenty				Gisborne / Hawkes Bay		
CR	Kta	Ker	Rsl	Dar	Whg	Man	Wark	Hen	MkH	Ak	Co	Tau	TeP	Wkt	Opo	Gis	Nap	Has
52%	70%	46%	59%	84%	58%	58%	69%	111%	87%	59%	75%	26%	14%	10%	7%	28%	32%	41%

Waikato				Thermal / Central Plateau				Taranaki / King Country				Lower North Island						
Hml	TeA	Mata	Ot	Rt	Taup	Chat	Waio	Taum	NP	Str	Wang	Pa	Dann	Lev	PN	Mast	UH	Wgt
60%	83%	33%	87%	22%	17%	84%	64%	49%	49%	19%	50%	48%	44%	116%	71%	40%	59%	19%

West Coast				Top South Island				Canterbury				South Canterbury				Coastal Otago		
West	Grey	Hok	BB	Ta	Mou	Nel	Ble	Dar	Ham	Kai	Ch	Rak	Meth	Ash	Tim	Oam	Dun	Bal
59%	48%	40%	36%	11%	8%	10%	18%	46%	38%	55%	58%	29%	43%	65%	58%	53%	40%	70%

S Hydrolakes		Inland Otago						Southland			
LP	LT	Twi	Alex	TeA	Crom	Qstn	Ran	Wtn	Gore	Inv	Puy Pt
35%	28%	104%	60%	76%	41%	30%	47%	89%	82%	95%	115%

North Is Overall	South Is Overall	New Zealand Overall
53%	49%	51%

DECEMBER 2025

Northland							Greater Auckland					Bay Of Plenty				Gisborne / Hawkes Bay		
CR	Kta	Ker	Rsl	Dar	Whg	Man	Wark	Hen	MkH	Ak	Co	Tau	TeP	Wkt	Opo	Gis	Nap	Has
279%	201%	232%	216%	156%	148%	128%	131%	115%	99%	85%	121%	92%	69%	35%	92%	216%	160%	233%

Waikato				Thermal / Central Plateau				Taranaki / King Country				Lower North Island						
Hml	TeA	Mata	Ot	Rt	Taup	Chat	Waio	Taum	NP	Str	Wang	Pa	Dann	Lev	PN	Mast	UH	Wgt
52%	48%	53%	85%	79%	105%	69%	131%	97%	38%	57%	95%	37%	85%	57%	54%	68%	41%	36%

West Coast				Top South Island				Canterbury				South Canterbury				Coastal Otago		
West	Grey	Hok	BB	Ta	Mou	Nel	Ble	Dar	Ham	Kai	Ch	Rak	Meth	Ash	Tim	Oam	Dun	Bal
57%	78%	115%	80%	69%	91%	130%	102%	90%	68%	43%	84%	39%	64%	68%	70%	59%	49%	64%

S Hydrolakes		Inland Otago						Southland			
LP	LT	Twi	Alex	TeA	Crom	Qstn	Ran	Wtn	Gore	Inv	Puy Pt
12%	22%	56%	57%	34%	31%	31%	112%	39%	33%	82%	100%

North Is Overall	South Is Overall	New Zealand Overall
100%	71%	87%

SUMMER 2025

Northland							Greater Auckland					Bay Of Plenty				Gisborne / Hawkes Bay		
CR	Kta	Ker	Rsl	Dar	Whg	Man	Wark	Hen	MkH	Ak	Co	Tau	TeP	Wkt	Opo	Gis	Nap	Has
136%	127%	99%	129%	115%	50%	54%	55%	51%	81%	72%	111%	99%	80%	67%	85%	80%	64%	98%

Waikato				Thermal / Central Plateau				Taranaki / King Country				Lower North Island						
Hml	TeA	Mata	Ot	Rt	Taup	Chat	Waio	Taum	NP	Str	Wang	Pa	Dann	Lev	PN	Mast	UH	Wgt
88%	101%	85%	76%	80%	93%	78%	94%	79%	72%	62%	104%	79%	57%	106%	71%	69%	98%	71%

West Coast				Top South Island				Canterbury				South Canterbury				Coastal Otago		
West	Grey	Hok	BB	Ta	Mou	Nel	Ble	Dar	Ham	Kai	Ch	Rak	Meth	Ash	Tim	Oam	Dun	Bal
107%	78%	76%	57%	76%	55%	77%	67%	108%	106%	78%	103%	89%	113%	150%	81%	84%	77%	97%

S Hydrolakes		Inland Otago						Southland			
LP	LT	Twi	Alex	TeA	Crom	Qstn	Ran	Wtn	Gore	Inv	Puy Pt
93%	100%	152%	79%	80%	78%	76%	117%	88%	91%	95%	109%

North Is Overall	South Is Overall	New Zealand Overall
83%	86%	84%

AUTUMN 2025

Northland							Greater Auckland					Bay Of Plenty				Gisborne / Hawkes Bay		
CR	Kta	Ker	Rsl	Dar	Whg	Man	Wark	Hen	MkH	Ak	Co	Tau	TeP	Wkt	Opo	Gis	Nap	Has
47%	58%	118%	128%	80%	71%	74%	84%	92%	82%	70%	153%	88%	73%	60%	51%	28%	31%	45%

Waikato				Thermal / Central Plateau				Taranaki / King Country				Lower North Island						
Hml	TeA	Mata	Ot	Rt	Taup	Chat	Waio	Taum	NP	Str	Wang	Pa	Dann	Lev	PN	Mast	UH	Wgt
67%	75%	52%	79%	73%	68%	77%	72%	61%	56%	39%	70%	38%	45%	73%	51%	35%	51%	52%

West Coast				Top South Island				Canterbury				South Canterbury				Coastal Otago		
West	Grey	Hok	BB	Ta	Mou	Nel	Ble	Dar	Ham	Kai	Ch	Rak	Meth	Ash	Tim	Oam	Dun	Bal
88%	90%	82%	56%	53%	47%	63%	68%	75%	44%	36%	70%	62%	66%	81%	105%	99%	32%	55%

S Hydrolakes		Inland Otago						Southland			
LP	LT	Twi	Alex	TeA	Crom	Qstn	Ran	Wtn	Gore	Inv	Puy Pt
61%	59%	90%	63%	83%	51%	68%	55%	83%	83%	95%	115%

North Is Overall	South Is Overall	New Zealand Overall
69%	72%	70%

WINTER 2025

Northland							Greater Auckland					Bay Of Plenty				Gisborne / Hawkes Bay		
CR	Kta	Ker	Rsl	Dar	Whg	Man	Wark	Hen	MkH	Ak	Co	Tau	TeP	Wkt	Opo	Gis	Nap	Has
108%	122%	132%	135%	126%	114%	113%	112%	126%	134%	92%	196%	108%	105%	96%	68%	96%	145%	183%

Waikato				Thermal / Central Plateau				Taranaki / King Country				Lower North Island						
Hml	TeA	Mata	Ot	Rt	Taup	Chat	Waio	Taum	NP	Str	Wang	Pa	Dann	Lev	PN	Mast	UH	Wgt
131%	111%	93%	115%	110%	79%	125%	119%	130%	109%	116%	137%	77%	109%	87%	102%	68%	80%	73%

West Coast				Top South Island				Canterbury				South Canterbury				Coastal Otago		
West	Grey	Hok	BB	Ta	Mou	Nel	Ble	Dar	Ham	Kai	Ch	Rak	Meth	Ash	Tim	Oam	Dun	Bal
102%	118%	100%	72%	84%	96%	97%	82%	60%	42%	67%	94%	58%	71%	89%	103%	151%	94%	137%

S Hydrolakes		Inland Otago						Southland			
LP	LT	Twi	Alex	TeA	Crom	Qstn	Ran	Wtn	Gore	Inv	Puy Pt
98%	64%	185%	144%	92%	95%	103%	91%	110%	111%	129%	125%

North Is Overall	South Is Overall	New Zealand Overall
113%	95%	106%

SPRING 2025

Northland							Greater Auckland					Bay Of Plenty				Gisborne / Hawkes Bay		
CR	Kta	Ker	Rsl	Dar	Whg	Man	Wark	Hen	MkH	Ak	Co	Tau	TeP	Wkt	Opo	Gis	Nap	Has
120%	73%	71%	70%	86%	78%	95%	93%	116%	105%	72%	129%	55%	56%	54%	60%	129%	78%	110%

Waikato				Thermal / Central Plateau				Taranaki / King Country				Lower North Island						
Hml	TeA	Mata	Ot	Rt	Taup	Chat	Waio	Taum	NP	Str	Wang	Pa	Dann	Lev	PN	Mast	UH	Wgt
81%	96%	53%	104%	57%	45%	96%	120%	94%	54%	67%	88%	83%	86%	93%	75%	101%	102%	72%

West Coast				Top South Island				Canterbury				South Canterbury				Coastal Otago		
West	Grey	Hok	BB	Ta	Mou	Nel	Ble	Dar	Ham	Kai	Ch	Rak	Meth	Ash	Tim	Oam	Dun	Ba
116%	89%	85%	54%	78%	88%	72%	73%	86%	92%	69%	94%	62%	59%	88%	70%	61%	74%	75%

S Hydrolakes		Inland Otago						Southland			
LP	LT	Twi	Alex	TeA	Crom	Qstn	Ran	Wtn	Gore	Inv	Puy Pt
122%	103%	181%	79%	105%	80%	109%	82%	113%	92%	106%	123%

North Is Overall	South Is Overall	New Zealand Overall
83%	87%	105%

YEAR OVERALL 2025

Northland							Greater Auckland					Bay Of Plenty				Gisborne / Hawkes Bay		
CR	Kta	Ker	Rsl	Dar	Whg	Man	Wark	Hen	MkH	Ak	Co	Tau	TeP	Wkt	Opo	Gis	Nap	Has
112%	103%	119%	128%	109%	88%	91%	93%	104%	107%	78%	153%	91%	82%	67%	64%	84%	87%	120%

Waikato				Thermal / Central Plateau				Taranaki / King Country				Lower North Island						
Hml	TeA	Mata	Ot	Rt	Taup	Chat	Waio	Taum	NP	Str	Wang	Pa	Dann	Lev	PN	Mast	UH	Wgt
92%	94%	68%	96%	80%	70%	93%	102%	93%	71%	72%	99%	66%	76%	84%	71%	65%	76%	65%

West Coast				Top South Island				Canterbury				South Canterbury				Coastal Otago		
West	Grey	Hok	BB	Ta	Mou	Nel	Ble	Dar	Ham	Kai	Ch	Rak	Meth	Ash	Tim	Oam	Dun	Bal
100%	94%	88%	61%	73%	77%	84%	76%	74%	63%	57%	82%	59%	68%	92%	83%	85%	65%	88%

S Hydrolakes		Inland Otago						Southland			
LP	LT	Twi	Alex	TeA	Crom	Qstn	Ran	Wtn	Gore	Inv	Puy Pt
86%	68%	135%	78%	83%	65%	80%	85%	90%	85%	102%	113%

North Is Overall	South Is Overall	New Zealand Overall
99%	82%	92%

DAYS OF MOST FREQUENT RAIN IN EACH MONTH

BY SAMPLE PROPORTION IN MM, USING TOTALS FOR 26 TOWNS IN EACH ISLAND

	Jan		Feb		Mar		Apr		May		June	
	NI	SI	NI	SI	NI	SI	NI	SI	NI	SI	NI	SI
1st	117	56	14	77	0	2	32	32	0	4	110	20
2nd	33	35	1	178	0	0	8	0	0	8	85	422
3rd	67	9	32	10	12	32	5	0	43	3	430	65
4th	79	23	49	11	28	29	1	0	20	94	13	110
5th	30	121	82	55	36	89	4	16	15	32	9	92
6th	1	322	119	69	80	68	5	0	1	48	8	42
7th	112	261	27	15	75	1	107	3	15	211	0	5
8th	6	61	480	157	26	52	976	45	421	130	0	0
9th	47	463	231	192	33	24	1591	51	236	48	0	2
10th	29	100	5	61	8	20	206	126	94	1	1	23
11th	123	27	1	197	86	3	31	140	664	226	1	156
12th	142	39	95	61	24	0	81	31	103	309	158	348
13th	30	37	13	105	19	6	1	24	180	296	183	152
14th	0	7	60	5	77	0	5	35	5	35	53	39
15th	0	1	94	1	14	2	88	30	5	52	40	100
16th	0	10	134	0	1	10	60	19	1	59	135	108
17th	1	201	729	1	14	224	1	63	18	66	61	73
18th	12	11	244	4	70	308	8	40	1	21	80	197
19th	107	5	17	123	24	27	18	9	11	12	56	93
20th	612	287	62	127	4	0	7	12	0	218	137	39
21st	126	197	79	4	0	0	73	91	45	137	555	3
22nd	163	44	4	9	17	1	228	252	74	176	150	23
23rd	439	34	14	100	742	498	273	165	4	21	25	6
24th	470	190	0	13	221	232	72	69	3	8	69	20
25th	15	81	21	3	539	279	24	11	6	21	110	15
26th	69	3	9	17	31	77	23	1	3	68	45	28
27th	32	13	16	2	13	298	59	105	28	28	98	43
28th	8	12	0	0	175	176	14	41	4	12	46	11
29th	1	9			184	97	6	3	0	6	12	11
30th	2	49			0	8	1	0	22	15	55	6
31st	10	163			0	2			105	206		
est	2882	2866	2634	1595	2553	2566	4006	1414	2124	2566	2724	2253
av	3356	3214	3405	2637	3714	3044	4061	2974	4741	3120	5148	2984
month	drier	drier	drier	drier	drier	drier	average	drier	drier	drier	drier	drier

	July		Aug		Sep		Oct		Nov		Dec	
	NI	SI	NI	SI	NI	SI	NI	SI	NI	SI	NI	SI
1st	534	48	380	27	16	32	142	0	12	90	1	0
2nd	213	247	17	20	19	15	27	4	15	105	0	13
3rd	273	280	21	95	4	12	62	243	23	57	0	12
4th	106	130	17	67	4	10	385	305	23	142	200	55
5th	64	100	34	73	8	26	246	61	0	32	27	24
6th	76	23	24	30	141	1	145	50	17	45	0	10
7th	3	3	30	41	462	27	15	16	88	18	0	26
8th	1	0	424	79	72	20	7	0	7	1	1	56
9th	9	134	931	331	51	124	0	2	0	4	43	8
10th	1034	1139	230	313	53	82	18	252	0	69	35	59
11th	480	167	292	68	1	33	259	301	21	153	40	6
12th	622	112	65	21	94	114	417	228	38	10	32	24
13th	210	305	142	24	8	49	319	272	30	2	13	5
14th	75	37	115	18	28	383	71	25	57	42	0	2
15th	355	160	584	385	399	164	217	323	642	80	18	0
16th	375	172	126	121	154	11	14	284	131	35	643	11
17th	532	34	281	57	22	9	8	259	218	15	532	7
18th	112	8	243	6	39	11	109	284	64	7	118	0
19th	22	17	28	50	139	18	134	61	5	3	241	0
20th	345	33	28	59	335	16	164	127	0	3	301	114
21st	940	20	75	125	46	13	433	382	12	8	81	175
22nd	142	34	378	380	10	4	261	99	27	8	9	244
23rd	41	4	128	121	0	1	55	10	17	161	4	388
24th	76	1	42	50	17	223	468	433	288	279	11	199
25th	74	2	17	5	91	80	111	108	58	31	18	38
26th	492	2	106	1	133	98	175	124	91	49	4	75
27th	367	5	989	166	164	44	85	302	62	6	7	103
28th	235	11	145	100	35	10	459	181	8	0	29	312
29th	100	1	283	11	41	0	250	33	0	1	834	251
30th	309	35	27	5	293	1	10	62	0	0	427	75
31st	132	16	22	8			41	152			83	23
est	8348	3277	6223	2853	2878	1631	5108	4981	1951	1456	3748	2314
av	5247	2895	4912	2967	4154	2899	4163	3388	3667	2967	3732	3256
month	wetter	wetter	wetter	drier	drier	drier	wetter	wetter	drier	drier	average	drier

ESTIMATED SUNSHINE HOURS SUMMARY

SELECTED TOWNS ESTIMATED BY MONTHLY TRENDS

	North Island 2025																				
	Kta	Whg	Dar	Ak	Hml	Tau	Rt	Taum	NP	Wang	Taup	Gis	Nap	Wai	Dann	PN	Mast	Pa	UH	Wgt	OVERALL
Jan	snnr	av	av	cldr	cldr	av	cldr	av	cldr	cldr	av	av	av	cldr	av	cldr	av	cldr	cldr	cldr	cldr
Feb	snnr	snnr	snnr	snnr	av	snnr	snnr	snnr	snnr	av	snnr	av	av	av	snnr	av	av	snnr	av	snnr	snnr
Mar	snnr	snnr	snnr	snnr	av	snnr	av	snnr	av	av	snnr	snnr	snnr	snnr	snnr	snnr	snnr	snnr	snnr	snnr	snnr
Apr	av	av	snnr	av	av	snnr	snnr	snnr	snnr	av	snnr	snnr	snnr	av	snnr	av	snnr	snnr	av	snnr	snnr
May	av	av	snnr	snnr	snnr	snnr	snnr	snnr	snnr	av	snnr	snnr	av	snnr	snnr	av	av	av	av	av	snnr
June	snnr	snnr	snnr	snnr	snnr	av	av	snnr	snnr	av	snnr	av	snnr	av	snnr	snnr	snnr	snnr	av	snnr	snnr
July	cldr	cldr	av	cldr	cldr	cldr	cldr	av	cldr	cldr	cldr	cldr	cldr	cldr	cldr	av	cldr	av	cldr	cldr	cldr
Aug	av	av	snnr	av	av	av	av	av	av	cldr	av	av	av	av	av	av	av	av	cldr	av	av
Sept	av	cldr	av	av	av	snnr	av	snnr	av	cldr	av	snnr	av	snnr	snnr	av	snnr	snnr	av	snnr	snnr
Oct	snnr	snnr	snnr	av	av	av	av	av	cldr	cldr	av	av	av	av	av	av	cldr	av	cldr	cldr	av
Nov	cldr	av	av	av	snnr	snnr	snnr	snnr	snnr	av	snnr	snnr	snnr	snnr	snnr	snnr	av	snnr	av	snnr	snnr
Dec	cldr	cldr	cldr	cldr	cldr	cldr	cldr	cldr	av	av	av	av	av	av	av	snnr	cldr	av	cldr	cldr	cldr
Year	av	av	snnr	a	av	snnr	av	snnr	av	cldr	snnr	snnr	snnr	av	snnr	av	av	snnr	cldr	snnr	av

	South Island 2025																				
	Ta	Nel	Ble	Mou	West	Grey	Hok	Ham	Kai	Ch	Ash	Tim	Crom	Qstn	Bal	Dun	Wnt	Gore	Inv	Alex	OVERALL
Jan	cldr	cldr	cldr	cldr	cldr	cldr	cldr	cldr	cldr	cldr	cldr	cldr	cldr	cldr	cldr	cldr	cldr	cldr	av	av	cldr
Feb	cldr	av	av	av	av	snnr	snnr	cldr	cldr	av	av	av	snnr	snnr	av	snnr	av	av	snnr	snnr	snnr
Mar	snnr	snnr	snnr	snnr	snnr	snnr	snnr	snnr	snnr	snnr	snnr	snnr	av	snnr	snnr	snnr	av	av	av	snnr	snnr
Apr	av	snnr	snnr	snnr	snnr	snnr	snnr	snnr	av	snnr	snnr	snnr	av	snnr	av	snnr	av	av	snnr	snnr	snnr
May	av	av	av	av	av	av	av	snnr	av	av	snnr	av	snnr	av	av	av	snnr	snnr	snnr	snnr	snnr
June	snnr	snnr	snnr	snnr	cldr	av	av	snnr	snnr	snnr	snnr	av	av	av	av	snnr	av	av	snnr	snnr	snnr
July	av	cldr	cldr	cldr	cldr	snnr	snnr	av	cldr	av	av	av	cldr	av	av	av	snnr	snnr	snnr	cldr	cldr
Aug	av	av	av	av	cldr	cldr	cldr	snnr	av	av	av	cldr	cldr	cldr	cldr	av	av	av	av	av	cldr
Sept	snnr	snnr	snnr	snnr	av	snnr	snnr	snnr	snnr	snnr	snnr	snnr	snnr	av	snnr	snnr	snnr	av	snnr	snnr	snnr
Oct	cldr	av	av	cldr	cldr	av	av	av	cldr	av	av	av	cldr	cldr	cldr	av	av	av	av	av	cldr
Nov	snnr	snnr	snnr	snnr	snnr	snnr	snnr	snnr	snnr	snnr	snnr	snnr	snnr	snnr	av	snnr	av	av	snnr	snnr	snnr
Dec	cldr	cldr	av	av	av	av	snnr	snnr	av	av	av	av	snnr	snnr	snnr	snnr	snnr	snnr	snnr	snnr	snnr
Year	av	av	snnr	av	cldr	snnr	snnr	snnr	av	av	snnr	av	snnr	snnr	av	snnr	av	av	snnr	snnr	snnr

MONTHLY AVERAGE TEMPERATURE SUMMARY

ESTIMATES OF SELECTED TOWNS VS AVERAGE MAXIMUM TEMPERATURES

NORTH ISLAND 2025

	Kta	Ker	Whg	Dar	War	Ak	M Hds	Coro	Hml	Taur	Te P	Rt	Wkt	Opt	Taum	Tur	Chat
jan	clr	clr	a-c	a-c	clr	clr	clr	a-c	a-c	clr	clr	clr	clr	clr	clr	clr	clr
feb	clr	clr	clr	clr	av	a-c	a-w	a-c	av	a-w	av	a-c	a-w	wmr	clr	av	a-w
mar	a-w	a-w	a-w	a-w	a-w	wmr	wmr	av	wmr	wmr	wmr	wmr	a-w	wmr	av	wmr	wmr
apr	av	av	av	clr	a-w	av	wmr	av	wmr	wmr	wmr	a-w	av	wmr	a-w	wmr	wmr
may	wmr	wmr	wmr	wmr	wmr	wmr	wmr	wmr	wmr	wmr	wmr	wmr	wmr	wmr	wmr	wmr	wmr
jun	wmr	wmr	a-w	a-w	wmr	av	wmr	a-w	wmr	wmr	wmr	a-w	a-w	wmr	wmr	wmr	wmr
jul	a-w	a-w	a-w	a-w	av	a-w	wmr	av	wmr	av	wmr	a-w	a-w	wmr	av	av	a-w
aug	a-w	a-w	av	a-w	av	av	wmr	av	av	a-w	a-w	av	av	wmr	av	av	av
sep	av	av	av	av	av	av	a-w	av	av	av	a-w	av	av	a-w	a-c	av	av
oct	wmr	a-w	wmr	wmr	av	a-w	a-w	a-w	a-w	av	av	av	a-c	wmr	clr	clr	clr
nov	a-w	av	av	av	clr	av	av	av	av	av	wmr	av	a-w	wmr	clr	clr	av
dec	wmr	av	a-w	wmr	clr	wmr	wmr	wmr	wmr	a-w	wmr	wmr	av	wmr	av	av	wmr
yr	a-w	av	av	av	av	av	wmr	av	a-w	av	a-w	av	av	wmr	a-c	av	a-w

	Waio	Tai	N P	Wang	Taup	Gis	Nap	Has	Waip	Lev	Dann	P N	Mast	Para	Wgt	Overall Trend
jan	clr	clr	clr	clr	clr	clr	clr	clr	clr	clr	clr	clr	av	clr	a-c	clr
feb	a-c	clr	av	clr	clr	clr	clr	clr	clr	a-c	clr	av	wmr	av	a-w	a-c
mar	wmr	wmr	wmr	wmr	wmr	wmr	wmr	wmr	av	a-w	wmr	wmr	wmr	a-w	wmr	wmr
apr	wmr	av	a-w	av	av	wmr	av	av	clr	a-w	wmr	a-w	wmr	wmr	wmr	a-w
may	wmr	wmr	wmr	wmr	wmr	wmr	wmr	wmr	a-w	wmr	wmr	wmr	wmr	wmr	wmr	wmr
jun	wmr	wmr	wmr	a-w	a-w	wmr	wmr	wmr	clr	wmr	wmr	wmr	wmr	wmr	wmr	wmr
jul	av	a-w	av	av	a-w	wmr	av	a-w	av	wmr	clr	a-w	av	av	wmr	a-w
aug	av	a-w	a-w	av	av	av	a-c	av	av	a-w	av	av	wmr	av	a-w	av
sep	a-c	clr	av	clr	a-c	av	a-c	av	a-w	a-c	wmr	a-c	wmr	av	av	av
oct	clr	av	a-w	av	av	av	av	av	a-c	a-w	a-c	wmr	a-c	a-c	av	av
nov	clr	a-w	a-w	av	av	wmr	wmr	wmr	av	av	a-w	a-w	a-w	a-c	wmr	av
dec	wmr	wmr	wmr	wmr	wmr	wmr	wmr	wmr	av	wmr	wmr	wmr	wmr	a-w	wmr	wmr
yr	av	a-w	a-w	av	av	a-w	av	a-w	a-c	a-w	av	wmr	wmr	av	wmr	a-w

SOUTH ISLAND 2025

	Tak	Nel	Ble	Mout	West	Grey	Hok	Darf	Hamn	L Col	Kaik	Ch	Ash	Rang
jan	clr	clr	clr	clr	av	a-w	wmr	clr	clr	clr	clr	clr	clr	clr
feb	av	av	clr	av	av	wmr	a-w	clr	a-c	clr	a-c	clr	clr	clr
mar	wmr	wmr	wmr	wmr	a-w	a-w	a-w	wmr	wmr	a-w	a-w	wmr	wmr	wmr
apr	wmr	av	av	wmr	av	av	av	wmr	wmr	a-w	wmr	av	av	wmr
may	wmr	wmr	wmr	wmr	wmr	a-w	wmr	wmr	wmr	a-w	wmr	wmr	wmr	wmr
jun	wmr	a-w	wmr	wmr	a-w	av	a-w	wmr	wmr	wmr	wmr	wmr	wmr	wmr
jul	clr	a-w	av	clr	av	a-w	av	clr	clr	clr	clr	av	av	clr
aug	av	av	av	av	wmr	wmr	wmr	av	wmr	clr	av	clr	clr	av
sep	av	a-w	wmr	a-w	a-c	a-c	a-c	a-w	wmr	a-c	av	av	a-w	a-w
oct	clr	a-c	clr	clr	a-w	wmr	wmr	clr	clr	clr	a-c	a-c	a-c	clr
nov	a-c	a-w	wmr	a-c	av	av	av	a-c	av	clr	a-c	a-w	wmr	av
dec	av	wmr	wmr	wmr	wmr	wmr	wmr	av	wmr	av	av	wmr	wmr	av
yr	av	a-w	av	av	a-w	a-w	a-w	av	a-w	clr	av	av	av	av

	Tim	Twiz	Oam	L Tek	Crom	Qu	Balc	Dun	Win	Te A	Gore	Inv	Alx	Puys	Overall Trend
jan	clr	clr	clr	clr	clr	a-c	clr	clr	a-c	a-c	clr	a-c	a-c	clr	av
feb	av	clr	clr	av	av	av	a-c	a-c	av	av	a-c	a-c	av	clr	a-c
mar	wmr	wmr	wmr	wmr	wmr	a-w	wmr	wmr	wmr	wmr	a-w	a-w	wmr	av	wmr
apr	wmr	wmr	av	wmr	wmr	av	wmr	a-w	wmr	wmr	av	av	wmr	av	wmr
may	wmr	wmr	wmr	wmr	wmr	wmr	wmr	wmr	wmr	wmr	wmr	wmr	wmr	wmr	wmr
jun	wmr	wmr	a-w	wmr	wmr	a-w	wmr	wmr	wmr	wmr	wmr	wmr	wmr	a-w	wmr
jul	av	clr	a-c	clr	clr	clr	clr	av	clr	clr	a-c	a-c	clr	clr	clr
aug	clr	av	clr	av	a-c	av	av	av	av	av	av	av	av	av	av
sep	wmr	av	a-w	a-w	wmr	av	a-w	wmr	a-w	wmr	av	av	wmr	av	av
oct	a-c	clr	clr	clr	clr	clr	clr	a-c	clr	clr	av	av	a-c	a-c	clr
nov	wmr	clr	wmr	a-c	a-c	av	clr	wmr	clr	av	av	av	wmr	clr	av
dec	wmr	a-w	wmr	wmr	wmr	wmr	a-w	wmr	wmr	wmr	wmr	wmr	wmr	av	wmr
yr	a-w	av	av	av	av	av	av	a-w	av	a-w	av	av	wmr	av	av

Key:

- clr cooler
- a-c av to clr
- av average
- a-w av to wmr
- wmr warmer

SEASONAL TEMPERATURE TRENDS AT A GLANCE

NORTH ISLAND 2025

	Kta	Ker	Whg	Dar	War	Ak	M Hds	Coro	Hml	Taur	Te P	Rt	Wkt	Opt	Taum	Tur	Chat	Waio	Tai	N P	Wang	Taup	Gis	Nap	Has	Waip	Lev	Dann	P N	Mast	Para	Wgt	Overall Trend
Summer	clr	a-c	av	av	clr	a-c	av	av	av	av	av	clr	a-c	av	clr	clr	a-c	clr	clr	av	clr	clr	a-c	a-c	clr	clr	a-c	clr	av	av	clr	av	a-c
Autumn	a-w	a-w	a-w	av	wmr	wmr	wmr	a-w	wmr	wmr	wmr	wmr	a-w	wmr	a-w	wmr	wmr	wmr	wmr	wmr	a-w	a-w	wmr	a-w	wmr	av	wmr	wmr	wmr	wmr	wmr	wmr	wmr
Winter	a-w	a-w	a-w	a-w	a-w	a-w	wmr	av	a-w	a-w	wmr	av	av	wmr	av	wmr	a-w	av	a-w	a-w	av	av	wmr	av	wmr	a-c	wmr	a-c	a-w	wmr	av	wmr	a-w
Spring	a-w	av	av	a-w	a-c	av	a-w	av	av	av	a-w	av	av	a-w	clr	a-c	av	a-c	av	av	av	av	av	av	a-w	av	av	a-w	av	av	a-c	av	av
December	wmr	av	a-w	wmr	clr	wmr	wmr	wmr	wmr	a-w	wmr	wmr	av	wmr	av	av	wmr	wmr	wmr	wmr	wmr	wmr	wmr	wmr	wmr	av	wmr	wmr	wmr	wmr	a-w	wmr	wmr

SOUTH ISLAND 2025

	Tak	Nel	Ble	Mout	West	Grey	Hok	Darf	Hamn	L Col	Kaik	Ch	Ash	Rang	Tim	Twiz	Oam	L Tek	Crom	Qu	Balc	Dun	Win	Te A	Gore	Inv	Alx	Puys	Overall Trend
Summer	clr	av	clr	a-c	av	a-w	a-w	clr	clr	clr	clr	clr	clr	clr	av	clr	clr	clr	clr	a-c	clr	av	a-c	clr	a-c	a-c	av	clr	clr
Autumn	wmr	wmr	wmr	wmr	av	a-w	a-w	wmr	wmr	a-w	wmr	wmr	wmr	wmr	wmr	wmr	a-w	wmr	wmr	wmr	wmr	wmr	wmr	wmr	wmr	a-w	wmr	a-w	wmr
Winter	av	a-w	av	av	a-w	a-w	a-w	av	wmr	a-c	av	av	av	av	av	av	av	av	av	av	av	a-w	av	av	a-w	av	a-c	av	av
Spring	clr	av	a-w	av	av	av	av	a-c	av	clr	a-c	av	av	av	wmr	clr	av	a-c	a-c	av	a-c	a-w	a-c	av	av	av	av	a-c	av
December	av	wmr	wmr	wmr	wmr	wmr	wmr	av	wmr	av	av	wmr	wmr	av	wmr	a-w	wmr	wmr	wmr	wmr	a-w	wmr	wmr	wmr	wmr	wmr	wmr	av	wmr

Key:

clr	cooler
a-c	av to clr
av	average
a-w	av to wmr
wmr	warmer

GARDENING GUIDE 2025

REST PERIOD:
The Rest Period in the garden is as important as the activity times. It's a time when the Moon is Void of Course, that is, traveling between one sign and another and during this time the Moon is deemed to have "no energizing power", also Rest Periods on the day of the Full Moon, New Moon and when the moon is at Southern and Northern Declinations. Hence it is a good time to sit and enjoy your garden or to attend to non-plant activities such as tidy your shed, repair or maintain outdoor structures or even to sit and plan your next "to do" list when the Moon is once more energized. Eclipse days and the day after the eclipse are also deemed days to rest in the garden.

ALL ABOVE-GROUND ACTIVITY SHOULD TAKE PLACE when the Moon is Waxing – i.e. building from New Moon through the 1st Quarter and up to the Full Moon. This is a time of increasing light and with the Moon pulling the sap upwards into the leaves and flowers.

ALL BELOW-GROUND ACTIVITY SHOULD TAKE PLACE when the Moon is Waning – i.e. from the Full Moon through the 3rd Quarter and up to the New Moon. This is a phase of decreasing light and the general trend for sap to be pulled downwards and into the root and tuber systems.

THE NEW MOON PHASE:
Activities can begin in the garden once the crescent of the New Moon begins to be visible.

At this time:

DO start sowing, transplanting, grafting or strike cuttings of all ABOVE-GROUND producers. Flowering annuals, brassicas (cabbage, broccoli etc) cover crops (mustard, lupins etc) and annual fruiting crops (cucumber, pumpkin, tomatoes etc) and house-plants all benefit from attention during this phase. Also, mow lawns should you wish to encourage growth or sow any large areas such as lawns.

DON'T plant root crops (as they will only produce plenty of leaves and little roots, or go to seed too early), nor prune at this time – your plants, and trees in particular, will be prone to excessive sap loss and suffer from die-back.

THE 1ST QUARTER PHASE:
Continue on with the New Moon phase activities and especially sow any seeds two days before the Full Moon to yield optimum results. During drought times, sow during this phase when the Moon is in Capricorn or Taurus for the strongest of survival qualities.

At this time:

DO sow and transplant all legume, grain, flowering plants (especially roses) and pot plants and start harvesting any above-ground produce, with 2-3 days before the Full Moon being the most beneficial time (especially for herbs and grape harvesting).

DO give plants a food boost if required, but do water it down and apply as close to the Full Moon as possible, particularly for those plants that require phosphorus.

DON'T harvest with the Moon in a water sign (Cancer, Scorpio or Pisces) unless for immediate use, nor prune or plant out root crops.

THE FULL MOON PHASE:
The Moon will have reached its full potency by the time of the Full Moon, and from this point to the New Moon, there is a gradual decrease in its vitality and the sap flow is now downwards into the roots – it is a time when the plants build up their root systems and store their goodness. However, you can continue to harvest above-ground producers for a few more days after the Full Moon. Leeks, onions, garlic etc are at their best potency when harvested with the Full Moon in Leo.

At this time:

DO sow or plant out any root crops (potatoes, kumaras, carrots etc) especially during the two days in the middle of the Full Moon phase. Also sow or plant out fruting plants (e.g. strawberries, passion fruit etc) as well as perennial flowering plants, bulbs and tubers (e.g. daffodils, dahlias, crocus etc) and this is also a good time to spray, prune and trim hedges.

DO perform drying activities of herbs, flowers and fruits.

DO dig herb roots or harvest leaves and bark intended for medicinal teas.

DO spray fruit trees, mulch your garden and kill weeds.

THE 3RD QUARTER PHASE:
This is considered to be the resting period of the Lunar cycle, the time when the sap in the plants runs at its slowest.

DO harvest all root crops that you wish to keep for long term storage (e.g. potatoes, garlic, onions and root herbs etc), especially if the Moon is in Taurus or Capricorn. Also lift your bulbs and tubers for their seasonal storage. You can also use this time to build up your compost, especially when the Moon is in Scorpio and attend to maintenance chores such as spreading compost, mulching, staking taller plants etc, as well as attend to all pruning and trimming of trees, vines and bushes, as there is less chance of die back and excessive sap bleeding. Cut your lawns during this phase to discourage growth.

DO plant trees and saplings during this phase for good strong foundations of root growth and bark development as well as plant strawberries or their runners so their growth is well developed before fruiting.

DO fertilize your crops for potassium if needed, as potassium absorption is at its peak at this phase.

When the Moon is in Cancer, Scorpio or Pisces:
This is a beneficial time for attending to vegetables such as lettuce, cabbage, herbs, broccoli, leafy pot plants, etc. Avoid any harvesting except for immediate use during these Moon signs as anything harvested for long term storage will deteriorate faster and be prone to mildew, except of course, unless you are attempting to build compost – choose when the Moon is in Scorpio for best results, as vegetation decomposes fastest under this influence.

When the Moon is in Taurus, Virgo or Capricorn:
This is a beneficial time for attending to vegetables such as potatoes, carrots, leeks etc. as well as to non-flowering shrubs and trees and ground covers such as lawns all respond best attended to under these signs. A Virgo Moon (particularly in the 3rd Quarter phase) is best for harvesting any medicinal leafy herbs due to the usual desire for long term storage.

When the Moon is in Gemini, Libra, Aquarius:
This is a beneficial time to attend to vegetables such as globe artichokes, flowers for medicinal purposes, and flowers in general. (Brocolli and cauliflower etc are considered "leafy" as if planted during a Flower phase they are more likely to bolt straight to seed). The Moon's focus is more directed towards developing perfume, fragrance, flower vibrancy and beauty as well as medicinal properties with the Full Moon in one of these signs.

When the Moon is in Aries, Leo or Sagittarius:
This is a beneficial time to attend to any fruit, berry, seed or grain crops such as tomatoes, peas, beans, peppers, wheat or corn, pumpkin, apples, strawberries, raspberries etc. Any seeds you wish to store for next year's planting, or for long term storage are best attended to during these Moon signs, as well as harvesting crops such as onions and garlic for their most potent flavour.

Above-ground - Leafy, Fruits and Seeds examples:

Vegetables	beans, broccoli, brussels sprouts, cabbages, cauliflower, chicory, celery, cress, cucumber, endive, lettuces, marrow, mustard, parsley, rhubarb, sage, silverbeet, corn, etc.
Flowers	cornflowers, forget-me-nots, marigolds, pansies, poppies, roses etc.
Fruits and seeds	apples, blue-berries, grapes, passion-fruit, pears, peas, peppers, raspberries, strawberries, tomatoes, wheat

Below-ground - Roots, Bulbs and Non-Flowering shrubs examples:

Vegetables	beetroot, carrots, garlic, leeks, onions, parsnips, radish, salsify, shallots, swedes, turnips etc.
Flowers	agapanthus, anemone, crocus, daffodils, dahlias, freesias, gladioli, hyacinth, ranunculus, red-hot pokers, tuberose, tulips
Fruits and Seeds	garlic, onions, peanuts

BASIC MOON IN SIGN GUIDE

Moon in Aries – Harvest root and fruit for storage. Good for destroying weeds and pests.

Moon in Taurus – Excellent time for planting and transplanting, particularly root crops. Also good for leaf crops.

Moon in Gemini – Good for harvesting root and fruit for storage. Good for destroying weeds and pests. Melon seeds respond well.

Moon in Cancer – Best time for all planting and transplanting. Good for grafting and irrigation.

Moon in Leo – Good for harvesting root and fruit for storage. Excellent for destroying weeds and pest.

Moon in Virgo – Good for destroying weeds and pests. Vines and some flowers respond well.

Moon in Libra – Best time for planting for fragrance and beauty of flowers, vines and herbs.

Moon in Scorpio – Best time for planting sturdy plants and vines. Especially favoured by tomatoes, corn and squash. Good for irrigation and transplanting.

Moon in Sagittarius – Good for harvesting roots. Especially favoured by onions and fruit trees. Good time to cultivate the soil.

Moon in Capricorn – Plant potatoes and root crops and encouraging hardy growth. Good for grafting and pruning to promote healing and applying organic fertilizers.

Moon in Aquarius – Harvest root and fruit for storage. Good for destroying weeds and pest. Good for planting onions.

Moon in Pisces – Good for planting and transplanting, especially for root growth. Good for irrigation.

JANUARY	2025		Hour:	1am 2am 3am 4am 5am 6am 7am 8am 9am 10am 11am Noon 1pm 2pm 3pm 4pm 5pm 6pm 7pm 8pm 9pm 10pm 11pm Midnight		
	1st	Wed	Above	Capricorn	REST	
	2nd	Thur	Above	Aquarius		
	3rd	Fri	Above	Aquarius	REST	
	4th	Sat	Above	REST	Pisces	
	5th	Sun	Above	Pisces		
XhN	6th	Mon	Above	Pisces	REST	Aries
1Q	7th	Tue	Above	Aries		
P13	8th	Wed	Above	Aries	REST	Taurus
	9th	Thur	Above	Taurus		
	10th	Fri	Above	Taurus	REST	Gemini
	11th	Sat	Above	Gemini		
^	12th	Sun	Above	REST DAY - Northern Declination		
	13th	Mon	Above	Cancer		
F	14th	Tue	Below	REST DAY - Full Moon		
	15th	Wed	Below	Leo		
	16th	Thur	Below	Leo	REST	
	17th	Fri	Below	REST	Virgo	
	18th	Sat	Below	Virgo		
XhS	19th	Sun	Below	Virgo	REST	Libra
	20th	Mon	Below	Libra		
A	21st	Tue	Below	Libra	REST	
3Q	22nd	Wed	Below	REST	Scorpio	
	23rd	Thur	Below	Scorpio		
	24th	Fri	Below	Scorpio	REST	Sagittarius
	25th	Sat	Below	Sagittarius		
	26th	Sun	Below	Sagittarius	REST	
V	27th	Mon	Below	REST DAY - Southern Declination		
	28th	Tue	Below	Capricorn		
	29th	Wed	Below	Capricorn	REST	Aquarius
N	30th	Thur	Abo	REST DAY - New Moon		
	31st	Fri	Abo	Aq	REST	Pisces

FEBRUARY	2025		Hour:	1am 2am 3am 4am 5am 6am 7am 8am 9am 10am 11am Noon 1pm 2pm 3pm 4pm 5pm 6pm 7pm 8pm 9pm 10pm 11pm Midnight		
	1st	Sat	Above	Pisces		
P10 XhN	2nd	Sun	Above	Pisces	REST	Aries
	3rd	Mon	Above	Aries	REST	
	4th	Tue	Above	REST	Taurus	
1Q	5th	Wed	Above	Taurus		
	6th	Thur	Above	Taurus	REST	Gemini
	7th	Fri	Above	Gemini		
^	8th	Sat	Above	REST DAY - Northern Declination		
	9th	Sun	Above	Cancer		
	10th	Mon	Above	Cancer		
	11th	Tue	Above	Cancer	REST	Leo
	12th	Wed	Above	Leo		
F	13th	Thur	Below	REST DAY - Full Moon		
	14th	Fri	Below	Virgo		
	15th	Sat	Below	Virgo	REST	

FEBRUARY	2025		Hour:	1am – Midnight
XhS	16th	Sun	Below	REST · Libra
	17th	Mon	Below	Libra
A	18th	Tue	Below	Libra · REST · Scorpio
	19th	Wed	Below	Scorpio
	20th	Thur	Below	Scorpio · REST
3Q	21st	Fri	Below	REST · Sagittarius
	22nd	Sat	Below	Sagittarius
V	23rd	Sun	Below	REST DAY - Southern Declination
	24th	Mon	Below	Capricorn
	25th	Tue	Below	Capricorn · REST · Aquarius
	26th	Wed	Below	Aquarius
	27th	Thur	Below	Aquarius · REST · Pisces
N	28th	Fri	A	REST DAY - New Moon

MARCH	2025		Hour:	1am – Midnight
XhN	1st	Sat	Above	Pisces · REST · Aries
P7	2nd	Sun	Above	Aries
	3rd	Mon	Above	Aries · REST · Ta
	4th	Tue	Above	Taurus
	5th	Wed	Above	Taurus · RE
	6th	Thur	Above	REST · Gemini
1Q	7th	Fri	Above	Gemini
^	8th	Sat	Above	REST DAY - Northern Declination
	9th	Sun	Above	Cancer
	10th	Mon	Above	Cancer · REST · Leo
	11th	Tue	Above	Leo
	12th	Wed	Above	Leo · REST · Virgo
	13th	Thur	Above	Virgo
F	14th	Fri	Below	REST DAY - Full Moon - Lunar Total Eclipse
XhS	15th	Sat	Below	REST DAY - Day after Eclipse Day
	16th	Sun	Below	Libra · REST
	17th	Mon	Below	REST · Scorpio
A	18th	Tue	Below	Scorpio
	19th	Wed	Below	Scorpio
	20th	Thur	Below	Scorpio · RE · Sagittarius
	21st	Fri	Below	Sagittarius
V	22nd	Sat	Below	REST DAY - Southern Declination
3Q	23rd	Sun	Below	Capricorn
	24th	Mon	Below	Capricorn
	25th	Tue	Below	Capricorn · RE · Aquarius
	26th	Wed	Below	Aquarius · RE
	27th	Thur	Below	REST · Pisces
	28th	Fri	Below	Pisces
XhN	29th	Sat	Below	Pisces · RE · Aries
N P4	30th	Sun	Abo	REST DAY - New Moon
	31st	Mon	Abo	REST DAY - Day after Eclipse Day

APRIL		2025	Hour:	1am – Midnight
	1st	Tue	Above	Taurus
	2nd	Wed	Above	Taurus / REST / Gemini
	3rd	Thur	Above	Gemini
^	4th	Fri	Above	REST DAY - Northern Declination
1Q	5th	Sat	Above	Cancer
	6th	Sun	Above	Cancer / REST / Leo
	7th	Mon	Above	Leo
	8th	Tue	Above	Leo / REST
	9th	Wed	Above	REST / Virgo
	10th	Thur	Above	Virgo
XhS	11th	Fri	Above	Virgo / REST / Libra
	12th	Sat	Above	Libra
F	13th	Sun	Below	REST DAY - Full Moon
A	14th	Mon	Below	REST / Scorpio
	15th	Tue	Below	Scorpio
	16th	Wed	Below	Scorpio / REST / Sagittarius
	17th	Thur	Below	Sagittarius
	18th	Fri	Below	Sagittarius / REST
V	19th	Sat	Below	REST DAY - Southern Declination
	20th	Sun	Below	Capricorn
3Q	21st	Mon	Below	Capricorn / REST / Aquarius
	22nd	Tue	Below	Aquarius
	23rd	Wed	Below	Aquarius / REST / Pisces
	24th	Thur	Below	Pisces
XhN	25th	Fri	Below	Pisces / REST / Aries
	26th	Sat	Below	Aries
	27th	Sun	Below	Aries / REST / Taurus
N P3	28th	Mon	Above	REST DAY - New Moon
	29th	Tue	Above	Taurus / REST / Gemini
	30th	Wed	Above	Gemini

MAY		2025	Hour:	1am – Midnight
^	1st	Thur	Above	REST DAY - Northern Declination
	2nd	Fri	Above	Cancer
	3rd	Sat	Above	Cancer / REST / Le
	4th	Sun	Above	Leo
1Q	5th	Mon	Above	Leo
	6th	Tue	Above	Leo / REST / Virgo
	7th	Wed	Above	Virgo
XhS	8th	Thur	Above	Virgo / REST / Libra
	9th	Fri	Above	Libra
	10th	Sat	Above	Libra / REST
A	11th	Sun	Above	FEST / Scorpio
	12th	Mon	Above	Scorpio
F	13th	Tue	Below	REST DAY - Full Moon
	14th	Wed	Below	Sagittarius
	15th	Thur	Below	Sagittarius
V	16th	Fri	Below	REST DAY - Southern Declination

MAY		2025	Hour:	1am – Midnight
	17th	Sat	Below	Capricorn
	18th	Sun	Below	Capricorn · REST · Aquarius
	19th	Mon	Below	Aquarius
	20th	Tue	Below	Aquarius
3Q	21st	Wed	Below	RE · Pisces
	22nd	Thur	Below	Pisces
XhN	23rd	Fri	Below	Pisces · RE · Aries
	24th	Sat	Below	Aries · RE
	25th	Sun	Below	REST · Taurus
P5	26th	Mon	Below	Taurus
N	27th	Tue	Above	REST DAY - New Moon
	28th	Wed	Above	Gemini
^	29th	Thur	Above	REST DAY - Northern Declination
	30th	Fri	Above	Cancer
	31st	Sat	Above	Cancer · REST · Leo

JUNE		2025	Hour:	1am – Midnight
	1st	Sun	Above	Leo
	2nd	Mon	Above	Leo · REST · Virgo
1Q	3rd	Tue	Above	Virgo
	4th	Wed	Above	Virgo · REST
XhS	5th	Thur	Above	REST · Libra
	6th	Fri	Above	Libra
A	7th	Sat	Above	Libra · REST · Scorpio
	8th	Sun	Above	Scorpio
	9th	Mon	Above	Scorpio
	10th	Tue	Above	REST · Sagittarius
F	11th	Wed	Below	REST DAY - Full Moon
V	12th	Thur	Below	REST DAY - Southern Declination
	13th	Fri	Below	Capricorn
	14th	Sat	Below	Capricorn · REST · Aquar
	15th	Sun	Below	Aquarius
	16th	Mon	Below	Aquarius
	17th	Tue	Below	Aquarius · RE · Pisces
	18th	Wed	Below	Pisces
3Q XhN	19th	Thur	Below	Pisces · REST · Aries
	20th	Fri	Below	Aries
	21st	Sat	Below	Aries · RE · Taurus
	22nd	Sun	Below	Taurus
P8	23rd	Mon	Below	Taurus · REST · Gemini
	24th	Tue	Below	Gemini
N ^	25th	Wed	Above	REST DAY - New Moon / Northern Declination
	26th	Thur	Above	Cancer
	27th	Fri	Above	Cancer · REST · Leo
	28th	Sat	Above	Leo
	29th	Sun	Above	Leo · REST
	30th	Mon	Above	Virgo

JULY 2025

	Date	Day	Position	Hour: 1am – Midnight
	1st	Tue	Above	Virgo
XhS	2nd	Wed	Above	Virgo / REST / Libra
1Q	3rd	Thur	Above	Libra / REST
	4th	Fri	Above	REST / Scorpio
A	5th	Sat	Above	Scorpio
	6th	Sun	Above	Scorpio
	7th	Mon	Above	Scorpio / RE / Sagittarius
	8th	Tue	Above	Sagittarius / REST
V	9th	Wed	Above	REST DAY - Southern Declination
	10th	Thur	Above	Capricorn
F	11th	Fri	Below	REST DAY - Full Moon
	12th	Sat	Below	REST / Aquarius
	13th	Sun	Below	Aquarius / REST
	14th	Mon	Below	REST / Pisces
	15th	Tue	Below	Pisces
XhN	16th	Wed	Below	Pisces / REST / Aries
	17th	Thur	Below	Aries
3Q	18th	Fri	Below	Aries / REST / Taurus
	19th	Sat	Below	Taurus
	20th	Sun	Below	Taurus / REST / Gem ni
P11	21st	Mon	Below	Gemini
^	22nd	Tue	Below	REST DAY - Northern Declination
	23rd	Wed	Below	RE / Cancer
	24th	Thur	Below	Cancer / REST
N	25th	Fri	Above	REST DAY - New Moon
	26th	Sat	Above	Leo / REST
	27th	Sun	Above	REST / Virgo
	28th	Mon	Above	Virgo
XhS	29th	Tue	Above	Virgo / REST / Libra
	30th	Wed	Above	Libra / REST
	31st	Thur	Above	REST DAY - Moon Void of Course all day

AUGUST 2025

	Date	Day	Position	Hour: 1am – Midnight
	1st	Fri	Above	REST / Scorpio
1Q A	2nd	Sat	Above	Scorpio
	3rd	Sun	Above	Scorpio / REST / Sagittarius
	4th	Mon	Above	Sagittarius
	5th	Tue	Above	Sagittarius
V	6th	Wed	Above	REST DAY - Southern Declination
	7th	Thur	Above	Capricorn / REST
	8th	Fri	Above	REST / Aquarius
F	9th	Sat	Below	REST DAY - Full Moon
	10th	Sun	Below	REST / Pisces
	11th	Mon	Below	Pisces / REST
XhN	12th	Tue	Below	REST / Aries
	13th	Wed	Below	Aries
	14th	Thur	Below	Aries / REST
P12	15th	Fri	Below	REST / Taurus
3Q	16th	Sat	Below	Taurus / REST

AUGUST		2025	Hour:	1am – Midnight
	17th	Sun	Below	REST / Gemini
	18th	Mon		Gemini / RE
^	19th	Tue		REST DAY - Northern Declination
	20th	Wed		Cancer
	21st	Thur		Ca / REST / Leo
	22nd	Fri		Leo / REST
N	23rd	Sat	Above	REST DAY - New Moon
	24th	Sun		Virgo
	25th	Mon		Virgo
XhS	26th	Tue		Virgo / RE / Libra
	27th	Wed		Libra / REST
	28th	Thur		REST / Scorpio
	29th	Fri		Scorpio
A	30th	Sat		Scorpio / REST
1Q	31st	Sun		REST / Sagittarius

SEPTEMBER		2025	Hour:	1am – Midnight
	1st	Mon	Above	Sagittarius
V	2nd	Tue		REST DAY - Southern Declination
	3rd	Wed		Capricorn
	4th	Thur		Capricorn / RE / Aquarius
	5th	Fri		uarius
	6th	Sat		Aquarius / REST
	7th	Sun		REST / Pisces
F	8th	Mon	Below	REST DAY - Full Moon - Lunar Total Eclipse
XhN	9th	Tue		y after Eclipse Day
	10th	Wed		Aries / REST
P9	11th	Thur		REST / Taurus
	12th	Fri		Taurus
	13th	Sat		Taurus / REST / Gemini
3Q	14th	Sun		Gemini
^	15th	Mon		REST DAY - Northern Declination
	16th	Tue		Cancer
	17th	Wed		Cancer / REST / Leo
	18th	Thur		Leo
	19th	Fri		Leo
	20th	Sat		RE / Virgo
	21st	Sun		Virgo
N XhS	22nd	Mon	Above	REST DAY - New Moon - Partial Solar Eclipse
	23rd	Tue		REST DAY - Day after Eclipse Day
	24th	Wed		Libra / REST / Scorpio
	25th	Thur		Scorpio
A	26th	Fri		Scorpio
	27th	Sat		Scorpio / REST / Sagittarius
	28th	Sun		Sagittarius
V	29th	Mon		REST DAY - Southern Declination
1Q	30th	Tue		Capricorn

OCTOBER	2025		Hour:	1am – Midnight
	1st	Wed	Above	Capricorn
	2nd	Thur	Above	Capricorn / REST / Aquarius
	3rd	Fri	Above	Aquarius
	4th	Sat	Above	Aquarius / REST / Pisces
	5th	Sun	Above	Pisces
XhN	6th	Mon	Above	Pisces / REST / Aries
F	7th	Tue	Below	REST DAY - Full Moon
	8th	Wed	Below	Aries / REST / Taurus
P6	9th	Thur	Below	Taurus
	10th	Fri	Below	Taurus / REST / Gemini
	11th	Sat	Below	Gemini / REST / Cancer
^	12th	Sun	Below	REST DAY - Northern Declination
	13th	Mon	Below	Cancer
3Q	14th	Tue	Below	Cancer / REST / Le
	15th	Wed	Below	Leo
	16th	Thur	Below	Leo / REST
	17th	Fri	Below	REST / Virgo
	18th	Sat	Below	Virgo
XhS	19th	Sun	Below	Virgo / REST / Libra
	20th	Mon	Below	Libra
	21st	Tue	Below	Libra
N	22nd	Wed	Above	REST DAY - New Moon
	23rd	Thur	Above	Scorpio
A	24th	Fri	Above	Scorpio / RE / Sagittarius
	25th	Sat	Above	Sagittarius
	26th	Sun	Above	Sagittarius
V	27th	Mon	Above	REST DAY - Southern Declination
	28th	Tue	Above	Capricorn
	29th	Wed	Above	Capricorn / RE / Aquarius
1Q	30th	Thur	Above	Aquarius
	31st	Fri	Above	Aquarius / REST

NOVEMBER	2025		Hour:	1am – Midnight
	1st	Sat	Above	RE / Pisces
	2nd	Sun	Above	Pisces
XhN	3rd	Mon	Above	Pisces / REST / Aries
	4th	Tue	Above	Aries
	5th	Wed	Above	Ar / REST / Taurus
F P1	6th	Thur	Below	REST DAY - Full Moon
	7th	Fri	Below	Taurus / REST / Gemini
^	8th	Sat	Below	REST DAY - Northern Declination
	9th	Sun	Below	Gemini / RE / Cancer
	10th	Mon	Below	Cancer
	11th	Tue	Below	Cancer / RE / Leo
3Q	12th	Wed	Below	Leo
	13th	Thur	Below	Leo / RE / Virgo
	14th	Fri	Below	Virgo
XhS	15th	Sat	Below	Virgo / REST / Libra
	16th	Sun	Below	Libra

NOVEMBER		2025	Hour: 1am 2am 3am 4am 5am 6am 7am 8am 9am 10am 11am Noon 1pm 2pm 3pm 4pm 5pm 6pm 7pm 8pm 9pm 10pm 11pm Midnight			
	17th	Mon	Below	Libra		
	18th	Tue	Below	Lib	REST	Scorpio
	19th	Wed	Below	Scorpio		
N A	20th	Thur	Above	REST DAY - New Moon		
	21st	Fri	Above	Sagittarius		
	22nd	Sat	Above	Sagittarius		
V	23rd	Sun	Above	REST DAY - Southern Declination		
	24th	Mon	Above	Capricorn		
	25th	Tue	Above	Capricorn	REST	Aquar
	26th	Wed	Above	Aquarius		
	27th	Thur	Above	Aquarius		
1Q	28th	Fri	Above	Aquarius	REST	Pisces
	29th	Sat	Above	Pisces		
XhN	30th	Sun	Above	Pisces	REST	Aries

DECEMBER		2025	Hour: 1am 2am 3am 4am 5am 6am 7am 8am 9am 10am 11am Noon 1pm 2pm 3pm 4pm 5pm 6pm 7pm 8pm 9pm 10pm 11pm Midnight			
	1st	Mon	Above	Aries		
	2nd	Tue	Above	Aries	REST	Taurus
	3rd	Wed	Above	Taurus		
	4th	Thur	Above	Taurus	REST	Gemini
F P2	5th	Fri	Below	REST DAY - Full Moon		
^	6th	Sat	Below	REST DAY - Northern Declination		
	7th	Sun	Below	Cancer		
	8th	Mon	Below	Cancer	REST	Leo
	9th	Tue	Below	Leo		
	10th	Wed	Below	Leo	REST	Virgo
	11th	Thur	Below	Virgo		
3Q XhS	12th	Fri	Below	Virgo		
	13th	Sat	Below	Virgo	REST	Libra
	14th	Sun	Below	Libra		
	15th	Mon	Below	Libra	RE	Scorpio
	16th	Tue	Below	Scorpio		
A	17th	Wed	Below	Scorpio		
	18th	Thur	Below	Scorpio	REST	Sagittarius
	19th	Fri	Below	Sagittarius		
N V	20th	Sat	Above	REST DAY - New Moon / Southern Declination		
	21st	Sun	Above	Capricorn		
	22nd	Mon	Above	Capricorn		
	23rd	Tue	Above	Capricorn	REST	Aquarius
	24th	Wed	Above	Aquarius		
	25th	Thur	Above	Aquarius	REST	Pisces
	26th	Fri	Above	Pisces		
XhN	27th	Sat	Above	Pisces	REST	Aries
1Q	28th	Sun	Above	Aries		
	29th	Mon	Above	Aries	REST	
	30th	Tue	Above	REST	Taurus	
	31st	Wed	Above	Taurus		

FISHING CALENDAR 2025

AV	VG	EX
1 fish	2 fish	3 fish

JANUARY	1st	2nd	3rd	4th	5th	6th	7th	8th	9th	10th	11th	12th	13th	14th	15th	16th	17th	18th	19th	20th	21st	22nd	23rd	24th	25th	26th	27th	28th	29th	30th	31st
MOON							First Q	P13						Full Moon							Ap	Third Q								New Moon	
Best Bite Time am/pm	1.00	2.00	3.00	4.00	4.30	5.00	6.00	7.00	8.00	9.00	10.00	10.30	11.00	12.00	1.00	2.00	2.30	3.00	4.00	5.00	5.30	6.00	7.00	8.00	9.00	10.00	10.30	11.00	11.30	12.00	1.00
Next Best Bite Time am/pm	7.00	8.00	9.00	10.00	10.30	11.00	12.00	1.00	2.00	3.00	4.00	4.30	5.00	6.00	7.00	8.00	8.30	9.00	10.00	11.00	11.30	12.00	1.00	2.00	3.00	4.00	4.30	5.00	5.30	6.00	7.00
RATING	EX	AV	AV	AV	AV	VG	VG	VG	AV	AV	AV	AV	EX	EX	EX	AV	AV	AV	AV	AV	VG	VG	VG	AV	AV	AV	AV	AV	EX	EX	EX

FEBRUARY	1st	2nd	3rd	4th	5th	6th	7th	8th	9th	10th	11th	12th	13th	14th	15th	16th	17th	18th	19th	20th	21st	22nd	23rd	24th	25th	26th	27th	28th
MOON		P10			First Q								Full Moon					Ap			Third Q							New Moon
Best Bite Time am/pm	2.00	2.30	3.00	4.00	5.00	6.00	7.00	8.00	9.00	10.00	11.00	11.30	12.00	12.30	1.00	2.00	3.00	4.00	4.30	5.00	6.00	7.00	8.00	9.00	10.00	11.00	11.30	12.00
Next Best Bite Time am/pm	8.00	8.30	9.00	10.00	11.00	12.00	1.00	2.00	3.00	4.00	5.00	5.30	6.00	6.30	7.00	8.00	9.00	10.00	10.30	11.00	12.00	1.00	2.00	3.00	4.00	5.00	5.30	6.00
RATING	AV	AV	AV	VG	VG	VG	AV	AV	AV	AV	AV	EX	EX	EX	AV	AV	AV	AV	AV	VG	VG	VG	AV	AV	AV	AV	EX	EX

MARCH	1st	2nd	3rd	4th	5th	6th	7th	8th	9th	10th	11th	12th	13th	14th	15th	16th	17th	18th	19th	20th	21st	22nd	23rd	24th	25th	26th	27th	28th	29th	30th	31st
MOON		P7					First Q							Full Moon				Ap					Third Q						New Moon	P4	
Best Bite Time am/pm	1.00	2.00	3.00	4.00	5.00	5.30	6.00	6.00	7.00	8.00	9.00	10.00	11.00	12.00	1.00	2.00	3.00	4.00	4.30	5.00	5.30	6.00	7.00	8.00	9.00	10.00	10.30	11.00	11.30	12.00	1.00
Next Best Bite Time am/pm	7.00	8.00	9.00	10.00	11.00	11.30	12.00	12.00	1.00	2.00	3.00	4.00	5.00	6.00	7.00	8.00	9.00	10.00	10.30	11.00	11.30	12.00	1.00	2.00	3.00	4.00	4.30	5.00	5.30	6.00	7.00
RATING	EX	AV	AV	AV	AV	VG	VG	VG	AV	AV	AV	AV	EX	EX	EX	AV	AV	AV	AV	AV	AV	VG	VG	VG	AV	AV	AV	EX	EX	EX	AV

APRIL	1st	2nd	3rd	4th	5th	6th	7th	8th	9th	10th	11th	12th	13th	14th	15th	16th	17th	18th	19th	20th	21st	22nd	23rd	24th	25th	26th	27th	28th	29th	30th
MOON					First Q								Full Moon	Ap							Third Q							P3 New Moon		
Best Bite Time am/pm	2.00	3.00	4.00	5.00	6.00	7.00	8.00	9.00	10.00	10.30	11.00	11.30	12.00	12.30	1.00	2.00	3.00	4.00	5.00	5.30	6.00	7.00	8.00	9.00	10.00	11.00	11.30	12.00	1.00	2.00
Next Best Bite Time am/pm	8.00	9.00	10.00	11.00	12.00	1.00	2.00	3.00	4.00	4.30	5.00	5.30	6.00	6.30	7.00	8.00	9.00	10.00	11.00	11.30	12.00	1.00	2.00	3.00	4.00	5.00	5.30	6.00	7.00	8.00
RATING	AV	AV	AV	VG	VG	VG	AV	AV	AV	AV	AV	EX	EX	EX	AV	AV	AV	AV	AV	VG	VG	VG	AV	AV	AV	AV	EX	EX	EX	AV

MAY	1st	2nd	3rd	4th	5th	6th	7th	8th	9th	10th	11th	12th	13th	14th	15th	16th	17th	18th	19th	20th	21st	22nd	23rd	24th	25th	26th	27th	28th	29th	30th	31st
MOON					First Q						Ap		Full Moon							Third Q						P5	New Moon				
Best Bite Time am/pm	3.00	4.00	5.00	5.30	6.00	6.30	7.00	8.00	9.00	10.00	11.00	11.30	12.00	1.00	2.00	3.00	4.00	5.00	5.30	6.00	7.00	8.00	9.00	10.00	11.00	11.30	12.00	1.00	2.00	3.00	4.00
Next Best Bite Time am/pm	9.00	10.00	11.00	11.30	12.00	12.00	1.00	2.00	3.00	4.00	5.00	5.30	6.00	7.00	8.00	9.00	10.00	11.00	11.30	12.00	1.00	2.00	3.00	4.00	5.00	5.30	6.00	7.00	8.00	9.00	10.00
RATING	AV	AV	AV	VG	VG	VG	AV	AV	AV	AV	AV	EX	EX	EX	AV	AV	AV	AV	VG	VG	VG	AV	AV	AV	AV	EX	EX	EX	AV	AV	AV

JUNE	1st	2nd	3rd	4th	5th	6th	7th	8th	9th	10th	11th	12th	13th	14th	15th	16th	17th	18th	19th	20th	21st	22nd	23rd	24th	25th	26th	27th	28th	29th	30th
MOON			First Q				Ap				Full Moon								Third Q				P8		New Moon					
Best Bite Time am/pm	5.00	5.30	6.00	7.00	8.00	9.00	10.00	10.30	11.00	11.30	12.00	1.00	2.00	3.00	4.00	5.00	5.30	6.00	7.00	8.00	9.00	10.00	10.30	11.00	12.00	1.00	2.00	3.00	4.00	4.30
Next Best Bite Time am/pm	11.00	11.30	12.00	1.00	2.00	3.00	4.00	4.30	5.00	5.30	6.00	7.00	8.00	9.00	10.00	11.00	11.30	12.00	1.00	2.00	3.00	4.00	4.30	5.00	6.00	7.00	8.00	9.00	10.00	10.30
RATING	AV	VG	VG	VG	AV	AV	AV	AV	AV	EX	EX	EX	AV	AV	AV	AV	AV	VG	VG	VG	AV	AV	AV	EX	EX	EX	AV	AV	AV	AV

AV	VG	EX
1 fish	2 fish	3 fish

JULY	1st	2nd	3rd	4th	5th	6th	7th	8th	9th	10th	11th	12th	13th	14th	15th	16th	17th	18th	19th	20th	21st	22nd	23rd	24th	25th	26th	27th	28th	29th	30th	31st
MOON			First Q		Ap						Full Moon							Third Q			P11				New Moon						
Best Bite Time am/pm	5.00	5.30	6.00	7.00	8.00	9.00	10.00	10.30	11.00	11.30	12.00	1.00	2.00	3.00	4.00	4.30	5.00	6.00	7.00	8.00	9.00	10.00	11.00	11.30	12.00	1.00	2.00	3.00	4.00	5.00	5.30
Next Best Bite Time am/pm	11.00	11.30	12.00	1.00	2.00	3.00	4.00	4.30	5.00	5.30	6.00	7.00	8.00	9.00	10.00	10.30	11.00	12.00	1.00	2.00	3.00	4.00	5.00	5.30	6.00	7.00	8.00	9.00	10.00	11.00	11.30
RATING	AV	VG	VG	VG	AV	AV	AV	AV	AV	EX	EX	EX	AV	AV	AV	AV	VG	VG	VG	AV	AV	AV	AV	EX	EX	EX	AV	AV	AV	AV	AV

AUGUST	1st	2nd	3rd	4th	5th	6th	7th	8th	9th	10th	11th	12th	13th	14th	15th	16th	17th	18th	19th	20th	21st	22nd	23rd	24th	25th	26th	27th	28th	29th	30th	31st
MOON		First Q Ap							Full Moon						P12	Third Q							New Moon							Ap	First Q
Best Bite Time am/pm	6.00	7.00	8.00	9.00	10.00	10.30	11.00	11.30	12.00	1.00	2.00	3.00	3.30	4.00	5.00	6.00	7.00	8.00	9.00	10.00	11.00	11.30	12.00	1.00	2.00	3.00	4.00	4.30	5.00	5.30	6.00
Next Best Bite Time am/pm	12.00	1.00	2.00	3.00	4.00	4.30	5.00	5.30	6.00	7.00	8.00	9.00	9.30	10.00	11.00	12.00	1.00	2.00	3.00	4.00	5.00	5.30	6.00	7.00	8.00	9.00	10.00	10.30	11.00	11.30	12.00
RATING	VG	VG	VG	AV	AV	AV	AV	EX	EX	EX	AV	AV	AV	AV	VG	VG	VG	AV	AV	AV	AV	EX	EX	EX	AV	AV	AV	AV	AV	VG	VG

SEPTEMBER	1st	2nd	3rd	4th	5th	6th	7th	8th	9th	10th	11th	12th	13th	14th	15th	16th	17th	18th	19th	20th	21st	22nd	23rd	24th	25th	26th	27th	28th	29th	30th
MOON								Full Moon			P9			Third Q								New Moon				Ap				First Q
Best Bite Time am/pm	6.30	7.00	8.00	9.00	10.00	11.00	11.30	12.00	1.00	2.00	3.00	4.00	5.00	6.00	7.00	8.00	9.00	10.00	10.30	11.00	11.30	12.00	1.00	2.00	3.00	4.00	4.30	5.00	5.30	6.00
Next Best Bite Time am/pm	12.30	1.00	2.00	3.00	4.00	5.00	5.30	6.00	7.00	8.00	9.00	10.00	11.00	12.00	1.00	2.00	3.00	4.00	4.30	5.00	5.30	6.00	7.00	8.00	9.00	10.00	10.30	11.00	11.30	12.00
RATING	VG	AV	AV	AV	AV	AV	EX	EX	EX	AV	AV	AV	VG	VG	VG	AV	AV	AV	AV	AV	EX	EX	EX	AV	AV	AV	AV	AV	VG	VG

OCTOBER	1st	2nd	3rd	4th	5th	6th	7th	8th	9th	10th	11th	12th	13th	14th	15th	16th	17th	18th	19th	20th	21st	22nd	23rd	24th	25th	26th	27th	28th	29th	30th	31st
MOON							Full Moon		P6					Third Q								New Moon		Ap						First Q	
Best Bite Time am/pm	7.00	8.00	9.00	10.00	10.30	11.00	12.00	1.00	2.00	3.00	4.00	5.00	5.30	6.00	7.00	8.00	9.00	9.30	10.00	10.30	11.00	12.00	1.00	2.00	3.00	4.00	5.00	5.30	6.00	7.00	8.00
Next Best Bite Time am/pm	1.00	2.00	3.00	4.00	4.30	5.00	6.00	7.00	8.00	9.00	10.00	11.00	11.30	12.00	1.00	2.00	3.00	3.30	4.00	4.30	5.00	6.00	7.00	8.00	9.00	10.00	11.00	11.30	12.00	1.00	2.00
RATING	VG	AV	AV	AV	AV	EX	EX	EX	AV	AV	AV	AV	VG	VG	VG	AV	AV	AV	AV	AV	EX	EX	EX	AV	AV	AV	AV	AV	VG	VG	VG

NOVEMBER	1st	2nd	3rd	4th	5th	6th	7th	8th	9th	10th	11th	12th	13th	14th	15th	16th	17th	18th	19th	20th	21st	22nd	23rd	24th	25th	26th	27th	28th	29th	30th
MOON						P1 Full Moon						Third Q								New Moon Ap								First Q		
Best Bite Time am/pm	9.00	9.30	10.00	10.30	11.00	12.00	1.00	2.00	3.00	4.00	5.00	5.30	6.00	7.00	8.00	9.00	10.00	10.30	11.00	12.00	1.00	2.00	3.00	4.00	4.30	5.00	6.00	7.00	8.00	9.00
Next Best Bite Time am/pm	3.00	3.30	4.00	4.30	5.00	6.00	7.00	8.00	9.00	10.00	11.00	11.30	12.00	1.00	2.00	3.00	4.00	4.30	5.00	6.00	7.00	8.00	9.00	10.00	10.30	11.00	12.00	1.00	2.00	3.00
RATING	AV	AV	AV	AV	EX	EX	EX	AV	AV	AV	VG	VG	VG	AV	AV	AV	AV	AV	EX	EX	EX	AV	AV	AV	AV	AV	VG	VG	VG	AV

DECEMBER	1st	2nd	3rd	4th	5th	6th	7th	8th	9th	10th	11th	12th	13th	14th	15th	16th	17th	18th	19th	20th	21st	22nd	23rd	24th	25th	26th	27th	28th	29th	30th	31st
MOON				P2	Full Moon							Third Q					Ap			New Moon								First Q			
Best Bite Time am/pm	10.00	10.30	11.00	11.30	12.00	1.00	2.00	3.00	4.00	4.30	5.00	6.00	6.30	7.00	8.00	9.00	10.00	10.30	11.00	12.00	1.00	2.00	3.00	4.00	4.30	5.00	6.00	6.30	7.00	8.00	9.00
Next Best Bite Time am/pm	4.00	4.30	5.00	5.30	6.00	7.00	8.00	9.00	10.00	10.30	11.00	12.00	12.00	1.00	2.00	3.00	4.00	4.30	5.00	6.00	7.00	8.00	9.00	10.00	10.30	11.00	12.00	12.00	1.00	2.00	3.00
RATING	AV	AV	AV	EX	EX	EX	AV	AV	AV	AV	VG	VG	VG	AV	AV	AV	AV	AV	EX	EX	EX	AV	AV	AV	AV	AV	VG	VG	VG	AV	AV

TIDES 2025

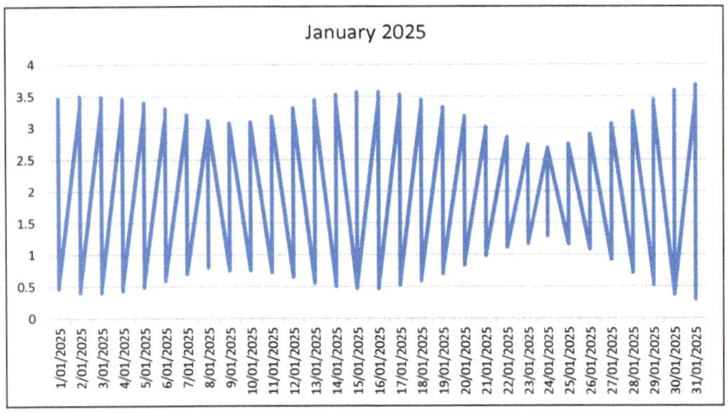
January 2025

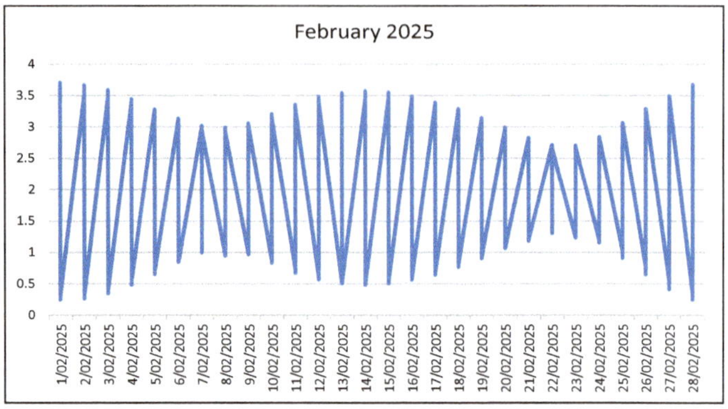
February 2025

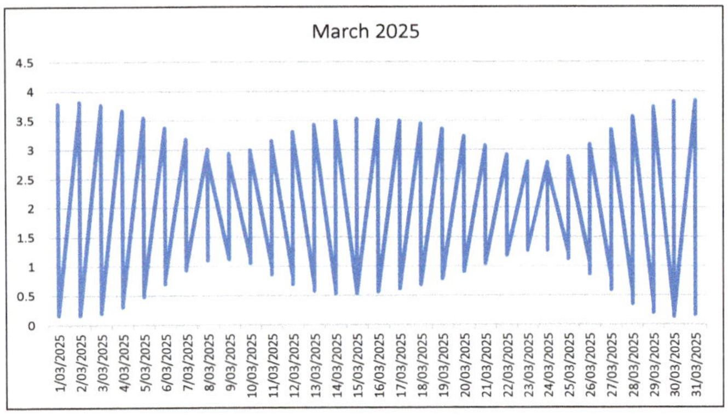
March 2025

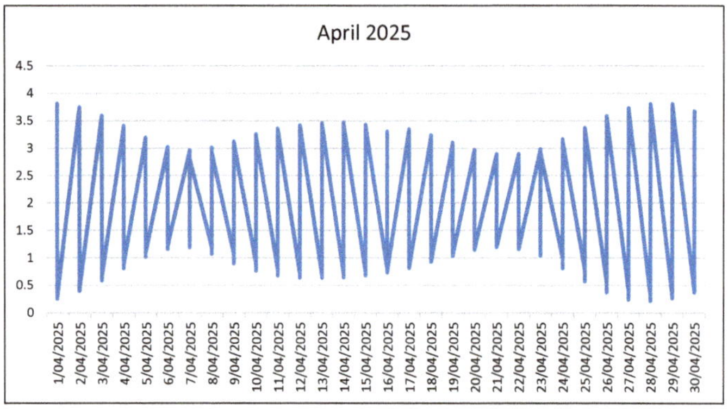
April 2025

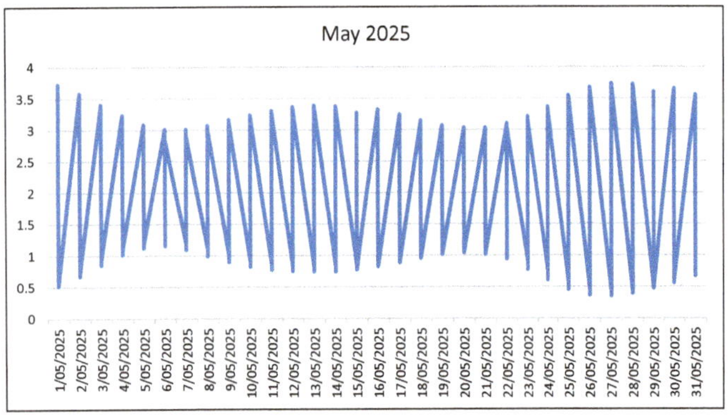
May 2025

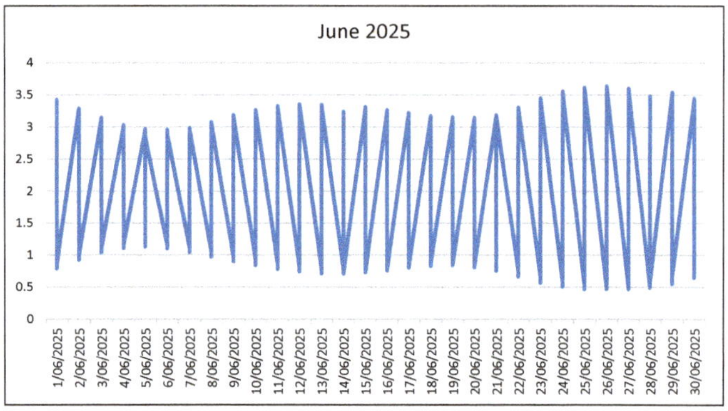
June 2025

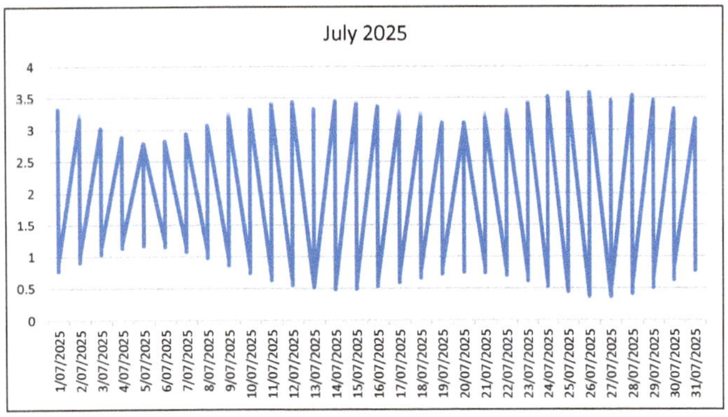
July 2025

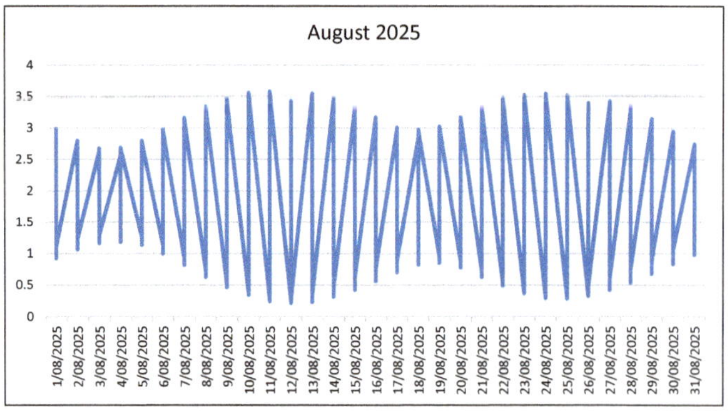
August 2025

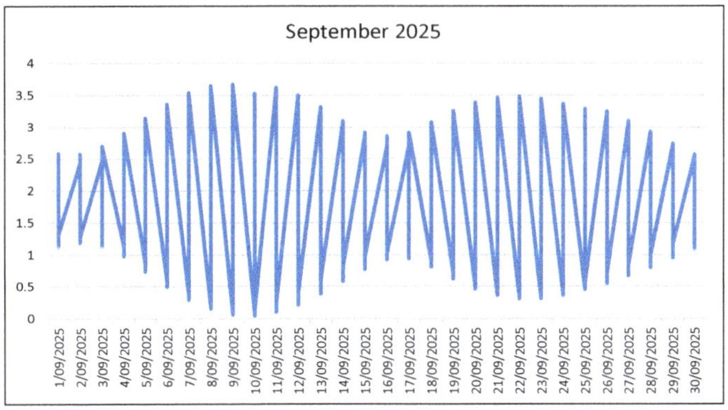

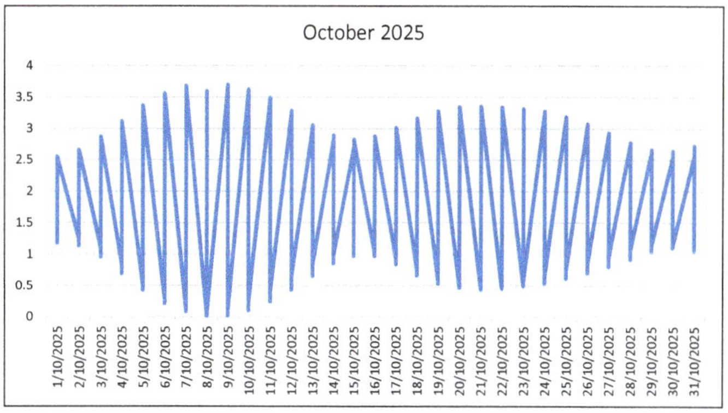

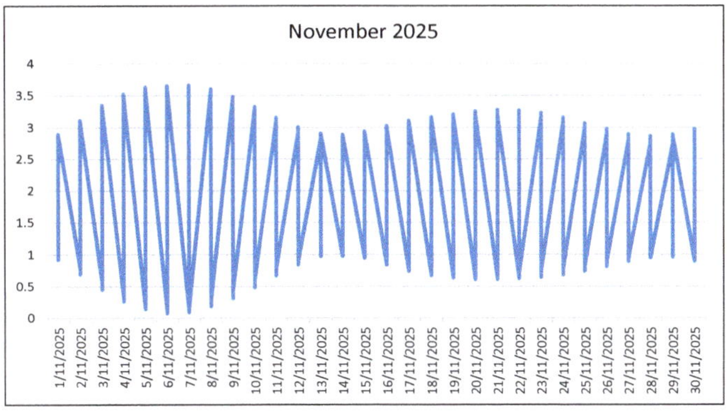

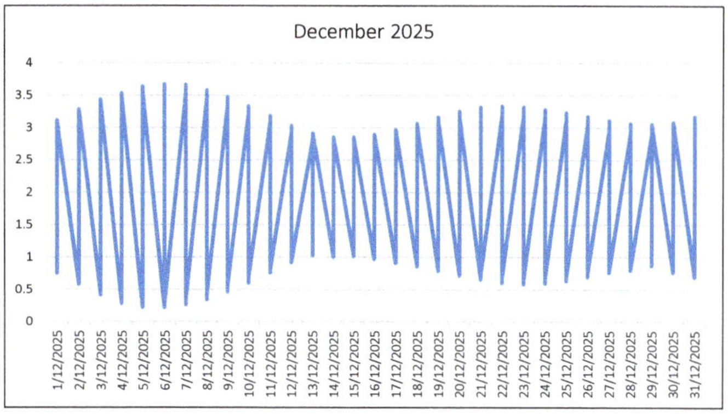

KING-TIDES IN ORDER OF HIGHEST

HIGHEST	**MARCH**	31/03/2025
	APRIL	01/4/2025
	MAY	27/05/2025
	FEBRUARY	01/02/2025
	OCTOBER	09/10/2025
	JANUARY	31/01/2025
	DECEMBER	06/12/2025
	SEPTEMBER	09/09/2025
	NOVEMBER	07/11/2025
	JUNE	26/06/2025
	JULY	25/07/2025
LOWEST	**AUGUST**	11/08/2025

SNOW DIARY 2025

January

8th-9th: The passage of a wet front crosses southern parts of Canterbury, which may cause some localised flooding, including in the Ashburton district. Colder temperatures may also generate some out-of-season sleet or light snowfalls around the Porter's Pass area.

February

24th-25th: A depression south of New Zealand may bring brief snow pellets onto the higher reaches of the west and south of the South Island as well as overnight in the higher reaches of the central North Island.

March

23rd-24th: A band of thunderstorms crossing central New Zealand may see lightning strikes and hail. Chance of light snow in the ranges around Wellington including on the Rimutaka Hills.

24th-25th: Very cold overnight with the chance of sub-zero frosts overnight in the central parts of the South Island including around the Mt Cook region and possible snowfall around Lake Rotoiti and frosts and snow in the central plateau districts of the North Island. Chance of snowflurries in the Christchurch area.

April

23rd-24th: An intense Southern Ocean storm flows up through the country bringing the chance of snow onto the higher reaches of the South Island, including in the Lake Rotoiti region, and later into the higher ranges from the Bay of Plenty to the Central Plateau.

May

4th: Colder temperatures in Otago may see wintry showers in the region, with snow likely in the surrounding higher ranges.

June

13th: Chance of snow in the higher ranges in the South Island.

17th-19th: A strong southwesterly flow over the South Island brings snow to low levels in Southland and Otago, and including in the high country passes. Some snowfall likely to reach the Dunedin hill suburbs and in Queenstown. Chance that snowstorms may cause disruptions to power services.

21st: A depression crossing the North Island brings snow on the Central Plateau.

24th-25th: Chance snowfall in the ranges around Wellington and on the Rimutaka Hills as well as in the west and south of the South Island.

29th-30th: Chance of snow in the Reefton area of the West Coast as well as snow flurries in Southland.

30th June-4th July: A strong southwesterly returns over the South Island bringing further snow to low levels in Southland and Otago, and including in the high country passes. Snow also in the Central North Island, possibly causing the closure of the Desert Road for a time, as well as snow in the Reefton and Hanmer Forest areas. More settled conditions in the east of the country, however, widespread severe frosts are likely to prevail. Snow may reach sea level in some coastal parts of Southland and Otago.

July

30th June-4th July: A strong southwesterly returns over the South Island bringing further snow to low levels in Southland and Otago, and including in the high country passes. Snow also in the Central North Island, possibly causing the closure of the Desert Road for a time, as well as snow in the Reefton and Hanmer Forest areas. More settled conditions in the east of the country, however, widespread severe frosts are likely to prevail. Snow may reach sea level in some coastal parts of Southland and Otago.

1st: Snow likely in the Central Plateau and southern ranges of the North Island, as well as in Southland and Otago to low levels.

2nd-3rd: Snow likely to low levels in the South island including snow may reach sea level in Otago, and accumulate on the hills around Dunedin.

5th: Stormy air with strong gusty winds in the Wellington region and chance of snow in the surrounding hills.

6th: Further snow likely in the Dunedin area.

9th-11th: Wintry depression brings further snowfalls in the south of the South Island and high country and around the Central Plateau.

19th-22nd: Chance of further snow in Canterbury to Otago high country, and incuding the chance of snowflurries Christchurch and Dunedin.

August

1st: Chance of snowflurries in the hills surrounding Dunedin.

10th: A depression over the South Island may bring snowfalls into the south of the South Island and up along the Southern Alps.

15th-18th: An intense wintry depression crosses the South Island bringing snowfalls once more onto the higher reaches of the southern regions of the South Island, including around the Dunedin hills. As the depression moves up through the country, snowfalls are likely in the central North Island districts and on the Central Plateau.

22nd-24th: An unstable cold front rolls up through the country bringing further snow into the south and east of the South Island as well as up onto the North Island Central Plateau. Snow flurries may reach into the major centres of Christchurch and Dunedin.

28th: Chance of light snow in the Dunedin hills.

September

14th-16th: Chance of snow in the east of the South Island, possibly reaching low-lying parts of Canterbury, including around Ashburton, Methven, Twizel, Tekapo, Darfield and Geraldine areas and snow flurries may be observed in Christchurch. The Lindis, Crown Range and other Passes may be closed for a time. Cold conditions may cause stock losses, particularly of new-born lambs. Snow falls may also occur during this spell in the central regions of the North Island.

October

12th-13th: Chance of hail or light snow and strong wind gusts in Christchurch.

12th-17th: Chance of snowfall in many high country areas, including in the Mt Cook district, Arthurs Pass and Tekapo districts, as well as in the North Island Central Plateau region around the Desert Road.

20th-21st: Chance of heavy rain in the top of the South Island and into Canterbury as well as snowfall in many high country areas of the South Island, including the Arthurs Pass area.

24th-25th: Chance of snowfall in many high country areas of the South Island. More snowfall likely in the Arthurs Pass area again.

26th-29th: Chance of snowfall in the Southern Alps and southern Canterbury high country including around Tekapo and Arthurs Pass again, with possible flurries as far to the east as Fairlie. Snow may also fall later in the North Island Central Plateau around the Desert Road.

EARTHQUAKE NOTES

Earthquakes occur at solar and lunar-influenced times. We are coming up to 2026 which is the solar maximum. There is increased earthquake activity during and after solar maximum years. This will encompass years from 2025-2028, which is in line also with the lunar periodicity, particularly 2028. This is the next likely year to have a series of big events, although they may not last longer than a few months.

2020 was the last solar minimum year and a relatively less active year for earthquake activity. The solar minimum means a weaker sun putting out less flares and solar disturbances. 2024 was drier, and with more highs, that led to higher barometric pressures..

Watch for road-kill

This has been found to increase during earthquake activity, with open roads in the country offering wild creatures relative safety from ground disturbance and surface disruption. It is hence a sign that earthquakes may be imminent.

Watch atmospheric pressure

The barometer is the warning tool for seismic activity, because land, sea and air are connected, and when air pressure changes by at least 12mbs per day, it may herald a seismic event, providing that the moon is also in line. Past examples can be seen taken from the 2011 Christchurch series. The day before the 4 September 2010, 7.1mag earthquake, the barometer leapt up by 18mbs.

3 September, 2010

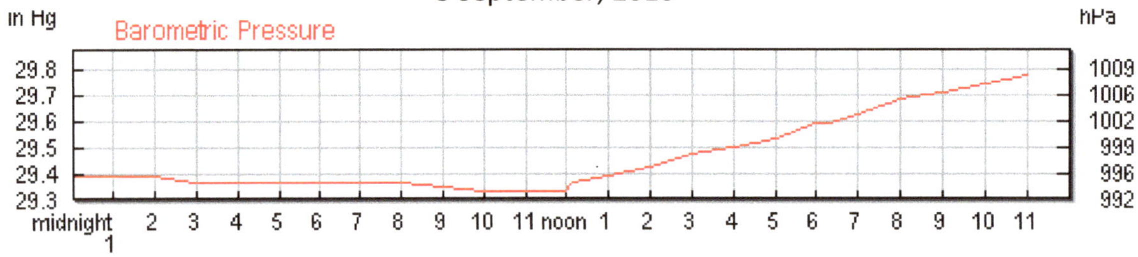

Atmospheric pressure change of about 12mbs is enough to warn of big earthquakes, usually within two days

Watch tide height

Around 3000kms depth of land may be on the move compared to 2-3kms depth of ocean, if averaged over the globe. It is **perigee** that causes kingtides, but the kingtide of the **land** is paramount. It is why large earthquakes invariably occur around kingtide days and most often the seismic action is either before or just after the kingtide, see pic below.

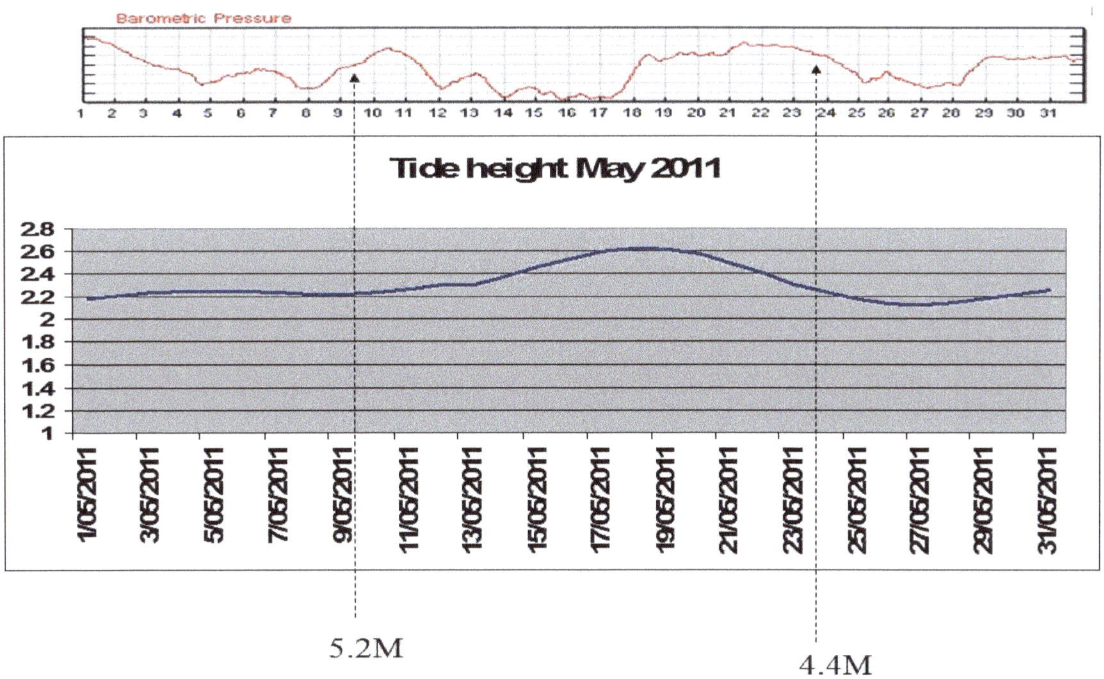

During high Earth (also called Land or Solid) tide, the land expands, and like the sides of a bathtub getting wider, the water level goes down, and we call it the Low (sea) Tide. Because high Land Tide (earthquake) = low sea tide, larger earthquakes can be seen to occur most often around or coming up to, the low (sea) tide. At 4.30am on 4 September, 2010, the tide needle shook itself off, coming up to low (sea) tide.

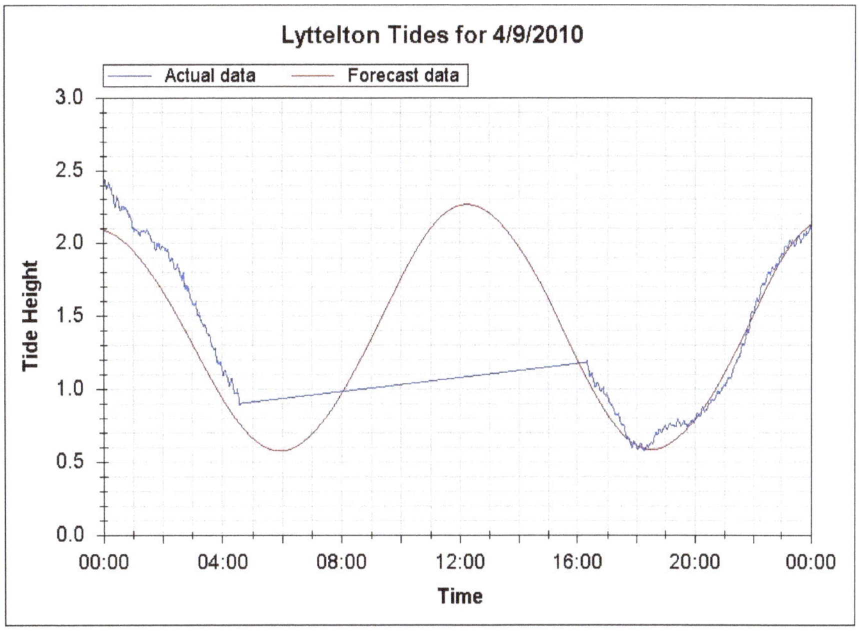

The needle just seemed to stop before low (sea) tide on 22 February 2011, suggesting the force of the event made the tide recorder malfunction. The same scenario repeated with other large 2011 Christchurch earthquakes.

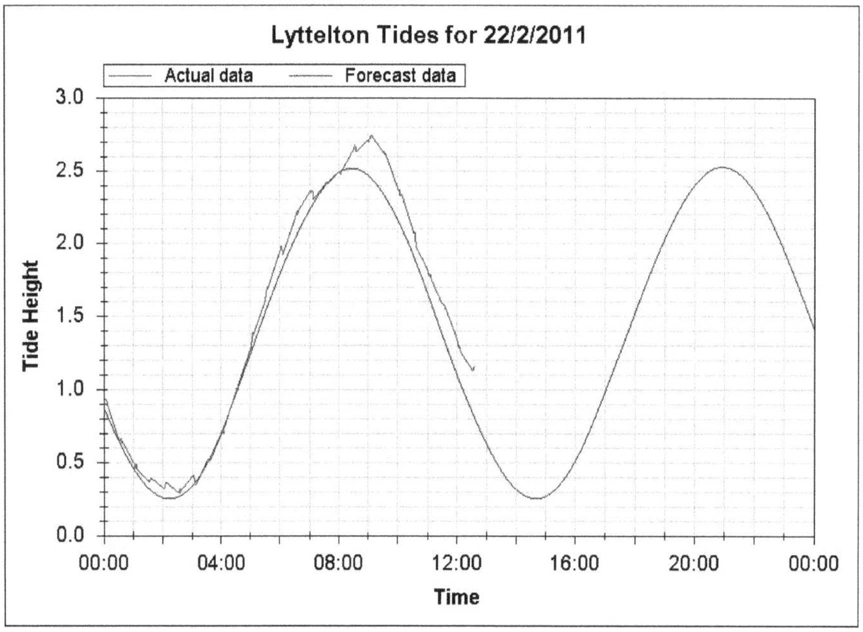

Every low (sea) tide produces a land expansion, which most often is not picked up by earthquake sensors. The land is in continual vertical displacement, and over hundreds of years, grinds shells into fine sand, and rocks into soil.

Comparison with known earthquakes (Christchurch 2010 and 2011).

Lunar perigee is the main factor, and most earthquakes occur in the perigee week. Various factors come together. The full moon synching with low (sea) tide, and planetary movements, notably Mercury and Pluto. See my book "How To Predict Earthquakes" (www.predictweather.com) for more detail.
Mercury features in NZ earthquakes. Mercury passed between Earth and the Sun between 1-4 September 2010. For every earthquake, from September through to the following June, a Mercury conjunction was the key.

In September 2010, ^+3rdQ=2nd, P2+NM=8th, XhS=9th, 1st+V=15th, A=21st, XhN=22nd, F=23rd, ^=30th.
Mercury was in front of the Sun 1-4 September 2010.
The 7.4mag earthquake occurred on the 4th.

Earthquake predictions, tweeted beforehand

On 8 September 2010, four days after the first large shake, we issued this tweet.

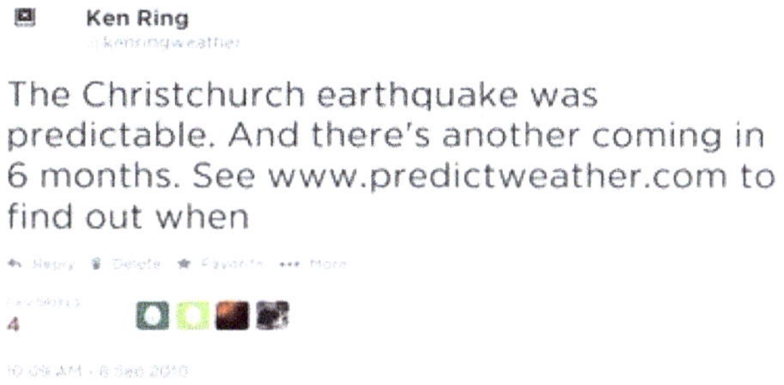

https://twitter.com/kenringweather/status/23856729753

We then tweeted a reminder on 14 February 2011, lest the public forgot the first tweet.

https://twitter.com/kenringweather/status/36763786 807345152

For February 2011, N=3rd, A+XhN=6th, 1st=11th, ^=13th, F=18th, P5=19th, XhS=20th, V+3rdQ=25th, Mercury - Sun 11th-23rd.

On the 18 February 2011, FM, Earth, Sun, Mercury, Mars and Neptune were in an alignment.

The historic 6.3mag earthquake occurred on the 22nd. It was just three days past the perigee, when the seismic movements started to build up..

Moon cycle

The lunar cycles are around 17-19 years. The solar, or sunspot cycles average 11-12 years and 23-24 years.

Looking at lunar cycles first, 2010/11 back to 1992 (2011-19), saw earthquake intense activity only 70 km from the CBD in the Arthurs Pass area. There were more and larger shakes during these years than was experienced in 2010/11. But because there was no loss of life in 1992, unlike the 185 in 2011 who were trapped in a building condemned by inspectors before any earthquakes happened, the activity in 1992/3 hardly received media attention, and the few houses in the Arthurs Pass area were all wooden railway cottages, which do not crumble like stone buildings. They simply tip over as the land drops away on one side and can be quickly jacked back into position.

In 1972/3 (1992-19) a 7mag event was experienced in Taumaranui. Then in 1953/54 (1973-20), there was a 7.2mag event in Bay of Plenty. 1934 (1953-19) saw a 7.2 earthquake in Pahiatua. Beyond that, pre-around 1935, the Richter Scale was not invented, and so we can only guess at the size. Relying on hearsay oral stories is not science, and so can be discarded, although they are intriguing in the telling.

Taking 18 years as the average of 17 and 19, around 2028 (2010/11+17), there may be significant activity again in the Christchurch area, based on lunar cycles. Close perigees are the key to the lunar cycle. In 2028-30, perigees (average) will be closer than in any year since 2011.

Sun cycle

Solar cycles occur at 11-12 and 22-24 years. Solar maximums have been 1956-60, 1968-72, 1979-83, 1989-92, 2000-02, 2013-15, 2025-28, and 2037-39. Solar minimums have been between these. Every 200 years there is a general lack of sunspots and that has been seen in the 1200s, 1400s, 1600s, etc.

Observing the cycles of solar maxima, we can go 11 years back from 2011 and see that in 2000/1 there were some 7mag events, especially in East Cape. Going back even further, to 1977/78 (2000-23) shows a 6mag event that occurred in Cook Strait, felt widely in both islands. Then we can observe that 1955/56 (1977-21) experienced a 6.3mag shake in Gisborne.

Tsunamis – not likely in NZ

The universal fear about tsunamis is overblown by local media. There is never a tsunami threat to NZ, as the waters around the country are too shallow. In the 84 years between 1929-2013 there have been only 25 registered tsunamis in the world. In the first 40 years, to 1960, 9 or 10 had occurred. In the next 40 years to 2000, another 9 or 10. During the 13 year gaps between 1933-1946 and between 1964-1976 no tsunamis were recorded.

The few-cm wave that arrived on NZ shores from Chile in 2013 was called a "small tsunami". But there is no such thing as a small tsunami. That is simply called a "wave." By definition a tsunami is a HUGE destructive tidal event.

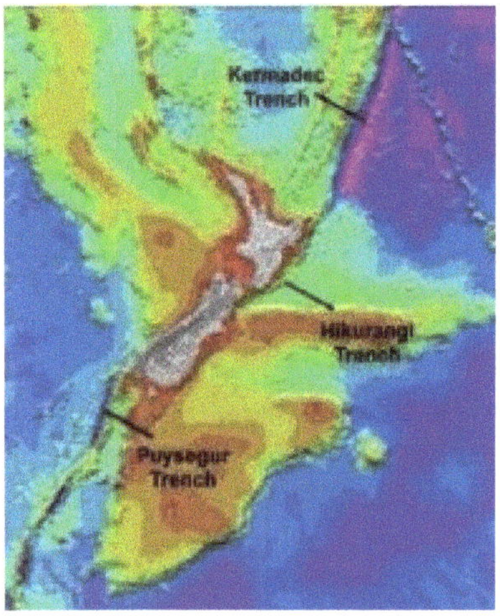

There has been ZERO people killed by tsunamis in NZ in all of our history. After 250 years of settlement, 2-3 million recorded earthquake events, and plenty of tidal and rogue waves, not one name can be attributed to a tsunami misfortune.

Our shallow coastline and therefore our geography forbids it. Luckily for NZ, only 7% of the country's mass sits above the waves. That means 93% sits underwater. NZ is therefore like a submarine, with only a small part poking out. It has relatively shallow water around it. You need deep water, to supply the destructive wave in enough quantity, and to keep 'loading' it.

With over 90% of the land beneath the waves, we are unique in the world, when it comes to the larger countries. To escape waves a surfer on a board dives beneath them. Our country is like that. It is as if tsunamis are constantly going over but not affecting us.

For a tsunami to occur, an earthquake must exceed a certain threshold. The 8.9-magnitude (on the Richter Scale) of Japan's earthquake was enough to trigger a tsunami, and the magnitude-7.7 earthquake that struck Indonesia in October 2010 only just surpassed the threshold.

A shake must be 7.8mag or higher. Earthquakes below 7.5 or 7.0 usually do not trigger them. A 7.1 or a 7.4 will not do it. Two earthquakes have exceeded 7.7mag, in 1855, in Lake Wairarapa, a 8.2mag and 2016, in Kaikoura. Neither saw any sign of a destructive tsunami and certainly no loss of life. (In 1855 a hotel was carried away when a higher tide entered an inlet, but no population was endangered). If one was going to occur, it would have been at Kaikoura.

There is a website about tsunamis put up by the earthquake commission. It bears the title of Tsunami but is and mentions the 1855 event, the 1868, the 1947, and the 1960 as extra-high tidal systems that *can* cause tsunamis. Of deaths it says “Only one tsunami death has been officially recorded – on the Chatham Islands in 1868”. If you read on it says this poor chap went into the water to try to retrieve a boat. That's asking trouble in a storm. Then we read; " .. the 1960 event was not a major disaster in New Zealand. No lives were lost or even threatened. So when we invited stories about natural disasters, the 1960 tsunami accounts tended to be more amusing than terrifying. "

But the Richter Scale was only developed in 1935. So magnitude of seismic events prior to that is only estimation. The fact remains that in our nearly 200 year history of settlement, over 2 million earthquakes have been recorded, and from them less than a single handful of tsunami scares.

Compare that to the annual tragic road toll of several hundred lives lost. We have extensive tsunami alarm systems in place costing ratepayers thousands of dollars e.g along the Kaikoura coastline, all for something that won't happen, but few road risk alarm systems in place for something that kills people every day. Some might say it is disproportionate public funding.

NZ is unlike Japan, which drops deeply into the Pacific. Resting on a plate that extends 200 miles east of the Wairarapa, NZ gets relatively shallow waves from the east. (See the picture at the top of this article). In the Pacific Islands the currents go east, so higher waves from a Samoa tsunami could not threaten NZ. They just dissipate further east. Earthquakes that occur around our coast are not able to threaten to travel inland with any destructive force.

To get some perspective on this, consider that the 2004 Indian Ocean Asian earthquake was only 10kms down, the Samoan tsunami (magnitude 8.0) was at a depth of 18 km, and the 11 March 2011 Japanese tsunami was an 9-mag at the depth of 32kms. And directly beneath each of those was a deep sea. Nothing comes up from the south. Australia shields from the west.

The reason we do not get tsunamis here is the same reason why the climate is so mild and Mediterranean-like. Yet our latitude is more like that of Switzerland. The shallow coast ensures that the sun warms most of the land, too (see picture).

Tsunami scares are part of the alarmist funding industry often called the fake news. They are alongside other dubious and unproven beat-ups intended for sensationalist value, so the media can stay relevant. There is now the incorrect belief that every earthquake is a potential tsunami.

Sadly, children have now been whipped into a state of permanent fear. You can see it in the schools as regular 'tsunami drill'. If a large earthquake hits us, making for higher ground which may open up, is a rather silly thing to recommend. Do not head for any hills. I believe the sand hills at the beach is a much safer place.

No animal or insect obeys the "drop, crouch, and hold" mantra. I believe this to be erroneous advice. I have never seen a horse with hooves over its eyes, nor ants gathering underneath a table. They all want to break out, and move quickly to open ground.It is a well-known fact that horses and snakes will violently break free of their enclosures, and head into the open.

At the time of the Asian earthquake of Christmas Eve, 2004, which generated a tsunami, many lives were saved by following elephants who seemed to know instinctively where to go. We should follow their example.

EARTHQUAKE DIARY 2025

Abbreviations

FM=full moon, NM=new moon, Pn=perigee (the number refers to priority of closeness for the year, e.g. P1 means the shortest earth-moon distance), A=apogee, ^=northern declination, V=southern declination, XhS=moon crossing equator heading south, XhN=moon crossing equator heading north, Mercury+Sun=Mercury crossing in front of sun, Mercury-Sun=Mercury crossing behind sun. (Numbers on *left* may be disregarded, being for purposes of model only).

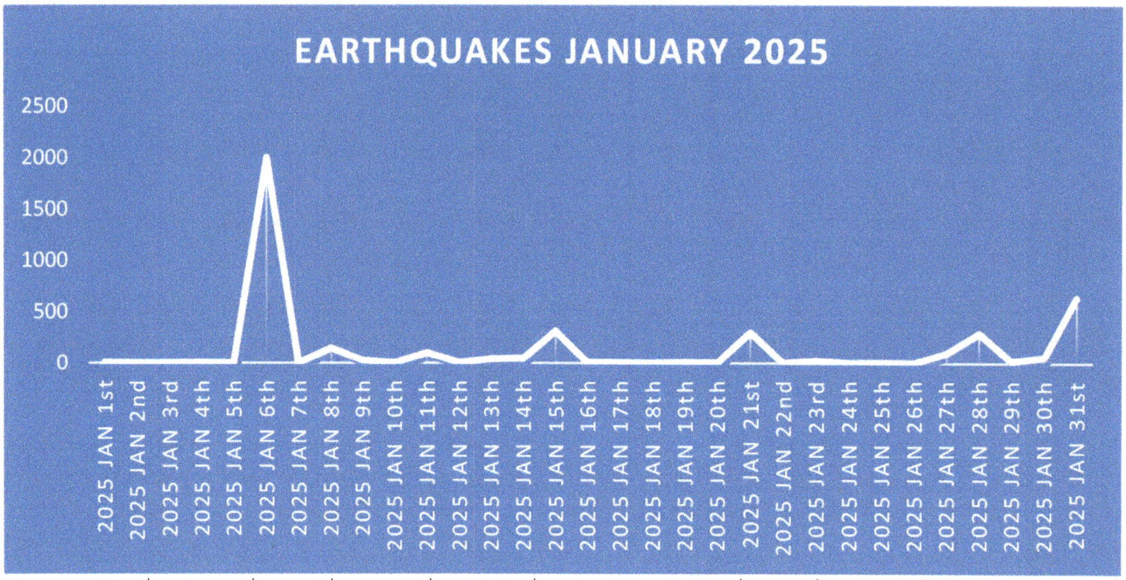

XhN, 1stQ,=7th, P#13=8th, ^=13th, FM=14th, XhS=20th, A=21st, 3rdQ=22nd, V=27th, NM=30th, Merc behind Sun 29-31st. Moon square Merc=5th, 20th, .

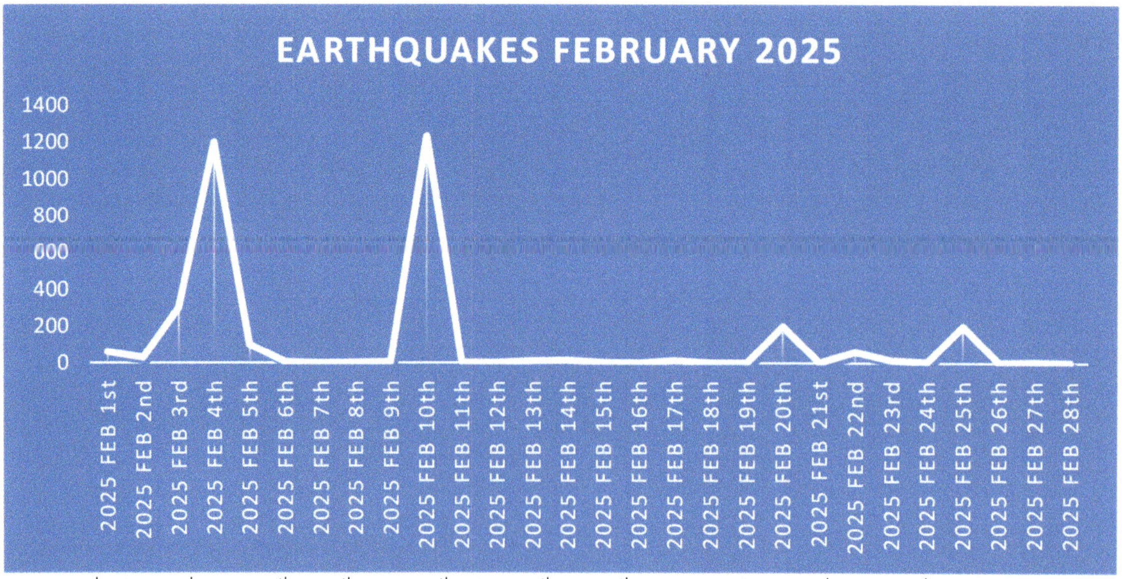

P#10=2nd, XhN=3rd, 1stQ=5th, ^=9th, FM=13th, XhS=16th, A=18th, 3rdQ=21st, V=23rd, NM=28th, Merc behind sun 1-6th. Moon square Merc=5th, 21st.

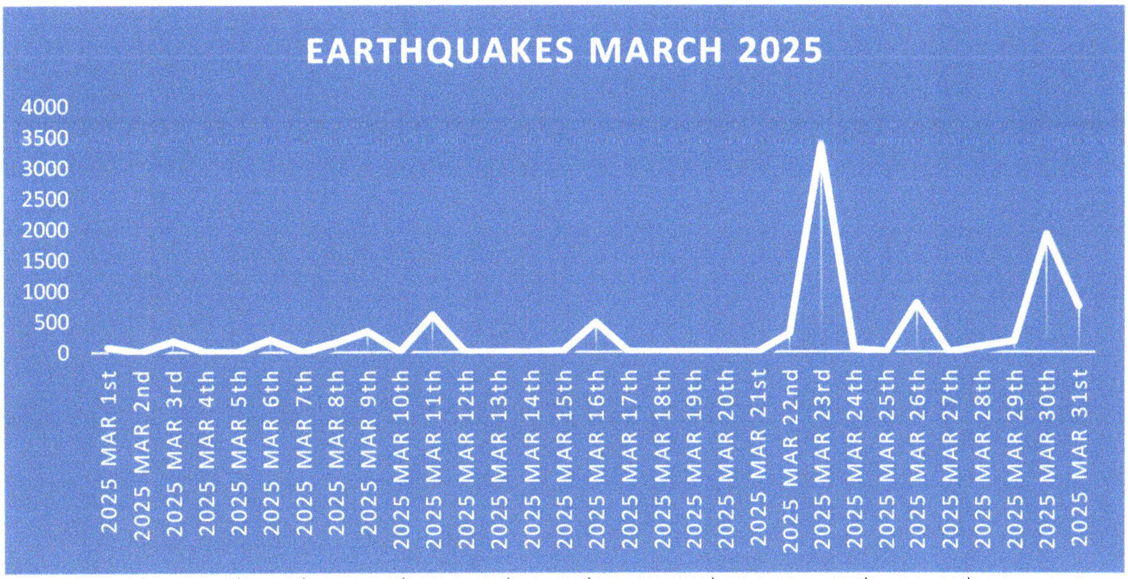

P#7, XhN=2nd, 1stQ=7th, ^=8th, FM=14th, XhS=15th, A=18th, 3rdQ=23rd, NM, XhN=29th, P#4=30th, Merc in front of sun 28th-31st. Moon square Merc=8th, 23rd.

^=4th, 1stQ=5th, XhS=11th, FM=13th, A=14th, V=19th, 3rdQ=21st, XhN=26th, P#3, NM=28th, Merc in front of sun 1st-5th. Moon square Merc=4th, 19th,

^=2nd, 1stQ=6th, XhS=9th, A=11th, FM=13th, V=16th, 3rdQ=21st, XhN=23rd, P#5=26th, NM=27th, ^=29th, Merc behind sun 23rd-31st. Moon square Merc=2nd, 19th.

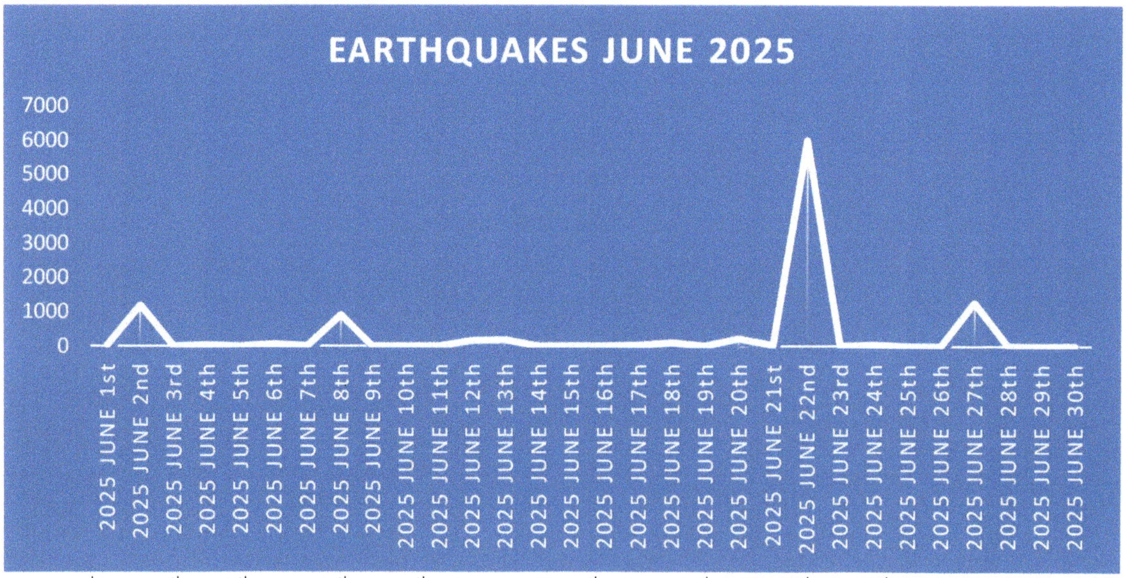

1stQ=3rd, XhS=5th, A=7th, FM=11th, V=12th, XhN, 3rdQ=19th, P#8=23rd, NM=25th, ^=26th, Merc behind sun 1st-4th. Moon square Merc=3rd, 20th.

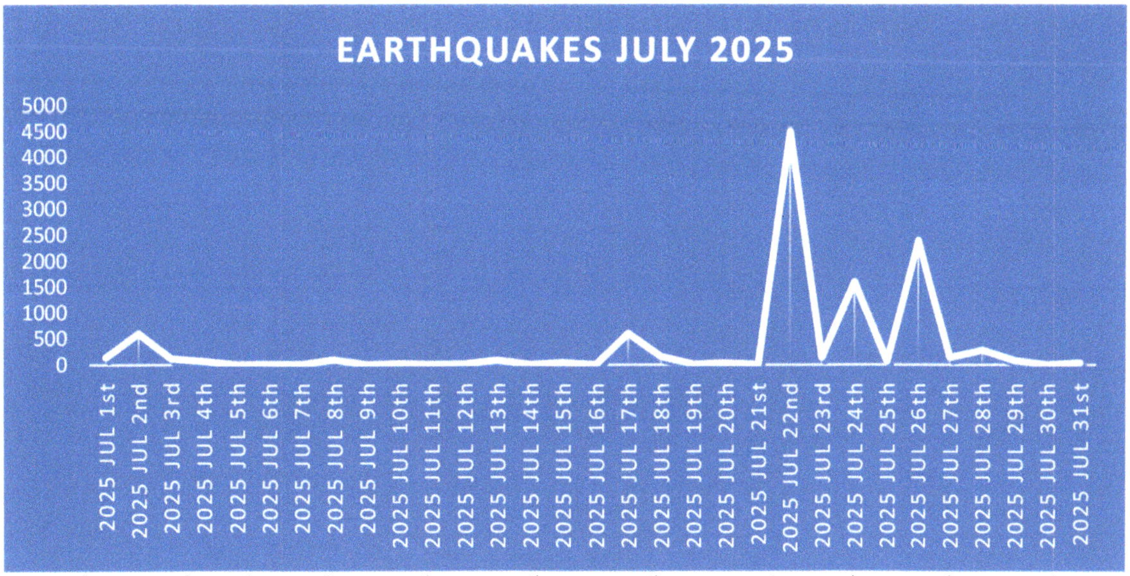

XhS=2nd, 1stQ=3rd, A=5th, V=10th, FM=11th, XhN=17th, 3rdQ=18th, P#11=21st, ^=23rd, NM=25th, XhS=30th, Merc in front of sun 20th-29th. Moon square Merc= 5th, 19th.

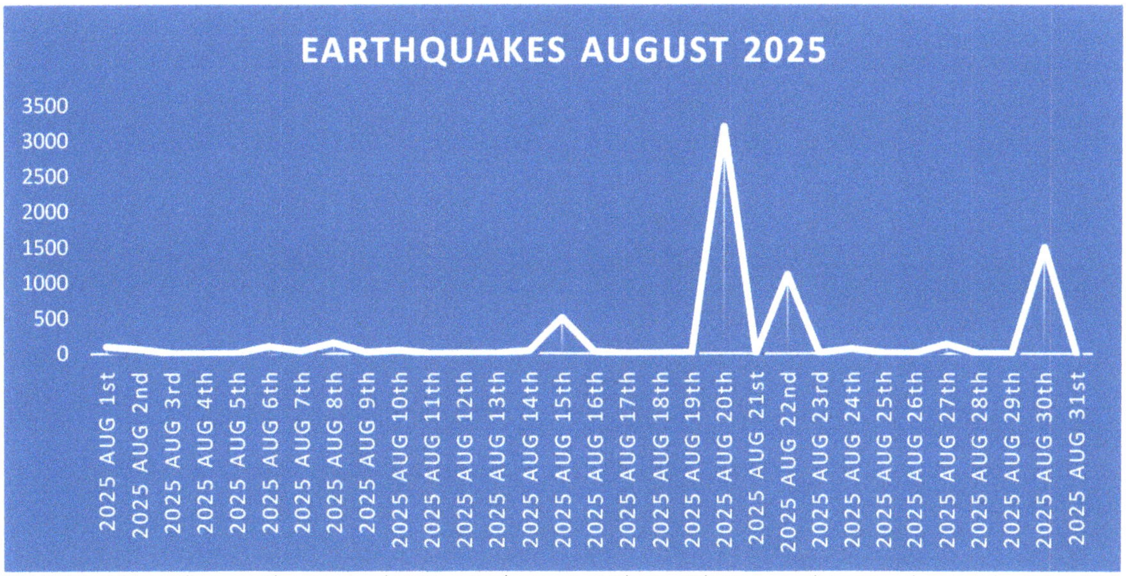

A, 1stQ=2nd, V=6th, FM=9th, XhN=13th, P#12=15th, 3rdQ=16th, ^=19th, NM=23rd, XhS=26th, A=30th, 1stQ=31st, Merc absent. Moon square Merc=1st, 15th, 30th.

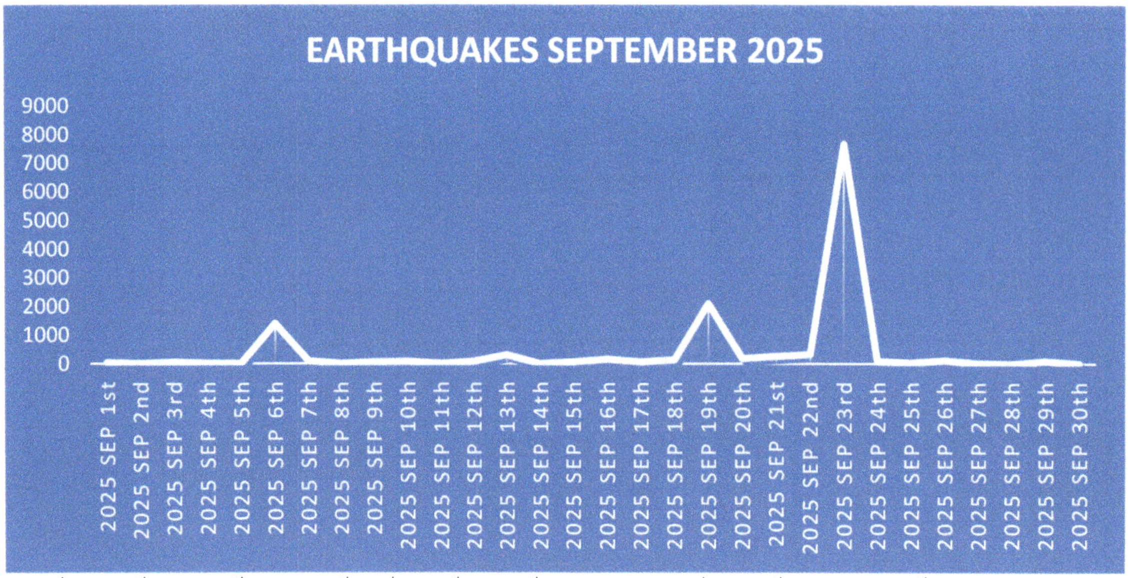

V=2nd, FM=8th, XhN=9th, P#9=11th, 3rdQ=14th, ^=15th, NM, XhS=22nd, A=26th, V, 1stQ=30th, Merc behind sun 17th-29th. Moon square Merc=14th.

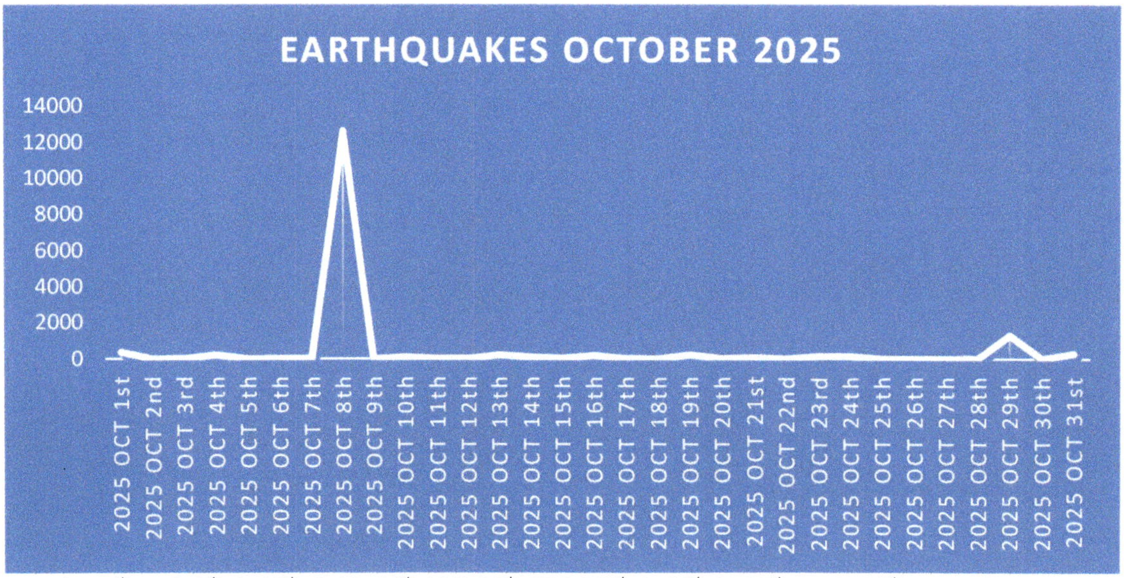

FM, XhS=7th, P#6=9th, ^=13th, 3rdQ=14th, XhS=20th, NM=22nd, A=24th, V=27th, 1stQ=30th, Merc absent. Moon square Merc=1st, 15th, 31st.